Fundamental Statistics
for Business and Economics

Fundamental Statistics for Business and Economics

Thomas R. Dyckman
Graduate School of Business and Public Administration
Cornell University

L. Joseph Thomas
Graduate School of Business and Public Administration
Cornell University

PRENTICE-HALL, INC., Englewood Cliffs, New Jersey 07632

Library of Congress Cataloging in Publication Data

DYCKMAN, THOMAS R
 Fundamental statistics for business and economics.

 Includes index.
 1. Statistics. I. Thomas, L. Joseph,
joint author. II. Title.
HA29.D92 519.5 76-28383
ISBN 0-13-344523-2

© 1977 by Prentice-Hall, Inc., Englewood Cliffs, New Jersey 07632

All rights reserved. No part of this book
may be reproduced in any form or by any means
without permission in writing from the publisher.

Printed in the United States of America

10 9 8 7 6 5 4 3 2 1

PRENTICE-HALL INTERNATIONAL, INC., *London*
PRENTICE-HALL OF AUSTRALIA PTY. LIMITED, *Sydney*
PRENTICE-HALL OF CANADA, LTD., *Toronto*
PRENTICE-HALL OF INDIA PRIVATE LIMITED, *New Delhi*
PRENTICE-HALL OF JAPAN, INC., *Tokyo*
PRENTICE-HALL OF SOUTHEAST ASIA PTE. LTD., *Singapore*
WHITEHALL BOOKS LIMITED, WELLINGTON, *New Zealand*

This book is dedicated to

Louis W. Thomas

Clovis E. Dyckman

Wallace Gardner

For those who know them, no words are necessary
and for those who are not so privileged,
no words would be sufficient.

The purpose of studying statistics is not to make you a statistician but to prevent you from being deceived by one.

With apologies to Joan Robinson

Contents

Preface *xiii*

Chapter **1**

Introduction *1*

1-1 Descriptive Statistics.   1-2 Decision Making—Statistical Inference.
1-3 Measurement Scales.   1-4 Problem Analysis.
1-5 Decision Problems and This Book.   1-6 A Warning.

Chapter **2**

Summarizing Data: Frequency Distribution *19*

2-1 Frequency Distributions.   2-2 Graphical Presentation.
2-3 Tabular Presentation.   2-4 A Warning.

Chapter 3

Summarizing Data: Statistical Descriptions *49*

3-1 Samples and Populations.   3-2 Measures of Location.
3-3 Measures of Variation.   3-4 Describing Grouped Data.
3-5 Skewness and Kurtosis.   3-6 A Warning.

Chapter 4

Index Numbers *85*

4-1 Nature of Index Numbers.   4-2 Types of Index Numbers.
4-3 Several Important Indexes.
4-4 Some Special Problems in Index-Number Construction.   4-5 A Warning.

Chapter 5

Uncertainty, Sample Spaces and Probabilities *115*

5-1 Uncertainty.   5-2 Sample Spaces.   5-3 Events.   5-4 Probability.
5-5 Addition Rule of Probability.   5-6 A Warning.

Chapter 6

Counting, Probability Distributions and Conditional Probability *154*

6-1 Counting Techniques.   6-2 Discrete and Continuous Probability Distributions.
6-3 Conditional Probability.   6-4 Revising Probabilities: Bayes' Theorem.
6-5 A Warning.

Chapter 7

Expectations *200*

7-1 Expectations: The Mean.   7-2 Expectations: Variance and Standard Deviation.
7-3 Expectations of Linear Functions.
7-4 Expectations: Expected Payoffs and Other Functions of a Random Process.
7-5 A Warning.

CONTENTS   ix

Chapter 8

Probability Distributions   *231*

8-1 Random Variables.   8-2 Some Probability Mass Functions.
8-3 Some Probability Density Functions.   8-4 A Warning.

Chapter 9

Sampling   *283*

9-1 Introduction to Sample Design: Definitions and Potential Problems.
9-2 Random Sampling and Judgment Sampling.   9-3 Types of Random Sampling.
9-4 A Warning.

Chapter 10

Sampling Distributions   *308*

10-1 Data-Generating Process.   10-2 Distribution of Sample Means.
10-3 Distribution of Sample Means When σ Is Unknown.   10-4 A Warning.

Chapter 11

Decision Making: Estimation   *334*

11-1 Characteristics of Estimators.   11-2 Interval Estimation—Confidence Intervals.
11-3 Confidence Intervals for the Variance and Standard Deviation.
11-4 Determining the Sample Size.   11-5 A Warning.

Chapter 12

Decision Making: Tests of Hypotheses   *366*

12-1 Introduction to Hypothesis Testing.   12-2 Selecting the Null Hypothesis.
12-3 Computing $\bar{x}_c$ Values: An Example.
12-4 Probability of Making Incorrect Decisions.   12-5 Hypothesis Tests.
12-6 A Warning.

Chapter 13

Statistical Decision Theory   409

13-1 An Urn Problem and a Managerial Problem.   13-2 Decision-Theory Model.
13-3 Making an Action Choice: The Bayesian Decision Criterion.
13-4 Opportunity Loss.   13-5 Multiple Decisions Over Time.
13-6 Decision To Obtain More Information.   13-7 Evaluating Payoff Values.
13-8 A Warning.

Chapter 14

Analysis of Variance   472

14-1 One-Factor Analysis of Variance.   14-2 Two-Factor Analysis of Variance.
14-3 A Warning.

Chapter 15

Regression Analysis   521

15-1 Objectives of Regression Analysis.   15-2 Regression Population.
15-3 Procedures for Estimating the Population Regression Line.
15-4 Using the Regression Equation To Predict.
15-5 Placing Confidence Intervals Around Predictions.   15-6 Testing Hypotheses.
15-7 Cross-Section and Time-Series Data in Regression Analysis.
15-8 Limitation of Linearity.   15-9 Multiple Relationships.   15-10 A Warning.

Chapter 16

Correlation Analysis   563

16-1 Correlation Analysis in a Regression Problem.
16-2 Correlation Analysis Without Regression.
16-3 Tests of Significance on the Correlation Coefficient.
16-4 Correlation and Statistical Independence.   16-5 A Warning.

Chapter 17

Time-Series Analysis *588*

17-1 General Time-Series Models. 17-2 Estimating the Trend Line.
17-3 Estimating the Seasonal Effect. 17-4 Exponential Smoothing Methods.
17-5 A Warning.

Chapter 18

Nonparametric Methods *626*

18-1 Estimating Methods. 18-2 Chi-Square Tests for Independence.
18-3 Goodness-of-Fit Tests. 18-4 The Runs Test for Randomness.
18-5 The Signs Test of Significance. 18-6 Rank Tests of Significance.
18-7 Nonparametric Measures of Association. 18-8 A Warning.

Appendix: Tables *667*

Solutions to Starred Problems *694*

Index *707*

Preface

This text is designed to introduce students of business and economics to the methods and applications of statistics. Our purpose is to give potential managers enough knowledge to be intelligent consumers of statistics. To accomplish our objective, we shall introduce applications of each statistical topic in several managerial settings including accounting, finance, marketing, production, and quality control. To emphasize the usefulness of statistical techniques, the development of theory is sacrificed in favor of an understanding as to how statistics can be properly and improperly used. Examples of both good and bad statistical approaches are given, to help the student select the proper data and analysis in a decision situation (considering cost and time limitations).

The proper application of statistical methods requires common sense. This book attempts to tie good statistical theory to common sense and intuition.

A Word to the Instructor

This text should accommodate a number of alternative statistical course offerings at the undergraduate and graduate levels. We suggest the following topics for consideration:

Length of Course	Chapters
One semester or one quarter	1, 2, 3, 5, 8 (normal distribution only), 9, 10, 11, 12, 16, 17
Two semesters or two quarters	First half of course: 1, 2, 3, 4, 5, 6, 7, 8, 9, 10, 11, 12
	Second half of course: 13, 14, 15, 16, 17, 18

Although the text is specifically designed for undergraduates, we have used it successfully at the graduate level for students who lack a strong background in mathematics. We have found it possible to cover 14 chapters in a 15-week semester course. Two weeks each were allocated to Chapters 12 and 13, and Chapters 1 and 2 were covered in a single week. Chapters 4, 14, 17, and 18 were omitted from our single-semester coverage.

Considerable flexibility is also possible. For example, Chapters 15 and 16, covering regression and correlation analysis, can be moved up in the sequencing to any point after Chapter 12. Chapters 4, 13, 14, 15, 17, and 18 can be used to supplement any course based on the introductory chapters, depending on the coverage desired.

The chapters proper do not make use of the calculus. However, some of the technical notes at the ends of the chapters apply the concepts developed in the chapter to the continuous-variable case using the calculus. Additional technical notes are used to introduce valuable but nonessential topics and, occasionally, to derive a statistical result when the derivation is enlightening.

Certain sections and subsections are starred. These sections can be omitted without loss of continuity or of any ideas essential to the understanding of later material. The topics are included at their particular location because it is the logical position for the topic, if it is to be covered.

A Word to the Student

This book is written using the word "we." The choice was made deliberately to emphasize the joint learning experience—student and instructor working together.

We suggest a quick first reading to obtain an overview of the chapter. At this time do not stop to underline, work through examples, attempt the solved problems at the end of each major section (called Student Shoulds and indicated by the symbol S.S.), or puzzle over difficult passages. Follow this quick reading immediately, if possible, with a slow and careful second reading. Keep a pencil and pad close at hand to work out difficulties. The Student Shoulds serve as a good review. Try to work each one during the second reading. Successful completion of the Student Shoulds is one good indication that the section material has been mastered.

Preface

After completing a chapter, try to work the problems assigned by the instructor. Answers to the starred problems appear at the end of the book. It will be necessary at times to return to the text material to work some of the assigned problems. This should be expected. We have designed many of the problems to cause just this result. Most important ideas do not crystallize until one has worked with them awhile. We cannot overemphasize the centrality of a conscientious effort on the problem material to the understanding of statistics.

One of the major difficulties we will experience involves notation. Care has been taken in this book to note and identify new symbols where they are introduced and to summarize them at the end of each chapter. The key formulas have also been summarized at the chapter's end. Once mastered, mathematical notation offers an efficient means of expressing complex ideas in an unambiguous way, and a more effective way of gaining new insights than if verbal reasoning alone were used. To tide us over this important but difficult learning experience, we will continue to express the important ideas in verbal as well as in mathematical form.

We have used these materials successfully and owe our past students a substantial debt for the present improved status. We wish to acknowledge the extensive problem work by Tim Critchfield of Cornell, the extensive and able typing of Jean Tubbs, and the technical assistance of Elva Lovell. We also wish to thank the reviewers of the manuscript: Robert L. Childress, University of Southern California and Brian Kritt, University of Baltimore. Once again, our foremost debt is to our wives, who continue to tolerate, if not encourage, these endeavors that their husbands persistently pursue in hopes that the revenues will justify the expenditures.

Fundamental Statistics
for Business and Economics

Chapter 1

Introduction

Stephen Butler Leacock, the Canadian economist and humorist, once wrote: "In earlier times they had no statistics, and so they had to fall back on lies." Carrying this situation one step further, Disraeli is reported to have observed that there are three kinds of lies: lies, damned lies, and statistics. It is our contention that the state of affairs is not so hopeless and that if one is careful, useful insights can be obtained and helpful conclusions reached using statistical methodology. (Of course, had we agreed with these distinguished gentlemen, this would have been a very short book.)

Too many, if not most of us, use the term *statistics* to conjure up images of numbers and figures: the crime rate, population data, the stock reports, baseball batting averages, and so on. Indeed, governmental and private organizations together produce enough data to bury us all. Yet properly selected, sorted, classified, summarized, and presented, these data can be extremely valuable in formulating, understanding, and even evaluating problem situations. For example, a pictorial presentation of the distribution of family incomes within a given geographical area is relevant to estimating the potential product appeal for a new, income-sensitive item such as diamond-studded grapefruit spoons. We refer to a use of data such as this as descriptive statistics. Chapters 2 through 4 deal largely with descriptive statistics and analysis based thereon.

But in the last 50 or so years, statistics has come to have a much more important function. The word "statistics" is used to describe a set of methods for making decisions when uncertainty is present. This use of statistics is known as statistical inference or, more recently, statistical decision theory. The latter, being the more erudite title, is preferred, naturally, by your authors.

The growth in the relevance of statistics to management decisions is due to a number of factors, including the increased educational background of managers and improved statistical methodology. But the most important factor for this growth is undoubtedly the dramatic increase in our ability to process data faster and more cheaply. This, in turn, is due to the advent of the computer. Because of the importance to the data-processing and computational powers of the computer, we will from time to time turn our attention to what computer programs are available to relieve the drudgery of statistical problem solving.

Because statistical decision making has its origin in uncertainty and uncertainty can be quantified using the theory of probability, we will, in Chapters 5 through 10, spend some time on the topic of probability before turning to statistical inference and other snazzier decision-making methods.

1-1 Descriptive Statistics

Lord Kelvin, the British physicist, has said: "When you can measure what you are speaking about, and express it in numbers, you know something about it; but when you cannot measure it, when you cannot express it in numbers, your knowledge is of a meager and unsatisfactory kind."

We believe that the greater measurement precision attainable in the pure sciences, such as physics, astronomy, and chemistry, is in large part responsible for the more rapid development of these areas in comparison to the social sciences. The question of measurement and summary description of what we have learned constitute the first major section of the book: descriptive statistics.

Among the more important reasons to summarize data is to improve one's ability to communicate with others. In Chapter 2, we examine the uses of graphs, tables, frequency distributions, and charts to accomplish this objective. Chapter 3 concentrates on the use of one or more numbers to characterize a much larger data set. And in Chapter 4 we look at a particularly important area in economics, index numbers, where such descriptive measures have had an important impact on governmental policy and management decision.

The reader of such summarized data must be ever alert to its misuse to suggest unsupportable conclusions. This is most likely to occur when the one preparing the display has something to be gained from the interpretation made. Further, the analyst must be careful to provide reliable data that adequately support the point being made. All too often summarized data are used for the same purpose that a drunk uses a lamppost: for support rather than for light. Yet the unorganized and unsummarized data alone are usually unintelligible. The data do not speak for themselves. Intelligence and insight

Introduction 3

are required if the figures are to be marshalled in justifiable support of a chosen objective. We also note in passing that the area of descriptive statistics, or summarized data, must not be separated from the decision-making process. A good analysis without adequate communication is, like a schoolboy with a slingshot but no spitball, unlikely to produce the desired effect.

Fortunately, most decisions in the ordinary course of affairs can be made without a great deal of fuss. The appropriate choice is clear and a formal analysis is not worth the bother. Statistical methods are not warranted in these situations. But this is not always the case and, unfortunately, it is sometimes true that seemingly simple decision situations are deceptively complex. This complexity may unfold when the manager attempts to formulate his problem, solve it or implement a solution. This should not surprise the reader, since a great deal of the art of decision making lies in understanding just what the problem is and searching for meaningful alternatives. Knowing the right question to ask is at least half the battle.

Once in a while, then, managers should be willing to juggle a few figures when they think it might help. Such situations must involve alternative consequences sufficiently serious to justify the time and other resources spent. For example, a rather extensive study might be undertaken by an automobile producer concerning the relative markets for large versus economy automobiles. On the other hand, the decision to carpet the car's interior may be made with little or no formal analysis at all.

The type of decisions we are concerned with in this book involve a choice among actions with important consequences. The manager's concern with the choice arises because events will occur over which the manager has no control or which cannot be predicted with certainty. These future or uncertain events will interact with the action choice to produce outcomes, and these outcomes affect the manager's ultimate well-being.

Some problems will be sufficiently complex that the manager will require the assistance of an expert statistician. In this event, the manager must be careful not to relinquish the decision-making responsibility. The useful resolution of important and complex problems requires the joint attention of both the manager and the statistical expert. Only when these two work jointly and both understand and appreciate the other's contribution are useful results likely. Yet the manager must never forget that the ultimate responsibility is his and his alone. One is reminded of the statement that President Truman kept on his desk: "The Buck Stops Here."

1-2 Decision Making—Statistical Inference

Statistical inference can be described as the process of drawing conclusions from incomplete data. The reliance is on sample data and hence on only a portion of the total information set. Yet the manager is interested in what is true in the larger set from which the sample observations were taken. An example is using available historic data (the sample) to predict (make inferences about) the future (the larger set). The methodology of statistical inference addresses this issue. Chapters 11 through 18 treat the topic of statistical inference.

We might ask why an exhaustive analysis is not made. Why not observe every item in the larger set? There are several reasons. First, it may simply be impossible to do so. Making inferences about the future is a good example. We cannot observe what has not yet occurred. It may also be impossible to observe the larger set since, before the task could be completed, the larger set's composition will have changed. An example would be to attempt an exhaustive census for any major geographic area. Before this task could be completed, births, deaths, and population relocation will have altered the composition of the larger set. Second, although a complete enumeration could be made, such a task is impractical. For example, the measurement or observation task might destroy the item, or sufficient time before a decision is required may not be available, or the cost of the complete enumeration activity may be too expensive compared either to the available funds or the expected benefits. Finally, measurement error, caused by fatigue for example, may increase over the data-gathering process leading to noncomparable and useless data.

Even if reliable data are available for analysis, care must be taken in generalizing from such data. Conclusions based, for example, on student opinions or behavior cannot safely be extended to the public at large. As another example, the estimated relation of a product's unit cost to the number of units produced may change abruptly when substantially larger levels of productive activity are encountered than those which were used to develop the estimated relationship. Moreover, the manager must always keep in mind the alternative of continued analysis. Based on the data available, a selection among alternatives may not be justified. Some alternatives may be rejected but new ones may be suggested. Indeed, a better problem formulation may be required and it may be appropriate to gather additional, perhaps even different, data. The manager should attempt to balance the costs of additional investigation against the anticipated additional benefits. An example of the idea of continued investigation is found in studying new products. At any point in time the possible decisions can be characterized as GO, NO, and ON. GO implies "market the product vigorously," NO implies "discard the product," and ON implies "continue the investigation."

Statistical inference attempts not only to allow the manager to draw conclusions but, equally important, it provides a means by which the probabilities of making wrong decisions or incorrect generalizations can be

INTRODUCTION

determined. Hence, using statistical methods, the decision maker can make reasoned decisions from incomplete data and calculate the likelihood that the decision is in error. An even more useful measure to the manager than the likelihood of error is the expected cost of the error. Indeed, a manager would be well advised to select the action with the lowest expected cost or largest expected benefit. When such costs or benefits can be determined, statistical methods are available which incorporate them. These methods are more powerful than traditional techniques, which attempt to establish at most the likelihood of being wrong. Often, unfortunately, it is not possible or feasible to measure these costs or benefits with sufficient accuracy to justify the more powerful (and more expensive) methodology.

Perhaps at this time a few definitions would be helpful. Statistical inference is characterized in this chapter as the process of drawing conclusions from an incomplete set of data called a *sample*. The sample consists of a set of measurements made on some variable of interest to the decision maker. The sample is drawn from the total set of possible observations, called the *universe*.

For example, suppose that we wish to study the income levels of households in a given city. The variable of interest is the level of income. The household becomes the basic unit sampled. The set of all such households in the city we will call the *population*. A subset of households is selected for observation, and the data on income are gathered for each. This set of income measures constitutes the sample of measurements, the data. The sample measurement data are then subjected to statistical procedures which permit one to infer something about the true income levels in the population. The set of all possible income measures from which the sample was drawn is the universe.

> A *population* is the set of units of observation in a statistical study. (Example: households)
>
> A *variable* is the particular property of each unit in the population that is of interest. (Example: income)
>
> A *universe* is the set of all values of the variable in the population, one for each unit. (Example: the income of each household in the area under investigation)
>
> The term *sample* is used in two ways (the particular meaning should be clear from the context of discussion):
>
> 1. The subset of the population for which measurements are obtained. (Example: the subset of households selected)
>
> 2. The subset of values obtained from the universe, sometimes also called the sample data. The term is used in this sense in the previous discussion. (Example: the incomes of the subset of households selected)

S.S. An inspector checks incoming washers to ascertain whether their diameters meet specifications. He does this by checking 100 washers in each shipment of 10,000.

(a) What is the population?

(b) What is the universe?

(c) What is the sample data?

Solution:

(a) The population is the 10,000 washers in each shipment.

(b) The universe is the 10,000 diameters, one for each washer.

(c) The sample data or simply the sample would be the 100 washer diameters checked by the inspector. (Sometimes it will be convenient not to distinguish between the population and the universe.)

1-3 Measurement Scales

Once the population has been identified and the variable of interest established, it is necessary to determine the appropriate type of measurement to use. Sometimes the nature of the problem dictates the type of measurement that will be appropriate. In other cases a conscious choice among several possible alternatives is required. Four levels or scales of measurement are described in this section: (1) nominal, (2) ordinal, (3) interval, and (4) ratio.

NOMINAL MEASUREMENT

Nominal measurement involves classification of the items to be measured. The only operational characteristic of nominal measurement is that the classes can be told apart or distinguished from one another. It is not necessary to be able to form a rank order of the classes, and it may even be impossible. All that is required is that the classes be different. Examples include classifying people by sex, or political belief; classifying units of a product by color, or by whether they are defective or not; and classifying costs as fixed, semifixed, semivariable, or variable.

> *Nominal measurements* only allow us to determine that the objects measured differ.

Numbers are often used in classification tests where the final result is at most of a nominal nature. For example, people may be classified as tall if they are over 6 feet, short if under 5 feet, and medium otherwise. If the data

INTRODUCTION   7

we are given only tell us how many people are in each category (that is, the original measurement on each individual's height is not available), then the data are nominal.

In some cases numbers are used to represent classes. A common approach in classifying items where only two categories are involved is to use the numbers 0 and 1 to stand for the classes. For example, suppose that we were to stand on the corner each day at high noon and note whether a woolly bear was in sight. The day is considered successful if one is sighted and the day is a failure otherwise. Five days of observation might produce the following sequence:

Day 1	Day 2	Day 3	Day 4	Day 5
woolly bear	no woolly bear	woolly bear	woolly bear	no woolly bear

This sequence could also be coded as follows:

	success	failure	success	success	failure
or:	1	0	1	1	0

where the number 1 stands for success and the number 0 for failure. A similar series could reflect the successive tosses of a coin, the ordered observations on tests of whether newly made electric motors function properly, or the results of successive tests of a new drug on patients. It is worth noting that although this approach is most common when only two categories are involved, it is extendable to three or more categories. Numbers used in this way yield nominal data.

ORDINAL MEASUREMENT

Many measurements permit us to ascertain more than just whether two items differ. Using ordinal measurement it is possible to rank-order the various measurements. Many measures in the social sciences are of this variety. Examples include preference measures among brands of a product, of political candidates, academic and military ranks, and the order of arrival for service. The numbers used in ordering the items measured is arbitrary in one sense. The preference order for four job offers might assign the number 4 to either the most or least desirable job. But once this is done, the number sequence is meaningful.

> *Ordinal measurement* allows us to determine that the items measured differ; it allows us to establish an ordering among the items.

We must be careful not to ascribe too much to ordinal data. A potential customer may rate three brands of coffee as 1 (the first choice), 2 (the second choice), and 3 (the third choice). But this does not mean that brand 1 is preferred over brand 2 by the same degree as brand 2 is preferred over brand 3. Similarly, if three individuals have IQ's of 140, 120, and 100, it means at most that they are ordered in intelligence by their scores. If the test is valid, it may be stated that the first individual is more intelligent than the second, but it is not possible to say by how much. Nor can it be said that the first individual exceeds the second in intelligence by the same degree as the second exceeds the third. We can only state that the second individual is between the other two in terms of the characteristic measured. Statements based on the intervals between scores require a stronger level of measurement.

INTERVAL MEASUREMENT

Measures, principally those found in the physical sciences, permit more precise conclusions to be reached. When interval measurements are available, the distance between any two measures is meaningful. For example, if three identical objects have temperatures of 30°, 20°, and 10° centigrade, the first is as much hotter than the second as the second is hotter than the third. The intervals between the measurements are meaningful.

Interval measurement allows us to make meaningful statements about the distance between measurements.

When interval measurement is used, the zero point on the scale may be set arbitrarily. The zero on the centigrade temperature scale is set at the freezing point of water, but it could be set anywhere. For example, the boiling point of water, 100° on the normal centigrade scale, could be used just as easily (−100 would then be the freezing point). Water at 50° centigrade would have a temperature of −50° if the boiling point of water was used as the origin. Such water is midway between freezing and boiling on either temperature scale.

RATIO MEASUREMENT

The strongest form of measurement in terms of the information it provides is called ratio measurement. With ratio measurement, statements concerning the relative magnitudes of the measures can be made. Such measurement is possible with weight, height, dollars, percent of impurities, and many other variables common to the physical sciences. For example, if three objects weigh 60, 40, and 20 pounds, respectively, we can say the first is three times as heavy as the third and the weight of the second and third together equals that of the first. Such statements are not possible with centigrade temperatures or other data that are measured at most on an interval scale.

Ratio measurement allows us to make meaningful statements about the relative magnitudes of the various measurements.

When measurement is of a ratio nature, the zero point on the scale signifies the absence of the property. A height of zero feet or a weight of zero pounds, for example, implies that the object under observation does not possess the property at all. The zero point, then, cannot be established arbitrarily on a ratio scale.

Of course, statements permitted by weaker measurement scales can also be made. Hence the first object weighing 60 pounds weighs as much more than the 40-pound object as the latter exceeds the 20-pound object—an interval measurement statement. We could also say the first object weighs the most and the third object the least—an ordinal measurement statement. We may also state that the objects' weights differ—a nominal measurement statement.

The stronger the measurement scale used, the more information that is contained in the measures. Statistical techniques always make some assumptions, often implicitly rather than explicitly, concerning the level of measurement attained. Those procedures which allow the most to be concluded from a given sample are those that assume the strongest measurement scale—ratio measurement. Most of the statistical tests discussed in this book assume ratio measurement. (Chapter 18 is an exception, as is the binomial distribution introduced in Chapter 8.)

The manager must be careful that the data used by the resident statistician to make any tests satisfy the required measurement level. Otherwise, the manager may be guilty of allowing the expert to conduct a mathematically precise analysis on irrelevant data. The well-known phrase "garbage in–garbage out" (meaning that bad data produce bad results even if the method of analysis is good) is appropriate. The manager must be careful not to ascribe properties to measurements they do not possess. The very ability to manipulate numerical values can lead to trouble. An almost unattainable degree of self-control is required at times not to endow data with properties they do not enjoy.

S.S. Indicate the highest level of measurement possible in each of the following cases:
(a) Time
Temperature:
(b) Kelvin (absolute)
(c) Fahrenheit

(d) Sales orders

(e) Job satisfaction

(f) Race

Solution: Items (a) and (c) are interval; items (b) and (d) are ratio (the Kelvin absolute temperature scale has an absolute zero that is not arbitrarily defined since it signifies the absence of molecular motion); item (e) is most likely ordinal; and item (f) is nominal.

1-4 Problem Analysis

Decision problems do not arrive on the manager's desk neatly wrapped, tied, and ready for routine solution. In fact, the most creative aspect of decision making is in deciding that a problem exists and electing to consider it seriously. Once this critical step has been taken, we can begin the necessary steps to find and implement a solution. We use the words "necessary" and "a" advisedly. The steps described below may help but will not assure a solution, and a given problem may have several solutions or none at all.

The steps involved include:
1. Formulating the problem.
2. Gathering the necessary data.
3. Analysis and testing of the data.
4. Implementation of a solution.

The process may repeat some steps, and it may do so even before all four steps have been completed. An example is when the data analysis suggests that the problem was initially formulated incorrectly.

Unfortunately, a text such as this one will give you little help in determining that an important problem is at hand. This is an art that comes with time and practice. We also have little to say concerning implementation. Implementation involves people and requires communication as well as leadership. These two tasks, implementation and detection, are best learned on the job. On the other hand, we will have a good deal to say concerning problem formulation, data gathering, and analysis. These are crucial tasks in making good decisions, and they lend themselves to formal study.

The reader will recognize that in trying to concentrate on problem formulation and analysis, the exercises at the end of each chapter are written to (1) be as clear as possible, and (2) to contain all the data needed for a solution without the difficulties introduced by extraneous data. Further, the solution technique the reader is expected to use is contained in the material

INTRODUCTION

just read. In essence, we will often act as though the problem to be solved is clear, the data gathered, and the range of decision techniques substantively reduced. Although we defend this approach pedagogically, it may give the reader a mistaken impression of problem solving at the acting manager's level.

FORMULATING THE PROBLEM

If the problem is incorrectly conceptualized, there is little chance of obtaining a useful solution since the manager will be addressing the wrong issue. A useful example, although not a statistical one, is provided by the manager selecting between two products to produce. The first had a contribution margin (price minus variable costs) of $7 per unit, and the second had a margin of $5. Since the manager was restricted to produce only one of the two items, the product with the greater unit margin was selected. The manager failed to consider that 3 units of the second product could be sold for every 2 units of the first product. Thus $15 = $5(3) in total margin would be obtained from the sale of the second as compared to $14 = $7(2) for the first. The manager determined the product with the larger margin but failed to examine the other critical variable: demand.

Of equal importance to correct problem identification is the need to formulate the problem so that a timely solution can be obtained. Solving the right problem too late to be of use is just as disastrous as formulating the wrong problem at the start.

The ability to simplify a problem to the point where it yields meaningfully to analysis is an essential managerial skill second only to problem recognition. The successful manager must have or develop the ability to abstract a problem from a complex environment and eliminate the unimportant elements without losing the problem's essence. Managers must be able to define or search for alternative solutions and methods of analysis while balancing the associated search costs with the expected benefits.

THE DECISION MAKER

An early issue to resolve is who is the decision maker. A useful question to ask in this regard is whether we or our immediate superior makes the decision. In the latter case, it is necessary to analyze the problem from the superior's point of view. And, of course, the final decision may be made even further up the chain of command: our immediate superior may only recommend. In the latter case, it is necessary to consider both individuals, the recommender and the decision maker. It may also be useful to consider the uncertainty in the reactions of the decision maker to the recommendation. When no one seems willing to assume responsibility or, alternatively, the decision is made in a meeting or by a group, the process is much more hazardous for the junior executive. As Howard Raiffa observes in *Decision Analysis* (Reading, Mass.: Addison-Wesley Publishing Company, Inc., 1968), "In some mysterious way a decision will eventually be made, and *after* everybody learns how it has turned out the identity of the decision maker (and the recommender) will suddenly come to the surface. In such a situation the analyst himself may be the 'fall guy' if his recommended strategy turns out to be a poor choice after the fact; and equally, the analyst may not gather any glory even if his strategy turns out well."

The analyst, of course, may actually play a significant role in the decision process in, as Raiffa notes, "a myriad of subtle ways: by what he chooses to incorporate in the analysis, how he phrases questions, the grimaces he makes in dialogue . . . the tone of voice he uses in an oral presentation, and the issues he may conceal behind a barrage of mathematical mumbo-jumbo."

IMPLEMENTATION

Determining who makes the decision is also important in the implementation process. In most organizations a substantial gap exists in the understanding of the problem, the knowledge required to find a solution, and the authority to accept and implement the solution. This gap necessitates communication among the executives concerned as well as between them and the statistical expert. But, more importantly, it means that those members of management who will of necessity implement a decision must be substantively involved in the various stages of problem solving. As many studies fail for lack of this vital ingredient as for any other.

QUALITATIVE FACTORS

Many who disparage the attempt of the measurer argue that the approach ignores the host of qualitative factors attendant to any real problem. Indeed, it is easy to believe in "the numbers," to put more faith in them than is warranted. We should always remember to question the numbers. The wise manager considers the source of the data and why it is valid and relevant to the problem. But if this task is done well, much of the complexity is removed and the impact of qualitative factors can be more easily seen. Moreover, if the qualitative factors favor the quantitatively superior conclusion, the decision is clear. If, on the other hand, they on balance favor a course of action that is inferior to another based solely on the quantitative analysis, it may be possible to ask whether the qualitative difference is "worth" the quantitative superiority of the other action. Quantitative analysis, we maintain, makes it easier to properly evaluate the importance of the qualitative factors that may initially be put aside.

INTRODUCTION

S.S. Indicate (using the numbers 1–7) the natural sequence and, separately, the order of importance of the following tasks in decision making:
- (a) Data analysis and testing
- (b) Data collection
- (c) Establishing conclusions and solutions
- (d) Implementation
- (e) Problem formulation
- (f) Problem recognition
- (g) Selection of analysis technique

Solution: The natural sequence is (f), (e), (g), (b), (a), (c), (d). The relative importance of the tasks is a matter of opinion, and there is no single best answer. We believe that task (f) is paramount. Tasks (e) and (d) are of next importance, followed by tasks (g), (b), and (a). If those are done well, task (c) should be easy. It might be useful for you to return to this issue of relative task importance at the end of your study.

1-5 Decision Problems and This Book

In this section, we briefly describe several decision situations similar to ones that you will learn how to analyze in this book. The chapters where the relevant materials are discussed are noted at the end of each brief problem statement.

PROBLEM 1

An organization has data on the costs and production levels in one of its divisions for several periods. It wants to know whether these data could be used to predict the future level of costs and, if so, just how this could be accomplished. (Chapter 15) It would also like to know whether the level of production explains a significant amount of the variability in costs from period to period. (Chapter 16)

PROBLEM 2

The director of a local welfare office would like to know whether certain characteristics such as whether individuals own their home and the size of their fixed expenditures on items such as food and clothing affect their presence on the list of welfare recipients. (Chapters 14 and 18)

PROBLEM 3

The manager of a regional hospital has data on the amount of type O-positive blood needed in the hospital in any given week. The manager would like to know the chance that stocks of various levels of blood will be sufficient. (Chapters 8 and 10)

If the cost of a rush order to meet an emergency and the cost of having blood left over which must be discarded are known, how much should be stocked? (Chapter 13)

PROBLEM 4

The manager of a productive activity wishes to control the quality of the product. Suppose the manager is concerned with the percentage of impurities in each finished unit. How many units should be sampled before drawing conclusions? Given a set of sample data, how does the manager conclude whether the process is performing as desired? How can the likely range within which the process is producing be estimated? (Chapters 9, 10, 11, and 12)

PROBLEM 5

A union desires information on the cost of living to use in wage negotiations. How can useful information for this purpose be obtained and what limitations do these data have? How might the union describe the data to management to present an accurate and forceful case? (Chapters 2, 3, and 4)

PROBLEM 6

The space program is considering backup devices for a critical life-support system. The system either works adequately or it fails. Given some data on past performance over time, how can the expected time the system will function and the increase that can be obtained in this time for each backup system added be estimated. (Chapters 5, 6, 7, and 10)

PROBLEM 7

An organization is concerned about planning and forecasting cash needs for a growing business subject to substantial seasonal influences. How can the available historical data be marshalled toward providing some insights into the problem? (Chapter 17)

INTRODUCTION   **15**

The list of problems is only illustrative of those to which the methods of this book may, if used appropriately, provide solutions. Each chapter provides a number of additional examples. Some of these are discussed in detail while others are left for the reader to resolve.

S.S. An electric utility is considering building a nuclear plant on a local lake. The lake's waters will be used to cool the reactor. This will raise the temperature of the water near the plant, with resulting environmental effects. Does this problem involve statistics?

Solution: Since moral issues are raised as well as the needs of a large user group in comparison to the disutility to local citizens, the final decision will turn not only on quantitative considerations in general or statistical issues in particular. However, the presence of uncertainty in needs, costs, environmental impact, and so on means that data gathered and analyzed according to statistical methods will play an important role in the decision. Often the position that supplies the better documentation (usually statistical) will prevail.

1-6 A Warning

The individual may often feel overwhelmed when faced with a plethora of data and statistical tests. The figures and commentary appear compelling yet this may not be the case. W. Allen Wallis provides this useful example in commenting on the Kinsey Report in the *Journal of the American Statistical Association*, 44 (1949), 466:

> When I first examined the volume, paying attention mostly to its fascinating substantive findings and scarcely at all to its methods, I was very favorably impressed indeed. When I diverted my attention to the general methods I began to note shortcomings; but I felt that these were technicalities—mere blemishes on the surface of the monument, which might modify some of the findings in detail but surely would not affect the broad conclusions. After all, many of (the) figures would still be important and interesting even if we had to allow for an error factor as large as two or even three. But when I spent some time studying the statistical methods in detail, I realized that my confidence in the basic significance of the findings cannot be securely buttressed by factual material included in the volume. In fact, it now seems to me that the inadequacies in the statistics are such that it is impossible to say that the book has much value beyond its role in opening a broad and important field.

The reader is urged to practice questioning answers rather than simply answering questions.

PROBLEMS

1-1. Distinguish between a universe and a sample.

1-2.* Why should a sample be taken in most situations even when it is possible to examine every item in the population?

1-3. How does a population differ from a universe?

1-4.* What are the more important steps in problem solving?

1-5.* A well-known saying named Murphy's Law states that, "If something bad can happen, it will." What should someone with even a limited knowledge of probability and statistics respond to someone advocating this law as a decision-making criterion?

1-6. Suggest several problem areas in production where statistical methods would be likely to arise.

1-7. What is wrong with the following statement? "Since there are more automobile accidents during the day, it is safer to drive at night."

1-8. A city once claimed to be the most healthy city since its death rate was the lowest in the country. What problems do you see with such a statement?

1-9.* What is ignored in the following statement? "Recent statistics show that the average income per person in a certain area is $2,000; thus, the average income for a family of five is $10,000."

1-10. The following quotation appeared in a weekly periodical. Do you agree with the statement or not and why?

> *We are all, God knows, inundated by statistics of doubtful validity and uncertain import, and anyone who wants to challenge the statistics can usually make a persuasive case that something is wrong with them. Perhaps businessmen should challenge more of the statistics around them. But let us venture one small suggestion: that businessmen not issue any such challenge at times when the figures happen to make them look good.*

1-11.* Procrustes, a legendary highwayman of Attica, tied his victims to an iron bed and, as the case required, either stretched them or cut off their legs to make sure they would fit the bed. What can we learn from Procrustes regarding the simplification of real-world problems?

1-12. Georgi Malenkov, a member of the Soviet Politburo, once said there were 14 million unemployed in the United States in 1949, and thus the United States was in a serious depression. A better estimate would have been about 4 million. Malenkov used United States data. Can you guess why his figure was so high?

INTRODUCTION 17

1-13. Studies of personal income based on census data usually underestimate aggregate personal income for the United States by more than 5 percent. Why?

1-14.* The "concentration ratio" in an industry is defined as the percentage of total industry sales made by the four largest companies. Do you see any difficulties in computing this ratio?

1-15.* "Wage rates are rising and yet take-home pay is declining." How could this happen?

1-16. One hospital claimed a need for more money than another because, it claimed, the number of severe cases treated were more numerous in every category than in a second hospital. Further, treatment of more severe cases was more costly. What might you wish to know before allocating funds in this case?

1-17. It has been said that the quantitative drives out the qualitative in analysis. What does this mean?

1-18. The production foreman in an automobile manufacturing plant reported that production of automobiles in the plant fell off 120 percent from last year. Interpret his report.

1-19.* The manager of a tool and die plant hired a safety director to initiate a program of safety consciousness. Soon after the inception of the program, the accident rate in the plant dropped sharply. Would you conclude that the special program was responsible for the decrease in the accident rate?

1-20.* An advertising firm has just presented the sales manager of a ballpoint-pen manufacturer with the layout of a new advertisement. The advertisement asserts that the pen is "guaranteed to write five times longer." What does this mean?

1-21. What are the population, statistical universe, and level of measurement implied in each of the following:
   (a) An accounts receivable manager for a parachute manufacturer attempted to estimate the average age of the 1,000 accounts with nonzero balances on December 31, by examining a random sample of 150.
   (b) The sales manager for a cereal manufacturer attempted to determine the number of raisins that a competitor puts in its 15-ounce box of Raisin Bran. He purchased 100 15-ounce boxes from various retailers and counted the number of raisins in each box.

1-22. The manager of a clothing manufacturing plant had assigned one inspector to each of three groups of sewing machine operators. At the end of each month, each inspector is required to report the number of defectives produced by the group. After one month, the inspectors reported that groups A, B, and C had produced 450, 500, and 700 defectives, respectively. The manager concluded that the groups could be ranked according to their efficiency with A best, followed by B, and then C. Do you agree? Why or why not?

1-23. The personnel director at United Radio Company is collecting data on employee attitudes. She provides the following statement to 150 of the 750 employees of the firm: "I feel that I am performing a valuable service for society when I do my job well."

1	2	3	4	5
strongly agree	agree	no opinion	disagree	strongly disagree

(a) What is the population of interest?
(b) What is the statistical universe?
(c) What are the sample data?
(d) What is the level of measurement?

1-24. What is the population and the universe in each of the following:
(a) A quality control inspector is sampling 1 day's production of 15-ounce boxes of cereal to determine their actual weight.
(b) The sales manager of a wholesale grocery company samples 100 orders in process to support his contention that all orders received now average about $100 each.
(c) The Updown investment firm claims that it can predict with 90 percent accuracy whether the Dow Jones Industrial Average will go up or down from one week to the next. To test its claim, a competing firm employed Updown to make predictions for the 4 weeks within the next month.
(d) The sales manager of Hotshot Gun Company conducted a survey of New York residents to determine how many favored a proposed gun-control law.

1-25.* Indicate the level of measurement involved in each of the following:
(a) A credit card company draws a random sample of 100 from its 100,000 cardholders to estimate the average income of its cardholders.
(b) An advertising firm engaged in research for its client surveyed consumers to determine what brands of soft drink they consume.
(c) In an investment research office, 5 of the 10 staff members favored purchasing Buyrite stock while the other 5 were against it.
(d) A clock manufacturer tests a sample of clocks to see how many minutes each registers in 1 hour as measured by a control clock.
(e) A department in United Radio Company produces dials for radios. The dials for AM radios are calibrated from 500 to 1600.

Chapter 2

Summarizing Data: Frequency Distribution

In many parts of the world, including the area where the authors live, one can hear the lament, "It is always cloudy here." This is an example of a summary of some data. It is, however, both incorrect and imprecise. It is incorrect because it is not cloudy all the time (although it may seem that way); it is imprecise because it is not clear what quantity is being measured. The quantity of interest may be the fraction of time between sunrise and sunset when there are any clouds in the sky. It may be the fraction of days during which blue sky is never visible or the proportion of the day when there is no direct sunlight. Clearly these two measures would be drastically different, but neither one would be 100 percent.

In Chapter 1, we spoke of problem-solving approaches and discussed the idea of choosing measures relevant to the decision process. An individual planning a picnic might want to define cloudy (bad picnic weather) as a day in which there is total overcast for all but at most 2 hours of the time between sunrise and sunset. A clear day (good picnic weather) is then one in which there are at least 2 hours of sunlight or partial cloudiness. This individual is using nominal measurement and might say, "It is cloudy 30 percent of the time." A commercial farmer, on the other hand, is interested in the amount of sunlight during the day and would define cloudiness as the absence of direct sunlight on the crops. For the same data a farmer, using a ratio scale, might say, "It is cloudy 60 percent of the time."

Both of the decision makers in the paragraph above are describing the relative frequency of an event. They use different events because of their different information needs, but they both have summarized a large amount of data into a very brief statement. The purpose of this chapter is to study methods of summarizing data. In summarizing data we will nearly always lose some information. For example, when we know that 30 percent of the days are cloudy, by one definition, we do not know whether those days typically come in groups of two or more. We also do not know how likely it is to be cloudy tomorrow. (It is very likely to be cloudy tomorrow if it is cloudy today.)

Why, then, do we summarize data at all? One reason is that summarized data are sufficient for some purposes. For example, the farmer may be concerned about total sun during the growing season, and knowing that the sun will be out 40 percent of that time may be all the farmer needs to know. Another reason for summarizing data is that we are incapable of dealing with the data in their raw form, because, typically, there are just too many. Knowing that on June 1 the sun was visible from 6:02 to 6:45, 7:45 to 8:14, 12:02 to 2:14, and 4:45 to 7:02 may be interesting, but 365 such scenarios would be very hard to deal with. Thus summarization is necessary. The manager must be careful in dealing with summarized data. The manager must remember that some information has been lost and consider the relevance of the data in summarized form to the decision at hand.

2-1 Frequency Distributions

A *frequency distribution* is a summary of some sample data. It tells us how often each possible outcome occurred.

A frequency distribution can be given for the cloudy/sunny example above, using the picnic planner's definition. Suppose the statement was based on a sample of 30 days.

Weather	Number of Days
Cloudy	9
Sunny	21
Total	30

The frequency distribution tells us how often each of the two possible outcomes occurred. (Note that the summary only makes sense to you because you know the definitions of cloudy and sunny, that is, because you have seen the prior discussion. We will return to tabular representation, and improve

Summarizing Data: Frequency Distribution

upon this table, in Section 2-3.) In most cases, of course, there are more than two possible categories, and choosing the categories is often the first problem to be faced. To illustrate a more complex situation, let us consider the problem of the manager of a hospital blood bank in a large city. He has data for the last 52 weeks on how many units (pints) of type O-positive blood were needed. The manager can partially control his supply by calling in volunteer donors, drawing from a regional supply, or paying for blood. (The average adult has 8 to 10 pints of blood. Certain types sell, in large cities, for up to $50 per pint, or as one wag put it, $500 for all of it.) Thus, the manager is interested in examining the data on blood usage to help him formulate an inventory policy. Table 2-1 gives the details.

Table 2-1 Weekly Blood Usage (Type O Positive) at Central Hospital, 1975[a]

Week No.	Usage	Week No.	Usage	Week No.	Usage	Week No.	Usage
1	104	14	89	27	68	40	104
2	61	15	76	28	97	41	98
3	72	16	79	29	91	42	86
4	85	17	85	30	86	43	81
5	92	18	82	31	92	44	67
6	76	19	81	32	89	45	86
7	88	20	78	33	65	46	87
8	81	21	84	34	77	47	90
9	96	22	80	35	83	48	85
10	98	23	87	36	83	49	88
11	112	24	94	37	89	50	94
12	91	25	54	38	92	51	74
13	82	26	115	39	90	52	91

[a] Usage is given in units (pints).

Table 2-1 is difficult to understand because of the large number of data. However, a frequency distribution using all possible outcomes will not be of much help, since most of the outcomes occur only once and no outcome occurs more than three times. If there were relatively few possible occurrences, we could simply write the number of times each occurred, and thereby produce a frequency distribution. However, in this case writing the number of times each blood usage occurred results in essentially no data summarization, even though it would be a frequency distribution. To obtain a more useful frequency distribution we group events into *classes*. To illustrate what is meant by classes, three different frequency distributions for our data, using three different sets of classes, are shown in Table 2-2.

Table 2-2 Three Frequency Distributions for Weekly Blood Usage Using Three Different Classifications[a]

Classification 1		Classification 2		Classification 3	
Blood Usage	*Frequency*	*Blood Usage*	*Frequency*	*Blood Usage*	*Frequency*
Less than 75	7	50–54	1	50–59	1
75 to 99	41	55–59	0	60–69	4
100 or more	4	60–64	1	70–79	7
Total	52	65–69	3	80–89	22
		70–74	2	90–99	14
		75–79	5	100–109	2
		80–84	9	110–119	2
		85–89	13	Total	52
		90–94	10		
		95–99	4		
		100–104	2		
		105–109	0		
		110–114	1		
		115–119	1		
		Total	52		

[a] Usage is given in units (pints).

Classification 1 includes *open-ended classes*, such as "100 or more." The other classifications are *closed-ended* or *bounded classes*, and each class includes the same number of possible outcomes. Bounded classes provide a preferable method of assigning classes, but it may not be possible to use bounded classes if there are a few data points distant from the other data. For example, if we had a usage of 12 pints during one period, many classes with zero frequency would be needed between 12 and 54. In this case we would use an open-ended class such as "less than 50." In addition, if the blood usage had included several readings above 115, we might use a few categories of larger size above 100, such as "100–119, 120–139, and 140 or above." Unfortunately, there is an art to choosing the classification to use, and there is no one correct way to do it. This is especially unfortunate since the choice of classes determines what the frequency distribution will look like and, thus, what conclusions will be drawn from it. We will return to this point in the "Student Should" at the end of this section.

Another characteristic of a classification scheme is the *class interval*, or how "wide" each class is. For example, classification 2 has a *class interval* of 5, since 5 possible outcomes are included in each class. We can also define *class limits* and the *midvalue* for each class. One of the classes, 60–64 for example, has a *nominal lower class limit* of 60, a *nominal upper class limit* of 64, and a *midvalue* of 62.

Summarizing Data: Frequency Distribution

A classification scheme can be used for either discrete or continuous data. In the blood-usage case we have only a few possible outcomes, and the classification uses discrete classes. In some situations, there are many possibilities in each class, but the terms above are still used in the same way. For example, the daily usage of water by a small city can take on any value in a class. One class might be "between 2 and $2\frac{1}{2}$ million gallons," and 2.18742 million gallons is a possible outcome. In such a case we say that the classes are *continuous*; the precision with which we can distinguish outcomes is limited only by the precision of measurement. For the class "between 2 and $2\frac{1}{2}$," the lower class limit is 2, the upper class limit is $2\frac{1}{2}$, and the midvalue is $2\frac{1}{4}$.

In practice, the distinction between continuous and discrete classes is frequently unclear. For example, water may be measured only to the nearest gallon, then the class 2 to $2\frac{1}{2}$ million gallons really should be 2,000,000 to 2,499,999. However, no one cares whether 2,499,999 gallons is mistaken for $2\frac{1}{2}$ million or vice versa. Thus we use 2 to $2\frac{1}{2}$ million for convenience. (When to use continuous classes for convenience is up to the individual making the summary.) One last notion regarding discrete classes is that the class 60–64 has *nominal class limits* of 60 and 64, as noted above. For graphical purposes we would like to use all real numbers, not just the integers. To do this, the *true class limits* are defined by 59.5 and 64.5. That is, when discrete classes are used, *true class limits* are the limits containing the numbers that when rounded off would be in the class. In a continuous classification there is no distinction to be made.

But we still have not chosen a classification scheme for the hospital manager to use in choosing a blood inventory policy. The issue hinges on how much summarization to use. If the data are summarized too much, as in classification 1, the information loss is too great, but if the summarization is insufficient, as, for example, if each possible outcome is treated as a class, it will be hard to use the data to reach any conclusions. Once again, there is no single answer as to how much summarization to use, but there is an heuristic. (An heuristic is a fancy word for "rule of thumb" or "good guess."One function of any book should be to introduce a few complicated words for easy ideas. We will try to limit this practice, and to warn you when we engage in it.) A rule of thumb for the number of classes to use is given by *Sturges' rule*.

Sturges' Rule
The number of classes to use = $1 + 3.3 \log(n)$, where n is the number of data points you have (52 in the example).

The symbol "$\log(n)$" refers to the logarithm, base 10, of the number n. Logarithms are explained in algebra books, and the value can be found in tables. A few values for Sturges' rule are as follows:

Number of Data Points (n)	log (n)	Number of Classes	
10	1.0	1 + 3.3(1) =	4.3 (use 4)
25	1.398	1 + 3.3(1.398) =	5.6 (use 6)
50	1.699	1 + 3.3(1.699) =	6.6 (use 7)
100	2.0	1 + 3.3(2) =	7.6 (use 8)
10,000	4.0	1 + 3.3(4) =	14.2 (use 14)
1,000,000	6.0	1 + 3.3(6) =	20.8 (use 21)

You will notice that the number of classes reaches 21 only when there are 1,000,000 data points. In our example, where $n = 52$, Sturges' rule implies seven classes, indicating that classification 3 should be chosen unless there is some strong reason to use another number. (At last, our manager has something to look at.) Classification 3 is restated in Table 2-3, and we use it now to introduce the notion of *relative frequency*.

Table 2-3 Frequency and Relative Frequency of Blood Usage (Type O Positive) at Central Hospital, 1975[a]

Blood Usage	Frequency	Relative Frequency
50–59	1	$\frac{1}{52}$ = 0.01923
60–69	4	$\frac{4}{52}$ = 0.07692
70–79	7	$\frac{7}{52}$ = 0.13462
80–89	22	$\frac{22}{52}$ = 0.42308
90–99	14	$\frac{14}{52}$ = 0.26923
100–109	2	$\frac{2}{52}$ = 0.03846
110–119	2	$\frac{2}{52}$ = 0.03846
Total	52	$\frac{52}{52}$ = 1.00000

[a] Usage is given in units (pints).

Relative frequencies are obtained for each class (or for each possible occurrence if classes are not used) by dividing the frequency for that class by the total number of data points. Relative frequencies tell us what fraction of the data points fell in a particular class. For example, 80–89 pints was the usage for 42 percent (0.42308) of the weeks during 1975. The notion of relative frequency is intuitively appealing, and we will return to it when we study probability.

We will conclude this section by giving an example of a frequency distribution where the classes use continuous intervals. The manager of a specialty-

Summarizing Data: Frequency Distribution

products machine shop knows that different tasks can take drastically different amounts of time. He wants to study the labor hours required using some past data, as follows:

Job No.	Labor Hours	Job No.	Labor Hours	Job No.	Labor Hours
1	1.46	9	1.40	17	0.94
2	1.92	10	0.71	18	1.10
3	0.95	11	8.51	19	1.19
4	0.96	12	2.14	20	0.61
5	0.64	13	0.81	21	0.97
6	1.12	14	1.23	22	1.42
7	4.89	15	1.65	23	0.80
8	1.85	16	0.80	24	1.82

Using Sturges' rule he decides to use six categories. Since there are some data points that are much higher than the average, he will probably use either unequal class intervals or open-ended classes or both. One reasonable classification and the resulting frequency distribution is as follows:

Task Time (hours)	Frequency	Relative Frequency
From 0.6 to less than 0.9	6	$\frac{6}{24} = 0.250$
From 0.9 to less than 1.2	7	$\frac{7}{24} = 0.292$
From 1.2 to less than 1.5	4	$\frac{4}{24} = 0.167$
From 1.5 to less than 1.9	3	$\frac{3}{24} = 0.125$
From 1.9 to less than 2.3	2	$\frac{2}{24} = 0.083$
2.3 or greater	2	$\frac{2}{24} = 0.083$
Total	24	$\frac{24}{24} = 1.000$

The classes above have unequal class intervals, and the last class is an open-ended class. The first class (0.6 to less than 0.9) means that all two-digit decimals from 0.60 to 0.89 are included. This class has a lower limit of 0.6, an upper limit of 0.9, and a midvalue of 0.75. (If we are certain to deal only in two-digit numbers, we might say that these values are 0.60, 0.89, and 0.745, respectively. This distinction is usually unimportant.) One last point to make regarding this classification is that we have lost some information, particularly the range of values. We should, perhaps, report, along with the frequency distribution, that 8.51 was the maximum value. (We have also lost other information—the order in which the numbers occurred. This is critical if, for example, the task is one on which the worker improves through time.)

S.S. The Tomdike corporation is having a labor dispute. Labor and management both have access to salaries of all 30 employees (including management salaries), and they both have prepared frequency distributions of the data. The data and both frequency distributions are shown below. As an arbitrator, guess which frequency distribution was made by labor and which by management, and prepare another one that you feel is less biased. Comment on the difficulties involved.

Salary data (in numerical order):

$4,800	$4,950	$7,320	$7,900	$8,140	$12,110
$4,800	$5,250	$7,320	$7,900	$8,260	$20,000
$4,800	$5,250	$7,650	$7,950	$8,260	$21,000
$4,800	$5,300	$7,650	$8,000	$12,110	$24,000
$4,850	$5,300	$7,900	$8,000	$12,110	$25,000

Frequency Distributions

	I			II	
Salary	Frequency	Relative Frequency	Salary	Frequency	Relative Frequency
$2,301–$5,300	10	$\frac{10}{30}$	$0–$4,799	0	$\frac{0}{30}$
5,301–8,300	13	$\frac{13}{30}$	4,800–7,300	10	$\frac{10}{30}$
8,301–12,110	3	$\frac{3}{30}$	7,301–11,300	13	$\frac{13}{30}$
12,111–16,000	0	$\frac{0}{30}$	11,301–25,000	7	$\frac{7}{30}$
16,001–19,999	0	$\frac{0}{30}$	Over 25,000	0	$\frac{0}{30}$
20,000–39,999	4	$\frac{4}{30}$			

Solution: Classification I was prepared by labor and II was prepared by management. Neither one gives a clear picture of the salary distribution. Actually, with only 30 data points, a clear picture can be obtained from the original data, but that would not be possible in a larger organization. It is very difficult here to pick an unbiased classification. Sturges' rule tells us to use six classifications, so we might divide the range of salaries into six intervals of roughly equal size. To be fair, the clusters (for example, around 5,000) should be in the middle of a class. A classification as follows might be the result. This form gives a truer picture, but it is certainly not a perfect representation of the salary distribution.

Salary	Frequency	Relative Frequency
$3,001–$7,000	10	$\frac{10}{30}$
7,001–11,000	13	$\frac{13}{30}$
11,001–15,000	3	$\frac{3}{30}$
15,001–19,000	0	$\frac{0}{30}$
19,001–23,000	2	$\frac{2}{30}$
23,001–27,000	2	$\frac{2}{30}$

2-2 Graphical Presentation

As Artemus Ward once said, "It ain't so much the things we don't know that get us in trouble. It's the things we know that ain't so." One form of information that makes us "know" things that are not so is graphical representation of statistical data. In the previous section we saw that frequency distributions can be designed to give faulty impressions. However, numbers do not evoke images as strongly as do charts, pictures, and graphs. Yet it is this very image formation that makes graphical representation very useful in many situations. Thus, one must be very careful when faced with graphical representation of statistical information and even more careful when preparing such a presentation. This section will concentrate first on four simple methods of graphically presenting statistical information: the histogram, the frequency polygon, the frequency curve, and the ogive (cumulative frequency diagram). Second, this section will briefly discuss examples of poor statistical representation, to give some idea of how not to do it.

The histogram, frequency polygon, and ogive all give a pictorial representation of a frequency distribution. The most common of these is the histogram.

> A *histogram* is a diagram consisting of one vertical bar for each class in a frequency distribution. The vertical axis of the diagram (and the height of each bar) gives frequency and/or relative frequency. The horizontal axis represents the value of the class; each bar is centered on the midvalue for that class.

For example, a histogram can be drawn to represent the Table 2-3 frequency distribution for blood usage (Figure 2-1). The highest bar in the histogram has a height of 22 (frequency) or $22/52 = 0.42308$ (relative frequency). The midvalue of the 80–89 class, for example, is 84.5. (In the diagram, the boundary values for this class would be shown as 79.5 and 89.5, so the horizontal axis includes all the real numbers in the relevant range and appears as an unbroken line.) The reader should try to match the height of each bar to a frequency and relative frequency from Table 2-3.

A frequency polygon is another representation of the same information as is shown in a histogram. There is no reason to prefer one over the other in general, but the frequency polygon will lead us to frequency curves, which are used in later chapters. The definition of a frequency polygon is given next, followed by an example using the data from Table 2-3.

> A *frequency polygon* is drawn by plotting points defined by the midvalue of each class and the frequency and/or relative frequency of that class. The frequency polygon is the curve drawn by connecting those points. The axes are the same as for a histogram. An

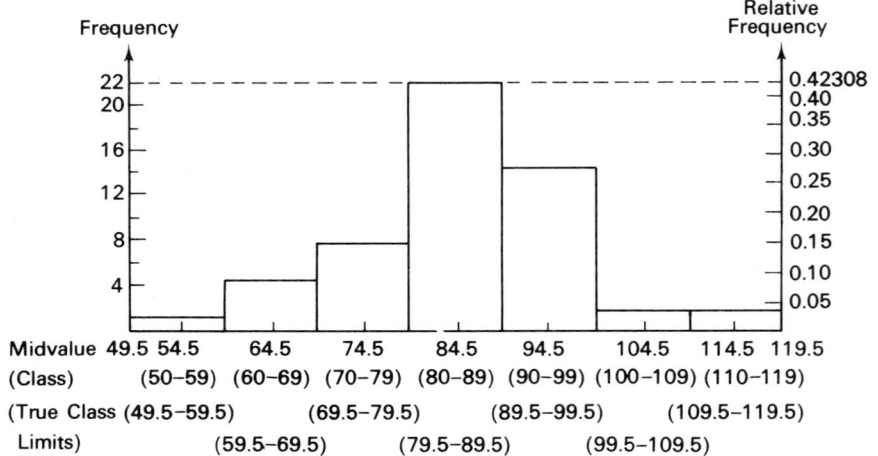

Figure 2-1 Histogram of Weekly Blood Usage (Type O Positive) at Central Hospital, 1975

extra class with zero frequency is usually added so that the polygon begins and ends on the horizontal axis.

In Figure 2-2, an example is given using the data from Table 2-3. An extra class, with zero frequency, has been added on each end to complete the frequency polygon. (If the first or last class is an open-ended class, the frequency polygon cannot be drawn in that class.) The frequency polygon in Figure 2-2 indicates that all occurrences fall in the range 50–120. It also indicates that the most likely occurrence is around 85 and that most values fall in the 80–100 range. We can return to the original data shown in Table 2-1 to see that these statements are correct.

The frequency polygon is characterized by "sharp corners," where the curve bends. It is often helpful, for reasons that will be clearer as we get further into the book, to draw the graph showing frequency and/or relative frequency as a smooth curve. This is done by drawing freehand a smooth curve along the frequency polygon, staying as close to the polygon as is consistent with having a smooth curve. Such a curve is called a *frequency curve*, and an example, based on Figure 2-2, is shown in Figure 2-3. (The classicists in the group may want to think of a frequency curve as the Platonic form of a frequency polygon.)

The values on these axes remain as before only to facilitate drawing the curve. The axis values do not have the same meaning in the frequency curve as the frequency polygon. The important thing is the shape of the curve. Unfortunately, there is no unique way to draw a frequency curve from a

SUMMARIZING DATA: FREQUENCY DISTRIBUTION

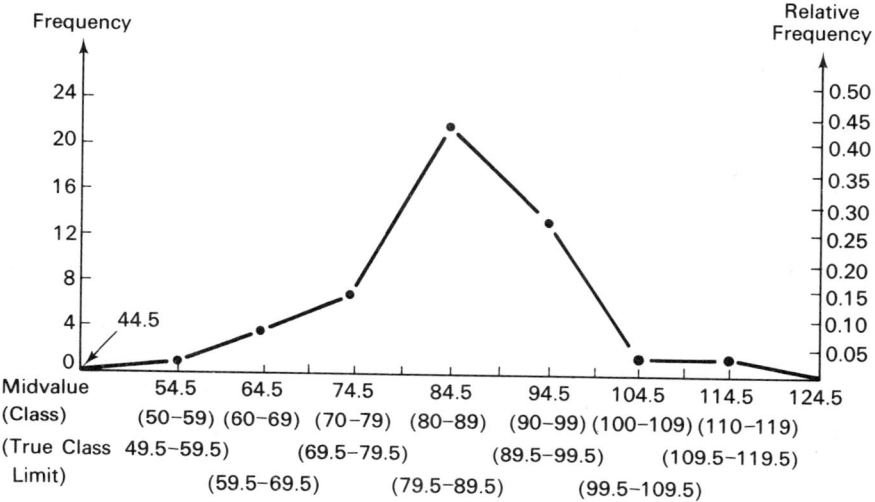

Figure 2-2 Frequency Polygon of Weekly Blood Usage (Type O Positive) at Central Hospital, 1975

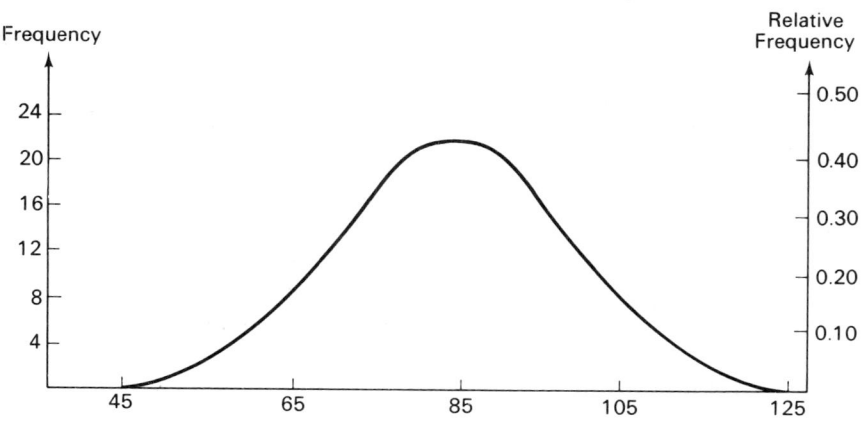

Figure 2-3 Frequency Curve for Weekly Blood Usage (Type O Positive) at Central Hospital, 1975

frequency polygon. Typically, we will try to draw a curve that resembles a curve type that we will study later. The curve in Figure 2-3 is approximately a "normal" curve. For now, you can just try to draw a smooth curve around the frequency polygon.

To help a little, four common curve types are shown in Figure 2-4 with their corresponding frequency polygons. The first curve represents a situation where the most common occurrence is a very small value (near zero), but much larger occurrences are possible. An example is the length of telephone calls (believe it or not). The second curve represents a situation wherein low values are more common than high values. An example might be weekly sales for a particular item. In the third curve, high values are more likely than low values; an example might be class attendance. The final curve is a symmetrical curve, where high and low values are equally common and both sides of the curve look the same. Curves such as this are used to represent errors in producing machined parts, scores on an IQ test, and a host of other things.

The final graphical representation we shall examine here is the ogive or cumulative frequency diagram. It graphically represents the frequency (or relative frequency) with which a value less than (or more than) a particular number occurred. To draw this representation we need to have a table showing both the cumulative frequency of less than each value and the cumulative frequency of more than each value. The definition of an ogive is given next, and Table 2-4 illustrates this idea using the data and class limits from Figure 2-1.

An *ogive* is drawn by connecting points defined by a true class limit value and either (1) the number of sample data values less than that value, for a *"less than"* ogive, or (2) the number of sample data values more than that value, for a *"more than" ogive*. The two types of ogives allow you to read the cumulative frequency up to a particular class limit (or the cumultiave frequency from that value on) directly from the ogive.

Table 2-4 "Less Than" and "More Than" Frequencies for Blood Usage

Blood Usage	Frequency	True Class Limits	"Less Than"	"More Than"
		49.5	0	52
50–59	1	59.5	1	51
60–69	4	69.5	5	47
70–79	7	79.5	12	40
80–89	22	89.5	34	18
90–99	14	99.5	48	4
100–109	2	109.5	50	2
110–119	2	119.5	52	0

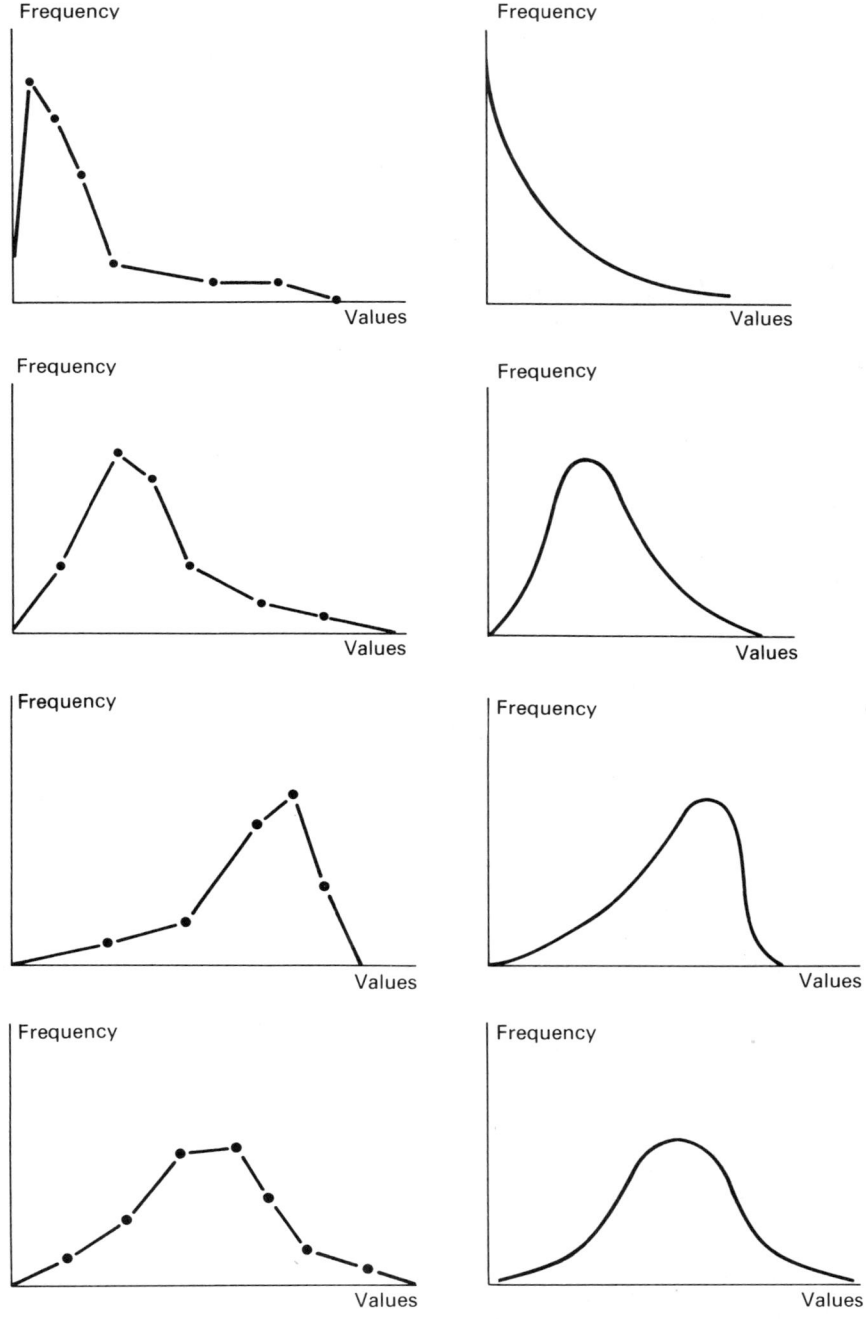

Figure 2-4 Four Common Frequency Curves and Corresponding Frequency Polygons

31

Table 2-4 shows (for example) that 5 weeks had blood usage less than 69.5 and that 18 weeks had blood usage more than 89.5. The true class limits are, for example, 49.5–59.5, with a midvalue of 54.5. The initial class limit, 49.5, must be included. (Class limits from 50 to 120 or 49 to 119 could be used, but using the true class limits of 49.5 to 119.5 is preferable.) The values for the "less than" column are computed by adding the frequency numbers as you proceed down the column. To obtain the numbers in the "more than" column, begin with 52 and subtract the frequency values as you proceed down the column. The corresponding "less than" and "more than" ogives are shown in Figure 2-5.

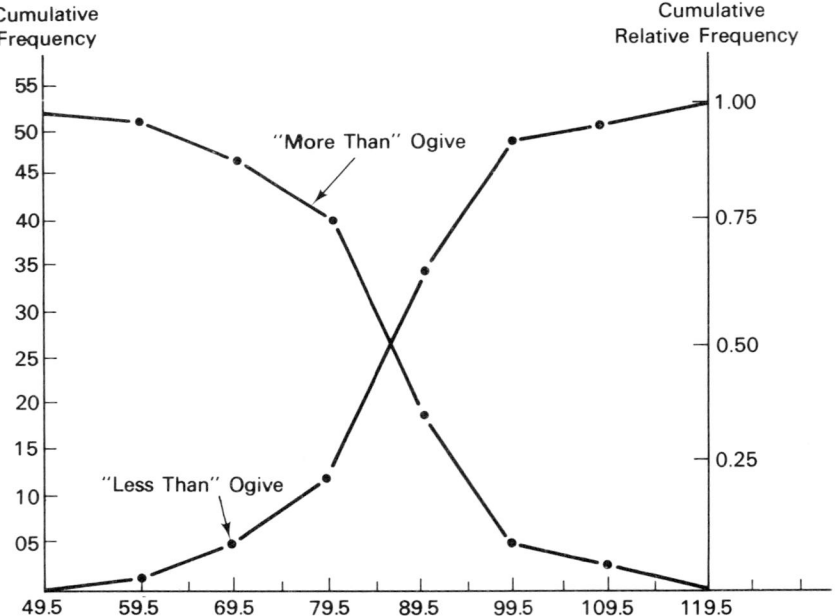

Figure 2-5 Ogives for Blood Usage

Cumulative relative frequency is defined in the same way as relative frequency, using cumulative frequency divided by the total number of data points, 52 in this case. Cumulative relative frequency can never be greater than 1. We can draw a cumulative frequency curve by drawing a smooth curve along the ogive as we did above to obtain a frequency curve.

At this point you might wonder why one would ever want an ogive. The manipulations are easy enough, and the picture does look pretty, but why bother? Rather than discussing the question at length, we will give one

Summarizing Data: Frequency Distribution

example of a problem where the ogive gives better information than any other summary technique discussed so far. Our hospital manager needs to know how many units of blood to stock. Looking at the ogive, he can see that if he had stocked 100 units at the beginning of each week (and made plasma with any leftover blood), he would have run out of blood 4 times (see the "more than" ogive), which means that he would have had a sufficient supply 48 times (see the "less than" ogive). If the future is like the past (an assumption we shall discuss further in Chapter 17), he runs a $\frac{4}{52}$, or 0.07, chance (see the cumulative frequency axis of the "more than" ogive) of running out of blood if he stocks up to 100 units. If he wants more or less safety, he can choose the stock level that yields the required safety level.

We conclude this section by giving one example of poor statistical information. It will be clear, we hope, that standard techniques, such as the methods of this chapter, would convey much better information. We hope that you will present data clearly when the occasion arises and that you will question statistical information that is presented to you. [For an extensive, and amusing, treatment of statistical misrepresentation, see Darrell Huff and Irving Geis, *How To Lie with Statistics* (New York: W. W. Norton & Company, Inc., 1954).]

The example comes from a car advertisement, which said, in effect, that of all the cars the company had produced from 1902 to 1974, 45 percent of them were still on the road in 1974. That statement is remarkably devoid of information. You might be tempted to think: (1974–1902)(0.45) = 32.4, so if I buy a car of this type it will last over 30 years. They neglect to mention that 1973 production is many times that of 1902, and 1963–1973 probably accounts for far more sales than 1902–1945. In any case a more useful representation of the data would be a histogram, showing the ages of cars scrapped in 1973. The information might look like that given in Figure 2-6.

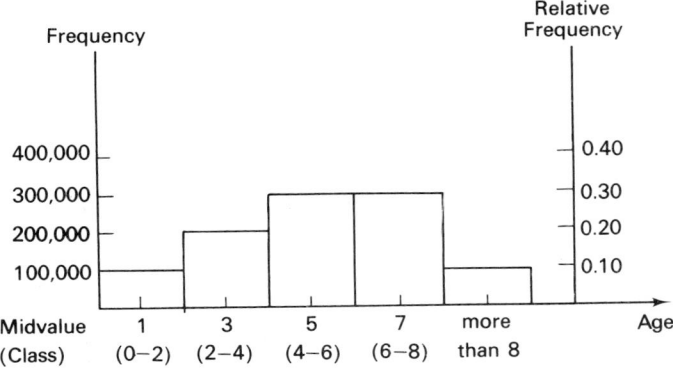

Figure 2-6 Ages of Cars Scrapped in 1973 (Fictitious)

(The values are fictitious.) The information here would imply that we could expect to have only a 10 percent chance of having the car last more than 8 years, a more reasonable expectation than was implied by the advertisement. This histogram still has the weakness that it does not reflect the life expectancy of cars currently on the road, a large fraction of which are new, including many improvements which may extend their life.

S.S. A large automotive-parts distributor is interested in studying the age of accounts that have not been paid (accounts receivable). They have taken a sample of 50 accounts, and they have the following data. You are to draw a histogram, a frequency polygon, a frequency curve, and an ogive (both "less than" and "more than"). The following numbers are the ages for each account, in weeks, to the nearest week:

1, 1, 0, 4, 0, 2, 2, 0, 1, 4, 7, 3, 2,
2, 11, 35, 1, 4, 4, 3, 5, 6, 1, 0, 0,
7, 16, 24, 8, 6, 1, 0, 0, 0, 3, 3, 2,
0, 8, 9, 0, 3, 5, 4, 2, 5, 1, 6, 1, 0.

Because of the high values (up to 35), an open-ended class will be necessary. You need not include that class in the pictorial representations.

Solution: There are 50 numbers, so Sturges' rule suggests that we use seven classes. The last class will be an open-ended class. One set of classes is: 0 and 1, 2 and 3, 4 and 5, 6 and 7, 8 and 9, 10 and 11, and more than 11.

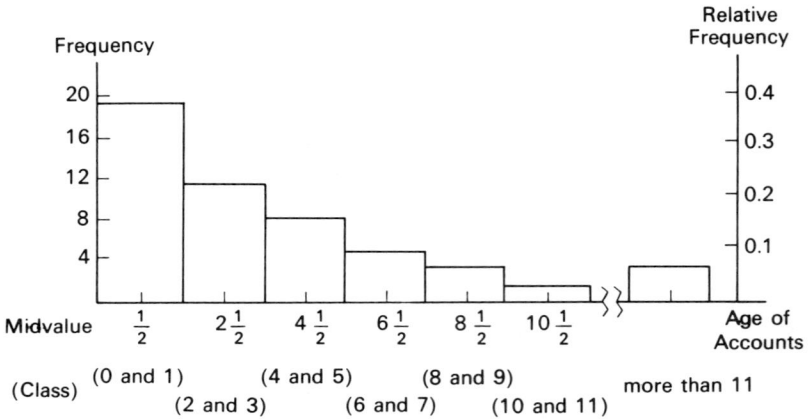

Figure 2-7 Histogram Showing Age of Accounts for an Automotive-Parts Distributor

SUMMARIZING DATA: FREQUENCY DISTRIBUTION   35

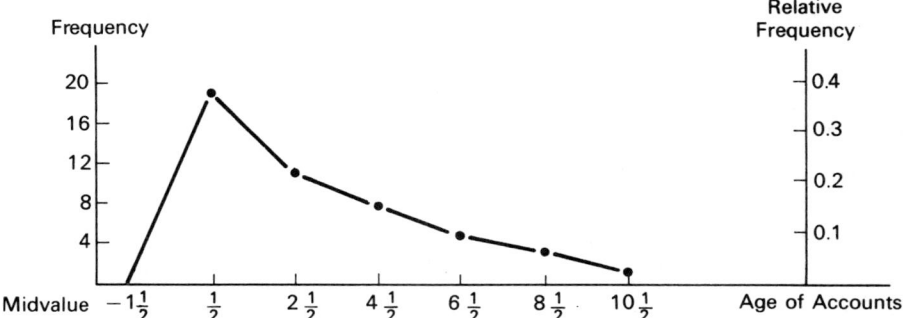

Figure 2-8 Frequency Polygon Showing Age of Accounts for an Automotive-Parts Distributor

Although we can add a category "more than 11" to the histogram (Figure 2-7) by showing a break in the axis, we cannot show that on the frequency polygon (Figure 2-8). The frequency curve and ogive are shown in Figures 2-9 and 2-10. The ogives never reach a relative frequency of 1 (for the "less than" ogive) or zero (for the "more than" ogive) because of the open-ended class.

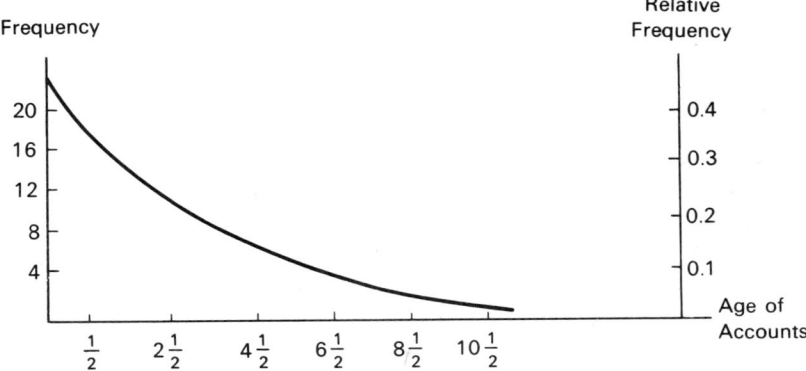

Figure 2-9 Frequency Curve Showing Age of Accounts for an Automotive-Parts Distributor

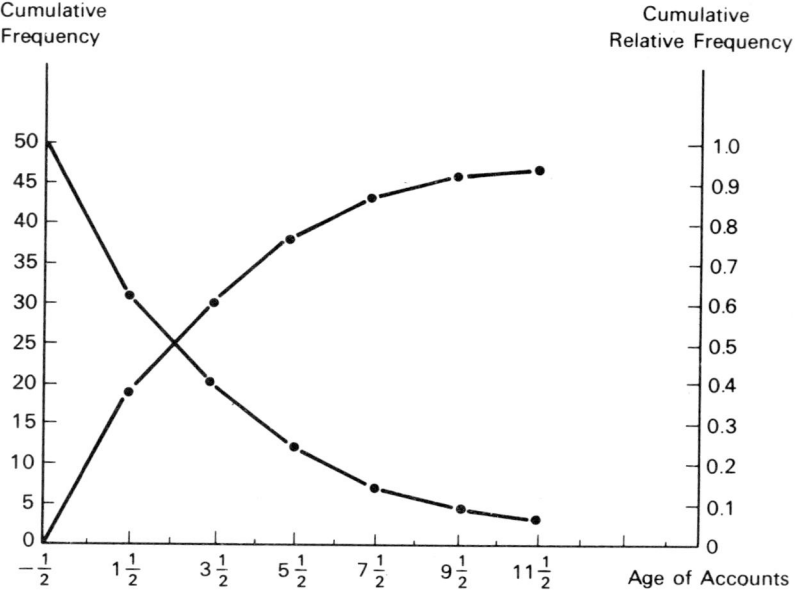

Figure 2-10 "More Than" and "Less Than" Ogives, Showing Age of Accounts for an Automotive-Parts Distributor

2-3 Tabular Presentation

We are all familiar with tabular presentation of statistical data, and the tables used so far in this book illustrate their use. Tabular presentation is, of course, subject to misuse in the same way that graphical presentation is. Hence, the general warning of taking statistical information with a grain of salt applies here. In this section we will, very briefly, discuss formats for tabular presentation of statistical data and, at the same time, we will discuss things to look for in interpreting tables. Several statements can be made about tables as a means of conveying information.

The table should be readable (or as nearly so as possible) even by a person who has not read the report or article (or has skimmed through it quickly). This requirement is often hard to fulfill completely. An example is Table 2-2, which is understandable only to someone who has read the preceding paragraph and knows what the three classifications mean. The reader might be able to figure it out but would have to notice that the three sets of frequencies are consistent before being able to conclude that the same data are used to produce all three classification schemes. It is not necessary to

Summarizing Data: Frequency Distribution

define terms such as frequency in a table; knowledge of common statistical terms can be assumed (which means that you should learn them).

The basic parts of a table include the title, which carries the burden of explanation with it, and various headings attached to the rows and/or columns. The table should be numbered so that the information can be easily referenced in the text. In addition, there may be footnotes, giving further explanatory remarks. In Table 2-2, the footnote indicates the units of quantity in which the numbers of the table are given. Units are usually indicated with the headings; the footnote form is used here so that we have a footnote to discuss. An important footnote that is missing in Table 2-2 is the source of the information. It is of paramount importance to know the source of information that you are using to form judgments. In some situations, for example in company reports, the source of information is obvious and can be omitted.

Finally, the numbers in the table are in the center, surrounded by headings, explanatory titles, and footnotes. Without the headings, titles, and footnotes, the numbers themselves are meaningless. We must know precisely what data are being presented, in what units they are given, and what the source is. Only then can we obtain useful information from the table. As an example, consider Table 2-5.

Table 2-5 Average Hourly Rate for Union and Nonunion Painters

	Area					
	Northeast	*Mideast*	*South*	*Southwest*	*Midwest*	*West*
Union	7.86	7.21	5.89	5.41	6.65	7.02
Nonunion	5.12	4.88	4.58	4.41	5.02	5.12

The weaknesses in the table are almost too numerous to recount. The title is unclear because it leaves out the period of time for which the averages apply and because the meaning of the word "rate" is not well defined. It is important to know whether the rate includes insurance (typically carried by union contractors but not by nonunion painters) and other such items if the rates are to be compared. We also need to know the definition of "union" (does it include apprentices?). Although it is a minor point, the units of the numbers are not given, but we can probably assume that they are dollars. Other important points include the definitions of the headings (Mideast, for example) and the source of the data. It is hard to imagine the use for which these data were prepared. Perhaps someone is making an argument related to the relative wage level of union and nonunion members. In that case, however, it would be necessary to know the relative experience of the two groups, whether insurance and other extras are included in the rate, and the period of time to which the data apply, as well as the all-important source of the data, before drawing any conclusions.

S.S. Discuss Table 2-6.

Table 2-6 Average Hourly Gross Pay for Union and Nonunion Painters, as of June 1975[a]

	Region of the United States					
	Northeast[b]	Mideast	South	Southwest	Midwest	West
Union[c]	$7.86	$7.21	$5.89	$5.41	$6.65	$7.02
Nonunion[d]	5.12	4.88	4.58	4.41	5.02	5.12

[a] Source: Informal guesses made by Joe the painter.
[b] Each region was represented by a single city of size less than 50,000.
[c] Union workers include apprentices.
[d] Nonunion workers include all persons, part-time and full-time, working as painters (including college and high school students).

Solution: First, this solution is really just a list of points to consider. Those points are:

1. This table is much improved over Table 2-5.

2. The table may be less trustworthy now that a source is named. Joe is an honest fellow and a reasonable painter, but there is no reason to trust his informal guesses.

3. If an argument for or against unions is to be made, you must consider whether small cities are typical of the area and whether the sample size is sufficiently large.

4. The percentage of high school and college students represented would be nice to know.

5. It is still not clear whether the same level of service is supplied by both groups. Specifically this might mean: Are the painters self-insured or is the owner liable? The quality and quantity of work may be different due to equipment differences, as well as for qualitative reasons.

The major point is that even after a table looks very precise, there are still many questions that remain.

2-4 A Warning

We have introduced many different methods of summarizing data so that the data can be more useful for decision making. Among these are: frequency distributions, histograms, frequency polygons, frequency curves, ogives, and tables. Each of these can be very useful when used properly. The basic warning that applies to any of these is that in summarizing data, some information has been suppressed. In many cases this is an absolute necessity if a decision maker is to be able to understand the data, so the loss of information is not an evil thing; it is simply something to be aware of.

Computer techniques are available that can cheaply print out histograms, ogives, or other statistical summaries for as many data points as you can imagine. Why on earth, with computers around, do we need to summarize (using computers or not, as the case may be)? Why not use the computers to deal with the original data, in all its complexity? The reason is that computers are not good at dealing with complex decisions in a real environment. It is the computer's role to assist the decision maker as much as possible, by providing understandable information to the decision maker in a timely fashion. That role is also the role we accept for statistics. That is the basis for the second warning: In summarizing data, keep in mind the decision(s) for which the data are needed and the decision maker who will use them. Summaries are useful only if they lead to better decisions.

PROBLEMS

2-1. Why do we need methods for summarizing data? What is gained and lost in the summarization process?

2-2. What is an open-ended class? When is it useful (or necessary)?

2-3. Briefly define frequency and relative frequency. Which notion seems more useful to you?

2-4. Is Sturges' rule a rule that must be followed? Why or why not? Why do we need any such rule at all?

2-5.* Which graphical presentation technique would be most useful if the manager wanted to know
 (a) How likely an out-of-stock situation is?
 (b) What sales values are most likely?
 (c) What frequency curve might be used to describe the data?

2-6. Comment critically on each of the following sets of class intervals:

(a) 20–under 25
26–under 30
31–under 35
etc.

(b) 20–25
50–84
80–114
etc.

(c) 20–25
25–30
30–35
etc.

2-7.* The manager of a wood-crate-assembling plant wanted to find uses for the lumber scraps which the plant accumulates from the production of crates. In order to find viable uses for the scrap, she needed to know the lengths which remained after use of the good lumber in production. She and a group of workers gathered the scraps from 1 day's production and measured the lengths. They found the following lengths to the nearest foot:

Length	Frequency	Length	Frequency
0	4	8	12
1	10	9	26
2	2	10	11
3	25	11	15
4	19	12	2
5	33	13	21
6	25	14	8
7	14	15	3

From this information, the manager constructed the following frequency table:

Frequency of Scrap-Lumber Lengths

Length to Nearest Foot	Frequency
1–6	114
6–12	105
13–20	29
Total	248

Find at least three errors that were made in grouping these data.

2-8.* The management of a corporation claims that its stock is widely held because more than 90 percent is owned by stockholders having 100 shares or less. The following distribution was used to develop their contention:

Number of Shares Held	Percentage of Total Number of Stockholders
1–25	46.61
26–50	23.13
51–100	22.25
101–500	6.94
Over 500	1.07
Total	100.00

Do you agree with management's interpretation of the data? If so, explain how the data support the conclusion; if not, present your interpretation of the data.

2-9. The president of Coldbox Refrigerator Company received data on the number of accidents by employee age during the previous year among all the factory workers in the company's plants. From these data, which follow, she noted that employees 50 years or older seem to have many more accidents than any other age group, and perhaps these employees do not pay a high-enough premium for the company's group accident insurance program. Do you agree with the president's interpretation of the data? Support your position.

Age Group	Number of Accidents	Age Group	Number of Accidents
15–16	15	30–34	196
17–18	96	35–39	259
19–20	111	40–44	183
21–22	102	45–49	218
23–24	208	50 and over	433
25–29	276		

2-10.* A new-car salesman has maintained his daily sales records for the past 1,000 days. He has always sold either 0, 1, or 2 cars, never 3 or more, on a single day. How many classes does Sturges' rule say he should use? How many should he use in your opinion?

2-11. In each of the following situations, describe the kind of data that might be useful and outline the method for organizing the data:
(a) Play, Inc., a national toy manufacturer, is currently spending $20 million for advertising. The president of the company would like to know if the current expenditure is too much or too little.
(b) An investor wants to know to which business forecasting service to subscribe.

2-12. The personnel manager in an automobile plant conducted a study of employees' attitudes with respect to the speed of the assembly line. The results of the study were as follows:

Distribution of Respondents by Their Opinions on the Speed of the Assembly Line and Their Estimates of all Workers' Opinions[a]

	Individual's Estimate of All Worker's Opinions		
Individual's Opinion	Too Fast	Too Slow	Total
Too fast	21	0	21
Too slow	16	2	18
Total			39

[a] Source: Personnel manager's study.

Write a summary of the table's main conclusions and offer critical comments concerning the table.

2-13. The cost accountant for Milkmaid Dairy Company provided the manager of the main plant with the following estimates of the direct labor costs (dollars) of repairing a carton-filling machine for the last 50 breakdowns:

```
21.22  28.86  48.75  35.90  37.62
38.96  25.30  56.15  25.93  61.92
45.72  44.18  54.25  33.60  47.32
27.00  52.25  21.73  61.45  27.23
64.50  31.75  28.57  42.25  33.95
26.22  72.63  48.67  56.15  27.44
75.90  63.27  64.17  71.90  57.18
58.63  41.18  35.40  31.30  61.62
55.40  28.93  67.46  43.50  36.79
48.76  32.67  63.15  55.43  27.62.
```

(a) Form a frequency distribution and draw a frequency polygon using the following classes: $15.00–$24.99, $25.00–$34.99, $35.00–$44.99, $45.00–$54.99, $55.00–$64.99, $65.00–$74.99, and $75.00–$84.99.

(b) Form a frequency distribution and draw a frequency polygon using the following classes: $20.00–$29.99, $30.00–$39.99, $40.00–$49.99, $50.00–$59.99, $60.00–$69.99, and $70.00–$79.99.

SUMMARIZING DATA: FREQUENCY DISTRIBUTION

(c) Note the differences between the polygon in (a) and the polygon in (b).

2-14.* The manager of a small retail store wants to know the distribution of daily sales volume for his establishment. He has collected the daily sales volume (dollars) for 50 days that the store was open in a 2-month period:

115	131	129	96	114
95	111	118	119	97
86	92	101	106	122
152	127	109	134	131
129	126	112	146	134
123	121	98	138	158
99	132	156	119	108
101	107	125	109	116
113	104	117	125	109
136	98	108	132	134

Construct a frequency table, a relative frequency table, and a histogram.

2-15. In order to determine the number of crates of processed cheese to load on trucks, a shipping foreman wants to find the range of weights that he can expect for each crate of cheese. He weighs 100 crates at random and constructs the following frequency distribution:

Weight (pounds)	Frequency	Weight (pounds)	Frequency
20	5	24	4
21	10	25	2
22	42	26	3
23	33	27	1
		Total	100

(a) Assemble a relative frequency table.
(b) Graph the data as a histogram and as a frequency polygon.
(c) Construct ogives for the data. Now consider:
   (1) 22 pounds or less.
   (2) 27 pounds or less.
   (3) 23 pounds or more.
   (4) From 21 pounds to 24 pounds.
(d) Find these cumulative frequencies from the ogives.

2-16.* The production manager for a tire manufacturer has tabulated the daily output of tires for a 30-day period:

$$\begin{array}{cccccc}
100 & 90 & 85 & 84 & 94 & 81 \\
92 & 83 & 78 & 92 & 99 & 84 \\
83 & 102 & 82 & 88 & 82 & 82 \\
89 & 79 & 86 & 96 & 61 & 87 \\
56 & 78 & 83 & 82 & 73 & 95.
\end{array}$$

(a) Construct a frequency and relative frequency table.
(b) Construct a histogram.

2-17. An accounts receivable manager draws a sample of 100 outstanding accounts to determine the number of weeks the balances have been outstanding. The age (weeks) of the balances are as follows:

$$\begin{array}{cccccccccc}
2 & 1 & 1 & 4 & 1 & 2 & 2 & 3 & 1 & 1 \\
1 & 2 & 1 & 2 & 1 & 1 & 1 & 1 & 1 & 25 \\
5 & 3 & 3 & 1 & 5 & 6 & 4 & 2 & 11 & 1 \\
4 & 1 & 2 & 2 & 1 & 1 & 1 & 8 & 6 & 5 \\
1 & 2 & 9 & 1 & 1 & 4 & 17 & 1 & 19 & 1 \\
1 & 2 & 1 & 1 & 1 & 6 & 1 & 1 & 1 & 1 \\
1 & 1 & 1 & 5 & 8 & 1 & 1 & 4 & 1 & 7 \\
2 & 1 & 7 & 6 & 1 & 1 & 1 & 7 & 1 & 1 \\
13 & 10 & 2 & 1 & 1 & 1 & 12 & 8 & 4 & 1 \\
1 & 4 & 2 & 3 & 1 & 1 & 4 & 30 & 3 & 10.
\end{array}$$

(a) Construct a frequency and relative frequency table.
(b) Draw a histogram.

2-18. Financial analysts for Gobroke Associates had estimated for 1975 the cost per unit of output for 200 plants in the same industry:

Cost per Unit of Output	Number of Plants	Cost per Unit of Output	Number of Plants
$1.00–$1.04	8	$1.20–$1.24	42
1.05–1.09	16	1.25–1.29	29
1.10–1.14	29	1.30–1.34	19
1.15–1.19	54	1.35–1.39	3
		Total	200

For 1965, they had estimated the cost per unit of output for 200 plants in this same industry as follows:

SUMMARIZING DATA: FREQUENCY DISTRIBUTION

Cost per Unit of Output	Number of Plants	Cost per Unit of Output	Number of Plants
$0.85–$0.89	11	$1.05–$1.09	33
0.90–0.94	18	1.10–1.14	22
0.95–0.99	41	1.15–1.19	9
1.00–1.04	62	1.20–1.24	4
		Total	200

(a) Construct frequency polygons on the same graph for each of the two frequency distributions.

(b) Combine the two frequency distributions into one frequency distribution and construct a frequency polygon for the combined data.

(c) Does the frequency polygon for the combined data resemble the polygons for each of the years 1975 and 1965?

(d) Of the graphs in parts (a) and (b), which would you believe provides more meaningful information for the analysts?

2-19. A quality control engineer in a light-bulb plant has attempted to determine the difference in the lives of the regular 100-watt bulbs compared to the soft-white 100-watt bulbs. During a previous month she had drawn samples of each bulb and had tested them for the length of life. The following table represents the data that she finally assembled:

Life (hours)	Regular	Soft White
Less than 50	6	4
From 50 to less than 100	18	9
From 100 to less than 150	25	16
From 150 to less than 200	40	29
From 200 to less than 250	5	10
From 250 to less than 300	2	6
From 300 to less than 350	1	3
From 350 to less than 400	1	2
400 and over	2	1
Total	100	80

(a) Construct a relative frequency table and graph frequency polygons on the same graph. What did you use for your vertical axis? Why?

(b) Construct ogives for each type of bulb. At approximately what life are 50 percent of the regular bulbs still burning? Fifty percent of the soft-white bulbs? How did you discover at what life 50 percent of the bulbs were burning?

2-20.* Owing to cooking errors, a number of batches of food prepared in a food-processing plant must be rejected. The plant recently instituted a brush-up course in food preparation in the hope that the program would reduce the number of batches rejected daily. In anticipation of the program's results, the production manager had gathered data for the number of batches rejected daily 30 days directly before the program was instituted. After the retraining program was completed, the production manager again gathered 30 days of data for the number of batches rejected daily. The data appear as follows:

Number of Batches Rejected Daily Prior to the Institution of the Course

29	36	43	42	48
47	42	45	38	46
42	55	21	17	19
51	54	58	21	42
48	57	23	32	51
43	49	19	25	36

Number of Batches Rejected Daily Subsequent to the Completion of the Course

18	11	24	27	8
25	27	14	29	14
26	24	19	16	25
14	21	31	11	21
33	26	14	22	29
51	16	17	26	13

(a) Form frequency and relative frequency tables for each data set.
(b) Draw ogives for the two sets of data.

2-21. A group of 100 students receiving education degrees received salaries on their first job as in the following frequency distribution:

Annual Salary	Frequency
$8,000–$8,499	1
8,500–8,999	3
9,000–9,499	14
9,500–9,999	22
10,000–10,499	32
10,500–10,999	16
11,000–11,499	8
11,500–11,999	4
Total	100

Construct a "less than" ogive from the frequency distribution. If you were going to use this information to describe beginning teachers' salaries, what else would you like to know? (List at least two types of information.)

2-22. Use the data in Problem 2-21 to construct a frequency polygon and draw an idealized frequency curve from that polygon. Why are frequency curves of importance?

2-23.* The sales manager of a small machined-parts manufacturing firm asked a market research group to prepare a summary of the sales to customers, broken down by dollar sales volume. The company has 30 customers, and last year's sales, to the nearest thousand dollars, were:

0, 0, 1, 1, 1, 1, 2, 2, 3, 4, 4, 4, 4, 5, 5,
6, 6, 7, 9, 9, 9, 12, 16, 19, 21, 26, 29, 48, 91, 155.

Sturges' rule suggests six classes. The market research group presents the following frequency distribution:

Sales (nearest thousand)	Frequency
$0–$4	13
5–9	8
10–14	1
15–19	2
20–24	1
25–29	2
30 or more	3

Is this an appropriate summary of the sales to customers for this firm? Why or why not?

2-24. In order to adjust its service charges, a telephone-answering service selected 80 of its calls at random and measured the time that elapsed for each of the calls. The data (minutes) appears as follows:

2.3	3.6	2.4	9.8	6.2	3.1	4.0	2.5
3.7	2.8	6.9	7.1	10.3	10.4	7.0	4.8
2.7	8.0	4.7	2.9	5.3	4.1	3.3	5.1
5.8	6.0	9.5	15.4	3.4	14.2	2.5	8.3
4.5	2.6	3.0	6.7	2.7	2.4	16.1	3.1
1.4	4.5	5.6	1.6	15.3	5.4	6.4	1.9
5.4	3.0	2.3	8.5	10.7	14.1	5.0	4.6
3.2	2.0	7.3	3.4	5.1	9.6	2.6	6.3
4.2	7.2	2.1	8.7	10.9	7.6	3.6	9.2
2.5	6.2	5.9	4.0	2.0	5.5	4.5	2.8

(a) Construct a frequency and relative frequency table.
(b) Construct a histogram.
(c) Construct a frequency polygon and a frequency curve.
(d) Construct ogives.
(e) About what percentage of calls were:
   (1) Less than 5 minutes?
   (2) 9 minutes or more?

2-25.* Critique the following table using the ideas from Section 2-4.

Expectation of Life (years) at Various Ages in the United States, 1972

Age	White		All Other		Total	
	Male	Female	Male	Female	Male	Female
0	68.3	76.0	61.3	69.9	67.4	75.2
20	50.5	57.6	44.4	52.6	49.8	57.0
40	32.1	38.5	28.4	34.6	31.7	38.1
45	27.7	33.9	24.8	30.5	27.4	33.5
50	23.5	29.5	21.4	26.6	23.3	29.2
55	19.7	25.2	18.3	23.0	19.5	25.0
60	16.2	21.1	15.6	19.6	16.1	20.9
65	13.1	17.3	13.3	16.6	13.1	17.2
70	10.5	13.8	11.3	13.9	10.5	13.7

Source: National Centre for Health Statistics, U.S. Department of Health, Education, and Welfare. The data are provisional.

Chapter 3

Summarizing Data: Statistical Descriptions

The Stumptown Office Supply Company (SOS Company) is considering producing a new kind of business form. Before entering that market, they would like to know how many units of this product are sold in their area of distribution. If the total usage is sufficiently large, they will produce the item. They assume that they will be able to win a reasonable share of the potential customers over to the SOS Company. Accordingly, the president requested that a study be made of the current usage of this item.

Mr. Eljay Samoht, head of SOS Company's market research department, investigated and discovered that 6,000 offices in their distribution area use the type of form in question. He decided to call 60 of these offices and see how many of the forms were used by each of the 60 offices. (The 60 were selected "randomly" and he instructed his assistants to try to get a response from each one and to call a substitute office only as a last resort. Sampling techniques will not be discussed here, since Chapter 9 deals extensively with that topic.) Mr. Samoht reported the information he obtained to the president in the form shown in Figure 3-1.

Now the president, unaccustomed to such mystical displays (having not read Chapter 2 of this book), said: "Samoht, I do not understand this mess. First of all, I am interested in future usage, not past usage. Second, I wish you would just give me one or two numbers, something that would be easier to understand than this picture."

Perplexed, Mr. Samoht went back to the basic data, shown in Table 3-1. He knows that the data must be summarized, but in a manner other than a histogram. In addition, he would like to relate these data to the problem at

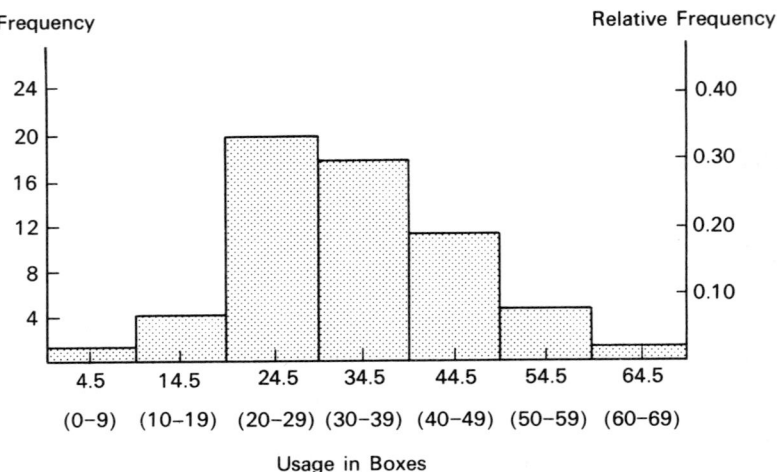

Figure 3-1 Business-Form Usage (in Boxes of 1 Gross Each) During 1975

Table 3-1 Business-Form Usage (in Boxes of 1 Gross Each) During 1975, for a Sample of 60 Firms (usage is shown in ascending order)

8, 12, 13, 17, 19, 20, 20, 21, 21, 21, 21, 21, 22, 22, 22,
24, 25, 25, 25, 26, 27, 28, 28, 28, 29, 31, 31, 32, 32, 32,
34, 34, 34, 34, 35, 36, 37, 37, 37, 37, 38, 39, 39, 40, 41,
42, 43, 45, 45, 45, 46, 47, 47, 48, 51, 53, 54, 57, 59, 63

hand: whether to enter the market for this product or not. This latter question is more complex than finding methods of summarization and is dealt with in later chapters. Statistical descriptions (using a few numbers to summarize many) is the subject of this chapter.

As Mr. Samoht knows, the most common way to summarize data is to give the average value. For example, he knows that the average rainfall in his area is 40 inches per year. For some purposes, however, he would rather know the extreme value rather than the average. For example, the lowest temperature in his area in 1975 was $-12°$ Fahrenheit, and the highest temperature was 102° Fahrenheit. For other purposes, he would like to know the most likely value. For example, the most likely amount of rain on a given day is zero inches. These data-summarization concepts, and several others, will be studied in Sections 3-2 and 3-3, at which time we will be able to help Mr. Samoht summarize his data. Before beginning that study, the ideas of populations and samples must be introduced.

3-1 Samples and Populations

In the SOS Company example, a sample of 60 firms was used, out of a total population of 6,000 firms. Mr. Samoht could not conceivably contact all 6,000 firms during the brief time he had to complete the study, so he had to be satisfied with a sample; this is often the case. If he could have contacted the entire population of 6,000, and if every firm gave him accurate data, he would have known exactly what the total business-form usage in the company's distribution area was during 1975. The qualifying statement "during 1975" is critical, since these were the data he had, not data from 1976 or 1974. We shall discuss this issue further after giving definitions of the terms "sample" and "population."

> A *population* consists of every unit of observation that might be obtained in a particular situation. The set of values of the variable of interest for all members of the population is called the *universe*.
>
> A *sample* is referred to both as the portion of the population selected as data in a particular situation, and the values the variable of interest takes on for those units of observation.

For example, the population in the SOS Company case is the 6,000 firms they could have contacted, and the universe is the 6,000 usage figures they might have obtained. The words "population" and "universe" are frequently used interchangeably; we will make a distinction in this book. But even when the words are used interchangeably, the meaning usually will be clear from the context of the discussion. The sample is the 60 usage figures they did obtain or the 60 firms contacted.

One important fact to note here is that in using the 6,000 usage figures for 1975 as the universe, Mr. Samoht has made a decision to study usage during 1975. When he is done, he will know something about usage during 1975, which is, per se, of no immediate value to him. Since he cannot obtain data from 1976 (the data do not exist as of January 1, 1976), it may seem that he cannot conclude anything about 1976. In defining a population, we must be very precise, and we can only draw valid conclusions about the population defined. However, all is not lost; Mr. Samoht and the president will undoubtedly use their conclusions about 1975 usage to predict usage in 1976 and perhaps beyond. There is nothing wrong with this, and we will study statistical methods for prediction in Chapters 15, 16, and 17. The important point is that a separate leap of faith is necessary to extrapolate from 1975 to 1976. Mr. Samoht must believe that the future will be similar to the past. This is what is needed to answer the president's first objection to Mr. Samoht's first presentation, the histogram. But we have not yet helped Mr. Samoht summarize the sample data in better form.

The purpose of the summary of the sample data is to draw conclusions about the population. This conclusion-drawing procedure is called *inferential statistics* since we seek to infer something about a larger set of things (the population) by examining a summary of a smaller set of things (the sample). (The summary measures based on the sample data are called *statistics*.) Inferential statistics will be discussed extensively in several chapters, including Chapters 11 and 12. In this chapter we are concerned only with methods of describing and summarizing the data in the sample and with further refining the notions of population and sample.

Samples are easy to identify, since they consist of the data that we can see, chew on, and manipulate. Populations are not always so easy to identify. In the above example, the population is a *finite population* in that there is a finite number of data points in the population; that is, it would be possible to obtain all of the data in the population, even though in our example Mr. Samoht chose not to do so. Alternatively, if we are interested in the average temperature during a year, the population is all temperature readings during the year. Since there is an infinite number of times during the year, the population is called an infinite population. There are 31,536,000 seconds in the year, and each second could be subdivided to obtain even more readings. We could not, therefore, write down every bit of data in the population as we could do for a finite population.

Other examples of samples and populations include a sample of 100 parts chosen for inspection from a production lot of 10,000 items and a sample of 500,000 returns to be audited from a population of 50,000,000 submitted tax returns. Sampling is widespread in business and government, and the populations are usually well defined.

S.S. 1. Give two examples of a sample and a population, one with a finite population and one with an infinite population.

2. If a firm is interested in the average time it takes to make a particular handmade item, what sample and population are appropriate?

Solution:
1. If we are concerned about the average temperature in a steam line in a plant at a particular time, we could sample the temperature at five points on the pipe. The population is all the points at which we might sample, which is an infinite population.

If we are concerned about the number of persons watching a particular TV program, the (finite) population is all persons with access to a TV set, and the sample is either the people we contact or their responses.

2. The population is all the items that we could observe to measure the time input. The sample is the items or the times we actually observe.

3-2 Measures of Location

But we have not yet helped Mr. Samoht summarize his data. And what on earth is this "measures of location" stuff? He knows where the data are; he wants them summarized, not located.

Measures of location in statistics are used to indicate what values the data most commonly have, or to characterize where the data tend to concentrate. That is, they indicate (in some sense) where the values of the data are located. The three measures we will examine, to aid Mr. Samoht, are the arithmetic average value (mean), the middle value (the median), and the most common value (the mode). Each of these is relatively easy to compute and understand.

THE MEAN

The most common summary measure for data is the arithmetic mean or average. For example, "the average per capita income in the United States is $5,000" or "a student's cumulative grade-point average is 2.84" are statements involving a mean. To obtain the mean of a sample, we add all the values in the sample and divide by the number of values.

If we have sample data points $x_1, x_2, \ldots, x_n$, where the dots indicate that not all terms in the series are written out, the *mean* is

$$\bar{x} = \sum_{i=1}^{n} x_i/n, \quad \text{where } \bar{x} \text{ denotes the arithmetic mean.} \quad (3\text{-}1)$$

New Symbol	Meaning
$\bar{x}$	The arithmetic mean of a sample.

For an example, suppose that the data points are 4, 5, 5, and 7. (Then $x_1 = 4, x_2 = 5, x_3 = 5, x_4 = 7$, and $n = 4$.) The mean is

$$\sum_{i=1}^{n} \frac{x_i}{n} = \frac{4+5+5+7}{4} = 5.25.$$

As a final example, we will find the mean of Mr. Samoht's sample data (which are shown in Table 3-1):

$$\bar{x} = 8 + 12 + 13 + 17 + \cdots + 57 + 59 + 63 = \frac{2{,}000}{60} = 33.33.$$

THE MEDIAN

The middle value, or median, is typically easier to compute than the mean. It is commonly used, but it should never be mistaken for the mean. Mistaken impressions can be given and received if we do not know which measure of location is being used. This is discussed further in the Student Should at the end of this section. The median is defined to be the data point that has at least half of the data values less than or equal to itself and at least half of the values greater than or equal to itself. It is found by arranging the data in ascending order and counting until at least half of the data points are less than or equal to the value at hand. (If there is an even number of data points, the median is between this data point and the next one.) Formally:

A number, M, is a *median* of a sample if:

1. At least half the sample values are greater than or equal to M.
2. At least half the sample values are less than or equal to M.

If there are two medians with different values, we define M to be the average of the two for convenience. The two values and any number between them would actually satisfy the definition.

New Symbol	Meaning
M	The median value of a sample

We will do several examples.

Sample Values (in ascending order)	Median
1, 2, 3, 4, 5	3
1, 2, 3, 4	$2\frac{1}{2}$ (or any number between 2 and 3 inclusive)
1, 5, 5, 8	5
1, 5, 5, 6, 8	5
1, 5, 5, 7, 8, 9	6 (or any number between 5 and 7 inclusive)
1, 2, 3, 102, 117	3

The last example shows the basic characteristic of the median that differentiates it from the mean. It is unaffected by the size of an extremely large (or small) value. The first and the last samples have identical medians; their means are 3 and 45, respectively. Unfortunately, we cannot say which measure of location is better; it depends on the particular situation.

SUMMARIZING DATA: STATISTICAL DESCRIPTIONS

Finally, we will find the median for business-form usage for Mr. Samoht, using his data from Table 3-1:

median = 33 (the average of 32 and 34).

THE MODE

The last measure of location discussed here is the mode. The mode is the most common value (or values). For the mode to be a sensible measure, there must be enough data that some values are repeated several times. If two, three, or more values occur an equal number of times, and no value occurs more frequently, each value with the maximum number of occurrences is a mode. However, the usefulness of the mode as a measure of location is severely limited in such cases. In fact, the mode is most useful when there is only one mode.

> The *mode*, denoted M_0, is the sample value (or values) that occurs at least as often as any other sample value.

New Symbol	Meaning
M_0	The modal value of a sample.

For example:

Sample Values (in ascending order)	Mode(s)
1, 2, 3, 4, 4, 5	4
1, 2, 3, 4, 5, 5	5
1, 1, 2, 3, 4, 4	1, 4
1, 2, 3, 4, 5, 6	1, 2, 3, 4, 5, 6
1, 2, 2, 3, 3, 3, 4, 4, 5,	3

Finally, for Mr. Samoht, we use the usage data from Table 3-1:

mode = M_0 = 21 (occurs 5 times).

Thus, Mr. Samoht can present a mean (33.33), a median (33), or a mode (21) as a summary or partial summary of his data. Since there are several values that occur 4 times and they are spread through the range of the data, the mode is suspect. Thus, Mr. Samoht considered presenting either the mean or the median (both 33 to the nearest integer) to the president. Since it is more often used, he decided, finally, to report the mean usage of $33\frac{1}{3}$.

S.S. In the Student Should for Section 2-1 we examined salary data for the Tomdike Corporation and found that there are biased ways of presenting data. We will continue that example here. The data for all 30 employees are as follows:

4,800	4,950	7,320	7,900	8,140	12,110
4,800	5,250	7,320	7,900	8,260	20,000
4,800	5,250	7,650	7,950	8,260	21,000
4,800	5,300	7,650	8,000	12,110	24,000
4,850	5,300	7,900	8,000	12,110	25,000

Find the mean, median, and mode of these values. (The sum of the 30 numbers is 278,680.) Comment on the representativeness of the three measures of location for this problem.

Solution:

$$\text{mean} = \bar{x} = \sum_{i=1}^{30} \frac{x_i}{30} = 9{,}289.33$$

$$\text{median} = M = 7{,}900$$

$$\text{mode} = M_0 = 4{,}800 \text{ (occurs 4 times)}.$$

The mode is not representative. It might, however, be a very pertinent statistic for the union to use in bargaining. The median and mean are representative, and we cannot choose the better measure. The amounts of the management salaries have a greater impact on the mean than on the median. A clear bias is introduced by reporting either measure. Labor might want to use the median so that wages will look small. Management might want to use the mean at the bargaining table. We cannot say which value is better, only that more information will be conveyed if we are precise about reporting and careful about reading statistical reports.

3-3 Measures of Variation

Mr. Samoht went to the president and said, "$33\frac{1}{3}$." The president said, "$33\frac{1}{3}$ what?" Mr. Samoht said, more precisely, "The average usage of the business forms in 1975 was $33\frac{1}{3}$ gross in the sample firms." The president was heartened by this news since 6,000 firms times $33\frac{1}{3}$ gross each was 200,000 gross of business forms. The president knew that total market sales of 150,000, or an average of 25 gross for each customer, was sufficient to make him want to enter the market. The average usage in the sample was larger than 25 gross.

At this point, however, the president became worried and asked Mr. Samoht how he could be sure that $33\frac{1}{3}$ was a good guess as to the average

SUMMARIZING DATA: STATISTICAL DESCRIPTIONS

usage in the entire population. He also wondered how likely it is that next year's average will be below 25. The summary is too terse; the president needs to have some measure of how variable these data are.

Measures of variability are usually used to indicate how tightly bunched the sample values are around the mean. A high degree of variability indicates that we might find a very different average if we examined a different or larger sample. (This question will be examined in more depth in Chapter 10.) The amount of variability in a sample is also of value in simply describing the sample. In particular, the shape of the frequency distribution is partially described by measures of variability.

The simplest way to give a measure of variation is to state the range of values (highest and lowest values) that occurred. Another way is to state the value below which (or above which) only 10 percent (or x percent) of the data values fell. In this section we will examine these two measures (range and fractiles, respectively) as well as the mean absolute deviation and the standard deviation. The last two measures of variability are more complicated (but unfortunately more useful) measures of variability.

THE RANGE

The *range* is a measure of variability which displays the lowest and highest values in the sample data.

For example:

Sample Values (in ascending order)	Range
1, 2, 3, 4	1–4
2, 6, 10, 15, 20, 23	2–23
2, 2, 3, 4, 4, 4, 5, 23	2–23

The weakness of the range is shown by the second and third examples. Both have identical ranges, but they are not equally variable. The range is strongly affected by the outlier, 23, in the third set of sample data. Moreover, it is based on only two of the sample observations and thereby ignores much of the sample information.

For the data in Table 3-1, Mr. Samoht might report the mean and the range as a summary of the data.

mean: $33\frac{1}{3}$

range: 8–63.

FRACTILES

Fractiles (and percentiles) are used to indicate what fraction (percent) of the data are below a certain value. Statements such as: "a grade-point average in the top 10 percent" (90th percentile or above) or "an income in the lower ¼ of the population" (25th percentile or below) involve fractiles or percentiles. The median is the 50th percentile or 0.5 fractile, which means that 50 percent of the data values are less than or equal to the median and 50 percent of the data values are greater than or equal to the median. This leads us to a definition of fractiles and percentiles.

The ath *fractile*, denoted F_a (100a percentile, denoted P_{100a}), of a sample, where $0 < a < 1$, is a sample data value such that (1) at least 100a percent of the data values are less than or equal to the ath fractile, and (2) at least $100(1 - a)$ percent of the data values are greater than or equal to the ath fractile. If there are two such data values, the ath fractile will be defined, for convenience, as the average of the two values.

New Symbol	Meaning
F_a, P_{100a}	The ath fractile, $0 < a < 1$, and the 100ath percentile, of a sample; $F_a = P_{100a}$.

For example:

Sample Values	Selected Fractiles and Percentiles
1, 2, 3	$F_{0.10} = 1$, $P_{10} = 1$
	$F_{0.30} = 1$, $P_{30} = 1$
	$F_{0.40} = 2$, $P_{40} = 2$
	$F_{0.50} = 2$, $P_{50} = 2$
1, 2, 4, 6, 12, 17	$F_{0.25} = 2$, $P_{25} = 2$
	$F_{0.75} = 12$, $P_{75} = 12$
	$F_{0.50} = 5$, $P_{50} = 5$

Fractiles are used to measure variability by giving one fractile above and one below the median, indicating how tightly the data values are centered around the median. Quartiles ($F_{0.25}$ and $F_{0.75}$) are the most commonly

Summarizing Data: Statistical Descriptions

used fractiles for this purpose, and $F_{0.25}$ to $F_{0.75}$ is defined to be the *interquartile range*. The interquartile range contains at least 50 percent of the sample data values. For example, in the second sample above, $F_{0.25}$ to $F_{0.75}$ or 2 to 12 is the interquartile range. Thus, a measure of variability in that example is the interquartile range, 2–12.

Using the data in Table 3-1, the interquartile range is 23–41½. Thus, Mr. Samoht could summarize his data as follows:

median: 33 (either the median or the mean can be used)

interquartile range: 23–41½.

Other fractiles could be used for a measure of variability as well as $F_{0.25}$ to $F_{0.75}$.

MEAN ABSOLUTE DEVIATION

Instead of examining how wide the range of the data is, or what range encompasses the middle 50 percent of the data values, the variability of a sample can be expressed using averages. This can be done by computing either the mean absolute deviation (MAD) or the variance (s^2). The mean absolute deviation is a computation of the average difference between the data values and the mean. It is called the absolute deviation because all differences are considered positive; that is, 3 and 7 are both treated as being +2 units away from 5.

The *mean absolute deviation* for a sample $x_1, x_2, \ldots, x_n$, with mean = $\bar{x}$ is computed as

$$\text{MAD} = \sum_{i=1}^{n} \frac{|x_i - \bar{x}|}{n}, \qquad (3\text{-}2)$$

where $|x_i - \bar{x}|$ is the absolute value of the difference between x_i and $\bar{x}$.

(The absolute value means that all differences are considered positive.)

New Symbol	Meaning
MAD	The mean absolute deviation of a sample; the average deviation from the sample $\bar{x}$.

For example:

Sample Values (in ascending order)	MAD Computation
1, 2, 3	$\bar{x} = 2$, so $$\text{MAD} = \frac{|1-2| + |2-2| + |3-2|}{3} = \frac{2}{3}$$
1, 2, 4, 6, 12, 17	$\bar{x} = 7$, so $$\text{MAD} = \frac{|1-7| + |2-7| + |4-7| + |6-7| + |12-7| + |17-7|}{6}$$ $$= \frac{6+5+3+1+5+10}{6} = 5$$

Mean absolute deviation, standard deviation and variance are not intuitively easy to understand. In fact, you may wonder why you are subjected to them at all. The reason is that they are much more important in applications than ranges or fractiles, and this will become clear later in the book.

To obtain MAD for the sample data in Table 3-1 is tedious, but possible.

$$\bar{x} = 33\tfrac{1}{3}$$

$$\text{MAD} = \frac{|8 - 33\tfrac{1}{3}| + |12 - 33\tfrac{1}{3}| + \cdots + |63 - 33\tfrac{1}{3}|}{60}$$

$$= \frac{594}{60} = 9.90.$$

Thus, Mr. Samoht could summarize his data as follows:

$$\bar{x} = \text{mean} = 33.33$$

$$\text{MAD} = \text{mean absolute deviation} = 9.90.$$

VARIANCE AND STANDARD DEVIATION

We can measure variability by averaging the absolute deviations. We cannot average the deviations without considering all of them to be positive, since values below the mean would cancel values above the mean, and we would always obtain an average deviation of zero. (For example, the samples 0, 5, 10 and 4, 5, 6 both have average deviations from the mean of zero, but the MAD values would be $3\tfrac{1}{3}$ and $\tfrac{2}{3}$, respectively.)

Summarizing Data: Statistical Descriptions

Another measure of variability that can be (and is) used, is the average squared deviation, which is called the variance. Squaring each deviation from the mean results in positive numbers, and these numbers are then averaged. (This avoids the problem of average deviations, which equal zero.) The standard deviation, an extremely important measure of variation, is the square root of the variance. Both of these measures are more common and easier to deal with mathematically than MAD.

The variance of a sample data set, denoted by s^2, is computed as

$$s^2 = \sum_{i=1}^{n} \frac{(x_i - \bar{x})^2}{n}, \qquad (3\text{-}3)$$

where $x_1, \ldots, x_n$ are the n sample values and $\bar{x}$ is the sample mean.

The standard deviation of a sample data set, denoted by s, is computed as

$$s = \sqrt{s^2} = \sqrt{\sum_{i=1}^{n} \frac{(x_i - \bar{x})^2}{n}} \qquad (3\text{-}4)$$

New Symbol	Meaning
s^2	The variance of a sample; the average squared deviation from the sample mean.
s	The standard deviation of a sample. The square root of s^2.

For example:

Sample Values (in ascending order)	Variance and Standard Deviation
1, 2, 3	$\bar{x} = 2$, so $$s^2 = \frac{(1-2)^2 + (2-2)^2 + (3-2)^2}{3} = \frac{2}{3}$$ $$= 0.6667$$ $$s = \sqrt{0.6667} = 0.8165$$
1, 2, 4, 6, 12, 17	$\bar{x} = 7$, so $$s^2 = \frac{6^2 + 5^2 + 3^2 + 1^2 + 5^2 + 10^2}{6}$$ $$= \frac{196}{6} = 32.667$$ $$s = \sqrt{32.667} = 5.715$$

To indicate the type of use to which the standard deviation can be put, for the normal distribution (a special animal we will study extensively in Chapter 8 and beyond), the range from the mean minus 1 standard deviation to the mean plus 1 standard deviation contains slightly more than 68 percent of the values. This type of statement will be refined later and used extensively. For now we simply state that the standard deviation is frequently the best summary value to use in describing the variability of sample data.

There is another formula for finding the variance (and thus the standard deviation) of a sample data set that is easier to use for hand computation. It is equivalent to formula (3-3). Frequently, computers will do the calculations for you, and you will not need this formula; we provide this formula only for your use in hand calculation of s^2 or s.

$$s^2 = \left(\sum_{i=1}^{n} \frac{x_i^2}{n} \right) - \bar{x}^2 \qquad (3\text{-}5)$$

and

$$s = \sqrt{s^2} = \sqrt{\left(\sum_{i=1}^{n} \frac{x_i^2}{n} \right) - \bar{x}^2}. \qquad (3\text{-}6)$$

Using the data in Table 3-1 as an example:

$$\bar{x} = 33.33 \quad \text{and} \quad \bar{x}^2 = 1{,}111.11$$

$$\sum_{i=1}^{60} \frac{x_i^2}{n} = \frac{8^2 + 12^2 + \cdots + 63^2}{60} = \frac{75{,}466}{60} = 1{,}257.767.$$

Thus,

$$s^2 = 1{,}257.767 - 1{,}111.111 = 146.656$$

and

$$s = \sqrt{146.656} = 12.11.$$

Mr. Samoht can now summarize his data using the sample mean and the standard deviation. This is a very common and useful method of summarizing sample data.

$$\text{mean} = 33.33$$

$$\text{standard deviation} = 12.11.$$

At this point, the president of SOS has a measure (indeed, several measures) of both location and variability. They tell him that a value of the population mean less than 25 is certainly possible. We do not know how, yet, to be more precise about this statement, but this type of question will occupy much of our time in subsequent chapters.

One final measure of variability that we will introduce here is really just a modification of the standard deviation. The *coefficient of variation*, $V = s/\bar{x}$, is the standard deviation divided by the mean. It shows how large the variability is relative to the average value. It is important since an s value of 2 is huge if $\bar{x} = 3$, important if $\bar{x} = 10$, and trivial if $\bar{x} = 4{,}000$.

New Symbol	Meaning
V	The coefficient of variation; $V = s/\bar{x}$. A measure of relative variability.

If, at this point, you think there are so many measures of variability that you cannot remember them all, much less understand them (the president of SOS, while reaching for an aspirin, would certainly agree), take heart. You will remember them and understand them better after seeing them in use and computing a few. You are not expected to understand exactly how to interpret them at this time; we will discuss them further in succeeding chapters.

S.S. Compute the mean, median, mode, range, interquartile range, mean absolute deviation, standard deviation, and variance of the following sample data set. The data represent the hours per week spent watching TV in a small sample of individuals taken by a telephone survey.

$$12,\ 0,\ 7,\ 4,\ 20,\ 12,\ 8.$$

Solution: First, place the data values in ascending order:

$$0,\ 4,\ 7,\ 8,\ 12,\ 12,\ 20.$$

Then,

$$\text{mean} \equiv \bar{x} = \frac{0 + 4 + 7 + 8 + 12 + 12 + 20}{7} = \frac{63}{7} = 9$$

$$\text{median} \equiv M = 8$$

$$\text{mode} \equiv M_0 = 12 \text{ (the mode is not very useful here)}$$

range: 0–20

interquartile range: 4–12

mean absolute deviation

$$\equiv \text{MAD} = \frac{|0-9| + |4-9| + |7-9| + |8-9| + |12-9| + |12-9| + |20-9|}{7}$$

$$= \frac{34}{7} = 4.857$$

$$\text{variance} \equiv s^2 = \left(\sum_{i=1}^{7} \frac{x_i^2}{n}\right) - \bar{x}^2$$

$$= \frac{0^2 + 4^2 + 7^2 + 8^2 + 12^2 + 12^2 + 20^2}{7} - 9^2$$

$$= \frac{817}{7} - 81 = 116.71 - 81 = 35.71$$

$$\text{standard deviation} \equiv s = \sqrt{s^2} = \sqrt{35.71} = 5.98.$$

3-4 Describing Grouped Data*

Frequently, data are only available in the form of a frequency distribution where the values have been grouped into categories. In that event we cannot simply add all the values and divide by the number of values, since we do not have the individual values. Instead, we know that there are four values in the 10 to 19 class in Mr. Samoht's data (see Figure 3-1). Since the data are grouped, we have no choice but to assume they average to the midpoint of the class, 14.5 in this example. That is the basic approach used in dealing with grouped data. In fact (see Table 3-1), the four numbers between 10 and 19 inclusive (12, 13, 17, 19) average 15.25, but if we only have the grouped data we would not know that.

Since we have discussed the measures of location and variability, only the formulas for dealing with grouped data and any potential problems will be given in this section. The ideas are the same, and the basic approximation made (which you must keep in mind) is discussed in the preceding paragraph. Figure 3-1 will be used as an example so that you can see the error introduced by the approximation. Before beginning, we need some symbols (some of which we have seen before).

New Symbol	Meaning
n	The number of data values, $\sum f_i$, in the notation below.
N	The number of classes in the frequency distribution.
m_i	The midvalue of class i, $i = 1, \ldots, N$.
f_i	The frequency of class i, $i = 1, \ldots, N$.
L_i	The lower (true) limit of class i, $i = 1, \ldots, N$.
U_i	The upper (true) limit of class i, $i = 1, \ldots, N$ (L_i and U_i are $29\frac{1}{2}$ to $39\frac{1}{2}$ if, for example, the interval contains the integers from 30 to 39).
C_i	The cumulative frequency up to and including class i.
I_i	The width of class i, equal to $U_i - U_{i-1}$.

---

* This section may be skipped without loss of continuity.

MEAN, GROUPED DATA

The *mean for grouped data* can only be computed if there are no open-ended classes. It is found as follows:

$$\bar{x} = \sum_{i=1}^{N} \frac{f_i m_i}{n}. \qquad (3\text{-}7)$$

For example, using Figure 3-1:

$$\bar{x} = \frac{1(4.5) + 4(14.5) + 20(24.5) + 18(34.5) + 11(44.5) + 5(54.5) + 1(64.5)}{60}$$

$$= \frac{4.5 + 58.0 + 490 + 621 + 489.5 + 272.5 + 64.5}{60}$$

$$= 33.33.$$

The mean computed using grouped data is only an estimate of the true $\bar{x}$, since m_i is assumed to be the mean of each class's values, and this may not be true. In this case they are, coincidentally, equal.

MEDIAN, GROUPED DATA

Once again, we can only estimate the median for grouped data, since we cannot know how values are arranged within a class. The formula used is logical, but it only yields an approximate answer. The logic (which is also used for fractiles below) is that we are after the 50 percent value of the data. Thus, we say that the median would occur halfway through the data, at the $(n/2)$th value $[(n + 1)/2$ would be more correct, but the analogy to 50 percent does not hold then]. For example, $n/2 = 60/2 = 30$ in our example. The basic idea, then, is that if the class contains 18 data values, and there are 25 values below it, we need to place the median a fraction equal to $(30 - 25)/18$ into the class interval. The class interval is taken to be $29\frac{1}{2}$–$39\frac{1}{2}$, using the true class limits.

The formula for the *median for grouped data* is

$$\text{median} \equiv M = L_j + \frac{(n/2) - C_{j-1}}{f_j} I_j. \qquad (3\text{-}8)$$

where j is the median class and $j - 1$ is the class below that.

For example, using the data from Figure 3-1:

$$\text{median} \equiv M = 29.5 + \frac{30 - 25}{18}(10) = 32.28.$$

MODE, GROUPED DATA

The *modal class* is the class with the largest number of values.

Since the mode itself is of less importance than the other measures of location, we will not pursue the mode further here. The modal class in Figure 3-1 is 20–29.

RANGE, GROUPED DATA

The range of grouped data cannot be given exactly, and it cannot be given at all if there are open-ended classes. For closed-ended classes, the upper limit of the highest class and the lower limit of the lowest class can be used to give a range that is guaranteed to be at least large enough to contain all data values. For example, 0–69 in the case of the Figure 3-1 data, does contain all the data.

INTERQUARTILE RANGE, GROUPED DATA

In the same manner in which we estimated the median, which is the 50th percentile, we can estimate the 25th and 75th percentile. For example, using Figure 3-1, the 20–29 ($19\frac{1}{2}$–$29\frac{1}{2}$) interval has $5/60 = 0.083$ of the values below its lower class limit, and $25/60 = 0.417$ of the values less than or equal to its upper class limit. The 25th percentile is estimated to be halfway into the interval since 0.25 is halfway between 0.083 and 0.417. So $24\frac{1}{2}$ is our estimate of the 25th percentile, the lower limit of the interquartile range. We could similarly estimate the upper limit (75th percentile) to be between $39\frac{1}{2}$ (with 17 values above this lower class limit) and $49\frac{1}{2}$ (with 6 values above this upper class limit). The 75th percentile is approximately $41\frac{1}{2}$. This is a process that can be understood intuitively (find the right class interval and then add the correct fraction of the interval to the true lower class limit), and, since it is not as important a measure of variation as the ones that follow, we will not formalize the process.

MEAN ABSOLUTE DEVIATION, GROUPED DATA

As with the mean, the MAD can only be calculated for grouped data where all classes are closed-ended classes. Then

$$\text{MAD} = \sum_{i=1}^{N} \frac{f_i |m_i - \bar{x}|}{n}. \tag{3-9}$$

SUMMARIZING DATA: STATISTICAL DESCRIPTIONS

For example, using the data in Table 3-1, with $\bar{x} = 33.33$ from above:

MAD

$$= \frac{1|4.5 - 33.33| + 4|14.5 - 33.33| + 20|24.5 - 33.33| + 18|34.5 - 33.33| + 11|44.5 - 33.33| + 5|54.5 - 33.33| + 1|64.5 - 33.33|}{60}$$

$$= \frac{28.83 + 75.32 + 176.6 + 21.06 + 122.87 + 105.85 + 31.17}{60}$$

$$= \frac{561.7}{60} = 9.36$$

VARIANCE AND STANDARD DEVIATION, GROUPED DATA

These measures cannot be calculated if any classes are open-ended, since no midvalue exists for the open-ended classes.

The formulas for s and s^2 for grouped data are

$$s^2 = \sum_{i=1}^{N} \frac{f_i(m_i - \bar{x})^2}{n} \qquad (3\text{-}10)$$

$$s = \sqrt{s^2} = \sqrt{\sum_{i=1}^{N} \frac{f_i(m_i - \bar{x})^2}{n}}. \qquad (3\text{-}11)$$

For example, tediously using Figure 3-1 data:

$$s^2 = \frac{1(4.5 - 33.33)^2 + 4(14.5 - 33.33)^2 + 20(24.5 - 33.33)^2 + 18(34.5 - 33.33)^2 + 11(44.5 - 33.33)^2 + 5(54.5 - 33.33)^2 + 1(64.5 - 33.33)^2}{60}$$

$$= \frac{831.17 + 1418.28 + 1559.88 + 24.64 + 1372.45 + 2240.84 + 971.57}{60}$$

$$= \frac{8{,}418.83}{60} = 140.3$$

$$s = \sqrt{140.3} = 11.85.$$

A summary of all the calculations regarding SOS Company's data is as follows:

	Ungrouped Data	Grouped Data
Mean, $\bar{x}$	33.33	33.33
Median, M	33	32.28
Interquartile range	23–41½	24½–41½
Mean absolute deviation, MAD	9.40	9.36
Standard deviation, s	12.11	11.85

As you can see, the differences are not large; further, the direction of the difference is not always the same and cannot be predicted. The more data classes there are, the smaller the difference between the ungrouped and grouped calculations will be. The grouped data calculations are necessary if further summarization of a frequency distribution is desired.

S.S. 1. In the previous section we considered the following data for TV watching, based on a telephone survey: 0, 4, 7, 8, 12, 12, 20. Using as classes 0–5, 6–11, 12–17, and 18–23, form a frequency distribution and use it to compute the mean, median, mean absolute deviation, and standard deviation.

2. Given your frequency distribution in part (1), find data values that fit into that distribution but yield an $\bar{x}$ estimate as far from the $\bar{x}$ value computed in part (1) as possible.

Solution:

1.

Class	Frequency
0–5	2
6–11	2
12–17	2
18–23	1

$$\text{mean} \equiv \bar{x} = \frac{(2.5)(2) + (8.5)(2) + (14.5)(2) + (20.5)(1)}{7} = \frac{71.5}{7} = 10.2$$

$$\text{median} \equiv M = 5.5 + \frac{3.5 - 2}{2} 6 = 10.0$$

Summarizing Data: Statistical Descriptions

mean absolute deviation $\equiv$ MAD

$$= \frac{2|(2.5 - 10.2)| + 2|8.5 - 10.2| + 2|14.5 - 10.2| + 1|20.5 - 10.2|}{7}$$

$$= \frac{37.7}{7} = 5.4$$

variance $\equiv s^2 = \dfrac{2(2.5 - 10.2)^2 + 2(8.5 - 10.2)^2 + 2(14.5 - 10.2)^2 + 1(20.5 - 10.2)^2}{7}$

$$= \frac{118.6 + 5.8 + 37.0 + 106.1}{7} = \frac{267.5}{7} = 38.2$$

standard deviation

$$\equiv s = \sqrt{38.2} = 6.2.$$

2. The $\bar{x}$ would be as far from 10.2 as possible if every value were at the lower class limit (or if every value were at the upper class limit). For example: 0, 0, 6, 6, 12, 12, 18 and 5, 5, 11, 11, 17, 17, 23 both fit the frequency distribution. These two data sets have $\bar{x}$ of 7.71 and 12.71, respectively. The low values, with $\bar{x} = 7.71$, show how wrong the grouped data $\bar{x}$ calculation can be $(10.2 - 7.71 = 2.49)$.

3-5 Skewness and Kurtosis*

Every book (and course) should have a few fancy words tossed in to mystify the masses. Skewness and kurtosis seem to fit this requirement, but, as we will see, they can be helpful in describing sample data. Measures of location indicate the number around which the sample data are centered; measures of variability indicate how close the sample data are, on the average, to the mean or median. Skewness indicates whether the histogram "leans to the left" or "leans to the right," and kurtosis indicates how peaked it is. They are calculated in a manner similar to s^2, as we will see, but they are not as easily understood. They are also not as useful or as common, so the discussion will be brief. To introduce skewness and kurtosis, it is helpful to introduce the concept of the moments of a sample.

---

* This section may be skipped without loss of continuity.

The mean is the *first moment* of the data, in that it measures the average distance from zero. The variance is the *second moment around the mean*, in that it measures the average squared deviation from the mean and can be written as

$$M_2 = \sum_{i=1}^{n} \frac{(x_i - \bar{x})^2}{n} \quad \text{for ungrouped data} \quad (3\text{-}12)$$

or as

$$M_2 = \sum_{i=1}^{N} \frac{f_i(m_i - \bar{x})^2}{n} \quad \text{for grouped data.} \quad (3\text{-}13)$$

There are also third, fourth, fifth, and higher moments around the mean. They are denoted and computed as follows.

The *k*th moment about the mean of a set of sample data is computed as

$$M_k = \sum_{i=1}^{n} \frac{(x_i - \bar{x})^k}{n} \quad (3\text{-}14)$$

for ungrouped data (*n* values) or as

$$M_k = \sum_{i=1}^{N} \frac{f_i(m_i - \bar{x})^k}{n} \quad (3\text{-}15)$$

for grouped data (*N* classes, *n* total values).

New Symbol	Meaning
M_k	The *k*th moment of a set of sample data.

We have use in this section for M_3 and M_4, and s^2 is the same as M_2. We will have no use for $M_5, M_6, \ldots$, and, in fact, very little use for M_4.

SKEWNESS

Figure 3-2 shows three frequency curves, each with different skewnesses. Figure 3-2a is symmetric. It is not skewed (skewness = zero). Figure 3-2b is skewed to the right in that there are more values to the right of the mode than to the left. It is positively skewed. Figure 3-2c is skewed to the left or negatively skewed. If we compute a positive skewness for a sample, we know that the histogram (or frequency polygon) is similar to the curve in Figure 3-2b. Similarly, a negative skewness value implies that the sample histogram

Summarizing Data: Statistical Descriptions

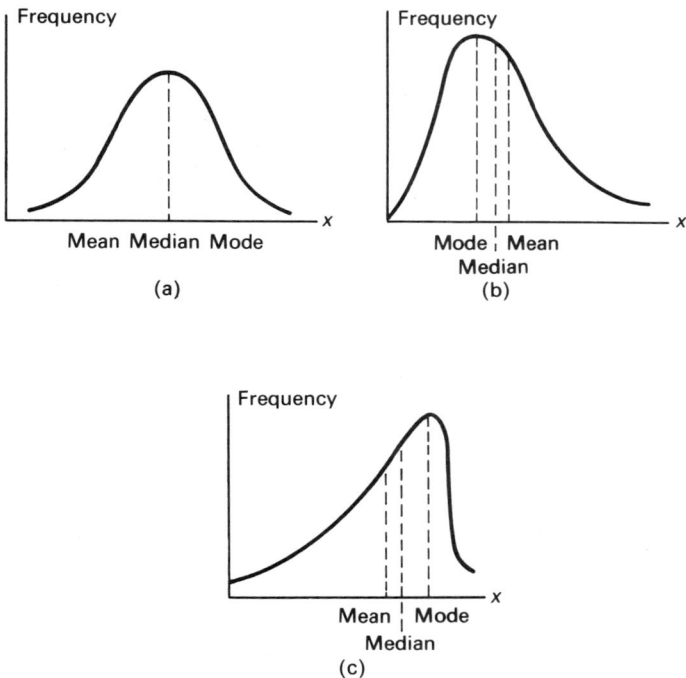

Figure 3-2 Frequency Curves with Zero, Positive, and Negative Skewness

(or frequency polygon) is similar to the curve in Figure 3-2c. The relation of the mean, median, and mode in the three situations in Figure 3-2 is shown in each figure. In Figure 3-2a, all three location measures are at the same point. In the positively skewed frequency curve in Figure 3-2b, mean > median > mode. In Figure 3-2c, mean < median < mode. These relationships always hold for symmetrical, positively skewed, and negatively skewed frequency curves, respectively.

Skewness can be measured in several ways. The easiest method does not require the calculation of the third moment.

$$\text{Sk} \equiv \textit{skewness} = \frac{3(\bar{x} - M)}{s}, \tag{3-16}$$

where $\bar{x}$, M, and s are the mean, median, and standard deviation, respectively, of the sample.

For example, we found that $\bar{x} = 33.33$, $M = 33$, and $s = 12.11$ for Mr. Samoht's data.

Thus,

$$\text{Sk} \equiv \text{skewness} = \frac{3(33.33 - 33)}{12.11} = 0.082.$$

That distribution, as you can see from looking at Figure 3-1, is very slightly skewed to the right. If the grouped data figures are used the skewness is larger, $\text{Sk} = 0.26$.

New Symbol	Meaning
Sk	A measure of skewness in a sample; it indicates the direction in which a frequency distribution (or frequency curve or frequency polygon) leans.

A measure of *relative skewness* involves a ratio of the third and second moments. Where M_3 and M_2 are the third and second moments about the mean of the sample, the following are measures of relative skewness in the sample data:

$$b_1 = \frac{M_3^2}{M_2^3} \quad \text{or} \quad a_3 = \sqrt{b_1} = \frac{M_3}{\sqrt{M_2^3}}. \qquad (3\text{-}17)$$

As before, zero for either b_1 or a_3 implies a symmetric distribution; a negative third moment implies negative skewness, and a positive third moment implies positive skewness. (We must remember to use this test involving M_3 since both a_3 and b_1 are always positive.)

New Symbol	Meaning
b_1, a_3	Measures of relative skewness; $\sqrt{b_1} = a_3$.

For example, using the grouped data formula, and data from Figure 3-1:

$$M_3 = \frac{\begin{array}{c} 1(4.5 - 33.33)^3 + 4(14.5 - 33.33)^3 + 20(24.5 - 33.33)^3 \\ + 18(34.5 - 33.33)^3 + 11(44.5 - 33.33)^3 \\ + 5(54.5 - 33.33)^3 + 1(64.5 - 33.33)^3 \end{array}}{60}$$

$$= \frac{-23{,}902.5 - 26{,}706.1 - 13{,}769.3 + 28.83 + 15{,}330.3 + 47{,}438.65 + 30{,}287.7}{60}$$

$$= \frac{28{,}648}{60} = 477.5.$$

Summarizing Data: Statistical Descriptions

Then (using $s^2 = 140.3 = M_2$ from the grouped data calculation)

$$b_1 = \frac{M_3^2}{M_2^3} = \frac{(477.5)^2}{(s^2)^3} = \frac{(477.5)^2}{(140.3)^3} = 0.083.$$

Also,

$$a_3 = \sqrt{b_1} = 0.287.$$

All three measures of skewness for Mr. Samoht's data indicate positive skewness. (It would be possible for a grouped data measure and an ungrouped data measure to have different signs.) They have different numerical values because they measure skewness in different ways. The easiest way to measure skewness is with the $3(\bar{x} - M)/s$ formula.

With kurtosis, we have no correspondingly easy formula. Kurtosis measures a distribution's peakedness, the degree to which one narrow range of values contains a large fraction of the sample data. Although it may seem that this is what variance measures, distributions with the same variance can have different measures of kurtosis and look different. This is illustrated in Figure 3-3. Figure 3-3a has "normal" kurtosis, while Figure 3-3b and c have low (not peaked) and high (peaked) values, respectively. [Just to be sure that you get your money's worth in big words, these situations are called mesokurtic, platykurtic, and leptokurtic, respectively. As far as the

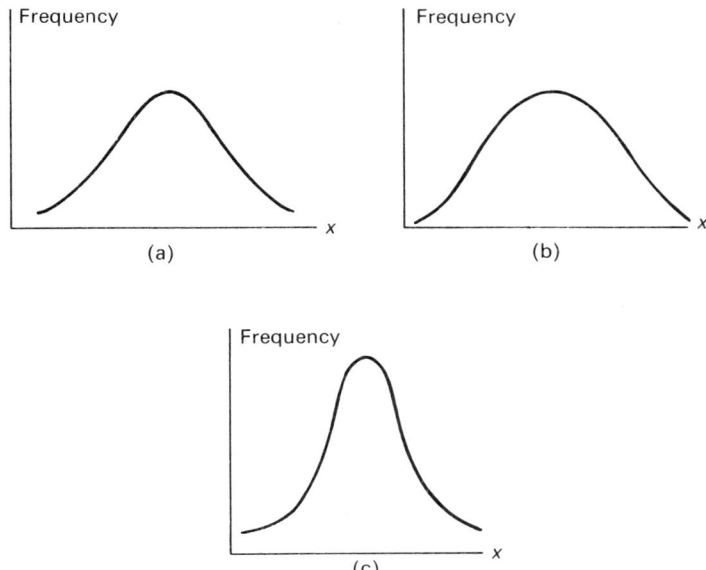

Figure 3-3 Three Frequency Distributions with the Same Mean and Variance but with Different Measures of Kurtosis

authors know (1) the rumor that there is a foundation to help stamp out leptokurtosis is not true, and (2) there probably should be, since the study of kurtosis is not all that useful.]

A measure of *relative kurtosis* can be computed as follows:

$$a_4 = \frac{M_4}{M_2^2} \qquad (3\text{-}18)$$

$a_4 = 3$ corresponds to a curve such as Figure 3-3a, while $a_4 < 3$ and $a_4 > 3$ correspond to Figure 3-3b and c, respectively.

New Symbol	Meaning
a_4	A measure of relative kurtosis; it indicates the degree of peakedness for a frequency distribution, frequency curve, or frequency polygon.

For example, using Figure 3-1 data:

$$M_4 = \frac{\begin{array}{c}1(4.5 - 33.33)^4 + 4(14.5 - 33.33)^4 + 20(24.5 - 33.33)^4 \\ + 18(34.5 - 33.33)^4 + 11(44.5 - 33.33)^4 \\ + 5(54.5 - 33.33)^4 + 1(64.5 - 33.33)^4\end{array}}{60}$$

$$= \frac{690{,}841 + 502{,}875 + 121{,}583 + 34 + 171{,}240 + 1{,}004{,}272 + 943{,}948}{60}$$

$$= \frac{3{,}434{,}793}{60} = 57{,}247.$$

Then

$$a_4 = \frac{M_4}{M_2^2} = \frac{57{,}247}{(140.3)^2} = \frac{57{,}247}{19{,}684} = 2.91.$$

Mr. Samoht's data is nearly mesokurtic (he must be very happy).

S.S. The previous two sections used data for TV watching that can be given in grouped form as follows:

Class	Frequency
0–5	2
6–11	2
12–17	2
18–23	1

Summarizing Data: Statistical Descriptions

We know that $s^2 = M_2 = 38.2$. Also, $\bar{x} = 10.2$ and $M = 10.0$ using the grouped data. Compute a measure of skewness (the easy one) and a measure of kurtosis. Comment on the degree of peakedness in the grouped data and on the direction in which the distribution leans, using the skewness value found.

Solution:

$$Sk \equiv \text{skewness} = \frac{3(10.2 - 10.0)}{6.2} = \frac{0.6}{6.2} = 0.097.$$

The distribution is skewed slightly to the right.

$$M_4 = \frac{2(2.5 - 10.2)^4 + 2(8.5 - 10.2)^4 + 2(14.5 - 10.2)^4 + 1(20.5 - 10.2)^4}{7}$$

$$= \frac{7{,}031 + 17 + 684 + 11{,}255}{7} = \frac{18{,}987}{7} = 2{,}712.43.$$

Thus,

$$a_4 = \frac{M_4}{M_2^2} = \frac{2{,}712.43}{(38.2)^2} = \frac{2{,}712.43}{1{,}459} = 1.86.$$

The value indicates, and the data show, that the sample values are not peaked, but spread out.

3-6 A Warning

This chapter discusses statistical methods of summarizing data, using a few numbers. The most useful of these are the mean (a measure of location) and the standard deviation (a measure of variation). The median, range, and mean absolute deviation will be useful in many situations as well.

The concepts of population and sample are discussed in this chapter. The warning that applies here is that you must try to avoid generalizing from a sample taken from one population to a second population; you must be careful even about generalizing from a sample to the correct population. Sometimes generalizations to a second population are necessary (you cannot have a sample from the future, for example), but extreme caution should be exercised in making these generalizations.

As with Chapter 2, when using a few numbers to summarize a large amount of data, some information is lost. This must be done because people (like the president of SOS Company) cannot deal effectively with a large amount of raw data. You should always be aware of possible biases introduced by the summary, however, and be prepared to report (or ask for) the original data if necessary. Care must be exercised in selecting the summary measures of location and variability, and there are no hard and fast rules for this selection. A manager must obtain enough information to deal with the decision under consideration. Summaries, as well as the original data, should be designed to aid the decision maker.

New Symbols

Symbol	Meaning
M_0	The most common value in a sample, the mode. If there are ties, each such value is a mode.
F_a, P_{100a}	The ath fractile, $0 < a < 1$, and $100a$th percentile, $F_a = P_{100a}$, of a frequency distribution; at least a of the values, $100a$ percent, are less than or equal to F_a, and at least $1 - a$ ($100 - 100a$ percent) are greater than or equal to F_a.

Key Formulas

Formula	Used to Compute
$\bar{x} = \sum_{i=1}^{n} \dfrac{x_i}{n}$	The arithmetic mean for ungrouped data.
$\text{MAD} = \sum_{i=1}^{n} \dfrac{|x_i - \bar{x}|}{n}$	The mean absolute deviation, from the sample mean, for ungrouped data.
$s^2 = \sum_{i=1}^{n} \dfrac{(x_i - \bar{x})^2}{n}$	The variance of a sample of ungrouped data; the average squared deviation from the mean.
$s^2 = \left(\sum_{i=1}^{n} \dfrac{x_i^2}{n} \right) - \bar{x}^2$	An alternative formula for computing s^2.
$s = \sqrt{s^2}$	The standard deviation of a sample.
$V = \dfrac{s}{\bar{x}}$	The coefficient of variation of a sample.
$\bar{x} = \sum_{i=1}^{N} \dfrac{f_i m_i}{n}$	The arithmetic mean for grouped data.
$M = L_j + \dfrac{(n/2) - C_{j-1}}{f_j} I_j$	The median for grouped data.

Summarizing Data: Statistical Descriptions

$$\text{MAD} = \sum_{i=1}^{N} \frac{f_i |m_i - \bar{x}|}{n}$$ The mean absolute deviation, from the sample mean, for grouped data.

$$s^2 = \sum_{i=1}^{N} \frac{f_i (m_i - \bar{x})^2}{n}$$ The variance of a sample of grouped data; $s = \sqrt{s^2}$.

$$M_k = \sum_{i=1}^{n} \frac{(x_i - \bar{x})^k}{n}$$ The kth moment of a sample of ungrouped data; $M_2 = s^2$.

$$M_k = \sum_{i=1}^{N} \frac{f_i (m_i - \bar{x})^k}{n}$$ The kth moment of a sample of grouped data.

$$\text{Sk} = \frac{3(\bar{x} - M)}{s}$$ A measure of skewness in a sample.

$$b_1 = \frac{M_3^2}{M_2^3}$$ A measure of relative skewness in a sample.

$$a_3 = \sqrt{b_1}$$ A measure of relative skewness in a sample. (Both a_3 and b_1 are always positive. The direction of the skewness is indicated by the sign of M_3.)

$$a_4 = \frac{M_4}{M_2^2}$$ A measure of relative kurtosis in a sample.

PROBLEMS

3-1.* A firm is test marketing a new brand of coffee. (The firm is going to give free jars to some individuals and obtain their reaction to it.) Management wants to decide whether or not to market the coffee in the entire United States market. What is the population of interest, and what is the sample?

3-2. A firm is considering producing a new soft drink, and it is interested in obtaining buying patterns of individuals who buy the particular type of soft drink. In particular, management wants to know how many bottles an individual consumer drinks per week. They are going to take a sample during one particular week. What is the population and what is the universe?

3-3. What is inferential statistics and why is it important?

3-4.* A new car salesman recorded the number of sales per day for 20 days. The sales were: zero, 11 times; one, 7 times; two, 2 times. Compute the mean, median, and mode for the number of sales per day. Are all three measures useful?

3-5. A clothing store sells 10 items, whose prices in dollars are: 1, 100, 100, 100, 100, 100, 110, 120, 120, and 140. They advertise that, while their median price is $100, the range is from $1 to $140. Is this a fair summary? What summary statistics would give a fairer picture? Why?

3-6. A frequency distribution has a mean of 10, a median of 12, and a range of 0 to 20. Very roughly sketch a frequency curve that might fit that frequency distribution. Does the frequency distribution have positive or negative skewness?

3-7.* During the first 5 years of existence, the Terrific Toys for Tots company sold, in sequence, 10, 14, 16, 17, and 18 thousand dollars worth of a particular line of toys, to the nearest thousand. Compute the mean, median, MAD, and standard deviation of sales per year. Are these summary statistics a useful representation of sales during the 5-year period? If not, what is missing?

3-8.* A group of persons who received master's degrees in business 2 years ago were asked their current annual salary. To the nearest thousand, their responses were: 17, 19, 19, 20, 20, 21, 24, 29, 30, 30. These numbers were grouped into a frequency distribution as follows:

Annual Salary (nearest thousand)	Frequency
$11–$15	0
16–20	5
21–25	2
26–30	3
Over 30	0

Compute the mean using the original data and the mean using the grouped data. Explain the reason for the difference.

3-9. If summary statistics computed using grouped data can be incorrect (when compared against the statistics from the ungrouped data calculations), why would we ever make such calculations?

3-10.* A publishing company is considering whether or not to reduce the number of books they publish in a particular topical area. Last year's sales, to the nearest thousand, were: 2, 2, 3, 3, 3, 4, 4, 6, 10, 14, 15, 24. The firm wants to know the mean, median, mode, range, mean absolute deviation, standard deviation, skewness (the easy way), and kurtosis. Before computing the skewness, look at the data and guess whether the skewness will be positive or negative. If the firm feels that a book must sell 5,000 copies per year to break even, do you think they should drop the books below that level from their product line? Why or why not?

3-11. From a survey used to determine some characteristics of employees, the personnel manager of a sporting goods manufacturer found that

SUMMARIZING DATA: STATISTICAL DESCRIPTIONS

the distribution of all employees by age had a mean of 37.3 years and a standard deviation of 21.5 years. The distribution of all employees by annual earnings had a mean of $6,849 and a standard deviation of $6,356. Which of the two distributions contains greater variability? Explain.

3-12.* The production manager in a soft drink bottling plant has drawn a random sample of six filled bottles, which he empties into a calibrated tube to determine the number of ounces in each bottle. He found that the six bottles contained 16.0, 15.9, 16.3, 16.1, 15.4, and 15.7 ounces.
(a) What is the average number of ounces?
(b) What is the median number of ounces?
(c) What is the variance of the sample data?

3-13. An investment company is considering the purchase of two stocks traded on the New York Stock Exchange, and the management wants information concerning the recent price movements of the two stocks. The two stocks are Growell Company and Dividend, Inc., and their price movements for last week were as follows:

	Growell Company	Dividend, Inc.
Monday	$52	$56
Tuesday	64	54
Wednesday	56	54
Thursday	53	55
Friday	50	56

(a) Compare the stocks by means of location measures.
(b) Compare the standard deviations for the two stocks and explain what information this provides about the price movements of the stocks.

3-14. A clothing manufacturer is considering the purchase of a subsidiary. The management is considering two companies, Steady Company and Highfly Company. Sales for the companies for the last 5 months were as follows:

Steady Company	Highfly Company
$26,000	$40,000
20,000	5,000
19,000	3,000
16,000	50,000
19,000	2,000
100,000	100,000

(a) Compute the average monthly sales for each of the companies.
(b) Compute the range, variance, standard deviation, and mean absolute deviation for each set of data.

3-15. Cowbell milk producers have two half-gallon production lines working 6 days per week. In order to discover which line is more efficient, the production manager examined the production records for the lines for 2 weeks in July. The number of cases produced by each line was:

Line 1: 1,250, 1,495, 1,316, 1,377, 1,544, 1,456, 1,433, 1,429, 1,375, 1,461, 1,456, 1,380

Line 2: 1,525, 1,486, 1,479, 1,491, 1,452, 1,512, 1,397, 1,483, 1,426, 1,485, 1,486, 1,438.

(a) For the sample from each line, find:

(1) $F_{0.80}$ (2) $F_{0.10}$ (3) $F_{0.90}$ (4) $F_{0.65}$ (5) $F_{0.30}$
(6) P_{100} (7) P_{99} (8) P_{20} (9) P_{45} (10) P_{50}

(b) Find the interquartile range for the sample from each line.

3-16.* An automotive magazine is testing two car models, one foreign and one American, to determine the mileage per gallon for each car. Engineers tested seven cars of each model and found the following mileage results:

(1) Foreign: 24, 28, 22, 33, 26, 17, 25
(2) American: 12, 13, 12, 14, 18, 16, 13.

(a) Which of the two models exhibits more variability for gasoline mileage?
(b) Add a constant $k = 10$ to each of the mileage results for the seven American cars. Compute the variance before adding the constant and compare it to the variance computed after adding the constant.
(c) Multiply each of the mileage results for the seven American cars by a constant $k = 10$ and compute the variance. What has happened to the variance?

3-17.* The Blackgold Oil Company owns and operates a small bulk plant from which it sells wholesale gasoline to independent retailers. The sales manager is worried about the current slump in sales, which she suspects results from increased conservation efforts on the part of consumers. Rather than attempt to increase sales to current customers, she is considering sending out a force of salesmen to win new customers. The firm wonders what sales a new customer might provide. The distribution of sales, to current customers for the previous month, was as follows:

Summarizing Data: Statistical Descriptions

Gallons of Gasoline (thousands)	Number of Sales
From 0 to less than 10	60
From 10 to less than 20	52
From 20 to less than 30	83
From 30 to less than 40	102
From 40 to less than 50	96
From 50 to less than 60	41
From 60 to less than 70	29
From 70 to less than 80	37
Total	500

(a) Estimate the total number of gallons sold last month.
(b) Compute the mean, median, and the mode of the distribution.

3-18. The purchasing agent for a lawn furniture manufacturer ordered samples of incandescent lamps for plant lighting from two suppliers. The agent tested the two samples and found the following results:

	Frequency	
Length of Life (hours)	Supplier A	Supplier B
700 and under 900	12	2
900 and under 1,100	20	35
1,100 and under 1,300	15	11
1,300 and under 1,500	3	2
Total	50	50

(a) Which supplier's lamps have the greater average life?
(b) Which supplier's lamps are more uniform?

3-19.* The office manager for a plumbing supply warehouse examined all past-due accounts to see how many weeks each account has been in default. The results are:

Weeks Past Due	Number of Accounts	Weeks Past Due	Number of Accounts
1–2	48	9–10	7
3–4	13	11–12	3
5–6	6	13–14	1
7–8	4	Total	82

(a) Find the mean number of weeks the past-due accounts have been in default.
(b) Find the 75th percentile and interpret this number.

3-20. In response to several complaints from office workers, the management of Twilight Radio Company decided to study the difference in hourly wages between factory workers and office workers. They pulled the employee records for 100 factory workers and 50 office workers and found the following distribution:

Hourly Wages	Factory	Office
$2.70–$2.89		3
2.90– 3.09		10
3.10– 3.29	13	11
3.30– 3.49	26	26
3.50– 3.69	29	
3.70– 3.89	17	
3.90– 4.09	8	
4.10– 4.29	3	
4.30– 4.49	4	
Total	100	50

(a) For each group of workers calculate the

    (1) Mean hourly wage.
    (2) Median hourly wage.

(b) By comparing the mean and median hourly wage for each group of employees, what can you say about the skewness of the distributions?

3-21. Before offering her restaurant for sale, the owner tabulated the number of customers per day for the 100 days directly prior to the offering. She felt that this information would enable her to make knowledgeable statements to potential purchasers. She found the following distribution of daily customers:

Number of Customers	Number of Days	Number of Customers	Number of Days
0–19	1	80–99	28
20–39	4	100–119	6
40–59	12	120–139	2
60–79	45	140–159	2
		Total	100

(a) Calculate the mean, median, and mode of the frequency distribution.
(b) Compute the interquartile range.

3-22. In order to determine the reasonableness of the accounts receivable balance which a client company reported in a Statement of Financial Position, auditors drew a random sample of 150 customer accounts and found the following distribution:

Account Balance	Number of Customers	Account Balance	Number of Customers
$0–$19.99	29	$60.00–$79.99	14
20.00–39.99	33	80.00– 99.99	11
40.00–59.99	56	100.00 and over	7
		Total	150

(a) Find the median of the above distribution.
(b) Why cannot you find the mean of the distribution?
(c) Can you find the relative skewness, b_1, for the sample? Explain.

3-23.* A food distributor wants to make sure that he receives the proper weight in fruits that he orders. In order to determine if he needs to investigate further, he weighs 100 crates of oranges. The 100 crates have weights as follows:

Weight (pounds)	Number of Crates
20–21	15
22–23	40
24–25	25
26–27	20
Total	100

(a) Find the mean, median, and mode of the frequency distribution.
(b) Compute the variance for the distribution.

3-24. The controller for a mail-order department store is concerned about the clerical expenses involved in processing the large number of small orders. He is considering offering a discount for orders greater than $50 in order to encourage customers to place large orders. His chief account clerk developed a table which shows the average dollar magnitude per transaction for each of 500 customers from last year:

Magnitude for Average Order Last Year	Number of Customers	Magnitude for Average Order Last Year	Number of Customers
$0–$9.99	301	$50.00–$59.99	11
10.00–19.99	82	60.00–69.99	4
20.00–29.99	45	70.00–79.99	2
30.00–39.99	33	80.00–89.99	2
40.00–49.99	20	Total	500

(a) Compute the mean, median, and mode for the distribution.
(b) Compute the mean absolute deviation and the standard deviation for the data.
(c) If the discount policy is successful:
   (1) Will the mean increase, decrease, or be unaffected?
   (2) Will the standard deviation increase, decrease, or be unaffected?

3-25. The manager of a manufacturing plant wants to determine the level of employee absenteeism which occurred this year in the first full week in January. She decides to determine this by drawing a random sample of 100 time cards for her full-time employees for that week and computing the hours worked for each of the 100 employees. She found the hours to be distributed as follows:

Hours Worked per Week	Number of Employees
31–33	3
34–36	10
37–39	25
40–42	32
43–45	16
46–48	14
Total	100

(a) What is the population? Is it finite?
(b) Compute the skewness for the distribution.
(c) Compute the relative kurtosis for the distribution.
(d) Using the results of (b) and (c), sketch the frequency curve for the sample.

Chapter 4

Index Numbers

For several days the United Pipefitters (UP) have been deadlocked with representatives of the Dundee Ohio Warm Nightware (DOWN) Company concerning the relation of wage rates to changes in the cost of living. The UP wishes to incorporate an escalator clause providing for a given cents-per-hour across-the-board raise tied to a given point change in the Consumer Price Index (CPI). They argue that only in such a way can their purchasing power keep pace with the cost of living. Further such contracts are in use by the UAW with General Motors and in the railroad industry.

The DOWN Company argues that the CPI does not reflect the change in the cost of living to members of the union and that the change in hourly wage rate increase should be by the same percentage as the increase in a more appropriate index. Finally, they point out that if the appropriate index were to decline, the wage rate should also decline.

At this point the negotiations are recessed to prepare for binding arbitration under Mr. I. B. Fare. I.B. believes it would be a good idea if both management and labor better understood index numbers before he renders a decision in the UP–DOWN wage negotiations. The same concerns lead us to a general examination of the topic.

4-1 Nature of Index Numbers

Index numbers, as their name suggests, indicate how a quantity has changed. They are used to make comparisons. For example, if there were 1,000,000 gallons of domestic gin sold in June and 1,200,000 in July, sales in

July were 120 percent of sales in June. We could also say that sales increased by 20 percent.

At the same time we must be careful to note that this does not necessarily mean that people are consuming more alcoholic beverages (only gin sales are covered) or even drinking more gin (they may be replenishing or increasing inventories). For these reasons and others, such as seasonal factors and possible changes in consumption habits, we could certainly not present these data as proof of increased imbibing on the part of the public.

Such simple indexes compare only two items and are quite useful if we are careful not to impute too much to them. More complex index numbers also exist which attempt to describe how more complex phenomena change. The CPI is one such index and a very important one. It attempts to measure the change in cost of a specific market basket of items purchased by a specific type of wage earner. Indeed, this index is important enough to many facets of our life that we will spend some time on it alone in Section 4-3.

Even if we know the type of indicator desired, it may not be clear which index to use. There are several indexes of price changes, for example, including the CPI, the Wholesale Prices Index (WPI), and the Gross National Product Price Deflator (GNPPD), to mention three. And things can be even more complex. There are, for instance, at least 16 indicators of stock conditions on the New York Stock Exchange as well as additional indicators for other stock markets. We have heard about the Dow Jones Index and the New York Stock Exchange Index. But do we know the differences between them? And what about the other indexes?

Indexes in business relate to several areas. One of these is prices. The CPI and the WPI are good examples. Another price index of importance is the Parity Index, computed by the federal government. This index has been central to setting the level of farm subsidies, and in doing so it has supported the level of farm income.

A second important area of business indexes is found in the production area. The most important of these indexes is the Federal Reserve Board's Index of Industrial Production. We will describe this important index in more detail in Section 4-3.

A third important area of business in which index numbers find substantial use is in the interpretation and evaluation of an organization's financial situation. This is accomplished by analysts through an evaluation of the data appearing in the organization's published financial statements. Analysts compute ratios (or indexes) from these data which measure solvency, earning power, and financial strength. For example, the ratio of cash plus cash-like assets (such as short-run investments in government securities) to current liabilities provides what is called the quick ratio. This ratio measures the liquidity of the organization and its ability to meet short-term obligations and take advantage of early-payment discounts. Ratios such as these are also watched by the organization's management to see if their financial affairs are in order. A low quick ratio may, for example, suggest that the organization delay capital expenditures, turn some of its other assets into cash, or borrow. The quick ratio helps the organization to better manage its short-run cash position. (There are also other financial ratios management uses to help it control its activities.)

S.S. Index numbers play an important role in many areas. Can you think of a use of index numbers in the fields of business, health, and public administration that are presently computed and published and to which we have not yet referred?

Solution: We cannot begin to suggest all the possibilities here, but some examples are given. In the health area several organizations construct indexes which measure or attempt to measure the effectiveness of hospital and mental health facilities. The federal government as well as many state governments compute measures to indicate the level of welfare activities. A very important index used in part to evaluate general economic conditions can be constructed from unemployment statistics at both the regional and national level. Indexes of trade provide still other examples.

On the business side an important index that is watched by many firms and others is the Index of Consumer Confidence, computed by the Survey Research Center at the University of Michigan. By scientific interviewing procedures, the Center attempts to capture consumer sentiment about the state of their own economic health and translate this sentiment into an index that will help organizations predict the effects of their own actions.

4-2 Types of Index Numbers

Index numbers can be classified in several different ways. One way is by *what* changes they portend to measure: prices, quantities, confidence, values, and so on. An alternative way to classify index numbers is by *how* they measure the change: weighted or unweighted. For pedagogical reasons, we will use the second classification scheme.

UNWEIGHTED INDEX NUMBERS

Unweighted index numbers are not of great practical importance, but they do provide a useful way of introducing the more common weighted index numbers. We will use prices to illustrate how an unweighted index number is obtained.

Since an index measures relative change, we are then interested in measuring the price change or the change in some other variable between two periods of time (typically years) or between two geographic areas (states, for example) or between different population groups (with and without a college education) or between two goods. To do so we need the prices for the two situations being compared. Since the most common of these comparisons involves the comparison of price changes over a period of time, consider a price index which reflects the cost of building a bridge.

First, we need to know all the materials and types of labor that are involved in bridge construction. Prices for some items are relatively easy to obtain: including prices for structural steel, concrete, and cable. Others are more difficult to obtain: such as the value of management time, the cost of equipment, and other elements of overhead that contribute indirectly to the total task. Some items may be considered small enough that they can be ignored, but the task remains a formidable one. Let us simplify our problem for expositional purposes to consider only an index of the cost of the major materials included in construction. Specifically, we will use only steel, concrete, and cable. Suppose that we have the (fictitious) price information for 1970 and 1975 given in Table 4-1. What can we do with these data? The

Table 4-1 Price Data for Three Items: Bridge Construction

Item	1970 Price	1975 Price
Steel	$100 per ton	$120 per ton
Concrete	$0.05 per pound	$0.04 per pound
Cable	$0.20 per foot	$0.24 per foot

answer is many things, but we should not use it to produce any price indexes because the prices are quoted on different bases (tons, pounds, and feet). The price index will be a function of the units in which the prices are quoted. In the present case the index would essentially be a function of the price of steel. First, then, we must convert these prices into prices measured in the same physical units (if possible). We shall elect to use tons of material (Table 4-2), although any common measure would do.

Table 4-2 Comparable Price Data on Three Items: Bridge Construction

Item	1970 Price/Ton	1975 Price/Ton
Steel	$100	$120
Concrete	$100	$80
Cable	$150	$180

Index Numbers

Now the simple aggregate of these prices, all measured in dollars per ton, can be used to measure the price change between 1970 and 1975. We obtain

$$\frac{120 + 80 + 180}{100 + 100 + 150} 100 = 108.57.$$

In symbolic form,

$$I_U = 100 \frac{\sum p_n}{\sum p_0}, \qquad (4\text{-}1)$$

where the summation is over all prices and

$$p_n \equiv \text{price in the present year}$$
$$p_0 \equiv \text{price in the base year}.$$

The figure 108.57 suggests that bridge material prices are about 8.5 percent higher in 1975 than they were in 1970, if the data are accurate. If we had used the original quoted units (see Table 4-1), the result would have been

$$I_U = 100 \left[\frac{120 + 0.04 + 0.24}{100 + 0.05 + 0.20}\right] = 100 \left[\frac{120.28}{100.25}\right] = 119.98$$

which is essentially the price increase index in basic structural steel.

The simple aggregative index given by (4-1) is easy to compute and understand. But does it tell an accurate story? It may not because it fails to reflect the relative magnitudes of the raw materials that go into the construction of a bridge. The price of concrete has gone down. If bridges are primarily concrete, this price change should receive more weight in the index than the price changes for structural steel and cable. The simple aggregate index implicitly assumes that all items in the index are equally important. In this case that means the same amount of each is used, and this may not be true.

Before we examine means of correcting for such biases if they exist, we illustrate one other simple index construction technique. This technique computes the simple index, the price relative, for each item and then averages them. It is called the *arithmetic mean of the price relatives*. For our data the calculations are as given in Table 4-3. We obtain $\frac{320}{3} = 106.67$. In symbolic

Table 4-3 Computations of Price Relatives: Bridge Construction

Item	1970 Price/Ton	1975 Price/Ton	Price Relative
Steel	$100	$120	120
Concrete	$100	$80	80
Cable	$150	$180	120
Total			320

form,

$$I' = 100 \frac{\sum (p_n/p_0)}{N}, \qquad (4\text{-}2)$$

where $N \equiv$ the number of price relatives averaged.

The arithmetic mean of the price relatives has all the problems possessed by the simple aggregate index. Its only redeeming feature is that it simplifies data collection. We will also use it in discussing weighted indexes in the next subsection.

Examples of unweighted indexes are rare in practice. Among government-produced indexes, only the Index of Spot Market Prices is of this form. It is computed as the geometric mean (rather than the arithmetic mean) of the price relatives. Symbolically,

$$I^* = 100 \cdot \sqrt[N]{\prod_{i=1}^{N} \frac{p_{in}}{p_{i0}}} \qquad (4\text{-}3)$$

where p_{in}/p_{i0} is the price relative between year n and the base year (indicated by the subscript zero) and $\prod_{i=1}^{N}$ directs the computation of the product of the N price relatives. Finally, the Nth root of this product is taken and multiplied by 100.

As a final point in this subsection it is perhaps worth noting that while our discussion has involved the computation of price indexes, the technique is general. We could just as easily have computed quantity indexes if the data were given. Such indexes are used to answer different questions, however.

WEIGHTED INDEX NUMBERS

When we began the computation of simple price indexes, we encountered a problem because the items were quoted in different units. At that time, we altered the basic price quotes to the same unit, namely tons. This is not always so easy (or even possible) to do as it was in our example. (For example, how many hours of an engineer's time is equivalent to 1 ton?) An alternative means of obtaining common units is to weight each price change by the quantities used and thereby transform the data to value terms. In addition, the weighting of price changes by the quantity of the item used provides a measure of the relative importance of that price change. For example, in establishing an index of the change in an individual family's purchasing power, the price changes would be weighted by the amount of each item purchased by that family over some time period.

The appropriate time period to be used in determining the quantities is not obvious. Three possibilities suggest themselves. We could use the quan-

INDEX NUMBERS

tities required (or purchased) in the base period, denoted by the subscript zero in equations (4-1), (4-2), and (4-3). We could use the quantities purchased in the current period, denoted by the subscript n above. Still a third alternative is to use a different period which is felt to be typical in some sense, such as being devoid of artificial scarcities, to develop the appropriate quantity weights. Both the first and last methods have the advantage that the quantity weights need not be reestablished each time that the index is to be computed.

The three choices of quantity years lead to three separate indexes. They can be calculated directly from equation (4-4) by a suitable choice of a.

$$I_W = \frac{\sum (p_n/p_0 \times 100)p_a q_a}{\sum p_a q_a}. \tag{4-4}$$

Equation (4-4) is a weighted average of the price relatives where the value weights ($p_a q_a$) relate to some (hopefully representative) base year denoted by the subscript a. When the subscript a is set equal to zero, that is, when base-year weights are used; equation (4-4) yields a base-year index called the *Laspeyres index*:

$$I_L = 100 \frac{\sum p_n q_0}{\sum p_0 q_0}, \tag{4-5}$$

where the sum is over all products included in the index.

If, on the other hand, we let $a = n$, then a present-year index called the *Paasche index* is obtained from equation (4-4):

$$I_P = 100 \frac{\sum p_n q_n}{\sum p_0 q_n}. \tag{4-6}$$

A related index that has received a good deal of recent support has an equation quite similar to equations (4-5) and (4-6). However, it cannot be obtained from equation (4-4). This index is called the *fixed-weight aggregative index*:

$$I_F = 100 \frac{\sum p_n q_a}{\sum p_0 q_a}. \tag{4-7}$$

Such an index allows the quantity weights to be determined (and hence updated) in a period other than the base period ($a = 0$), but then to be held constant for a time (unlike $a = n$). This updating can also be accomplished by changing the base year occasionally in the Laspeyres index.

We point out that the Paasche index is seldom used because of the need to continually reestablish the weights. This is true even though this index reflects the new quantity relationships resulting from changes in composition over the period from year zero to year n. The choice of some representative (arbitrary) year, as is done in the fixed-weight index and in the Laspeyres index, is an attempt to avoid changes in the quantity composition. This advantage holds whether an index of construction prices or consumer purchase costs is under consideration. For example, the Consumer Price Index is based on the fixed-weight index with the weights revised about every 10 years to reflect changes in taste.

The data from our continuing example can be used to illustrate the calculation of the weighted indexes described in this subsection. Let us make these calculations for the Paasche and Laspeyres indexes, and in doing so we will return to the originally quoted price data to illustrate that the conversion to price per ton is no longer required. All data (Table 4-4) are for a specific

Table 4-4 Cost Figures for the Components of Bridge Construction

Item	Price		Quantity		Value	
	$p_0 = p_{1970}$	$p_n = p_{1975}$	$q_0 = q_{1970}$	$q_n = q_{1975}$	$p_0 q_0$	$p_n q_0$
Steel	$100/ton	$120/ton	2,000 tons	2,200 tons	20×10^4	24×10^4
Concrete	$0.05/lb	$0.04/lb	(3×10^6) lb	(4.0×10^6) lb	15×10^4	12×10^4
Cable	$0.20/ft	$0.24/ft	(0.1×10^6) ft	(0.1×10^6) ft	2×10^4	2.4×10^4
Total					37×10^4	38.4×10^4

type of bridge. Other types of bridges may have other components and, hence, costs.

$$I_L = 100 \frac{38.4}{37.0} = 103.78.$$

The calculations for the Paasche index, not given in Table 4-4, yield

$$I_P = 100 \frac{(0.12 \times 10^3)(2.2 \times 10^3) + (0.04)(4 \times 10^6) + (0.24)(0.1 \times 10^6)}{(0.10 \times 10^3)(2.2 \times 10^3) + (0.05)(4 \times 10^6) + (0.20)(0.1 \times 10^6)}$$

$$= 101.82.$$

We note that these indexes, which are now weighted to reflect the composition of the individual items in the final package, yield different answers from the unweighted indexes. The cost of the components of bridge construction has risen, but not as much as the unweighted average, which assumes that

INDEX NUMBERS

all items are required in equal amounts. However, the cost of constructing a bridge has risen much more than any of the indexes, owing to increased use of both steel and concrete. (This may be due to new state or federal safety standards.) A 1975 bridge would have cost $448,000 (the numerator of the Paasche index), while in 1970 a bridge would have cost $370,000 (the denominator of the Laspeyres index). Thus, the cost of building a bridge has an index of (448,000 ÷ 370,000) = 121.08. The product is no longer the same, so the cost of components does not present a complete picture of the cost of bridge building. As with all statistics, index numbers can lead to erroneous conclusions if we are not careful with regard both to what the index represents and what we want to know.

A further observation can be made. The cost of steel has risen while the cost of concrete has fallen. The index also reflects any adjustment of construction inputs to these price changes. If concrete can be substituted for steel, it will be as its relative price compared to steel declines. This substitution effect may be involved with or without changes in safety requirements. Substitution changes are, perhaps, more common in the case of consumer price indexes. In any case, substitution will tend to increase the difference between the Laspeyres and the Paasche indexes.

S.S. Given the data on butter and margarine in Table 4-5, construct

Table 4-5 Butter–Margarine Data

Item	Price		Quantity/Week		Value	
	1974	1975	1974	1975	1974	1975
Butter	$1.00/lb	$1.30/lb	1 lb	0.8 lb	___	___
Margarine	0.80/lb	0.90/lb	1 lb	0.9 lb	___	___

the unweighted price index, and a present year's (Paasche) weighted index. Has there been a substitution effect and is it in the expected direction? Explain.

Solution:

$$I_U = 100 \frac{2.20}{1.80} = 122.2$$

The items are already in the same units, but this computation ignores the change in usage.

$$I_P = 100 \frac{1.30(0.8) + 0.90(0.9)}{1.00(0.8) + 0.80(0.9)} = 121.7$$

There has been a substitution effect. The use of butter has declined more proportionally. Since its price went up more proportionally, this is expected. Apparently, the use of both products has declined. This may reflect the use of other substitute products in cooking or for the table or, perhaps, a lower level of fat consumption.

4-3 Several Important Indexes

In this section we will briefly examine several indexes that find extensive use in government and business. They include the Consumer Price Index, the Wholesale Price Index, and the Index of Industrial Production.

CONSUMER PRICE INDEX

Because all of us are consumers, the impact of prices on consumption is important to us. We wish to know if our wages are increasing relative to what it costs us to live. Furthermore, many labor contracts now have provisions whereby increases in the Consumer Price Index (CPI) automatically lead to increases in wages.

The CPI is calculated to measure the effect of price changes on a "market basket" of some 400 items purchased by a typical urban wage earner. The list includes food (consumed at home and away), housing (shelter, utilities, furnishings, and operations), apparel, transportation, health, and recreation (including medical and personal care). Consumer expenditures must be surveyed periodically to establish the relative importance (weight) of each type of item included in the market basket.

Owing to the prohibitive expense of including all items bought by urban wage earners, only a representative sample is included. Prices are obtained by selectively sampling 56 metropolitan areas and cities selected to be representative of such geographic areas with populations over 2,500. Hence sampling error is present in computing this index.

The type of index used is a fixed-weight one, as in equation (4-7). Separate indexes are calculated for the different geographic areas and then combined using weights which are intended to reflect the relative importance of the area in terms of the proportion of urban wage earners it represents. A description of the data-collection procedures is too elaborate to be given here. (The interested reader should consult Doris P. Rothwell's "The Consumer Price Index Pricing and Calculation Procedures" published by the U.S. Bureau of Labor Statistics.)

One of the major problems with this index, as we saw with the cost of bridge components and which is true with any price index, relates to changes in quality. An increase in quality given a fixed price is equivalent to a price decline. Attempts are made to specify products included in the index in some detail so that changes in quality are covered. Comparisons are usually also kept fairly close in time to one another, which helps to minimize quality changes which usually take place over extended periods of time. Neither of these procedures is totally successful, however.

A second major problem is the change in tastes which occurs over time. The periodic analysis of consumer expenditures is designed to update the composition of the "market basket" of goods and services to reflect such changes. A major updating in 1964 produced substantive changes for the weights that had been used since 1953. For example, health and recreation increased in relative importance. The 1964 change also involved procedural changes, including a general increase in the number of items included and prices sampled. Another updating took place in 1967.

WHOLESALE PRICE INDEX

The Wholesale Price Index (WPI), like the CPI, is produced by the Bureau of Labor Statistics. The index consists of prices quoted on organized commodity markets and received by manufacturers and producers. It is, then, not a measure of the prices received by wholesalers and other middlemen, but a measure of prices received by large-scale users of the products.

The WPI is also based on a sample. However, around 2,300 items carefully selected by experts as important and representative are involved. Subsets of these items are used to construct price indexes under various classification schemes. Some of the classification breakdowns used include: stages of processing (raw materials, intermediate goods, and finished products), types of farm production (fruits, grains, poultry, and livestock), and industrial products classified into 13 separate groups (including wood products, textiles, and so on). These classifications and other subdivisions of these categories are used to calculate many separate indexes of wholesale prices.

The Wholesale Price Index is computed by a formula consistent with equation (4-4) and is, therefore, a weighted average of the price relatives.

INDEX OF INDUSTRIAL PRODUCTION

The Index of Industrial Production (IIP), published by the Federal Reserve Board, is intended to measure business activity. This index is intended to measure changes in the physical volume of activity in factories, mines, and utilities. The index is not totally comprehensive since directly it covers substantially less than half of the total value of the country's goods and services. However, if we include those activities reflected indirectly by the index, such as the use of these products in construction, the percentage is nearer 60. Nevertheless, production within the construction industry, on farms, in the transportation industry, and in service organizations is excluded from this index.

The basic data are also used to form subindexes, which apply to industry categories. Three major groups—manufacturing, mining, and utilities—are recognized. Further breakdowns by durable versus nondurable, and into smaller industry groupings, are made. Since 1959 a second combination of the data is used to obtain indexes by market categories. Three primary markets are distinguished: consumer goods, equipment, and materials. Further subdivisions are also used, as is the case for the industry groups. The added data on the individual market groupings is useful in economic analyses, which should precede policy formulation.

The index is constructed using methods consistent with the fixed-weight aggregative index given by formula (4-7). Currently the index takes the following form:

$$\frac{\sum p_{1957} q_n}{\sum p_{1957} q_{1957-1959}}, \qquad (4\text{-}8)$$

where the summation is taken over all commodities. In a quantity index, it is the prices that are the weights. The base period is the average quantity produced over the 1957–1959 period.

The purpose of this index is to measure and monitor changes in economic activity. The movements of this index are used in part to gauge the health of the economy and as an indicator of business conditions. Together with other published data on economic conditions (such as plans for capital expenditures and business sales) and psychological indicators (including the Index of Consumer Confidence), the data on industrial production help managers and public officials to plan their actions.

SELECTED ADDITIONAL APPLICATIONS

Although we have concentrated on value indexes so far, the concept is much more general in its use. For example, indexes can be constructed to reflect changes in sales for a region, a line, a salesman, or a division. Comparisons among different series and of a single series over time may provide management with useful insights. Similar series could be constructed to reflect changing inputs, service demands, and so on.

In the areas of accounting and financial analysis, ratio analysis is common. Tests of solvency, liquidity, and financial strength are often made using ratios of values taken from an organization's financial statements. The quick ratio was mentioned previously. A similar ratio, the current ratio, reflects the ratio of all current assets (including inventories and accounts receivable as well as cash) to current liabilities. Banks and other financial institutions use such ratios together with other data on the organization and the economy to determine whether they will loan money to the organization. Ratios and the trends in these ratios are used in bond ratings, mergers, acquisitions, pricing new stock issues, and by management in selecting among alternative courses of action.

A major use of index numbers in economic analysis is in deflating dollar value series to constant amounts. If the country's gross national product were to increase from $447.3 billion in 1958 to $865.7 billion in 1968 (as it did), we might conclude that the value of goods and services produced had nearly doubled. But this is not true if the level of prices also increased. During the same period, for example, the Consumer Price Index increased from about 100 to 121, a 21 percent rise. In order to obtain the increase in gross national product in constant (sometimes called real) dollars, we must deflate it for this price change. The results are as follows:

Year	GNP (billions of current dollars)	Consumer Price Index (1957 − 1959 = 100)	GNP (billions of 1957–1959 dollars)
1958	$447.3	100.7	$444.2
1968	$865.7	121.2	$714.3

Once the series is expressed in constant dollars, we see that rather than doubling, the increase was about 61 percent (714.3 − 444.2) ÷ 444.2.

Gross national product is an attempt to measure economic growth. Changes due purely to price changes should be excluded. Hence gross national product can rise in current-dollar terms when the basic cause is inflation. Once the series is adjusted for inflation, gross national product in real terms may fall. We would not wish to conclude that economic growth was occurring if the rise in gross national product were caused by inflation. Similar calculations are appropriate to measuring changes in personal income, wages, the profitability of business organizations, government expenditures, and so on. Indeed, deflation is a key step taken in economic studies involving dollar series over time. Labor negotiations are now strongly influenced by the notion of real wages. Contracts involving adjustments for changes in prices, typically measured by the Consumer Price Index, are common. Accountants also continue to debate the merits of requiring that price-level adjusted data be reported in published financial statements of earnings and financial condition.

A related index that has had an important bearing on the welfare of the farm community for many years is the *parity ratio*. This ratio relates the index of prices received by farmers for all farm products to the index of prices paid by farmers for commodities, interest, labor, and taxes. A base period of 1910–1914 is used, when prices received by farmers were in the most favorable position relative to prices paid. When this ratio exceeds 100, prices received by farmers are higher, relative to the base period, than those paid. (We must remember that this is an average and need not be true for all farmers). The parity ratio is an input to the U.S. government's agricultural policy as it is made in Washington. The indexes of prices paid and received by farmers also has been used to calculate parity prices of individual agricultural commodities. These prices represent price supports and hence yield insurance through subsidies against falling prices.

S.S. Suppose that we were presented with an index of efficiency for a particular organization. Before using this index as a relevant measure, what information pertaining to this index should be obtained? List five items.

Solution: In determining whether the index is relevant, we must decide whether the index is appropriate. It is important to ask whether the definition of efficiency is a useful one. Should the measure be supplemented? We would also like to know how the index is compiled. What items are used to make it up? How are the data obtained? What base period is used and is it a normal or typical period? Is the index up to date? What assumptions are made by those constructing the index? What manipulations are made on the data? Finally, we may wish to ask those who have been involved with the existing index what problems they have discovered.

4-4 Some Special Problems in Index-Number Construction

This section examines briefly four problems in index-number construction. They are

1. Selecting the base period.
2. Shifting the base period.
3. Splicing two series covering different time periods.
4. Sampling.

SELECTING THE BASE PERIOD

Changes are measured from the base period. Hence, it is useful to have as a base period a fairly recent time when conditions are similar to those being measured. The parity ratio is an example where the base period is not a

INDEX NUMBERS

recent one. (Other considerations have prevailed.) Comparisons lose their relevance when large time (or geographic) separations are involved due to changing efficiency, quality, objectives, and techniques of measurement. In practice, when the items covered by the index or the weights used are revised, most indexes are updated.

Comparisons, which are the essence of index numbers, should reflect comparisons with a normal or typical base period. For example, an index of automobile sales using a war year as a base would give misleading ideas about the relative level of such sales. The same would be true of price indexes using a war-year base since prices may be controlled and goods may be unavailable. For this reason, the base period for an index should be a time period, or an average of several time periods, when conditions for the variable measured are normal. Major strikes, restrictions in the availability of an important material, or natural disasters can make a given time period inappropriate for a base if they affect the economic activity variable being measured.

It is common for averages over several years to be used, to avoid giving undue importance to any short period that may reflect the effect of atypical events. Yet in a period where substantial changes are occurring, say in tastes, or where there is a pronounced and continuous drift in a variable, say the general price level, the base period for an index may require rapid updating.

As a final observation, the base period may be used to obtain the weights or the weights may be obtained from some other period. For example, in equation (4-7) the base period is represented by the subscript 0 on the p term in the denominator. The present period is period n. But the quantity weights are for time period a. There is no reason why the year denoted by the subscript a and the year denoted by the subscript zero need be the same. The CPI is an example where different values are used. The IIP is another example.

SHIFTING THE BASE PERIOD

The shifting of the base period is, as we have seen, designed to make comparisons more meaningful. When the base is moved (updated), the new index series is constructed by simply dividing all the index values by the index for the new base. Consider the following values of the Consumer Price Index:

Year	Consumer Price Index, 1957–1959 = 100	Consumer Price Index, 1962 = 100
1958	100.7	95.5
1962	105.4	100.0
1966	113.1	107.3

To shift the base period to 1962, we divide all the indexes in the second column by 105.4.

The new index series still uses the same quantities and hence has not been completely updated. But, as we noted in the previous subsection, changing tastes and new products may make the weighting system obsolete as well. If the weighting system becomes outdated, new weights should be devised. Since the new weights usually reflect new commodities, the index is not usually carried back in time. Instead, the index for the present and future years is computed using new weights.

SPLICING INDEX-NUMBER SERIES

Sometimes one index-number series does not go far enough back into time for the purpose at hand. If a second series can be found that does so, the two series may be combined as long as there is some overlap in them. The same can be done when a new series is started if the calculations for the old series are also completed for the same year, thus providing a year of overlap. This is often the case when an organization wishes to revise the construction of the index. For an example, assume that a firm revises its selling prices and also shifts its base for a price index of its products. The old and the new index are given in Table 4-6. The values in the spliced series from 1975 on

Table 4-6 Splicing Two Index-Number Series

Year	Old Index, 1970 = 100	Revised Index, 1975 = 100	Spliced Index, 1975 = 100
1970	100.0		92.3
1971	101.3		93.5
1972	103.3		95.3
1973	104.5		96.4
1974	107.6		99.3
1975	108.4	100.0	100.0
1976		103.2	103.2
1977		104.1	104.1

are taken from the revised index series. The values in the spliced index series for 1974 and earlier are obtained by dividing each value in the old index series by 108.4. Dividing by 108.4 also puts the old index series at 100 in 1975. This is common procedure when an index presently being computed is revised.

INDEX NUMBERS

When two series already exist, there may be more than one year of overlap. In this case an average of the ratios of the two indexes in each of the overlap years can be used. It is appropriate to use a geometric average for averaging ratios. The *geometric average* is the nth root of the product of the n quantities to be averaged. For an example, we will use the data in Table 4-6 and assume that, in addition, an index of 99.1 is available for the revised index series for the year 1974. Then in 1974 the two indexes have values of 107.6 and 99.1; in 1975 the values are 108.4 and 100.0. Taking the geometric mean of the 2 ratios, we obtain

$$100 \sqrt{\frac{108.4}{100.0} \frac{107.6}{99.1}} = 108.5.$$

The spliced index series with 1975 = 100 is as follows:

Year	Spliced Index	Year	Spliced Index
1970	92.2	1974	99.1
1971	93.4	1975	100.0
1972	95.2	1976	103.2
1973	96.3	1977	104.1

The figures from 1974 on are obtained from the revised index series. The values for 1970 through 1973 are obtained by dividing the index for that year in the old index series by 108.5.

SAMPLING

Index numbers are usually based upon only a subset of all the possible commodities or other items to which they relate. This is done for cost reasons and so that data can be gathered and processed quickly. Timeliness of index numbers is important if they are to be used in decision making.

The items involved in most index series are chosen to be representative. Sampling, then, is judgmental and not random in the selection of items to be included in the index. Items in the Consumer Price Index, for example, are chosen because (1) they are typically bought by wage earners and clerical workers, (2) they represent significant fractions of total expenditures, (3) price quotations are easy to obtain, (4) the item is a good indicator of price movements for a subclass of items, or (5) the item's price movements are distinctive and not likely to be reflected by other items.

It is also the case with the Consumer Price Index that prices are gathered only in certain cities. Here, again, sampling is judgmental. The problem with judgment sampling is that error measures cannot be calculated and hence probabilistic statements cannot be made. Sampling is also used to obtain price quotations and from time to time to obtain information on consumer spending habits. This information is used to determine whether new weights are needed and, if so, what they should be.

In some cases, such as within organizations, complete enumerations may be possible as a means of establishing proper weights if the organization's records are sufficient. This is also true of the Federal Reserve Board's indexes of department store sales and industrial production which are obtained from a periodic census. However, sampling is used to obtain current sales and production data.

S.S. Splice the two economic index series given in the following table:

Year	Old Index, 1970 = 100	Revised Index, 1975 = 100	Spliced Index, 1975 = 100
1970	100.0		
1971	98.0		
1972	100.0		
1973	102.0		
1974	107.0	95.0	
1975	110.0	100.0	
1976		103.2	

Solution: The factor to use on the old index series is

$$\sqrt{(110 \div 100)(107 \div 95)} = 1.113.$$

The values in the new spliced series from 1970 to 1973 are 89.8, 88.1, 89.8, and 91.6. They are obtained by dividing each of the corresponding values in the old index series by 1.113. The spliced series from 1974 to 1976 are those of the revised index series, namely 95.0, 100.0, and 103.2.

4-5 A Warning

Index numbers are often interpreted as averages. We must recall that averages may not provide a complete picture of the facts. This is particularly true of index numbers where wide changes occur in habits, tastes, productive activities, and so on over time. Further, subcomponents of some index may move quite differently from the overall index.

Data on which index numbers are based may be difficult and costly to collect. Biases and measurement errors easily arise in collecting data on prices, quantities, costs, and so on. Judgmental mistakes can occur in selecting the items to include in the index, the normal period in which to gather data on the weights, and the choice of a base period.

Perhaps most serious is the problem of allowing for changes in quality of the items included in the index. Consider the following example. The Consumer Price Index in 1969 was about 128 while in 1946 it was about 67 (based on 1957–1959 = 100): prices had about doubled. Suppose now that we must buy strictly from the world of goods available in 1946. Would you be satisfied with an income half the size of our 1969 income? If not, then there are some quality differences (including new goods) which indicate that the real price rise is something less than the doubling suggested by the index.

Index numbers ought to behave according to logic. By this we mean that a reasonable index ought to obey some expected behavior patterns. For example, consider the following propositions for a single good and its price index:

1. If the good's price doubles, and there is no change in quantity, the index should double.

2. The index between any two dates should not change if the base period of the index is changed from one date to another (called the time-reversal test).

3. A dimensional change in the unit of measurement of the good, say from quarts to gallons, should not change the index.

4. A change in the monetary unit, say from pounds to dollars, should not affect the index.

5. The product of the price index (p_1/p_0) and a quantity index (q_1/q_0) should equal the ratio of the values (p_1/p_0) ÷ (q_1/q_0) between any two comparable dates (called the factor-reversal test).

Ragnar Frisch (a Nobel laureate in economics) proved long ago that when the number of goods exceeds 1, it is not possible to find index numbers that satisfy all these criteria. Indeed, the index numbers we have derived in this chapter do not satisfy many of these logical characteristics. For example, the Laspeyres and the Paasche price indexes fail to satisfy the time- and factor-reversal tests, items 2 and 5 above. Nevertheless, that does not mean that such indexes are not useful. To be useful the index must be simple to understand, easy to calculate, available on a timely basis, and appropriate for the purpose to which it is applied.

Comparisons should be made using recent data to avoid changes in quality and tastes. Base periods should be selected with care to avoid atypical conditions. Linking systems should be used to obtain consistency when new weights are required for the series. Care should be used in selecting items to be included in the index to assure it is representative and in gathering data to avoid biases and measurement errors.

TECHNICAL NOTE 4-1: PRICE INDEXES AND SATISFACTION

This note links the concepts of economic utility to index numbers. It also helps us to see just what types of questions the Laspeyres and Paasche indexes answer, and it introduces a new index number.

The Laspeyres index uses base-year weights. It measures how we live today versus in the base period if we buy *now* what we bought *then*. It measures the cost of living now as we lived then. But if prices have changed, it is likely that relative prices will have changed, too, and our consumption patterns with them. That is, we will not live today as we did in the base year. Given an income today that would permit us to live precisely as we did in the base period, that is, to buy the same goods in the same amounts we did then, we would allocate this income somewhat differently today. By this reallocation we would achieve a higher level of satisfaction. Otherwise, there would be no reallocation. This means that we could reach the same level of satisfaction with a smaller real income today than that required to buy the base period quantities. Using the Laspeyres index (4-5), this means that the numerator value need be no larger and it could be smaller to reach the same level of satisfaction. Thus, Laspeyres index is an upper bound to the effective price change based on the satisfaction level (utility) achieved in the base period. (Note that tastes are held constant in this argument.)

The Paasche index, on the other hand, uses present-year weights. It measures the price change to live now as we lived in the base period if we bought *then* what we buy *now*. But again there will have been relative price changes and we would not have lived this way; we would have bought somewhat differently, allocating expenditures generally toward relatively less expensive items. Given precisely enough money in the base period to live as we live today, we would elect to buy a slightly different market basket and thereby reach a somewhat higher level of satisfaction. Hence, a smaller real income level in the base period would allow today's satisfaction level to be reached. Using the Paasche index (4-6), this means the denominator need be no larger and it could be smaller to reach the same level of satisfaction. Thus, the Paasche index gives a lower bound to the effect price change based on the satisfaction level (utility) achieved in the present period. This satisfaction level is different than that achieved in the base period and discussed in connection with the Laspeyres index.

If we can assume that the level of satisfaction reached in the present period under present prices is S_t, we can construct the following picture. We assume

INDEX NUMBERS 105

that good B is relatively less expensive in the base period. (Note that since S_t is further from the origin than S_b, S_t represents a higher level of satisfaction.)

$B_b \equiv$ budget line base period
$B_t \equiv$ budget line today
$S_b \equiv$ satisfaction level reached in base period
$S_t \equiv$ satisfaction level reached today.

Figure 4-1A applies to both the Paasche and the Laspeyres indexes. But the budget lines, using the same total income, will not be the same in the two periods if the relative prices have changed. Then the satisfaction levels reached are different. In this situation the Laspeyres index gives an upper bound

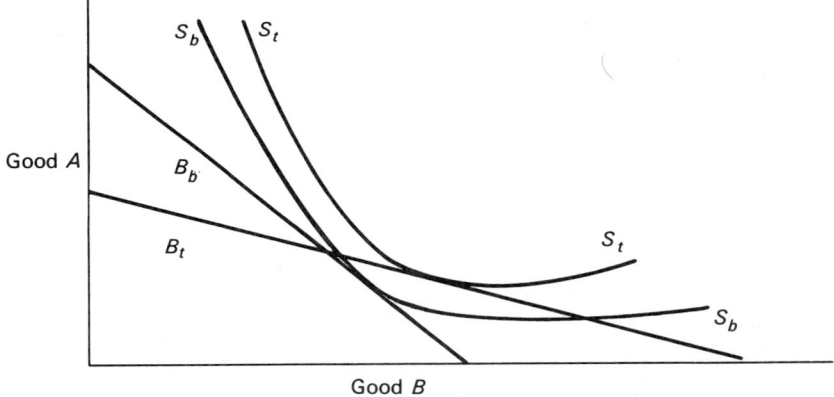

Figure 4-1A Index Numbers and Satisfaction Levels

and the Paasche index a lower bound to the effective price change. It we can assume that the difference in the two satisfaction levels is reasonably close—probably a reasonable assumption if comparisons are kept close in time—then an average of the two measures may be a reasonable measure of the price change. Using the appropriate geometric average, we obtain

$$I_I = \sqrt{I_P I_L}, \qquad (4\text{-}1\text{A})$$

where

$I_P \equiv$ Paasche index
$I_L \equiv$ Laspeyres index.

This index is known as *Fisher's ideal index*. It was devised by Fisher because of the failure of other indexes to satisfy the time- and factor-reversal tests described by items 2 and 5 in Section 4-5. Although Fisher's index does satisfy these conditions, it requires more data and is harder to calculate than the Laspeyres index. It also lacks the simple verbal interpretation of either the Laspeyres or the Paasche index. Fisher's ideal index is not widely used.

Key Formulas

Formula	Used to Compute:
$I_L = 100 \dfrac{\sum p_n q_0}{\sum p_0 q_0}$	The Laspeyres index
$I_P = 100 \dfrac{\sum p_n q_n}{\sum p_0 q_n}$	The Paasche index.
$I_F = 100 \dfrac{\sum p_n q_a}{\sum p_0 q_a}$	Fixed-weight aggregative index.
$I_I = \sqrt{I_P I_L}$	Fisher's ideal index.

PROBLEMS

4-1. A recent teachers' contract calls for future salaries to be tied to the Consumer Price Index, both up and down. (If the CPI falls, so will salaries.) Why is such a contract useful, in general, and why may it not work as desired in individual cases?

4-2.* If you own a house during inflationary times, does the CPI overstate or understate the impact of inflation on you?

4-3. Some labor contracts have a portion of the salary determined by average productivity in the plant. Discuss the problems of measuring such an index.

4-4.* Discuss the difference between a weighted index and an unweighted index. Is the CPI a weighted or unweighted index?

4-5.* A government agency wants to demonstrate that its salaries have kept pace with inflation. They have four levels of employee: from level 1 (lowest salary level) to level 4 (highest salary level). During the last five years, the Consumer Price Index (CPI) has gone from 1.219 to 1.680. During the same period, average salaries have changed as follows:

INDEX NUMBERS

Level	Salary Five Years Ago	Salary Now
1	$12,468	$16,208
2	15,920	21,650
3	18,842	26,756
4	22,150	31,896

(a) Form an unweighted salary index, and compare it to the change in the CPI.

(b) Discuss the use of an unweighted index in this situation.

4-6. What are the advantages of a fixed-weight aggregative index? Why would it ever be used? Finally, discuss the three time periods used in the formula for a fixed-weight aggregative index and the data collected from each time period.

4-7. The Laspeyres index is used more frequently than the Paasche index. Why? If there were no data problems, which index would you use and why?

4-8.* Suppose that the Breakfast Eaters Index (BEI) is designed to show the price of breakfast. Some data are given below. Quantities are the amount used per week by an average family.

	1967		1972		1977	
Item	Price	Quantity	Price	Quantity	Price	Quantity
Eggs (dozen)	$0.77	2.4	$0.84	2.2	$0.98	2.0
Bacon (pound)	1.18	1.9	1.64	1.3	1.65	1.5
Bread (pound)	0.38	1.0	0.42	1.3	0.44	1.6
Coffee (pound)	0.82	0.5	0.91	0.5	1.04	0.5

Use 1967 as the base (zero) year in every calculation. Compute:

(a) The Laspeyres index values as it would be computed in 1972 and 1977.

(b) The Paasche index values as it would be computed in 1972 and 1977.

(c) The fixed-weight aggregative index for 1977, using 1972 as the period in which quantities are determined.

4-9. Problem 4-8 gives the data for the Breakfast Eaters Index. (The solutions will be found at the back of the book.) Examine the data and the answers. The Breakfast Eaters Lobbying Team (BELT) wants you to answer the following questions.

(a) Using only the Laspeyres index values, shift the base year to 1972. What are the new index values for 1967, 1972, and 1977? Are the new values completely updated? Why or why not?
(b) Briefly discuss the changes in the quantity data through this period. Why do you think some of the changes took place? Discuss the impact of the changes on the indexes.

4-10. How should a year be selected for use in determining quantity information in a Laspeyres or fixed-weight index?

4-11. What are the IIP and WPI? What do they measure, and how might they be used?

4-12.* The following are 1968, 1969, and 1970 prices and production data for three products manufactured by a synthetics company:

	Prices (cents per yard)			Quantities (hundred million yards)		
Product	1968	1969	1970	1968	1969	1970
1	11.3	11.6	11.9	1.94	2.03	2.10
2	10.3	10.1	10.8	8.41	8.48	8.97
3	11.3	11.1	11.8	1.02	1.03	0.96

(a) Using 1969 value weights, calculate a weighted mean of price relatives with 1968 = 100 for the year 1970.
(b) Using 1968 = 100, construct Paasche indexes.
(c) Using 1968 = 100, construct Laspeyres indexes.
(d) Using the 1969 quantities as weights and 1968 = 100, calculate fixed-weight aggregative indexes.

4-13.* The following data represent values of the 1975 Consumer Price Index (1967 = 100) for several regions of the United States:

Region	July	August	September
Chicago	158.3	159.1	159.6
Detroit	161.0	161.4	162.9
Los Angeles	158.1	158.8	160.4
New York City	166.6	167.5	169.3
Philadelphia	165.0	165.6	166.9

Source: *Monthly Labor Review*, November 1975, p. 120.

INDEX NUMBERS

(a) Do the data indicate that prices in New York City are higher than any of the other four cities? Explain.

(b) Which city experienced the greatest increase in the CPI between July and September? Which city experienced the smallest increase?

4-14. The branch manager for a large manufacturing plant has recently received word that he would be transferred from Portland, Oregon, to the corporate headquarters in New York City. His salary at the New York office would be $50,000 and would increase by 5 percent each year. The manager was concerned, however, that the Consumer Price Index for New York City had recently been increasing at an 8 percent annual rate. Based upon his information, the manager believed he could present a strong case for larger annual increments. Do you agree? Why or why not?

4-15.* Shift the base year in each of the following cases.

(a) Belgium's index of industrial production with 1965 = 100 was 98.2, 100.0, 101.8, 103.7, 109.2, and 121.1, respectively, for the years 1964 through 1969. Shift the base to 1968.

(b) The 1965 through 1971 values of an index of sales for retail stores in the United States with 1967 = 100 are, respectively, 90.5, 96.9, 100.0, 109.0, 114.1, 119.7, and 130.3. Shift the base so that 1970 = 100.

4-16. The average prices of kitchen chairs produced by a certain furniture manufacturer were $14.87, $15.50, $16.00, $16.95, and $17.75 in the years 1966 through 1970. Use the corresponding values of the Wholesale Price Index with 1967 = 100, which were, respectively, 99.8, 100.0, 102.5, 106.5, and 110.4 for the same years, to judge whether the increases in the prices of these kitchen chairs have fallen behind, kept pace with, or exceeded the general increase in wholesale prices.

4-17. The gross national product of the United States was $793.9, $864.2, $930.3, $976.4, and $1,050.4 billion, respectively, in the years 1967 through 1971. Given that the corresponding values of the Consumer Price Index (with 1967 = 100) are, respectively, 100.0, 104.2, 109.8, 116.3, and 121.3, express the 1967 through 1971 figures for the gross national product of the United States in terms of constant 1967 dollars. Also find the percentage increase from 1969 to 1971 using

(a) The actual dollar amounts.

(b) Constant 1967 dollars.

4-18. A businessmen's association in a large city publishes a paper which circulates among all small appliance retailers in the city. One feature of the paper is a section that is devoted to trends in prices within the city. A recent article written by one industrious retailer noted that, in the previous year, an index for wholesale prices of small appliances climbed to 141, while the index for consumer prices of small appliances reached the level of 150. The author noted that since the profit margin indicated by the indexes was only nine points (that is, 150 − 141), most small appliance dealers in the city would be forced to enact price increases for the coming year.

(a) Do you agree with the author's conclusion? Why or why not?
(b) How would you interpret the implications of the indexes if you knew that both were constructed using the same base year?

4-19.* As part of a company's review of its capital-expenditures program, the president gathered data representing the normal amount of annual expenditures for building construction undertaken by the company. The data appear below, together with an index series that represents construction costs applicable to the company's industry and geographical region:

Year	Construction Expenditures (millions)	Construction Costs Index
1970	$5.1	123
1971	5.5	156
1972	6.0	165
1973	6.2	178
1974	6.3	183
1975	6.4	195

Has the trend in "real" construction costs of physical facilities differed from what might be inferred from the expenditure data?

4-20.* In a bargaining session over future wage levels carried on in a local firm, the union leader noted that while wages had increased 20 percent in a 5-year period, consumer prices had increased by 50 percent, leaving the workers with 30 percent less in terms of real wages. Was the union leader correct? Explain.

4-21. A company has been preparing an index of the prices it pays for inputs to its manufacturing process. The company began compiling the index in 1959. In 1969 the company decided to reevaluate the index since the proportions of the different materials used in production had changed significantly from that used in 1959. The old and new series are as follows:

Year	Old Index, (1959 = 100)	New Index, (1970 = 100)
1959	100.0	
1960	121.6	
1961	133.7	
1962	142.5	
1963	139.6	
1964	143.7	
1965	146.5	
1966	152.3	
1967	151.1	
1968	162.0	
1969	164.3	101.6
1970	164.7	100.0
1971		107.1
1972		98.2
1973		99.3
1974		105.4
1975		110.8

(a) Splice the two indexes using 1970 as the new base year.
(b) What is the percentage increase in raw-material prices from 1959 to 1975?

4-22. Thread Bare Shirt Company produces three major lines of shirts: top, middle, and economy. Last year, Thread Bare significantly revamped its pricing policy for the three lines. The following data provide information on sales and volume for the last 2-year period:

	Average Selling Price		Sales (thousands of shirts)	
Line	1974	1975	1974	1975
Top	$14.00	$17.00	22.7	22.8
Middle	8.00	10.00	29.7	35.6
Economy	3.29	3.99	44.6	45.1

(a) Construct the Laspeyres index, the Paasche index, and Fisher's ideal index for 1975 prices with 1974 = 100.
(b) How does the use of indexes provide information not obtainable from an examination of the data above?

4-23. During selected months in 1975, the value of two farm-price indexes were as follows:

Month	Prices Received by Farmers, All Farm Products (1910–1914 = 100)	Prices Paid by Farmers, All Farm Products (1910–1914 = 100)
January	438	617
February	427	615
March	420	612
April	431	621
May	452	627
June	463	632
July	476	636

Source: *Survey of Current Business,* November 1975, pp. 5–8.

(a) Compute the parity ratio for each month.
(b) Would a declining parity ratio indicate that commodity prices are falling?
(c) Would a parity ratio of 100 today signify that the average farmer's income today was the same as the average farmer's income in the period 1910–1914? Explain.

4-24. The following data represent prices and quantities of a raw material purchased by a manufacturer for two different years:

	1970		1975	
Material	Price	Quantity	Price	Quantity
1	$0.75	5,000 pounds	$0.20	5,500 pounds
2	1.22	1,000 pounds	1.55	800 pounds
3	0.55	100 yards	0.70	50 yards
4	3.21	250 dozen	2.56	325 dozen
5	0.45	800 feet	1.10	400 feet

(a) Construct the Laspeyres index, the Paasche index, and Fisher's ideal index for both price and quantity with 1970 = 100.0.
(b) Apply the factor-reversal test to each of the indexes in part (a). The factor-reversal test is explained in Technical Note 4-1.

4-25.* High-Flier Mutual Fund is composed of a group of stocks whose past performances have been highly volatile, where volatility is defined as the relationship between the percentage change in stock price and the percentage change in a market index developed by the fund. For each stock held by the fund, the managers have estimated an equation of the following form:

$$R_{it} = a_i + b_i R_{mt},$$

Index Numbers

where

$R_{it} \equiv$ percentage change in the share price for company i between day $t - 1$ and day t (expressed in decimals rather than percentage points)

$R_{mt} \equiv$ percentage change in a market index between day $t - 1$ and day t (expressed in decimals rather than percentage points)

$a_i, b_i \equiv$ parameters estimated by the managers for company i.

The most important parameter in this relationship is b_i, since it describes how the price of a share will change as the market index changes. The relationship is not perfect, but the managers agree that it is sufficient for decision-making purposes, since errors among shares within the portfolio will tend to cancel one another. Using available historical data, the managers have estimated that $a = 0.005$ and $b = 1.322$ for one stock, Growth Company.

(a) If the market index has the following values for a period of 1 week, what price should Growth Company's stock be at the end of the week if the share price is $30.00 on Monday and the estimated relationship holds without error?

	Market Index
Monday	886.1
Tuesday	892.3
Wednesday	888.1
Thursday	895.8
Friday	896.9

(b) Can you see any danger in holding a portfolio composed of all stocks whose b_i parameters are greater than 1.00?

4-26. An accountant for Peebles Manufacturing Company believes that financial information compiled by the company for both internal decision making and external use by investors would be more useful if the present historical cost-accounting procedures were changed to a replacement-cost basis. Under historical cost-accounting procedures, assets are recorded by a company at a dollar amount equal to the purchase price, and adjustments are usually not made to reflect changes in prices for the assets. The accountant believes that by continually adjusting the recorded amount of assets to reflect changing prices, better information could be obtained concerning the real value of the assets to the company. The following data indicate the company's recent equipment purchases, together with an index of machinery and equipment prices applicable to the industry. (Assume that all equipment is purchased at the end of the year indicated.)

Year	Purchases (thousands)	Equipment Index (1960 = 100)
1970	$452.67	171.6
1971	463.81	182.3
1972	471.22	188.9
1973	466.70	186.7
1974	489.51	195.4
1975	421.62	198.2

(a) Approximately how much would the equipment purchased in 1970 cost if the company had bought the same equipment in 1975?
(b) The company is required by law to estimate a dollar value for all the equipment owned by the company, and this amount must be reported to the company's shareholders. In arriving at a total dollar amount, the company's accounting procedures require deducting 10 percent for depreciation each year from the purchase price of the equipment beginning in the year following the purchase. The purchase price less deductions is then totaled for all equipment, and this total amount is reported to shareholders. What amount would Peebles report for the equipment at the end of 1975? (Assume that the company began operations in 1970.)
(c) If the company supplied shareholders with replacement-cost information, the company would first compute the replacement cost at the reporting date for each piece of equipment. From this value, the company would deduct 10 percent of this replacement cost for each year in which the equipment has been used. This net amount would then be totaled and the total reported to shareholders. Using these procedures, how much larger would the amount be than the total reported in part (a)?

Chapter **5**

Uncertainty, Sample Spaces and Probabilities

Templeton P. Lerr, Tipler to his friends, is the only salesman for the Bubs Beer Company. His job entails convincing restaurants and other establishments to stock his company's product. The company needs to know something about the success of Tip's activities if it is to plan production to meet required deliveries. Because it is necessary to begin deliveries almost immediately after a new customer is signed, the beer must be produced and stocked before the new customers are known. This means that the production decision must be made under uncertainty.

Uncertainty means that the process being considered can lead to more than one outcome and there is no assurance which outcome will occur. In the present situation, the impact of Tip's sales ability on new orders represents the process whose outcomes are uncertain. We do not know how many new orders he will bring in. We will return to this problem from time to time throughout this chapter to illustrate some of the ideas developed. We will even try to help the company with its demand-estimation problem. First, however, we need some tools to assist us. These tools are provided by probability theory.

Probability theory is concerned with processes (or experiments as they are sometimes called) that are uncertain. Uncertainty, we recall, means that the process has more than one outcome. With the help of probability we can assign numbers to these outcomes that are useful to a manager in problem solving. These numbers are called the *probabilities of the outcomes*.

The manager plays a key role since a knowledge of the process is needed to specify what may happen; that is, he must specify the outcomes. For example, one must know the candidates before predicting who will win an election. In the Bubs Beer Company case it is necessary to know the order levels that are possible before demand can be estimated. Once the set of all possible outcomes is established, it may be possible to assign a number to each outcome that measures its likelihood of occurrence. These likelihood measures are called *probabilities*.

The manager will often play an important role in the process of establishing the outcome probabilities. Once these probabilities are determined they can be combined in accordance with the laws of probability (the more important of which you will encounter in this book) to give the probabilities of other events in which the manager is interested. Thus, if the Bubs Beer Company knows the probability that its ace salesman will obtain an order from each of its customers, it may then be able to determine the probability of any level of total demand.

Determining probabilities is not an end in itself, however. Probabilities are merely an input to the decision process. If the Bubs Beer Company knew the probabilities of various demand levels, it might consider changing its planned production levels or inventory policies. Management might reevaluate whether they have the right-sized sales force and whether that force is composed of the right sales personnel. The company might use the probabilistic information in a study of whether to build a larger plant or extend the geographical coverage of their distribution system. A systematic analysis of these and similar decisions requires the use of probability theory.

In this chapter and the next, we will explore the answers to three basic questions about probability numbers. They are:

1. What do we mean when we say that the probability of some event (say, a head on one flip of a coin) is a specific number (say, $\frac{1}{2}$)?

2. How can probability numbers be determined for a specific problem?

3. How do probability numbers behave? In other words, what mathematical rules do they obey?

It is the answer to the third question that allows probabilities to be combined in ways that assist decision making.

5-1 Uncertainty

Uncertainty exists if a process can lead to more than one outcome. Hence, we define an *uncertain process* as one with two or more outcomes. Not all processes involve uncertainty. A pot of water put over a sufficiently hot fire will eventually boil; a small boy left unattended in the presence of a mud puddle will ultimately make its acquaintance; a worker will eventually leave a given company; a machine run constantly without maintenance will eventually break down. These things will happen; there is no uncertainty about them. There typically is uncertainty, however, about the time at which these events occur, and this uncertainty could be important. For example, in funding a company pension plan for retired workers it is necessary to estimate both their length of service and the likelihood that they will reach retirement in the company's employ. Probability theory finds extensive use in such problems.

Still other processes exhibit negligible uncertainty, which we elect to ignore. For example, in tossing a coin we typically ignore (assign no probability to) the outcome described by the words "the coin landed on edge." As another example, we assume that the principal as well as the interest on government obligations will be paid when due. These cases are not the typical ones that cause managers to lose sleep. Moreover, managers like to believe that they can control or at least influence the final results of decisions they make. This often leads managers to express the position, "If only I knew enough about the situation, the uncertainty would go away." This reminds us of a poem by Hughes Mearns:

> *As I was walking up the stair*
> *I saw a man who wasn't there.*
> *He wasn't there again today.*
> *I wish, I wish he'd go away.*

But alas, this is not the case. The uncertainty remains. It will not go away. Greater knowledge will allow the manager to quantify his uncertainty, but it will not remove it. Uncertainty is inevitable when conclusions must be reached which go beyond the data at hand.

In addition to the problems of uncertainty, managers must also be alert to the possibility that the data at hand may be less than perfect for the purposes required. Wallis and Roberts in their book *Statistics: A New Approach* (Glencoe, Illinois: The Free Press, 1958), tell us of such a case reported in the April 17, 1948, issue of *This Week*.

> *A museum was proud of its amazing attendance.* [*Such data might be used, for example, in the justification of requests for foundation support.*] *Shortly thereafter a small building was erected nearby. Museum attendance fell off by 100,000 in the following year. The small building, you see, was a comfort station.*

Uncertainty pervades most managerial problems. The decision situations described at the end of Chapter 1 are typical examples. All involve uncertainty, as does the question concerning the level of orders for the Bubs Beer Company.

Yet, as Howard Raiffa declares in his book, *Decision Analysis* (Reading, Mass.: Addison-Wesley Publishing Company, Inc., 1968), "There are uncertainties and there are uncertainties." On the one hand, there is the uncertainty that attaches to the random flip of a perfectly balanced coin and the uncertainty of whether a light bulb produced from a stable production process will light. This type of uncertainty may yield to logical analysis or to probabilities based on accumulated data such as that given by a frequency distribution of bulb life. (We would be tempted to classify the manager who does not recognize this as a dim bulb.) On the other hand, how do we deal with the uncertainty in predicting either the demand for a new product or the impact of a modified farm subsidy program on agricultural production several years from now? These are more complex issues, and we must become more technically proficient before we can cope with them.

Before we get unavoidably more technical, there is an important point about uncertainty that needs to be made. Suppose we were to select a given light bulb just produced and ask the production manager for the likelihood that it will light. Now this bulb will either light or it will not. There is in this case no uncertainty about the fact of the bulb's ability to function: that is a fact and facts are not subject to uncertainty. However, our knowledge of the fact is subject to uncertainty. In the present case, to estimate the likelihood that the particular bulb in question will light, we might use data on the percentage of bulbs produced by this process that light. Thus, we use a probability that describes the process over a long period of time and is based on the process to answer our question concerning one item produced by that process. Probability statements that are attached to particular outcomes for decision purposes are derived from the process that produces the outcome and not from the outcome itself.

S.S. Which of the following situations describe or are based on an uncertain process?

(a) Drilling for oil
(b) The timing of tides
(c) The verdict in a jury trial of an innocent man
(d) The results of a tuberculosis test
(e) Agricultural production

Solution: The timing of tides is sufficiently accurate that we could say that this is a process without uncertainty. All the other processes involve uncertainty.

UNCERTAINTY, SAMPLE SPACES AND PROBABILITIES   119

5-2 Sample Spaces

Up to now we have relied heavily on our intuitive grasp of such terms as outcome and event. These ideas are important enough for us to spend some time on them. An *outcome* is a result of an uncertain process. If we flip a coin, the possible outcomes are head and tail. If we roll a die, the outcomes are the integers 1 through 6. The term "outcome" applies to the finest division that can be distinguished. Thus, the result "odd" when a die is rolled is not an outcome, since a finer distinction can be made—the integer value that occurs.

The set of all possible outcomes for some uncertain process is called a *sample space*. For example, the outcomes from tossing a coin once are

$$\{\text{head, tail}\} \text{ or simply } \{h, t\}.$$

The braces stand for the word "set," and the outcomes are separated by commas. It is often useful to let a capital letter denote the entire set. Thus, we might write

$$S \equiv \{1, 2, 3, 4, 5, 6\},$$

where S is the set of outcomes from rolling a die once. A discussion of the set terminology used in this chapter is given in Technical Note 5-1.

In the cases just described, the outcomes in the sample space are fairly easy to establish. But what about the sample space for our friendly salesman, Tipler, of the Bubs Beer Company? What sample space describes the uncertain order-obtaining process? Well (and one may not like this answer in a mathematics book), it depends. It depends on what we want to know. Several different sample spaces can be used to describe the order-obtaining process for a single customer. Some are as follows, in increasing complexity for a single customer:

$S_1 \equiv \{\text{order, no order}\}$

$S_2 \equiv \{\text{none, small, medium, large}\}$

$S_3 \equiv \{\text{integers representing the size in number of cases of an order received}\}$

$S_4 \equiv \{\text{numbers representing the size in dollars of an order received}\}$.

Several sample spaces, then, may be used to describe a single uncertain process. This may be somewhat disturbing. How are we to know which sample space to select? One possibility is to pick the simplest sample space that will allow us to answer the questions we want to know. For the Bubs Beer Company, interested in establishing production levels, this might be the sample space denoted by S_3. Another useful criterion is to use a sample space which is composed of outcomes that cannot be further divided or distinguished for the purposes at hand. This allows more complex events to be built up from the basic ones. (It is not possible to work in the opposite direction.) Using this criterion, sample space S_2 may not be adequate in the Bubs Beer Company case. In fact, sample space S_3 contains all the information that S_2 contains and more. Unfortunately, the specification of the best sample space in a given situation is an art one acquires with experience and not a science for which definitive rules can be given.

Sometimes the manager is interested in the joint results of two or more uncertain processes or with successive trials of the same process. Examples of each of these possibilities are given next.

The joint result of the order process and whether a visit was made to the customer as described separately by sample spaces S_1 and $Z \equiv$ {visit, no visit}, respectively, is

$S_5 \equiv$ {(no order, visit), (no order, no visit), (order, visit), (order, no visit)}.

The successive results of visits to two customers where each visit is described by sample space S_1 is

$S_6 \equiv$ {(order, order), (order, no order),
(no order, order), (no order, no order)}.

It is worth noting that in sample space S_5, some possible combinations may not occur. For example, the pair (order, no visit) may not be a reasonable possibility. However, it does not hurt to include it in the sample space.

The practice of indicating the successive results in S_5 and in S_6 by ordering the outcomes in parentheses is common notation. This notation could be extended to trials of any number of processes or to any number of successive repetitions of the same process. For example, if Tipler visits three potentially new customers, the sample space

{(order, order, order), . . . , (no order, no order, no order)}

could be used to study the impact on total demand.

SAMPLE SPACES AND DATA

The data used to construct a sample space are the result of some measurement scale. Sample space S_1 represents nominal data and S_2 uses ordinal data.

Data at these measurement levels (nominal and ordinal) are often described as *qualitative* data. Certain statistical techniques have been developed which are of particular relevance to problems involving qualitative data. They include the binomial probability law which we will meet initially in Chapter 8, and a group of methods collected in Chapter 18 under the title "nonparametric statistics." Some additional examples of qualitative data include: brand preference, quality designations, bond ratings, letter grades representing course or exam performance, and categories of crimes committed.

Data involving numbers, that is, data involving measurement on either an interval or ratio scale, are called *quantitative* data. Examples include income, age, percentage of impurities, time, and number of occurrences of a particular outcome. Sample spaces S_3 and S_4 are based on quantitative data. Particular statistical methods are also available for problems that involve quantitative data.

It is easy to make a mistake as to just which type of data one has. This is particularly true of count data such as the number of people in favor of a particular proposal. Is this qualitative or quantitative data? The best way to resolve this question is to consider the basic unit of observation (the basic sample unit) and ask whether this unit is only categorized or whether it has some one of its properties measured. Thus, in the case of the number of people in favor of a particular proposal, the sample unit is the person whose opinion is sought. The person's opinion is used to categorize that person as in favor, opposed, or perhaps neutral on the issue at hand. We have, therefore, qualitative (nominal) data in this case.

In considering the problem of ordering the week's supply of **Bubs Beer** we would be interested in measuring the number of bottles or cases demanded, quantitative data. Another example of quantitative data would be to measure the revenue earned by some sales unit (possibly a salesman like T. P. Lerr) over a given period of time.

The distinction between qualitative and quantitative data is of central importance and should be determined in any problem situation. However, within the category of quantitative measurement a further distinction is worth our attention. This is the distinction between discrete and continuous measurement. When a variable is subject to continuous measurement, any real number (such as 1, $\sqrt{2}$, e, 6.847) may be the measurement result. Theoretically—a word we promise to use sparingly—it must be possible to obtain as fine a measurement in terms of accuracy as is desired. On the other hand, in discrete measurement not all real values are possible. Typically, values with a constant interval between them, such as the integers, are used. Examples of discrete measurement include measuring costs to the penny or height to the nearest inch.

Now, in fact, all available measuring instruments are capable of at most discrete measurement even where the unit involved could be measured more accurately given a more precise measuring device. Thus, a table may be longer than 2 feet 6¾ inches but less than 2 feet 7 inches, which is the best our yardstick can do. Nevertheless, it is often convenient to assume continuity in the measurement process for purposes of solving problems expeditiously. Hence, we will often assume continuity, both when it is reasonable theoretically (the table example), and even sometimes when it is not (for example, the number of bottles of Bubs Beer ordered).

S.S. Indicate whether the following represent qualitative or quantitative data:
   (a) Number of defective items produced by a process
   (b) Rate of motor vehicle accidents at different intersections on a given day
   (c) Sex of purchasers of a particular product
   (d) Length of time it takes to process a complaint
   (e) Percentage of weed seed in a box of grass seed

Solution: Items (b) and (d) involve quantitative data. In the other cases the basic unit of analysis is categorized. For example, in (a), the units are categorized as defective or not defective. In (e), we need more information. A seed is either a weed seed or not, and if this is the classification method, the data are qualitative. However, if a physical measure such as weight is used, the data are quantitative.

NUMBERS AND REAL-WORLD PHENOMENA

Statisticians do not deal directly with the phenomena of the real world. Instead, they work with numbers that result from some measurement process. Hence, strictly speaking, statistical analysis applies to these numbers, and the conclusions of the analysis apply to the processes that produced the numbers rather than directly to the physical phenomena they purport to reflect. It is in the province of the manager (with, of course, the assistance of a statistician) to make the transition from the statistical conclusions to their real-world implications. And it is the manager's ultimate responsibility to determine if the data obtained and statistical conclusions reached can be extrapolated to the issues that are the manager's concern.

For one example, the count of the number of new housing starts is often quoted as an index of economic activity. But do we know how this figure (or statistic) is obtained? Are all housing starts counted in the tally? What constitutes a housing start; how is one defined? What about housing stops?

What does this statistic tell us about economic activity in the building industry, let alone in the general economy? Housing is not the only part of the construction industry nor does it measure either separable activities, such as cement, electrical, and plumbing work, or production for use in the construction industry.

It is imperative that the manager take care in accepting figures for what they purport to measure but may not. The number of crimes reported may not be a good indication of the level of crime. Recent studies suggest that less than half of all crimes are reported. The number of those admitted to mental institutions may not provide an accurate measure of mental illness. (Do you see why?) In evaluating any set of figures, the decision maker should remain skeptical until satisfied that the data measure what they claim. In this regard, we mean more than that there may be variability in the reported data. Variability can and should be reported using measures such as the range or standard deviation (see Chapter 3). We are concerned here instead with the relevance of the data.

S.S. 1. Not too long ago a graduate student from a Western institution of higher learning questioned a sample of Japanese-Americans about their treatment by the U.S. government during World War II. To his surprise the overwhelming response was favorable. Yet it turned out that the data, although accurately reported, were invalid. Can you guess why?

2. List three reasons why the number of persons admitted to mental institutions may not be an accurate measure of mental illness.

Solution:

1. There are several possibilities, including asking the wrong people or the wrong questions. The problem in this case was that the data on treatment of Japanese-Americans was collected by a Caucasian interviewer. The Japanese were apparently reluctant to be honest with the interviewer. This was verified when the task was repeated by a Japanese-American interviewer.

2. Several reasons why the admissions figure might not be a useful indicator of mental illness are:

(a) Inadequate facilities to admit all cases, or new facilities may allow patients to be treated who were not treated previously.

(b) The reluctance to report some cases.

(c) The difficulty of establishing the presence of mental illness in specific cases.

(d) Not all cases diagnosed are appropriate for institutional care.

If there were no dramatic changes in items (b), (c), and (d), admissions might serve as valid surrogates for mental illness. However, inadequate facilities will not respond to changes in such a way so as to yield useful data.

5-3 Events

Events are the phenomena managers wish to explain or predict. They are the substance of decisions. We need, then, to be able to find the probabilities of events.

An event is described using either a single outcome or a group of outcomes. Events that are described by a single outcome are called *fundamental events*. Using the sample space $S_7 \equiv \{1, 2, 3, 4, 5, 6\}$ for the roll of a die, six fundamental events can be described. One of them is the event "five," which is written $\{5\}$. Another is the event "three," written $\{3\}$. *Compound events* are those which must be described using more than one outcome. An example of a compound event is suggested by the words "an odd number." This event would be written $\{1, 3, 5\}$. Any event can be written as a subset of the sample space.

A decision maker may be interested in the likelihood of several possible events. It is thus useful to specify a sample space that will allow him to combine fundamental events (the outcomes), whose likelihoods are known or can be estimated, into the more complex events of interest. We will learn in this and the next chapter how the likelihoods (probabilities, as they are known in the trade) of the fundamental events can be combined to yield the probabilities (likelihoods) of compound events.

S.S. Using the sample space $V \equiv \{(h, h), (h, t), (t, h), (t, t)\}$, which represents the flipping of one coin twice (or two coins once), write down all the events possible and, if you can, give each a name. There are 16 events. Five of the possibilities are given below.

Event	*Description*
1. $\{(h, h)\}$	The fundamental event "two heads."
2. $\{(h, h), (h, t)\}$	The compound event "a head on the first flip."
3. $\{(h, h), (t, t)\}$	The compound event "like sides."
4. $\{(h, t), (t, h), (t, t)\}$	The compound event "at least one tail."
5. $\{\ \}$	The null event that "nothing happens."

Solution: The first four events are the fundamental events. Those not given above are

$$\{(h, t)\},$$
$$\{(t, h)\},$$

and

$$\{(t, t)\}.$$

Uncertainty, Sample Spaces and Probabilities

The next 11 events are compound events. The ones not given with some (but not all) names, are:

$$\{(h, h), (t, h)\},$$

head on second flip,

$$\{(h, t), (t, h)\},$$

$$\{(h, t), (t, t)\},$$

$$\{(t, h), (t, t)\},$$

$$\{(h, h), (h, t), (t, h)\},$$

at least one head,

$$\{(h, h), (h, t), (t, t)\},$$

$$\{(h, h), (t, h), (t, t)\},$$

and the sample space itself,

$$\{(h, h), (h, t), (t, h), (t, t)\},$$

the somewhat uninteresting event that something happened. The sixteenth and last event is called the *null* or *empty* event. It implies that nothing happened and thus it, too, is of little practical interest.

OCCURRENCE OF EVENTS

An event is said to occur if the outcome that describes the process belongs to the event. Hence, if a die is rolled and a "three" results, the event {3} has occurred. In addition, the event "odd," written {1, 3, 5}, has occurred. And so has any other event that has "three" listed in it.

S.S. Consider again the sample space $V \equiv \{(h, h), (h, t), (t, h), (t, t)\}$. If two heads are flipped, which of the following events occurred?

Event	Description
(a) {(h, t)}	A head followed by a tail
(b) {(h, h), (t, t)}	Like sides
(c) {(h, t), (t, h), (t, t)}	At least one tail

Solution: Only the event described in part (b) occurs, since the outcome "two heads," written (h, h), belongs only to the event described in part (b).

VENN DIAGRAMS, SAMPLE SPACES, AND EVENTS

Pictures help us to visualize ideas. Venn diagrams, named after the nineteenth-century English logician John Venn, provide one means of visualizing sample spaces and events. Consider first the single roll of a die described by the sample space $S_7 \equiv \{1, 2, 3, 4, 5, 6\}$. This sample space can be pictured as in Figure 5-1. The flags name the points.

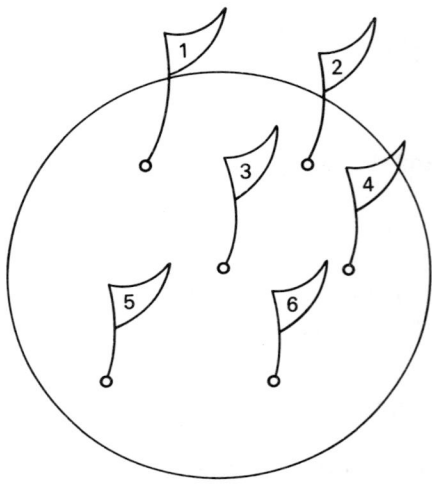

Figure 5-1 Venn Diagram for One Roll of One Die

Using this illustration, any fundamental event is represented by one of the six points indicated by the flags. The event six, $\{6\}$, is represented by the lower-right-hand dot, for example. Complex events can be represented by sets of the dots in the illustration. Thus, the event "odd" can be denoted by the dots in the shaded area of Figure 5-2. The important idea about the picture is that the three points of the event "odd" lie in the shaded area. The size of the shaded area is not significant in this illustration.

Venn diagrams can be used to extend our understanding of events and relationships among events. In Figure 5-3 three events in particular are defined, so we may develop some new ideas. These events are:

The event "odd" $\equiv \{1, 3, 5\}$, denoted by all points in the area shaded (including the area that is both shaded and crosshatched).

The event "even" $\equiv \{2, 4, 6\}$, denoted by the points in the clear or unshaded area (including points in the area crosshatched but not shaded).

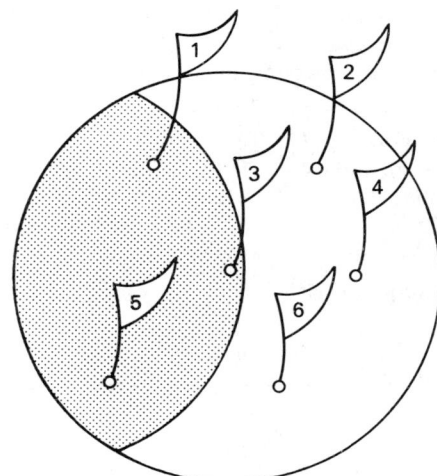

Figure 5-2 Event "An Odd-Numbered Face Is Rolled"

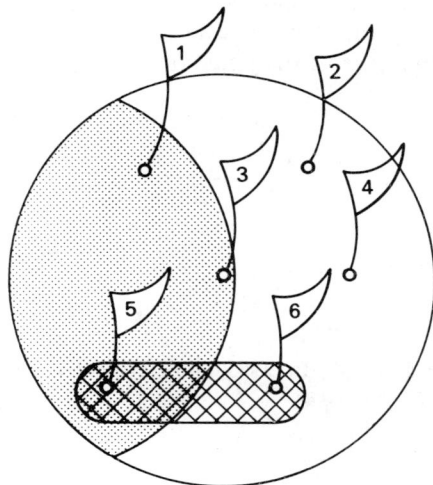

Figure 5-3 Three Specific Events

The event "at least a five was rolled" ≡ {5, 6}, indicated by the points in the crosshatched area.

When two or more events have no common points, we say they are *mutually exclusive (or disjoint)* events. The events "odd" ≡ {1, 3, 5} and "even" ≡ {2, 4, 6} provide an example. Note that on a single roll only one of a set of mutually exclusive events can occur. The event "odd" may occur or the event "even" may occur on one roll, but not both.

When two events have one or more points in common, we say they *overlap*, and this overlap is called the *intersection* of the two events. The events "even" and "at least a five" overlap at the point (or outcome) 6. This is written {2, 4, 6} ∩ {5, 6} = {6}, where the symbol ∩ means overlap or intersection. It can also be read as "and." Six is the only such integer in {2, 4, 6} and {5, 6}. The intersection of two or more events consists of only those outcomes common to all the events. If a 6 is rolled, both events occur simultaneously.

New Symbol	Meaning
∩	Intersection: the elements simultaneously in all sets under consideration.

The intersection of the events "even" and "at least 5" occurs in the area that is crosshatched but not shaded in Figure 5-3, that is, the area containing only the point 6. The intersection (the overlap) of several events, three in this case, is suggested in general by the shaded area in Figure 5-4.

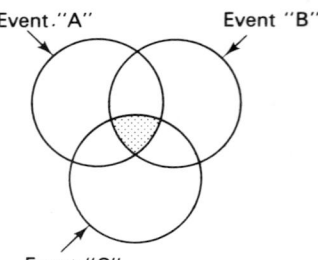

Figure 5-4 Intersection of Three Events: Shaded Area

Sometimes we may need to ask a question such as the following: If a 6 is rolled, did either the event "odd" or "at least 5" occur? The answer is yes, since the event "at least 5" occurred. The event either "odd" or "at least 5" is written {1, 3, 5} ∪ {5, 6} = {1, 3, 5, 6}, where the symbol ∪ denotes the word "union." The word "or" could also be used for the symbol ∪. The union of two events consists of the outcomes in either or both of the separate events. The union of several events consists of all outcomes that appear in any one (or more) of the several events.

UNCERTAINTY, SAMPLE SPACES AND PROBABILITIES

New Symbol	Meaning
∪	Union: the elements in each set under consideration.

The union of the events "odd" and "at least 5" is represented in Figure 5-3 by the area that is only shaded plus the area that is only crosshatched, plus the area that is both shaded and crosshatched. It is the area in Figure 5-3 containing the points 1, 3, 5, and 6. More generally, the union of several events, three in this case, is suggested by the shaded area in Figure 5-5.

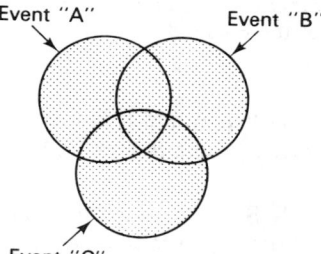

Figure 5-5 Union of Three Events: Shaded Area

One other type of event is of interest at this point. We shall sometimes find it useful to consider an event that did not occur. Given that an event, say, "even," occurred, the event that did not occur is the event "odd." It includes all the points in the sample space not included in the event that occurred. Such an event is called the *complement* event. The event "odd" is the complement of the event "even" (and vice versa). This event is written using a prime symbol. Hence we write A', which is read "not A." For the sample space in our example, the event "not odd," that is, the complement of the event "odd" is written

$$\{1, 3, 5\}' = \{2, 4, 6\}.$$

Using the Venn diagram technique, the complement of the event A is illustrated by the shaded area in Figure 5-6. The notion of a complementary event is helpful, as we shall see later, in simplifying the calculation of the probabilities of certain compound events. Thus, suppose that a sample space is divided (some would say *partitioned*) into two events, call them the events A and $B = A'$. Assume that A is a compound event but that B is a fundamental event. If we wish to find the probability of the compound event A, it may

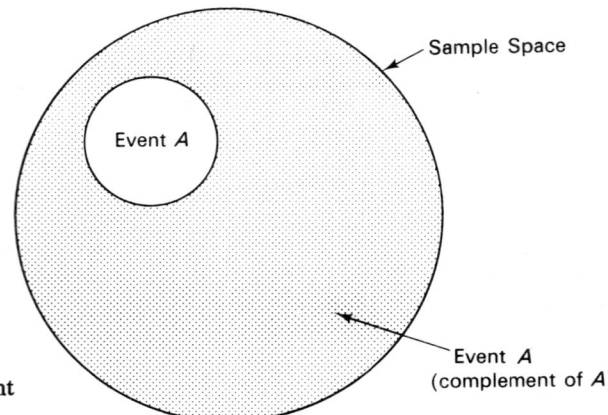

Figure 5-6 Complement of the Event *A*

be easier to find it indirectly by first calculating the probability of the fundamental event $B = A'$. Then $P(A) = 1 - P(B)$.

S.S. Consider the sample space for two rolls of a symmetrical die where the outcomes are represented by numbers giving the sum of the two faces. One way to write this sample space is

$$\{2, 3, 4, 5, 6, 7, 8, 9, 10, 11, 12\}.$$

(a) Write the simple event "snake eyes"; that is, the total was 2.

(b) If two sixes were rolled ("boxcars"), did the event "at most 8" occur?

(c) Write the event "7 or 11." Is it a fundamental event?

(d) In the game of craps if "my point" is four, and I roll a four, then we say the event "my point" . . . .

(e) Write the complement event to the event "four."

Solution:

(a) $\{2\}$.

(b) No, the event "at most 8" $\equiv \{2, 3, 4, 5, 6, 7, 8\}$.

(c) $\{7, 11\}$. It is a compound event since it consists of two outcomes.

(d) Occurred.

(e) $\{2, 3, 5, 6, 7, 8, 9, 10, 11, 12\}$ or $\{4\}'$.

DETERMINING THE POSSIBLE OUTCOMES

For the examples we have used so far, it is a relatively easy task to write down all the fundamental events. In a more complex problem situation, this may not always be true. One technique that is useful in this regard is called the *tree diagram*. Figure 5-7 provides a tree diagram for one roll of a die.

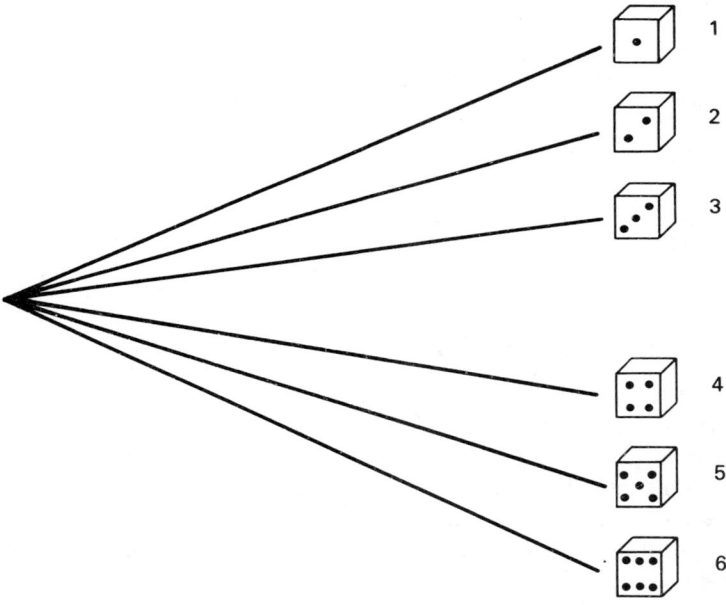

Figure 5-7 Tree Diagram for One Roll of a Die

One particularly useful result from this technique is that it helps one count the total possible outcomes from some uncertain (random) process. For example, suppose that we tossed a coin three times; how many fundamental events are there? Guess the answer before looking at Figure 5-8.

The answer, as we can see by examining Figure 5-8, is that there are 8 possible outcomes from tossing a coin three times. (If you wrote down 6 because you reasoned that 2 outcomes, h and t, were possible on each of the 3 tosses, hence $3 \times 2 = 6$, don't be worried. Probability often seems at first to contradict one's intuition. That is why we study it formally and learn rules and techniques to help us avoid these pitfalls.)

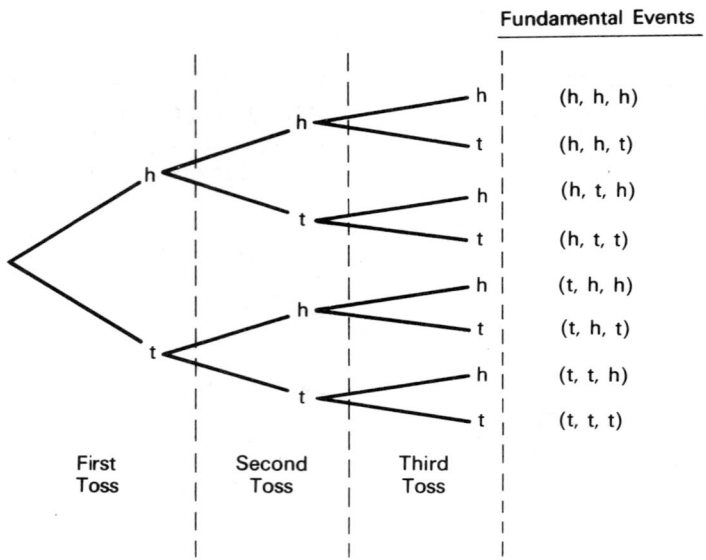

Figure 5-8 Tree Diagram for Three Tosses of a Coin

The concept of a tree diagram is useful not only in establishing outcomes but also in searching for alternatives in a decision problem, as we shall see in Chapter 13. One must be sure, however, that the outcomes at the end of the tree are mutually exclusive and exhaust the possible results for the random process. In the next section, we will show how tree diagrams play a role in helping to calculate probabilities in many common situations.

S.S. Construct a tree diagram for the results from two consecutive rolls of a die. How many fundamental events are there if we keep track of the results of both rolls?

Solution: The results for a single roll are given in Figure 5-7. For two rolls, each of the six branch endings of Figure 5-7 would have an additional six paths (or branches) leading from it. This is suggested for only the first branch in Figure 5-9. The total number of fundamental events is 36, one for each branch ending.

5-4 Probability

Up to now we have studied only what is possible. A coin flipped once lands either heads or tails. We now propose to ask what is probable and what

Uncertainty, Sample Spaces and Probabilities

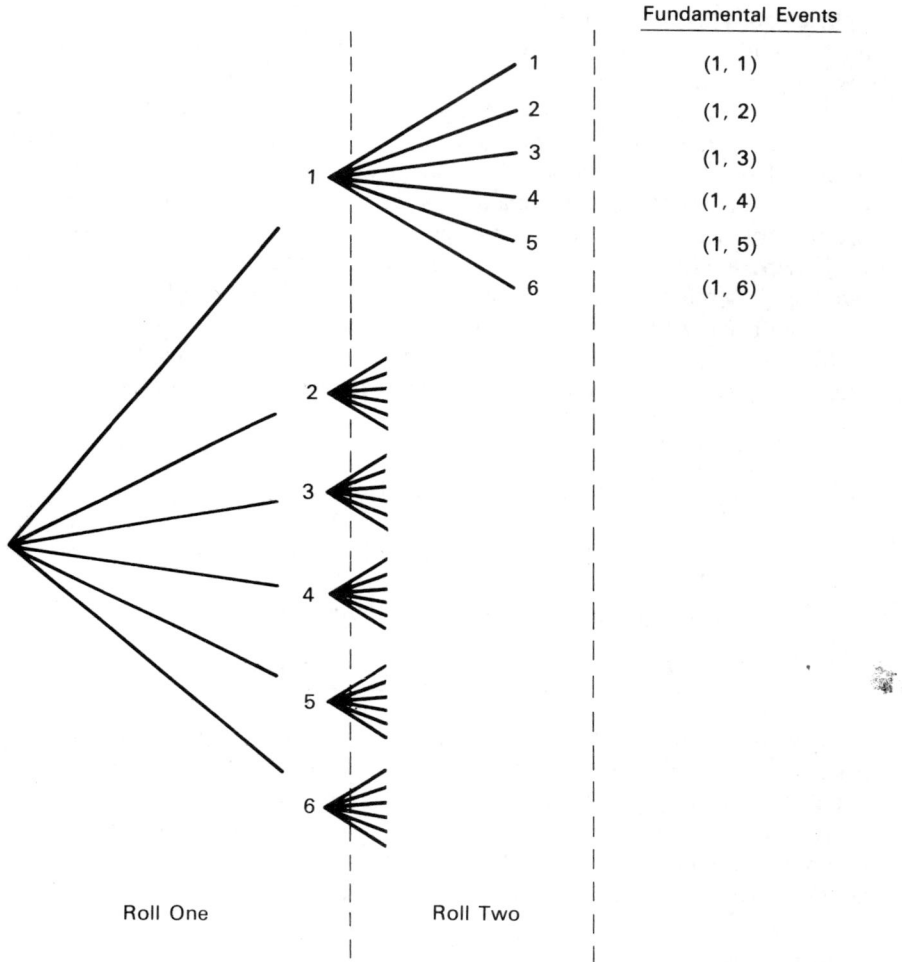

Figure 5-9 Tree Diagram for Two Rolls of a Die

is improbable. We wish to establish a numerical measure for each outcome that tells us how likely that outcome is to occur. We suspect that you already have a good idea of what values would be useful in describing the likelihood of the fundamental events "head" and "tail" for one flip of a coin. One obvious choice is to assign a probability of $\frac{1}{2}$ to each event. Notationally, where $P(\cdot)$ means "the probability of what appears within the parentheses in place of the dot,"

$$P(h) = P(t) = \tfrac{1}{2} = 0.5.$$

The equal probabilities assigned to the two fundamental events are based on the understanding that the coin is symmetrical and that no tendency toward one side or the other is introduced by the flipping process.

Before we explore just where these numerical measures come from and what they mean, let's observe some of the properties that a probability number has. First, the probability number attaching to any event is a non-negative number. (Can you imagine what it would mean to say that the probability of rain is $-\frac{1}{2}$ or -3?) Second, the sum of probabilities assigned to the separate fundamental events for a random process must equal 1. These two ideas are reflected by our coin-tossing example. First, $P(h) = P(t) = 0.5$. This number is nonnegative and $P(h) + P(t) = 0.5 + 0.5 = 1.0$.

More formally, if we are given a sample space, call it S, with n fundamental outcomes, $O_1, O_2, \ldots, O_i, \ldots, O_n$, then (where O_i is any outcome):

$$S \equiv \{O_1, O_2, \ldots, O_i, \ldots, O_n\} \equiv \{O_i\}, \quad i = 1, 2, \ldots, n$$

$$P(O_i) \geq 0, \quad i = 1, 2, \ldots, n$$

$$\sum_{i=1}^{n} P(O_i) = 1.$$

New Symbol	Meaning
$P(O_i)$	Probability of the outcome O_i.

In words, a probability is a number between zero and 1 inclusive and is assigned to an event. The sum of the probabilities assigned to the fundamental events in any sample space must equal 1.

We have not yet studied just how to compute the probability numbers for compound events. That is done later in this subsection. We turn next to how probabilities are determined for fundamental events.

S.S. Given the sample space $S_7 \equiv \{1, 2, 3, 4, 5, 6\}$ for one roll of a "fair" die and the sample space $V \equiv \{(h, h), (h, t), (t, h), (t, t)\}$ for two tosses of a "fair" coin, answer the following questions:

(a) How many fundamental events are in S_7 and in V?

(b) What probability number indicates the likelihood of the event $\{3\}$ in S_7 and the event $\{(t, h)\}$ in V?

Solution:

(a) There are 6 fundamental events in S_7 and 4 in V.

(b) The probabilities are

$P(\{3\}) \equiv P(3) = \frac{1}{6}; \quad P(\{t, h\}) \equiv P(t, h) = \frac{1}{4} \quad$ if the die and coin are "fair."

WHERE DO PROBABILITIES COME FROM?

The probability numbers we used for tossing a coin once came from our experience with symmetrical objects in general and coins in particular. We did not even need to see this coin or try it out before assigning probabilities to the fundamental events. The reasoning goes as follows. A symmetrical coin is as likely to fall heads as tails. Therefore, the total probability of 1 should be divided equally over the fundamental events. Extending this logic, if we have an n-sided symmetrical object, each fundamental event would be assigned a probability of 1 divided by n: each of the 6 fundamental events from rolling a die once (see S_7) is assigned a probability of $\frac{1}{6}$.

When we want to indicate that an object is symmetrical and that all fundamental events should be assigned the same probability, we will refer to the item as "fair": for example, a "fair" coin or a "fair" die.

Of course, if the coin is not perfectly symmetrical or if the flipping process favors one side, the probability numbers assigned to the fundamental events will not be equal. Since most random processes are of this type, the reader may wonder why we worry about symmetrical objects at all. There are two reasons. First, although not precisely accurate, the assumption of symmetry and equal probabilities is often sufficiently accurate to be useful. Second, it provides a base from which we can measure deviations from "fairness."

When a coin is not "fair," we can no longer rely on our experience with similar objects or situations to help establish reasonable probabilities. An alternative is to experiment with the object by flipping it many times. Given a sufficiently large number of flips, we can then use the relative frequency of occurrence of each outcome as its probability. (So that is the reason for all that relative frequency stuff in Chapter 3, you say.) Thus, if 1,000 flips yield 520 heads and 480 tails, we might let $P(h) = 0.52$ and $P(t) = 0.48$. We should be aware that 1,000 flips is not sufficient to establish the probability for certain. Indeed, that cannot be done; certainty is not possible. But the relative frequencies may still be the most useful data that we have. We should also note that the probability of a head might still be one-half even if 1,000 flips produced 520 heads. Even with a "fair" coin, we would not expect to obtain exactly half heads. Try it on 10 flips and see! (If you obtained exactly 5 heads, try 11 flips.)

The relative frequency concept of probability just described is the most common. Under this interpretation the probability of an event is considered to be the proportion of times that the event will occur over the long run, a notion proposed by the French mathematician Siméon D. Poisson in 1837. Using the relative frequency notion, an event's probability is based on the relative frequency with which the event *has* occurred in the past. Thus, if T. P. Lerr (remember him?) of the Bubs Beer Company has received an order from the Tipsy Tavern in 7 of the last 10 visits, he might assign a probability of 0.7 to the event "order" on his next visit.

Of course, Mr. Lerr should be careful that nothing has changed to make his historical data no longer relevant. Remember: he wants a measure of how often the event will occur in the future. If there is a new owner at the Tipsy Tavern or if hard times have hit the area, the historically established probability may be inaccurate.

What could one, namely T. P. Lerr, do if things have changed? Many of those steeped in the mysteries of probability would say there is no probability number that can now be applied to the event "order." These statisticians would only use probabilities based on objective evidence, such as historical frequencies.

Others, however, would argue that the decision maker might be able to use knowledge of the situation (or similar situations) to establish probabilities that are useful in making decisions. These statisticians favor using what they term *subjective, intuitive,* or *personal probabilities.* Thus, if the new owner of Tipsy Tavern favors Bubs beer and is more easily swayed by Mr. Lerr's loquaciousness than the prior owner, the probability of an order may even exceed 0.7.

The notion of personal or subjective probability was suggested by James Bernoulli, who wrote in 1713 that probability is no more than the "degree of confidence" one has that something will happen. DeMorgan in 1847 and more recent writers have substituted "degree of belief." We need not take a position that accepts one of these positions (objective vs. subjective) to the exclusion of the other. The relevance of long-run historically or experimentally determined probabilities to many problems has been demonstrated. However, managers and other decision makers will continue to modify conclusions based on objective data using their understanding of the problem at hand. For this reason alone, it is useful to understand how subjective probabilities are presently obtained and used. We shall pursue this issue in Chapter 13.

S.S. A machine turns out castings. The relative frequency of defective castings has been 1 out of 100. List five reasons why this relative frequency might not be a useful measure of the probability of a bad casting in the future.

Solution: Some reasons are:

(a) An increased time rate of production, causing more defective items.

(b) Fewer defective castings, a result of workers learning how to do the job better.

(c) A new machine.

(d) A new operator.

(e) A new inspector or different definition of defective.

(f) New raw materials.

Many other possibilities exist. We have only listed a few.

PROBABILITIES AND ODDS

The concern with uncertainty and hence with probability goes way back in time, no doubt to before recorded history. We know that the Chinese writer Sun-Tze discussed probability in the first or third century B.C., that backgammon was played by the Pharaohs, that dice have been found in the excavations of prehistoric civilizations, and, as Raiffa writes, "perhaps even the cavemen tossed a bone to decide minor matters." But it seems that the notions were first formalized during the sixteenth century by Gerolamo Cardano, and others, in relation to gambling.

Cardano and others of his time and before used the idea of "odds." The probability of an outcome was considered to be the ratio of the ways the outcome could occur to the total number of possible outcomes. In effect, all possible outcomes were considered to be equally likely. This is an approach consistent with the concept of long-run frequency discussed earlier. Moreover, the use of this concept requires that each of the outcomes have the same probability.

Consider, for example, the roll of a single die. There are 6 outcomes or fundamental events, each of which may be assumed to have the same probability, if the die is fair. They are the elements in the sample space $S_7 \equiv \{1, 2, 3, 4, 5, 6\}$. Hence, the probability of a 6, as we know from our earlier study of symmetrical objects, is

$$\frac{\text{number of outcomes labeled 6}}{\text{number of possible outcomes}} = \tfrac{1}{6}.$$

The concept of "odds" is embodied in these considerations. You have no doubt heard it said that the odds on a horse in a particular race are 3 to 1. Odds in horse races are the odds against the horse winning. This means that if the total number of equally likely outcomes (or fundamental events) were four, three of these would be the case where the horse loses, and in the one other case, the horse would win. Hence, odds of 3 to 1 that an event will occur (the horse loses here) imply a probability of $\tfrac{3}{4}$ for that event. (Someone seems to be forever overestimating these odds, at least for your authors' race selections.) Note that in this case the fundamental events are not defined to be that a given horse loses, for all horses; the odds on a given horse can be 1 to 4 even if there are, say, 10 horses in the race. The events are more complex in the case of horse racing.

What are the odds of a 6 in one roll of a die? Since a 6 represents one of 6 equally likely fundamental events, the odds are 1 to 5. In general:

> If the *odds* in favor of an event are b to c, then the *probability* of the event is $b/(b + c)$.

In 1926 the English mathematician Frank Ramsey used the notion of odds and betting behavior to operationalize subjective probability. Suppose there are two equally attractive outcomes, call them E and F. In essence he argued that if someone is willing to bet b dollars that event E will occur versus c dollars that the equally attractive event F will occur, then $P(E) = b/(b + c)$. The individual's betting behavior is used to establish his personal or subjective probabilities.

The words "equally attractive" in the statement are important. Consider again the roll of a single die. Assume that we may bet on the event $\{6\}$ or the event $\{1, 2, 3, 4, 5\}$, that is, on the event "six" or the event "nonsix." The die is rolled and we are paid $6 if the event we select occurs. If we would bet $1 on the event "six," then we should be willing to bet $5 on the event "nonsix." The betting odds on the event "six" are 1 to 5 and the probability of a six is $1/(1 + 5) = \frac{1}{6}$.

S.S. Answer the following questions.

(a) If the odds favoring the event E are 2 to 1, what is the value of $P(E)$?

(b) Since the sample space for two tosses of a fair coin can be described by $W \equiv$ {two heads, two tails, one head and one tail}, is the probability of two heads in two tosses $\frac{1}{3}$?

Solution:

(a) $P(E) = 2/(2 + 1) = \frac{2}{3}$.

(b) No. The probability $\frac{1}{3}$ was formed by taking the ratio of outcomes in W where two heads occurred, to the total number of possible outcomes. But this only works when all outcomes are equally likely, and this is not true here. A sample space with equally likely events is given by

$$V \equiv \{(h, h), (h, t), (t, h), (t, t)\},$$

and the correct probability is given by $\frac{1}{4}$. (The practical gambler knows this result without a formal knowledge of probability theory, as you would soon find out if you were to "odd-man" with him.)

PROBABILITIES AND UNIQUE CASES

When T. P. Lerr calls on a potential customer, it is a unique event. It is not exactly like calls on other customers or even previous calls on that particular customer. Are probabilities meaningful in the case of such unique events? As another example, of what relevance would it be to know that the probability of an accident during a normal driving day on the freeway is less than 1 chance in 1,000? After all, each trip is unique. The individual who makes the trip may have even less patience with such probabilities quoted to him by a friend from the foot of his hospital bed.

Although it may seem reasonable to apply probabilities only to a long series of identical events, the last event in any series (after the earlier ones have been completed) is a unique event: the 1,000th toss of the same coin is a unique event. But if probabilities are applicable to each trial in a series, they are also applicable to the last unique trial. Consider the probability of rain on a specific day or the probability that an investment will break even. There is only one day for which the weather forecast is made, and there is only a single investment that will or will not break even. They are not repeated processes; these things either happen or they do not. Yet it is reasonable to consider such probability statements as reflecting the long-run frequency of occurrence of similar events under identical (or nearly identical) circumstances.

Of course, if there are reasons why some of the data used to estimate a desired probability are inappropriate, the offending data should be excluded. But the expeditious use of available information can be valuable in planning space travel, introducing a new federal program, using a new drug or therapy, or planning for new products and productive methods, although the precise activities have not been done before.

S.S. A manager of a toy company is considering the introduction of an electric soccer game to the market. To estimate the growth potential for this game, the manager is considering use of data on the company's electric hockey game. Is this wise?

Solution: The data on sales growth for the hockey game may be useful if the manager is careful to adjust or at least recognize the tentative limitations of this historical evidence. Problems can arise because of competition, less interest by the public in soccer, lack of interest in a second and similar game, better-quality production, and improved distribution or sales. These are only a few of the possibilities.

PROBABILITIES ZERO AND 1

What does it mean to assign a probability of zero or of 1 to some event from an uncertain process? Our intuition tells us that if the probability given an event is zero, then it will not occur. And a probability of 1 implies that the event must occur. But this removes the uncertainty. There is nothing uncertain about an event that cannot occur or that must occur. Sometimes we hear on the radio that the probability of rain is 100 percent. This simply means that it is raining, has rained, or the forecaster believes it is certain to rain in the region covered by the forecast. Again, there is no uncertainty.

We, and most others, use the probability number zero to represent the likelihood of events that are too unusual to worry about. We wish to omit them from further consideration. For example, a possible sample space for one flip of a coin could be written

{head, tail, on edge, a bird got it, down a hole}.

For most normal considerations, we would assign zero probabilities to the last three fundamental events and thereby reduce the sample space to {head, tail} $\equiv$ {h, t}.

Still another example is provided by the monkey placed at a typewriter and asked to type Lincoln's Gettysburg Address, word for word, without a single mistake. A sample space for this process is {success, failure} $\equiv$ {s, f}. We feel reasonably comfortable with the subjective probability assignment $P(s) = 0$ and $P(f) = 1$, even without specifying the number of monkeys involved, the available time, or observing any data that might be available. (One recent experiment reported to your authors, produced "Four score and seven years ago our fathers brought forth on this continent a new gazorgenplex....")

In a similar vein, we reserve the probability one for those events that are so nearly certain that we are willing to assign zero probability to all other possibilities. We are willing to assign a probability of 1 to the event that it will rain in Seattle sometime during the next year. (Indeed, some would not require such a long period of time in Seattle or even in San Francisco, for that matter.) Note that the process is uncertain and the events take place in the future, yet a probability of 1 may be still appropriate.

S.S. Suppose that we flip a coin and note the following sequence on the first four flips: head, head, head, head. What probability would you assign to the event head for the next flip?

Solution: Although you may not be happy with this answer, we would pick "it depends." If we originally believed the coin was a "fair" coin and that the flipping process does not favor one side over the other, we might be willing to assume that the coin is "fair" and the results above, somewhat unusual. Under these assumptions we might set $P(h) = 0.5$. The evidence is not considered sufficient to alter our prior beliefs based on coin flipping and similar experiences with symmetrical objects. By the way, four heads in a row will occur one time in 16 on the average even with a "fair" coin (we will learn how to calculate such astounding probabilities shortly).

Uncertainty, Sample Spaces and Probabilities

On the other hand, if we observed 10 heads in a row, we might begin to question the fairness of the coin-flipping task. And, we might even do so after four flips. In this case we might conclude that $P(h) \neq 0.5$. Yet it still seems unreasonable to set $P(h) = 1.0$, which implies that $P(t) = 0$. Do we really believe that a tail is impossible?

Twenty heads in a row might lead us to let $P(h) = 1.0$, since a fair coin will land heads 20 times in a row one time in slightly over 1 million (1 in 1,048,576 if you like to be precise about such matters) on the average.

5-5 Addition Rule of Probability

We have already calculated a number of probabilities. The rules presented in this section will allow us to combine the probabilities of some events to obtain the probabilities of other events. This is one of those tasks pleasurable to a statistician and necessary in problem solving.

In Section 5-3 we studied the sample space $S_7 \equiv \{1, 2, 3, 4, 5, 6\}$ for one roll of a die. If the die is fair, we know that each fundamental event (outcome) has the same probability, and this probability is given by the number of ways the event can occur divided by the total number of equally likely fundamental events. Hence,

$$P(1) = P(2) = P(3) = P(4) = P(5) = P(6) = \tfrac{1}{6}.$$

Suppose, now, that we request the probability of the event "odd." A sample space that might be relevant is {odd, even} or {{1, 3, 5}, {2, 4, 6}}.

The event "odd" is given by {1, 3, 5}. Since the events "odd" and "even" are equally likely, $P(\text{odd}) = \tfrac{1}{2}$. But this probability is also equal to the sum of the probabilities of the (mutually exclusive) fundamental events that comprise it.

$$P(\text{odd}) = \tfrac{1}{2} = \tfrac{3}{6} = \tfrac{1}{6} + \tfrac{1}{6} + \tfrac{1}{6} = P(1) + P(3) + P(5).$$

This suggests our first rule for combining probabilities.

> Rule 1: *Addition Rule for Mutually Exclusive Events*
> Let an event A be the union of a number of mutually exclusive events $B_1, B_2, \ldots, B_i, \ldots, B_n$. Then the probability of the event A is the sum of the probabilities of the mutually exclusive events. Symbolically, if
>
> $$A = B_1 \cup B_2 \cup \cdots \cup B_i \cup \cdots \cup B_n,$$

then

$$P(A) = P(B_1) + P(B_2) + \cdots + P(B_i) + \cdots + P(B_n)$$
$$= \sum_{i=1}^{n} P(B_i). \tag{5-1}$$

Rule 1 is satisfactory as long as the component events are mutually exclusive. But it must be generalized before it can be used in situations where the events are not mutually exclusive. For an example of such a case, consider the sample space for two tosses of a fair coin,

$$V \equiv \{(h, h), (h, t), (t, h), (t, t)\}.$$

Each outcome in V has a probability of $\frac{1}{4}$. Now suppose that we wish to know the probability of tossing either a head on the first toss or like sides on two tosses. Call this event M. The three outcomes that satisfy this specification are (h, h), (h, t), and (t, t). These are the three outcomes in the event M. $M \equiv \{(h, t), (h, h), (t, t)\}$. Since the three outcomes are mutually exclusive (they cannot occur simultaneously), we can use rule 1 to obtain the probability of the event we desire, M. This probability is

$$\tfrac{1}{4} + \tfrac{1}{4} + \tfrac{1}{4} = \tfrac{3}{4} = P(M).$$

But there is a more general method for finding these probabilities that will be useful to know. The event "a head on the first toss" is given by

$$\{(h, h), (h, t)\} \equiv A$$

and the event like sides is given by

$$\{(h, h), (t, t)\} \equiv B.$$

The probability of each event is $\frac{1}{2}$, using rule 1. If we were told to add the probabilities of the events A and B as a possible way of obtaining the probability of M, we would obtain $P(A) + P(B) = \frac{1}{2} + \frac{1}{2} = 1$. But this answer exceeds the correct probability of $\frac{3}{4}$. What have we done wrong? Perhaps the picture in Figure 5-10 will suggest our mistake. See if you can spot the problem before we describe it.

The problem is that in adding the probabilities for the two events, we have added the probability of the outcome in their intersection, namely, $P(h, h)$,

Uncertainty, Sample Spaces and Probabilities

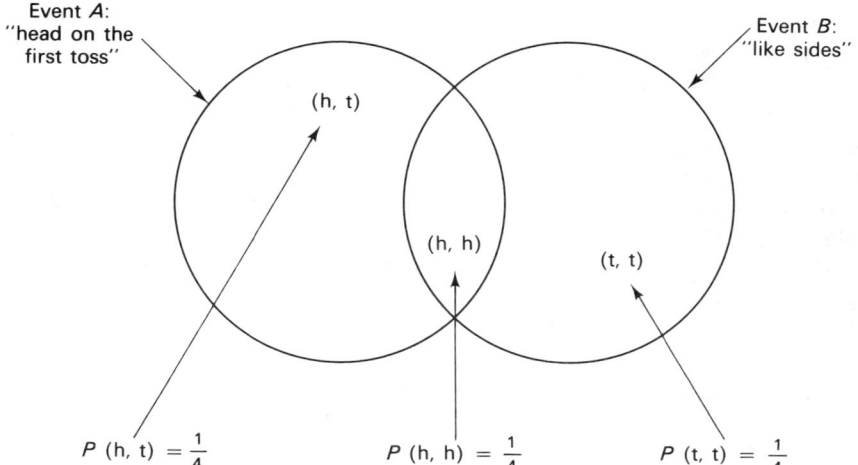

Figure 5-10 Addition Rule When Events Intersect (Events Are Not Mutually Exclusive)

twice. But it should be included only once. The means used to correct for this is simply to subtract the probability of the intersection. Thus,

$$\begin{aligned} P(M) &= P\{(h, t), (h, h)\} + P\{(h, h), (t, t)\} - P(h, h) \\ &= P(A) + P(B) - P(A \cap B) \\ &= \tfrac{1}{2} + \tfrac{1}{2} - \tfrac{1}{4} \\ &= \tfrac{3}{4}. \end{aligned}$$

This example suggests the general way in which probabilities can be added.

Rule 2: *General Addition Rule for Two Events*

Let A and B be two events. Their intersection is given by $A \cap B$, which may be empty. Then the probability that either A or B occurs is given by the probability of A plus the probability of B, less the probability of their intersection.

Symbolically, if A and B are two events, then

$$P(A \text{ or } B) = P(A \cup B) = P(A) + P(B) - P(A \cap B). \tag{5-2}$$

Several comments concerning this rule are in order. First, note the stress on the word "or." It is useful to verbalize any probability calculation to see if we wish to find the probability that *either* one event *or* some other event will occur. If so, the general addition rule will help us. Later we shall learn what to do if we want the probability that one event *and* some other event *both* occur.

Second, the general addition rule can be used when the two events are mutually exclusive. When this is the case, the probability of the intersection is zero and the last term in rule 2 disappears: $P(A \cup B) = P(A) + P(B)$ when A and B are mutually exclusive. This is the special case given by rule 1.

Finally, the general addition rule has only been defined for two events. It is really not so general after all. What do we do when there are three or more events, all of which may overlap? Well, this is a sticky wicket. Fortunately, most managerial situations can be handled considering two events at a time, and hence we elect to omit a formal discussion of this problem. For those not faint of heart, we examine this more general case in Problem 5-25.

S.S. Consider the case of Mr. T. P. Lerr making two calls on two separate customers. A relevant sample space is:

$$S_6 \equiv \{(\text{order, order}), (\text{order, no order}), (\text{no order, order}), (\text{no order, no order})\}.$$

If the probabilities of the four fundamental events in S_6 are 0.04, 0.16, 0.16, and 0.64, respectively, select the correct probability for each of the following events:

	Probabilities
(a) The first customer orders	0.04, 0.16, 0.20, 0.32, 0.40
(b) At least one order	0.04, 0.20, 0.32, 0.36, 0.40
(c) Either the first customer ordered, or both customers ordered	0.04, 0.20, 0.32, 0.36, 0.40

Solution: The event the first customer orders is given by the event {(order, no order), (order, order)}. The probability of this event is $0.04 + 0.16 = 0.20$. For part (b), the event is given by

{(order, order), (order, no order), (no order, order)}.

Using rule 1, the probability is obtained by adding the probabilities of the individual events,

$$0.04 + 0.16 + 0.16 = 0.36.$$

Two events are described in part (c). The first customer ordered is the event {(order, order), (order, no order)} $\equiv A$, with a probability given by rule 1 of $0.04 + 0.16 = 0.20$. Both customers ordered is the event {(order, order)} $\equiv$

UNCERTAINTY, SAMPLE SPACES AND PROBABILITIES   **145**

B, with probability 0.04. Since these two events intersect at the point (order, order), we can use rule 2 to obtain

$$P(A \text{ or } B) = P(A) + P(B) - P(A \cap B)$$
$$= 0.20 + 0.04 - 0.04$$
$$= 0.20.$$

This is the same answer that we obtained in part (a). We should not be surprised, however, since the two events are the same, only described differently. This shows you how tricky we can be!

5-6 A Warning

Up to this point we have typically been given the probabilities of some events, often the fundamental events, and asked to find the probabilities of others. In real-life problems, we will be forced first to establish the probabilities of the fundamental events. Even with historical data this may not be easy to do because conditions are continually changing.

A related problem is to establish an appropriate sample space that will help us answer the questions of interest. This sample space must include all the fundamental events that may occur (those with nonzero probability). These fundamental events must not overlap, and they must exhaust all the possibilities. Finally, the fundamental events need to be defined in such a way that the more complex events of interest to the manager can be expressed in terms of the union of two or more of the fundamental events.

Performing these tasks well is an art that only experience will provide. Some of that experience can be obtained by problem solving, but much of it awaits us after graduation.

Finally, we have not yet earned our fee consulting for the Bubs Beer Company. We have told them relatively little about what they should expect demand to be. In Chapter 6 we will learn some additional rules of probability which, together with what we already know, will permit us to do a somewhat better, if not complete, job. Our friend T. P. Lerr must bear with us a while longer.

TECHNICAL NOTE 5-1: SETS

The algebra of sets is used extensively in this chapter. This technical note discusses a few of the essential elements. The basic notion of a set is intuitive. It is just a well-defined group of elements. Examples include the present Congress, the set of items you are wearing now, the currently outstanding stock certificates of a particular company, the positive integers less than 5, the countries belonging to the United Nations, the assets of some organization, and so on.

Although the notion of a set is not defined, it has two important characteristics or properties. They are:

1. The set must be "well defined." That is, we must be able to determine whether any element belongs to the set or not.

2. The elements in the set must be distinct. That is, we must be able to tell them apart; the same element can be counted as a member of a given set once, at most.

We use the symbol $\in$ to indicate that an item belongs to a set and the symbol $\notin$ to indicate that the item does not belong to the set. Elements of the set are represented in general by lowercase letters, and the set itself (usually but not invariably) by a capital letter, by braces around the elements, or by some defining property. For example, consider the set of positive integers less than 6. We could write

$$I \equiv \{1, 2, 3, 4, 5\} = \{4, 3, 1, 5, 2\} = \{x : 0 < x \leq 5 \text{ and } x \text{ an integer}\}.$$

In this case we can write further that $3 \in I$ but $6 \notin I$. The letter x is a variable over the set, and the elements are separated by commas when listed separately. Note that the order in which the elements are listed is not important.

New Symbol	Meaning
$I, M, R, \ldots$	Capital letters stand for a set.
$x, y, z, \ldots$	A variable that can be used to represent any element of a set.
$\{\ \}$	Braces used to enclose the members of a set or the defining property of a set. Sets are often written by listing the elements within braces.
$\in$	Set inclusion symbol.
$\notin$	Set exclusion symbol.

An ordered set is a set in which the characteristic of order is important. Ordered sets are written using parentheses. The ordered set of positive integers less than 6 is written (1, 2, 3, 4, 5). This set is not the same as the set (4, 3, 1, 5, 2). Sometimes order is an important characteristic, say, of a set of observations and sometimes it is not. Hence, it is useful to have a means of indicating order.

UNCERTAINTY, SAMPLE SPACES AND PROBABILITIES

New Symbol	Meaning
(,)	The ordered set of elements appears within the parentheses separated by commas.

Several special sets are worth defining. First there is the empty, sometimes called the null, set. The *null* set has no elements. It is written $\emptyset$ or { }.

New Symbol	Meaning
$\emptyset$ or { }	The null or empty set. A set with no elements.

Second, the *universe set* is the set of all possible elements under consideration. If we wish to deal with the positive integers less than 6, then I is the universe set. Usually, the universe set is quite large, and it can be infinitely large: for example, if the set of possible elements under consideration were to be all the positive integers or the real numbers between 1 and 2. Statistics often involves looking at a portion of some universe set, or simply universe, and trying to infer what is true in the universe from the portion (sample) examined. Hence, it is important that we agree on the appropriate universe set as a first step in any analysis.

Third, a subset of some other set is itself a set. A *subset*, call it B, is a subset of a set A if every element of B is an element of A. The subsets of the set {1, 2, 3} are: {1}, {2}, {3}, {1, 2}, {1, 3}, {2, 3}, {1, 2, 3}, and $\emptyset$. It turns out to be useful to define $\emptyset$ as a subset of any set. The notion of a subset is not useful for ordered sets. We write $B \subseteq A$ to denote that B is a subset of A.

New Symbol	Meaning
$\subseteq$	The set on the left of the symbol is a subset of the set on the right.

We conclude this discussion of sets with a look at several operations that can be performed on sets. The *union* of two (or more) sets, call them A and B, is the set of elements in A or in B or in both. If $A \equiv \{1, 2\}$ and $B \equiv \{2, 3\}$, then the set union, written $A \cup B$, is the set {1, 2, 3}. The *intersection* of two (or more) sets is the set of elements simultaneously common to each. For the sets A and B defined in this paragraph, we write $A \cap B = \{2\}$. Finally, the difference of two sets is the set of all elements in the first which are not in the second. Again, for the sets A and B, we write $A - B = \{1\}$.

New Symbol	Meaning
$A \cup B$	The union of the sets A and B. The set of all elements in A or in B or in both.
$A \cap B$	The intersection of the sets A and B. The set of all elements that are in A and B simultaneously.
$A - B$	The difference of the sets A and B. The set of elements in A but not in B.

These three concepts are illustrated in Figure 5-1A.

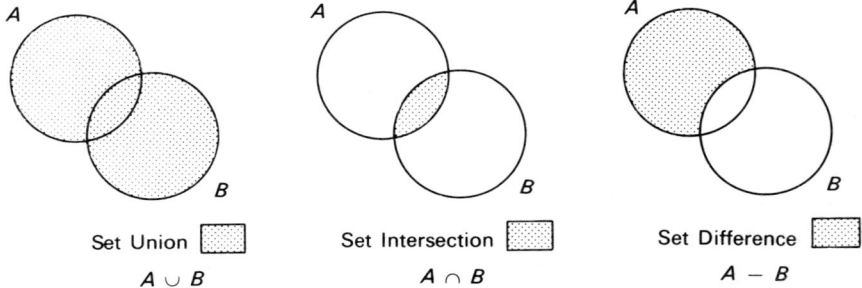

Figure 5-1A Set Union, Intersection, and Difference: Illustrated for Two Sets

New Symbols

Symbol	Meaning
$\cap$	Intersection: the elements simultaneously in all sets under consideration.
$\cup$	Union: the elements in each set under consideration.
$P(O_i)$	Probability of the outcome O_i.

Other set symbols are defined in Technical Note 5.1.

Key Formulas

Formula	Used to Compute:
$P(A) = \sum_{i=1}^{n} P(B_i)$	Probability of an event that can be written as the union of n mutually exclusive events.
$P(A \cup B) = P(A) + P(B) - P(A \cap B)$	The probability that either the event A or the event B or both occur.

UNCERTAINTY, SAMPLE SPACES AND PROBABILITIES

PROBLEMS

5-1. What is a probability?

5-2. What probability would you assign to a die landing on a corner or on an edge and remaining there? What does this probability mean?

5-3.* Provide a sample space for the inspection of a part using a go–no go gauge.

5-4.* What is the major flaw in the following statement? "Since there were fewer airplane accidents in 1940 than in 1960, it was safer to fly in the forties."

The following description holds for Problems 5-5 to 5-8.
A sample space for an opinion poll is given by

$$S \equiv \{\text{in favor, opposed, no opinion}\} \equiv \{f, o, n\}.$$

5-5.* Define all possible events for this sample space.

5-6.* In Problem 5-5, how many fundamental events are there? How many compound events?

5-7.* For Problem 5-5, define symbolically the event "I have an opinion."

5-8. Show the event "I have an opinion" using a Venn diagram. Are the fundamental events based on S mutually exclusive?

5-9.* If $P(A) = \frac{1}{3}$, $P(B) = \frac{1}{5}$, and $P(A \cap B) = \frac{1}{4}$, find $P(A \cup B)$. Recall that $(A \cap B)$ means A and B, while $(A \cup B)$ means A or B.

5-10.* Consider the "tossing" of a tack. Assume that only two positions, called position A and position B, will occur.

Position A   Position B

We suspect you have had little experience with "tossing" tacks, yet we ask you to give us an estimate of the probability that tacks X and Y below, if tossed, would land in position A. If you find this difficult, can you estimate for which tack this probability would be smaller?

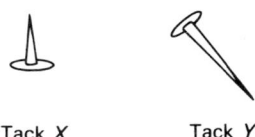

Tack X   Tack Y

5-11.* A national franchising company is interviewing prospective buyers in Indiana. The odds that an interviewee in the Muncie area will buy a franchise are 2 to 1. The odds that either an interviewee in the Fort Wayne area or an interviewee in the Muncie area will buy a franchise are 3 to 1. Consistent with this information, what are the smallest and largest possible values for the probability that an interviewee in the Fort Wayne area will buy a franchise? Is the event that an interviewee in the Fort Wayne area purchases a franchise a fundamental event?

5-12. The promotional manager for Book-of-the-Month Club is attempting to decide from which categories to select books as featured selections for July. For the categories fiction, biography, and history (which may have books in common), draw Venn diagrams and shade in the areas that represent each of the following decisions:
(a) He will offer only fiction and no books in other categories.
(b) He will offer either fiction or biography, but not both and not history.
(c) He will offer no books from those categories, but he will offer books from other categories.
(d) He will offer books from all three categories only.

5-13. A manufacturer of ice cream products has three machines: A, B, and C, which produce ice cream novelties. Let

$A \equiv$ event that machine A breaks down on a particular day

$B \equiv$ event that machine B breaks down on the same day

$C \equiv$ event that machine C breaks down on the same day.

Using only the symbols $\cap$, $\cup$, $'$, A, B, C, and parentheses, write expressions for the events:
(a) At least one occurs.
(b) Only A occurs.
(c) A and B occur, but not C.
(d) All three occur.
(e) None occurs.
(f) Exactly one occurs.
(g) Exactly two occur.
(h) At most two occur.

5-14. An inspector in a light-bulb-manufacturing firm selects bulbs from a lot just produced until she finds a defective bulb. Lot 20 contains 4 good bulbs and 1 defective bulb. Draw a tree diagram to represent the possible outcomes of the inspector's examination of lot 20.

5-15.* Which of the following pairs of events are mutually exclusive?
(a) In a shipment of light bulbs:
(1) Exactly one is defective.
(2) Two are defective.
(b) (1) The prime rate of interest in Chicago rose to $11\frac{3}{4}$ percent on a day in June.

Uncertainty, Sample Spaces and Probabilities

  (2) The prime rate of interest in Chicago dropped to $11\frac{1}{4}$ percent on a day in June.
  (c) A refrigerator manufacturer wants to build two new plants and is considering three cities: Cincinnati, Atlanta, and Philadelphia.
  (1) Cincinnati is selected.
  (2) Atlanta is selected.
  (d) In a bin of ballpoint pens just produced:
  (1) Five are defective.
  (2) At least one pen is defective.

5-16.* Britelite Company is considering the purchase of Buyme Corporation. Analysts estimate that the probability that the purchase will occur is 0.6. What are the odds that Britelite will not purchase Buyme?

5-17. An electronics firm wants to develop a new component for its electronic calculators. It assigns the development task to its own research team. It knows that a competing firm is trying to develop the same component. The company believes that there is a 0.6 probability that the company's own research staff will develop the component and that there is a 0.8 probability that the competing firm will develop the component. It also believes that there is a 0.5 probability that both of them will develop the component. What is, then, the probability that the needed component will be developed?

5-18. A clothing manufacturer produces three garments: shirts, pants, and coats. She can produce only one product on any given day. Let X, Y, and Z be shirts, pants, and coats, respectively, and let (X, Y) represent the fact that she has produced X on the first day and Y on the second day. Define the events:

$A \equiv$ manufacturer produces the same product on both the days

$B \equiv$ manufacturer produces product X in at least one of the 2 days

$C \equiv$ manufacturer produces Z on the first day.

(a) Describe the sample space for the manufacturing process for 2 days.
(b) Let all the outcomes in the sample space be equally likely and compute the following probabilities:
  (1) $P(A)$
  (2) $P(B)$
  (3) $P(C)$
  (4) $P(A \cap B)$
  (5) $P(A \cap C)$
  (6) $P(A' \cap C)$
  (7) $P(A \cup B \cup C)$
  (8) $P(B \cup C)$
  (9) $P(B \cap A')$
  (10) $P(A \cap B \cap C)$

5-19. A machine in a manufacturing plant consists of three major components. The machine is designed in such a way that it will function effectively as long as two of the three components are operative. The three components will be designated A, B, and C. The components are subject to completely different stresses, and, as a result, the fact that one component is damaged will in no way influence the likelihood that damage has been suffered by the other two.
  (a) Draw a Venn diagram showing the interrelationship of the three subsets of interest.
  (b) Shade in the portion of the Venn diagram corresponding to the inability of the machine to function.

5-20. Explain why each of the following claims must contain a mistake:
  (a) A quality control engineer claims that the probabilities that a large shipment of plate glass will contain 0, 1, 2, 3, 4, or 5 defectives are 0.11, 0.24, 0.37, 0.16, 0.09, and 0.05, respectively.
  (b) The probability that there will be 0, 1, 2, or 3 or more fires in a company's warehouse during a particular year are 0.47, 0.25, 0.13, and 0.05, respectively.
  (c) The probability that during a policy year an insurance company will receive one claim from a married male over 25 years of age is 0.17, and the probability that during the same policy year the company will receive one or more claims from a married male over 25 years of age is 0.11.
  (d) The probability that a company will have a fire in warehouse A during a particular year is 0.25, and the probability that the company will have a fire in warehouse A as well as warehouse B in that same year is 0.35.

5-21. An employee for a ballpoint-pen-manufacturing firm tests each pen as it is produced and continues testing until he finds one that is defective. Describe a sample space of possible outcomes for the testing procedure.

5-22. An investment company plans to add to its portfolio two securities from a list of five recommended stocks: Britelite, Leanto, Funtoy, Eatmore, and Claybird. Describe the sample space that represents the possible choices.

5-23. A firm in Dayton, Ohio, ordered a specialized machinery component from a London firm. The company in London can fly the part to New York, Philadelphia, or Chicago. Once it reaches one of these cities, it can be sent by either plane, train, or truck to Dayton. Use a tree diagram to describe all possible shipping routes.

Uncertainty, Sample Spaces and Probabilities

5-24. A food processor is attempting to decide from which firm to buy its packaging boxes. The plant manager tests three boxes for strength. The test subjects the box to a given pressure. One box is from firm A, one from firm B, and one from firm C. State whether each of the following is a fundamental event, and if the event is not a fundamental event, list the fundamental events of which it is composed.
   (a) Only manufacturer A's box is crushed.
   (b) Manufacturer A's box is crushed.
   (c) Two of the three boxes are crushed.
   (d) None of the boxes is crushed.

5-25. In this chapter we discussed the general addition rule for the probability that either of two events would occur.
   (a) What would the general addition rule be for three events? Use Figure 5-4 to help you.
   (b) Can you write the general addition rule for n events?

Chapter 6

Counting, Probability Distributions, and Conditional Probabilities

The topic of probability is just too large for a single chapter. In this chapter we continue the discussion of probabilistic topics that are central to the resolution of statistical questions. Specifically, we will examine first the question of counting the number of outcomes from an uncertain or random process. This task can be useful in establishing the probabilities of events based on a random process. We will then turn to the question of what can be done when there are too many outcomes to count.

Sometimes the probabilities of certain events are altered by the occurrence of other events. For example, the likelihood of a misstatement increases with the number of words we write. The probability of stubbing one's toe decreases if shoes are worn. The chance that a machine will function properly is increased if it is serviced regularly. The odds that an employee will leave an organization rise if the employee is unhappy. Such probabilities are called conditional probabilities. A substantial portion of this chapter is devoted to this topic.

6-1 Counting Techniques

To this point in our discussion we have dealt with relatively simple sample spaces. First, these sample spaces have contained only a few outcomes.

COUNTING, PROBABILITY DISTRIBUTIONS, AND CONDITIONAL PROBABILITIES

Tossing a single coin once led to a sample space with only two outcomes, head and tail. Rolling a die once produced a six-outcome sample space.

Second, and of equal importance, we dealt with processes in which each outcome in the sample space was as likely as any other. Thus, each face on a die is equally likely to turn up on one roll of a fair die. Given equally likely outcomes, we saw in Chapter 5 that the probability of any event, E, defined over a sample space, S, could be calculated as

$$P(E) = \frac{\text{number of outcomes leading to } E}{\text{number of outcomes in } S}. \qquad (6\text{-}1)$$

Thus, if E is the event "odd face on one roll of a fair die," then

$$E \equiv \{1, 3, 5\}, \qquad S \equiv \{1, 2, 3, 4, 5, 6\},$$

and

$$P(E) = \tfrac{3}{6} = \tfrac{1}{2}.$$

If the outcomes are equally likely and we can count them, formula (6-1) can be used to obtain the probability of interest.

When there are only a few outcomes in E and S, determining the probability of the event E is easy. But when the sample space, S, contains many elements, it is often a difficult task to count them. The purpose of this section is to provide several methods of counting the outcomes in a large (but finite) sample space. We will learn to count the number of elements in S so that we can answer three questions. Suppose that there are n outcomes in S and let r be an integer equal to or less than n; $r \leq n$.

Question 1: How many ordered sets of size r can be obtained from S? Each ordered set of r outcomes (sometimes called an r-tuple) is known as a *permutation* of the r outcomes taken from S.

Question 2: How many sets of size r can be obtained from S? Note that this counting task does not involve order. Each set of r outcomes (a subset of S) is known as a *combination* of the r outcomes taken from S.

Question 3: How many ways can the n outcomes in S be placed into k classes (or categories) where there are n_1 outcomes in class 1, n_2 outcomes in class 2, ..., n_i outcomes in class i, ..., and n_k outcomes in class k? We require

$$n_1 + n_2 + \cdots + n_i + \cdots + n_k = n.$$

Soon we will illustrate each of these questions with an example. Before doing so, however, we introduce the basic technique by which such counting problems can be solved. This technique is called the *fundamental principle of counting*.

156 CHAPTER 6

FUNDAMENTAL PRINCIPLE OF COUNTING

Suppose that the coach of a little league baseball team must complete two tasks. For the first task, he must select a team name from the possibilities submitted by the team members. The list of suggested names is

$$\text{names} \equiv \{\text{streakers, old toads, hotdogs}\}.$$

He must, as his second task, select a color for the team's uniform from the following set of colors:

$$\text{colors} \equiv \{\text{red, puce, purple, chartreuse}\}.$$

We wish to know how many possible ways there are to complete the two tasks taken together, which we will call the job. Figure 6-1 illustrates all the

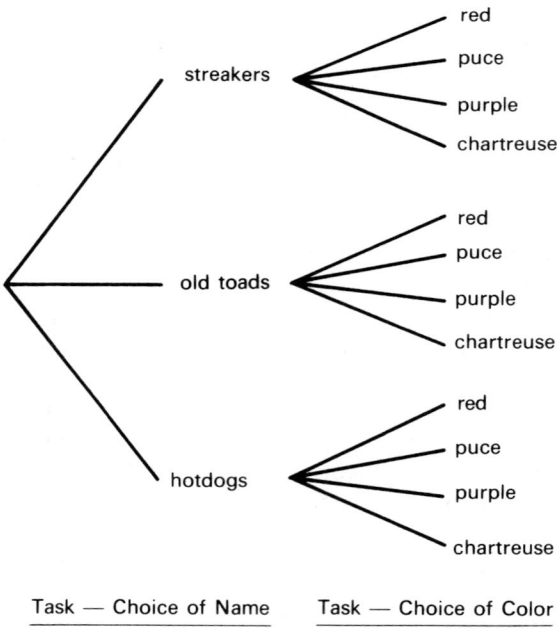

Figure 6-1 Selection of a Name and Uniform Color

possibilities using the tree-diagram technique introduced in Chapter 5. The tree diagram has 12 separate paths or end points. Each represents a different way of completing the coach's job. Do you think that if we had first selected a color and then a name, we would still have 12 paths? What does your intuition tell you? Are there fewer than 12 paths, exactly 12 paths, or more than 12 paths? (Draw the appropriate tree diagram and see that there are still exactly 12 paths.)

But surely there must be an easier way! Tree diagrams can become messy bushes very quickly as the number of tasks and the ways in which each task can be completed are increased. Well, we are in luck! There is an easier way. Recall that there were 3 possible names and hence 3 ways to complete the name-selection task. There were 4 colors and hence 4 ways to complete the color-selection task. The fundamental principle of counting says there will be $(3)(4) = 12$ ways, then, of doing the entire job.

More formally, if there are k tasks to be accomplished and if there are n_1 ways of doing the first, n_2 ways of doing the second, and so on down to n_k ways of doing the last task, then there are

$$(n_1)(n_2) \cdots (n_i) \cdots (n_k)$$

ways of doing the entire job.

Fundamental Principle of Counting
Given k tasks, where n_i is the number of ways of doing the ith task, then there are

$$(n_1)(n_2) \ldots (n_i) \ldots (n_k) = \prod_{i=1}^{k} n_i \qquad (6\text{-}2)$$

ways of performing all the tasks. (This principle can be proved by mathematical induction, but we will not do so here.)

It is important to note that the number of ways in which any one of the tasks can be performed is dependent neither on the number of ways that any other task can be performed nor on the order in which prior tasks are completed. (More complex counting procedures, which we will not study, consider situations where the ways task i can be done depend on which tasks have been completed already. See Problem 6-11.)

PERMUTATIONS AND COMBINATIONS

We are now in a position to illustrate the three questions referred to at the start of Section 6-1. In this subsection, we will treat the first two questions. To do so we will use the following example. Suppose that the marketing faculty at the Better Business School consists of the following five forlorn faculty members: Chase, Adams, Morton, Esmon, and Love. In more concise form:

$$\text{faculty} \equiv F \equiv \{C, A, M, E, L\}.$$

Suppose that we wish to select a committee of two members, a chairperson and one other person. (A camel, it will be recalled, is a horse designed by a committee.) Suppose that we define the tasks as follows:

Task 1: Select a chairperson.
Task 2: Select the committee member.

There are five possible choices for chairperson; $n_1 = 5$. There are four possible choices remaining for the committee member; $n_2 = 4$. Using the fundamental principle of counting there are $5(4) = 20$ possible committees. We could also have found this number by listing all the possible ordered pairs that could be formed from the five letters, such as (C, A), (A, C), and so on. We note that the ordered pair (C, A) differs from the ordered pair (A, C) since the first member of each pair is the chairperson. We could list all such pairs, perhaps using a tree diagram, to see that there are indeed 20.

But again there is an easier way to find the number of ordered pairs (or r-tuples for more than two elements) that can be formed from n things. In our committee example the number of ordered arrangements or permutations of 2 people, $r = 2$, taken from 5 people, $n = 5$, was 20. Symbolically,

$$5(4) = 20 = {}_5P_2,$$

where ${}_5P_2$ is the symbol for the number of permutations of 5 things taken 2 at a time. Rewriting, using the fact that $n = 5$,

$$n(n - 1) = {}_nP_2.$$

More generally,

$$n(n - 1) \cdots (n - r + 1) = {}_nP_r.$$

For our example,

$$(n - r + 1) = (n - 2 + 1) = (n - 1) = (5 - 1) = 4.$$

In practice, the formula is further simplified as follows:

$$[n(n - 1) \cdots (n - r + 1)] \frac{(n - r)(n - r - 1) \cdots 1}{(n - r)(n - r - 1) \cdots 1} = {}_nP_r.$$

Letting $n! = n(n - 1)(n - 2) \cdots 1$, this simplifies to

$$\frac{n!}{(n - r)!} = {}_nP_r.$$

Counting, Probability Distributions, and Conditional Probabilities

The number of *permutations* of n things taken r at a time is given by

$$_nP_r = \frac{n!}{(n-r)!}. \qquad (6\text{-}3)$$

For our committee example,

$$_nP_r = {_5P_2} = \frac{5!}{(5-2)!} = \frac{5!}{3!} = \frac{5 \cdot 4 \cdot 3 \cdot 2 \cdot 1}{3 \cdot 2 \cdot 1} = 20.$$

The exclamation point is notation for convenience only. It tells us to multiply together all integers from 1 up to and including the integer preceeding the exclamation symbol. We read $n!$ as "n factorial."

New Symbol	Meaning
$n!$	Multiply together all integers from 1 to n inclusive: $\prod_{j=1}^{n} j$.

Suppose that we were interested in the number of committees that we could form from our five friendly faculty members, where the first selected is chairperson, the second vice-chairperson, the third secretary, the fourth treasurer, and the last official custodian of the keg. Inspection tells us there are several possibilities, but using the permutation formula yields

$$_5P_5 = \frac{5!}{(5-5)!} = \frac{5!}{0!}$$

and it is not clear what $0!$ means. We shall define $0!$ to be 1. This we do because it works, always. Could anyone find a better reason than that? Hence, $_5P_5 = 5!/1 = 120$. You may count them if you wish, to see that we are right.

We shall now investigate a slightly different question. Suppose that we wish to select a committee of two from our five felicitous faculty but that we do not wish to designate a chairperson. How many such committees are possible? In this case the order of choice is irrelevant. The 20 committees we found earlier when a chairperson was designated included pairs distinguishable by order such as (C, A) and (A, C). If order is no longer important, only one of each two such committees counts. Only the top line, 10 of the 20 ordered pairs in the following array, can be counted:

(C, A) (A, E) (C, M) (A, L) (C, E) (M, E) (C, L) (M, L) (A, M) (E, L)
(A, C) (E, A) (M, C) (L, A) (E, C) (E, M) (L, C) (L, M) (M, A) (L, E)

The second line gives the same committees. There are 10 such committees. Another way to answer this question is to say that there are 20 permutations of 5 things taken 2 at a time, but only half of these count because each of these committees can be ordered in $_2P_2 = 2$ ways.

The number of unordered sets of r things taken from n is called the number of *combinations*, $_nC_r$. Using the data from the example,

$$\frac{(5)(4)}{2} = 10 = {_5C_2},$$

or rewriting using permutation symbols,

$$\frac{_nP_r}{_rP_r} = {_nC_r}.$$

Use of formula (6-3) for $_nP_r$ yields

$$_nC_r = \frac{n!/(n-r)!}{r!/(r-r)!} = \frac{n!}{(n-r)!\,r!}\frac{0!}{} = \frac{n!}{(n-r)!\,r!}.$$

The number of *combinations*, sets without regard to order, of n things taken r at a time is given by

$$_nC_r = \binom{n}{r} = \frac{n!}{(n-r)!\,r!}. \tag{6-4}$$

For our committee-without-chairperson example,

$$_nC_r = {_5C_2} = \frac{5!}{(5-2)!\,2!} = \frac{5 \cdot 4 \cdot 3 \cdot 2 \cdot 1}{(3 \cdot 2 \cdot 1)(2 \cdot 1)} = 10.$$

There is a close relation between combinations and permutations, as our committee example suggests. In fact, we can use the idea of a combination in answering a question that requires permutations. Consider again the job of selecting a committee of 2 with a chairperson from our 5 frolicsome faculty. Define the tasks as:

Task 1: Select the 2 faculty members to be on the committee.
Task 2: Select a chairperson from the 2 committee members.

Now task 1 asks for the number of combinations of 2 taken from 5. Using formula (6-4) for $_nC_r$, $_5C_2 = 5!/(5-2)!\,2! = 10$. Task 2 asks us to order the 2 committee members, the first being the chairperson. This ordering is a permutation of 2 things taken 2 at a time. Using formula (6-3), $_2P_2 = 2!/0! = 2$. Hence, using the fundamental principle of counting, the job of selecting a committee of 2 with a chairperson can be accomplished in $10(2) = 20$ ways. Symbolically,

$$_nP_r = (_nC_r)(_rP_r).$$

WAYS OF CLASSIFYING A SET OF OBJECTS

We are now ready for the third counting question. Let us suppose that faculty member Adams dies. (We did in one faculty member in order to make the problem small enough so that we could list all the alternatives.) The remaining four frugal faculty are arranging rides to his funeral in three available cars which can accommodate two, one, and one individuals, respectively. How many possible ways (arrangements) are there for the four remaining faculty to ride to this happy occasion in the three cars? The possibilities are listed below, where car 1 can accommodate two people and cars 2 and 3, one person each.

Car 1:	CM	CM	CE	CE	CL	CL	ME	ME	ML	ML	EL	EL
Car 2:	E	L	M	L	M	E	C	L	C	E	C	M
Car 3:	L	E	L	M	E	M	L	C	E	C	M	C
Way No.:	1	2	3	4	5	6	7	8	9	10	11	12

By simply listing the possibilities, we find that there are 12 possibilities. However, again there is an easier way. (Yes, we can use the fundamental principle of counting.) Let us define the tasks as follows:

Task 1: Select 2 people from 4 to ride in the first car:

$$_4C_2 = \frac{4!}{(2!)(2!)} = 6.$$

Task 2: Select 1 person from the remaining 2 to go in the second car:

$$_2C_1 = \frac{2!}{(1!)(1!)} = 2.$$

Task 3: Select 1 person from 1 to ride in the third car:

$$_1C_1 = \frac{1!}{(0!)(1!)} = 1.$$

Using the fundamental principle of counting there are $(_4C_2)(_2C_1)(_1C_1) = 6(2)(1) = 12$ ways. Using the combination formula and letting

$$n_1 = \text{number to ride in car } 1 = 2$$
$$n_2 = \text{number to ride in car } 2 = 1$$
$$n_3 = \text{number to ride in car } 3 = 1,$$

where $n_1 + n_2 + n_3 = n$, we have

$$_nC_{n_1}(_{n-n_1}C_{n_2})(_{n-n_1-n_2}C_{n_3})$$

or

$$\frac{n!}{(n-n_1)!\,n_1!} \cdot \frac{(n-n_1)!}{(n-n_1-n_2)!\,n_2!} \cdot \frac{(n-n_1-n_2)!}{(n-n_1-n_2-n_3)!\,n_3!}.$$

This equals

$$\frac{n!}{n_1!\,n_2!\,n_3!} \quad \text{since } (n-n_1-n_2-n_3)! = 0! = 1.$$

Ways to Place n Items in k Classes (Also the Number of Distinguishable Permutations of the n Items)
The number of ways *n* items can be placed in *k* classes so that there are n_1 in the first class, n_2 in the second class, and so on is

$$\binom{n}{n_1, n_2, \ldots, n_k} = \frac{n!}{n_1!\,n_2!\ldots n_k!}, \tag{6-5}$$

where $n_1 + n_2 + \cdots + n_k = n$.

USING COUNTING FORMULAS IN PROBABILITY

Let us try our wings a little to see how these formulas can be used to establish probabilities. Suppose that a committee consisting of a chairperson, a vice chairperson, and one additional member is drawn by lot from a group of 10 people that includes the authors of this book. What is the probability that your authors will fill the chairperson and the vice-chairperson positions? The probability of this unhappy event is given by the ratio of the number of committees with the authors in officer roles to the total number of three-person committees. The number of different three-person committees con-

COUNTING, PROBABILITY DISTRIBUTIONS, AND CONDITIONAL PROBABILITIES

sisting of a chairperson, a vice chairperson, and one other member that can be drawn from a group of 10 is given by

$$_{10}P_3 = \frac{10!}{7!} = 720.$$

The number of committees with the authors as officers is found using the following tasks and the fundamental principle of counting.

Task 1: Select the two authors. This can only be done in $_2C_2 = 1$ way.

Task 2: Select one other committee member from the eight remaining people. This can be done in $_8C_1 = 8$ ways.

Task 3: Order the set of authors with the first being the chairman and the second the vice chairperson. This task can be done in $_2P_2 = 2$ ways.

Now using the fundamental principle of counting, the selection of a committee with your authors as officers from a total of 10 people can be accomplished in $(_2C_2)(_8C_1)(_2P_2) = 1(8)(2) = 16$ ways. Hence the probability of such a committee is $\frac{16}{720} = \frac{1}{45}$, happily a relatively small number.

This is an example of using formula (6-1) to obtain a probability, since each committee has the same probability of selection.

So that you can claim to have learned something practical from this book, consider the game of poker. What is the probability of exactly one pair in a 5-card poker hand if no wild cards are allowed? The number of possible 5-card poker hands is given by $_{52}C_5 = 2,598,960$, since a deck of cards contains 52 cards. (If you do not believe it and you have a few hours of free time to waste, work it out!) The number of hands having exactly one pair is found using the following tasks.

Task 1: Select a face value from the 13 available for the pair:

$$_{13}C_1 = 13.$$

Task 2: Select the 2 cards to make the pair from the 4 with this face value:

$$_4C_2 = 6.$$

Task 3: Select the 3 remaining face values for the hand from the 12 remaining face values:

$$_{12}C_3 = 220.$$

Task 4: Select 1 card from each of the 4 cards with this face value for each of the 3 face values determined in the solution to task 3:

$$(_4C_1)(_4C_1)(_4C_1) = (_4C_1)^3 = 64.$$

Using the fundamental principle of counting yields

$$13(6)(220)(64) = 1{,}098{,}240.$$

The probability of a poker hand with exactly one pair is

$$\frac{1{,}098{,}240}{2{,}598{,}960} \approx 0.42.$$

As the average poker player knows, a hand with a pair is quite common on the deal; and hence equal or better hands are even more likely, particularly if the player is allowed to discard and draw additional cards in an attempt to improve the hand.

For those of you who prefer the sedate game of bridge, what is the probability that each of the 4 players is dealt a single ace? The total number of bridge hands is given by the number of ways that 52 cards can be placed into 4 hands (classes). This is given by

$$\binom{52}{13,\,13,\,13,\,13} = \frac{52!}{(13!)^4}.$$

To find the ways to divide the aces, one to a player, we use the following tasks:

Task 1: Distribute 4 aces, one to each player:

$$\binom{4}{1,\,1,\,1,\,1} = 4! = 24.$$

Task 2: Distribute the remaining 48 cards to the 4 players so each player receives 12:

$$\binom{48}{12,\,12,\,12,\,12} = \frac{48!}{(12!)^4}.$$

The probability we desire is now given by

$$24\,\frac{48!}{(12!)^4} \div \frac{52!}{(13!)^4} = \frac{24(48!)(13!)^4}{(12!)^4 (52!)} = \frac{24(13)^4}{52(51)(50)(49)} \approx 0.11.$$

Each player receives exactly one ace 1 deal out of every 9, approximately.

It is important to emphasize that solving probability problems using these counting methods only works when each outcome has the same probability. Only in such cases is the answer given by the ratio of the number of ways an event can occur to the total number of possible outcomes. For the majority of probability problems, the outcomes have different probabilities. However,

COUNTING, PROBABILITY DISTRIBUTIONS, AND CONDITIONAL PROBABILITIES **165**

as we shall see, several important problem types, including certain sampling techniques and problems involving the binominal probability function, involve equally likely outcomes. Finally, we have discussed the subject in terms of games rather than managerial problems since we wished to draw on your intuition. Managerial applications are suggested in succeeding sections and in the problems.

S.S. Try to answer the following questions.

(a) What is the probability of a full house in a 5-card poker hand with no wild cards? (A full house is 3 cards of one denomination and 2 of another.)

(b) If 5 items are removed from a group of 10 of which half are defective, what is the probability that all 5 removed are defective?

(c) In sampling it is typical to include an item in a sample only once. This is called sampling without replacement. If a population is of size 25, how many samples are there of size 4, (a) with and (b) without replacement? Solve where the order of draw is relevant and solve where order is not relevant for (b) only.

Solution:

(a) Set up the following tasks to count the number of full-house hands:

Task 1: Select the two denominations.
Task 2: Select the one to be represented by 3 cards.
Task 3: Draw 3 of 4 cards from the first denomination selected in task 2.
Task 4: Select the 2 cards of the second denomination.

$$\frac{(_{13}C_2)(_2C_1)(_4C_3)(_4C_2)}{_{52}C_5} = \frac{3{,}744}{2{,}598{,}960} \approx 0.0014.$$

(b) The probability is

$$\frac{_5C_5}{_{10}C_5} = \frac{1}{10!/5!\,5!} = \frac{1}{252} \approx 0.004.$$

(c) With replacement; order relevant: $(25)^4 = 390{,}625$
Without replacement; order relevant; $_{25}P_4 = 303{,}600$
Without replacement; order irrelevant: $_{25}C_3 = 12{,}650.$

6-2 Discrete and Continuous Probability Distributions

In our initial discussion of probability in Chapter 5, we dealt with sample spaces where we could list, and hence count, the outcomes. Sample spaces where the outcomes can be counted are called *discrete*. At the time we introduced discrete sample spaces we noted that additional outcomes could be added as long as no probability is assigned to them. Thus, if we are measuring a table and the outcomes are listed to the nearest inch, we could include all the real numbers in the sample space as long as no probability is assigned to those not used.

An obvious question is why we should do this given that our measuring instruments are capable of only discrete measurement and life is already complicated enough. Well, we are not merely diabolical by nature. It turns out that putting all the real numbers (typically in some interval) into the sample space often makes our task easier. We do it, then, because it simplifies calculations in many otherwise difficult problems. When all the real numbers (often in some interval) are included in the sample space, the sample space is said to be *continuous*. The outcomes can no longer be counted.

The interpretation of what a probability is in such a case requires some explanation. To accomplish this we will first introduce some additional notation as it relates to discrete sample spaces and then expand this notation to the case of continuous sample spaces.

PROBABILITY MASS FUNCTIONS

Suppose that the Cleansweep Company manufactures superduper pooper scoopers for sale to hardware and garden stores. The historical frequency of weekly material costs per 100 scoopers is given in Table 6-1.

Table 6-1 Frequency Distribution of Weekly Material Cost for the Past 5 Years

Cost (x)	Number of Occurrences	Relative Frequency
$100	26	0.1
110	78	0.3
120	78	0.3
130	52	0.2
140	26	0.1
	260	1.0

We use the historical data to determine the relative frequencies of each cost level. The relative frequencies can then be used to establish the probabilities of each cost level. This gives us the probability mass function (abbreviated pmf) graphed in Figure 6-2. The probabilities are given by the five dots in the graph.

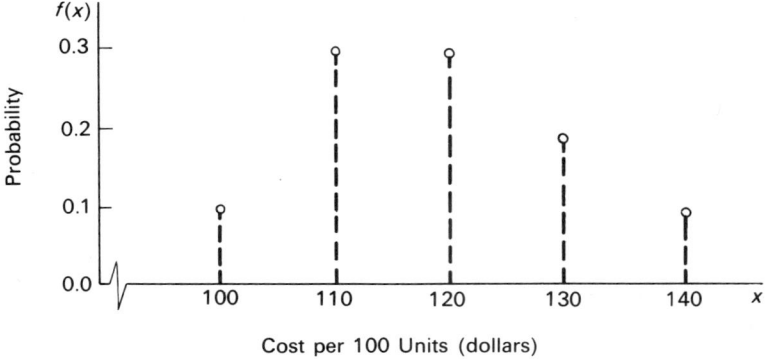

Figure 6-2 Probability Mass Function for Material Costs

The probability mass function can also be written in formula form, where we now use $f(x)$ in place of $P(x)$ to stand for probability.

$$f(x) = \begin{cases} 0.1 & \text{for } x = 100 \\ 0.3 & \text{for } x = 110 \\ 0.3 & \text{for } x = 120 \\ 0.2 & \text{for } x = 130 \\ 0.1 & \text{for } x = 140. \end{cases} \quad (6\text{-}6)$$

Here x stands for the specific values (namely, 100, 110, 120, 130, and 140) that the cost variable can assume, and $f(x)$ stands for the probability of that cost. For example, the probability of a cost level of 100 is 0.1. Note the similarity between the graph of a pmf and the graph of a relative frequency distribution. They are similar because the latter is often used to obtain the former.

> When the sample space is discrete, that is, when the outcomes can be counted, the function, $f(x)$, that gives the probability of each outcome, x, is called a *probability mass function*, or pmf.

The function $f(x)$ given by (6-6) is a probability mass function. Each value of $f(x)$ is the probability of that x; thus, 0.2 is the probability that the weekly cost will be $130.

The probability mass function can be used to calculate the probability of any event desired. For example, the probability that cost per 100 units will exceed $110 is obtained using the general rule for addition of probabilities:

$$f(x = 120) + f(x = 130) + f(x = 140),$$

or, more simply,

$$f(120) + f(130) + f(140) = 0.3 + 0.2 + 0.1 = 0.6.$$

PROBABILITY DENSITY FUNCTIONS

To illustrate probability density functions, we can consider the material-cost problem from a different point of view. For the problem we are considering, it may be reasonable to consider cost as a continuous variable which may assume any value between, say, $100 and $150. In this case there are so many outcomes, they can no longer be counted. Suppose management believes that any cost interval of a given size is no more or less likely than any other cost interval of the same size. For example, suppose that cost is as likely to be between $100 and $105 as between $105 and $110 as between $110 and $115, and so on. Then we could graph the appropriate probability function as in Figure 6-3. This function is called a *probability density function*.

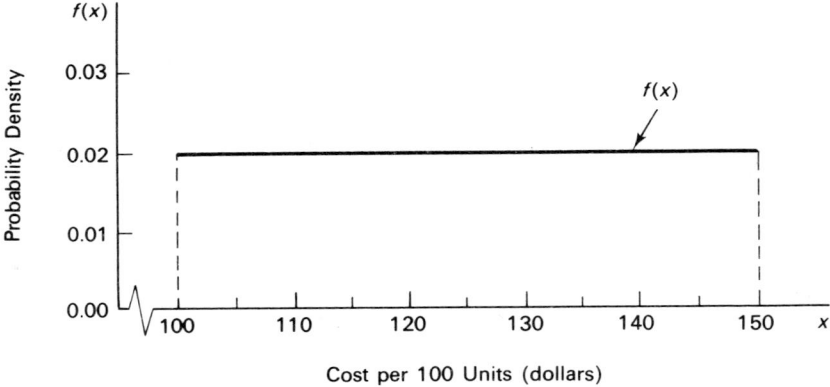

Figure 6-3 Probability Density Function of Material Costs (New Problem Assumptions)

(Note that this graph and the assumptions underlying it as to the proper probabilities are different from those suggested by the data in Table 6-1 and graphed in Figure 6-2. The concept is the same as with the frequency curve discussed in Chapter 2.)

The probability density function (abbreviated pdf) can also be written in formula form, using the functional notation $f(x)$:

$$f(x) = \begin{cases} 0.02 & \text{for } 100 \leq x \leq 150 \\ 0 & \text{otherwise.} \end{cases} \tag{6-7}$$

When the sample space is continuous, that is, the outcomes cannot be counted, the function, $f(x)$, that gives the probability density of each outcome, x, is called a *probability density function*, or pdf.

There are several important differences between a probability density function and a probability mass function. The first is suggested by the words "probability density." The height of a probability mass function (see Figure 6-2) at a value of x is the probability of that value of x. The height of a probability density function (see Figure 6-3) at a value of x is the probability density of that value of x. Not very helpful, you say? Well, try this. For probability density functions, probabilities are given by areas between the function and the horizontal axis. Hence probabilities apply only to ranges of x values. The area for a single x value would be zero. For example, the probability that cost per 100 units will be between $100 and $150 is given by the area in the rectangle bounded by the x axis, the horizontal line labeled $f(x)$, and the two vertical dashed lines. This area is given by the product of the rectangle's height by its length or $0.02(150 - 100) = 1$. Since this is the total probability, an answer of 1 makes sense. Indeed that is why we draw the horizontal line for $f(x)$ at a height of 0.02. The problem stipulates that the cost must be in the range 100 to 150.

Similarly, the probability of any event is defined by the area above that event and under the function $f(x)$. Thus, the probability that the cost exceeds $125 is the area under $f(x)$ in the interval 125 to 150. This area is $0.02(150 - 125) = 0.5$. It represents one half of the total rectangle. Now observe what happens when we ask for the probability that cost is exactly $125. The appropriate rectangle has no length, and hence there is no area. The probability is zero. This is true of any single point. It is also the reason that we obtain the same probability for any interval whether or not we include its end points.

The probability of any event for a random process described by a probability density function is given by the area under the function and above the interval defining the event on the x axis. Since some specific cost will, in fact, occur, we characterize this result by saying that so many possible costs could occur (indeed any one of an infinite number) that any one cost has a very small, "essentially zero," probability. Operationally, "essentially zero" is converted to zero probability.

The height of the graph at any point represents the height of a rectangle that has been squeezed until no area is left. In Figure 6-3 all rectangles have the same height. Thus, two events defined by the same size interval on the x axis and between 100 and 150 have the same probability. We say the probabilities are equally dense everywhere, and this is suggested by the fact that the function $f(x)$ has the same value (height) for all x. A pdf of constant height is called a *uniform probability density function*.

In contrast, if the appropriate probability density function took on a triangular shape, as does the one in Figure 6-4, this is no longer the case.

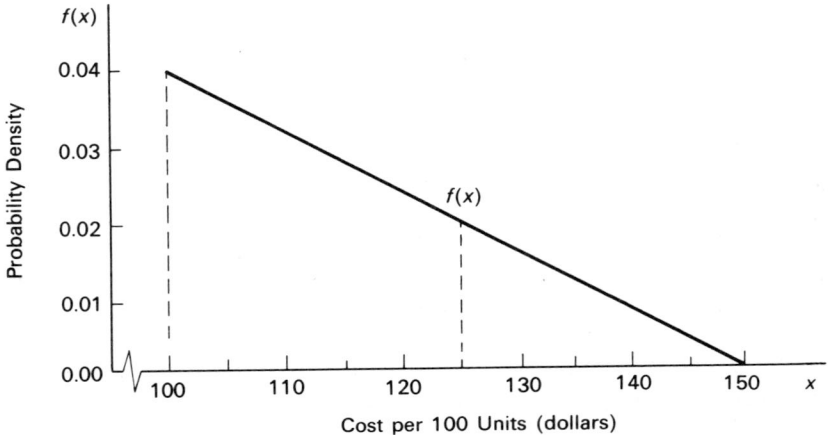

Figure 6-4 Alternative Probability Density Function of Material Costs

The probability is bunched toward the lower cost levels.

In formula form this new pdf, called a *triangular pdf*, is written

$$f(x) = \begin{cases} 0.12 - 0.0008x & 100 \leq x \leq 150 \\ 0 & \text{otherwise.} \end{cases} \quad (6\text{-}8)$$

[The equation for the straight line given by $0.12 - 0.0008x$ was found by inspection. The slope is $-0.04 \div 50$, and the intercept is found by solving $f(x) = 0$ when $x = 150$.]

We can see by inspecting the graph as well as by using our area procedure for finding probabilities that, for this pdf, the probability that cost exceeds $125 is much less than with the uniform pdf of Figure 6-3. Specifically, the probability that cost exceeds $125 equals the area under a triangle with height 0.02 and base $150 - 125$. This area is given by the expression $\frac{1}{2}(0.02)(150 - 125) = 0.25$. Inspection, or the value of $f(x)$, tells us that there is more probability, probabilities are more dense, at the lower cost levels in the interval 100 to 150. The fact that the graph is higher, $f(x)$ is greater, for costs near (but above) 100 reflects this fact.

Summarizing what we know about pmf's (probability mass functions) and pdf's (probability density functions):

PMF's:

1. The probability of any outcome is a nonnegative number.
2. The sum of the probabilities assigned to all outcomes is 1.

PDF's:

1. The probability density of an outcome is a nonnegative number. (It may exceed 1.)
2. The area under the pdf and above the horizontal axis is 1.

New Symbol	Meaning
$f(x)$	pmf: the probability of x. pdf: the density of probability at x.

One problem still remains. What do we do when the pdf is not a simple linear function? How can we find areas when our simple geometric figures, such as the rectangle and triangle, cannot be used? The solution requires the mathematical technique of integration. The procedures are omitted here and covered in Technical Note 6-1 at the conclusion of the chapter.

CUMULATIVE PROBABILITY FUNCTIONS

Cumulative probability functions, called *cdf's*, which are sometimes also called probability distribution functions, answer the question: What is the probability that x is less than or equal to some number, say, b? The cdf adds up the probability for all x less than or equal to the value b, hence the adjective "cumulative" in the name.

For example, using the probability mass function given in Figure 6-2 or by equation (6-6), the cumulative probability function is illustrated in Figure 6-5 and given by equation (6-9). The symbol used to indicate the cumulative probability function is $F(x)$. A capital F replaces the small f.

$$F(x) = \begin{cases} 0 & \text{for } x < 100 \\ 0.1 & \text{for } 100 \leq x < 110 \\ 0.4 & \text{for } 110 \leq x < 120 \\ 0.7 & \text{for } 120 \leq x < 130 \\ 0.9 & \text{for } 130 \leq x < 140 \\ 1.0 & \text{for } 140 \leq x. \end{cases} \quad (6\text{-}9)$$

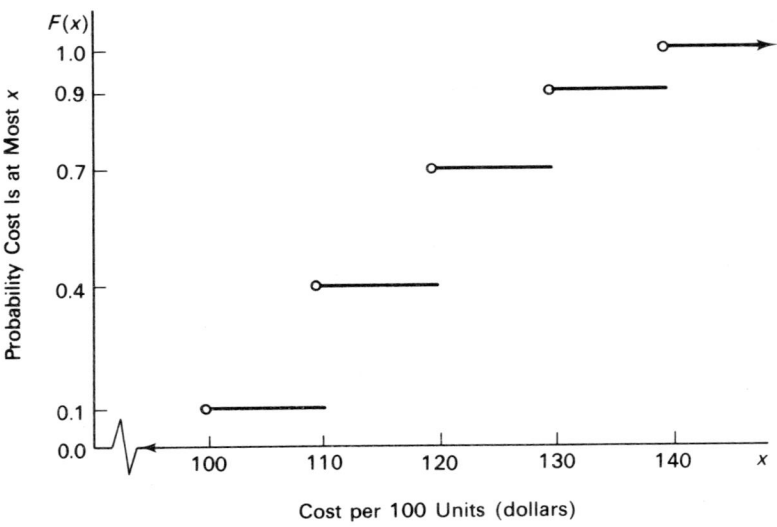

Figure 6-5 Cumulative Probability Function for the Probability Mass Function Illustrated in Figure 6-2

New Symbol	Meaning
$F(x)$	The probability of being equal to or less than x.

The height of this function at any point is obtained by summing the probabilities of all values up to and including the point. Probabilities such as the likelihood that the cost is less than or equal to $130, less than $120, between $120 and $130 inclusive, or greater than $117 can all be answered using the cumulative probability function. We obtain

$$P(x \leq 130) = F(x = 130) = 0.9$$
$$P(x < 120) = F(x = 110) = 0.4 \text{ (read directly from the graph)}$$
$$P(120 \leq x \leq 130) = F(x = 130) - F(x = 110) = 0.9 - 0.4 = 0.5$$
$$P(x > 117) = 1 - F(x = 117) = 1 - F(x = 110) = 1 - 0.4 = 0.6.$$

[These probabilities could also be obtained using the pmf given by equation (6-6).] Since it is easier to see the answers using the pmf, we will typically do so for discrete sample spaces. The dots in Figure 6-5 indicate, for example, that at $x = 110$, $F(x) = 0.4$ and not 0.1.

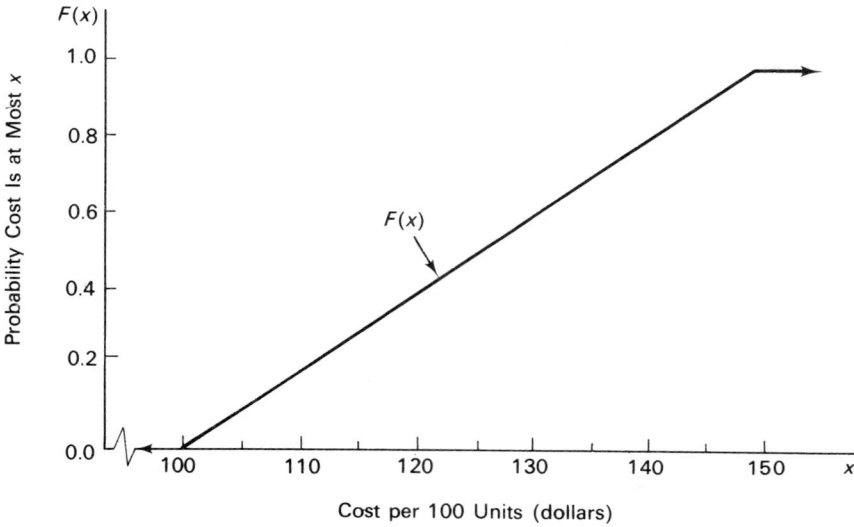

Figure 6-6 Cumulative Probability Function for the Probability Density Function Illustrated in Figure 6-3

For an example using a pdf, we give the cumulative probability function for the uniform cost function of Figure 6-3 and equation (6-7). The cumulative probability function is given in Figure 6-6. The cumulative probability function in this case is written as follows:

$$F(x) = \begin{cases} 0 & \text{for} \quad x < 100 \\ -2 + 0.02x & \text{for } 100 \leq x < 150 \\ 1.0 & \text{for } 150 \leq x. \end{cases} \qquad (6\text{-}10)$$

Technical Note 6-1 explains what must be done in the case of nonlinear pdf's.

There is no probability until a cost of $100 is reached. Because the pdf is uniform, the probability accumulates uniformly as x increases until the cost reaches $150. At this point the total probability of 1 has been accumulated. For this reason the cumulative probability function is linear and increasing between $100 and $150. The slope of this line is given by the ratio 1.0/50, and the intercept, a, is obtained by solving $F(x) = 0 = a + (1.0/50)x$ when $x = 100$.

The probability of the event "cost is at most $120," which is the same as requesting the probability that cost is equal to or less than $120, is given by the height of the function $F(x)$ at $120. Using the graph in Figure 6-6 or equation (6-10), $F(120) = 0.4$. This answer could also have been obtained by finding the area under the pdf.

S.S. Let us return to our friend T. P. Lerr. Remember him? Suppose that the dollar value of his weekly sales activity is given by (where $s \equiv$ sales dollars in thousands)

$$f(s) = \begin{cases} \frac{1}{2} & \text{for } 1 \leq s \leq 3 \\ 0 & \text{otherwise.} \end{cases}$$

(a) Does $f(s)$ describe a pmf or a pdf? Graph it.

(b) What is the probability that sales will exceed $2,500?

(c) Draw and write the cumulative probability function.

(d) Use the answer in part (c) to determine the probability that sales exceed $2,500 and check your answer using part (b).

Solution:

(a) It gives a pdf, since the number of possible cost values between 1 and 3 cannot be counted. The necessary graph (Figure 6-7) is the horizontal line labeled $f(s)$. Note that $f(s) \geq 0$ and the area under $f(s)$ between 1 and 3 is unity.

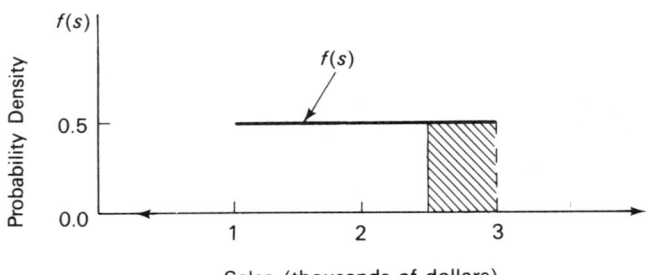

Figure 6-7 Probability Density Function of Sales (Thousands of Dollars)

(b) The probability is given by the area of the crosshatched rectangle in Figure 6-8. Probability $= 0.5(3 - 2.5) = 0.25$.

(c) $$F(s) = \begin{cases} 0 & \text{for } s < 1 \\ -0.5 + 0.5s & \text{for } 1 \leq s < 3 \\ 1 & \text{for } 3 \leq s. \end{cases}$$

(d) $F(s = 2.5) = -0.5 + 0.5(2.5) = 0.75$.

Probability $s > 2.5 = 1 - F(s = 2.5) = 1 - 0.75 = 0.25$. This checks with part (b). (And that is all we wish to do for T.P. at this time.)

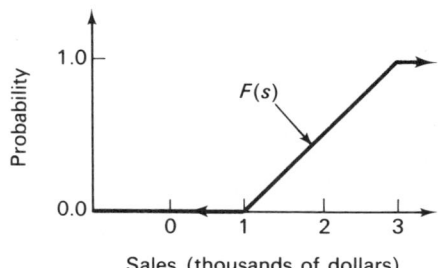

Figure 6-8 Cumulative Probability Density Function of Sales (Thousands of Dollars)

6-3 Conditional Probability

In many situations a decision maker is interested in several events and their effect on one another. For example, is the sale of a product affected by the weather? Does the origin of supplied raw material (the supplier) affect the probability of acceptable final product quality? Will training improve job performance? And so on. To put it another way, can decision makers use knowledge about one event to improve their ability to estimate the probability of some other event? The answer is that "it depends."

We shall first provide a simple example of a case where knowledge that one event has happened will affect the estimate of the probability of another event. Suppose that we roll one die once and ask for the probability of an odd face. The probability is $\frac{3}{6} = \frac{1}{2}$, as we know from our previous work. Now suppose we are told that the result was a number greater than 3 and asked for the probability that the result was also an odd face. We now know that the result was a 4, 5, or 6 and the probability of the event "odd face" is reduced to $\frac{1}{3}$, conditional on the knowledge that the result exceeded 3. Our knowledge that one event, {4, 5, 6}, occurred decreased the likelihood of the event "odd face" to $\frac{1}{3}$, since only 1 of the 3 remaining outcomes satisfies the event description.

On the other hand, if we were told instead that the roll produced a number greater than 2, that is, the event {3, 4, 5, 6} occurred, the probability of the event "odd face" remains $\frac{2}{4} = \frac{1}{2}$. The probability of the event of interest is not changed, since 2 of the remaining 4 possibilities satisfy the event description.

Now consider a somewhat more realistic problem. The manager of an electrical equipment manufacturing firm purchases switches from two outside suppliers; denote them A and B. She has recently received a report from the production manager recommending that they cease purchasing from supplier B. The production manager bases his recommendation on the data in Table 6-2.

Table 6-2 Frequency of Defective Switches by Manufacturer

Supplier	Number Defective	Relative Frequency
A	200	0.25
B	600	0.75
Total	800	1.00

The recommendation surprised the manager, who had the impression that supplier B produced a higher-quality product. Hence, she decided to gather further information on the problem before making a final decision. In particular, she obtained information on total usage. The results of her investigations are given in Table 6-3.

Table 6-3 Frequency of Use and Defective Switches by Supplier

Supplier	Condition of Switch		Total
	Defective	Not Defective	
A	200	1,800	2,000
B	600	7,400	8,000
Total	800	9,200	10,000

Based on the data in Table 6-3, the manager concluded that she should not discontinue purchasing switches from supplier B as recommended by the production manager. Her reasoning was as follows. The relative frequency of defective switches from supplier A was 200/2,000, or 10 percent. The percentage was $600/8,000 = 7.5$ percent for supplier B. Hence, the frequency of defective switches is less for supplier B than A. Added to the quantitative results was the fact that the manager did not want to be dependent on a single source of supply.

Let us see what this problem has to say about using probabilities if we let the historical relative frequencies be the probabilities of the events of interest. The following probabilities are determined directly from the data in Table 6-3.

Counting, Probability Distributions, and Conditional Probabilities

$$P(\text{defective}) \equiv P(d) = \frac{800}{10,000} = 0.08$$

$$P(\text{not defective}) \equiv P(n) = \frac{9,200}{10,000} = 0.92$$

$$P(\text{supplier } A) \equiv P(A) = \frac{2,000}{10,000} = 0.20$$

$$P(\text{supplier } B) \equiv P(B) = \frac{8,000}{10,000} = 0.80.$$

These probabilities are sometimes called *marginal probabilities* because they give the probabilities of the events listed in the margins of the table.

Now let us expand our list of probabilities somewhat:

$$P(\text{defective from supplier } A) = P(d \text{ and } A) \equiv P(d \cap A) = \frac{200}{10,000} = 0.02$$

$$P(\text{defective from supplier } B) = P(d \text{ and } B) \equiv P(d \cap B) = \frac{600}{10,000} = 0.06.$$

These probabilities are called *joint probabilities* since they yield the probability of the simultaneous occurrence of both events included within the parentheses. They are the probabilities of the events indicated by the numbers in the body of the table. Table 6-4 suggests the ideas of marginal and joint probabilities.

Table 6-4 Switches Example: Joint and Marginal Probabilities

Supplier	Condition of Switch		Total
	Defective	Not Defective	
A	$P(A \cap d) = \frac{200}{10,000}$	$P(A \cap n) = \frac{1,800}{10,000}$	$P(A) = \frac{2,000}{10,000}$
B	$P(B \cap d) = \frac{600}{10,000}$	$P(B \cap n) = \frac{7,400}{10,000}$	$P(B) = \frac{8,000}{10,000}$
Total	$P(d) = \frac{800}{10,000}$	$P(n) = \frac{9,200}{10,000}$	$P(S) = 1$

The probabilities of interest to the manager and those which she used in her decision are not in this table. The probabilities that she is interested in are: the probability that the switch is defective given (conditional on the fact) that A is the supplier, and the probability that the switch is defective given that B is the supplier. These probabilities are written

$$P(\text{defective given } A) \equiv P(d \mid A) = \frac{200}{2,000} = 0.10$$

$$P(\text{defective given } B) \equiv P(d \mid B) = \frac{600}{8,000} = 0.075.$$

These probabilities are called *conditional probabilities*.

New Symbol	Meaning
$P(E \mid F)$	Probability of the event E given (or conditional on) the occurrence of the event F.

Such probabilities direct our attention to a subset of the original sample space. If we are told that A is the supplier, then only the 2,000 switches supplied by A are relevant for further consideration. Of these 2,000 switches, 200 were defective. Hence, the conditional probability of a defective switch given it is produced by supplier A is $200/2,000 = 0.10$. We note that this probability is different from the unconditional probability of a defective switch, namely $P(d) = 800/10,000 = 0.08$. Hence, knowledge that the event A occurs gives information about the event d. This causes the manager to adjust her probabilities. Good managers use such information when it is available. Knowledge that A is the supplier affects the probability that the switch is defective.

FINDING CONDITIONAL PROBABILITIES

We have just seen that $P(d \mid A) = 200/2,000$. Dividing both the numerator and denominator by 10,000 gives

$$P(d \mid A) = \frac{200/10,000}{2,000/10,000} = \frac{P(d \cap A)}{P(A)}.$$

COUNTING, PROBABILITY DISTRIBUTIONS, AND CONDITIONAL PROBABILITIES

In words, the conditional probability of a defective switch given supplier A is:

1. The ratio of the number of defective switches supplied by A, 200, to the total number of switches supplied by A, 2,000; or
2. The proportion of defective switches supplied by A of the total switches, 200/10,000, as a ratio to the proportion of all switches supplied by A, 2,000/10,000.

In general:

> Rule 3: *Conditional Probability of an Event E Given an Event F*
> Given two events, E and F, defined over a sample space S, the *conditional probability* of the event E given the event F [provided that $P(F) \neq 0$] is
> $$P(E \mid F) = \frac{P(E \cap F)}{P(F)}. \qquad (6\text{-}11)$$

In fancy terms, the conditional probability of the event E given the event F is defined as the joint probability of E and F divided by the marginal probability of F.

MULTIPLICATION LAW OF PROBABILITY

From formula (6-11) we obtain formula (6-12), the *multiplication law* for the probability of the joint occurrence of two events, by cross multiplying.

> Rule 4: *General Multiplication Law for the Probability for Two Events*
> $$P(E \text{ and } F) \equiv P(E \cap F) = P(F)P(E \mid F). \qquad (6\text{-}12)$$

This formula allows us to compute the probability that two events both occur. For example, the probability that a switch drawn at random is defective and supplied by B is given by

$$P(d \cap B) = P(B)P(d \mid B)$$
$$= (0.80)(0.075)$$
$$= 0.06.$$

This probability can be verified directly from Table 6-4 by the ratio 600/10,000.

It is also useful to note that the event (E and F) is the same as the event (F and E). That is,
$$(E \cap F) = (F \cap E)$$
and hence
$$P(E \cap F) = P(F \cap E)$$
or, using equation (6-12),
$$P(F)P(E \mid F) = P(E \cap F) = P(F \cap E) = P(E)P(F \mid E). \qquad (6\text{-}13)$$

Thus, the probability that the switch is defective and supplied by B is the same as the probability that the switch is produced by B and is defective; they are the same event. Equation (6-13) will be used in Section 6-4 in deriving Bayes' theorem.

The multiplication law for more than two events is more complicated. Since we shall have little use for it, we omit it here. (See Problem 6-12.)

STATISTICAL INDEPENDENCE

In the first portion of this section we asked whether knowledge of the occurrence of one event causes a change in the probability of another event. When a die is rolled, does the knowledge that the event "over 2" occurred alter the probability of the event "odd face"? What about the event "over 3"? Does the fact that the switch was supplied by B change the likelihood that it is defective? Let us look at the probabilities for rolling a die once.

Event	Definition	Symbol	Event Probability
Odd face	$\equiv \{1, 3, 5\}$	G	$P(G) = \frac{1}{2}$
Over 2	$\equiv \{3, 4, 5, 6\}$	H	$P(H) = \frac{2}{3}$
Odd face given over 2	$\equiv \{3, 5\}$ given $\{3, 4, 5, 6\}$	$(G \mid H)$	$P(G \mid H) = \frac{1}{2}$
Over 3	$\equiv \{4, 5, 6\}$	J	$P(J) = \frac{1}{2}$
Odd face given over 3	$\equiv \{5\}$ given $\{4, 5, 6\}$	$(G \mid J)$	$P(G \mid J) = \frac{1}{3}$

We see that knowledge that the event "over 2" occurred did not change the probability of the event "odd face." The probability of the event G is not altered by knowing that the event H occurred. In other words, $P(G \mid H) = P(G) = \frac{1}{2}$; the event H gives no information about the event G. This was not the case for the events G and J; the probability of an "odd face" is changed from $\frac{1}{2}$ to $\frac{1}{3}$ if we are told that the result of the toss is "over 3."

If E and F are two events, $P(F) \neq 0$, then we say that the two events are *statistically independent* if and only if $P(E \mid F) = P(E)$, and we say that the events are *statistically dependent* if and only if $P(E \mid F) \neq P(E)$.

Using the above statement, we can say that the events G and H are statistically independent while the events G and J are statistically dependent. Statistically independent events give no information about each other that would cause us to change our probability assignments. Statistically dependent events do yield such information. [Hence, the events "defective switch" and "supplied by B" are statistically dependent since $P(d \mid B) = 0.075 \neq 0.08 = P(d)$.]

Using equation (6-12) and the events that we have defined for the roll of a die:

$$P(G \cap H) = P(H)P(G \mid H) = P(H)P(G)$$

$$\tfrac{1}{3} = \tfrac{2}{3}(\tfrac{1}{2}) = \tfrac{2}{3}(\tfrac{1}{2})$$

but

$$P(G \cap J) = P(J)P(G \mid J) \neq P(J)P(G)$$

$$\tfrac{1}{6} = \tfrac{1}{2}(\tfrac{1}{3}) \neq \tfrac{1}{2}(\tfrac{1}{2}).$$

This relationship can also be used to define statistical independence.

> If E and F are two events defined over the same sample space, then E and F are *statistically independent events* if and only if
> $$P(E \cap F) = P(F)P(E)$$
> $$= P(E)P(F).$$

The concept of statistical independence allows us to modify the multiplication law for the case of statistically independent events.

Rule 5a: *Multiplication Law for Two Statistically Independent Events*

$$P(E \text{ and } F) \equiv P(E \cap F) = P(F)P(E). \quad (6\text{-}14a)$$

Rule 5b: *Multiplication Law for n Statistically Independent Events*

$$P(E_1 \text{ and } E_2 \text{ and } \ldots E_n) \equiv P(E_1 \cap E_2 \cap \cdots \cap E_n)$$
$$= P(E_1)P(E_2) \cdots P(E_n)$$
$$= \prod_{i=1}^{n} P(E_i). \quad (6\text{-}14b)$$

The multiplication law for more than two events is easily written if the events are independent. It is more difficult to write and use where the events are dependent. (See Problem 6-12.)

S.S. Answer the following questions.

(a) If two events are mutually exclusive, are they statistically independent?

(b) Is the probability that an individual will live 10 more years conditional on his present age?

(c) The number of people in the population with tuberculosis is, say, 1 in 1,000. The likelihood that a person with tuberculosis will test positive on X rays is, say, 0.90. Therefore, the likelihood of someone having tuberculosis and testing positive on X rays is?

(d) Is the probability that a crime will be reported a function of the type of crime?

Solution:

(a) If two events are mutually exclusive, then one or the other, but not both, can happen on a single trial. The event "head" and the event "tail" on one flip of a coin are examples of two mutually exclusive events. Thus, if one of these events occurs, say heads, the probability of the other is zero.

$$P(h) = \tfrac{1}{2} \quad \text{but} \quad P(h \mid t) = 0.$$

(b) and (d) Life expectancy has been shown by actuarial studies to depend on age. Crimes against one's person, for example, are less likely to be reported, according to recent studies.

(c)
$$P(\text{tb and } + \text{ X ray}) = P(\text{tb})P(+ \text{ X ray} \mid \text{tb})$$
$$= 0.001(0.90) = 0.0009 = \frac{9}{10,000}.$$

6-4 Revising Probabilities: Bayes' Theorem

Events can occur under different conditions. A defective switch may be due to supplier B or supplier A. A poor examination score may indicate a student's lack of understanding, but it may be due to other factors as well. A positive tuberculosis test may not mean the patient has the illness and vice versa. Nevertheless, if we have some reasonable estimates of the conditional probabilities involved, perhaps using historical frequencies or subjective evaluations, the likelihoods of the various conditions which might have produced the sample results can be estimated. The probabilities are initially assigned, so the underlying conditions can be altered or modified by new knowledge or events. A positive tuberculosis test will cause us to revise our probability that someone has tuberculosis. A series of out-of-tolerance parts leads a manager to revise beliefs concerning the likelihood that the machine producing them is functioning properly.

The revision of probabilities is a common and very important task for a manager and stands at the heart of the approach described in Chapter 13.

For one additional example, consider the manager who must decide whether or not to introduce a new consumer product to the market. Among other relevant factors is the level of demand that is forecast. The level of demand may, in turn, be closely related to the state of the economy: Are we in an expansion or recessionary period for example? The decision may depend on an evaluation of the likelihood of these two future conditions. Now we can never (well, hardly ever) be certain about the future, but we frequently encounter relevant evidence from the stockmarket, GNP statistics, statements about monetary policy from the Federal Reserve Board or the banking community, and other similar sources that may lead to a revision of previous probabilistic assessments of the future.

Such revision is not limited to managerial problems. We all do this, at least informally, every day. The likelihood of a happy evening for your authors is altered by the behavior of our children during the day. (The same events may also cause us to revise the probability of eating dinner out, as does the lack of delicious odors emanating from the kitchen upon arriving home.) A new, or even used, car is bought with the expectation that it will be mechanically sound. But, as we all know, a car may turn out to be a "lemon," a condition that a series of minor difficulties tends to confirm. And when one morning you awake to find 4,000 pounds of scrap metal in the garage, you know.

The purpose of this subsection is to provide a logical and systematic method by which new information can be used to revise our beliefs about the conditions producing those events. In this task we will make extensive use of the concept of conditional probability. We will first show how the procedure works, second we will interpret our results, and finally we will attempt to use our intuition to see why the technique works.

Let us return to the example introduced earlier in this chapter involving defective switches. The problem we will examine here is determining the likelihood that a switch was supplied by B, given that it tested defective. Symbolically, this is $P(B \mid d)$, a conditional probability that we did not calculate in Section 6-3. (We will assume that the test always tells us the actual condition of the switch.)

Perhaps the easiest method to establish $P(B \mid d)$ empirically is to use the equations developed in Section 6-3. Using formula (6-11),

$$P(B \mid d) = \frac{P(B \cap d)}{P(d)} = \frac{P(d \cap B)}{P(d)}. \qquad (6\text{-}15)$$

Now we already know these probabilities for our problem. Hence,

$$P(B \mid d) = \frac{0.06}{0.08} = \frac{3}{4}. \qquad (6\text{-}16)$$

However, in many cases, we will not know the probabilities $P(d \cap B)$ and $P(d)$ directly. A more complex approach is required. Using formulas (6-13) and (6-15),

$$P(B \mid d) = \frac{P(B \cap d)}{P(d)} = \frac{P(B)P(d \mid B)}{P(d)}. \tag{6-16}$$

In equation (6-16) a very significant change has been made. We have written $P(B \cap d)$ in terms of $P(d \mid B)$. Managers may often experience the condition of knowing some conditional probabilities [such as the probability of a defective switch given it is supplied by B, namely $P(d \mid B)$], but needing others [such as the probability of supplier B given the switch is defective, $P(B \mid d)$]. We are nearly ready to help such a manager. First, we need one more step.

The event "defective" can occur in two ways. The item may be defective and supplied by B, or it may be defective and supplied by A. Writing this gobbledygook symbolically,

$$\text{Event: } d = \underbrace{(B \cap d)}_{\text{Event } d_1} \underset{\text{or}}{\cup} \underbrace{(A \cap d)}_{\text{Event } d_2} = \underbrace{(B \cap d)}_{\substack{\text{(supplied by } B \\ \text{and defective)} \\ \text{Event } d_1}} \underset{\text{or}}{\cup} \underbrace{(A \cap d)}_{\substack{\text{(supplied by } A \\ \text{and defective)} \\ \text{Event } d_2}}$$

Since the events d_1 and d_2 are mutually exclusive, we can use rule 1, the addition rule of probability, to write

$$P(d) = P(d_1) + P(d_2) = P(B \cap d) + P(A \cap d)$$

or using formula (6-13), we may write

$$P(d) = P(d \cap B) + P(d \cap A)$$

and then, using formula (6-12),

$$P(d) = P(B)P(d \mid B) + P(A)P(d \mid A).$$

Now, using this last result for $P(d)$ we may rewrite equation (6-16) in the form

$$P(B \mid d) = \frac{P(B)P(d \mid B)}{P(B)P(d \mid B) + P(A)P(d \mid A)}. \tag{6-17}$$

We urge you not to try to be able to reproduce all the steps used to obtain equation (6-17) from equation (6-15). You should, however, follow each step of the derivation to satisfy yourself that we have not pulled any rabbits out of the hat, so to speak. Accept, then, the conclusion, equation (6-17), and you can then forget the derivation.

The formula given by equation (6-17) is known as Bayes' theorem or Bayes' law, after an eighteenth-century clergyman and mathematician, Thomas Bayes, who apparently found his clerical duties at most a part-time vocation. The particular form given by equation (6-17) involves two events. Bayes' theorem can be generalized to more than two events and our statement of the theorem does so.

Rule 6: *Bayes' Theorem*
If S is a sample space and if the set of events $F_1, F_2, \ldots, F_n$ are mutually exclusive such that $F_1 \cup F_2 \cup \cdots \cup F_n = S$, and if E is any other event defined over S such that $P(E) \neq 0$, then

$$P(F_i \mid E) = \frac{P(F_i)P(E \mid F_i)}{P(E)}. \qquad (6\text{-}18)$$

where

$$P(E) = \sum_{i=1}^{n} P(F_i)P(E \mid F_i).$$

For our example, the events F are:

$$F_1 \equiv \text{switch is supplied by supplier } B$$
$$F_2 \equiv \text{switch is supplied by supplier } A.$$

These two events are mutually exclusive and their union gives the sample space S, which, for this problem, can be written
$S \equiv \{Ad, An, Bd, Bn\}$ or $\{(A \cap d), (A \cap n), (B \cap d), (B \cap n)\}$.
The event E for the example can be written

$$\{Bd\} \quad \text{or} \quad (B \cap d).$$

Using the probabilities given in Section 6-3 and Bayes' theorem [for our example, equation (6-17)],

$$P(B \mid d) = \frac{P(B)P(d \mid B)}{P(B)P(d \mid B) + P(A)P(d \mid A)} = \frac{0.80(0.075)}{0.80(0.075) + 0.20(0.10)}$$
$$= \tfrac{6}{8} = \tfrac{3}{4}.$$

Using the notion of the complementary event, the probability that the defective switch was supplied by A is $1 - \tfrac{3}{4} = \tfrac{1}{4}$. Alternatively, we could use Bayes' theorem again, this time using the second term of the denominator in the numerator. This would give

$$P(A \mid d) = \frac{P(A)P(d \mid A)}{P(B)P(d \mid B) + P(A)P(d \mid A)} = \frac{0.20(0.10)}{0.80(0.075) + 0.20(0.10)} = \tfrac{1}{4}.$$

It is worth reiterating that some conditional probabilities may be desired when others involving the same events are known. In such cases, Bayes' theorem can prove to be quite helpful.

Finally, we ask whether our intuition can be of any help in understanding why Bayes' theorem works and thereby ease our acceptance of it as non-magical. Looking at the monster given by formula (6-18) provides little encouragement, nor is it settling to our lunch. But be of brave heart and try this on for size. Suppose that we took 100 switches at random. Then, on the average, 20 would be supplied by A and 80 by B. Of the 20 supplied by A, 10 percent—2 switches—are defective (again on the average) and 18 are acceptable. Of the 80 supplied by B, $7\tfrac{1}{2}$ percent—6 switches—are defective and 74 are acceptable. These data are illustrated in the tree diagram given in Figure 6-9.

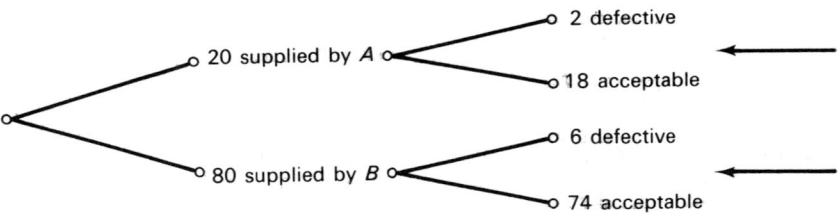

Figure 6-9 Results of 100 Switches, on the Average

Now we know, we are given, that the switch is defective. Hence, our concern is only with those switches that are defective. These two results, the 2 defective switches supplied by A and the 6 supplied by B, are indicated by the arrows in Figure 6-9. The defective switch must be one of the 8 defective switches indicated in the last column and by the arrows. And, because 6 of the 8 possible defective switches were supplied by B, the likelihood that the one we are given came from B is 6 out of 8 or $\tfrac{6}{8} = \tfrac{3}{4}$. The probability that the defective switch came from supplier B is the ratio of defective switches supplied by B to the total number of defective switches. This is $\tfrac{6}{8}$ or $\tfrac{3}{4}$. It is the same result that we found using Bayes' theorem.

S.S. Suppose that 30 percent of the population over 30 (and, hence, "over-the-hill") suffer from Dunlap's Condition. A new visual diagnostic test gives a positive indication 90 percent of the time when an individual has the condition and a negative indication 80 percent of the time when the condition is absent. An individual selected at random from the population over 30 is given the test, and the test is positive. What is the approximate chance that he suffers from Dunlap's Condition? (Dunlap's Condition is when one's stomach done-laps over one's belt.)

Solution: Let

$$D \equiv \text{has condition}$$
$$D' \equiv \text{does not have condition}$$
$$+ \equiv \text{test positive}$$
$$- \equiv \text{test negative}$$

$$P(D \mid +) \equiv \frac{P(D)P(+ \mid D)}{P(D)P(+ \mid D) + P(D')P(+ \mid D')} = \frac{0.30(0.90)}{0.30(0.90) + 0.70(0.20)}$$
$$\approx 0.659.$$

6-5 A Warning

It is always useful to estimate in advance what a reasonable answer to any probabilistic question is. Then check the calculated answer with the estimate as one means of verifying your work. One of the authors was once required in a civil engineering exam to calculate the diameter of a railroad-engine wheel required on a new diesel locomotive. After several minutes of frantic effort, the answer 120 feet emerged from his scribblings. No credit was received despite use of the proper method. The following note was appended to the exam by the instructor: "Presentation of such an answer to the management would quickly land you among the ranks of the unemployed. You should have realized that 120 feet is an unreasonable answer and checked your work. If you had done so, you would have quickly seen that the figures were in inches and the answer is 10 feet. In fact, half credit would have been forthcoming, given use of the correct method, if the unrealistic magnitude of the figure had even been noted."

Sometimes, however, our intuition may not work so well. In such cases only scrupulous attention to methodology will help. Consider the following classical paradox as an example. There are three boxes placed before a prisoner. The first box contains two gold coins, the second contains a gold coin and a brass coin, and the third contains two brass coins. A box is selected at random and a coin is drawn. It is gold. The prisoner is required to predict the coin remaining in the box. If he does so correctly, he will be freed. Otherwise, he will be dispatched to the next life. What should he guess, or does it matter? (Before reading on, estimate the probability that the second coin is also gold.)

A cursory examination of the problem might well lead us to conclude that the probability of a second gold coin is $\frac{1}{2}$, since the first coin, being gold, must have come from box 1 or box 2. Box 3 is ruled out and, since the box was selected at random, there is a 50–50 chance that the box selected is box 1 or 2. Thus, it does not matter what he guesses. There is a 50–50 chance that the second coin is also gold. But this analysis is not correct.

The appropriate way to solve this problem is to use Bayes' theorem or the equivalent analysis using a tree diagram similar to the one used in Figure 6-9. Either will work. Before looking at a coin:

$$P(\text{box 1}) = P(1) = \tfrac{1}{3}, \quad P(G \mid 1) = 1.0$$
$$P(\text{box 2}) = P(2) = \tfrac{1}{3}, \quad P(G \mid 2) = 0.5$$
$$P(\text{box 3}) = P(3) = \tfrac{1}{3}, \quad P(G \mid 3) = 0.0.$$

Let us find the probability of box 1 given that a gold coin is drawn. Using Bayes' theorem,

$$P(1 \mid G) = \frac{P(1)P(G \mid 1)}{P(1)P(G \mid 1) + P(2)P(G \mid 2) + P(3)P(G \mid 3)}.$$

In this case Bayes' theorem involves three events, whose union is the sample space rather than the two we have had in previous examples.

$$P(1 \mid G) = \frac{\tfrac{1}{3}(1.0)}{\tfrac{1}{3}(1.0) + \tfrac{1}{3}(0.5) + \tfrac{1}{3}(0)} = \frac{\tfrac{1}{3}}{\tfrac{1}{3} + \tfrac{1}{6}} = \tfrac{2}{3}.$$

Since the probability is $\tfrac{2}{3}$ that the box from which the coin selected was the first box, the prisoner should guess gold to maximize his chance of freedom. Did you guess $\tfrac{2}{3}$? We did not either the first time we tried this problem. (As an exercise, it is useful to work this problem using a tree diagram.)

TECHNICAL NOTE 6-1: NONLINEAR PROBABILITY FUNCTIONS

In this note we discuss how to find areas under pdf's described by curved functions such as in Figure 6-1A. We can no longer use simple geometric

Counting, Probability Distributions, and Conditional Probabilities

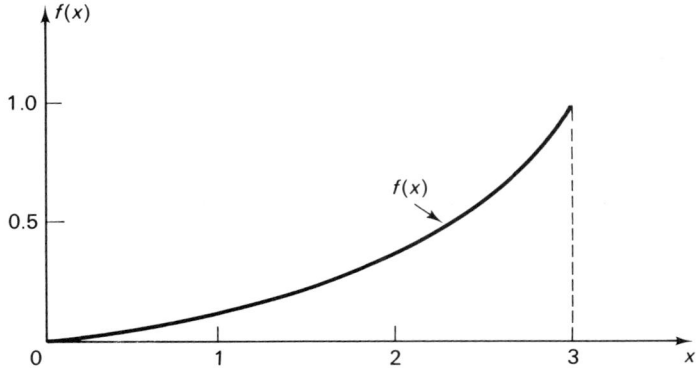

Figure 6-1A Probability Density Function Given by $f(x) = \frac{1}{9}x^2$ for $0 \le x \le 3$

relationships such as rectangles and triangles (except as approximations) to calculate the areas we seek. There are two ways to proceed, the direct method and the preferred method.

First, we can try the mathematical technique of integration to obtain the required area. This method will usually work for simple functions, including the one in Figure 6-1A. Although we will have relatively little need to do so, we will illustrate the technique of finding probabilities using integration for this example.

Probability x is between 2 and 3:

$$\int_2^3 \frac{1}{9} x^2 \, dx = \frac{1}{9} \int_2^3 x^2 \, dx = \frac{x^3}{3(9)} \bigg|_2^3 = 1 - \frac{8}{27} = \frac{19}{27}.$$

This procedure is quite general and could have been used for the pdf's in Figures 6-3 and 6-4. For example, using Figure 6-4: probability cost exceeds 125:

$$\int_{125}^{150} (0.12 - 0.0008x) \, dx = \int_{125}^{150} 0.12 \, dx - \int_{125}^{150} 0.0008x \, dx$$

$$= 0.12x \bigg|_{125}^{150} - \frac{0.0008}{2} x^2 \bigg|_{125}^{150}$$

$$= (18 - 15) - (9 - 6.25) = 0.25.$$

As one more example, suppose that we desire the cumulative probability function for the pdf given next.

$$f(x) = \begin{cases} e^{-x} & x \geq 0 \\ 0 & \text{otherwise.} \end{cases}$$

The cumulative probability function, $F(x)$, gives the probability up to and including a value of x, say $x = b$. Integrating e^{-x} we obtain

$$\int_{x=0}^{x=b} e^{-x}\, dx = -e^{-x}\Big|_0^b = -e^{-b} + 1.$$

Therefore,

$$F(x = b) = \begin{cases} 1 - e^{-b} & b \geq 0 \\ 0 & \text{otherwise.} \end{cases}$$

The probability that $x \leq 3$ is $1 - e^{-3}$, and so on.

But some problems, indeed most of the more interesting ones, do not yield to such relatively easy integration techniques. For example, if someone asked you to find the indefinite integral of the function e^{-x^2}, you would probably think he had lost his marbles. Well, he has not. We will find that a function very similar to this one is extensively used in statistics. The function to which we allude is called the normal probability function. (We apologize for using the word "normal" for such an apparently wierd fellow, but that is what it is called, and for a good reason, as we shall see in Chapter 8.) Anyway, this function does not have an indefinite integral. Numerical integration using a computer is used to obtain satisfactory approximations to the areas under this curve. Fortunately, this has been done for us, and we can simply look up the answer in a table. This is the second and preferred method for finding such probabilities.

New Symbols

Symbol	Meaning
$_nP_r$	The number of ordered arrangements (permutations) of n things taken r at a time.
$_nC_r$	The number of sets (combinations) of n things taken r at a time.
$n!$	$n(n-1)(n-2) \cdots 1.\quad 0! = 1.$
$f(x)$	The probability mass (or density) of x.
$F(x)$	The probability of a value equal to or less than x.

Key Formulas

Formula	Used to Compute:
$P(E \mid F) = \dfrac{P(E \cap F)}{P(F)}$	The conditional probability of the event E given the event F.
$P(E \cap F) = P(E)P(F)$	The joint probability of the events E and F if the events E and F are independent.
$P(F_i \mid E) = \dfrac{P(F_i)P(E \mid F_i)}{\sum_{i=1}^{n} P(F_i)P(E \mid F_i)}$ $= \dfrac{P(F_i)P(E \mid F_i)}{P(E)}$	Bayes' theorem, to find the probability of the event F_i given that the event E has occurred.

PROBLEMS

6-1.* The following brief piece appeared in *Newsweek*, May 6, 1968.

How to Win at Wordsmanship

After years of hacking through etymological thickets at the U.S. Public Health Service, a 63-year-old official named Philip Broughton hit upon a sure-fire method for converting frustration into fulfillment (jargonwise). Euphemistically called the Systematic Buzz Phrase Projector, Broughton's system employs a lexicon of 30 carefully chosen "buzzwords":

Column 1	Column 2	Column 3
0. integrated	0. management	0. options
1. total	1. organizational	1. flexibility
2. systematized	2. monitored	2. capability
3. parallel	3. reciprocal	3. mobility
4. functional	4. digital	4. programming
5. responsive	5. logistical	5. concept
6. optional	6. transitional	6. time-phase
7. synchronized	7. incremental	7. projection
8. compatible	8. third-generation	8. hardware
9. balanced	9. policy	9. contingency

The procedure is simple. Think of any three-digit number, then select the corresponding buzzword from each column. For instance, number 257 produces "systematized logistical projection," a phrase that can be dropped into virtually any report with that ring of decisive, knowledgeable authority. "No one will have the remotest idea of what you're talking about," says Broughton, "but the important thing is that they're not about to admit it."

Required: How many buzzwords (or three-digit numbers) are possible using this list? What technique did you use to calculate the answer?

6-2.* Uri Geller, a young Israeli psychic, in one of several tests conducted at the Stanford Research Institute, correctly called the upper face of a covered die eight times in eight tries. (*Time*, March 4, 1975, p. 69.) What is the probability that this would be done by chance?

6-3. Suppose that a sample space, S, consists of 5 outcomes.
(a) How many events can be defined over S?
(b) Use an argument other than counting the events in part (a) to show that the number of events is given by 2^5 in this problem. (In general, 2^n, where n is the number of outcomes in S.)
(*Hint:* Define 5 tasks, one each to include or exclude each outcome, and then use the fundamental principle of counting.)

6-4.* In poker, find the probability of being dealt the following hands without using wild cards.
(a) Two pairs (the last card to be of a different denomination).
(b) Three of a kind (three cards of the same denomination and two of different face values).
(c) Four of a kind.

6-5. In bridge (where 13 cards are dealt to each of 4 players) your partner holds 9 spades, including the ace and king. The defenders hold 4 spades, including the queen, between them. What is the probability that the queen will fall to successive leads of the ace and king? (That is, what is the probability that the 4 remaining spades will be split 2–2 or 3–1 with the queen alone?)

6-6.* A woman was mugged in the suburbs of a large city. A couple was convicted of the crime; the evidence upon which the prosecution rested its case was largely circumstantial. Probability was used to demonstrate that an extremely low probability existed that any specific couple could have committed the crime. The probability value was determined, using the multiplication law for independent events. The events were the characteristics ascribed by witnesses to the couple who actually did the deed. The characteristics, together with the assumed probabilities, are as follows:

Characteristic Event	Assumed Probability
Drives yellow car	1/10
Interracial couple	1/1,000
Blonde girl	1/4
Girl wears hair in ponytail	1/10
Man bearded	1/10
Man black	1/3

Source: *Time*, April 26, 1968, p. 41.

(a) Assuming independence, what is the probability of a couple with the identified characteristics?
(b) Evaluate the assumption of independence here.

6-7. Solve the gold-coin example in Section 6-5 using the tree-diagram technique. Show the probability that the box selected is box 1 is $\frac{2}{3}$, and hence a guess of "gold coin" has a $\frac{2}{3}$ probability of being correct.

6-8. A marketing executive is planning the introductory promotion for a new product. She has decided to use two television specials, t_1 and t_2; three full-page newspaper ads, n_1, n_2, and n_3; and one radio broadcaster, b_1. The question remaining is to establish the order in which the promotional events should appear.
  (a) If all 6 events are treated as separate promotions, how many alternatives are there?
  (b) If the two television specials are identical and the three newspaper ads are identical, how many distinguishable possibilities are there?

6-9.* An interesting and surprising probability is given by the probability that in a group of r people, no two have their birthday on the same day of the year. Assuming 365 days in a year and that each person's birthday is equally likely to be on each day, write an expression for this probability. How large do you think r must be before this probability is less than 0.5? (*Hint:* Find the probability of no matches.)

6-10. Consider the problem of sampling with replacement when order is not relevant. How many possible samples of size 4 are there from a population of 25 items?

6-11.* Consider picking a vacation. A vacation plan is defined as a destination and a method of travel. Our vacationer starts from New York and we will allow him to go to either San Francisco or London. Travel to San Francisco may be by car or plane. The trip to London must be made by ship or plane.
  (a) Does the number of vacation plans depend on whether the destination or means of transportation is selected first?

(b) How many vacation plans are there?
(c) If we want to use the fundamental principle of counting, does it matter whether the destination or means of travel is considered as the first task?

6-12. This problem considers the conditions for independence among several events. Basically the question is whether pairwise independence is sufficient to guarantee the independence of the events taken together. Consider the sample space for the roll of two dice once and define the following events:

$$E \equiv \text{odd total}$$
$$F \equiv \text{odd face on first die}$$
$$G \equiv \text{odd face on second die.}$$

(a) Show that E, F, and G are independent if taken in pairs but not all taken together.
(b) Write an equation that generalizes the multiplication law for probabilities to three events that are not statistically independent. Show that it works on this problem.

6-13. In the manufacture of fluorescent lights, a plant manager has decided to test the quality of the manufacturing process by randomly selecting tubes and testing them. He has decided to test at random tubes from a large lot of newly produced tubes and note whether the tube is defective or nondefective. He will continue to test until he finds two consecutive defective tubes or until he has tested four tubes, whichever occurs first. The result of a test does not affect the result of any subsequent test.

(a) Describe a sample space for the testing procedure.
(b) For each test, the probability of a tube's being nondefective is $\frac{3}{4}$ and the probability of its being defective is $\frac{1}{4}$. Compute the probability of each point in the sample space.
(c) Compute and graph the probability mass function for the number of defective tubes found from the testing procedure.

6-14.* Clever Associates, an advertising firm, has conducted a market survey in Cincinnati which shows that of three leading brands of aspirin, A, B, and C, 30 percent of the aspirin consumers use brand A, 20 percent use brand B, 15 percent use brand C, 4 percent use brands A and B, 6 percent use brands B and C, 2 percent use brands A and C, and 1 percent use all three brands. The uses are not mutually exclusive. A user of brand B may also use C or A. If an aspirin consumer in Cincinnati is selected at random, what is the probability that
(a) He uses none of the brands A, B, and C?
(b) He uses either brand A or B or both?
(c) He uses brand A if it is known that he uses A or B?

6-15. The distribution center for a catalogue company has received a defective toaster. The manager of the center had just received shipments from two warehouses, A and B, so she felt certain that the toaster had come from one of them. Each of the warehouses receives toasters from the same three manufacturers, X, Y, and Z, and the shipments to the distribution center had been part of large shipments to a number of distribution centers. Data for the shipments of toasters are as follows:

	Manufacturer			
Warehouse	X	Y	Z	Total
A	20	25	40	85
B	80	75	160	315
Total	100	100	200	400

What is the probability that a toaster selected at random
(a) Is from warehouse A?
(b) Is from manufacturer X?
(c) Is from manufacturer Z?
(d) Is from manufacturer Y if it is known to be from warehouse A?
(e) Are warehouse and manufacturer independent? Support your answer. If they are not, what adjustments to the numbers in the table would render the two factors independent?

6-16. The output from a machine that produces nails is known to contain 10 percent defectives, and the other 90 percent are satisfactory. From the large lot of nails produced by the machine, an inspector draws two nails at random.
(a) Define a sample space for this experiment and, by using a tree diagram, assign reasonable probabilities to its fundamental events.
(b) Find the probability that at least one of the nails is defective.

6-17. The inspectors in a manufacturing plant examined a lot of 5,000 bolts produced on four machines and classified them according to three grades:

	Machine				
Grade	W	X	Y	Z	Total
A: good	1,200	800	600	1,000	3,600
B: rework	400	300	200	300	1,200
C: scrap	100	30	40	30	200
Total	1,700	1,130	840	1,330	5,000

The plant manager was concerned with these figures and was considering halting production on any machines that had too high a probability of producing defective bolts.

(a) What is the probability that a bolt selected at random both needs to be reworked and was produced by machine W?

(b) Provide symbolic notation for the probability that a bolt selected at random:
   (1) Was produced by machine X.
   (2) Was produced by machine Z and is unsatisfactory.
   (3) Was produced by machine Y and should be scrapped.
   (4) Needs to be reworked.
   (5) Needs to be scrapped, given that it was produced by machine W.

(c) Find each of the probabilities in (b).

(d) Which of the probabilities in (c) are marginal probabilities?

(e) Are the two variables (machine and grade) statistically independent in the population of 5,000 bolts? Explain how you arrived at your answer.

6-18. In a dairy products plant, the production foreman must halt production whenever she wishes to change from one size of milk container to a different size. Let $x \equiv$ the number of minutes it takes the crews to make the necessary adjustments. The probability mass function for x is

$$f(x) = \begin{cases} x/15 & \text{for } x = 1, 2, \ldots, 5 \\ 0 & \text{otherwise.} \end{cases}$$

(a) Prove that $f(x)$ is a probability mass function.

(b) What is the probability that it will take exactly 3 minutes to make the adjustments?

(c) What is the probability that it will take at least 2 minutes but not more than 4 minutes?

(d) Find the cumulative probability function in formula form.

(e) Graph the probability mass function and the cumulative probability function.

(f) What is the probability that it will take at most 3 minutes?

(g) What is the probability that it will take more than 2 minutes?

6-19. The trust department of First National Bank is considering adding a stock to its list of acceptable purchases, and the officers in the department decided to chart its price movements. Each day the share price moves up one point with probability $\frac{1}{4}$ and down one point (not both) with probability $\frac{3}{4}$. After 4 days the analyst charting the price presented her first report to the officers. What is the probability that the stock had returned to its original price after 4 days? If the analyst's report indicated that the stock had returned to its original price after

4 days, what is the probability that the price had gone up on the first day?

6-20.* A refrigeration firm needs to build seven new plants in several portions of the United States to efficiently satisfy the growing demand for its new portable refrigerators. It is considering building one plant in California, three in the 13 southern states, one in either New York, Massachusetts, or Rhode Island, and two in any of five Midwestern states: Ohio, Indiana, Illinois, Nebraska, and Iowa. How many combinations will the firm have to examine if it wants to study the advantages of each possible combination?

6-21. Cando Company is in the process of storing its accounts receivable records in a computer file. The controller has decided that each account will be assigned a four-character identification code which will be punched in the first four columns of all data cards pertaining to the particular account.
(a) If each character in the code can only be one of the digits from 0 to 9, how many accounts can be stored?
(b) If the first character is one of the letters of the alphabet and the remaining characters are numbers, how many accounts can be stored?
(c) If each character can be either a letter or a number, how many accounts can be stored?

6-22.* John Marshall, office manager for a branch office of a large corporation, always uses the following technique to determine whether to write letters to customers whose accounts are past due. He draws a random sample of 10 accounts receivable records. If he finds seven or more accounts in current status, he will not write letters. If he finds more than three balances overdue, he will examine all account balances for the office, and write letters to all customers whose balances are delinquent. In the past, he has found three or fewer accounts past due only 20 percent of the time. The probabilities that he will collect the last overdue balance in a certain week depends on whether he writes letters:

If He Writes Letters

Last Past-Due Account Will Be Collected	Probability
In first week	0.3
In second week	0.4
In third week	0.2
In fourth week	0.1
Later than fourth week	0
	1.0

If He Does Not Write Letters

Last Past-Due Account Will Be Collected	Probability
In first week	0
In second week	0
In third week	0.2
In fourth week	0.3
Later than fourth week	0.5
	1.0

(a) Mr. Marshall is currently collecting a sample of 10 customer accounts. What is the probability that he writes letters?

(b) The last past-due account was collected in the fourth week. Using this new information, revise your estimate of the probability that he wrote letters.

6-23. An advertising firm is conducting an experiment to test the ability of consumers to taste the difference between its client's instant coffee and a leading brand of percolator coffee. In random order, six cups of coffee, three instant and three percolated, are presented to a subject. The subject is told that there are three cups of each kind, and he is asked to identify the contents of each cup. Find the probability mass function for the number of cups identified correctly for a subject who cannot discriminate and only guesses. Assume that he guesses three of each kind.

6-24. Wearwell shirt manufacturers has established three inspection points throughout the production of its shirts. The material that is used for the shirts is first inspected to determine any defects. Material accepted by the inspector is then cut to conform to the shirts being produced. Another inspector examines the cutting to see that the specifications are met and then forwards all cut material that has passed inspection to the sewers, who assemble the shirts. Once the shirts have been sewn, an inspector examines the completed shirts and rejects any that have defects. The inspectors are not always correct and the probability that each of the inspectors will incorrectly accept a faulty item or reject a good item is 0.05.

(a) What is the probability that
  (1) A shirt with good material but with faulty cutting will be sent to the sewers?
  (2) A shirt with defective material and faulty cutting will be sewn?

(b) If material that is actually good has passed inspection, and the cutting is actually good, what is the probability that the cut material will eventually be sewn?

6-25. In the manufacture of a specialized tool, the production system is composed of two machines, A and B. The system functions only if both machines A and B operate properly. Each machine A and B is composed of three components. The operation of either machine is assured if any one of its components is functioning. The operation of each component is independent of all other components. All components are operating at the start of a run. (We are studying breakdowns.) In order to schedule maintenance crews, the production manager wants to determine the probability that the system will fail to function properly. If she knows that a component of machine A fails with probability 0.01 and a component of machine B fails with probability 0.05, what is the probability that the system fails to function? Even if the system fails, the production manager could save time if she could guess which component to repair. If the system fails, what is the probability that the defective component is in machine A?

6-26.* Let
$$f(x) = \begin{cases} \frac{1}{2}x & 0 \leq x \leq 2 \\ 0 & \text{otherwise.} \end{cases}$$
(a) Show that $f(x)$ is a pdf.
(b) Find $P(x < 1)$, $P(x \leq 1)$, $P(x > 1)$, and $P(x = 1)$.

6-27. The productive output from an operation is given by the following pdf:
$$f(x) = \begin{cases} 0.5 & 1 \leq x \leq 2 \\ 1.0 & 2 < x \leq 2.5 \\ 0 & \text{elsewhere,} \end{cases}$$
where x is in thousands of units.
(a) If the firm fails to break even at less than 1,500 units, find the probability that it will fail to break even. Graph it.
(b) What is the probability that it will make a profit?
(c) Write the cumulative probability density function. Graph it.

Problems 6-28 and 6-29 require Technical Note 6-1.

6-28. Given the exponential pdf, $f(t) = \lambda e^{-\lambda t}$ for $t \geq 0$, $\lambda > 0$, find the cumulative probability density function.

6-29. Given $f(x) = 3x^2$ for $0 \leq x \leq 1$, show that $f(x)$ is a pdf and find $P(x \geq \frac{1}{2})$.

Chapter 7

Expectations

Having patiently read Chapters 1 to 6, the president of the Brighten-Up Roof Paint Company, Will Fixit, wanted to know if statistics would help him with his immediate problem. Brighten-Up Roof Paint Company (Burp Company) is considering renting a special roof-spraying machine to spray a latex compound on certain kinds of roofs, to act both as a sealer and as decoration. The machine that Burp Company owned broke last week. They have no capital to purchase one, and they wonder if renting one is a good idea. Mr. Fixit has done a market survey and has obtained a probability mass function for the number of customers who will use this service per week. This probability mass function (pmf) was based on the last 2 years' sales data, obtained from a frequency distribution on the 100 one-week periods in those 2 years. (They always have 2 weeks of vacation.) The frequency distribution has $\bar{x} = 2.86$ and $s = \sqrt{3.56} = 1.89$. The pmf is as follows:

$$f(x) = \begin{cases} 0.08 & x = 0 \\ 0.18 & x = 1 \\ 0.24 & x = 2 \\ 0.16 & x = 3 \\ 0.14 & x = 4 \\ 0.10 & x = 5 \\ 0.06 & x = 6 \\ 0.02 & x = 7 \\ 0.02 & x = 8 \\ 0 & \text{otherwise.} \end{cases}$$

In using a pmf based on the frequency distribution from the last 2 years, Mr. Fixit is implying that he thinks this year's demand, on the average, will be the same as before. That is, the last 2 years' demands and this year's demands are drawn from the same population. This is an important assumption that should be stated explicitly. (There are ways to avoid this assumption, as we will see in Chapter 17.) Since Mr. Fixit believes that last year's data are relevant here, he expects average demand to be 2.86 customers. He also expects the standard deviation of demand to be 1.89 customers. In other words, his *expectation* is that average demand will be 2.86, and that the standard deviation will be 1.89.

The universe mean (the average of all the demands that might occur) is also estimated to be 2.86. The universe mean will never be known exactly, but that is the value Mr. Fixit would like to know. Since all possible demands cannot be observed, Mr. Fixit will have to be content with an estimate of the population mean. (The difference between the value of a universe parameter and an estimate of that value is a difficult concept. We will discuss that difference in this chapter.)

The company has the following additional data regarding the new sprayer, as well as the pmf given above:

$$\begin{aligned}
\text{rental cost per week} &= \$100 \\
\text{revenue per customer} &= \$150 \\
\text{labor cost per customer} &= \$84 \\
\text{material cost per customer} &= \$30 \\
\text{maximum number of customers serviced} & \\
\text{in 1 week by one machine} &= 5.
\end{aligned}$$

(All customers above that number go to a competitor and are lost to Burp Company.)

Mr. Fixit would like to investigate his demand distribution further. However, the question he would really like answered is: "Should I rent the spraying machine?" We will, eventually, get to that question, but first we must learn more about expectations, one of the more important ideas in statistics.

7-1 Expectations: The Mean

As an example of expectations, if we toss a coin, say, 100 times, we expect to see (about) half heads. That does not mean that we will see exactly half heads in a particular run of 100 tosses. Suppose, for example, that we toss a coin 100 times. Each time that we get a head we write 1; each time that we get a tail we write 0. Then the average of those numbers will be the average number of heads per toss. If the coin is fair (that is, if it is balanced and there is no bias in the tossing process), we expect that average to be 0.5. This means that 0.5 is the true underlying mean of that process (the universe mean). On a particular sample we might obtain an average equal to 0.43 (based on 43 heads out of 100 tosses). If we did not believe that 0.5 were the universe mean, we might use $\bar{x} = 0.43$ as our estimate of it. (If we use 0.43 as the estimate of the universe mean and as the probability of a head on any toss, we are implying that the coin is not fair.)

The difference between the universe mean and the sample mean is that the universe mean is a single, usually unknown, number. The sample mean is known, and each time we take a sample we may observe a different sample mean. We may use the sample mean as an estimate of the universe mean. The universe mean can also be thought of as the mean of the probability mass function (pmf) or probability density function (pdf) that describes a random process. There is a special symbol for the universe mean. It is the lowercase Greek letter μ.

New Symbol	Meaning
μ	The universe mean; the mean of the pmf or pdf which describes the random process.
$\bar{x}$	The average of the values in a sample. This value is used as an estimate of μ, if μ is unknown.

The universe mean is the value that we would expect to obtain for the average of a sample. Yet it usually will not be the result obtained for a particular sample. The universe mean is also the expected value of a sample of size 1. In coin flipping, the expected number of heads is $\frac{1}{2}$ each time a coin is tossed. The coin will either have one head or no heads, but the expected number of heads is $\frac{1}{2}$. In this case the expected value cannot occur, yet the concept of expected value remains. (If you think this is a difficult notion, you are right.)

An expected value can be obtained from a probability mass function for a random process if all possible outcomes have numerical values. If we know the probability of each possible event, we can find the expected value of the variable. (Expected values can also be obtained for probability density functions in a similar way, as is shown in Technical Note 7-2 at the end of the chapter.) The formula and symbol definitions are given below.

EXPECTATIONS

Given a probability mass function with possible outcomes $x_1, x_2, \ldots, x_n$ (all numerically valued) and probabilities $f(x_1), f(x_2), \ldots, f(x_n)$, the *expected value* is denoted $E(X)$ and is given by

$$E(X) = \sum_{i=1}^{n} x_i f(x_i). \tag{7-1}$$

New Symbol	Meaning
$E(X)$	The expectation of the variable, X.
X	Capital letters are used to denote the name of a (random) variable. Lowercase letters are used to denote particular values of the variable. For example, $X \equiv$ sales, and x might equal 10,000 or some other value of X. The concept of a random variable is discussed in Technical Note 7-1 and Chapter 8.

As an example, suppose that $X \equiv$ daily sales of encyclopedias, and $x = 0, 1, 2,$ or 3. Suppose further that the pmf is

$$f(x) = \begin{cases} 0.2 & \text{for } x = 0 \\ 0.4 & \text{for } x = 1 \\ 0.3 & \text{for } x = 2 \\ 0.1 & \text{for } x = 3 \\ 0 & \text{otherwise.} \end{cases}$$

Then
$$E(X) = 0(0.2) + 1(0.4) + 2(0.3) + 3(0.1) = 1.3.$$

In Mr. Fixit's problem, where $X \equiv$ the number of customers, with values of $0, 1, 2, 3, 4, 5, 6, 7,$ and 8, the pmf values are as follows:

$$f(0) = 0.08, \quad f(3) = 0.16, \quad f(6) = 0.06$$
$$f(1) = 0.18, \quad f(4) = 0.14, \quad f(7) = 0.02$$
$$f(2) = 0.24, \quad f(5) = 0.10, \quad f(8) = 0.02.$$

We obtain the expected value:

$$\begin{aligned} E(X) &= 0(0.08) + 1(0.18) + 2(0.24) + 3(0.16) + 4(0.14) + 5(0.10) \\ &\quad + 6(0.06) + 7(0.02) + 8(0.02) \\ &= 0 + 0.18 + 0.48 + 0.48 + 0.56 + 0.50 + 0.36 + 0.14 + 0.16 \\ &= 2.86. \end{aligned}$$

But 2.86 is the $\bar{x}$ value that was given earlier! Why bother with $E(X)$ if $\bar{x}$ will do? Several observations are in order here:

1. If a pmf is based directly on a frequency distribution, $E(X)$ computed using the pmf will equal $\bar{x}$ [and, thus, $E(X)$ yields no information beyond that given by $\bar{x}$].

2. If the pmf is known to be the probability mass function that exactly describes a random process (either for theoretical reasons, as in coin tossing or because a large amount of data has been taken) then $E(X) = \mu$, the universe mean. In that case the expected value of the pmf for a random process equals the universe mean, μ. If no evidence to the contrary is available, we will assume that a pmf or pdf exactly describes the random process. (In Technical Note 7-2, we will deal with probability density functions.)

3. The concept of an expectation is a more general notion than $\bar{x}$ is. While $\bar{x}$ is an estimate of $E(X) = \mu$, expectations can be taken of things other than the process mean. We will discuss several other expectations in this chapter.

We have helped Burp Company and Mr. Fixit a little in that he now knows that $E(X) = 2.86$ is his best guess of the average number of customers per week. However, if he is to use 2.86 as his forecast, he is assuming, like it or not, that the upcoming year's demand is drawn from the same universe as the last 2 years' demand. However, while Mr. Fixit is willing to assume that next year's data is drawn from the same universe as the last 2 years, he is unimpressed with the information that he had his best guess about $E(X)$ all along. For the price of this book, he expects something more—some measure of variability, or even a direct answer to whether or not he should rent the new machine. We will take up the variability question in the next section, so Mr. Fixit will have to wait awhile for a direct answer.

S.S. 1. A quality control manager is investigating the number of defectives in a batch of 100 units of a machined part. The manager has the following data from 25 batches. There were:

0 defectives	14 times
1 defective	7 times
2 defectives	3 times
3 defectives	1 time.

Form a probability mass function using these data, and use the pmf to find $E(X)$. Is this $E(X)$ the universe mean of the random process?

EXPECTATIONS

2. The pmf for the number of heads on two tosses of a coin can be given by $f(0) = \frac{1}{4}$, $f(1) = \frac{1}{2}$, and $f(2) = \frac{1}{4}$. Find $E(X)$ for this pmf. Is $E(X)$ the universal mean for this random process?

Solution:

1.
$$f(x) = \begin{cases} \frac{14}{25} & \text{for } x = 0 \\ \frac{7}{25} & \text{for } x = 1 \\ \frac{3}{25} & \text{for } x = 2 \\ \frac{1}{25} & \text{for } x = 3 \\ 0 & \text{otherwise.} \end{cases}$$

Thus,
$$E(X) = 0(\tfrac{14}{25}) + 1(\tfrac{7}{25}) + 2(\tfrac{3}{25}) + 3(\tfrac{1}{25}) = \tfrac{16}{25} = 0.64.$$

This value is $\bar{x}$, not the universe mean, μ.

2. $E(X) = 0(\frac{1}{4}) + 1(\frac{1}{2}) + 2(\frac{1}{4}) = 1$. This value is the universe mean; $E(X) = \mu = 1$.

7-2 Expectations: Variance and Standard Deviation

Just as we can find the expected value of a variable, we can find the expected squared deviation (variance) of a variable. The same puzzling distinction, between the sample variance and the universe variance, is faced here also. For the universe standard deviation and variance we use the Greek letter σ (standard deviation) and σ^2 (variance).

New Symbol	Meaning
σ, σ^2 [σ^2 is written sometimes as Variance (X) or Var (X)]	The universe standard deviation (σ) and variance (σ^2). The standard deviation of a pmf or pdf. The average squared deviation from μ is σ^2; $\sigma = \sqrt{\sigma^2}$.
s, s^2	The average squared deviation from $\bar{x}$ of a set of sample values is s^2; $s = \sqrt{s^2}$.

A random process will always have some variability. If we take a sample, we will not get all values the same; they will have some variance, s^2. If we took all values in the universe (if that is possible), the average squared deviation from the mean will be σ^2, the universe variance. In any sample, the expected value of s^2 is σ^2. If we have only sample data, the s^2 value is our best guess as to the value of σ^2. Probability mass functions and probability density functions (Technical Note 7-2) can be used to find σ^2 and σ in a manner similar to that used to find $E(X) = \mu$.

Given a probability mass function for a random process with possible outcomes $x_1, x_2, \ldots, x_n$, with $E(X) = \mu$, and with probabilities $f(x_1), f(x_2), \ldots, f(x_n)$, the *variance* of the variable, X, is given by

$$\text{variance of } X \equiv \sigma^2 = \sum_{i=1}^{n} (x_i - \mu)^2 f(x_i)$$

and

$$\text{standard deviation of } X \equiv \sigma = \sqrt{\sigma^2}.$$

The variance is the expected squared deviation from the mean, that is,

$$\sigma^2 \equiv E[(X - \mu)^2] = E[(X - E(X))^2]. \qquad (7\text{-}2)$$

For example, using the same data we used above to find $E(X) = \mu = 1.3$:

$$f(x) = \begin{cases} 0.2 & \text{for } x = 0 \\ 0.4 & \text{for } x = 1 \\ 0.3 & \text{for } x = 2 \\ 0.1 & \text{for } x = 3 \\ 0 & \text{otherwise.} \end{cases}$$

Thus,

$$\sigma^2 = E[(X - E(X))^2] = \sum_{i=1}^{n} (x_i - \mu)^2 f(x_i)$$
$$= (1.3)^2(0.2) + (0.3)^2(0.4) + (0.7)^2(0.3) + (1.7)^2(0.1)$$
$$= 0.338 + 0.036 + 0.147 + 0.289 = 0.810.$$

So

$$\sigma^2 = 0.810 \quad \text{and} \quad \sigma = \sqrt{0.810} = 0.90.$$

In the Burp Company example,

$$\sigma^2 = E[(X - E(X))^2] = (2.86)^2(0.08) + (1.86)^2(0.18) + (0.86)^2(0.24)$$
$$+ (0.14)^2(0.16) + (1.14)^2(0.14) + (2.14)^2(0.10) + (3.14)^2(0.06)$$
$$+ (4.14)^2(0.02) + (5.14)^2(0.02)$$
$$= 0.654 + 0.622 + 0.178 + 0.003 + 0.182 + 0.458 + 0.592 + 0.343$$
$$+ 0.528$$
$$= 3.56$$

$\sigma = \text{standard deviation} = \sqrt{3.56} = 1.89.$

EXPECTATIONS

Once again, this is the same as the sample value, and for the same reason—the pmf was derived from the sample data. If a pmf results directly from a frequency distribution, the variance and standard deviation computed from that pmf are the sample variance (s^2) and standard deviation (s). If the pmf is the pmf that exactly describes the random process, we obtain σ^2 and σ when we perform the indicated calculations. We use μ and σ for the universe parameters; $\bar{x}$ and s are the sample values, which are the best guesses we have as to the universe parameters when the universe parameter values are unknown.

Once again we have helped Mr. Fixit by telling him that the value he already knew was his best guess as to the standard deviation of demand. He understands the concepts of expected value and expected squared deviation now, but he wishes we would get on to his real problem of whether to rent the machine or not. In fact, he has a proposed solution, which we will consider in the next section.

S.S. 1. The quality control manager in the problem at the end of Section 7-1 formed a pmf from 25 batches of empirical data.

$$f(x) = \begin{cases} \frac{14}{25} & \text{for } x = 0, \\ \frac{7}{25} & \text{for } x = 1, \\ \frac{3}{25} & \text{for } x = 2, \\ \frac{1}{25} & \text{for } x = 3, \\ 0 & \text{otherwise.} \end{cases} \qquad \begin{aligned} f(0) &= \tfrac{14}{25} = 0.56, \\ f(1) &= \tfrac{7}{25} = 0.28, \\ f(2) &= \tfrac{3}{25} = 0.12, \\ f(3) &= \tfrac{1}{25} = 0.04, \end{aligned}$$

Find the variance and standard deviation of this pmf. Are these values the universe variance and standard deviation of the random process?

2. The pmf for the number of heads on two tosses of a coin can be given by

$$f(0) = \tfrac{1}{4}, \qquad f(1) = \tfrac{1}{2}, \qquad f(2) = \tfrac{1}{4}.$$

Find the variance and standard deviation of this pmf. Are these values the true variance and standard deviation of this process?

Solution:

1. $\bar{x} = 0.64 = E(X)$:

$$E[(X - E(X))^2] = \sum_{i=1}^{4} f(x_i)[x_i - E(X)]^2$$

$$= (0.64)^2(0.56) + (0.36)^2(0.28) + (1.36)^2(0.12)$$
$$+ (2.36)^2(0.04)$$

$$= 0.229 + 0.036 + 0.222 + 0.223 = 0.710.$$

This value is s^2, not σ^2. Thus, $s^2 = 0.710$ and $s = \sqrt{0.710} = 0.843$.

2. $E(X) = 1.0 = \mu$. Thus, the universe variance,

$$\sigma^2 = 1^2(0.25) + 0^2(0.50) + 1^2(0.25) = 0.50 \quad \text{and} \quad \sigma = \sqrt{0.50} = 0.707.$$

7-3 Expectations of Linear Functions

At this point Mr. Fixit became tired of waiting. He performed the following calculations using the data introduced in the introductory section.

$$E(X) = \bar{x} = 2.86$$

each customer brings in $(150 - 84 - 30) = \$36$ net contribution

$$\text{expected net contribution} = (2.86)(36)$$
$$= \$102.96 \text{ per week}$$
$$\text{rental cost per week} = \$100.$$

Burp Company will be $2.96 better off per week if the machine is rented (according to Mr. Fixit's calculation). He estimates profit to be $2.96 per week. (Profit = net contribution − fixed costs.)

Mr. Fixit believes he has learned enough statistics to solve his problem. He has both learned some things and intuited some new ideas correctly. This is good news. However, there is also some bad news. First, let us review the good news.

Mr. Fixit has correctly surmised that if X is a variable with expectation $E(X)$, then if X is multiplied by a constant, so is the expectation. Further, if a constant is added to X, the same constant is added to $E(X)$. In Mr. Fixit's case, he wants to consider 36 times the number of customers, since his net contribution per customer (selling price less variable cost) is $36. Further, he wants to subtract fixed costs (rental cost) of $100, which is the same as adding -100. Therefore, if X is the number of customers, he will make $36X - 100$. He would like to find the expected value of that quantity. He wants $E(36X - 100)$, and he thinks that

$$E(36X - 100) = E(36X) - E(100) = 36E(X) - 100.$$

The good news is that he is right.
In general:

If X is the variable of interest, with expected value $E(X)$, then

$$E(aX + b) = aE(X) + b, \qquad (7\text{-}3)$$

where a and b are constants. [$E(b) = b$ for any constant b.]

For example,

$$E(X) = 2.86, \quad a = 36, \quad b = 100$$

and

$$E(36X - 100) = 36(2.86) - 100 = \$2.96.$$

The $2.96 value is the value Mr. Fixit had computed. The bad news is that while $2.96 is $E(36X - 100)$, it is not the expected profit. (The expected profit is complicated by the fact, given earlier, that Burp Company can only accept five or fewer customers per week. After that, potential customers are lost to the competition.) We will (finally) resolve Burp Company's problem in the next section. Before doing so, however, some other facts about expectations are discussed. The following statements are true.

Four Rules for Expectations
For variables X and Y with $E(X)$, $E(Y)$, Var (X) and Var (Y):
(a) $E(X + Y) = E(X) + E(Y)$. (7-4)
(b) Var $(aX + b) = a^2$ Var (X). (7-5)
(c) Var $(X + Y) =$ Var $(X) +$ Var (Y) only if X and Y are independent. (7-6)
(d) Var $(X - Y) =$ Var $(X) +$ Var (Y) only if X and Y are independent. (7-7)

For example, using the Burp Company data, $X \equiv$ demand, Var $(X) = 3.56$, $E(X) = 2.86$, $a = 36$, and $b = 100$. Then Var $(36X - 100) = 36^2[\text{Var}(X)] = 36^2(3.56) = 4{,}614$.

While that number may seem large, you should remember that the standard deviation of $36X - 100$ is $\sqrt{4{,}614} = 67.93$, which is the same proportion of $E(36X) = \$102.96$ as the standard deviation of X, 1.89, was of $E(X) = 2.86$.

As another example, consider $E(X) = 10$, Var $(X) = 4$, $E(Y) = 20$, and Var $(Y) = 9$, where X and Y are independent. Then

$$E(X + Y) = E(X) + E(Y) = 10 + 20 = 30$$

$$\text{Var}(X + Y) = \text{Var}(X) + \text{Var}(Y) = 4 + 9 = 13.$$

Also, using X and Y:

$$\text{Var}(2X + 3Y + 15) = 2^2 \text{Var}(X) + 3^2 \text{Var}(Y) = 4(4) + 9(9) = 97.$$

This example shows that we can combine the two facts, Var $(aX + b) = a^2$ Var (X) and Var $(X + Y) =$ Var $(X) +$ Var (Y) when X and Y are independent, into one rule:

Rules (b) and (c) combined:

$$\text{Var}(aX + bY + c) = a^2 \text{Var}(X) + b^2 \text{Var}(Y), \quad (7\text{-}8)$$

only if X and Y are independent.

We can also deal with linear functions of three or more variables by extending the rules. For example:

$$E(X + Y + Z) = E(X) + E(Y) + E(Z).$$

$$\text{Var}(X + Y + Z) = \text{Var}(X) + \text{Var}(Y) + \text{Var}(Z)$$

if X, Y, and Z are independent. Some managerial problems and hence some of the problems at the end of the chapter deal with three or more variables.

To solidify these important facts, let us try another example. First, to illustrate $E(aX + b)$, we can find $E(3X + 5)$ for the pmf given by the first two columns in the following table:

x	$f(x)$	$3x + 5$	$f(3x + 5) = f(x)$
1	0.2	3 + 5 = 8	0.2
2	0.5	6 + 5 = 11	0.5
3	0.3	9 + 5 = 14	0.3

$$E(X) = 1(0.2) + 2(0.5) + 3(0.3) = 2.1$$

$$E(3X + 5) = 8(0.2) + 11(0.5) + 14(0.3) = 11.3$$

$$E(3X + 5) = 3E(X) + 5 = 3(2.1) + 5 = 11.3.$$

$E(3X + 5)$ is found in two ways, using the pmf for $3X + 5$ and using the $E(aX + b)$ result. The fourth column above, $f(3x + 5) = f(x)$, is really unnecessary since it is the same as the second column. It is included to drive home the point that, for example, the probability that $x = 2$ is the same as the probability that $3x + 5 = 11$.

We can also use these data to illustrate the use of the formula $\text{Var}(aX + b) = a^2 \text{Var}(X)$:

$$\text{Var}(X) = \sum_{i=1}^{3}[x_i - E(X)]^2 f(x_i)$$

$$= 0.2(1 - 2.1)^2 + 0.5(2 - 2.1)^2 + 0.3(3 - 2.1)^2$$

$$= 0.49$$

$$\text{Var } (3X + 5) = 0.2(8 - 11.3)^2 + 0.5(11 - 11.3)^2 + 0.3(14 - 11.3)^2$$
$$= 4.41$$

$$\text{Var } (3X + 5) = (3^2) \text{ Var } (X) = 9(0.49) = 4.41.$$

Before concluding this section some facts that are not true in general, although they may be true in some instances, are given.

Six Formulas Involving Expectations That Do Not Hold in General
For variables X and Y with $E(X)$, $E(Y)$, Var (X), and Var (Y), the following statements do *not* hold in general.

(a) $E(XY) = E(X) \cdot E(Y)$.

(b) $E\left(\dfrac{1}{X}\right) = \dfrac{1}{E(X)}$.

(c) $E\left(\dfrac{X}{Y}\right) = \dfrac{E(X)}{E(Y)}$.

(d) Var $(X + Y) =$ Var $(X) +$ Var (Y). This does hold if X and Y are independent.

(e) Var $\left(\dfrac{1}{X}\right) = \dfrac{1}{\text{Var }(X)}$.

(f) Var $(XY) =$ Var $(X) \cdot$ Var (Y).

These statements are given because important managerial mistakes are made sometimes by assuming that they are correct. For example, if both selling price and unit sales are random, we cannot find the expected revenue by simply multiplying expected price by expected sales. That is, $E(XY) \neq E(X)E(Y)$. Nor could we obtain expected unit sales by dividing revenue by average price. That is, $E(X/Y) \neq E(X)/E(Y)$. In many situations this would not be a bad guess, especially if no other information is available, but it is not statistically correct.

S.S. 1. A nationally known furniture manufacturer has monthly sales data, in units sold, of a particular item by region. It has divided the country into four regions, and management believes it knows that $\mu = E(X)$ and $\sigma^2 =$ Var (X) in each of the regions. The information is as follows:

Region	$E(X)$	Var (X)
1	$10 = E(X_1)$	$4 =$ Var (X_1)
2	$12 = E(X_2)$	$4 =$ Var (X_2)
3	$6 = E(X_3)$	$3 =$ Var (X_3)
4	$7 = E(X_4)$	$3 =$ Var (X_4)

(a) Find the mean and variance (if possible) of total nationwide sales.

(b) Use the answers to part (a) to find the mean of total revenue if each unit sells for $100.

2. Let X be a variable with the following pmf:
$$f(10) = 0.5, \quad f(20) = 0.4, \quad f(30) = 0.1$$
To find the mean and variance of a new variable, $Y = 1/X$, we know that the pmf is
$$f(\tfrac{1}{10}) = 0.5, \quad f(\tfrac{1}{20}) = 0.4, \quad f(\tfrac{1}{30}) = 0.1$$
Show that $E(1/X) \neq 1/E(X)$ and $\text{Var}(1/X) \neq 1/\text{Var}(X)$ for this case.

Solution:

1.

(a) The nationwide mean is the sum of the regional means. Thus,
$$\begin{aligned}\text{nationwide mean} &= E(X_1 + X_2 + X_3 + X_4) \\ &= E(X_1) + E(X_2) + E(X_3) + E(X_4) \\ &= 10 + 12 + 6 + 7 \\ &= 35.\end{aligned}$$

We cannot find the variance of total nationwide sales unless the demands are independent, and they probably are dependent. (Can you think of some reasons why the demands might be dependent?)

(b) The mean of total revenue = 100 times expected nationwide sales = $100(35)$ = \$3,500 from this particular item. If we had a variance from part (a), the variance for total nationwide dollar sales would be $10,000 = 100(100)$ times that answer.

2.
$$E(X) = 0.5(10) + 0.4(20) + 0.1(30) = 16$$
$$\begin{aligned}\text{Var}(X) &= E[(X - E(X))^2] \\ &= 0.5(10 - 16)^2 + 0.4(20 - 16)^2 + 0.1(30 - 16)^2 \\ &= 18 + 6.4 + 19.6 = 44.\end{aligned}$$

Now to find $E(1/X)$ and $\text{Var}(1/X)$.
$$E\left(\frac{1}{X}\right) = E(Y) = 0.5(0.1) + 0.4(0.05) + 0.1(0.033)$$
$$= 0.05 + 0.02 + 0.0033 = 0.0733$$

EXPECTATIONS

$$\text{Var}\left(\frac{1}{X}\right) = \text{Var}(Y) = E[(Y - E(Y))^2] = 0.5(0.0733 - 0.1)^2$$
$$+ 0.4(0.0733 - 0.05)^2 + 0.1(0.0733 - 0.033)^2$$
$$= 0.000356 + 0.000217 + 0.000162 = 0.000735$$

$$E\left(\frac{1}{X}\right) = 0.0733, \quad \text{where } \frac{1}{E(X)} = \frac{1}{16} = 0.0625$$

$$\text{Var}\left(\frac{1}{X}\right) = 0.000735, \quad \text{where } \frac{1}{\text{Var}(X)} = \frac{1}{44} = 0.0227.$$

7-4 Expectations: Expected Payoffs and Other Functions of a Random Process

The reason Mr. Fixit's solution in Section 7-3 is incorrect is that he found $E(36X - 100)$, not expected profit. When demand is 6, 7, or 8, the profit to Burp Company is not $36x_i - 100$, since they only service five customers; the profit is $36(5) - 100$ for 5, 6, 7, or 8 customers. The basic outcomes, the profit figures, and the probabilities are as follows:

Outcome (Demand)	Probability of Outcome	Profit (dollars)
0	0.08	$36(0) - 100 = -100$
1	0.18	$36(1) - 100 = -64$
2	0.24	$36(2) - 100 = -28$
3	0.16	$36(3) - 100 = 8$
4	0.14	$36(4) - 100 = 44$
5	0.10	$36(5) - 100 = 80$
6	0.06	$36(5) - 100 = 80$
7	0.02	$36(5) - 100 = 80$
8	0.02	$36(5) - 100 = 80$

The last two columns contain profit and probability of profit figures, and the expected profit can be obtained.

$$\text{expected profit} = 0.08(-100) + 0.18(-64) + 0.24(-28) + 0.16(8)$$
$$+ 0.14(44) + 0.10(80) + 0.06(80) + 0.02(80)$$
$$+ 0.02(80)$$
$$= -\$2.80.$$

The expected profit is negative, and he should not rent the machine. The more careful analysis included specifically noting how much payoff (profit in this case) occurred for each possible outcome. Then the expected value of the payoff (profit) is found by multiplying each payoff by its associated probability and summing.

Finding expectations is a very general operation. Anytime we have possible values of something and can find the probability of each value, we can compute the expected value of that "something." This sounds easy stated this way, but it causes difficulty for many students of probability and statistics. The complication arises because, as in Burp Company's case, there is a basic variable, demand, and the payoff is a function of demand. Thus, we want the expected value of a second variable that is a function of the variable for which the probabilities are known. This leads us to the following statement.

Expectation of a Function of a Variable

Given a variable, X, with values $x_1, x_2, \ldots, x_n$, and with probability mass function values $f(x_1), f(x_2), \ldots, f(x_n)$, and given a second variable Y, with values $g(x_1), g(x_2), \ldots, g(x_n)$, the *expected value* of Y is given by

$$E(Y) = E[g(x)] = \sum_{i=1}^{n} g(x_i) f(x_i). \qquad (7\text{-}9)$$

New Symbol	Meaning
$E(Y) = E[g(x)]$	Expected value of the variable Y, which is a function of the variable X, with values given by $g(x)$.

The preceding definition and formula give the general form of expectation. We can compute the expected value of any variable for which the values and probabilities are known. Expected payoff, expected lost sales, expected number of jobs completed per week, and several other expected values are all possible using the probabilities for the same basic variable, such as demand. Several examples are given below, based on the Burp Company's data.

Probability	Demand (X)	Lost Sales	Jobs Completed	X^2
0.08	0	0	0	0
0.18	1	0	1	1
0.24	2	0	2	4
0.16	3	0	3	9
0.14	4	0	4	16
0.10	5	0	5	25
0.06	6	1	5	36
0.02	7	2	5	49
0.02	8	3	5	64

EXPECTATIONS

The expected value of any of the last three columns is found by multiplying the probabilities by the value in the appropriate column and summing. For example:

$$\text{expected lost sales} = 0(0.08) + 0(0.18) + 0(0.24) + 0(0.16) + 0(0.14)$$
$$+ 0(0.10) + 1(0.06) + 2(0.02) + 3(0.02)$$
$$= 0.16$$

$$\text{expected jobs completed} = 0(0.08) + 1(0.18) + 2(0.24) + 3(0.16) + 4(0.14)$$
$$+ 5(0.10) + 5(0.06) + 5(0.02) + 5(0.02)$$
$$= 2.7.$$

We notice that these two expected values add up to 2.86, which is the expected demand. This is true since a demand for a job results in either a completed job or a lost sale in this situation. There is no other possibility. As another example:

$$E(X^2) = 0.08(0) + 0.18(1) + 0.24(4) + 0.16(9) + 0.14(16) + 0.10(25)$$
$$+ 0.06(36) + 0.02(49) + 0.02(64)$$
$$= 11.74.$$

Lost sales and X^2 both have direct application. The expected lost sales, for example, would be of interest to an inventory control manager or a marketing manager, in that the firm cannot continue to lose large numbers of sales and maintain its market share. Nor can they afford, typically, to act so as to assure that no lost sales will occur.

$E(X^2)$ can be used for computing Var (X). The formula, which we will not derive (see Problem 7-26), is

$$\text{Var}(X) = E(X^2) - [E(X)]^2. \tag{7-10}$$

To check that formula using Burp Company's data, where $E(X^2) = 11.74$, $E(X) = 2.86$, and Var $(X) = 3.56$,

$$\text{Var}(X) = 11.74 - (2.86)^2 = 11.74 - 8.18 = 3.56.$$

(By golly, it works!)

The point of these examples and of this section is that expectations of interest other than $E(X)$ can be computed. Further, such expectations may be important in many situations. $E(X^2)$ and expected lost sales are two examples.

The most important example of expectations other than $E(X)$ is expected payoff. Expected payoffs form the basis for decision making, a subject studied extensively in Chapter 13. In decision-making problems, it is of paramount importance to properly represent the payoffs. Mr. Fixit of Burp Company reached an incorrect conclusion because he did not represent payoffs properly. In that case, the expected payoff from renting the machine changed from positive to negative after the proper representation was made. (Mr. Fixit should not rent the machine. We finally solved his problem.)

S.S. The Fancy Fresh Fruit Firm sells fresh corn by the bushel to individuals at a roadside stand. They pick once a day, and any corn not sold must be thrown away. (Corn more than 1 day old cannot be sold as "fresh" corn.) The pmf of daily demand, X, is

$$f(x) = \begin{cases} 0.2 & \text{for } x = 6 \\ 0.2 & \text{for } x = 7 \\ 0.3 & \text{for } x = 8 \\ 0.2 & \text{for } x = 9 \\ 0.1 & \text{for } x = 10 \\ 0 & \text{otherwise.} \end{cases}$$

The firm is considering a policy of picking 8 bushels every morning. They want to know:

(a) $E(X)$ and Var (X). Calculate Var (X) using equation (7-10).

(b) The expected number of lost sales if they pick 8 bushels each morning.

(c) The expected number of bushels that must be thrown away if they pick 8 bushels each morning.

Solution:

Probability	X	X^2	Lost Sales	Throw-Aways
0.2	6	36	0	2
0.2	7	49	0	1
0.3	8	64	0	0
0.2	9	81	1	0
0.1	10	100	2	0

The last two columns are based on a stock level of 8.

(a) $E(X) = 0.2(6) + 0.2(7) + 0.3(8) + 0.2(9) + 0.1(10) = 7.8$

$E(X^2) = 0.2(36) + 0.2(49) + 0.3(64) + 0.2(81) + 0.1(100) = 62.4$

Var $(X) = E(X^2) - [E(X)]^2 = 62.4 - (7.8)^2 = 1.56$

(b) Expected lost sales = (0.2)(0) + (0.2)(0) + (0.3)(0) + (0.2)(1) + (0.1)(2) = 0.4

(c) Expected throw-aways = 0.2(2) + 0.2(1) + 0.3(0) + 0.2(0) + 0.1(0) = 0.6.

The optimal number of bushels to pick will depend on the cost of throwing away a bushel and the cost of a lost sale. This type of decision is discussed further in Chapter 13.

7-5 A Warning

Expectations are of primary importance in statistics and in everyday life. We base many decisions on what we expect to happen. In statistics, expected value has a very precise meaning, namely the average value we think we would see if we obtained a very large number of sample values. This chapter has discussed the expected value of a variable, $E(X)$, the expected squared deviation from $E(X)$, called the variance, as well as expected lost sales, expected payoffs, and several other expectations.

There are two basic warnings regarding these expectations. The first is that care must be taken to determine whether $E(X) = \mu$ or $E(X) = \bar{x}$. That is, are we dealing with the universe value or the exact pmf or pdf of the random process, or is the expectation based on a small number of sample data? (If there is a large number of data, using the frequency distribution to estimate the pmf for the random process results in very little error.) The reason for concern is that we have more accurate information if we have $E(X) = \mu$ than if we have only $\bar{x}$, an estimate of μ.

The second, and more important, warning is that expected values (or statistics in general) do not protect a decision maker from misspecifying the problem. When we are dealing with expected payoffs, the payoffs must be correctly specified or the expected value will be incorrect. The manager must know what information is needed and be sure that the proper quantities are being considered. Expectations can assist in making decisions, but only if the problem formulation has been properly specified.

TECHNICAL NOTE 7-1: RANDOM VARIABLES

Throughout this chapter the term "variable" is used when the basic random process, such as demand, produces numerically valued outcomes. Such numerically valued variables are called *random variables*. The term "random variable" has a precise statistical definition, which is the subject in this note. When we think of a random variable, we think about a quantity whose value is unknown ahead of time. The word "quantity" is significant. We speak only of numerically valued random variables (demand is 0, 1, 2, or 3); other types of random processes (the weather is rainy, sunny, cloudy, or snowy, for example) do not define random variables. Since the idea is intuitive, and since the precise formulation may be confusing, the definition is relegated to this technical note. In Chapter 8 we will discuss the intuitive notion of random variables at more length.

> A *random variable* is a function that assigns a real number to every possible outcome of a random process. Random variables are denoted by capital letters, such as X, and the values that the random variable may take on are denoted by lowercase letters, such as x. (In mathematical terminology, the domain of the random variable is the set of fundamental outcomes in the sample space; the range of the random variable is the set of real numbers to which the outcomes are matched.)

As one example, consider a sample space for tossing two coins.

$$S \equiv \{(h, h), (h, t), (t, h), (t, t)\}.$$

We can define a random variable, X, as the number of heads that occurred. Then the four outcomes would be matched to 2, 1, 1, and 0, respectively. The random variable is formally the set of all pairs, one from the sample space and the number to which it is matched. Thus,

$$X \equiv \{((h, h), 2), ((h, t), 1), ((t, h), 1), ((t, t), 0)\}.$$

We could define $Y \equiv$ the number of tails; then

$$Y \equiv \{((h, h), 0), ((h, t), 1), ((t, h), 1), ((t, t), 2)\}.$$

The important characteristic is that we finally obtain only numerical values; we will not, as a rule, write the random variable out as shown here.

For another example, we might describe a random variable for demand as follows:

Outcome	Value of the Random Variable
1. One customer arrived at 10:02 A.M. Another customer arrived at 1:43 P.M.	2
2. One customer arrived at 11:43 A.M. Another customer arrived at 2:14 P.M. Another customer arrived at 4:12 P.M.	3

The random variable describes the entire outcome by a single number. The random variable is a complete list of all possible outcomes, each paired

EXPECTATIONS

with a real number. This is necessary because outcomes of the type described on the left above are hard to analyze. For example, we cannot compute an expected value for these outcomes.

Probability density functions can only be used with numerically valued outcomes and, hence, with random variables. Probability mass functions or density functions defined for a random variable are often indicated using a capital-letter subscript: $f_X(x)$ or $f_X(x_i)$. Then the expected value of a random variable is given as shown in Section 7-1 for a pmf or as shown below for a pdf.

$$E(X) = \sum_{i=1}^{n} x_i f_X(x_i) \qquad \text{for a pmf with values } x_1, \ldots, x_n$$

or (7-11)

$$E(X) = \int_a^b x f_X(x)\, dx \qquad \text{for a pdf defined on } a \leq x \leq b.$$

The subscript X may be omitted when the random variable under consideration is clear.

Probability density functions are discussed further in Technical Note 7-2. As one more example, consider the random phenomenon of weather in March (in a certain area of the country). We have the following probability mass function for the random process:

Outcomes	Probability
Rain	0.3
Snow	0.1
Cloudy	0.4
Sunny	0.2

The outcomes for this example are not numerically valued. But do not let that worry you; we can establish a rule for assigning a number to each outcome. In the demand case the rule was that the random-variable value equals the number of customers. Here, one rule that can be used is: Assign 1 if there is any precipitation, assign 0 if there is no precipitation. (The best rule to use depends on what we want to know.) Then the random variable can be written:

Outcome	Value of the Random Variable
Rain	1
Snow	1
Cloudy	0
Sunny	0

The expected value of this random variable is the expected number of days of precipitation. To find that value, we must find the pmf for the random variable. This can be done using the addition rule for probabilities discussed in Chapter 5. (The four outcomes do not, by definition, occur on the same day. If there is any snow, the day is classified as snowy. If there is no snow, but it rains, the day is classified as rainy, and so on.) Thus, the probability mass function is given by

$$f(1) = f(\text{rain}) + f(\text{snow}) = 0.3 + 0.1 = 0.4$$
$$f(0) = f(\text{cloudy}) + f(\text{sunny}) = 0.4 + 0.2 = 0.6.$$

The expected value of the random variable is

$$E(X) = 0.4(1) + 0.6(0) = 0.4.$$

S.S. In the weather example above, form a different random variable designed to study the number of days of sun (versus no sun). Then the outcome "sun" would be matched with 1. Find the expected value of this random variable.

Solution:

Outcome	Value of the Random Variable
Rain	0
Snow	0
Cloudy	0
Sunny	1

$$f(1) = f(\text{sunny}) = 0.2$$
$$f(0) = f(\text{rain}) + f(\text{snow}) + f(\text{cloudy}) = 0.8$$
$$E(X) = 0.2(1) + 0.8(0) = 0.2.$$

TECHNICAL NOTE 7-2: EXPECTATIONS INVOLVING PROBABILITY DENSITY FUNCTIONS

The concept of expectation translates directly from mass functions to density functions. All the important notions discussed in this chapter for mass functions hold for density functions as well. The difficulty is that integration is

EXPECTATIONS

required, in place of summation, to obtain the expectation. (Integration is a continuous summing-up process.) This note gives the integral form of the formulas in the chapter and provides a few examples. Often the integration itself is the hardest part of a problem. The best way to handle an unfamiliar integral is to look it up in a calculus book or a table of integrals.

Expected Value, Variance, and Expected Value of a Function for a Probability Density Function

Given a probability density function $f_X(x)$, defined on a random variable X, with values $a \leq x \leq b$, the *expected value* and *variance* of X are found as follows:

$$E(X) = \int_a^b x f_X(x)\, dx$$

$$\begin{aligned} \text{Var}(X) &= \int_a^b [x - E(X)]^2 f_X(x)\, dx \\ &= \int_a^b x^2 f_X(x)\, dx - \left[\int_a^b x f_X(x)\, dx\right]^2 \\ &= \left[\int_a^b x^2 f_X(x)\, dx\right] - [E(X)]^2 = E(X^2) - [E(X)]^2. \end{aligned}$$
(7-12)

[As before, $f_X(x)$ is the pdf for the random process with $E(X) = \mu$ and Var$(X) = \sigma^2$.] If, further, Y is a function of X with values given by $g(x)$, defined for $a \leq x \leq b$, then the *expected value* of Y is

$$E(Y) = E[g(X)] = \int_a^b g(x) f_X(x)\, dx. \qquad (7\text{-}13)$$

For example, suppose that the probability density function for labor hours required for a particular unit is given by

$$f_X(x) = \begin{cases} 2x & 0 \leq x \leq 1 \\ 0 & \text{otherwise.} \end{cases}$$

The job takes a maximum of 1 hour since the maximum value of x defined by the function is 1.

$$\mu = E(X) = \int_0^1 x f_X(x)\, dx = \int_0^1 x(2x)\, dx = \tfrac{2}{3} x^3 \big|_0^1 = \tfrac{2}{3}$$

$$\text{Var}(X) = \int_0^1 (x-\mu)^2 f_X(x)\, dx = \int_0^1 (x-\tfrac{2}{3})^2 2x\, dx$$
$$= \int_0^1 (x^2 - \tfrac{4}{3}x + \tfrac{4}{9}) 2x\, dx = \int_0^1 (2x^3 - \tfrac{8}{3}x^2 + \tfrac{8}{9}x)\, dx$$
$$= (\tfrac{2}{4}x^4 - \tfrac{8}{9}x^3 + \tfrac{8}{18}x^2)\big|_0^1 = \tfrac{2}{4} - \tfrac{8}{9} + \tfrac{8}{18} = \tfrac{1}{18}$$
$$= 0.0556.$$

If each hour costs \$5, we could write the hourly cost as $g(x) = 5x$. The expected value and variance of $g(x)$ could be found as

$$E[g(x)] = \int_0^1 g(x) f_X(x)\, dx$$

$$\text{Var}[g(x)] = \int_0^1 [g(x) - E[g(x)]]^2 f_X(x)\, dx.$$

However, the results of Section 7-3 can be used to obtain

$$E(5X) = E[g(x)] = (5)E(X) = \tfrac{10}{3} = 3.333$$
$$\text{Var}(5X) = \text{Var}[g(x)] = (5^2)\text{Var}(X) = \tfrac{25}{18} = 1.389.$$

As a final example, consider the exponential distribution, which is used to describe the time between customer arrivals in a waiting line, the length of time for telephone calls, and several other phenomena. This example requires some relatively complex integration techniques.

$$f_X(x) = \begin{cases} ae^{-ax} & 0 \le x < \infty,\ a > 0 \\ 0 & \text{otherwise} \end{cases}$$

$$E(X) = \int_0^\infty x(ae^{-ax})\, dx = \left(-xe^{-ax}\big|_0^\infty + \int_0^\infty e^{-ax}\, dx \right)$$
$$= \left(-xe^{-ax} - \tfrac{1}{a}e^{-ax} \right)\bigg|_0^\infty = +\tfrac{1}{a}.$$

[The integration uses "integration by parts" and the fact that $(-\infty)(e^{-\infty}) = 0$.] Then

$$\text{Var}(X) = \int_0^\infty [x - E(x)]^2 f_X(x)\, dx$$
$$= \int_0^\infty \left(x - \tfrac{1}{a} \right)^2 ae^{-ax}\, dx = \left(\tfrac{1}{a}\right)^2.$$

EXPECTATIONS

This integral requires both repeated applications of integration by parts and patience.

S.S. Suppose that a service time is described by

$$f_X(x) = \begin{cases} 0.2e^{-0.2x} & 0 \le x < \infty \\ 0 & \text{otherwise.} \end{cases}$$

Using the results above for any $a > 0$, what are the expected values and variance?

Further, if service time exceeds 8 hours, overtime must be paid. Find the expected amount of overtime. (This is a partial expectation, analogous to the lost sales example in the chapter. The integral will be taken only over the appropriate range, $x > 8$.)

Solution:

$$E(X) = \frac{1}{a} = \frac{1}{0.2} = 5$$

$$\text{Var}(X) = \left(\frac{1}{a}\right)^2 = \left(\frac{1}{0.2}\right)^2 = 25$$

$$\text{expected overtime} = \int_8^\infty (x - 8)(0.2)e^{-0.2x}\, dx$$

$$= \int_8^\infty x(0.2)e^{-0.2x}\, dx - 8\int_8^\infty (0.2)e^{-0.2x}\, dx$$

$$= [-e^{-0.2x}(x+5) + 8e^{-0.2x}]\Big|_8^\infty$$

$$= 0 + e^{-1.6}(13) - (8)e^{-1.6} = 5e^{-1.6}.$$

Using $e^{-1.6} = 0.2019$, expected overtime $= 5(0.2019) = 1.01$.

New Symbols

Symbol	Meaning
μ	The true mean value of a random process.
$\sigma, \sigma^2 = \text{Var}(X)$	The true standard deviation and variance of a process.
$E[g(x)]$	Expected value of a random variable which has values given by a function, $g(x)$, of the random variable X.

Key Formulas

Formula	Used to Compute:
$E(X) = \sum_{i=1}^{1} x_i f(x_i)$	The expected value of a variable X with probability mass function values given by $f(x_i)$.
$\sigma^2 = E[(x - E(X))^2]$	The variance of the variable X.
$E(aX + b) = aE(X) + b$	The expected value of a linear function of X.
$E(X + Y) = E(X) + E(Y)$	The expected value of a sum of two variables.
$\text{Var}(aX + b) = a^2 \text{Var}(X)$	The variance of a linear function of X.
$\text{Var}(aX + bY) = a^2 \text{Var}(X) + b^2 \text{Var}(Y)$	The variance of the sum of a constant times X and a constant times Y; this holds only if X and Y are independent.
$E(Y) = E[g(x)] = \sum_{i=1}^{1} g(x_i) f(x_i)$	The expected value of Y, whose values are given by $g(x)$, a function defined on X, when the probability values for X are given by a pmf.
$\text{Var}(X) = E(X^2) - [E(X)]^2$	An alternative formula for computing the variance of the variable X.

PROBLEMS

7-1. Briefly describe the difference between $\bar{x}$ and μ.

7-2. What is the difference between the following formulas?

$$\sigma^2 = \sum_{i=1}^{n} (x_i - \mu)^2 f(x_i) \quad \text{and} \quad \sigma^2 = E(X^2) - [E(X)]^2.$$

7-3. Name at least three business situations in which expectations are important. How important is an estimate of variability around the expectation in each of those situations?

7-4.* If you roll a die, with 1 through 6 spots on the faces, what is the expected number of spots on any one roll? (Assume that all 6 values are equally likely.) Will you ever observe this expected value on any roll?

7-5.* If $E(X) = 2$ and $E(Y) = 3$, indicate what the following quantities are, if possible. If a specific answer cannot be given, say why.
 (a) $E(X + Y)$
 (b) $E(2X + 3Y)$
 (c) $\text{Var}(X + Y)$
 (d) $E(1/X)$

7-6. A gambler has lost one of his two dice, so he must play a game with one die. The game is as follows. If you roll a "2" on the first roll, you win $1. If you roll a "1" or a "6," you lose $2. If you roll a 3, 4, or 5, you win $0.75. What is the expected value of the payoff from playing this game?

7-7.* The Nadir Marketing Company is considering distributing a new item. They think there is a 20 percent chance that the item will "catch on," in which case they will make $1,000,000 minus $200,000 distribution expense. There is a 40 percent chance that the item will cover the expenses of $200,000 exactly, and there is a 40 percent chance that the item will make nothing and the $200,000 expense will be lost. What is the expected net revenue (after distribution expense) of this item? Should they distribute the new item?

7-8.* In Problem 7-7, the Nadir Marketing Company considered distributing an item which they estimate has a 20 percent chance of netting $800,000, a 40 percent chance of netting $0, and a 40 percent chance of netting $-$200,000. What is the variance of the return in this case? Is the variance (or standard deviation) a meaningful measure of risk here?

7-9. Let $x = 0$ if a radio produced on an assembly line is defective and let $x = 1$ if a radio is not defective. If 1 percent of all radios produced are defective, what is the expected value of X?

7-10. An insurance company offers a 1-year $15,000 fire insurance policy to a home owner for $36 but is considering raising the premium. Actuaries for the company have found that 2 out of 1,000 homes of this type will have fires in any year and that all fires will cause damages of at least $15,000. What is the expected gain in a year from this policy if each loss is assumed to be exactly $15,000?

7-11. A company is considering opening a new retail store in either Cleveland or Atlanta. The company estimates that if it opens the store in Cleveland, the store will achieve either a profit of $4,000 with probability of 0.8 or a loss of $10,000 with probability 0.2. If the company establishes the store in Atlanta, the probability is 0.4 that the store will be successful and earn $5,000, and the probability is 0.6 that the store will be unsuccessful and lose $1,000. Which city should the company select if it wishes to choose the city for which the expected profit is greater?

7-12.* A ballpoint-pen manufacturer offers a "double-your-money-back" guarantee for every pen that does not function for at least 6 months from the date of-purchase. The company sells the pens for $10 each and they cost $5 to manufacture. The probability mass function for the percentage of pens produced that will be returned under the guarantee is as follows:

$$\begin{array}{cccc} y & 5 & 10 & 15 \\ f(y) & 0.80 & 0.15 & 0.05 \end{array}$$

or

$$f(y) = \begin{cases} 0.80 & y = 5 \\ 0.15 & y = 10 \\ 0.05 & y = 15 \\ 0 & \text{elsewhere.} \end{cases}$$

What is the expected profit from each batch of 100 pens sold, assuming that the returned pens must be thrown away?

7-13. A camera store sells a special film which yields a profit of $2 per roll. If the film is out of stock, the customers buy at another store in the same city. The number of rolls which customers purchase on any day is a random variable Y, and $E(Y) = 10$. At the end of one day the manager noticed that only five rolls of film remained in stock, and there was no way of getting additional rolls for the next day. Let X denote the profit that was lost due to the manager's failure to reorder. Find $E(X)$ if demand always exceeds 5 rolls.

7-14. Brightlite Bulb Company has two plants, one of which produces an average of 5,000 bulbs per week with a standard deviation of 500. The other plant produces an average of 12,000 bulbs per week with a standard deviation of 1,000.
  (a) What are the mean and standard deviation for the total number of bulbs produced weekly by the two plants?
  (b) What are the mean and standard deviation for the difference between the weekly production of the larger plant and the smaller plant? Assume that the two levels of production are independent.

7-15.* Let X be a random variable denoting the number of times that a certain machine will break down during any week. The probability mass function for X is as follows:

$$f(x) = \begin{cases} \dfrac{x+1}{10} & \text{if } x = 0, 1, 2, 3 \\ 0 & \text{otherwise.} \end{cases}$$

Find the variance of this probability mass function by two different methods.

7-16.* Seventy percent of the consumers in a certain city use brand X coffee. An advertising company is conducting a survey to find why consumers use brand X instead of brand Y. A consumer is selected at random. Let $x = 0$ if he does not use brand X, and let $x = 1$ if he does use brand X. Find $E(X)$ and Var (X).

EXPECTATIONS

7-17.* A real estate agency has found the probability mass function for the number of houses it sells on any day as follows:

$$f(x) = \begin{cases} 0.4 & \text{for } x = 0 \\ 0.2 & \text{for } x = 1 \\ 0.1 & \text{for } x = 2, 3, 4, 5 \\ 0 & \text{otherwise.} \end{cases}$$

For 100 independent occurrences of this random variable, find the expected value and the standard deviation of the total number of houses sold. Assume that each day is independent from every other day.

7-18. The manufacture of a specialized machine part takes place in four stages, and each stage must be completed before the next stage can be initiated. A salesman has just sold the part and wants to know how long it will be before the part is delivered to the customer. Since the part is so specialized, its manufacture is not begun until an order is received, and it must proceed through the following stages:

Stage	Expected Time of Completion of Stage (weeks)	Standard Deviation (weeks)
I	6	2
II	5	3
III	7	1
IV	4	2

What is the expected value and standard deviation of completion time for the part, assuming that the completion times for the stages are independent?

7-19. A college campus bookstore is considering its policy on purchasing used books from students. They currently set a price, then buy a specified number of any one book. For example, they might buy up to 10 books at $4, and after 10 have been purchased they would buy no more. Based on expected class size, they have estimates of how many used books they will sell. As an example, we will consider a small class in which the probability of used book sales is as follows:

Sales	Probability
1	0.2
2	0.5
3	0.3

If they buy the book for $4 and sell it for $7, how many books should they buy if they want to make as much money as possible? (If they do not sell the book, they lose the $4; they cannot expect to sell it next year.) (*Hint:* If they purchase one book, the probability of selling the book is $0.2 + 0.5 + 0.3 = 1.0$. If they buy 2 books, the probability of selling both books is $0.5 + 0.3 = 0.8$. You must compute the expected profit of each of the three possible stock-level decisions.)

7-20. In Problem 7-19 a college campus store considered buying a used book where the probability of selling the book is as follows:

Sales	Probability
1	0.2
2	0.5
3	0.3

When they paid $4 for the book and sold it for $7, we found that the store should buy two books if they want to maximize profit.

(a) If they paid $2 for the book and sold it for $7, how many books should they buy?

(b) Briefly discuss how they should decide what price to pay and what price to charge for the used books. Which of the above data might change as they change the selling price, and what other factors should they consider in making the decision?

7-21.* Taco-Yaco holds franchises for 20 fast-food Mexican restaurants and earns $1,500 per month from each franchise. By the beginning of next year, the company would like to have 5 more franchises in full operation. The probability mass function for the number of franchises obtained for next year is

$$f(x) = \begin{cases} 0.01 & \text{for } x = 0 \\ 0.04 & \text{for } x = 1 \\ 0.05 & \text{for } x = 2 \\ 0.20 & \text{for } x = 3 \\ 0.30 & \text{for } x = 4 \\ 0.40 & \text{for } x = 5 \\ 0 & \text{otherwise.} \end{cases}$$

What are the expected monthly earnings and the variance of the monthly earnings from all franchising rights next year, after the new franchises are in operation?

7-22. The manager of Playtime Toys is considering introducing one of two new toys to the market for next year. Her market research team has provided her with the probability mass functions for next year's profits for each toy:

Toy A	Toy B
$f(-2{,}000) = 0.4$	$f(0) = 0.5$
$f(0) = 0.2$	$f(1{,}000) = 0.2$
$f(1{,}000) = 0.2$	$f(2{,}000) = 0.2$
$f(5{,}000) = 0.15$	$f(3{,}000) = 0.05$
$f(10{,}000) = 0.05$	$f(4{,}000) = 0.05$

Find the expected net profit and the variance of net profit for each toy. If only one toy must be chosen to be marketed, which one should the manager choose, assuming all other factors to be equal? Assume that the sales of the two products are independent, and that less variability is preferred, for a given level of profits.

7-23. The probability mass function for the number of customers entering a bank between 9:00 and 10:00 A.M. on a Monday is as follows:

$f(20) = 0.03,$ $f(23) = 0.06,$ $f(26) = 0.21,$ $f(29) = 0.02,$

$f(21) = 0.05,$ $f(24) = 0.15,$ $f(27) = 0.12,$ $f(30) = 0.01.$

$f(22) = 0.07,$ $f(25) = 0.18,$ $f(28) = 0.10,$

(a) What is the expected number of customers entering the bank during the period?
(b) What is the variance for the number of customers?

7-24. A sample of four employees from the personnel records of Driveway Auto Parts Company reveals the following information:

Employee	Marital Status	Total Income	Number of Children
A	Married	$10,000	3
B	Married	8,600	0
C	Single	5,000	0
D	Married	12,000	2

One of the four employees is to be chosen at random. Let X have the value 1 if the employee is married and 0 otherwise, let Y be the total income, and let Z be the number of children. Find the mean and variance of each of these random variables.

7-25.* A bakery produces a certain kind of cake which costs $1.00 to make and sells for $2.50 on the day it is baked. Any cakes of this kind that are not sold on the day they are baked are thrown away. Let X be the random variable which denotes the number of this type of cake that customers order on a randomly selected day. The baker has found that the probability mass function for X is given as follows:

$$f(x) = \begin{cases} 0.1 & \text{for } x = 0 \\ 0.2 & \text{for } x = 1 \\ 0.3 & \text{for } x = 2 \\ 0.2 & \text{for } x = 3 \\ 0.1 & \text{for } x = 4 \\ 0.1 & \text{for } x = 5 \\ 0 & \text{otherwise.} \end{cases}$$

(a) How many cakes should the baker make in order to maximize his expected profit?

(b) Suppose that the baker counts each customer turned away as a 50-cent loss of goodwill. How many cakes should he bake to maximize the sum of profit from part (a) minus the goodwill cost?

7-26. Show using algebra that

$$\text{Var}(X) = E(X^2) - [E(X)]^2$$

by squaring and collecting terms using equation (7-2):

$$\text{Var}(X) = E[X - E(X)]^2.$$

7-27. (a) Find the expected value and variance of the following probability density functions. (This problem requires a knowledge of the material in Technical Note 7-2.)
  (1) $f(x) = 1, 4 \leq x \leq 5$.
  (2) $f(x) = (x^2/3) + \frac{1}{18}, 0 \leq x \leq 2$.
  (3) $f(x) = \frac{5}{2} - x, 1 \leq x \leq 2$.

(b) Describe how you would verify that each function is a probability density function.

Chapter **8**

Probability Distributions

In a book titled *Future Shock*, Alvin Toffler described the explosion of knowledge. The individual gropes for ways of dealing with this phenomenon. One approach is for individuals to categorize and summarize the plethora of information they wish to retain. In this way order may be brought to what would otherwise be chaos.

In Chapters 2 and 3, for example, we learned how a complex set of observations could be described by frequency distributions and also summarized by descriptive measures such as the mean and variance. In Section 6-2, we studied ways of describing the phenomena reflected by the observations of a frequency distribution using probability functions. Such functions describe the functional form that we expect frequency distributions to approximate based on probability theory.

For example, Table 8-1 illustrates two frequency distributions based on rolling two dice 720 times. Each of these frequency distributions represents what actually happened when your authors, having nothing better to do at the time, performed this experiment. The next-to-last column illustrates the probability function (in this case a probability mass function) based upon probability theory as it applies to tossing two dice considered to be "fair" objects. We have also included the descriptive measures, the mean and variance, for each frequency distribution and for the probability function.

We do not expect the frequency distribution to exactly replicate the probability function on any given set of 720 rolls, but that does not make the probability function any less useful. In fact, it is perhaps the best way to describe the process of rolling two fair dice the given number of times. If a still more general, summarized, description is required, we might use the description of the probability function given by the last column.

Table 8-1 Comparison of Two Frequency Distributions with the Probability Function for 720 Rolls of a Pair of Dice

	Frequency Distribution		Probability Function	
Outcome	Trial 1 720 Rolls	Trial 2 720 Rolls	Expected Number	Probability
2	19	23	20	$\frac{1}{36}$
3	37	49	40	$\frac{2}{36}$
4	51	62	60	$\frac{3}{36}$
5	82	87	80	$\frac{4}{36}$
6	99	108	100	$\frac{5}{36}$
7	130	115	120	$\frac{6}{36}$
8	110	101	100	$\frac{5}{36}$
9	82	70	80	$\frac{4}{36}$
10	55	53	60	$\frac{3}{36}$
11	38	37	40	$\frac{2}{36}$
12	17	15	20	$\frac{1}{36}$
Total	720	720	720	1
Mean:	$\bar{x} = 7.02$	$\bar{x} = 6.78$	$\mu = E(X) = 7$	
Variance:	$s^2 = 5.44$	$s^2 = 5.80$	$\sigma^2(X) = 5.83$	

In this chapter we will learn additional ways to categorize knowledge, which are both numerical and probabilistic. In particular, we will learn that a number of phenomena have similarities that allow us to describe them, probabilistically, by the same probability function. For example, we will find that flipping coins, checking for defective items in a production process, examining opinions for and against a proposal, assessing the results of certain diagnostic tests, determining the possibilities that a new program will be successful or not, and many other apparently dissimilar processes can be described by a single type of probability function, the binomial pmf.

In this chapter we will examine five such probability functions. These five functions will provide most of the generality we will need in this book. Many probabilistic problems both here and in a manager's world can be adequately quantified using the five theoretical probability functions discussed in this chapter. These functions are divided into two types: discrete probability functions and continuous probability functions. They are, in the order we will meet them:

Discrete probability functions (probability mass functions or pmf's):

1. The binomial
2. The hypergeometric
3. The Poisson

Continuous probability functions (probability density functions or pdf's):

1. The exponential
2. The normal

Sometimes we cannot be sure which of the above probability functions is relevant to a specific problem. And it may be that the appropriate function to use is one of the functions introduced later in this book, for example the chi square distribution, discussed in Chapter 11, or the t distribution, which is discussed in Chapter 10. Or it may be that the appropriate distribution is one not studied here at all, such as the Cauchy distribution. This is one excellent reason to obtain the help of a statistician when you have a problem.

Another problem, to which we will return, includes not just picking the right type of distribution, but, further, determining how to describe it. Thus, we may know that a particular random phenomenon, such as the sex of a newborn child, can be described by a binomial probability function. But we may not know what numbers to put into the formula for this function in order to appropriately define it. For example, in the present case, the probability of a female child (or male child) is required. One solution is to use a figure based on a frequency distribution of the number of such children. This approach assumes that the future will be like the past.

In our discussion of probability functions, the concept of a random variable will prove useful. It is itself a generalized concept that will apply across all probability functions. Indeed, we have already used this concept in several earlier chapters, although not by name. Random variables were specifically introduced in Technical Note 7-1. The discussion in that note is more formal than is given here.

8-1 Random Variables

It is easiest to begin with a definition.

> A *random variable* is a numerical quantity whose value is determined by the outcome of a random process.

The values of the random variable are the numbers we assign to the outcomes of the random process. Why should we complicate our lives with this somewhat mystical notion? Well, there are two reasons. First, we often want to derive new insights from our data. To do so we must manipulate the data that we have obtained. This manipulation is most easily done using the rules of mathematics. The rules of mathematics, in turn, are designed to be used on numerical quantities, and random variables yield numerical quantities. Second, even if we wish only to describe the observations in hand, many useful descriptions are supplied by numerical representations. The mean provides a good example.

Let us see how we might go about defining a particular random variable for a particular random process. Consider tossing a coin once. The possible results are head and tail. These are not numerical quantities. But if we define a random variable so that it is 1 if a head occurs and zero if a tail occurs, then we have converted the outcomes from this random process to the numerical values of a random variable.

The following analog, which we will call a random-variable machine, may help in understanding this process. Each outcome from the sample space is fed into the random-variable machine (Figure 8-1). The machine, or rule, assigns a unique numerical value to each outcome. The two alternative ways to write the rule for the random variable we defined for the coin example are given in Figure 8-2. The second way is preferred by most writers and will be used hereafter.

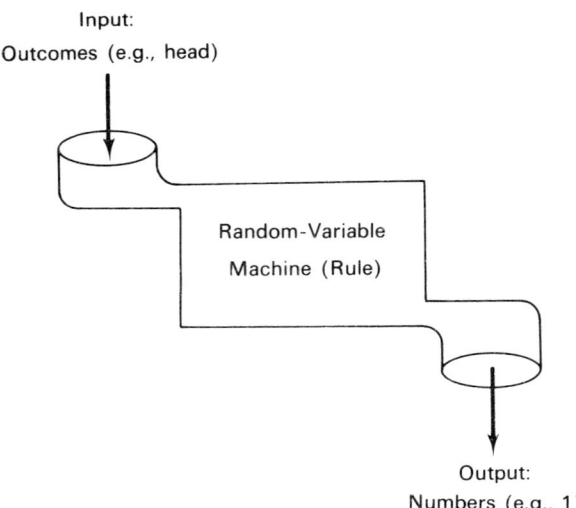

Figure 8-1 Random-Variable Machine

Probability Distributions

1.

Figure 8-2 Two Ways to Illustrate a Random Variable (Example: Toss One Coin Once)

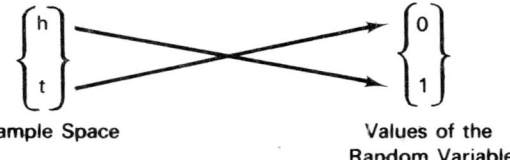

Sample Space

Values of the Random Variable

2. Give the values of the random variable the symbol (name) x, and let

$$S \equiv \{s\} \equiv \{h, t\}.$$

Then

$$x = \begin{cases} 1 & s = h \\ 0 & s = t. \end{cases}$$

(s stands for either of the two outcomes in S.)

Several important observations should be made concerning random variables. First, different random variables can be defined over the same sample space. For example, if one were to toss a coin and win $1 if the result were "head" but lose $1 if the outcome were "tail," a useful random variable would be

$$y = \begin{cases} 1 & s = h \\ -1 & s = t. \end{cases}$$

We use y for the values of this random variable to distinguish it from the previous random variable and to show the generality of the concept—not just to be difficult.

Second, some sample spaces are initially defined in numerical terms. The sample space for rolling two dice once given by the outcomes in Table 8-1 is an illustration. These numerical values can be thought of as the values of a random variable. Alternatively, we may elect to define a new random variable over this sample space. Suppose, for example, a bet of $10 returns $20 (including the initial bet, a gain of $10) if the outcome is 7 or 11. The same $10 bet is lost on the outcomes 6 and 8, and the $10 is simply returned if the outcome is any other value. Then a relevant random variable, called the player's gain (call its values w) would be

$$w = \begin{cases} +10 & x = 7 \text{ or } 11 \\ 0 & x = 2, 3, 4, 5, 9, 10, 12 \\ -10 & x = 6 \text{ or } 8, \end{cases}$$

where
$$S \equiv \{x\} \equiv \{2, 3, 4, 5, 6, 7, 8, 9, 10, 11, 12\}$$
x is defined as the sum of the two faces

w is defined as the amount of money won.

Third, the values of any random variable can be used as the elements in a new sample space for the same random process. The choice of the random variable to use is dictated by the nature of the questions we want to answer.

Basic Sample Space	Type of Question	Choice of Random Variable Sample Space
Toss of 1 coin once: {h, t}	Number of times a head occurs	{1, 0}
Toss of 1 coin once: {h, t}	Money won: +$1 for a head −$1 for a tail	{1, −1}
Roll of 2 dice once: {2, 3, 4, 5, 6, 7, 8, 9, 10, 11, 12}	Money won per trial in the above example	{−10, 0, +10}

The specific questions, however, will typically be probabilistic. For example, what is the chance we will win $1 in one toss of a fair coin? What are our expected winnings in the dice-rolling game? To answer these and similar questions we need a way of assigning probabilities to the values of the random variable based on the probabilities assigned to the outcomes in the basic sample space. How is this done? We will first state a rule that will work for the problems considered so far and then illustrate it for each of the previous examples.

Rule for Finding the Probability of an Event Defined Using the Values of a Random Variable

Let E be a value of a random variable defined over a sample space S. Let the outcomes in S that take on the value E have probabilities $p_1, p_2, \ldots, p_r$. Then

$$P(E) \equiv f(E) = p_1 + p_2 + \cdots + p_r.$$

In words, to find the probability of a particular value (or values) of the random variable, we add up the probabilities of all outcomes with that value (or values). This is illustrated in Tables 8-2 and 8-3 for the first and last of the examples we have worked. In the coin-tossing example, the calculations are relatively easy. Since, for example, only the outcome h leads to a value of 1 for the random variable,

$$f(1) = f(h) = 0.5.$$

Probability Distributions

Table 8-2 Values of the Random Variable and Their Probabilities for Tossing One Coin Once

Sample-Space Outcomes s	Probability of Outcome $f(s)$	Value of the Random Variable x	Probability of the Value of the Random Variable $f(x)$
h	0.5	1	0.5
t	0.5	0	0.5

Table 8-3 Values of the Random Variable and Their Probabilities for Rolling Two Dice Once

Sample-Space Outcomes x	Probability of Outcome $f(x)$	Value of the Random Variable w	Probability of the Value of the Random Variable $f(w)$
7	$\frac{6}{36}$	10 ⎫	
11	$\frac{2}{36}$	10 ⎭	$\frac{8}{36}$
6	$\frac{5}{36}$	−10 ⎫	
8	$\frac{5}{36}$	−10 ⎭	$\frac{10}{36}$
2	$\frac{1}{36}$	0 ⎫	
3	$\frac{2}{36}$	0	
4	$\frac{3}{36}$	0	
5	$\frac{4}{36}$	0	$\frac{18}{36}$
9	$\frac{4}{36}$	0	
10	$\frac{3}{36}$	0	
12	$\frac{1}{36}$	0 ⎭	

In the dice-rolling game, we can win $10 if either a 7 or 11 occurs. Since these outcomes are mutually exclusive, the addition rule for probability tells us

$$f(\$10) = f(7) + f(11) = \tfrac{6}{36} + \tfrac{2}{36} = \tfrac{8}{36}.$$

Table 8-3 also illustrates the calculations needed to obtain the probability of the remaining values of the random variable.

The outcomes for a random process are often numbers. When this is the case, we say that the outcomes are given by the values of a random variable. Statistical calculations are easier when the outcomes are already in numerical form. Even when the outcomes are not numbers it may be possible to change them into numbers, as was done in the coin example.

The procedure we have described is generally applicable and provides a useful way to think about such questions. The technique for finding the probability function for a random variable when the sample space is continuous can become quite involved. We examine this issue briefly in Technical Note 8-3. We close this section with a managerially oriented example for you to solve.

S.S. Suppose that a consumer may select between two brands at each time of purchase. Call them brand A and brand B. Consider the random process of three successive selections where we hypothesize that the consumer is equally likely to select brand A or brand B on each selection.

(a) Write the sample space for this process using A's and B's. We have started the process for you.

$$S \equiv \{s\} \equiv \{(A, A, A), \ldots\}$$

(b) Write the values of the random variable in formula form that assigns to each outcome a value equal to the number of brand switches observed.

$$t = \begin{cases} 0 & s = (A, A, A) \text{ or } (B, B, B) \\ s = \\ s = \end{cases}$$

(c) Write the probability mass function for the random variable in part (b).

$$f(t) = \begin{cases} \frac{2}{8} & t = 0 \\ & t = \\ & t = \end{cases}$$

(d) Suppose that a company makes \$1 profit on each A sold (B is a competitor's brand). Write the values of a new random variable, r, that gives this company's profit for each outcome in S and write the probability mass function. Again we have started the process for you.

$$r = \begin{cases} 0 & s = (B, B, B) \\ 1 & s = \\ & s = \\ & s = \end{cases}$$

$$f(r) = \begin{cases} \frac{1}{8} & r = 0 \\ & r = \\ & r = \\ & r = \end{cases}$$

Solution:

(a) The sample space consists of the different triples that can be formed using A's and B's:

$$S \equiv \{s\} \equiv \{(A, A, A), (A, A, B), (A, B, A), (B, A, A), (A, B, B),$$
$$(B, A, B), (B, B, A), (B, B, B)\}.$$

Probability Distributions

(b) The values of the random variable required by part (b) are

$$t = \begin{cases} 0 & s = (A, A, A) \text{ or } (B, B, B) \\ 1 & s = (A, A, B) \text{ or } (B, A, A) \text{ or } (A, B, B) \text{ or } (B, B, A) \\ 2 & s = (A, B, A) \text{ or } (B, A, B), \end{cases}$$

where s stands for any outcome in S.

(c) Since the choice of brand on each trial is as likely to be A as B, each of the eight outcomes in S has the same probability of $\frac{1}{8}$. Thus,

$$f(t) = \begin{cases} \frac{2}{8} & t = 0 \\ \frac{4}{8} & t = 1 \\ \frac{2}{8} & t = 2. \end{cases}$$

(d)

$$r = \begin{cases} 0 & s = (B, B, B) \\ 1 & s = (A, B, B) \text{ or } (B, A, B) \text{ or } (B, B, A) \\ 2 & s = (A, A, B) \text{ or } (A, B, A) \text{ or } (B, A, A) \\ 3 & s = (A, A, A) \end{cases}$$

$$f(r) = \begin{cases} \frac{1}{8} & r = 0 \\ \frac{3}{8} & r = 1 \\ \frac{3}{8} & r = 2 \\ \frac{1}{8} & r = 3. \end{cases}$$

8-2 Some Probability Mass Functions

When the number of outcomes in a sample space can be counted, a pmf is required. In most managerial applications where a pmf is useful, the number of outcomes in the sample space is finite, although no such limitation is placed on the functions discussed here. Three very common probability mass functions will be discussed in this section. They are, in order of presentation:

1. The binomial probability mass function.
2. The hypergeometric probability mass function.
3. The Poisson probability mass function.

Each of these functions is characterized by one or more parameters. A *parameter* is a descriptive measure of a probability function. In Chapter 3 we discussed descriptive measures of frequency distributions which are based on samples. Descriptive measures calculated from sample data (including frequency distributions) are called *statistics*. Statistics, we recall from Chapter 7, are often used as estimates of the parameter values we do not know.

We shall write probability functions using functional notation in the form $g(x; \Delta, \beta)$. The Greek letters Δ and β stand for the parameters of the distribution called g. In this book the parameters are always separated from the variable x by a semicolon. Some probability functions have one parameter (the Poisson is an example) and some have more. Sometimes the parameter values are known or given to us, and sometimes they are not. Quite often we must estimate them using the analogous sample statistic. The sample statistic is calculated from the sample observations. Statistics will be represented by English letters and never by Greek letters. Hence, when we use a Greek letter in this book, it *always* denotes a parameter. An English letter *may* be either a parameter or its sample estimate (a statistic); only the context of the discussion will tell us which.

BINOMIAL PROBABILITY MASS FUNCTION

When we are trying to establish which of several probability functions is appropriate to use for a specific problem, one useful way is to ask a set of questions related to a particular function. If the answer to each question is "yes," then that particular probability function is applicable. Ah, you say, but how are we to know the right set of questions to ask? We shall supply a list of such questions for each probability function discussed.

A *binomial process* is a set of trials for which the following questions can be answered yes.

1. Is the number of outcomes in the sample space finite?

2. At each trial of the random process are there two and only two possible outcomes?

3. If we call one of the two possible outcomes a success (the other, failure), is the probability of a success the same on every trial of the underlying random process?

If the answers to each of the three questions posed is yes, the random process is called a binomial random process and the binomial probability mass function applies. The binomial probability mass function allows us to calculate the probabilities we need more easily than is possible using the basic concepts of Chapters 5 and 6. Some examples of binomial processes include tossing a coin (whether it is "fair" or not), a series of phone calls where the line is either busy or not busy, and the inspection process for some product in which an item is classified as defective or not. One warning—question 3 requires that the probability of the outcome defined as success (say head) remain constant from trial to trial (toss to toss). In statistical

PROBABILITY DISTRIBUTIONS

language, this means that the trials must be independent. The result on one trial must not affect the result on any other trial.

Let us try an example that we are familiar with to see how this fellow we call the binomial pmf works. We will then generalize the result and try it out on a managerial problem. Consider three independent tosses of a coin. The sample space S is:

$$S \equiv \{s\} \equiv \{(h, h, h), (h, h, t), (h, t, h), (t, h, h), (h, t, t), (t, h, t), (t, t, h), (t, t, t)\}.$$

There are eight fundamental events. Hence, the number of outcomes is finite. The random process involves tossing a coin. We allow only two outcomes, "head" or "tail," on any trial of the random process (toss). Finally, the probability of a head (or a tail) remains constant from trial to trial since the tosses are given to be independent. All three questions have been answered in the affirmative. The process is therefore binomial.

We now make three arbitrary determinations to help us work on the problem. We shall first arbitrarily define a head as a success and a tail as a failure. (Although this first step is not essential, it is helpful.) Second, we shall define a random variable whose values, x, are the number of heads that occur on the three tosses. Finally, we will assume that the probability of a head on a single toss of this "unfair" coin is 0.6 and denote this value with a small p.

New Symbol	*Meaning*
p	The probability of a success on one trial of a binomial process (p is a parameter of the binomial).

Using these data, Figure 8-3 summarizes the resulting outcomes and their probabilities. The probability of each triple is the product of the probabilities of the three trial outcomes that comprise the triple. This is because the three trials are independent, and we can use the simplified multiplication law for probabilities. Thus, the probability of the first triple, (h, h, h), is the probability that the first toss is a head, and then the second is also a head, and finally the last toss is a head. This probability is $(0.6)(0.6)(0.6) = 0.216$.

In Table 8-4, the triples are converted to the values of the random variable. The probabilities of securing the various number of successes based on the probabilities in Figure 8-3 are indicated. This table is very important to understanding just how the binomial probability mass function can save us a great deal of drudgery. We are most concerned with the number of heads in three tosses and their probabilities. These two facts are given in columns 2 and 3 of Table 8-4. The probabilities of the number of heads on

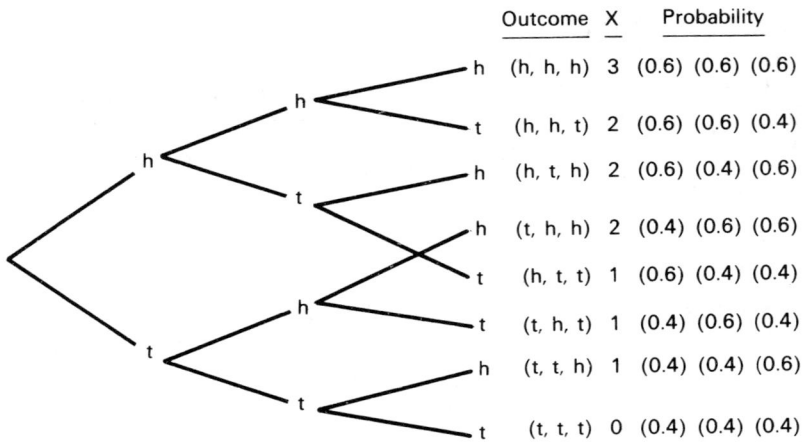

Figure 8-3 Three Tosses of a Coin Where $P(\text{Head}) = 0.6$

Table 8-4 Probability of the Number of Heads in Three Tosses of One Coin

Outcome Triple: Three Tosses of One Coin	Number of Heads on Three Tosses: x	Probability of That Number of Heads	Probability of the Outcome Triple (see Figure 8-3)
(h, h, h)	3	$(0.6)^3$	$(0.6)^3$
(h, h, t)			$(0.6)^2(0.4)$
(h, t, h)	2	$3(0.6)^2(0.4)$	$(0.6)^2(0.4)$
(t, h, h)			$(0.6)^2(0.4)$
(h, t, t)			$(0.6)(0.4)^2$
(t, h, t)	1	$3(0.6)(0.4)^2$	$(0.6)(0.4)^2$
(t, t, h)			$(0.6)(0.4)^2$
(t, t, t)	0	$(0.4)^3$	$(0.4)^3$

three tosses are obtained from the probabilities for the outcomes from the random process using the addition rule for probabilities. For example, the probability of two heads in three tosses is given by adding the probabilities of the three outcomes that produce two heads.

$$P(h, h, t) + P(h, t, h) + P(t, h, h)$$

or

$$P(1, 1, 0) + P(1, 0, 1) + P(0, 1, 1)$$

or
$$(0.6)^2(0.4) + (0.6)^2(0.4) + (0.6)^2(0.4)$$
or
$$3(0.6)^2(0.4).$$

The probabilities are added since these are three different and mutually exclusive ways of obtaining two heads in three tosses; only one of the three possible configurations of heads and tails can occur in any one set of three tosses.

Now grab your hat! We are about to generalize the results. We shall do so by giving the answer and then showing why it works.

Binomial Probability Mass Function

Given n trials of a binomial process, the probability of exactly x successes is given by

$$b(x; n, p) = \frac{n!}{x!\,(n-x)!} p^x (1-p)^{n-x}, \qquad (8\text{-}1)$$

where

$n \equiv$ number of trials,

$x \equiv$ number of successes,

$p \equiv$ probability of a success on a single trial.

Let us see how formula (8-1) applies to our coin-tossing problem. We tossed the coin three times, so $n = 3$. Defining, arbitrarily, a head to be a success, we were told that $p = 0.6$. Now suppose that we wish to know the probability of three heads in three tosses. We have $n = 3$, $x = 3$, and $p = 0.6$. Using formula (8-1), $P(3 \text{ heads in } 3 \text{ tosses}) = P(x = 3) = b(3; 3, 0.6) = 3!/3!\,0!\,(0.6)^3(0.4)^0 = (0.6)^3$. For two heads in three tosses: $n = 3$, $x = 2$, and $p = 0.6$. Thus,

$$P(x = 2) = \frac{3!}{2!\,1!}(0.6)^2(0.4)^1 = 3(0.6)^2(0.4).$$

It works! We know it works since the answer agrees with the one given in Table 8-4. Let us take just a moment to see why. The term $n!/x!\,(n-x)!$ represents the number of different ways of arranging n items of which x are of one kind and $n - x$ of another. This number is given by equation (6-5). It is

$$\binom{n}{x,\,n-x} = \frac{n!}{x!\,(n-x)!}.$$

If you find that difficult, think of this number as the number of ways of picking the x places to put the successes. This is the number of combinations of n things taken x at a time:

$$_nC_x = \frac{n!}{x!\,(n-x)!}. \tag{8-2}$$

For example, for two heads in three tosses, it is the number of ways of selecting the two spots for the two heads from the three available locations. So the constant given by $n!/x!\,(n-x)!$ gives the number of outcomes leading to x successes. And $p^x(1-p)^{n-x}$ is the probability of each mutually exclusive outcome involving exactly x successes and $n-x$ failures. Multiplying the number of outcomes yielding x successes by their common probability gives the probability of the event desired. Once n and p are known, we can compute the probability of any number of successes, x. Thus, n and p are the parameters of the binomial pmf.

A process in which the required probabilities can be calculated using the binomial pmf is called a *binomial process*. One such process that is very common in statistics is the selection of a sample where only two outcomes are possible for each item in the sample and the probabilities of each type of item remain constant from draw to draw. A sample selected by a binomial process is called a *binomial sample*.

New Symbol	Meaning
$b(x; n, p)$	The probability of x successes in n trials of a binomial process where the probability of a success on a single trial is p.

The binomial probability mass function is defined only for integer values. (Whoever heard of $2\frac{1}{2}$ heads on 3 tosses?) Since it is a probability function, it has a mean and a variance. The mean is the expected number of successes in n trials. If a coin with an 0.6 probability of a head is tossed 10 times we expect $6 = 10(0.6) = np$ heads. Using this as motivation:

The *mean of any binomial pmf* $= np$. (8-3)

No such simple motivation is available for the variance of the binomial pmf. Further, although formulas can be mathematically derived from the definitions for means and variances, little is gained for practical management purposes from doing so. Hence we will simply give the answer.

The *variance of any binomial pmf* $= np(1-p)$. (8-4)

In our coin-tossing example where the probability of a head on a single toss is 0.6:

$$\text{mean} = np = 3(0.6) = 1.8$$
$$\text{variance} = np(1 - p) = 3(0.6)(1 - 0.6) = 0.72.$$

We note that the mean (or expected value) is descriptive of the pmf, but it may not itself occur. We cannot get 1.8 heads on three tosses.

The binomial has the property that it becomes more and more symmetrical as the probability of a success on a single trial approaches 0.5, regardless of the sample size. This is illustrated in Figure 8-4. The binomial pmf also becomes more symmetrical as n increases for any probability of success, as is shown in Figure 8-5. This symmetrical behavior of the binomial is useful when n becomes large since it allows us to approximate the desired

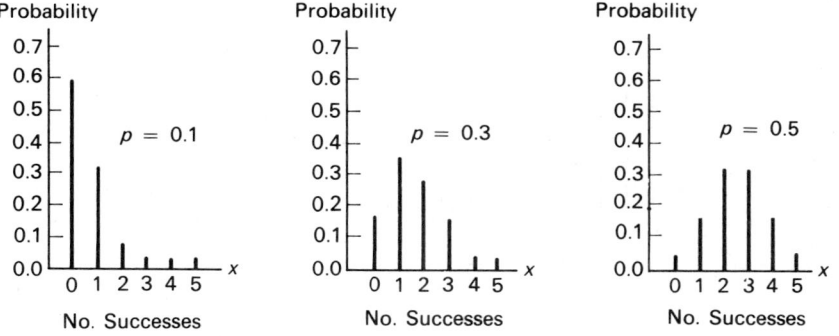

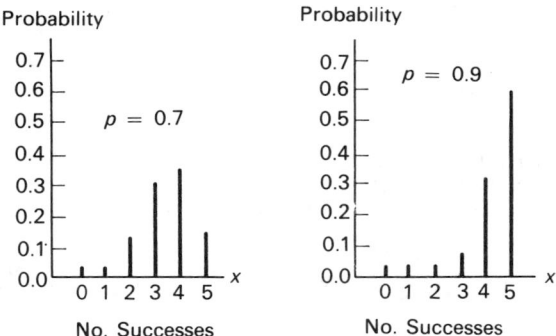

Figure 8-4 Binomial Probability Function for Several Values of p and $n = 5$

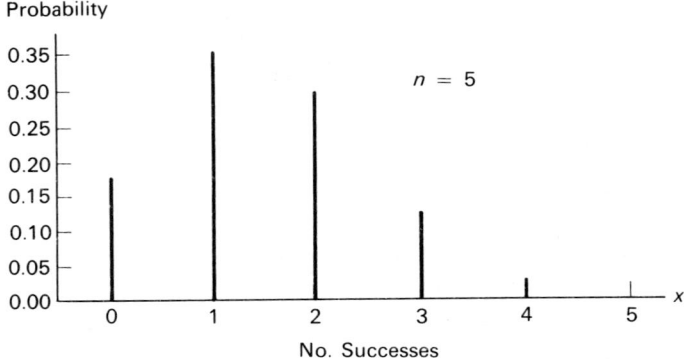

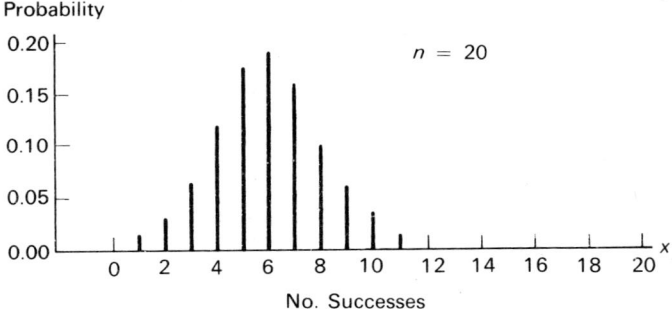

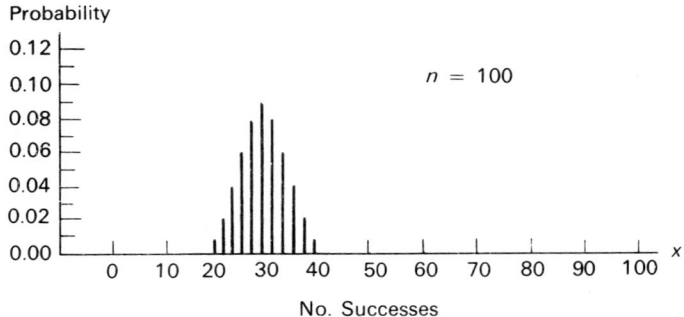

Figure 8-5 Binomial Probability Function for Several Values of n and $p = 0.3$

Probability Distributions

binomial probability using the normal probability function. We will show how this is done later in this chapter. Otherwise, the calculation of binomial probabilities for large sample sizes can be quite tedious.

To avoid such calculations even when n is relatively small, we have included a table of binomial probabilities at the end of this book. Since the table only goes up to 0.5 for p, a few examples are worked below to help you learn how to use the table. The table gives the probability of x or fewer successes; that is, it gives values of the cumulative probability function.

Event Description	Calculation of Event Probability
(a) 6 successes in 10 trials; $p = 0.4$. $b(6; 10, 0.4)$	$0.9452 = P(6 \text{ or fewer successes})$ $0.8337 = P(5 \text{ or fewer successes})$ $\overline{0.1115} = P(\text{exactly } 6 \text{ successes})$
(b) 6 or more successes in 10 trials; $p = 0.4$. $\sum_{x=6}^{10} b(x; 10, 0.4) = \sum_{x=0}^{4} b(x; 10, 0.4)$	$1.0000 = P(0 \text{ or more successes})$ $0.8337 = P(5 \text{ or fewer successes})$ $\overline{0.1663} = P(6 \text{ or more successes})$
(c) 6 or more successes in 10 trials; $p = 0.6$. $\sum_{x=6}^{10} b(x; 10, 0.6)$	$0.6331 = P(4 \text{ or fewer failures in } 10 \text{ trials; probability of a } failure = 0.4)$
(d) 6 successes in 10 trials; $p = 0.6$. $b(6; 10, 0.6) = b(4; 10, 0.4)$	$0.6331 = P(4 \text{ or fewer failures})$ $0.3822 = P(3 \text{ or fewer failures})$ $\overline{0.2509} = P(\text{exactly } 4 \text{ failures})$

In real problems it is not unusual for p to be unknown. When this is true, we may use a sample to estimate it. Thus, if we had no idea of the probability of a defective part being produced by a machine, we might take a sample of, say, 1,000 items, and use the ratio of the number of defectives observed, say 10, to the total of 1,000 as an estimate of p. Hence $p' = 10/1,000 = 0.01$ is a sample estimate of p.

New Symbol	Meaning
p'	An estimate of the binomial parameter based on sample data.

Of course, all situations do not result in a two-way classification. What do we do with replies to a questionnaire which permits the respondent to select in favor, against, or no opinion? How do we deal with forecasts of increase, decrease, or no change? These two examples involve three categories, but we could have used examples involving even more categories. If questions 1 and 3 given at the start of this subsection can be answered yes, then we may use the multinominal probability mass function to deal with problems involving more than two categories. We will not discuss this probability function, but Problem 8-11 deals with an example.

S.S. A mail-order house tests the quality of its order-filling operations by selecting a sample of 20 completed orders each day and checking to see if the orders contain exactly what the customer requested. The quality control manager sends a memorandum to the vice-president for operations any time that the sample contains one or more incorrectly filled orders. Suppose that 99 percent of all orders are being correctly filled; what is the probability a memorandum will be sent on a given day? How many memorandums would be expected over a year of 260 working days? What do you conclude based on this number, if anything? [In part (a) first write the probability, using formula (8-1).]

Solution: A memo will be sent if any order is incorrect. A memo will not be sent only if all orders are correct. The probability of this second event is $b(20; 20, 0.99)$, and the probability we seek is $1 - b(20; 20, 0.99)$. Using the binomial table at the back of the book the calculations are found by noting that 20 successes in 20 trials with an 0.99 probability of success is identical to 0 failures in 20 trials with an 0.01 probability of failure. Thus, we want to find $b(0; 20, 0.01) = 0.8179$ from the table.

1.0000	$P(20$ or fewer successes in 20 trials$)$
-0.8179	$P(20$ successes$) = P(0$ failures$) = b(0; 20, 0.01)$
0.1821	$P($less than 20 successes$)$.

The probability of a memo on a given day is 0.1821, and hence the vice-president should expect 260(0.1821) or about 48 such memos per year, nearly one per week, even when the process is working correctly 99 times in every 100. Unless incorrectly filled orders are very costly, a different quality control program may be in order. For example, if checking orders is not too expensive, a larger sample size could be used. If checking is costly, the memorandum might be sent only if, say, more than 1 order in 20 was incorrectly filled. This probability is equal to 1 minus the probability that either zero or one order is filled incorrectly:

$$1 - \sum_{x=0}^{1} b(x; 20, 0.01) = 1 - 0.9831 = 0.0169$$

PROBABILITY DISTRIBUTIONS

if 99 percent of the orders are being filled correctly. The vice-president would now expect only 260(0.0169), or about four memos per year when the order-filling process is working this well. This should give him more time for meetings and other important duties.

HYPERGEOMETRIC PROBABILITY MASS FUNCTION

As we have just seen in the example at the end of the last section, the binomial is useful in finding the probabilities of samples. But the binomial does not work well in all cases. If the population is finite and if an item once selected for the sample is not replaced prior to selecting the next item, the trials are no longer independent. For example, suppose a shipment of 10 parts contains 8 good parts and 2 defective parts. What is the probability that a sample of 2 will contain both bad parts? If we used the binomial, the answer would be $b(2; 2, 0.20) = 0.2(0.2) = 0.0400 = \frac{2}{10}(\frac{2}{10})$. But this answer assumes, incorrectly, that the probability of a bad part remains the same for each of the two sample draws.

The probability of a bad part on the first draw is $\frac{2}{10} = \frac{1}{5}$. But once a bad part is selected, the chance that the second part will also be bad is $\frac{1}{9}$. There are nine parts left, of which only one is bad. Hence if we do not replace the first part before selecting the second, the probability of our sample is given (using the multiplication law) by

$$P(2 \text{ bad in } 2) = \tfrac{2}{10}(\tfrac{1}{9}) = 0.0222.$$

This probability is smaller than suggested by the binomial since, once drawn, the first bad part is not available for the second draw.

There is another way of obtaining this probability and, although it is more difficult, it leads us to the general probability formula needed to solve this type of problem. This method uses the concepts of tasks and combinations. It works this way. There are 10 parts, from which we wish to select 2. This can be done in $_{10}C_2$ ways. So there are $_{10}C_2$ equally likely sample outcomes. How many of these lead to the result of 2 bad and no good parts? Well, from the 2 bad parts in the shipment we wish to select 2, and from the 8 good parts we wish to select zero. These two tasks can be accomplished in $_2C_2$ and $_8C_0$ ways, respectively. Using the fundamental principle of counting, a sample of 2 bad parts and no good ones can be obtained in $(_2C_2)(_8C_0)$ ways. Thus, the desired probability is found, using combination symbols, to be:

$$\frac{(_2C_2)(_8C_0)}{_{10}C_2} = \frac{\binom{2}{2}\binom{8}{0}}{\binom{10}{2}} = \frac{1}{45}.$$

Letting a population of N items have S successes (where a success is defined as a bad part to show our versatility), we can generalize what we have done. The appropriate pmf is called the *hypergeometric pmf*. The sampling process is described as being *hypergeometric* and the sample is called a *hypergeometric sample*.

Hypergeometric Probability Mass Function

Given n trials of a hypergeometric process, the probability of exactly x successes is given by

$$\frac{(_sC_x)(_{N-s}C_{n-x})}{_NC_n} = \frac{\binom{S}{x}\binom{N-S}{n-x}}{\binom{N}{n}}, \qquad (8\text{-}5)$$

where

$N \equiv$ size of the population
$n \equiv$ size of the sample
$S \equiv$ number of successes in the population
$x \equiv$ number of successes in the sample.

A *hypergeometric process* is a set of trials for which the following questions can be answered yes.

1. Is the number of outcomes in the sample space finite?
2. At each trial of the random process, are there 2 and only 2 possible outcomes?
3. Is it true that the item resulting from any one trial is not available on succeeding trials?

The symbolic representation of the hypergeometric is

$$h(x; n, N, S) \equiv \frac{\binom{S}{x}\binom{N-S}{n-x}}{\binom{N}{n}} \equiv \frac{\dfrac{S!}{x!(S-x)!} \dfrac{(N-S)!}{(n-x)!(N-S-n+x)!}}{\dfrac{N!}{n!(N-n)!}}. \qquad (8\text{-}6)$$

The mean and variance of the hypergeometric are given by equations (8-7) and (8-8).

The *mean of the hypergeometric pmf* $= n(S/N)$. $\qquad (8\text{-}7)$

The *variance of the hypergeometric pmf*
$\qquad\qquad = n(S)(N-S)(N-n)/N^2(N-1). \qquad (8\text{-}8)$

Probability Distributions

The most common hypergeometric process is sampling without replacement. (Sampling with replacement is a binomial process.) Items are typically not replaced because the one doing the sampling wishes each sampled item to constitute new information. Thus, in an opinion poll, he does not want the same individual's response twice.

For large samples, the computations required by the hypergeometric take a long time. Fortunately, if the sample is small relative to the population, the binomial pmf can yield an approximate probability that is often adequate for problem solving. For example, let $N = 100$, $S = 40$, $n = 3$, and $x = 2$. Then

$$b(2; 3, 0.4) = 0.2880 \quad \text{from the binomial table}$$

$$h(2; 3, 100, 40) = 0.2894 \quad \text{by hand calculation.}$$

These answers are close enough, for most practical purposes, to use interchangeably. Since binomial calculations are easier (nonbelievers are urged to try the last example), we will use the binomial probability if the sample size is less than 5 percent of the population, that is, if $n < 0.05N$. This is only a rule of thumb, however, and in an actual problem, such approximations may be inadequate if large losses can result from making a wrong decision.

S.S. In acceptance sampling of rocket components certain switches in each incoming batch are selected and tested using an excessively high voltage. The voltage destroys the switch, so only three are tested in each batch of 100. If all three switches tested are satisfactory, the incoming batch is accepted. Otherwise, the remaining items are returned to the supplier with full credit.

(a) Is testing here a hypergeometric process?

(b) Could the binomial be used? Why or why not?

(c) What is the probability of rejecting the batch if 1 of every 10 switches is defective?

Solution:

(a) Since an item once tested will not be tested again (it cannot be, since it is destroyed), the sampling process is hypergeometric.

(b) Since the sample is less than 5 percent of the population, we may use the binomial pmf.

(c) The probability is:

$$\text{exactly:} \quad 1 - h(3; 3, 100, 90) = 0.273$$

$$\text{approximately:} \quad 1 - b(3; 3, 0.9) = 0.271.$$

POISSON PROBABILITY MASS FUNCTION

A Poisson (pronounced Pwă · sōn, after the early-nineteenth-century French mathematician and physicist Siméon D. Poisson) process is similar to a binomial process. The binomial gives probabilities for the number of successes occurring in a fixed number of trials. The trials are discrete and therefore can be counted. In a Poisson process there is no counterpart to a fixed number of trials. Instead, the events occur randomly over time or space, both of which are continuous rather than finite. Hence, there is no way to count the number of trials. The Poisson pmf is concerned with the number of times an outcome occurs over a given length of time or within a given unit of space.

An example of a Poisson process is supplied by considering the occurrence of fires within a large plant. If we collected data on the number of separate fires starting in a 1-hour period and arrayed these data by number of fires reported together with the number of occurrences, the frequency distribution would be closely approximated by a Poisson. Table 8-5 is an example where the Poisson probabilities have been computed by a formula we will learn shortly.

Table 8-5 Number of Fires Starting in a Large Plant, by Hour

Number of Fires Starting Within Same Hour x	Number of Occurrences (number of hours in which exactly x fires started)	Poisson Probability (based on an average of 1)
0	36	0.368
1	38	0.368
2	18	0.184
3	6	0.061
4	2	0.015
5	0	0.003
Total	100	1.000

The occurrence of a fire is similar to the occurrence of a success for the binomial except there is no unique way to count the number of trials or the number of successes. If we were to try to count the trials in a 1-hour period, we might consider using the number of independent structures in the plant. But every separate location within each structure could also be treated as the place where a fire could originate and hence a trial. The same arbitrariness is also present in the time frame used. Thus, we could have used the number of fires per minute or per day just as easily as the number per hour.

PROBABILITY DISTRIBUTIONS

Fortunately, when events may be viewed as occurring over a continuous time or space dimension and, hence, the Poisson pmf is relevant, it is not necessary to specify the number of trials. All we need to know is the mean number of occurrences during the (arbitrarily) selected unit of time or space. For our fire example, this value can be calculated from the first two columns of Table 8-5. We let this value be denoted by m, which is a sample estimate of the single Poisson parameter λ (lambda). Lambda is the mean number of occurrences (or rate of occurrence) per unit of time (or space). For our example:

$$m = \frac{0(36) + 1(38) + 2(18) + 3(6) + 4(2) + 5(0)}{100} = 1.0.$$

New Symbol	Meaning
m	Mean number of occurrences in a sample from a Poisson process. An estimate of λ, the process average per unit of time or space.
λ	The process average per unit of time or space.

Using the value of λ or its estimate m, we can calculate the probability of any number of occurrences using the Poisson probability mass function.

Poisson Probability Mass Function
The probability of exactly x occurrences of an outcome described by a Poisson process is given by

$$j(x; \lambda) \equiv \frac{e^{-\lambda}\lambda^x}{x!}. \tag{8-9}$$

The Poisson, like the binomial, is defined only for integer values of x ($x = 0, 1, 2, \ldots$).

Values of the Poisson probability mass function are given in Table IV at the end of this book. In our fires example, we set $\lambda = m = 1$, and the probabilities in the last column of Table 8-5 are obtained directly from the Poisson table for $\lambda = 1$.

The mean and variance of the Poisson are given by expression (8-10) and (8-11), respectively.

The *mean of any Poisson pmf* $= \lambda$. (8-10)

The *variance of any Poisson pmf* $= \lambda$. (8-11)

Our intuition will help with the mean (but again not with the variance). If the mean number of occurrences of some event described by a Poisson random process is, say, 6 per hour, λ, then we expect 6 occurrences in each hour. (An hour is the unit time period here. If we change this time period to half-hours, then $\lambda = 3$ and we expect 3 occurrences per half-hour.) We do not derive these formulas, since few insights for operating managers are gained thereby.

A Poisson process can be described by the following questions:

1. Does the process occur over space or time in such a way that the event of interest could occur at any point in time or space?

2. Is the average number of occurrences that occur in a given interval proportional to the length or size of the interval?

3. Are the occurrences independent of each other?

Several conditions are implied by these questions. First, the process has no memory. The number of occurrences in one period does not influence the number in some other period. Second, the rate of occurrence, λ, remains constant over the duration of analysis. And third, for sufficiently small intervals of time or space, more than one occurrence would be extremely rare.

The second condition, which derives from the second question, means that we can estimate λ using one length of time (or space) and then convert it to answer probabilistic questions about other units of time or space. In our fires example we calculated $m = 1$, which means that one fire occurs per hour on the average. If we now wished to ask questions concerning the number of fires per half-hour we simply convert m to the new value, $m/2 = 0.5$.

The Poisson probability function has a number of applications. E. Parzen in his book entitled *Modern Probability Theory and Its Applications* (New York: John Wiley & Sons, Inc., 1960) writes:

> *This law arises frequently in the fields of operations research and management science since demands for service, whether upon cashiers or salesmen of a department store, the stock clerk of a factory, the runways of an airport, the cargo-handling facilities of a port, the maintenance man of a machine shop, and the trunk lines of a telephone exchange, and also the rate at which the service is rendered, often lead to random phenomena either exactly or approximately obeying a Poisson probability law. Such random phenomena also arise in connection with the occurrence of accidents, errors, breakdowns, and other similar calamities.*

We must be equally careful not to apply this probability function where it is inappropriate. Although the number of fires per hour may follow the Poisson, the number of structures catching fire would not be described by

PROBABILITY DISTRIBUTIONS 255

the Poisson, due to the likelihood of adjacent structures catching fire from the first. The occurrences would not be independent. Similarly, the number of accidents per hour on an assembly line or the number of errors in the accounts receivable record may follow a Poisson, but the number of people involved in the accidents or responsible for the account errors would not.

Another common error in using the Poisson is to assume that λ, the mean number of occurrences, holds over too long a period of time. Thus, the mean rate of airplane arrivals changes over the course of a typical day, and it is different on Fridays and weekends from what it is on other days of the week. This rate has seasonal patterns as well. A manager must be careful that the time periods used to estimate λ are similar to the time periods for which answers are needed. (Another way to say this is that the manager should draw all the data from a single, appropriate population.)

Furthermore, the Poisson does not put an upper limit on the number of occurrences. Theoretically, it is possible that (the pmf allows) 10,000 depositors may crowd into the offices of the Stumptown National Bank in a 10-minute interval, all interested in making a deposit. (This is so, even though there are not that many people in the whole county.) Of course, this probability is exceedingly small. But what it implies is that while the Poisson is only an approximation, it may be sufficiently accurate to provide useful data to the bank manager concerning how many tellers should be on duty at any time.

The Poisson pmf has one additional important use. The Poisson may be used to obtain reasonably accurate approximations to binomial probabilities as long as the number of trials is sufficiently large and the probability of a success (or failure) is sufficiently small. Unfortunately, it is not easy to define the phrase "sufficiently small." The accuracy of the approximation depends in rather complex ways on the values of n, p, and x. For our purposes, we shall assume that the approximation is adequate if $n \geq 50$, $p \leq 0.1$, and $np < 5$. (Note that this also works if $p \geq 0.90$, since we may then redefine a success.) Another common rule of thumb is to require that $np(1 - p) < 3$. As an example, suppose that we request the binomial probability of three successes in 50 trials when $p = 0.05$. Using the binomial,

$$b(3; 50, 0.05) = 0.2199.$$

Since $n \geq 50$, $p \leq 0.1$, and $np < 5$, we can use the Poisson to approximate the binomial probability. Setting the average number of successes $\lambda = np = 2.5$, $j(3; 2.5) = 0.2138$, using the Poisson table. The binomial probability was given in Table V in the Appendix, at the end of the book. In this case, there is no reason to use the Poisson approximation. But when such probabilities are not tabulated, the Poisson can be very helpful if the conditions for its use are met.

S.S. National studies indicate that the fatal accident rate among drivers and passengers on motorcycles and motor scooters is 1 per year per 1,000 registered vehicles. Suppose that 1,500 such vehicles are registered in a given location.

(a) What is the probability of no fatalities among such drivers and passengers in a 1-year period?

(b) What is the probability of one or more deaths in a 6-month period?

Solution:

(a) The best estimate of λ is given by the average number of fatal accidents per year, 1. Hence, set $\lambda = 1$ for a population of 1,000 or set $\lambda = 1.5$ for a population of 1,500. Thus, the correct probability is given by $e^{-1.5}(1.5)^0/0!$. The probability is 0.2231.

(b) We need first to adjust our estimate of λ to reflect the fact that the period of interest is $\frac{1}{2}$ year. To do so, we set $\lambda = 0.75$. Second, we note that a complementary probability is required:

$$P(x \geq 1) = 1 - P(x = 0).$$

The correct probability is given by $1 - e^{-0.75}(0.75)^0/0!$. This probability is not given precisely in the tables, but it is between the tabulated values of 0.5034 and 0.5507. The actual value is $1 - e^{-0.75} = 0.5276$.

8-3 Some Probability Density Functions

When the number of outcomes in a sample space is uncountable, the appropriate probability function is called a probability density function. An alternative way of remembering when a pdf is required is to ask whether the random variable used to represent the outcomes is a continuous variable or is being treated as a continuous variable. If so, a pdf is proper.

We have already become acquainted with two probability density functions in Chapter 6. These two functions were the uniform pdf and the triangular pdf. At that time we examined specific examples of each. That is, we examined one of each kind where the parameter values were given. Although the uniform and triangular pdf's are occasionally found in managerial applications, we used them in Chapter 6 primarily for their simplicity. (Surprised?) In this section we will introduce two more pdf's, which are more common in managerial applications. They are the exponential and the normal probability density functions. In particular, the normal pdf finds widespread application in managerial (as well as many other) applications. We will have reason to make extensive use of the normal pdf in the remainder of this book.

EXPONENTIAL PROBABILITY DENSITY FUNCTION*

In Section 8-2 we studied the Poisson pmf which was applied to counting the number of occurrences of an event that occurs over a continuum of time or space. In such problems the decision maker may also be interested in the time until the first occurrence or the time between occurrences. Time is a continuous variable, and the exponential pdf is frequently the appropriate probability function for studying such phenomena.

To a customer in line at the check-out counter, the important fact is how long he must wait for service, not how many precede him in line. The captain of a fishing vessel capable of handling a single whale is concerned with how long it takes to locate the first one rather than how many may be in the ship's cruising range.

The exponential distribution is very similar to the Poisson. The parameter λ is again defined as the number of occurrences per unit of time (or space): equivalently, the mean rate of occurrence over time (or space). For example, if, on the average, 60 cars arrive at a freeway exit per hour, then $\lambda = 60$ arrivals per hour or 30 per half-hour or 1 per minute. The exponential distribution differs from the Poisson in that the questions of interest concern the time or distance between occurrences rather than the number of occurrences. Let t be time until the next occurrence.

Cumulative Exponential Probability Function
   The probability that the time until the next occurrence in a Poisson process will be less than or equal to t is given by

$$F(t) = 1 - e^{-\lambda t} \quad \text{for } t \geq 0, \quad (8\text{-}12)$$

where

$\lambda \equiv$ mean rate of occurrence

$t \equiv$ time, measured in the same units as λ.

Formula (8-12) is the cumulative exponential probability function. It gives the probability that the time until the next occurrence is less than t. The cumulative exponential probability function is based on the exponential probability density function, which has the formula

$$f(t) = \lambda e^{-\lambda t} \quad \text{for } t \geq 0, \quad (8\text{-}13)$$

where λ and t are defined as before. We emphasize that formula (8-12) and not (8-13) provides the probabilities in which we will normally be interested. These two functions are graphed in Figure 8-6.

---

* This subsection can be omitted without loss of continuity.

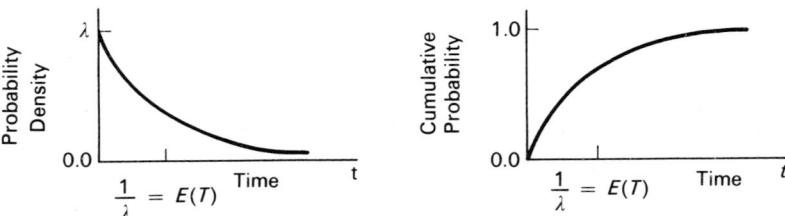

Figure 8-6 Exponential Probability Density Function (Left) and the Cumulative Exponential Probability Function (Right)

The exponential pdf has a mean and variance given by formulas (8-14) and (8-15), respectively.

$$\text{The {\it mean of any exponential pdf}} = \frac{1}{\lambda}. \tag{8-14}$$

$$\text{The {\it variance of any exponential pdf}} = \frac{1}{\lambda^2}. \tag{8-15}$$

Intuition is helpful in understanding the mean but not the variance. If the occurrences of some random process described by an exponential pdf occur at the rate of 2 per hour, λ, then we would expect to wait $\frac{1}{2}$ hour after an occurrence for the next one. The value $1/\lambda$ is the mean time between occurrences.

The exponential pdf is appropriate to our problem if

1. The process is Poisson, but

2. We are interested in the time between occurrences rather than in their number.

The nature of the process to which the exponential applies is such that small probabilities are assigned to long interevent times and larger probabilities to short interevent times. If a histogram of the data does not reflect this characteristic, then an exponential distribution is not useful for estimating interevent times.

One situation that commonly fits the exponential arises in establishing equipment reliability based on the mean time between failure. Suppose that a firm wishes to design its equipment so that no more than 10 percent of the machines produced will fail before 200 hours of operation. What mean failure rate, λ, measured in hours, must be achieved? Symbolically, we wish to solve

$$0.10 = 1 - e^{-\lambda(200)}$$

or

$$0.90 = e^{-\lambda(200)}.$$

PROBABILITY DISTRIBUTIONS 259

Since when $x = 0$, the Poisson probability $e^{-\lambda}(\lambda)^0/0! = e^{-\lambda}$, we can use the Poisson table and $x = 0$ to find the power of e needed to yield a probability of 0.90. The closest probability in the table is 0.9048 when $\lambda = 0.1$. Therefore, $e^{-200\lambda} = e^{-0.1}$ and $200\lambda = 0.1$. Thus, $\lambda = 0.0005$ failure per hour. This implies that 2,000 hours, 1/0.0005, is the mean time between failures.

S.S. A hospital is considering replacing its back-up power source used in emergency power failures. Suppose that the back-up generator has a mean time between failures of 100 hours and that the maximum time power is expected to be out is 10 hours before repair.

(a) What is the probability that the hospital's back-up generator will fail during the next power failure if the power failure were to last the full 10 hours?

(b) If three such blackouts are expected this year, give an expression for the chance that the back-up generator will not fail over the year.

Solution:

(a) This problem describes an exponential with a mean time between failure of $1/\lambda = 100$. This implies that $\lambda = 0.01$. Hence, the probability of no failures in 10 hours is given by $1 - e^{-0.01(10)}$. Using the Poisson table with $x = 0$, the probability is 0.0952.

(b) The probability of no failures in three tries can be approximated, using the binomial, to be

$$\binom{3}{0}(0.0952)^0(0.9048)^3 = (0.9048)^3 \approx 0.7407.$$

This is a reasonably high probability of need, considering the consequences. Perhaps additional back-up equipment should be obtained.

NORMAL PROBABILITY DENSITY FUNCTION

The normal probability density function is a symmetrical bell-shaped function. A graph of this function is given in Figure 8-7. The right and left tails continue without ever touching the x axis. This means that, conceptually speaking, an outcome from a normal random process can take on any value from positive to negative infinity. For this reason and because we must assume continuous measurement, the normal pdf is an approximation in any practical application. Very few random processes allow values from plus to minus infinity to occur. And no measuring system allows all the real numbers as potential outcomes. Further, very few processes in nature follow a normal pdf precisely.

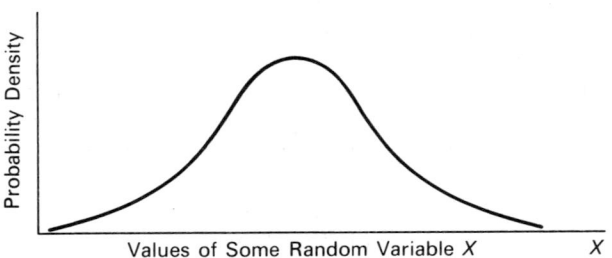

Figure 8-7 Normal Probability Density Function

If the normal pdf is only an approximation, what random processes does it approximate? Intelligence, weights, and certain other processes may be adequately approximated by a normal pdf. But the more important application for statistical purposes involves the distribution of averages. We shall learn in more detail in Chapter 10 that the averages of samples (of a given size), taken so that each sampled item is selected independently of the others, yield a distribution that is close to normal. Since we often wish to generalize about an actual random process from an average computed using sample data, this is an extremely useful fact.

If the following conditions hold, a normal random process is often relevant:

1. Are a "large" number of observations made on a continuous or near-continuous scale involved? (Continuity may be only approximate without requiring a "no" response to this question.)

2. Is each observation made independently of all the others?

3. Does each observation come from a single universe with a constant mean and variance?

Note that we did not require the underlying universe to be normal. Indeed, we placed few restrictions on the underlying universe from which the sample values come. The powerful conclusion is that the averages (or the totals) of the observations are approximately described by a normal pdf even if the underlying universe of values is not normal. We will return to the ideas in this paragraph again in Chapter 10 when we discuss the Central Limit Theorem.

In this subsection we will learn how to find probabilities when the normal pdf applies. But before doing so, we take much glee in giving the formula for the normal probability density function.

Normal Probability Density Function

The probability density for a random variable described by a normal process is given by

$$\frac{1}{\sigma\sqrt{2\pi}} \exp\left[-\frac{1}{2}\frac{(x-\mu)^2}{\sigma^2}\right], \qquad (8\text{-}16)$$

where

$\mu \equiv$ mean of the random process

$\sigma \equiv$ standard deviation.

Now that you have seen this rascal, you might believe it to be thoroughly unfriendly. But be of stout resolve. You will be overjoyed to learn that we will never need to use this messy-looking fellow. In fact, the reason for writing it at all was merely to introduce the two parameters of the normal pdf, μ and σ. The lowercase Greek letters μ and σ stand for the mean (or expected value) and standard deviation of the normal pdf, respectively.

The *mean of any normal pdf* $= \mu$. $\qquad (8\text{-}17)$

The *standard deviation of any normal pdf* $= \sigma$. $\qquad (8\text{-}18)$

As is indicated in Figure 8-8, the mean is located under the highest point of this density function. Because the normal pdf is symmetric, the mean, median, and mode are all identical. The standard deviation is a measure of the variability of the density function. The larger the standard deviation, the "fatter" the curve. This is also indicated in Figure 8-8.

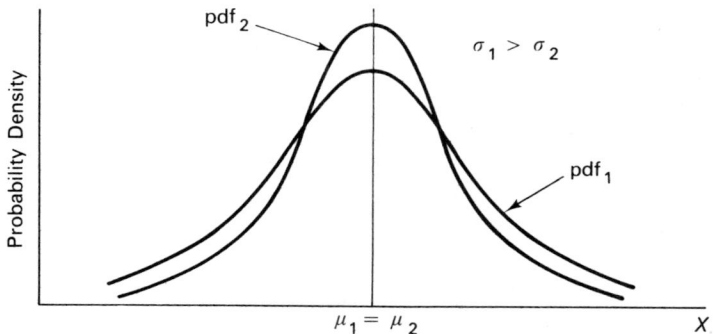

Figure 8-8 Two Normal Probability Density Functions with the Same Mean but Different Standard Deviations

The standard deviation of a normal pdf helps us determine probabilities of the events in which we are interested. Probabilities, we recall, are given by areas under the curve for continuous functions. If we include an area equal to ±1 standard deviation from the mean, approximately 68 percent of the total area and hence probability is included. This is illustrated in Figure 8-9. Figure 8-9 also shows that ±2 standard deviations includes about 95 percent of the area and ±3 standard deviations includes over 99 percent. Hence, the probability of being within one σ of the mean is 0.68; within two σ's the probability is 0.95, and the probability of being within three σ's of the mean is 0.99.

Using notation similar to that developed earlier, we will indicate that the

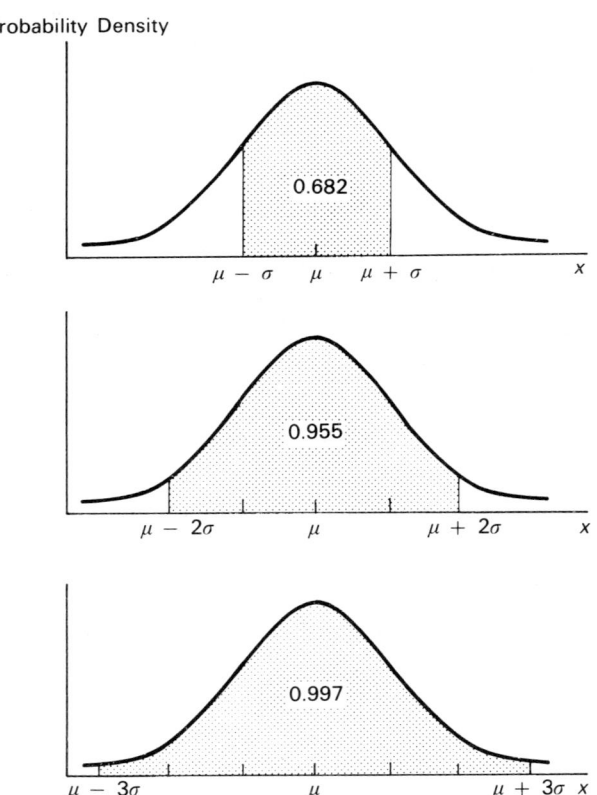

Figure 8-9 Probabilities Given by Intervals Under the Normal Curve and Based on Standard-Deviation Intervals

PROBABILITY DISTRIBUTIONS

values, x, of some random variable follow a normal pdf with a given mean and standard deviation by writing the normal probability density function as

$$n(x; \mu, \sigma). \qquad (8\text{-}19)$$

The cumulative normal probability function will be written

$$N(x; \mu, \sigma). \qquad (8\text{-}20)$$

New Symbol	Meaning
$n(x; \mu, \sigma)$	The probability density of the value x when the random variable follows a normal pdf with mean μ and standard deviation σ.
$N(x; \mu, \sigma)$	The probability of a value equal to or less than x when x follows a normal pdf with mean μ and standard deviation σ.

The notation $N(x; \mu, \sigma)$ can be related to the previous notation for cumulative probability functions, as is shown in the following examples.

$P(x \leq c)$
$\quad = P(x < c)$
$\quad = P(x < c \mid \mu, \sigma)$
$\quad = N(x = c; \mu, \sigma)$
$\quad = N(c; \mu, \sigma)$

The probability that x takes on a value less than or equal to c. Also, the probability that x is less than c when x follows a normal pdf with mean μ and standard deviation σ. These notations will be used interchangably.

The two probabilities, $P(x \leq c)$ and $P(x < c)$, are equal for a density function since there is no area and hence no probability associated with a point (such as c) on the x axis. This probability is illustrated for two cases by the shaded area in Figure 8-10.

$P(x \geq c) = P(x > c)$
$\quad = 1 - N(c; \mu, \sigma)$

The probability that x takes on a value equal to or greater than c. Also, the probability that x exceeds c.

$P(b \leq x \leq c)$
$\quad = N(c; \mu, \sigma) - N(b; \mu, \sigma)$

The probability that x takes on a value between b and c ($b \leq c$).

These last two probabilities are illustrated in Figure 8-11. The values of the constants b and c are completely general; these values could be anywhere

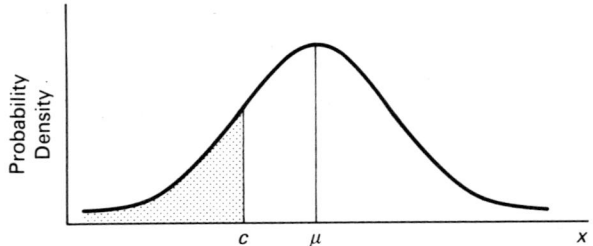

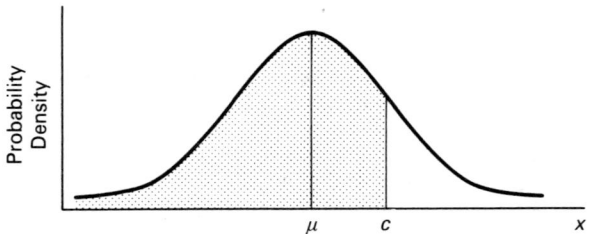

Figure 8-10 Normal Probability that $x \leq c$. $P(x \leq c) = P(x < c)$: Two Cases

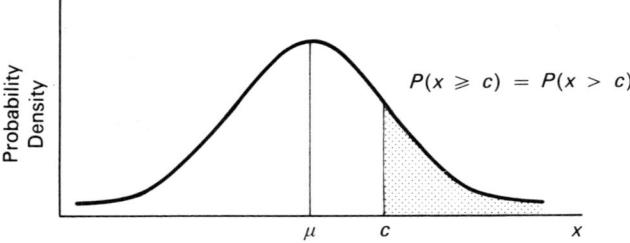

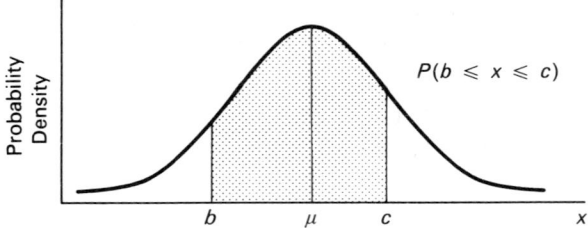

Figure 8-11 Normal Probablities of Two Events Illustrated

Probability Distributions

along the x axis. Also, the equal portion of the symbols $\leq$ and $\geq$ can be omitted without error since no area (probability) attaches to a point.

THE STANDARDIZED NORMAL

Finding probabilities (that is, areas under a curve) for a function like the normal requires numerical integration, a task usually done on a computer and not in one's head. This has been done for us for one normal pdf called the standardized normal. The results are given in Table I in the Appendix. Fortunately, it is a relatively simple task to convert any normal pdf into the standard normal pdf. First we will define the standard normal, and then we will give the steps necessary to convert any problem involving a normal probability to a problem of looking up an area given by the standard normal pdf.

The *standard normal pdf* has mean zero and variance 1. To convert any normal pdf with a mean other than zero to one with mean zero, it is sufficient to subtract the mean from all values. This has the effect of shifting the distribution so that it is centered over zero. Symbolically, we subtract the mean from each value, x, of the random variable, to obtain

$$x - \mu. \tag{8-21}$$

Thus, when the value of x is the mean value, we obtain

$$\mu - \mu = 0. \tag{8-22}$$

To reduce the variance (also the standard deviation) to 1 it is sufficient to divide expression (8-21) by the standard deviation, yielding a value denoted by z:

$$z \equiv \frac{x - \mu}{\sigma}. \tag{8-23}$$

(The logic of these steps is explained in Technical Note 8-1.) Thus, if $\mu = 17$, $x = 30$, and $\sigma = 13$, x is 1 standard deviation above the mean. Formula (8-23) reflects this fact by yielding $(30 - 17)/13 = 1$. The value of z is called the *standard normal deviate*.

New Symbol	Meaning
$z \equiv \dfrac{x - \mu}{\sigma}$	The number of standard deviations of x from the mean.

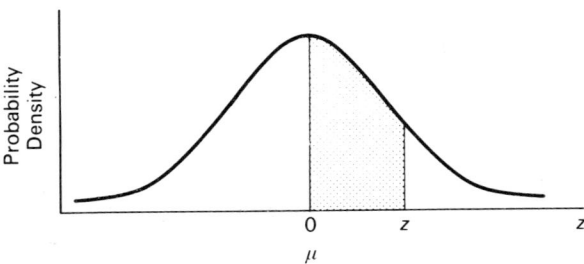

Figure 8-12 Tabulated Normal Probability

The value z is positive if x exceeds μ, and it is negative if x is less than μ. It is this value z that is tabulated in the normal curve given in Table I. In particular the shaded area between the mean and the positive value of z illustrated in Figure 8-12 is tabulated. This table is sufficient for any problem involving a normal pdf, because after converting to z the probabilities depend only on z and not on the μ and σ of the original distribution. (This simplification is not possible for most pdf's.)

Figure 8-13 illustrates the process of converting to the standardized normal.

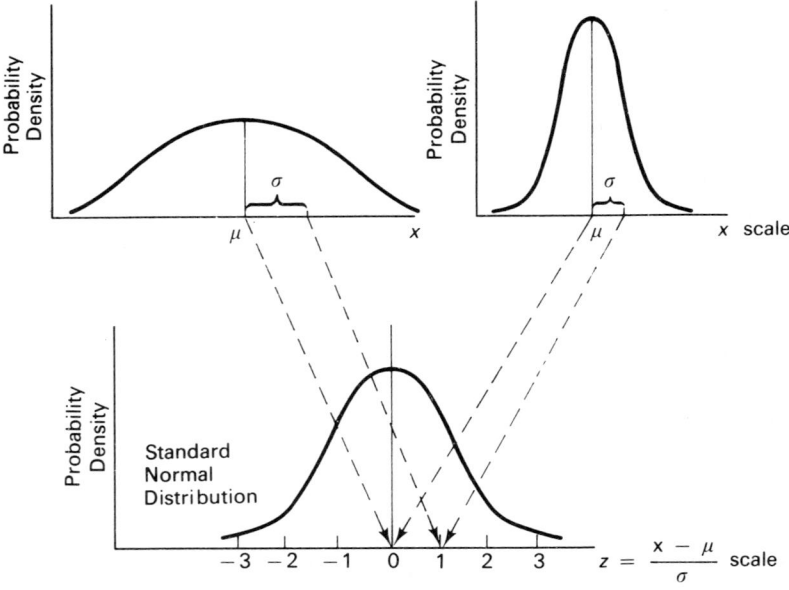

Figure 8-13 Converting to the Standard Normal

PROBABILITY DISTRIBUTIONS

CALCULATING PROBABILITIES

We are now in a position to try calculating the probabilities of several events using the standard normal. To do so we shall use an example involving a random variable whose values are described by a normal pdf with a mean of 80 and standard deviation of 20; $n(x; 80, 20)$. This function is illustrated in Figure 8-14. The lower (z) scale represents the conversion to the standardized normal. For illustrative purposes the probabilities requested in the left-hand column of the examples given in Table 8-6 are calculated by looking up the appropriate values of z in the middle column. The answers are given in the right-hand column and are illustrated in Figure 8-15.

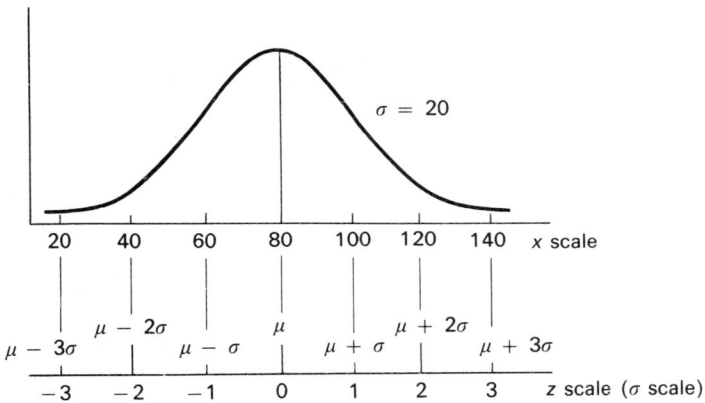

Figure 8-14 Normal Probability Function with Mean 80 and Standard Deviation 20

Table 8-6 Example of Some Normal Probability Calculations

Probability	Calculation of z	Numerical Answer
(a) $P(x \leq 60 \mid \mu = 80, \sigma = 20) =$	$P\left(z \leq \dfrac{60 - 80}{20} = -1\right) =$	$0.5 - 0.3413 = 0.1587$
(b) $P(x \leq 110 \mid \mu = 80, \sigma = 20) =$	$P\left(z \leq \dfrac{110 - 80}{20} = 1.5\right) =$	$0.5 + 0.4332 = 0.9332$
(c) $P(x > 130 \mid \mu = 80, \sigma = 20) =$	$P\left(z > \dfrac{130 - 80}{20} = 2.5\right) =$	$0.5 - 0.4938 = 0.0062$
(d) $P(110 < x \leq 130 \mid \mu = 80, \sigma = 20) =$	$P(1.5 < z \leq 2.5)$	$= 0.4938 - 0.4332 = 0.0606$

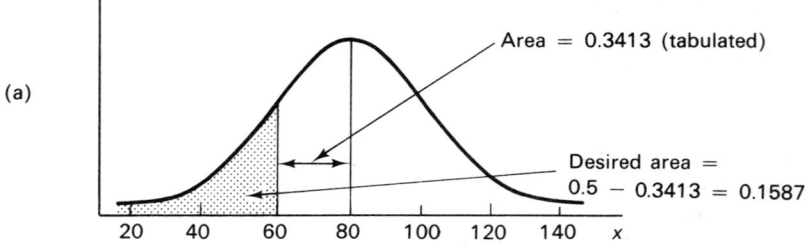
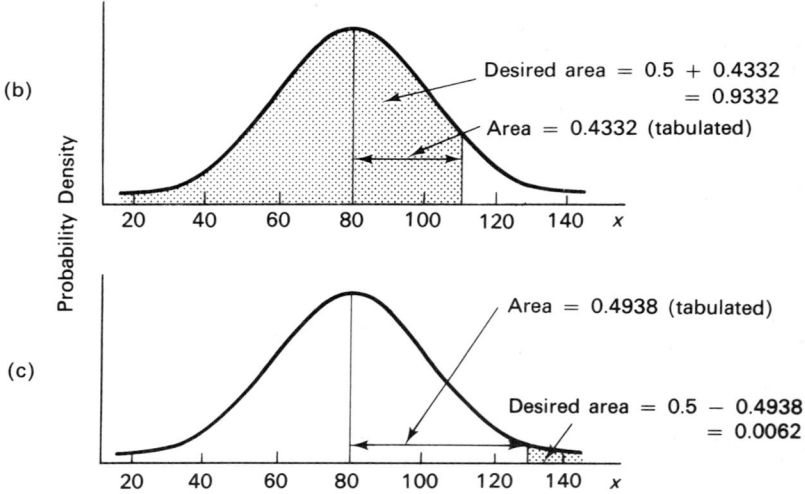
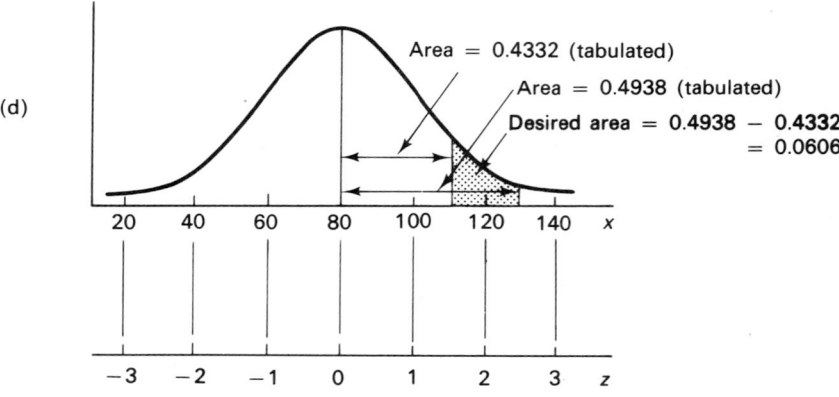

Figure 8-15 Probability Calculations Illustrated (Desired Area Shaded)

Probability Distributions

In computing these probabilities, we observe that the normal table only gives probabilities for values between the mean and $+z$. To obtain some of the probabilities in the example, we need to keep in mind that the normal pdf is symmetrical. Therefore, the following relationships hold for the normal:

$$P(0 \le x \le z) = P(-z \le x \le 0) \tag{8-24}$$

and

$$P(x \ge \mu) = P(z \ge 0) = 0.5$$
$$P(x \le \mu) = P(z \le 0) = 0.5. \tag{8-25}$$

In words, the probability that x is between the mean and plus some number of standard deviation units equals the probability that x is between the mean and minus the same number of standard deviation units [equation (8-24)]. Equation (8-25) says that 50 percent of the area is on each side of the mean.

In solving probability problems such as these, it is very useful to first draw a picture and locate the area required. This should help us judge the correctness of the answer. Remember that since no area is associated with a point, the equal portions of the $\le$ and $\ge$ symbols can be ignored.

S.S. For the normal pdf illustrated in Figure 8-14, find the probability that x is between 60 and 110.

Solution: First we note that omitting (or including) the word "inclusive" after the figure 110 in the statement of the problem is unnecessary. The answer is unchanged, since only the end points of the interval are involved and no area (hence probability) attaches to a point.

$$P(60 \le x \le 110) = P(60 \le x \le 80) + P(80 \le x \le 110)$$
$$= 0.3413 + 0.4332 = 0.7745.$$

THE NORMAL AS AN APPROXIMATION TO THE BINOMIAL

In discussing Figures 8-4 and 8-5, we observed that the binomial tends to approximate a normal probability function as p approaches 0.5 and/or as n gets large. See the bottom graph in Figure 8-5, in particular. As a rule of thumb, we will use the normal approximation to the binomial when

1. The specific binomial probabilities are not readily available.
2. The value of np and $n(1-p)$ are both at least 30. (An alternate, less stringent, rule of thumb is $np(1-p) \ge 3$.)

We must always remember that this is just an approximation. Let us try an example. Suppose that in sampling to obtain worker's views concerning a closed shop in a particular plant, an independent survey organization hired by both the union and management randomly selects a sample of 100 from 10,000 workers. If 60 percent of the workers are union members, what is the probability that the majority of those sampled will not belong to the union? First, we note that the sample is taken without replacement from a finite population. Hence, the hypergeometric applies. However, because the sample is less than 5 percent of the population, we may exercise the option of using the binomial. Second, we note that the binomial probability for a sample size of 100 is in our tables. Hence, we can and should solve our problem by looking up the answer in the binomial table. We obtain, letting $X \equiv$ the number of nonunion members in the sample,

$$p(x > 50) = \sum_{x=51}^{100} b(x; 100, 0.4)$$

$$= 1 - \sum_{x=0}^{50} b(x; 100, 0.4)$$

$$= 0.0168.$$

Let us use the normal pdf as an approximation, even though we already have an answer. Since $np = 100(0.4) = 40$ and $n(1 - p) = 100(0.6) = 60$, condition 2 for using the normal is satisfied.

The first step in the solution is to compute the mean and standard deviation using formulas (8-3) and (8-4), respectively.

$$\text{mean} = \mu = np = 100(0.4) = 40 \quad (8\text{-}26)$$

$$\text{standard deviation} = \sigma = \sqrt{np(1 - p)} = \sqrt{100(0.4)(0.6)} \approx 4.9. \quad (8\text{-}27)$$

The values for μ and σ are calculated using the formulas for the binomial mean and variance, but they are used as the normal parameters to approximate the required probabilities.

The second step is to solve for z. We find, using formula (8-23) and using $x = 51$,

$$z = \frac{51 - 40}{4.9} \approx 2.24.$$

The probability we seek is

$$P(z > 2.24) = 0.50 - 0.4875 = 0.0125.$$

Probability Distributions

This is a reasonably close approximation to a very small probability. And it is in the tail areas of the normal pdf where the probability estimation will always be the poorest. [Because the binomial is discrete, a more accurate answer would be given by

$$P\left(z > \frac{50.5 - 40}{4.9}\right) = P(z > 2.14) = 0.0162,$$

which is very close to the answer given by the binomial pmf. See Technical Note 8-2.]

S.S. Shipments of a small part are received in batches of 10,000. These parts are tested when they arrive using a randomly selected sample of 400 items. Suppose that the shipment is rejected if more than 50 defective parts are found. Suppose that the quality level of incoming shipments is guaranteed to be at least 90 percent good. If the fraction of good parts equals the guaranteed level, what is the probability that the shipment will be rejected?

Solution: Since the sample is of size 400, the binomial probability requested cannot be found in the binomial tables. Checking, we find that $np = 40$ and $n(1 - p) = 360$, where p is the probability of a defective. The value of $\sigma = \sqrt{400(0.1)(0.9)} = 6$. The probability we seek is

$$P\left(z > \frac{50 - 40}{6}\right) = P(z > 1.67) = 0.5 - 0.4525 = 0.0475.$$

More accurately,

$$P\left(z > \frac{50.5 - 40}{6}\right) = P(z > 1.75) = 0.5 - 0.4599 = 0.0401.$$

8-4 A Warning

When it comes time in the problem-solving process to select a probability function, the proper choice is critical. If a Poisson pmf is appropriate, using the normal pdf will give inaccurate and, hence, useless results. In the selection process it is necessary to assure ourselves that the assumptions underlying the method are satisfied. The assumptions are implicit in the questions we used to determine the nature of the random process. If we are uncertain, the help of a qualified statistician will prove essential.

Further, although we have discussed five probability functions in this chapter and several others in earlier chapters, we have far from exhausted the set of possibilities. For example, several other functions will be introduced in succeeding chapters. These include the t probability density function, the chi-square probability density function, and the F probability density function. Moreover, numerous other probability functions are not considered in this book.

TECHNICAL NOTE 8-1: DERIVATION OF THE MEAN AND VARIANCE OF THE STANDARDIZED NORMAL

Using the following facts involving expectations, we can show mathematically that the mean and standard deviation of the standardized normal pdf are zero and 1, respectively. Expectation formulas tell us that

1. $E(aX + b) = E(aX) + E(b)$.
2. $\text{Var}(aX + b) = a^2 \text{Var}(X)$, where a and b are constants, $E(X) = \mu$, and $\text{Var}(X) = \sigma^2$.

Now since $z = (X - \mu)/\sigma$, then

$$E\left(\frac{X-\mu}{\sigma}\right) = E\left(\frac{1}{\sigma}X\right) - E\left(\frac{\mu}{\sigma}\right)$$

$$= \frac{1}{\sigma}E(X) - \frac{1}{\sigma}E(\mu)$$

$$= \frac{1}{\sigma}\mu - \frac{1}{\sigma}\mu$$

$$= 0.$$

Further,

$$\text{Var}\left(\frac{X-\mu}{\sigma}\right) = \text{Var}\left(\frac{1}{\sigma}X - \frac{\mu}{\sigma}\right)$$

$$= \frac{1}{\sigma^2}\text{Var}(X) - \text{Var}\left(\frac{\mu}{\sigma}\right)$$

$$= \frac{1}{\sigma^2}\sigma^2 - 0$$

$$= 1,$$

and since $\sigma = \sqrt{\sigma^2}$, the standard deviation is also equal to 1. These results hold after standardizing any normal pdf.

PROBABILITY DISTRIBUTIONS

TECHNICAL NOTE 8-2: CONTINUITY CORRECTION

When a continuous probability function such as the normal is used to approximate a discrete probability function such as the binomial, a correction for continuity is appropriate. Thus, if we are counting units in integers, we should add or subtract, as appropriate, $\frac{1}{2}$ unit. The following examples are illustrative:

Discrete		Continuous Approximation
$P(x > 10)$	$=$	$P(x \geq 10.5)$
$P(x \geq 10)$	$=$	$P(x \geq 9.5)$
$P(5 \leq x < 10)$	$=$	$P(4.5 \leq x \leq 9.5)$
$P(x = 10)$	$=$	$P(9.5 \leq x \leq 10.5)$

Furthermore, since all actual measurement is discrete by the nature of the process, a correction for continuity is always proper even when a continuous probability function is under consideration. For example, suppose we can measure to the nearest hundredth of an inch, and we desire the probability of a length greater than 10.36 inches. Using a correction for continuity, we would write this probability as $P(x \geq 10.365)$. (Again, the equal sign has no effect on the calculation.) Typically this refinement is unnecessary and can be ignored, but it is never incorrect to consider it.

TECHNICAL NOTE 8-3: FINDING THE PROBABILITY LAW OF A RANDOM VARIABLE DEFINED OVER A CONTINUOUS SAMPLE SPACE

OBTAINING A PROBABILITY MASS FUNCTION OF A RANDOM VARIABLE FROM THE PROBABILITY DENSITY FUNCTION OVER THE BASIC SAMPLE SPACE

The procedure here is similar to the summing procedure described in Chapter 6, where a discrete sample space and a discrete random variable are involved. Instead of summing, we integrate over the range of the basic sample space, which gives the value of the random variables for which the probability is required. Note that continuous sample spaces are already numerical, so we are concerned here only with the values—and their probabilities—for a new random variable defined over the same sample space.

Consider an example. Let the random variable W consist of the weights in pounds of a package after it has been filled with soap by an automatic machine. Suppose the pdf for W is

$$f(w) = \begin{cases} 1 & 1 \leq w \leq 2 \\ 0 & \text{otherwise.} \end{cases}$$

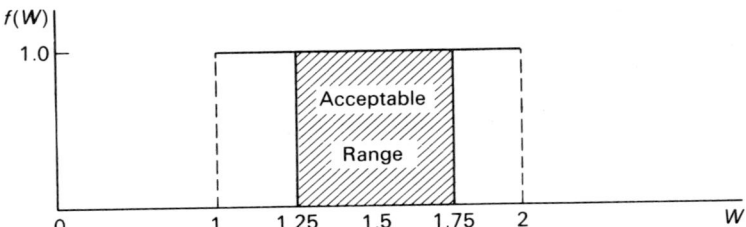

Figure 8-3A Soap Problem

Suppose further that the package is acceptable for shipment only if the weight is between 1.25 and 1.75 pounds. This is illustrated in Figure 8-3A.

Using integration techniques we can write

$$f(\text{acceptable}) = \int_{1.25}^{1.75} 1\ dx = x \Big|_{1.25}^{1.75} = 1.75 - 1.25 = 0.5.$$

Hence, $f(\text{unacceptable}) = 1 - 0.5 = 0.5$. Or using the random variable X with values

$$x = \begin{cases} 1 & 1.25 \leq w \leq 1.75 \\ 0 & \text{otherwise,} \end{cases}$$

then

$$f(x) = \begin{cases} 0.5 & x = 0 \text{ or } 1 \\ 0 & \text{otherwise.} \end{cases}$$

This case is relatively simple. We find the required probabilities by integrating over the appropriate range of the sample space. The next case is harder.

OBTAINING A PROBABILITY DENSITY FUNCTION OF A RANDOM VARIABLE FROM THE PROBABILITY DENSITY FUNCTION OVER THE BASIC SAMPLE SPACE

The procedure in this case is not difficult to state, but finding the proper ranges for the integration can become difficult. The steps are:

1. Compute the cumulative probability function for the density function defined for the new random variable.

2. Differentiate the cumulative probability function found in step 1.

For example, suppose that in our soap example the cost in dollars of a package of soap is given by twice its weight. That is,

$$c = C(w) = 2w.$$

Let us find $f(c)$ based on $f(w)$. Here we go. Using $f(w)$, we can find $F(c)$ by graphical analysis. For our problem, if the cost is to be less than some value, c, then the weight must be less than $w/2$, since cost is twice weight. Thus, if cost is to be less than 2 dollars then weight must be less than 1 pound, $c = 2w$ and $2 = 2(1)$. Therefore, we reason, the probability c is at most some value equals the probability w is at most $c/2$. Further, the smallest cost is $c = 2.00$, since $w \geq 1$ and the largest cost is $c = 4.00$, since $w \leq 2$. Therefore, we can write

$$F(c) = \begin{cases} 0 & c < 2 \\ \int_1^{c/2} 1 \, dw & 2 \leq c \leq 4 \\ 1 & c > 4, \end{cases}$$

or in simpler form since

$$\int_1^{c/2} 1 \, dw = w \Big|_1^{c/2} = \frac{c}{2} - 1.$$

This expression states that the probability that the cost, c, is at most some value (say 3) is equal to the probability that w is at most $c/2$ (which is $\frac{3}{2} = 1.50$).

Hence,

$$F(c) = \begin{cases} 0 & c < 2 \\ \dfrac{c}{2} - 1 & 2 \leq c \leq 4 \\ 1 & c \geq 4. \end{cases}$$

This completes step 1. We can use $F(c)$ to find any probability we desire. If we want, the density function for cost is found by differentiating $F(c)$.

$$f(c) = \frac{d}{dc}[F(c)] = \begin{cases} \dfrac{d}{dc}(0) = 0 & c < 2 \\ \dfrac{d}{dc}\left(\dfrac{c}{2} - 1\right) = \dfrac{1}{2} & 2 \leq c \leq 4 \\ \dfrac{d}{dc}(1) = 0 & c > 4. \end{cases}$$

Hence, $f(c)$ is a uniform pdf between 2 and 4 with height $\frac{1}{2}$ in this range.

Key Formulas

Formula	Used to Compute:
$b(x; n, p) = \dfrac{n!}{x!(n-x)!} p^x (1-p)^{n-x}$	The binomial probability of x successes in n trials where p is the probability of success on 1 trial.
np	Mean of the binomial.
$np(1-p)$	Variance of the binomial.
$h(x; n, N, S) = \dfrac{\dfrac{S!}{x!(S-x)!} \cdot \dfrac{(N-S)!}{(n-x)!(N-S-n+x)!}}{\dfrac{N!}{n!(N-n)!}}$	The hypergeometric probability of x successes in n trials where there are S successes in a population of size N.
$\dfrac{nS}{N}$	Mean of the hypergeometric.
$\dfrac{nS(N-S)(N-n)}{N^2(N-1)}$	Variance of the hypergeometric.
$j(x; \lambda) = \dfrac{e^{-\lambda} \lambda^x}{x!}$	The Poisson probability of x occurrences with an average rate of occurrence of λ. (The sample estimate of λ is represented by m.)
$z = \dfrac{\text{value} - \text{mean}}{\text{standard deviation}} = \dfrac{x - \mu}{\sigma}$	The standardized normal deviate.

PROBLEMS

8-1.* A number of financial analysts working in a brokerage firm are interested in their ability to predict each morning before the stock market opens whether the closing level of the Dow Jones average for that day will be above or below the previous closing average. One of the group claims that accurate short-run predictions of this type are possible. The others disagree. To settle the argument they let the claimant make predictions each day and record the results. After 2 weeks (10 days on which the market is open), the analyst who made the claim has predicted the direction of change correctly 8 of 10 times. What is the probability of being right *at least* 8 of 10 times if the true probability of making a correct prediction is 0.5 and if the results of each prediction are statistically independent?

PROBABILITY DISTRIBUTIONS 277

8-2. A production process produces parts which show a probability of 0.3 that a part will be defective. Find the probability that a lot of 100 parts will contain fewer than 20 defective parts (solve two ways). How many defectives should be expected in every 100 parts? What is the variance of the number of defective parts?

8-3.* Suppose that the number of significant new innovations in the Post Office follows the Poisson distribution. Past experience suggests that the average rate of occurrence is four per year. What is the probability no new innovations will occur in the next 6 months?

8-4. Errors in the accounts receivable of a large hospital occur at the rate of 20 per 1,000 accounts. If errors are made independently of one another, what is the probability that a drawer of 100 accounts selected for audit will have at most one error? Work this problem two ways and you may assume that the hospital has over 10,000 accounts.

8-5.* Suppose a random variable whose values, denoted by x, is normally distributed with a mean of 6 and a standard deviation of 3. Find the probabilities of the following events.

(a) $P(x \geq 6)$ (b) $P(x < 0)$ (c) $P(0 \leq x \leq 12)$

8-6. Suppose that an automatic soap-filling machine used to fill soap cartons operates so that the amount of soap in a given carton follows a normal probability law, with a mean of 10.0 pounds and a variance of 0.04 pound. Each trial (filling a carton) is a statistically independent event.
(a) What is the probability a single carton will contain less than 9.5 pounds?
(b) What is the probability that 4 cartons will contain less than 39 pounds?
(c) What is the probability that 100 cartons will *average* less than 9.95 pounds?

8-7. A production process yields metal rods whose length is normally distributed with a mean of 5.10 inches and a variance of 0.04 inch. The length of each rod is statistically independent of the lengths of preceding rods. A rod is acceptable if its length is between 4.84 and 5.20 inches.
(a) Find the proportion of rods that will be acceptable; too short; too long.
(b) What is the probability that a lot of 10 rods will contain less than half acceptable? Give a formulation that, if solved, would give the exact answer and also approximate the answer.

8-8. On Mondays, customers arrive at the Stumptown Farmer's Bank at the rate of 2 per minute and according to a Poisson distribution.
  (a) What is the probability that no customers arrive at the bank in the first 5 minutes that the bank is open?
  (b) What is the mean time between customer arrivals?
  (c) What is the probability that the time between successive arrivals will be shorter than 1 minute?

8-9.* A procurement analyst buys condensers in batches of 100. Before accepting delivery, 5 condensers are selected at random and tested. If none are found defective, the batch is accepted; otherwise, it is returned to the manufacturer for full credit. In the testing process the condenser is ruined. If 20 percent of the condensers are defective, what is the likelihood that a batch will be accepted? Compute the exact answer and an approximation using the binomial. Why is the binomial answer an approximation?

8-10. A company is considering two possible means of supplying specialized parts to machinists in its plant.
  Alternative 1: Use an experienced machinist at an hourly wage of $6 per hour to staff the parts cage.
  Alternative 2: Use a clerk to staff the parts cage at $3 per hour.
The experienced machinist services customers at twice the rate of a clerk. The expected service rate at the cage, λ_s, is 10 per hour for a clerk and the expected arrival rate is Poisson with mean rate $\lambda_A = 5$ machinists per hour. Mathematically, it is known that the expected waiting time, W, for a machinist is given by $E(W) = 1 \div (\lambda_s - \lambda_A)$. If lost time of machinists is valued at $6 per hour, which alternative is least expensive? Show your work. [*Hint:* A machinist is also unproductive during $1/\lambda_s$ hours, and λ_A machinists arrive per hour.] Why is the factor $1/\lambda_s$ added to the unproductive time?

8-11. A multinomial process is like a binomial process except there are more than two possible outcomes on each trial. Its formula is given by

$$m(x_1, x_2, \ldots, x_k; N, p_1, p_2, \ldots, p_k) = \frac{N!}{x_1! \, x_2! \cdots x_k!} p_1^{x_1} p_2^{x_2} \cdots p_k^{x_k},$$

where

$$\sum_{i=1}^{k} x_i = N \quad \text{and} \quad \sum_{i=1}^{k} p_i = 1.$$

Suppose that items produced by a process are classified as overweight, acceptable, or underweight, with the relative frequencies of occurrence of 0.2, 0.7, and 0.1, respectively, based on historical experience.

(a) In 10 trials, what is the probability of obtaining exactly 1 overweight, 6 acceptable, and 3 underweight items?
(b) Find this probability using the binomial pmf. (*Hint:* First find the probability of 6 acceptable items out of 10.)

8-12. The county safety director is not certain that the national average of 1 fatal accident per 1,000 vehicles registered is applicable in his county. Before examining the statistics for his county, he assigns a probability of 0.4 to the event that the national average given is applicable to his county, and a probability of 0.6 to the event that the correct fatal accident rate for his county is 0.5 fatal accident per year per 1,000 registered vehicles. After consulting his records for the most recent year, he discovers that 1,500 vehicles were registered in his county and that there were no fatal accidents involving such vehicles during the preceding year. Considering this evidence, and his prior judgments, what (revised) probabilities should he assign to the two accident rates that he considers possible?

8-13. A binomial process involving a single trial is known as a Bernoulli process. An example is a single toss of a fair coin.
(a) Is a pmf or pdf involved?
(b) Write the formula for this probability law in general form, where one denotes success and zero denotes failure.
(c) See if you can find its mean and variance.

8-14. Suppose that two single trial random processes follow one another with the probabilities shown.

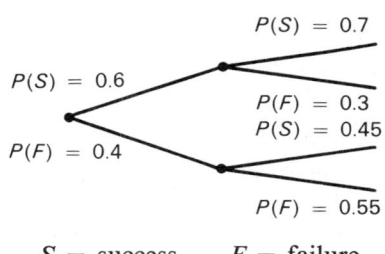

S = success F = failure

(a) Is the first trial a binomial process with $n = 1$?
(b) Does the joint process constitute a binomial process with $n = 2$?
(c) What probability would you assign to the event "success on the second trial"
　　(1) If the outcome from the first trial is a failure?
　　(2) If the outcome from the first trial is a success?
　　(3) If the outcome from the first trial is not known?

8-15.* In the manufacture of processed cheese, the measurements for the net weight of the cheese are not exact. If a manufacturer produces 12-ounce packages of sliced American cheese but the packages are actually normally distributed with $\mu = 12.2$ ounces and $\sigma = 0.50$ ounce, what proportion of the packages are underweight?

8-16. In a certain manufacturing plant, the machines break down for a number of reasons. If, on the average, 5 machines break down per day and the number of breakdowns per day are Poisson distributed, what is the probability that on any given day there will be
(a) No breakdowns?
(b) Exactly 2 breakdowns?
(c) At least 6 breakdowns?
(d) At most 4 breakdowns?
(e) Between 4 and 6 breakdowns inclusive?

8-17.* Before deciding whether to accept or reject a shipment of glass bottles, the production manager of Bubble-Up Soda Company selects a sample of 20 bottles randomly from each shipment. Each shipment contains a very large number of bottles. If she finds no more than one defective bottle in the sample, she accepts the entire shipment; otherwise, she sends the entire shipment back to the supplier. If a given shipment actually contains 20 percent defectives, what is the probability that the manager accepts the shipment?

8-18.* A quality control inspector for Manufacturers, Inc., wants to select a sample of items from among those produced by a machine that is known to yield 20 percent defectives. The inspector draws two samples, one of size 10 and one of size 30. Which of the following is the more likely result of the inspection?
(a) No defectives among the sample of size 10.
(b) At most one defective from the sample of size 30.

8-19. Owing to the failure of customers to pay their account balances, the credit manager of Nationwide Food Distributors has decided to be more selective in granting credit. He decided to ask the company's six salesmen their opinions on his proposal. Suppose that two of the salesmen favor the proposal and four oppose the change. If he selects three salesmen at random:
(a) What is the probability that exactly two of them are in favor of the policy change?
(b) What is the probability that one or more of them are in favor of the policy change?

8-20.* The owner of a small grocery store employs one cashier, and is trying to decide whether to add another checkout station. If customers

PROBABILITY DISTRIBUTIONS 281

arrive at the cashier according to a Poisson distribution with a mean arrival rate of one customer every 2 minutes:
(a) What is the probability that no customer will arrive at the cashier in a 10-minute period?
(b) What is the probability that no more than one customer arrives at the cashier in a 10-minute period?

8-21.* Widget Company manufactures a specialized electronic circuit that has a mean operating life of 5,000 hours. (Time between failures is distributed exponentially.) What is the probability that a circuit selected at random will operate effectively for at least 6,000 hours?

8-22.* A recording company uses mail solicitation as one means of obtaining customers. From past experience, the company has found that 1 percent of all mail offerings result in an order of records. If the company sends out 10,000 solicitations, what is the probability that:
(a) Fewer than 80 orders will be received?
(b) Between 80 and 100 orders inclusive will be received?

8-23. The diameter of a machine part produced by Hammer Hard Tool Company is normally distributed with a mean of 2 inches and a standard deviation of 0.002 inch.
(a) If the diameter of a part is less than 1.9944, the quality control inspectors will reject it. What percentage of the units produced would be expected to be less than 1.9944?
(b) The quality control inspectors will also reject any parts with a diameter greater than 2.0056. What percentage of the production would be expected to be greater than 2.0056?
(c) In redesigning the production process for the machine part, the engineers want a mean diameter of 2 inches for the parts produced. Furthermore, they want 99.8 percent of the parts produced by the process to have diameters between 1.9944 inches and 2.0056 inches. What must be the standard deviation of the diameters?

8-24. In the manufacture of aluminum lawn furniture, Easy Rest Company uses a certain drill to bore holes in aluminum tubing. Although the bits for the drill break randomly, it is known that on the average a bit will break once every 2 days. How many bits must the operator have on hand so that there is less than a 5 percent chance of having that number or more bits break than are available on any given day?

8-25. In an auto repair shop which employs one mechanic, an average of two customers per hour enter the shop to request some sort of service. The mechanic can repair each car at an average rate of three per hour. Customer arrivals at the shop are assumed to be distributed as Poisson and the mechanic's service time is assumed to be exponentially distributed. The results of this process are:

$$\text{expected length of customer waiting line} = \frac{a^2}{s(s-a)}$$

$$\text{expected percentage of the time the mechanic will be idle} = 1 - \frac{a}{s}$$

$$\text{expected waiting time of a customer before the mechanic begins work on his car} = \frac{a}{s(s-a)},$$

where

$a \equiv$ mean arrival rate

$s \equiv$ mean service rate.

(a) Compute the value for each of the above formulas.
(b) If a more efficient mechanic were found who could service cars at a mean rate of four per hour, what would be the expected length of the waiting line?

Chapter 9

Sampling

The director of a small executive placement service, The Dike Manpower Company, is interested in determining how well the people they place in jobs are doing. The service they provide is basically that of an intermediary between companies needing executive and technical personnel and persons with these skills who are interested in a new job. The fees for service are all paid by the company, not the individual. Business has been booming recently, and the firm has not had enough job applicants to match the job openings. This is particularly true since they entered the area of computer systems 2 years ago and that field has grown from 0 to 30 percent of their business in those 2 years.

The director, B. T. Drum, wants the sample to demonstrate the value of his firm's services to potential customers, but he wants to do so fairly, using income figures for previous clients. The firm has records showing how many persons they have placed in each of their 12 full years of operation. Those data are as follows (year 12 is the most recent year):

Year Number	Number of Individuals Placed	Year Number	Number of Individuals Placed
1	40	7	96
2	38	8	101
3	66	9	112
4	72	10	105
5	70	11	134
6	94	12	142

Mr. Drum has heard about sampling, and he believes he can use a sample to make his point. He assumes that a good sample here should involve taking some individuals from, for example, each year's placement. The figure Mr. Drum wants from the individuals is their average income, so that he can advertise that amount in the firm's promotional materials. Mr. Drum knows that some of the addresses in his file are out of date, so he has devised the following scheme to obtain a random sample; he will randomly choose three names from each year from among those that have an address that has been updated within the last 2 years.

Using this rule, 36 names were chosen, and letters were sent asking for salary information. Only one letter came back as undeliverable, so Mr. Drum was pleased with his sampling rule. Of the 35 letters delivered, 7 persons did not respond. The responses of the remaining 28 individuals are shown, by year, in Table 9-1. The figures in the table have been rounded to the nearest thousand. Mr. Drum averaged the numbers in Table 9-1 and produced the following advertisement, based on $\bar{x} = 46.46$:

Table 9-1 Sample Salary Values (thousands of dollars) for Persons Placed by Dike Manpower Company

Year Number	Income	Year Number	Income
1	110, 90, 45	7	44, 38, 27
2	100, 72, 48	8	40
3	90, 51	9	41, 28
4	48, 45	10	35
5	60, 47, 30	11	24, 17
6	41, 35, 28	12	28, 21, 18

Earn big money! Use the Dike Manpower Company Executive Employment Service. Our placements earn an average of $46.5 thousand per year. We specialize in computer-related positions, as well as general management positions.

Before printing the advertisement in newspapers and trade journals, Mr. Drum decided to hire a professional statistician to comment on his statistical analysis. The first statistician examined the methodology employed and issued a one-word verbal report—AARGGH!!—whereupon he was relieved of his duties. The second statistician, being more politic than the first, suggested that Mr. Drum read the remainder of this chapter before deciding on an advertisement.

9.1 Introduction to Sample Design: Definitions and Potential Problems

In taking a sample, we are always seeking to say something about a larger population. We wish to generalize from the sample to the population. Mr. Drum wants to be able to say something about the average salary of all of his firm's clients.

The reason for not sampling the entire population is that it often costs too much when compared to the benefit. This can be caused by many different factors, five of which are as follows:

1. A small percentage sample may give extremely good information. This is the case, for example, when a firm has 1,000,000 accounts outstanding, but the total amount of money owed the firm can be estimated within 1 or 2 percent using 1,000 accounts. Thus, a reasonably accurate picture of accounts receivable can be obtained, perhaps at more frequent intervals, at a much smaller cost than for complete enumeration.

2. The population may be so large (or inaccessible) that it is impossible to sample every item in the population. This occurs every night when a sample is taken of people who are watching various TV programs. The entire population numbers in the millions.

3. Even if the population is not very large, each unit sampled may be so costly that a small fraction of the population is all that can be taken as a sample. Destructive sampling, where the item must be destroyed to test it, is one example.

4. In addition to the above cost-benefit tradeoffs, there may be a "number of items sampled versus amount of information taken in each item" tradeoff. For example, should we interview 50 persons for 1 hour each, or have a telephone canvas of 2,000 persons where each person is asked only a few quick questions?

5. Finally, the information may be needed quickly, if it is to be of any value at all. Thus, there is a "number of units sampled versus time before the data is available" tradeoff.

For these reasons, most information for decision making is collected on a sample basis. The notion of a sample is intuitive, and it was discussed in Chapters 1 and 3. A definition is given below.

> A *sample* is a portion of the values in the universe. We draw a sample by selecting some of the items in a population, and observing the values of the variable of interest.

For example, the numbers in Table 9-1 make up a sample. The first statistician was dismayed because of the selection procedure used to obtain the sample. Mr. Drum does have a sample in Table 9-1. The problem is that the sample may not be representative of the population. The first reason for a sample to be nonrepresentative is the manner in which the sample is obtained. In order to take a sample, a physical representation of the population, called a *frame*, must be used. This *frame* may not exactly duplicate the population.

> A *frame* is the concrete form in which a population is represented so that a sample may be taken. In some cases, the frame and the population are identical; in others, a *frame* different from the population must be accepted.

For example, the phone book may be used as a representation of the households in a city. The sample households are drawn from the phone book. The phone-book list is not the same as the true population because of unlisted numbers and households without phones. The difference may not be serious today, but it certainly was in 1936, when a telephone sample predicted a landslide for Alf Landon over Franklin Roosevelt in the race for the U.S. presidency. In that election, the frame was drastically different from the population, in that it contained only persons who could pay for or had access to telephones. These people tended to be Republicans, thereby biasing the results. [This is the most famous case of poor choice of a frame; similar problems are discussed in Chapter 1 of Darrell Huff and Irving Geis, *How To Lie with Statistics* (New York: W. W. Norton & Company, Inc., 1954).]

Mr. Drum's frame is the portion of his ex-customers who have addresses that have been updated within the last 2 years. This frame is unacceptable because persons who stay in touch with Mr. Drum's firm are more likely to be individuals satisfied with the results. Thus, Mr. Drum is systematically excluding some persons whose careers are not what they expected, and thus the average income figures may be somewhat high. Unfortunately, we cannot be sure of the direction of the error. (We can test for the direction of the error by tracking down some of the people who are currently excluded.) In general, the choice of a frame is based on judgment. There are no rules for selecting a sampling frame, but careful inspection should help us avoid many potential sources of bias. In Mr. Drum's case, we have a feeling that an error has been made.

Having established a frame, a sampling scheme then selects some of the units from the frame to constitute the sample. Mr. Drum's selection scheme is complicated. First, he stratified his sample, choosing to select some individuals from each year. He chose the same number of clients from each year, even though the population does not consist of equal numbers in each year. Then he claims to have sampled three randomly chosen individuals from each year. However, he actually sampled only those individuals who responded to his letter. This is certainly not a random selection, since persons proud of their income are more likely to respond to the questionnaire. This type of error is called the *nonrespondent bias*, and it is very important in many situations.

It seems that Mr. Drum has biased the result (average income) in his favor on two accounts: when he chose updated addresses and when he accepted respondents based on their own initiative. There is also a third sampling problem introduced by the equal representation of each year, even though recent years contain more individuals. One would expect this to be another factor that causes the sample average to be too high, in that the recent years are underrepresented. All three of these factors cause a *nonsampling error* (which seems strange in that they all are attacks on Mr. Drum's sampling scheme). To clarify the notion of nonsampling error, two definitions are given below.

> *Nonsampling error* is the difference between the universe parameter and the sample value that is introduced by bias, conscious or unconscious, on the part of the researcher. This is due to improper sample selection, improper questionnaires, and so on.
>
> *Sampling error* is the difference between the universe parameter and the sample value that is caused purely by chance, since the entire population is not sampled.

Mr. Drum probably has some sampling error. We know that he has introduced some nonsampling error, but we do not know how much. Given the above definition, we can expand the discussion of possible sources of nonsampling error. This should help Mr. Drum both with his immediate problem and to avoid such problems in the future.

First of all, the questionnaire should be examined thoroughly. The questions should be very precisely worded and should relate to the desired information. Now, a questionnaire on income would seem to be well defined, but let us examine that issue further. First, does the question ask for total income, including interest, dividends, and other income not related to the job? If so, that income value is not relevant to the issue, which is: How much money do clients make from their job? (The firm may like the bias introduced, but, to be fair, it should be excluded.) Let us suppose, for the moment, that the questionnaire asked not for total income but for job-related income only. That seems simple enough—that equals salary plus any managerial bonus. Or does it also include gains made from stock options? (Stock options allow management to buy the company's stock at a specified price. If the specified price is below the market price, the manager can immediately make money.)

Suppose that income is defined to include gains made from stock options. Do the gains get counted when the individual buys stock at the lower price or when it is sold at a later date? Even more basic, is job-related income last year's income, or the rate at which the individual is earning money this year? This year, you say! But then the manager does not know what bonus will be received (and bonuses are highly variable in some industries).

After that long discussion, there must be a moral to the story. The moral is that questionnaire design and survey techniques in general require careful thought and pretesting. There is no one correct way in which to proceed, but we must try to be very precise. If a generally understood phrase such as "income" can have as many interpretations as those discussed above, how many more does a notion such as "level of satisfaction" with a product have? We shall not discuss this important area further, leaving it to common sense, rigorous thought, and more advanced books.

One final complaint about Mr. Drum's analysis is that the statements he makes about income have very little meaning, even if he is able to correct the above problems. He really wants to imply something about future clients. But his statistical population is his ex-clients. They do not represent next year's clients, for several reasons. First, the computer area is a larger fraction of the business than it was heretofore. Second, the average executive in the sample used the firm's service 6 years ago, and is further along in his career than a new client. It seems that Mr. Drum must rewrite his advertisement if he wants to obtain better communication. (Admittedly, that may not be his goal.) And he may have to redesign his sample if he wants to have more useful data to communicate.

Mr. Drum was correct that one key concept in sampling is the notion of random sampling. He was also correct that occasionally stratified sampling is helpful. The next two sections will examine these ideas, and others related to sampling techniques, so that we can help Mr. Drum conduct a better survey.

S.S. A business professor wants to have some feedback on her course in statistics. She teaches two sections, which she samples separately, using the same, rather lengthy, questionnaire in both sections. In the first section she mails a questionnaire to every student who is registered in the course and has an address on file. She asks that completed questionnaires be dropped in a box outside her office. In the other section, she hands the questionnaire out the last day of class and asks for it back during the same class period. Briefly discuss the population, frame, sampling, and non-sampling errors involved in the two procedures. In what direction do you believe the bias is, in both cases: for or against the course?

Solution: The mail sampling version has a frame of those students who have addresses on file, while the in-class version has a frame of those students who attend class on the last day. The population is composed of all students registered for the course (auditors may or may not be included), and the samples are the responses actually obtained.

Nonrespondent bias is the major problem for both sampling schemes, but the mail version will have more nonrespondents. There may be, of course, problems with the questionnaire, but we have no information on that. We cannot be sure in which direction the bias will occur in either case. The mail version will draw individuals with strong opinions more heavily than those with weakly held opinions, but that may either favor or not favor the course. In the in-class version the bias toward persons attending the class may eliminate more negative opinions than positive ones, but again we cannot be sure.

9-2 Random Sampling and Judgment Sampling

Mr. Drum wanted his sample to be a random sample. A random sample is one in which the researcher uses a selection rule that gives every unit in the population an equal chance of being included in the sample. Thus, the investigator's bias cannot creep in. Mr. Drum did not really select a random sample, as we saw. A judgment sample, on the other hand, is one in which the sample is chosen using someone's judgment as to what units are representative. The possibilities for introducing bias are obvious here, but a judgment sample may be the quickest way to get the necessary information. These notions are defined below.

> A *random sample* is a sample in which every unit in the population has the same probability of being included in the sample.
>
> A *judgment sample* is a sample in which the units of the sample are chosen according to the investigator's judgment. Special forms of judgment sampling include *quota sampling*, in which the population is stratified (separated into groups), then some units are judgmentally selected from each stratum (group), and *convenience sampling*, in which the units in one stratum (group) are selected as being representative of the population.

For example, suppose that Mr. Drum remembered 10 ex-clients whom he considered to be representative of his previous clients. If he sampled those 10 individuals, that would be a *judgment sample*. If he remembered one person from each year whom he considered to be representative, those individuals would constitute a *quota sample*. A *convenience sample* could be constituted, for example, of all the individuals in year 6, if he considered that group to be representative of the entire population.

On the other hand, if he put all of his ex-clients' names into a hat, mixed them well, and drew the sample from the hat, he would have a *random sample*. He would, of course, draw the second name without putting the first back into the hat, thus sampling *without replacement*. If he were to put the names back into the hat after each draw he would be sampling *with replacement*. In both cases the sampling would be random, but sampling without replacement is usually preferred in most situations, since each sampled unit then provides additional data. In both cases, the probability that a particular unit will be chosen on (say) the third draw (or any other draw) is 1 over the number of units remaining in the population at that time. Hence the probability of each possible sample is the same. (This fact will allow us to investigate the sampling error for a random sample. This is studied in Chapter 10.)

Random sampling has the advantage that the investigator (or manager, or whoever is collecting the data) cannot introduce bias into the choice of sample units, so it would seem to be preferable to judgment sampling. (The investigator still can introduce bias in the questionnaire design or the choice of a frame.) There are several forms of random sampling, and these are discussed in Section 9-3. However, before proceeding to that discussion, we need to point out that a manager's bias is not always bad.

Judgment sampling may be the best approach to use when a small sample is being taken. When a decision must be made quickly, a manager may need a "quick-and-dirty" sample. A carefully selected judgment sample may be the best way to obtain a small, useful sample. Since the sample is not random, we lose the ability to generalize statistically about the population. For example, the mean of our judgment sample may be the best guess as to the mean of the population, but the variability of the sample is not a useful estimate of the variability of the population. The loss of the ability to generalize may be offset by a gain in the increased speed (and reduced cost) with which a useful estimate of the average can be obtained. The validity of the estimate depends on good judgment by the manager, making this type of sampling a little more art and a little less science. We must remember, in this regard, that we are studying statistics because it can be useful in decision making, not so that we can marvel at its scientific beauty.

If we decided to take a random sample (or one of the variants of a simple random sample which we will discuss in the next section), a way must be found to choose the sample other than drawing the names (for example) out of a hat. Drawing names out of a hat can be a little cumbersome; instead, we draw numbers out of a table, in particular Table VII of Random Numbers, at the end of the book. The advantages of this procedure are that we do not have to cut pieces of paper (which may not be the same shape, thus destroying the randomness), the process is quicker, and we do not have to own a hat. The random number table takes the place of the paper and the hat.

Sampling

Here is how it works. Suppose that we had a population containing 79 elements, from which a sample of size 10 (without replacement) must be drawn. First, the population units are numbered from 1 to 79. Then we draw 10 two-digit numbers from the random number table. This is done by picking a starting point, for example by randomly placing your finger on the page, then reading two digits at a time successively, reading either across the page and continuing on the next line, or down the page and continuing on the next column. Suppose that the numbers drawn were 73, 90, 86, 09, 11, 44, 25, 31, 69, and 44. That means that our sample should include, in order, items number 73, 09, 11, 44, 25, 31, and 69. For 90 and 86 we must draw another number because they do not correspond to any sample value. For 44 we draw another number, since 44 has been previously drawn and sampling without replacement is being used. The next numbers drawn are 88 (which cannot be used), 69 (which also cannot be used), 15, 70, and 10. So the random sample without replacement consists of items 73, 09, 11, 44, 25, 31, 69, 15, 70, and 10.

If there were 45 items in the population, two-digit numbers would be used, but 1 and 46 would be matched with item 1. Similarly, 2 and 47 would be matched to item 2, 3 and 48 to 3, and so on. If there were 850 items in the population, three-digit numbers would be used. (If you ever run out of random numbers, or if you want some bedtime reading, try *One Million Random Digits*, put out by the Rand Corporation. It is not an exciting book, but the story line is consistent.)

S.S. 1. A manufacturer of consumer durables is interested in sampling user satisfaction on its new toaster oven. They sold 14,000 units during the preceding year and they have 8,400 returned warrantee cards, which is their only source of customer addresses. The 8,400 cards come from all 50 states, U.S. territories, Canada, and several other countries (which they group together for reports). Briefly discuss the population and frame for this sampling problem and discuss how you would obtain a random sample, a judgment sample, a quota sample, and a convenience sample from these data.

2. A standard deck of cards has 52 cards, 13 in each of 4 suits (clubs, diamonds, hearts, and spades). The 13 cards are ace, king, queen, jack, and the numbers 10, 9, 8, . . . , 2.

(a) Give an expression for the number of possible samples of size 5 there are, without regard to the order in which the individual cards appear. (You need not compute a final answer here.) What is the probability that any one set of 5 cards will occur in a sample?

(b) Number the possible outcomes of a sample from 1 to 52: 1 to 13 for clubs, ace, king, . . . , 2 and the same sequence for diamonds (14 to 26), hearts (27 to 39), and spades (40 to 52). Use the following random numbers to generate two samples of size 5:

55, 63, 14, 81, 15, 46, 82, 45, 61, 16, 69, 20,
09, 90, 20, 64, 05, 12, 95, 09, 03, 41, 81, 31.

In words, what are you doing in generating these samples?

Solution:

1. The population consists of the 14,000 individuals who purchased the toasters, and the frame consists of the 8,400 individuals who returned the warranty cards. It is not clear what bias, if any, is introduced by this choice of a frame, but we should try to find out before proceeding. A random sample would be drawn using random numbers from the 8,400 cards. A judgment sample might, for example, be based on choosing the 100 cards that have the most detailed information. This certainly is not random, but the firm might want the most critical customers. A quota sample might pick 5 persons from each state using the same rule. A convenience sample, a likely candidate if the investigation is going to thoroughly examine each case, would involve selecting one town and studying each purchaser in the town.

2. (a) The number of samples of size 5 (poker hands) is the number of combinations of 52 things taken 5 at a time, $_{52}C_5$, which is $52!/5!\,47!$. The probability of obtaining any one sample is $1/_{52}C_5$, assuming that the sampling is random. Random sampling implies that every possible sample is equally likely.

(b) The first 10 usable numbers are (14, 15, 46, 45, 16) and (20, 09, 05, 12, 03), excluding numbers over 52 and repeat numbers in the second sample of 5. Repeat numbers are excluded, since we are sampling without replacement; two players cannot have the same card. The two samples are:

1. Ace of diamonds, king of diamonds, 8 of spades, 9 of spades, queen of diamonds.

2. 8 of diamonds, 6 of clubs, 10 of clubs, 3 of clubs, queen of clubs.

In doing this you are *simulating* drawing 5 cards (a poker hand) out of a deck of 52 cards. Those of you who know poker, look at the next random number after the 03 to see if the second player would fill out the possible flush (draw a club). He does not; the 41 implies that he draws the king of spades, and if he bet the rent, he is in trouble.

9-3 Types of Random Sampling

For the most part we will deal in this book with random sampling, since only random sampling allows us to analyze the sampling errors. In fact, in later chapters we will deal almost exclusively with *simple random sampling*. However, there are several other types of sampling which are useful in many situations. They are examined in detail in books on sampling theory, and we will deal with them only very briefly here.

Simple random sampling involves choosing a random sample from the entire frame with which we are working. In some cases this procedure is too expensive, in that a large sample must be taken to give a representative sample (and reliable results). Other methods are required. They include stratified sampling, systematic sampling, cluster sampling, and multiple-stage sampling.

In *stratified sampling*, the population is divided into several groups, called strata, using some natural breakdown. A random sample is then taken from each stratum. The estimate of the population mean is found by multiplying the mean of the sample from a stratum by the fraction of the total population in that stratum, and adding over all strata. Symbolically,

$S \equiv$ number of strata

$\bar{x}_i \equiv$ mean of the n_i sample values taken from stratum i, which contains N_i units

$s_i^2 \equiv$ sample variance for stratum i.

Then

$$n = \sum_{i=1}^{S} n_i = \text{total sample size}$$

and

$$N = \sum_{i=1}^{S} N_i = \text{total number of items in the population.}$$

The estimate of the mean is given by

$$\bar{x} = \sum_{i=1}^{S} \frac{N_i}{N} \bar{x}_i. \qquad (9\text{-}1)$$

The estimate of the variance around $\bar{x}$ is

$$s_{\bar{x}}^2 = \sum_{i=1}^{S} \frac{N_i}{N} \frac{N_i - n_i}{N_i - 1} \frac{s_i}{n_i}. \qquad (9\text{-}2)$$

A discussion of how to choose the n_i values is given in books on sampling. Briefly, if we know nothing about the strata, n_i values proportional to N_i values make sense. If one strata is more variable than another, a larger fraction of the total sample should be drawn from this stratum.

In Mr. Drum's case, a stratified sample was taken, but he improperly weighted each sample value equally in computing a mean. The stratum means, computed using Table 9-1 data, are as follows:

Stratum (Year Number)	$\bar{x}_i$	Stratum (Year Number)	$\bar{x}_i$
1	81.67	7	36.33
2	73.33	8	40.00
3	71.50	9	34.50
4	46.50	10	35.00
5	45.67	11	20.50
6	34.67	12	22.33

The N_i values, which were previously given, are 40, 38, 66, 72, 70, 94, 96, 101, 112, 105, 134, and 142, which sum to $1{,}070 = N$. Thus, Mr. Drum should have estimated:

$$\bar{x} = (81.67)\frac{40}{1{,}070} + (73.33)\frac{38}{1{,}070} + (71.5)\frac{66}{1{,}070} + (46.5)\frac{72}{1{,}070}$$

$$+ (45.67)\frac{70}{1{,}070} + (34.67)\frac{94}{1{,}070} + (36.33)\frac{96}{1{,}070} + (40)\frac{101}{1{,}070}$$

$$+ (34.5)\frac{112}{1{,}070} + (35.0)\frac{105}{1{,}070} + (20.5)\frac{134}{1{,}070} + (22.23)\frac{142}{1{,}070}$$

$$= 39.03.$$

This average is drastically different from Mr. Drum's 46.46 value. If the selection procedure had not been biased, as discussed earlier, this value would be a good estimate of the average income of all the firm's ex-clients. The question of whether this value is a meaningful value to advertise is left to the next section.

In general, a stratified sample implies that the person taking the sample believes that the strata are different from each other, and that values from all strata should be taken in order to get a representative sample. Further, the units within a stratum are believed to be similar. Choosing the proper strata is important, but this is often easy to do. The strata should be chosen considering expected differences in the variable under study. Thus, the same population might be stratified differently when a different variable is investigated. The number of strata will sometimes be implied by the structure, as in Mr. Drum's case, where every year is a stratum. In any case, the number of strata should be chosen so that a significant sample is possible from each

Sampling

stratum. When stratified sampling is appropriate, it will yield estimates with less variability than those obtained using simple random sampling.

The next type of sampling to be discussed is systematic sampling. It is used because of the ease with which a sample can be drawn using it.

> In *systematic sampling*, the population (or the frame) is arrayed in order, and every kth unit is included in the sample. The first unit should be chosen randomly from among the first k units.

Systematic sampling is not really random. However, before the first unit is chosen, every unit has the same probability of being included in the sample. A problem arises only if for some reason a certain critical characteristic of the population is commonly held by individual units that are k units apart in the array. This usually does not happen. To be sure to avoid it, we must check to see that the choice of k does not coincide with some pattern in the array. For example, if a concert audience is arrayed according to seat numbers, and there are 10 seats in each row, choosing $k = 10$ will always get persons in the same part of the row. It may be, for example, that persons with center seats are more serious about music, and sampling their opinion regarding, for example, how many concerts should be held each year may not give a representative response. In choosing a systematic sampling plan, the researcher is stating a belief that there is no hidden pattern that occurs every kth unit.

If the above problem can be overcome, a representative sample should result. Further, systematic sampling then has the same advantage over simple random sampling as does stratified sampling—sample values from all portions of the array will be chosen, thus perhaps giving a clearer picture of the population. Random sampling may, on the other hand, occasionally result in a bunching of the units chosen for the sample. Systematic sampling is useful in dealing with any ordered lists, such as tax returns, sales records, and so on. It is easier to use a systematic sampling procedure than a random sampling procedure in those cases.

In some sampling situations the population units are not arrayed numerically but are clustered, for example, in towns and cities. *Cluster sampling* is used in such situations.

> In *cluster sampling*, the population is divided into groups, called clusters. (Typically these refer to geographic areas.) Each cluster should, as nearly as possible, be similar in composition to the entire population. Randomly, some of these clusters are chosen for the sample. Then, either (1) every unit in the chosen clusters is sampled, (2) a random sample is chosen from the cluster, or (3) a cluster sample is chosen from the cluster (this is called a two-stage cluster sample).

The reason for choosing a cluster sample is that it is often too expensive to sample in every cluster (geographic location). In stratified sampling we use groups, and sample from each group, so that the full spectrum of values within the population can be observed. In cluster sampling, we believe that if a few clusters are chosen, the entire spectrum will be represented; that is, each cluster is thought to be a miniature replication of the population. If this is not true, a cluster sample may give spurious results. In taking a cluster sample, the researcher is assuming that the clusters are representative of the population.

In Mr. Drum's case, a cluster sample is not a sensible idea, but such a sample could be performed. The clusters might be defined using the city (or geographic area) in which the individuals are currently working. A stratified sample will give much better results for the Dike Manpower Company.

The last type of sampling discussed in this section is multiple-stage sampling. In this approach, the investigator is allowed the option of taking an additional sample if the first sample is not convincing.

> In a *multiple-stage sampling* plan, the decision maker states that a sample will induce him to make a yes decision or a no decision or a decision to take a larger sample. (This may be stated as GO—NO—ON, where ON means that additional samples are to be taken.) Such a sampling plan, and decision system, is a multiple-stage sampling plan. If the second sample always forces a GO or NO decision, that is a *double sampling* plan. If the sample units are taken one at a time, and a GO—NO or ON decision is made after every unit is sampled, that is a *sequential sampling* plan.

Multiple sampling's basic advantage is that some sampling cost may be saved since an early decision is possible. Thus, it is used when sampling cost per unit is high. Examples include quality control sampling, where it is necessary to decide whether or not to accept a manufactured batch of items, and new product decisions, where we must decide whether enough customers will buy the new product to warrant its introduction into the market.

There must be a decision associated with the sample results if multiple sampling is to be appropriate. In Mr. Drum's case we can force multiple sampling to hold by introducing the decision to, say, run the advertisement if the average income exceeds $42,000, not run the advertisement if the average income is below $38,000, and to continue to sample in groups of 10 until either 100 individuals are sampled or the average income goes above $42,000 or below $38,000. (At 100 we might say, "advertise only if mean income is > $40,000.") This is not a well-designed multiple-sampling plan, but it does convey the form of such a plan. Double-sampling plans may be encountered in courses and books on production and operations management.

Sampling

S.S. The Tri-State Distribution Company is a distributor of food items to hotels, motels, and restaurants (the Hospitality Industry) in a three-state area. They have 24 customers, and last year's total sales to these customers is as follows (sales figures are in thousands of dollars):

Customer Number	State	Total Sales	Customer Number	State	Total Sales
1	1	8	13	2	12
2	1	22	14	2	40
3	1	6	15	3	21
4	1	10	16	3	18
5	1	12	17	3	16
6	1	45	18	3	48
7	2	6	19	3	82
8	2	6	20	3	12
9	2	14	21	3	35
10	2	28	22	3	29
11	2	30	23	3	14
12	2	6	24	3	26

The firm is considering adding a new product to their line, and they want to obtain an estimate of how well it will sell. They are going to ask some customers how much of this new product they will buy. They have hired us as statistical consultants, and they have asked us to do the following tasks.

1. Choose customers to sample using

   (a) Simple random sampling, sample size = 8, with random numbers: 14, 28, 27, 95, 97, 65, 21, 02, 84, 62, 47.

   (b) Systematic sampling with $k = 3$ and random number 2.

   (c) Stratified sampling with strata defined by
   1. Less than 10 (thousand) sales.
   2. 10 to 20 (thousand) sales.
   3. More than 20 (thousand) sales.

   Choose three from each strata, using the random numbers from part (a).

   (d) Cluster sampling, using the random number 6 to choose a cluster (state), and using every customer in that cluster.

2. Check the representativeness of each sample by computing the mean of last year's sales of the selected customers. Use the formula for stratified sampling in that case, and the arithmetic mean in all others.

3. How would you suggest they choose a sample?

4. What nonsampling problems do you foresee?

Solution:

1. (a) Customers number 14, 4, 3, 23, 17, 21, 2, and 12. (Number 97 cannot be used. The numbers recycle four times, so 28 implies customer 4, 27 implies customer 3, and 95 implies customer 23.)

   (b) Starting with customer 2, sample customers number 2, 5, 8, 11, 14, 17, 20, and 23.

   (c) The strata are

   Less than 10: 1, 3, 7, 8, 12.

   10 to 20: 4, 5, 9, 13, 16, 17, 20, 23.

   Over 20: 2, 6, 10, 11, 14, 15, 18, 19, 21, 22, 24.

   The random numbers 14, 28, and 27 imply, since the stratum has 5 members, a choice of the 4th, 3rd, and 2nd customers from stratum 1, which is customers number 8, 7, and 3. (This is, admittedly, a tricky procedure: for example, 14 matches with 4, since the matching is 1 to 1, 2 to 2, and so on; 6 to 1, and so on; and finally 11 to 1, 12 to 2, 13 to 3, and 14 to 4. The customers in this stratum are numbers 1, 3, 7, 8, and 12, and the fourth customer is number 8.) The next stratum has 8 members, so random numbers 95, 65 (97 still cannot be used), and 21 imply the 7th, 1st, and 5th members, or customers 20, 4, and 16. The final stratum has 11 members, so random numbers 2, 84, and 62 imply the 2nd, 7th, and 7th (whoops! We have to use the next random number, 47, which implies the 3rd customer in the stratum), so we sample customers 6, 18, and 10. The sample is: (8, 7, and 3), (20, 4, and 16), and (6, 18, and 10).

   (d) The random number 6 implies a choice of cluster 3. (1 is matched to 1, 2 to 2, 3 to 3, then 4 to 1, 5 to 2, 6 to 3, and so on.) Thus, the sample is customers 15, 16, 17, 18, 19, 20, 21, 22, 23, and 24, every customer from state 3.

2. (a) $(40 + 10 + 6 + 14 + 16 + 35 + 22 + 6)/8 = 18.625$.

   (b) $(22 + 12 + 6 + 30 + 40 + 16 + 12 + 14)/8 = 19.0$.

   (c) $\bar{x}_1 = (6 + 6 + 6)/3 = 6$, $\bar{x}_2 = (12 + 10 + 18)/3 = 13.33$, and $\bar{x}_3 = (45 + 48 + 28)/3 = 40.33$. Thus, since $N_1 = 5$, $N_2 = 8$, $N_3 = 11$, and $N = 24$, $\bar{x} = \frac{5}{24}(6) + \frac{8}{24}(13.33) + \frac{11}{24}(40.33) = 24.18$.

   (d) $(21 + 18 + 16 + 48 + 82 + 12 + 35 + 29 + 14 + 26)/10 = 30.1$. The average of all 24 values is 22.75. The purpose of showing the different values is to indicate the variability in a sampling process, not to discredit one of the methods. We should note, however, that cluster sampling does not

Sampling

work well here because the three states are not similar in composition, as assumed in a cluster sampling procedure.

3. A better suggestion, since in this type of operation each customer is visited regularly, is to ask each customer.

4. There are two main problems (not directly related to this chapter but definitely related to proper use of statistics), which are:

(a) What people say they will buy (or do) and what they actually buy (or do) are two different things.

(b) Even if the customers are correct, the firm has not studied whether to some extent the new product will substitute for another product and thus reduce sales for the other product.

9-4 A Warning

The warnings here are copious, but they all reduce to a general statement: "Make your data work for you." An effective manager must manage the statistical analysis if it is to be useful. Specifically related to this chapter, several points can be made about the two basic kinds of sampling discussed.

1. Judgment sampling and its variants are useful when a manager knows something about the data ahead of time. The manager must be aware of the possible bias introduced and the loss of the ability to generalize. This loss occurs because we cannot compute the sampling error using a judgment sample; this is critical if anyone else must be convinced based on the results.

2. Random sampling and its variants are useful when generalizations must be made that will convince someone else, or if the manager wants to be sure that an unbiased view is obtained. The manager must, unlike Mr. Drum, make sure that the sample is random. In using the variants, we must remember the implicit assumptions that are made. An example is the assumption that a cluster is representative of the entire population.

However, these points do not exhaust the potential problems. To manage data properly we must:

1. Be careful about finding explanations (rationalizations?) for data after the data have been seen.

2. Be sure, before the data are collected, that they will be useful for the decisions under consideration. To do this, for example, questions must be asked precisely and uniformly for all participants. Data collection is expensive, and it must be made to pay for itself.

On this last point, let us return to Mr. Drum's problem. Some time ago, we found that a proper stratified sample, with no built-in bias and a higher response rate, could give a good estimate of the average income of his ex-clients. The question is, though, who cares? Without knowing the average number of years since placement, the average income means very little. Also, as stated previously, the new population of clients is different than it was heretofore. A prospective client would be more interested, perhaps, in the current average salary of the clients who have used the service during the past 12 months rather than an amorphous "average income of all ex-clients." Perhaps the best information of all would be to know how much of an increase in income a client had when changing jobs to the job found by Mr. Drum's firm.

This leaves open the question of whether the data should be broken down into categories (computer-related, general management, financial analysis, and so on). Even if someone feels that it should be broken down, eventually the analysis has to cease because of cost and because 28 data points can only be broken down a few ways before all power to generalize is lost. Finally, the question of whether Mr. Drum wants to give a fair picture is not resolved. This is not a statistical question; statistics are meant to be used properly, but no revocable license is required to use them.

New Symbols

Symbol	Meaning
S	The number of strata in a stratified sample.
n_i	The sample size from strata i.
N_i	The numbers of items in strata i.
$\bar{x}_i$	The mean of the items in strata i.

Key Formulas

Formula	Used to Compute:
$\bar{x} = \sum_{i=1}^{S} \dfrac{N_i}{N} \bar{x}_i$	The sample mean for a stratified sample.
$s_{\bar{x}}^2 = \sum_{i=1}^{S} \dfrac{N_i}{N} \dfrac{N_i - n_i}{N_i - 1} \dfrac{s_i^2}{n_i}$	The variance around $\bar{x}$ in a stratified sample.

PROBLEMS

9-1.* An automotive company is interested in examining the quality of a particular portion of their assembly operation. The person doing the sampling is to carefully inspect 100 assemblies. The sample, a convenience sample, is chosen as the first 100 assemblies off the line on Monday morning. Discuss the validity of the sample. How might a better sample be obtained?

9-2.* What are the reasons for taking a sample rather than examining the entire population? Give an example in which a sample would be the only practical way to gain information and an example in which the entire population could be examined.

9-3. A computer manufacturer is interested in the extent to which scientific methods of inventory control are used by business firms. They decide to take a sample to find out, and they use the membership list of the American Production and Inventory Control Society (APICS) as the frame, to represent the population of individuals engaged in inventory control. Discuss their choice of a frame and how it may affect their results.

9-4. Describe a sampling situation in which nonrespondent bias might be a major factor. How might the sampling be done to avoid (or minimize) the problem?

9-5. Discuss the reasons for choosing a systematic sampling plan, and give an example of a situation in which it would be appropriate.

9-6. In both stratified and cluster sampling the population is broken down into subgroups. Also, we have different implied assumptions about the similarities or differences among the subgroups in the two cases. What are these assumptions, and why do they indicate the kind of sampling with which they are associated?

9-7.* If a population contains 10 individuals, is it possible to take a sample of size 25, with replacement? Is it intelligent to do so?

9-8. In each of the following situations, state whether the errors should be considered nonsampling error, sampling error, or both, and why.
(a) An inspector for a soft drink company selected fifty 12-ounce bottles of cola at random and poured the contents into a calibrated tube. The calibrated tube is incorrectly scaled and overestimates the contents by 0.05 ounce. The average contents of the fifty bottles was measured as 12.03 but the bottle-filling machine actually dispenses 12 ounces on average.

(b) To determine the effectiveness of its waitresses, a restaurant manager placed evaluation cards on each table for a 1-week period. Customers could rate the waitresses as either "very good," "good," "bad," or "very bad." Ten cards were returned, four of which rated the waitresses as "very bad."

(c) A consumer organization selected five Never-Die automobile batteries from five different stores in five different wholesale sales regions. The organization subjected the batteries to tests to determine the average life of the batteries.

(d) An advertising firm is conducting a survey for its client, Sky-High Department Store, in order to determine if most people in the community prefer Sky-High to its largest competitor. To conduct the survey, questioners stood outside Sky-High each day for a week and asked persons leaving the store to provide their opinions.

9-9. A town with a population of 50,000 residents is considering a $10,000,000 downtown renovation program. To study citizen opinion, the town board commissions a student to set up a stand on the busiest corner in town. The student speaks to everyone who stops at the stand, explains what renovations are being considered, and asks the individual if the renovations would improve, harm, or have no effect on the downtown area. The number of times each response is given is reported to the town board.

(a) Describe the population, frame, sample, nonrespondent bias, and other nonsampling errors in this case.

(b) Describe how a simple random sample, stratified sample, quota sample, and convenience sample might be taken to obtain the information they seek.

9-10.* A sample of 50 parts is to be taken from a production run containing 2,000 machined parts. Sampled parts are categorized as good or bad by the test.

(a) What probability distribution can be used to describe this process? Does the answer depend on whether the sampling is done with or without replacement?

(b) If there are 400 defectives among the 2,000 parts, what is the probability that there will be exactly 10 defectives in the sample, assuming that simple random sampling is used?

(c) Why might a systematic sample be considered in this case?

9-11.* The serial numbers for each of the transistor AM–FM radios produced by Ulisten Radio Company on a certain day are as follows:

Sampling

62675-1	62675-6	62675-11	62675-16	62675-21	62675-26
62675-2	62675-7	62675-12	62675-17	62675-22	62675-27
62675-3	62675-8	62675-13	62675-18	62675-23	62675-28
62675-4	62675-9	62675-14	62675-19	62675-24	62675-29
62675-5	62675-10	62675-15	62675-20	62675-25	62675-30

An inspector wants to draw a random sample of five radios from the day's production. Using the random numbers, 60, 08, 28, 28, 62, 72, 60, 81, 89, 57, 02, 99, and 20, draw a random sample of five radios.

9-12. The management of Morris Manufacturing Company wants to determine whether the employees in two of its buildings think their work areas are lighted brightly enough. One building consists solely of manufacturing processes while the second building consists of an office area and a section containing manufacturing processes. Each employee has a badge with a number on it and the distribution of employees is as follows:

Badge Numbers	Type of Job	Number of Employees
1–120	Office worker	120
300–499	Plant I worker	200
700–949	Plant II worker	250

Use stratified random sampling and the following random numbers to choose five employees from each group: 777, 570, 518, 383, 697, 634, 517, 213, 305, 165, 927, 892, 393, 473, 063, 652, 912, 607, 461, and 654.

9-13. What sampling method might be appropriate in each of the following situations? Why?
   (a) A firm that owns 400 typewriters wants to determine the average age of the typewriters. For each typewriter, the firm maintains an index card which contains the date of purchase, the purchase price, and the model specifications.
   (b) A television station in New York City wants to determine the proportion of viewers who watch a new show.
   (c) Buy-Lo Food Stores, Inc., has six supermarkets, one in each of six small towns. The manager of the company wants to determine the customer preferences toward shopping hours, which will be the same for each store.

9-14.* Using the numbering system introduced in the chapter for the 52 cards in a deck (01 to 13 assigned to clubs, ace, king, and so on down to 2, then 14 to 26 for diamonds, 27 to 39 for hearts, and 40 to 52 for spades), generate four samples of size 5 using the following random numbers:

09, 23, 72, 46, 07, 12, 81, 79, 81, 20, 30, 93, 77,
06, 66, 76, 05, 94, 44, 46, 94, 18, 81, 40, 60, 11,
00, 74, 54, 61, 24, 72, 50, 28, 77, 74, 66, 71, 84,
27, 45, 65, 84, 00, 44, 59, 48, 45, 99, 66, 60, 62,
27, 84, 44, 39, 58, 29, 19, 30, 97, 94, 81, 13, 22,
19, 92, 34, 99, 62, 92, 24, 30, 49, 24, 77, 99, 73,
43, 66, 81, 95, 99, 30, 02, 16, 85, 72, 31, 81, 63,
30, 33, 14, 59.

(a) If you know the rules of poker, which of the four hands would win? Remember that no card can be used twice.
(b) Discard the cards you would choose to discard from each hand and draw, using the remaining random numbers, to fill each hand to 5. Which hand is best now? (If you do not know the rules of poker, just draw 2 additional cards for each hand.)

9-15. A producer of a particular vitamin product (dietary supplement, including iron) is interested in knowing the "demographic" characteristics of its customers. In particular, the manager wants to know whether the users of its product are married or not, and their age, sex, family income, and type of occupation.
(a) Why might they want to make such a study?
(b) Briefly describe how you would conduct this study. (There are several ways to answer this part.)

9-16. Describe what factors you would consider in selecting strata (in stratified sampling) and clusters (in cluster sampling). Include the various costs of the sampling procedure in the discussion.

9-17. (a) A manufacturing firm is studying employee satisfaction and wants to take a simple random sample from among its 842 employees. They want to sample 6 individuals. What sample will they obtain if they use 391, 590, 479, 551, 163, 844, 756, 072, 235, and 268 as random numbers?
(b) If they, instead, want to sample 3 employees from each of their two plant locations, what sample will they obtain if

(1) The two plant locations have 612 and 230 employees, respectively?

(2) The same random numbers are used as in part (a)? Is this sample different from that taken in (a)?

9-18. A durable goods manufacturer is interested in knowing the average usage of electric range tops, in burner hours per day, in the United States. They are considering marketing a new, more efficient, electric range top, and they want to estimate average saving per household, for use in their promotional campaign. They believe that usage may be related to income, so they take a stratified sample. The data are as follows for the region in which the sample was taken:

Stratum	n_i	N_i	Average Usage ($\bar{x}_i$)	Sample Variance (s_i^2)
Less than $10,000 income	100	540,000	5.4	1.0
$10,000–$20,000 income	100	400,000	6.1	1.6
Greater than $20,000 income	100	150,000	5.9	1.9

Compute the sample mean and sample variance around $\bar{x}$ for the stratified sample.

9-19.* In Problem 9-18, data were given regarding a stratified sample taken on electric-range-top usage. The firm would like you to answer some questions about their sample.
(a) Is a stratified sample reasonable here? If so, was it taken properly? Suggest any changes that come to mind.
(b) If the $\bar{x}$ values are normally distributed, if $\sigma_{\bar{x}} = 0.01375$ (that is, the sample value is the parameter value), and if $\mu = 5.5$, how likely is an $\bar{x}$ value greater than or equal to 5.755 (the observed value)? This type of calculation will be very important to us in later chapters.

9-20. Discuss the main advantage and disadvantage of judgment sampling.

9-21. The managers of Sell-More Food Company want to determine the total amount of bonuses that their sales personnel will be paid. Since the bonuses will not be finally determined for several months, the managers are interested in estimating the total amount the company will need to include in this year's revised operating budget. The amount of bonus is computed as 1 percent of each salesman's total sales for the year. The company maintains five distribution centers, which employ 12, 21, 25, 15, and 30 salesmen. A stratified sample of salesmen's records was selected and the following results obtained:

Distribution Center	Total Number of Salesmen	Number Selected from Center	Sample Mean	Sample Variance
1	12	6	$75,185	$5,850
2	21	11	82,420	4,521
3	25	13	51,621	2,897
4	15	8	69,377	4,826
5	30	15	61,162	3,149

Find estimates for $\bar{x}$ and $s_{\bar{x}}^2$.

9-22. In testing attitudes toward a new inoculation program for rural areas, a public health official wants to use a cluster-sampling approach to choose a sample of individuals. The official wants to randomly choose 1 of the 48 contiguous states. Each state has had 20 rural townships designated for consideration, and one of those is to be chosen. Within the selected township, the agency will try to involve the entire population. What kind of sampling is this? Using the random numbers 47 and 22, what number state and township would be selected? Do you think they can safely generalize to the entire country based on their results?

9-23.* A market research organization is interested in sampling attitudes toward a potential new luxury car. In taking a stratified sample, what variable or variables might they use to form the strata, and why?

9-24.* In producing a forecast of next year's total sales, a toy manufacturing company asks some of its sales personnel to make a guess. They ask 4 persons, 2 of whom have less than 5 years of experience (there are 100 such persons with the company). The data are as follows:

Person	Years of Experience	Forecast of Annual Sales
A	Less than 5	$385,000,000
B	Less than 5	340,000,000
C	5 or more	305,000,000
D	5 or more	310,000,000

(a) Is the formula for computing the mean of a stratified sample appropriate here? Why or why not?
(b) What would you forecast for next year's sales, if you had to make a forecast based on these data?

9-25. A toy manufacturing firm is interested in knowing its daily sales performance during its own and its competitors' promotional periods.

Since they cannot receive and process orders on a daily basis, they perform a phone survey. On one day, the data obtained are as follows (showing sales for two groups of 3 salesmen):

Person	Years of Experience	Average Daily Sales	Sample Variance
A, B, C	Less than 5	$1,466.67	2,900,000
D, E, F	5 or more	2,166.67	1,120,000

(a) Is a stratified sample appropriate here? Why or why not?
(b) Without making any calculations, comment on the adequacy of the sample size.
(c) Compute $\bar{x}$ and $s_{\bar{x}}^2$ using the formulas for stratified sampling.

Chapter **10**

Sampling Distributions

The reason for taking a sample is to learn something about the population (or universe, the set of values the population takes on for the variable of interest). In particular, the universe mean is frequently the subject of concern in managerial problems. In previous chapters, there have been several examples where the sample was important because it implied something about the universe mean. However, we were never sure how good the sample estimate of the universe mean was.

If a manager is to know how accurate an estimate the sample mean is of the universe mean, the manager must know what values are likely to be observed if other samples of the same size were taken. The closer these values are to each other, the better the estimate is. That is, the manager needs to know the distribution of sample means for all samples of a given size. If all the sample means that might occur are expected to fall very close to the one obtained, that is, if they exhibit very little variability, then the manager's one sample mean value is a good estimate of the universe mean. In this chapter we investigate the distribution of sample means. In Chapter 11, information on the distribution of sample means will be used to study estimation problems in managerial settings.

Several facts must be kept in mind while studying sampling distributions. The universe mean that we want to estimate must be the mean of the universe from which each and every sample value was taken. If we extrapolate the results of a sample taken from one population to another population, as is often necessary, we must recognize the implied assumption that the two populations are identical or at least similar. A situation of this type occurs if

Sampling Distributions

the universe mean of a process changes while we are taking the sample; if early sample values are included in the sample mean, it cannot properly estimate the new universe mean. This may happen, for example, when sales data are gathered for several weeks and used to estimate average weekly sales. The average weekly sales may have changed during the data-gathering period. Any time a universe mean is estimated using sample data, the implicit assumption is made that all sample values are drawn from the same universe. (That is, we assume that all the sample values are generated by the same "data-generating process.")

Another fact that must be considered is the size of the sample as compared to the size of the population. Typically, a sample is taken without replacement, so a large-enough sample could eventually exhaust a finite population, and we could calculate the universe mean exactly. However, if the sample constitutes only a small fraction of the population, sampling with replacement is essentially the same as sampling without replacement. The two methods yield essentially the same results. In fact, our statistical methods will be based on sampling with replacement. (In sampling without replacement, the population is not identical for each draw.) Sampling without replacement from a finite universe is examined in more depth later in this chapter.

As an example of a sampling process, consider the problem of the Reliable American Tool and Foundry Incorporated (RATFINC). RATFINC produces therbligs in a continuous manufacturing process, at a rate of 10,000 units per 8-hour shift. There is one critical dimension in a therblig, the diameter of a hole that is machined during the process. This diameter must be within 0.010 inch of 0.500 inch. That is, the diameter must be within 0.500 inch $\pm$ 0.010 inch (0.490 to 0.510 inch). To ensure that this characteristic is maintained, the company is going to inspect some units every hour. They are considering two alternative methods of inspection. Method one uses a GO–NO GO gauge, whereby each unit's diameter is determined to be either acceptable or unacceptable. The number of acceptable units in the sample is then used to determine whether the production process is satisfactory or not. Method two uses a measurement, where the diameter of each unit is recorded and an average is taken. In this case, the average hole diameter of the sampled units is used to evaluate the production process. Both of these methods constitute sampling schemes, but they are treated differently, as we will see in the next section.

10-1 Data-Generating Process

The basic sample data are very different using the two inspection methods suggested above. In using the first method, each therblig is either acceptable or unacceptable, and the data are binomial in nature. In the second method, the data measurements are continuous, and they may be normally distributed, but we cannot be certain. In any case, we must be concerned about the form of the sample data obtained. The type of data obtained is said to result from the *data-generating process* used.

> The *data-generating process* is the process by which a sample is drawn. It determines the type of data that we obtain.

If the sampling process is random and with replacement, the individual sample values are distributed according to the same probability distribution which describes the universe. Random sampling is assumed for the remainder of the book, so the sample values have a distribution identical to that of the universe. For example, this means that if 40 percent of a population (of marbles) is blue and 60 percent is red, then on any one draw, the probability of a blue marble is 0.4.

In the RATFINC example, the population can be thought of as all the parts that are produced during 1 hour. If the diameters are measured, the universe is the set of all diameter measurements during the time period. If the GO–NO GO gauge is used, and if 0 is used to stand for acceptable and 1 is used to stand for unacceptable, the universe is the set of 0's and 1's corresponding to the units produced in the time interval. In both cases, the sample will be a subset of the values in the universe.

When the GO–NO GO gauge is used, a binomial sample results. The purpose of taking a binomial sample is to estimate the probability of "success" in the universe. (Success in this case is defined to be an unacceptable item.) In the RATFINC case, the reason for estimating the probability of an unacceptable item would be to see if the process is turning out too many defective units. In general, the reason for obtaining a binomial sample is that either the data are binomial in nature (two categories), or we choose to make the data binomial by categorizing the data into two classes, because of the simplicity introduced. An example of data that are originally binomial would be a survey asking if individuals are currently watching a particular TV show or not, and the RATFINC company problem is an example where binomial data generated using the GO–NO GO gauge may contain enough information to use in monitoring the production process.

If the hole diameters in the therbligs are measured and the measurements are used as the sample, we have more information; this information will be more costly. In the case of diameter measurements, the universe values may be approximately normally distributed. In other cases, the normal distribution may be a very poor approximation of the actual distribution. However, the next section of this chapter introduces a result that allows us to treat the sample mean as if it were generated by a normal process even if the universe data are not. (This is important because the normal distribution is tabulated on one page, and hence is easy to use.)

One other important question remains concerning the data-generating process in the RATFINC example. The company wants to sample some units every hour, from the hourly production of $10,000/8 = 1,250$ units. Why do they want to do that? Why not sample the entire 10,000 items during an 8-hour shift, obtain the mean diameter or fraction defective, and make a decision based on this value? A sample of 10,000 should be sufficient to obtain a very accurate estimate. But the company has elected to sample a few units every hour. Why?

If all units are from the same population and this will continue to be true in the future, the firm should take one large sample once and use it. The only reason for taking small samples periodically is to investigate the possibility that the data-generating process, and hence the population, is changing. The company samples to find out when the process has changed, so that it can react to the change. For example, the machine setting may have shifted so that the average diameter is no longer 0.500 but now is 0.520. Since the larger diameter is unacceptable, the manager may wish to stop the process and reset the machine. If the machine produced parts with a mean of 0.480 half the day and parts with a mean of 0.520 the other half, a day's sample would be close to 0.500, when in fact very few if any of the units are in the acceptable range (0.490 to 0.510). The point is that all units in a sample must be drawn from the same population if valid statistical generalizations are to be made. Thus, the firm chooses to sample a few units every so often and assumes that, although the data-generating process may be changing, it either does not change or changes an insignificant amount during the sampling procedure.

We might be tempted to think that a sample of size 1 would be ideal since no change can occur while that sample is being taken. However, most managers would not be willing to base decisions on a sample of size 1. They would feel more confident using a mean based on a larger sample. Thus, sampling procedures must balance the desire for greater statistical confidence with the desire to observe any changes in the process. Several other factors, including sampling cost and the cost of not detecting unacceptable units, must also be considered in designing a sampling plan.

S.S. A fountain pen company advertises that its new long-life pen will write a line 1 mile long on each refill. They want to test that assertion by periodically examining the length of the line that a refill will make.

(a) How might they take a binomial sample in this case?

(b) What does the periodicity of the sampling procedure imply about the firm's beliefs concerning process stability?

(c) They say that the length of the line is reasonably approximated by a normally distributed random variable. Do you believe this to be true? Why or why not?

Solution:

(a) If they take data on the length of the line, then assign 0 to lines less than 1 mile long and 1 to lines 1 mile long or more, they will have a binomial sample.

(b) Periodic sampling implies that they think the data-generating process might change.

(c) There is no a priori reason to question the assertion. In fact, a variable such as this would probably be close to normal.

10-2 Distribution of Sample Means

RATFINC's quality control manager, Mr. Jones, has decided to examine the hole diameter measurements on the therbligs rather than using a GO–NO GO gauge. He notes that a small sample will be more informative using this method because more of the information about each unit is preserved than with a GO–NO GO gauge. Mr. Jones took four samples of size 5, and the sample data are given in Table 10-1.

Table 10-1 Therblig Diameter Measurements (inches), RATFINC

Sample 1	Sample 2	Sample 3	Sample 4
0.493	0.500	0.499	0.500
0.498	0.499	0.499	0.500
0.502	0.497	0.500	0.501
0.508	0.502	0.496	0.497
0.501	0.495	0.503	0.494

Mr. Jones, being naturally inquisitive, made a few computations.

The mean of sample 1 is	0.5004
The mean of sample 2 is	0.4986
The mean of sample 3 is	0.4994
The mean of sample 4 is	0.4984
The overall mean is	0.4992
The basic data range from	0.493 to 0.508
	(a total range of 0.015)
The 4 sample means range from	0.4984 to 0.5004
	(a total range of 0.0020)

SAMPLING DISTRIBUTIONS

Mr. Jones observed that the sample means do not vary as much as the basic data. This is a very important observation, and we will verify it in the next paragraph. Before doing so, however, it is worth mentioning that this result (means are less variable than the sample or the universe) is the reason for taking averages. Since sample means are less variable, it is unlikely that a different sample would have a drastically different result and thus lead us to a different conclusion. The larger the sample, the better a sample mean performs as an estimator of the universe mean. This fact tends to make us want to take large samples. The countervailing reasons that hold the sample size down are: (1) the cost of the sample, (2) the population may change while the sample is being taken and we wish to minimize this effect, and (3) we may need to make a quick decision.

Looking at the data in Table 10-1, we can understand intuitively why means are less variable than the sample. To precisely specify the relationship between the variance of the sample mean and the variance of the basic data, the results from Section 7-4 must be used. Those results were:

1. For independent random variables X and Y,

$$\text{Var}(X + Y) = \text{Var}(X) + \text{Var}(Y).$$

2. $\text{Var}(aX) = a^2 \text{Var}(X)$.

In our case, the mean is calculated as $(1/n) \sum_{i=1}^{n} x_i$, a constant $1/n$ times a sum. If all the x_i values are drawn from the same population, they have the same variance, $\text{Var}(X)$ or σ^2. If we add n of them, the sum has a variance equal to $\sigma^2 + \sigma^2 + \cdots + \sigma^2 = n\sigma^2$ or $n \text{Var}(X)$. But then, we multiply the sum by a constant, and this multiplies the variance by the constant squared. Thus, the variance of the arithmetic mean is

$$\left(\frac{1}{n}\right)^2 [n \text{Var}(X)] = \left(\frac{1}{n}\right)^2 (n\sigma^2) = \frac{\sigma^2}{n} \quad \text{or} \quad \frac{\text{Var}(X)}{n}.$$

The variance of the arithmetic mean is denoted by $\sigma_{\bar{X}}^2$. Thus, we can show the above manipulations as

$$\sigma_{\bar{X}}^2 = \text{Var}\left(\frac{1}{n} \sum_{i=1}^{n} X_i\right) = \left(\frac{1}{n}\right)^2 \text{Var}(X_1 + X_2 + \cdots + X_n)$$

$$= \frac{1}{n^2} [\text{Var}(X_1) + \cdots + \text{Var}(X_n)]$$

$$= \frac{1}{n^2} [n \text{Var}(X)] = \frac{1}{n} \text{Var}(X) = \frac{\sigma^2}{n}. \tag{10-1}$$

Formula (10-1) is a precise statement of why an average has less variability than the universe data. The new symbol for the variance (and standard deviation) of the mean is:

New Symbol	Meaning
$\sigma_{\bar{X}}^2, \sigma_{\bar{X}}$	The true variance and standard deviation of the mean, where σ^2 is the variance of the universe, $\sigma_{\bar{X}}^2 = \sigma^2/n$, and $\sigma_{\bar{X}} = \sigma/\sqrt{n}$ (n is the sample size).

As mentioned previously, these formulas are based on an assumption of random sampling, with replacement. Sampling is typically done without replacement, but if the sample is a small fraction of the population, as is often the case, we can proceed to use the above formulas. However, if the sample constitutes a large fraction, say 5 percent or more, of the population, then a *finite population correction factor* should be used. Since this factor is usually unnecessary, we postpone its discussion until Technical Note 10-1.

Notice that up to this point we assume that σ^2, the universe variance, is known. This might be reasonable in the RATFINC example in that a machine's precision, its variability, may stay the same while the settings drift off their original values. Thus, a variance based on a large amount of historical data may be an accurate estimate of the process variance. When there is no such historical data, and all we have is a sample variance based on a relatively small amount of data, the problem must be handled differently. We will discuss that case in the next section.

As an example using our new formula, we can investigate Mr. Jones' data. In addition to the information in Table 10-1, suppose Mr. Jones says that $\sigma = 0.0032$. That is, based on a large amount of historical data, $\sigma = 0.0032$ is calculated as the universe standard deviation. This means that, if we consider a sample of size 5:

$$\sigma_{\bar{X}} = \frac{0.0032}{\sqrt{5}} = 0.0014$$

$$\sigma_{\bar{X}}^2 = \frac{(0.0032)^2}{5} = 0.000002.$$

If we want to consider a sample of size 20, which Mr. Jones can have by combining his four samples of size 5, then

$$\sigma_{\bar{X}} = \frac{0.0032}{\sqrt{20}} = 0.0007$$

$$\sigma_{\bar{X}}^2 = \frac{(0.0032)^2}{20} = 0.0000005.$$

Given values for $\sigma_{\bar{x}}$ and $\bar{x}$, Mr. Jones can try to decide if it is reasonable to assume that the machine is still producing items with a universe mean of $\mu = 0.500$. For example, using all his data, with $n = 20$ and $\bar{x} = 0.4992$, and using $\sigma_{\bar{x}} = 0.0007$, Mr. Jones might ask: If the mean is $\mu = 0.500$, how likely is a sample value as low as $\bar{x} = 0.4992$? For the moment, suppose that the $\bar{x}$ values are normally distributed. (This is true if the universe is normally distributed, and that is reasonable to assume for hole diameters.) Then we need to find the normal probability of $\bar{x} \leq 0.4992$ given $\mu = 0.500$ and $\sigma_{\bar{x}} = 0.0007$. That is, find

$$P(\bar{x} \leq 0.4992 \mid \mu = 0.500, \sigma_{\bar{x}} = 0.0007) = P\left(z \leq \frac{0.4992 - 0.500}{0.0007}\right)$$

$$= P(z \leq -1.14)$$

$$= 0.5 - 0.3729 = 0.1271,$$

using Table I.

A sample value as low as $\bar{x} = 0.4992$ with $n = 20$ is not very likely if $\mu = 0.500$. This may lead Mr. Jones to believe that the machine is no longer set at exactly 0.500. He may try to stop the process and adjust the settings, but since 0.4992 is relatively close to 0.500, he may not. The notion of deciding whether a mean is equal to a certain value (0.500 here) is a crucial idea, and such problems are investigated extensively in Chapter 12.

A normal probability distribution is used to calculate the probabilities above regarding $\bar{x}$. This was possible because the individual diameters were assumed to be normally distributed. Frequently, the individual measurements in the universe are not normally distributed, but we would like to use the normal distribution anyway. (After all, the table for normal probabilities fits on one page.) Fortunately, if we wish to generalize about the universe mean from a sample mean, it is possible, due to the Central Limit Theorem, to use the normal distribution.

Central Limit Theorem
If a random sample of size n is drawn from a universe with mean $= \mu$ and variance $= \sigma^2$, then the distribution of sample means, $\bar{X} = \sum_{i=1}^{n} x_i/n$, tends to be normally distributed, regardless of the distribution of the universe. The closeness to the normal improves as n gets large. Further, the mean of the distribution of $\bar{X}$ is μ, and the variance of the distribution of $\bar{X}$ is $\sigma_{\bar{X}}^2 = \sigma^2/n$. That is,

$$E(\bar{X}) = \mu$$

$$\sigma_{\bar{X}}^2 = \frac{\sigma^2}{n}.$$

The Central Limit Theorem allows us to treat the distribution of sample means as if it were normal. The importance of this result rests in the fact that we do not need to know the distribution of the individual items (the universe) to know the distribution of the sample means.

However, if we do know that the universe is normally distributed, then the distribution of sample means is exactly (not approximately) normal, regardless of the size of the sample. The $E(\bar{X}) = \mu$ and $\sigma_{\bar{X}}^2 = \sigma^2/n$ formulas still hold. The Central Limit Theorem also applies (as does the previous sentence) to sums as well as averages. We will deal with it most often as it applies to means.

The "tends to be normally distributed" phrase in the definition is a vague statement which should be further refined. How quickly the distribution of the sample mean approaches the normal distribution (as n increases) depends on how close the universe distribution is to the normal. If it is nearly normal, $n = 2$ or 3 may suffice. If all possible values in the universe are equally likely (that is, if the universe is described by a rectangular or uniform distribution), $n = 10$ is sufficient. If the original universe is distributed according to an exponential distribution (see Chapter 8), $n = 100$ may be necessary. The exponential is an extreme case, and in general $n = 20$ or so is sufficient.

Like most rules of thumb, using $n = 20$ as a required sample size is questionable. Most universes of interest in managerial problems will have one concentration of frequency around the mode, with the frequency diminishing in either direction away from the mode. (Such frequency curves are called unimodal; see Figure 2-4 for a picture.) If this is a reasonable assumption, and if values on both sides of the mode are likely (this excludes the exponential distribution), then a sample size of $n = 20$ is sufficiently large to assume that the sample means are normally distributed, and $n = 10$ is probably sufficient. The manager must use knowledge of the data-generating process to decide if it is safe to assume that the sample means are normally distributed, and the $n = 20$ rule of thumb should not be taken as an irrevocable truth.

The Central Limit Theorem applies to a discrete universe as well as to a continuous universe. In particular, when the sampling process is binomial, the sample proportion of successes (which is the same as the sample mean) will be approximately normally distributed for sufficiently large values of n. The approximation is better if the sample proportion is not close to either 0 or 1, but for n large enough, the Central Limit Theorem will hold in any case. (We used the Central Limit Theorem previously when saying that a normal distribution can approximate the binomial when np and $n(1 - p)$ are both at least 30.)

Based on the Central Limit Theorem, Mr. Jones can feel that his calculation of $P(\bar{x} \leq 0.4992 \mid \mu = 0.500, \sigma_{\bar{x}} = 0.0007) = 0.1271$ is proper. Exactly what this implies is left to Chapter 12. Here we must note that the probability value read from the normal table is exactly correct if the universe of diameter measurements is normally distributed, and the Central Limit Theorem says

Sampling Distributions

that the probability is approximately correct even if the universe is not normally distributed. For this reason, the Central Limit Theorem is extremely important in statistics, as are the $E(\bar{X}) = \mu$ and $\sigma_{\bar{X}} = \sigma/\sqrt{n}$ relationships.

One difficulty we may encounter is that σ may be unknown. Mr. Jones claimed above to know the universe variability for his machine based on a large amount of previous data. If he had only a small amount of sample data, and if the sample standard deviation, s, is used to estimate σ, then a different treatment is necessary. This is covered in the next section.

S.S. There are 1,000 employees in a lock factory, each of whom will work assembling the same new style of lock. The plant manager is interested in studying the number of locks they will be able to assemble per day. She decides to count the number of locks produced by 10 of the workers on their first day (after training). She says that $\sigma = 10$ is the universe standard deviation, based on information collected from other similar assembly jobs. The data are as follows:

$$45, \quad 49, \quad 53, \quad 64, \quad 64, \quad 64, \quad 64, \quad 64, \quad 71, \quad 74.$$

(a) What is $\bar{X}$ and $\sigma_{\bar{X}}$?
(b) If $\mu = 60$, what is the probability of observing a sample value as high as $\bar{X}$ or higher?
(c) Comment briefly on their sampling procedure and their results.

Solution:

(a) $\bar{x} = \dfrac{45 + 49 + 53 + 64 + 64 + 64 + 64 + 64 + 71 + 74}{10} = 61.2$

The basic formula is

$$\sigma_{\bar{X}} = \frac{\sigma}{\sqrt{10}} = \frac{10}{3.162} = 3.162.$$

(b) $P(\bar{x} \geq 62.1 \mid \mu = 60.0, \sigma_{\bar{X}} = 3.162) = P\left(z \geq \dfrac{61.2 - 60.0}{3.162}\right)$

$$= P(z \geq 0.380)$$

$$= 0.5 - 0.1480 = 0.3520.$$

(c) First, a sample of first-day performance will not be representative of the future. Second, the firm should, when it does sample, sample everyone, since it is easy to do so. Third, the σ value may be based on locks very different from the current one, so it may be incorrect. Finally, the five 64's lead us to believe that the data may not be normally distributed.

10-3 Distribution of Sample Means When σ Is Unknown

In the RATFINC example and the previous Student Should, the true value of σ, the universe standard deviation, was assumed known. The basis for the assumed value was previous data, presumably relevant to the case at hand. The relevancy of past data can always be challenged, and in some situations past data do not even exist. In that event, the only variability value that we have to work with is the standard deviation (or variance) computed from the sample.

Intuitively, this should make a decision maker less sure of the estimate, $\bar{x}$, of μ. There is more uncertainty (than if σ were known) associated with the estimated $\bar{x}$ value. For example, we found that $P(\bar{x} \leq 0.4992 \mid \mu = 0.500, \sigma_{\bar{x}} = 0.0007) = 0.1271$ in the RATFINC case. If we do not know $\sigma_{\bar{x}}$, but instead have only a sample estimate of it, $s_{\bar{x}}$, we would expect the probability of $\bar{x} \leq 0.4992$ to be greater than 0.1271. That is, the probability of low (or high) $\bar{x}$ values is increased due to the uncertainty about σ.

RATFINC can again be used as an example. Suppose that Mr. Jones has no information about σ; then he would compute s, the sample standard deviation, and use it as his estimate of σ. Suppose that Mr. Jones calculated, for the data listed in Table 10-1,

$$s = 0.003412$$
$$s^2 = 0.00001164$$
$$\bar{x} = 0.4992.$$

Now Mr. Jones really would like to have $s_{\bar{x}}^2$, the sample variance of the distribution of means. Luckily, the same relationship holds between the sample variance measures as holds between σ^2 and $\sigma_{\bar{x}}^2$.

The relationship between the sample variance (and standard deviation) and the variance (and standard deviation) of the sample mean is

$$s_{\bar{x}}^2 = \frac{s^2}{n} \tag{10-2}$$

$$s_{\bar{x}} = \frac{s}{\sqrt{n}}. \tag{10-3}$$

New Symbol	Meaning
$s_{\bar{x}}^2, s_{\bar{x}}$	The sample variance and standard deviation of the mean. They are based on s^2 and s, the sample variance and standard deviation. The relationships are given in equations (10-2) and (10-3).

Sampling Distributions

These relationships are logical since $s_{\bar{x}}$ is to be used as an estimate of $\sigma_{\bar{x}}$. Thus, we would expect the same relationships to hold. In the RATFINC example, Mr. Jones would obtain

$$s_{\bar{x}} = \frac{s}{\sqrt{n}} = \frac{0.003412}{\sqrt{20}} = 0.00076.$$

Using this value, Mr. Jones would like to know

$$P(\bar{x} \leq 0.4992 \mid \mu = 0.500, s_{\bar{x}} = 0.00076).$$

We would expect this probability to be somewhat larger than

$$P(\bar{x} \leq 0.4992 \mid \mu = 0.500, \sigma_{\bar{x}} = 0.0007),$$

because of the additional uncertainty caused by having $s_{\bar{x}}$ instead of $\sigma_{\bar{x}}$. That is, with everything else the same (ceteris paribus, if you took Latin in high school) we would expect the probability of values far below the mean (or values far above the mean) to be larger when $\sigma_{\bar{x}}$ is unknown. The $\bar{x}$ values do not cluster as closely around μ.

Through this discussion we assume that $\mu = 0.500$. The reason we can do this is that we want to test the idea (hypothesis) that $\mu = 0.500$ is true. To do so we first assume it is and then see how likely the $\bar{x}$ value is. Perhaps this sounds backwards, but that is the way it is done. This approach is discussed extensively in Chapter 12.

The notion of $\bar{x}$ values being less tightly bunched (more disperse) about the mean can be shown graphically. Figure 10-1 shows one probability

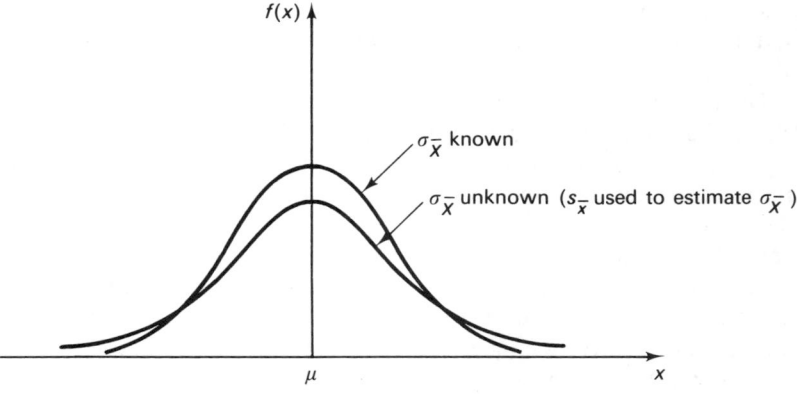

Figure 10-1 Two Probability Density Functions for the Distribution of Sample Means

density function for $\bar{x}$ values when $\sigma_{\bar{x}}$ is known and another for the case when it is unknown and only $s_{\bar{x}}$ is available. The curve with $\sigma_{\bar{x}}$ known has more area centered around the mean than the curve with $\sigma_{\bar{x}}$ unknown.

In both cases, for a given sample size, we expect to get values of the sample mean close to the universe mean, μ. As the sample size increases, $\sigma_{\bar{x}}$ and $s_{\bar{x}}$ decrease, so both curves in Figure 10-1 become more tightly concentrated around the universe mean. Moreover, as n increases, the two curves become closer and closer to one another. This implies that when n is large enough, we can use the two distributions interchangeably. Large enough will be defined more precisely after we examine the distribution of sample means when $\sigma_{\bar{x}}$ is unknown.

When $\sigma_{\bar{x}}$ is known, the Central Limit Theorem says that

$$z = \frac{\bar{x} - \mu}{\sigma_{\bar{x}}}$$

is approximately normally distributed with mean zero and standard deviation 1 regardless of the distribution of the universe. (z is exactly normal if $\bar{x}$ is normally distributed.) That is how we find probabilities such as

$$P(\bar{x} \leq 0.4992 \mid \mu = 0.500, \sigma_{\bar{x}} = 0.0007) = 0.1271.$$

When $\sigma_{\bar{x}}$ is unknown, we are dealing, instead, with $(\bar{x} - \mu)/s_{\bar{x}}$, which we will denote by t. That is,

$$t = \frac{\bar{x} - \mu}{s_{\bar{x}}}.$$

As we can infer from the fact that different pictures (see Figure 10-1) result for the $\sigma_{\bar{x}}$ known and unknown cases, $(\bar{x} - \mu)/s_{\bar{x}}$ does not follow the same distribution as $(\bar{x} - \mu)/\sigma_{\bar{x}}$ and, hence, is not normally distributed.

In fact, $(\bar{x} - \mu)/s_{\bar{x}}$ follows the *Student's t distribution*, or simply the *t distribution*, a distribution we have not yet studied. ("Student" was the name that W. S. Gosset used when introducing this distribution in 1908.) The t distribution, like the normal, is continuous and symmetrical. Unlike the normal, its shape depends on the number of data values used to compute the sample mean.

In particular, the lower curve in Figure 10-1 illustrates, in the $\sigma_{\bar{x}}$ unknown case, the t distribution for an $\bar{x}$ computed using 10 data points. If there were 20 data points, the t distribution would look more like the normal. The t distribution more nearly resembles the normal for larger numbers of data points. In the limit, with an infinitely large sample, the two distributions are identical. This is shown in Figure 10-2.

Sampling Distributions

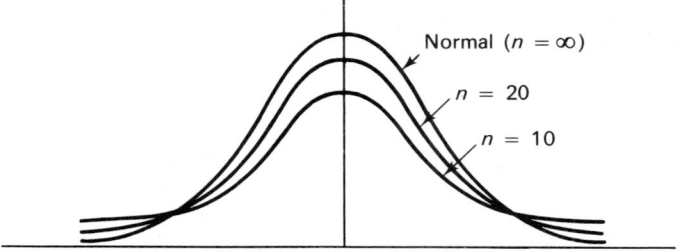

Figure 10-2 t distribution for Three Different Degrees of Freedom

Because the distribution differs for each sample size, a separate table is required for each value of n. However, so that the t table can be placed on one page, we choose to concentrate the many tables onto one page by giving only a few selected probability values for each sample size. For a very large sample size, the normal distribution can be used as an approximation to the t distribution, so we need not list high values of n in the table. The point at which the difference can be ignored and the normal used in place of the t distribution depends on the accuracy required in the problem at hand. As a rule of thumb, we will say that a sample size over 30 is sufficiently large that the normal distribution can be used to approximate the t distribution.

In order to compute probabilities using the t distribution, we need $\bar{x}$ and $s_{\bar{x}}$, and we need to consult Table II in the Appendix. This table gives the number of sample standard deviations, t values, away from the mean necessary to contain a given amount of probability between the mean and the critical value of $\bar{x}$. The tail probability (as with the normal) equals 0.5 minus the table reading. The probability in either tail is the same, since the t distribution is symmetric. Thus, we can find the probability under any portion of the curve from the table.

There is one t distribution for each sample size, and the t table has one line for each t distribution. However, the value $n - 1$ is listed instead of n, and this is called "degrees of freedom." The notion of degrees of freedom is difficult (although basically the degrees of freedom is the number of data values free to vary). We can proceed here without a detailed knowledge of the concept, and a brief discussion is given later in this section.

An abbreviated t table is given in Table 10-2 for easy reference. The degrees of freedom (d.f.) are shown on the left-hand side. For the purpose of studying the distribution of an arithmetic mean, $\bar{x}$, when σ_X is unknown, the degrees of freedom always equals $n - 1$. (The appropriate degrees of freedom does not equal $n - 1$ in some other situations involving the t distribution, however, so d.f. is listed instead of sample size.) Probability (between the mean and the stated number of t values) is given along the top of Table 10-2.

Using any table is tricky the first time; using any table is easy once we understand it. A few examples, using Table 10-2, may help to make the *t* table easy to use.

Table 10-2   Abbreviated *t* Table: The Area Between the Mean and +*t*

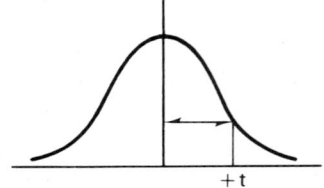

d.f.	Probability					
	0.25	0.40	0.45	0.475	0.49	0.495
1	1.000	3.078	6.314	12.706	31.821	63.657
2	0.816	1.886	2.920	4.303	6.965	9.925
⋮	⋮	⋮	⋮	⋮	⋮	⋮
19	0.688	1.328	1.729	2.093	2.539	2.861
⋮	⋮	⋮	⋮	⋮	⋮	⋮
30	0.683	1.310	1.697	2.042	2.457	2.750
∞	0.674	1.282	1.645	1.960	2.326	2.576

If $\bar{x} = 4.0$, $s_{\bar{x}} = 2.0$, and $n = 31$, then

1. Degrees of freedom = d.f. = 30.
2. The *t* table is applicable.

$$P(\bar{x} \geq 4.0 \mid \mu = 0, s_{\bar{x}} = 2.0) = P\left(t \geq \frac{4.0 - 0}{2.0} \mid \text{d.f.} = 30\right)$$

$$= P(t \geq 2.0 \mid \text{d.f.} = 30)$$

$$\approx 0.025.$$

The 0.025 probability is 0.5 − 0.475, and the 0.475 column is chosen because 2.042, the table reading for 0.475 and d.f. = 30, is close to the 2.0 value desired. Actually, we are only sure that the probability desired is between 0.025 and 0.05. A more complete set of tables would show that the

SAMPLING DISTRIBUTIONS

probability is, indeed, very close to 0.025. Thus, we are (roughly) $97\frac{1}{2}$ percent sure that the universe mean, μ, is greater than zero. The chance of obtaining an $\bar{x}$ value greater than or equal to 4.0 when $\mu = 0$ is only about $2\frac{1}{2}$ percent.

For some other examples, we can make the following statements based on Table 10-2. When d.f. = 1,

$$P(t \geq 1.0) = 0.50 - 0.25 = 0.25$$

$$P(t \geq 2.0) \text{ is between } 0.25 \text{ and } 0.10$$

$$P(t \geq 3.078) = 0.50 - 0.40 = 0.10.$$

If $\bar{x} = 3.5$, $s_{\bar{x}} = 0.5$, and d.f. = 1,

$$P(\bar{x} \geq 3.5 \mid \mu = 0, s_{\bar{x}} = 0.5 \text{ and d.f.} = 1) = P(t \geq 7 \mid \text{d.f.} = 1) < 0.05.$$

The last probability is between 0.05 and 0.025, and is much closer to 0.05. Using the table at hand, this is all that we can say.

Several important comments must be made about the t distribution and the t table if we are to use them correctly. First, interpolation between table values is often unnecessary, since we seek only an approximation to the probability. For example, knowing that the probability is ≤ 0.05 may be sufficient. If more accuracy is required, there are methods of (nonlinear) interpolation which we will not discuss, or a more complete set of tables can be consulted. The need for either of these will be rare.

Second, $(\bar{x} - \mu)/s_{\bar{x}}$ is theoretically a t distribution only if $\bar{x}$ is normally distributed. This is definitely true if the original universe from which the sample is drawn is normally distributed. It is approximately true, by the Central Limit Theorem, for any sufficiently large sample. In practice, this means that we can use the t distribution when σ_X is unknown for even small n values if the universe is unimodal and smooth, as described in Section 10-2.

Third, the difference between the t distribution and the normal can be seen in the table. When d.f. $= \infty$, the tabulated values shown are for the normal distribution. The differences between the tabulated values for d.f. $= 30$ and d.f. $= \infty$ (the normal) are small, although the difference is larger as we get farther out into the tail. (For example, examine 2.750 versus 2.576, using the 0.495 column, as opposed to 0.683 versus 0.674, using the 0.25 column.) The differences between the t value and the z value (normal) are quite large when the degrees of freedom for the t distribution is small.

Theoretically, the t distribution is the correct distribution to use any time $\bar{x}$ is normally distributed and we have only $s_{\bar{x}}$ instead of σ_X. Since the differences are slight when d.f. > 30 (or so), the normal distribution can be used as an approximation of the t distribution if $\bar{x}$ is normally distributed and d.f. > 30.

Finally, the notion of degrees of freedom, which we will see in several other parts of this book, refers to the effective size of the sample. It is based on the number of the sample values that are free to vary. If we know that $\bar{x} = 40$ and $n = 5$, only 4 of the x values are free to vary. The fifth value must cause the 5 values to average to 40. That is, there are 4 degrees of freedom in the data, not 5. If, for example, $x_1 = 30$, $x_2 = 40$, $x_3 = 50$, and $x_4 = 50$, then $x_5 = 30$ is required if the 5 values are to average to 40.

Another way to say this is that each calculation or conclusion based on the data leaves less freedom in the data; conclusions after the initial ones are based on a smaller set of information. When we calculate $\bar{x}$, some of the freedom is removed, and the effective size of the sample is reduced for any further calculations. Thus, we subtract one degree of freedom to obtain $n - 1$. In other situations, examined later in the book, it will be necessary to subtract more than one degree of freedom. For the purpose of obtaining a distribution of $\bar{x}$, $n - 1$ will always be the number of degrees of freedom.

As a final example of the t distribution, we return to the RATFINC data. With $n = 20$, they computed $\bar{x} = 0.4992$ and $s_{\bar{x}} = 0.00076$. We can now find:

$$P(\bar{x} \leq 0.4992 \mid \mu = 0.500, s_{\bar{x}} = 0.00076, \text{d.f.} = 19)$$

$$= P\left(t \leq \frac{0.4992 - 0.500}{0.00076} \,\middle|\, \text{d.f.} = 19\right)$$

$$= P(t \leq -1.05 \mid \text{d.f.} = 19)$$

$$= P(t \geq 1.05 \mid \text{d.f.} = 19).$$

This probability is somewhere between $0.50 - 0.25 = 0.25$ and $0.50 - 0.40 = 0.10$, since $t_{0.25} = 0.688$ and $t_{0.40} = 1.328$ for 19 degrees of freedom. Without a more complete table, we cannot obtain a more precise value. We can note that an $\bar{x}$ value this low is not a rare event with $\mu = 0.500$ and $s_{\bar{x}} = 0.00076$, since there is more than a 10 percent chance of such an event. This probability is larger than the corresponding normal probability. Even though we cannot obtain the precise value here, the fact that the tail probabilities are larger for the t distribution can be seen in general by looking at the t table and noting that the numbers in any column are all larger than the value for d.f. $= \infty$, which is the value for a normal distribution. That is, to leave any specified area in the tail, we must move more t values from the mean than the number of z's we would use if the normal were appropriate.

S.S. An advertising manager is interested in testing the effect of new advertising on sales, and has computed the average increase in sales from last month to this month in 25 cities. The manager has computed $s = 120.0$. How large must the sample mean be if the probability of an $\bar{x}$ value that large or larger, given $\mu = 0$, is to be equal to 0.05? Equal to 0.01?

Solution: First, $s_{\bar{x}} = s/\sqrt{n} = \frac{120}{5} = 24$. For d.f. $= 25 - 1 = 24$, $t = 1.711$ leaves 0.05 in one tail, and $t = 2.492$ leaves 0.01 in one tail. Thus, we need

$$\bar{x} = \mu + ts_{\bar{x}} = 0 + 1.711(24) = 41.06$$

in order to have

$$P(\bar{x} \geq 41.06 \mid \mu = 0, s_{\bar{x}} = 24, \text{d.f.} = 24) = 0.05.$$

This calculation is correct because we want $t = 1.711$ to obtain a 0.05 probability. Thus, we want to be $(1.711)s_{\bar{x}}$ above the mean; the mean is assumed to be zero. For a 0.01 probability,

$$\bar{x} = 0 + 2.492(24) = 59.81.$$

To check our understanding, we can ask the question in the reverse manner, finding

$$P(\bar{x} \geq 59.81 \mid \mu = 0, s_{\bar{x}} = 24, \text{d.f.} = 24).$$

This is

$$P\left(t \geq \frac{59.81 - 0}{24} \,\Big|\, \text{d.f.} = 24\right) = P(t \geq 2.492 \mid \text{d.f.} = 24)$$

$$= 0.5 - 0.49 = 0.01.$$

10-4 A Warning

Throughout this chapter we have dealt with approximations. The finite population correction factor (discussed in Technical Note 10-1) can be ignored if the sample constitutes a sufficiently small fraction of the population. The Central Limit Theorem states that the distribution of $\bar{x}$ tends to normality as n gets sufficiently large regardless of the universe distribution. The t distribution can be used as the distribution of $(\bar{x} - \mu)/s_{\bar{x}}$ when $\bar{x}$ is normally distributed and we have only $s_{\bar{x}}$, not $\sigma_{\bar{x}}$. Finally, for n sufficiently large, the normal distribution can be used as an approximation to the t distribution.

All these approximations may leave us a little uncomfortable, and that is good. We should be aware of an approximation when it is used, and we should know how far it can lead us astray. It is perfectly reasonable to use approximations, but if a decision is to be based on the statistics, we should be sure that the data strongly imply one course of action, or if not, study the issue more carefully.

The material in this chapter is used frequently in the next several chapters. To understand this new material, we must understand the Central Limit Theorem and what it tells us about the distribution of the sample means. We should know when the t distribution is appropriate, and we should feel comfortable with both the t table and the normal table.

Finally, all the information on sampling distributions is based on an assumption of random sampling from a single universe. That is, we must not take some readings from one universe and some from another. This, unfortunately, is a matter of art, not science. It requires practice or expert assistance. Moreover, a leap of faith is often needed, since all statistical conclusions are based on that assumption. If the assumptions about sampling do not hold, the conclusions *may* be false. (They may, of course, still be true, but we can have no confidence in them.) In the RATFINC example, introduced in Section 10-1, we saw that a sample should be taken in a relatively short period of time to avoid a population change. At the same time, we want to take enough data to reduce the $s_{\bar{x}}$ value. Those two conflicting goals must be intelligently balanced. In the RATFINC case, as in all use of statistics, intelligent data taking, as well as data analysis, is essential if the results and conclusions are to be valid.

TECHNICAL NOTE 10-1: FINITE POPULATION CORRECTION FACTOR

One modification of the $\sigma_{\bar{x}} = \sigma/\sqrt{n}$ result that should be introduced here involves the case of a finite and relatively small population. If a finite population is being sampled without replacement, *a finite population correction factor* may be required. This is needed, for example, when the entire population is sampled. In that case, $\bar{x}$ is the universe mean, and there is no variability in $\bar{x}$ ($\sigma_{\bar{x}} = 0$). The formula (10-1) does not yield $\sigma_{\bar{x}} = 0$ in that case. Thus, formula (10-1) overstates the variance when the population is finite.

When a sample of size n is taken from a population of size N, the relationship between $\sigma_{\bar{x}}$ and σ is

$$\sigma_{\bar{x}} = \frac{\sigma}{\sqrt{n}} \sqrt{\frac{N-n}{N-1}}. \qquad (10\text{-}4)$$

The factor $\sqrt{(N-n)/(N-1)}$ is called the *finite population correction factor*.

The finite population correction factor is approximately the square root of the fraction of the population that is not sampled (as can be seen by ignoring the -1 in the denominator). If, for example, 0.05 of the population is sampled, then the finite population correction factor is $\sqrt{0.95} = 0.9747$.

Sampling Distributions

As you can see, if 5 percent or less of the population is sampled, the finite population correction factor is very close to 1, and can safely be ignored. Theoretically, it should be used for any finite population, but it can, for practical purposes, be ignored if the sample is less than 5 percent of the population. (Hence, Mr. Jones, in the RATFINC example, can ignore it given a production rate of 1,250 per day and a sample size of only 20.)

Key Formulas

Formula	Used to Compute:
$\sigma_{\bar{x}}^2 = \dfrac{\sigma^2}{n}$	Universe variance of the sample means: $\sigma_{\bar{x}} = \dfrac{\sigma}{\sqrt{n}}.$
$s_{\bar{x}}^2 = \dfrac{s^2}{n}$	Variance of the sample means estimated from a sample $s_{\bar{x}} = \dfrac{s}{\sqrt{n}}.$
$t = \dfrac{\text{value} - \text{mean}}{\text{standard deviation}}$	The standardized t statistic, used when the sample estimate is used for the standard deviation.

PROBLEMS

10-1.* Even though the data-generating process produces data that are not normally distributed, we sometimes use the normal distribution in analyzing the data. Why is that acceptable?

10-2.* In taking a sample from a batch of items, a quality control inspector characterizes the items as good or bad. The data from a particular inspection of a sample of 100 units from a very large batch of items might be: 2 bad, 98 good.
(a) What kind of sample is being taken?
(b) If the batch really has 5 percent defective, how likely is a sample value of 2 or fewer defectives from a sample of 100 items? (This type of question will be asked in Chapter 12.)

10-3. How many data values are necessary before we can safely use the normal distribution to describe the distribution of sample means?

10-4. When is the t distribution the (exact) theoretically correct distribution to use? When we use the t distribution to describe the distribution of sample means, do we need to appeal to the Central Limit Theorem?

10-5.* The producer of "Wurst Hotdogs" is concerned about the weight of its 2-pound package of hotdogs. Each 2-pound package has 16 hotdogs in it, each supposedly weighing $\frac{1}{8} = 0.125$ pound. They know the stuffing machine has $\sigma = 0.005$ pound on each hotdog. They can set the mean value, and they have decided to use $\mu = 0.126$.
(a) What is the mean and standard deviation of the weight of one 2-pound package?
(b) What is the probability that one package will weigh less than 2 pounds?
(c) What is the probability that the mean of a sample of 25 packages will be less than 2 pounds? What if an average of four packages is taken?

10-6. A coal mining company is interested in estimating the total usage of coal by public-institution heating systems in their state. (Public institutions include schools, universities, courthouses, and so on.) There are currently 2,000 public-institution buildings heated by coal in their state. They obtained last year's usage for 400 of these users and obtained $\bar{x} = 127$ tons and $s = 30$.
(a) Compute $s_{\bar{x}}$.
(b) Find the expected value of total annual usage last year.
(c) If $\mu = 125$, how likely were they to see an $\bar{x} \geq 127$? What does this imply about $\mu = 125$?
(d) Is the data they have relevant to this year's usage?

10-7. In Problems 10-5 and 10-6, why did the solution find probabilities using the normal distribution?

10-8.* During an investigation of the cost of a particular type of service (such as TV or car repair), a public agency obtained the identical service from 36 firms and received 36 bills, where $\bar{x} = \$45.75$ and $s_{\bar{x}} = \$1.10$. The median bill was $\$42.00$.
(a) What is the probability of $\bar{x} \geq \$45.75$ if $\mu = \$45.00$?
(b) The highest bill was $\$70.00$. What is the probability, if it can be obtained, of a value that high when the mean is $\$45.75$? What assumption about the universe of bills is required to obtain an answer? Is the assumption reasonable?
(c) In which direction is the frequency distribution skewed?

10-9. Problem 10-8 can be solved if we assume that $(\bar{x} - \mu)/s_{\bar{x}}$ is normally distributed.

SAMPLING DISTRIBUTIONS

(a) What assumptions are necessary for that assumption to be valid?
(b) What assumption is unnecessary if the t distribution is used?

10-10. A paper cutting machine produces rolls of paper in 12-inch widths with a standard deviation of 1 inch. On a certain day, 225 rolls are cut.
   (a) What is the standard deviation of the total width cut during the day's production?
   (b) What is the standard deviation of the mean width cut during the day?
   (c) What assumption is needed in answering (a) and (b)? Is the assumption reasonable?

10-11.* A trucking company runs 500 trips per year between two major cities. In order to determine the average amount of travel time between the two cities for the most recent year, the manager of the company selected 100 driver logs from the 500 total. If the true average time on the 500 logs is 30 hours with a standard deviation of 2.5 hours, what is the probability that the sample average will be greater than 32 hours?

10-12.* The manager of a plant that manufactures yarn used for knitting sweaters has found an additive to the yarn which should increase the mean breaking strength by 50 percent without affecting the standard deviation. From past studies, the breaking strength of the untreated yarn can be well approximated by a normal distribution with mean of 10 pounds and a standard deviation of 2 pounds. A random sample of 100 lengths of yarn were treated with the new material and the mean breaking strength was found to be 13 pounds. What is the probability of finding a sample mean at least this low if the manager's beliefs are correct?

10-13. A quality control manager must devise a sampling procedure for production batches containing 25,000 items. She wants to sample 500 of these items and categorize each as good or bad. She has decided to accept the batch if 50 or fewer bad items are found. For one particular batch the manager found exactly 50 items, so $p' = 0.10$. (p' can be thought of as a sample mean as well, so $\bar{x} = 0.10$.)
   (a) Does the Central Limit Theorem apply here?
   (b) Given $p' = \bar{x} = 0.10$, and assuming that

$$\sigma_{\bar{x}} = \sigma_p = \sqrt{\frac{p'(1-p')}{n}} = 0.0134,$$

how likely is an $\bar{x} \geq 0.10$ if $\mu = p = 0.10$? If $\mu = p = 0.08$? If $\mu = p = 0.12$?

10-14. In order to determine the potential for increased sales, a marketing consultant for Flavorful Beverage Company wants to determine what percentage of a city's population drink Flavorful Soda.
  (a) What kind of sampling procedure would you suggest in order to determine this percentage?
  (b) Assuming that 40 percent of the city's population drink Flavorful Soda, what is the probability that from a random sample of 100, more than 50 drink the soda?

10-15.* A chain of retail stores consist of five establishments, one in each of five cities. Each store has 20 employees who are paid weekly salaries plus bonuses at year's end based upon the total dollar value of their sales. In order to plan effectively for this year's commission payments, the manager of the chain decided to select a sample of employees in a typical sales month to estimate the total annual sales. From previous years, the manager believes that the mean monthly sales for any employee is $5,000 and that the variance is the same for each employee. The manager selected a sample of 25 employees at random. The mean sales for the sample was $5,500, and the sample standard deviation was $1,000. What is the probability of obtaining a sample mean at least as great as $5,500 if the true mean is $5,000?

10-16. The mean number of bottles broken per day during filling in a soft drink bottling plant is 550 with a standard deviation of 75. Each bottle that is broken must be replaced by the company at a cost of $0.02. What is the probability that during a period of 100 working days the average replacement cost per day is more than $10.60?

10-17. A chemical company orders large quantities of XBT, which it uses as raw material for producing a variety of chemical compounds. The chemical is delivered in shipments composed of several crates of 100-cc flasks. The slightest variation in the chemical properties of XBT can cause disastrous results in the production process. Therefore, for each shipment, a random sample of flasks is drawn for testing. The effectiveness of the chemical is measured by the volume of precipitant that collects on the bottom of a calibrated tube when XBT is subjected to certain combinations with other substances. The volume of precipitant that results from a 100-cc volume of "good" XBT is known to have a mean of 4.5 cc and a standard deviation of 1.5 cc. Shipments of XBT are rejected in their entirety if a random sample of 100 flasks fails to produce a mean precipitant volume within the range 4.4 to 4.6 cc. What is the probability that a shipment of good XBT will be rejected?

Sampling Distributions

10-18. A machine used by Good-Grain Food Company fills 10-ounce boxes of cereal. The filling process averages 10 ounces, with a standard deviation of 2 ounces, and the distribution of the amount of cereal is approximately normal. What is the probability that from a random sample of n boxes drawn from a day's very large production, the average net weight is greater than 11 ounces if
(a) $n = 1$?
(b) $n = 9$?
(c) $n = 25$?
(d) $n = 36$?
(e) $n = 100$?

10-19.* In a cannery a brand of tomatoes is packed in 16-ounce cans for distribution to supermarkets. The filling machine must be periodically checked to make certain that cans are indeed filled with 16 ounces of tomatoes. When the filling machine is properly adjusted, the mean amount of tomatoes packed into cans is 16 ounces, and the standard deviation is 1 ounce. The plant manager has drawn a sample of size 36 to determine if the machine needs adjustment. If the machine is properly adjusted:
(a) What is the probability that the sample average will be greater than 16.25 ounces?
(b) What is the probability that the sample fails to yield an average of at least 15.9 ounces?
(c) What is the probability that the sample average is between 15.5 and 16.5 ounces?

10-20. A marketing manager for a furniture company has weekly sales records on one product line for a 10-week period. The data yield $\bar{x} = 111.7$ and $s = 25.68$. They would like to know if average weekly sales is above 100. They want to know how likely a value as high as the $\bar{x}$ value observed is if $\mu = 100$. Is there anything else they should consider here?

10-21. Consider the following two probabilities, using the t distribution:
(a) $P(t \leq -1 \mid \text{d.f.} = 4)$
(b) $P(t \leq 1 \mid \text{d.f.} = 6)$
For each of these, find the two probability values between which the true probability lies.

10-22. A consumer organization is testing the gasoline mileage obtained by a particular car. They drove 100 miles in each of 9 cars of this type and found that $\bar{x} = 16.8$ and $s = 1.35$. The organization knows that 16.8 is their best guess as to the true mean, but they would like to say more. For example, if $\mu = 16.8 + 1(1.35/\sqrt{9}) = 17.25$, then $P(\bar{x} \le 16.8 \mid \mu = 17.25, s_{\bar{x}} = 0.45) = P(t \le -1 \mid \text{d.f.} = 8)$. However, the research director does not like to interpolate, and so reads the table for d.f. $= 8$ and notices that $P(t \le -1.397 \mid \text{d.f.} = 8) = 0.50 - 0.40 = 0.10$. Thus, $P(\bar{x} \le 16.8 \mid \mu = 16.8 + 1.397(0.45) = 17.42, s_{\bar{x}} = 0.45, \text{d.f.} = 8) = 0.10$.

(a) What value of μ allows us to say: $P(\bar{x} \ge 16.8 \mid \mu, s_{\bar{x}} = 0.45, \text{d.f.} = 8) = 0.10$?

(b) Use the 1.860 table reading in the d.f. $= 8$ line of the t table to make statements as made above (using both $\bar{x} \ge 16.8$ and $\bar{x} \le 16.8$).

10-23. A paper manufacturer produces long rolls of paper and then cuts the rolls into desired widths. As a result of this process, a certain amount of each large roll becomes scrap. From a very large stock of 20-foot rolls, the scrap from a sample of 9 is drawn at random and measured. The average length for the sample is 1.1 inches and the sample standard deviation is 0.5 inch. What is the probability that the population mean differs from the sample mean by more than 0.5 inch if it is known that the sample is drawn from a normal distribution?

10-24.* In order to determine the relative effectiveness that certain materials contribute to durability, a tire manufacturer is conducting a durability test using 24 different types and quantities of materials. The durability tests are conducted using six different automobiles running over a test track. The tires are assigned to cars at random. Assume that types of material, quantities of material, and differences in cars have no effect on the life of the tires; that is, assume that all tires are drawn from the same population. If the life of each tire is independent of the lives of all other tires and the life of each tire is normally distributed with mean of 30,000 miles and standard deviation of 1,000 miles,

(a) What is the probability that the sample average tire life for two specific cars differs by more than 500 miles? (*Hint:* The difference between two normally distributed random variables is, again, a normally distributed random variable.)

(b) What is the probability that the lives of two specific tires differs by more than 500 miles?

10-25. From past experience, the auditors for Best Buy Company believe that the accounts receivable balances for the firm are normally distributed with a mean outstanding balance of $1,000 and a standard deviation of $400.

(a) If one account is selected at random, what is the probability that the outstanding balance is between $900 and $1,100?

(b) If a random sample of 100 accounts is selected at random, what is the probability that the sample mean is between $900 and $1,100?

(c) Determine A such that if 100 accounts are selected at random, the probability that the sample mean exceeds A is 0.025.

10-26. A random sample of 25 steel rods is drawn from a day's production. The average length of the rods is 16.5 inches and the standard deviation is 0.25 inch. Assuming a normal distribution, between what two values, symmetric about the sample mean, should the true mean for the production process lie with 0.90 probability?

10-27. An airline is conducting a test to determine the average length of telephone calls to the reservations desk. The sample average is to be within 0.5 minute of the population average with 95 percent probability. Previous studies have shown that the standard deviation of the length of calls is 1 minute and normality is a good approximation. What is the smallest sample that should be drawn to preserve the probability requirement? How might your answer change if it were not known that normality provided a good approximation?

10-28.* The output from a manufacturing process is normally distributed with a standard deviation of 10 inches.

(a) How large a sample should be selected in order for the sample mean to lie within 1 foot of the population mean with 80 percent probability?

(b) Repeat part (a) for 99 percent.

Chapter **11**

Decision Making: Estimation

In Chapter 10, we repeatedly spoke about $\bar{x}$ as the best estimate of the universe mean. We also spoke of s as being the best estimate of the universe standard deviation. In this chapter, we will be more precise about what "best" means. We will also discuss a method for placing an interval around the estimate, within which we can have a given degree of confidence that the universe value lies. In announcing an estimate, we have made the decision that the announced value (or interval) is our best estimate of an unknown universe parameter. For that reason, this chapter begins the section of this book explicitly devoted to decision making. This chapter is the first step toward obtaining methods for analyzing decision alternatives.

Estimating universe parameters is important in almost every type of organization, and examples have appeared in many of the foregoing chapters. For example, the proportion of persons with access to TV who are watching a particular program is a universe parameter that can be estimated by a telephone sample. This parameter would be useful to someone making advertising allocations. The average income of the members of a particular professional organization can be estimated by obtaining true income figures (if possible) from a random sample of the membership, and this average might be used to establish membership fees. The average time required to change the ball joints on an automobile can also be estimated using data from several such repairs, and this figure might be used to set a standard fee for that operation.

In all these cases, decisions are likely to be based on the estimate. Since that is so, we are interested in the characteristics of the estimate. As an

Decision Making: Estimation

example, one desirable characteristic is: as a larger and larger sample is taken, we get more confident that the estimate is close to the universe parameter. (One would certainly hope that this is so, and the estimates studied here behave in this manner.)

We will be interested in how close the estimate is to the true parameter value. Unless we sample the entire universe (which is usually impossible), we can never be sure of the universe parameter value, and, therefore, of how close some estimate is to that value. There always will be some uncertainty. However, we can be precise about how uncertain we are. (There's a good phrase for you.) An example of a statement regarding the uncertainty around an estimate of the sample mean is: We are 95 percent confident that the universe parameter value is between 8 and 12, and 10 is the best single-valued estimate.

In this chapter, we will examine both the characteristics of estimators and the uncertainty surrounding an estimate. Before we begin, a few preliminary definitions are necessary. The notion of a *statistic* has been used in previous chapters, but we have not defined the idea. The notions of *point estimates* and *interval estimates*, which are defined following the discussion of statistics, are contained in the preceding paragraph. Again, definitions will be helpful.

> A *statistic* is a value calculated from the values in a sample; it is used to summarize the data in the sample. A statistic is often used to estimate the related universe parameter.

The sample mean and the sample standard deviation are examples of statistics. In Chapter 3 these statistics (and others) are used to summarize the data in a sample. In this chapter we will use the values of these statistics (or *estimators* as they are sometimes called) to estimate the related universe parameters. In particular, the sample mean will be used to estimate μ, and the sample variance will be used to estimate σ^2. We will combine $\bar{x}$ and s values in a particular way to obtain an *interval estimate*, within which we have a stated degree of confidence that μ lies, whereas an $\bar{x}$ value by itself will be a *point estimate*. The arithmetic mean is an *estimator* of μ. (A particular $\bar{x}$ value gives an estimate of the particular μ.) The sample variance is an estimator of σ^2. (A particular s value gives an estimate of the particular σ.)

> A *point estimate* is a single-valued estimate of a universe parameter. It can be thought of as the best guess as to the universe parameter value.
>
> An *interval estimate* is a range of values for a universe parameter value within which we have a stated degree of confidence (say 90 percent) the universe parameter value lies.
>
> An *estimator* of a universe parameter is a statistic (such as the sample mean, $\bar{x}$) whose value is used to estimate a universe parameter (such as μ).

Thus, if $\bar{x} = 10$ and $\sigma_{\bar{x}} = 1$, we might say that $\bar{x} = 10$ is the point estimate of the universe mean. The interval from 8 to 12 ($\bar{x} \pm 2\sigma_{\bar{x}}$) is called an interval estimate for the mean, and, if the distribution of sample means is normal, we can be slightly over 95 percent sure that this interval contains μ. (The mean plus or minus 2 standard deviations contains 95.45 percent of the distribution.) Many different interval estimates, each with its own associated probability, can be given. That is, a different level of confidence (greater confidence for a larger interval) can be ascribed to different interval estimates. This notion is refined later in this chapter when *confidence intervals* are discussed. Before discussing confidence intervals, however, we will discuss the characteristics of estimators to see, for example, why $\bar{x}$ is chosen as the estimator of μ, the true universe mean.

11-1 Characteristics of Estimators

If you were an estimator, you certainly would want to be unbiased, efficient, and consistent. As a matter of fact, that sounds like a good list of characteristics for people to have. Unfortunately, the statistical definition of these terms is different from their common meanings. But they are as desirable for estimators as the common meanings of the words are for people. The statistical definitions are given in this section.

> A statistic is an *unbiased* estimator of a universe parameter if the expected value of the statistic is equal to the universe parameter.

For example, since $E(\bar{X}) = \mu$, the sample mean is an unbiased estimator of μ. A biased estimator is one that on the average will not equal the universe parameter. (This is different from the concept of sampling bias, which was introduced in Chapter 9 and which means that any estimate involves uncertainty.) The median, for example, is a biased estimator of the universe mean unless the universe distribution is symmetric. As discussed in Chapter 3, there still may be situations where the median is the preferred measure of location. In general, however, unbiased estimators will be used throughout the remainder of this book and in most applications of statistics to actual decision problems.

> A statistic, used as an estimator of a universe parameter, is more *efficient* than an alternative statistic, used to estimate the same parameter, if, for the same sample size, it has a smaller standard deviation around the parameter.

Decision Making: Estimation

For example, for a particular sample size, n, we could investigate the standard deviation of the arithmetic mean $\bar{x}$ and of the median M. We know that $\sigma_{\bar{x}} = \sigma/\sqrt{n}$, and it is also true that $\sigma_M = 1.2533(\sigma/\sqrt{n}) = 1.2533\sigma_{\bar{x}}$, although we will not demonstrate this here. Thus, we know that, since $\sigma_{\bar{x}} < \sigma_M$, the sample mean is more efficient as an estimator of μ than is M. Even for a normal distribution, in which case the median is an unbiased estimator of the universe mean, the sample mean is more efficient.

The concepts of bias and efficiency are illustrated in Figure 11-1, which shows sampling distributions for some estimators.

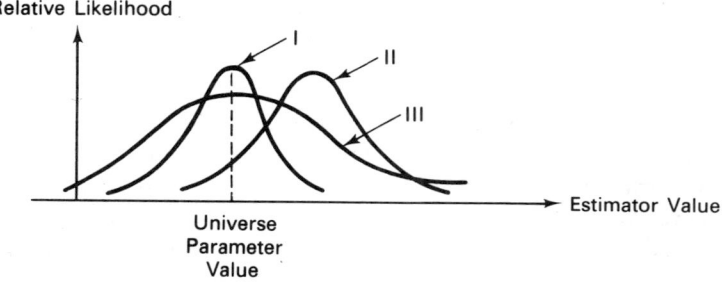

Figure 11-1 Sampling Distributions of Some Estimators

Estimator I is an unbiased estimator, and it is more efficient than estimator III, which is also unbiased. Estimator II is a biased estimator, since its expected value is not coincident with the parameter value.

> A statistic is a *consistent* estimator if, as the sample size increases, the probability that the sample estimate is very close to the universe parameter value approaches 1.

What this means is that the probability that the estimate is within any specified distance, say, 0.01 of the universe parameter value, approaches 1 as n approaches infinity. This statement is true for a consistent estimator even if 0.001 or 0.0001 (or any other number) is used in place of 0.01. However close we want to be, there is a sample large enough to get us that close with a probability of nearly 1 if the estimator is consistent. A consistent estimator, then, is one that gets better and better as the sample size gets larger and larger.

In the case of the sample mean, when σ is known, we can see that the sample mean is a consistent estimator of μ because $\sigma_{\bar{x}} = \sigma/\sqrt{n}$. As n gets large, $\sigma_{\bar{x}}$ gets very small, and we can be sure that $\bar{x}$ is close to μ. For example, if $\bar{x} = 10$ and $\sigma = 2$, then $n = 100$ implies that $\sigma_{\bar{X}} = 0.2$, while $n = 10{,}000$ implies that $\sigma_{\bar{X}} = 0.02$. By choosing n large enough, $\sigma_{\bar{X}}$ can be made as small as we wish. Since $E(\bar{X}) = \mu$, $\bar{x} - \mu$ will average zero, and we are 99.8 percent confident it will never be more than three $\sigma_{\bar{X}}$ away from zero. Three $\sigma_{\bar{X}}$ can be made as small as we wish simply by increasing n, so the sample mean is a consistent estimator.

The sample mean, as estimator of μ, has all three characteristics that we have described. Another sample statistic, the sample variance, is often used as an estimator of σ^2. The formula we have used to date (when a finite population correction factor is unnecessary) is

$$s^2 = \sum_{i=1}^{n} \frac{(x_i - \bar{x})^2}{n} = \frac{\sum_{i=1}^{n} x_i^2}{n} - \bar{x}^2. \qquad (11\text{-}1)$$

The second form of (11-1) is particularly useful in calculations.

This estimator is consistent, but it turns out to be biased. In particular, it tends to be too low; it is optimistic in that on the average it gives a low estimate of the variance. This formula was introduced since it is analogous to the σ^2 formula. It does have some other good properties, which we will not discuss here, and it is a perfectly good estimator to use. The reason the estimator in (11-1) is slightly biased low is that squared deviations are taken from the sample's mean, $\bar{x}$, rather than μ, thus measuring the variation around a statistic rather than the universe value. Further, the values in the sample are closer on the average to the sample mean than they are to μ. We can use a correction factor to remove the bias. The correction factor is easy to use, and if the logic eludes you, just remember what it is or where to find it. The unbiased estimator of σ^2, which we give as an alternative formula for computing a sample variance, is

$$s^2 = \sum_{i=1}^{n} \frac{(x_i - \bar{x})^2}{n-1} = \frac{n \sum_{i=1}^{n} x_i^2 - (\sum_{i=1}^{n} x_i)^2}{n(n-1)}. \qquad (11\text{-}2)$$

[Again the second form of (11-2) is useful in calculations.]

There is very little difference. The new formula is $n/(n-1)$ times the old formula, and this factor is unimportant when n is large. From now on, we will use the unbiased estimate of σ^2 given by (11-2) as our measure of the sample variance. As before, $s = \sqrt{s^2}$, and the $n/(n-1)$ factor is even less significant when its square root is taken. This formula for s is the appropriate formula to use in conjunction with the t table, where t distribution values are given by $t = (\bar{x} - \mu)/s_{\bar{x}}$. [In fact, we computed s from the Table 10-1 data in the RATFINC example using (11-2).] In Technical Note 11-1, a proof is given that s^2 computed using (11-2) is unbiased.

Decision Making: Estimation

One other statistic that is commonly used as an estimator is the sample proportion, p', from a binomial sample. It is the estimate of the universe proportion, p. That estimator is both an unbiased and consistent estimator of the true parameter. Actually, p' is just an arithmetic average, since the basic data values are 0 or 1, and p' is the average of those values. Thus, p' has all the characteristics that the sample mean has.

One final desirable characteristic for estimators is that of *sufficiency*. In this case, the common meaning of the word and the statistical meaning are very close. Something is sufficient if it is all that is needed. A statistic is sufficent if it contains all the information necessary for the question at hand.

> A statistic is a *sufficient* estimator of a universe parameter if it contains all the information about the parameter that is contained in the sample.

A sufficient estimator contains all the information contained in the sample, for the purpose at hand. A *minimal sufficient statistic* reduces the sample as far as possible without losing any pertinent information. The sample mean is an example of a *minimal sufficient statistic*. It contains all the information in the sample about the mean, μ. One can speak of a set of sufficient statistics as being a few statistics, computed from a sample, that contain all pertinent information. For example, the sample mean and standard deviation are sufficient to describe an entire sample if we are sure the sample is drawn randomly from a normally distributed universe. On the other hand, the first value in a sample of 10 is not a sufficient estimate of μ since it omits data in the sample relevant to estimating μ.

S.S. To show that the sample mean is unbiased and consistent, we proceed as follows.

$$\bar{x} = \frac{x_1 + x_2 + \cdots + x_n}{n}$$

$$= \frac{1}{n}(x_1 + x_2 + \cdots + x_n).$$

But from the properties of expectations discussed in Section 7-3, we know that

$$E\left(\frac{X_1}{n}\right) = \frac{1}{n}E(X_1)$$

and

$$E(X_1 + X_2) = E(X_1) + E(X_2).$$

Thus,

$$E\left(\frac{X_1}{n} + \frac{X_2}{n} + \cdots + \frac{X_n}{n}\right) = \frac{1}{n}E(X_1 + X_2 + \cdots + x_n)$$

$$= \frac{1}{n}[E(X_1) + E(X_2) + \cdots + E(X_n)].$$

But each sample is drawn from a universe with mean $= \mu$, $E(X_i) = \mu$ in each case. Hence

$$\frac{1}{n}(\mu + \mu + \cdots + \mu) = \frac{1}{n}(n\mu) = \mu.$$

Now, believe it or not, that proves $E(X) = \mu$, when each sample is drawn randomly from the same universe with mean $= \mu$. However, an odd professor we know likes to use only odd-numbered sample items in computing his average, to use as an estimate of μ. That is, he might collect 10 data points, and average readings number 1, 3, 5, 7, and 9 to use as an estimator of μ. (That is, he discards readings number 2, 4, 6, 8, and 10.) He wants to know:

(a) Is his estimator unbiased?

(b) Is his estimator more efficient than the sample mean calculated using all the values?

(c) Is his estimator consistent?

(d) Is his estimator a sufficient estimator of μ? Discuss your reasons for each answer.

Solution:
(a) Yes; if we compute $(x_1 + x_3 + x_5)/3$ for example, we can prove unbiasedness the same way as we did for the sample mean.

(b) $\sigma_{\bar{x}} = \sigma/\sqrt{n}$. If only half the readings are used, the standard deviation would be $\sigma/\sqrt{n/2}$, which is larger than $\sigma_{\bar{x}}$. Thus, his estimator is less efficient than the sample mean. Any average based on less than all the data will be less efficient than the sample mean.

(c) His estimator is consistent because eventually $\sigma/\sqrt{n/2}$ will get very small. It just does not get small as quickly as does $\sigma/\sqrt{n}$.

(d) His estimator is not a sufficient estimator because there is information about μ in the sample that is not included in the estimator.

11-2 Interval Estimation—Confidence Intervals

The previous section gave a method of obtaining point estimates for three important universe parameters: the mean, the proportion, and the standard deviation (or variance). In particular, we used the sample mean, the sample proportion, and the sample standard deviation [where $s = \sqrt{\sum (x_i - \bar{x})^2/(n-1)}$] as estimators. This section discusses interval estimates for the first two of those parameters, and the next (optional) section discusses interval estimates for the standard deviation. In addition to finding interval estimates, we will ascribe a level of confidence to the interval. In fact,

DECISION MAKING: ESTIMATION

an interval, associated with a specified level of confidence, will be called a *confidence interval*.

A *confidence interval* with a level of confidence, C, is an interval estimate such that C is the probability that the interval contains the universe parameter.

New Symbol	Meaning
C	The level of confidence for an interval estimate. Frequently, C will be stated in percent terms, such as 95 percent. It is also written as 0.95 or $C_{0.95}$.

In the introductory portion of this chapter, the statement, "We are 95 percent sure that the interval between 8 and 12 contains the universe parameter value" means that 8 to 12 is the 0.95 confidence interval based on the sample we had. We will hereafter speak of confidence intervals rather than interval estimates; the difference is that the confidence interval is an interval estimate with a specified confidence level associated with it.

In this chapter, a confidence interval with a level of confidence (say, 0.95), will always mean that there is a $(1 - C)/2$ (or 0.025 for a 0.95 confidence interval) chance that the universe parameter value is higher than the upper limit of the confidence interval, and the same probability that the universe parameter value is lower than the lower limit of the interval. That is, we will design confidence intervals that have the same probability of having too low an upper limit as of having too high a lower limit. This is done because it is common practice, and no other choice is easy to justify. In the case of a symmetric distribution of the statistic (for example, the normal or t distribution for the sample mean), the confidence interval will be centered around the sample statistic. For example, the 8 to 12 confidence interval given above is centered around $\bar{x} = 10$, and is from $10 - 2$ to $10 + 2$. For $\bar{x}$ and p' (point estimates of μ and p, respectively), confidence intervals will be designed in this manner. Section 11-3 will deal with confidence intervals for the standard deviation, where asymmetric confidence intervals are obtained.

In this section, we will discuss confidence intervals for the mean and proportion. We will begin with the mean.

CONFIDENCE INTERVALS FOR μ, THE UNIVERSE MEAN

As an example, we will use the data from RATFINC's quality control problem, introduced in Table 10-1. At that time we learned that

$$\bar{x} = 0.4992, \quad n = 20, \quad s = 0.003412.$$

Hence

$$s_{\bar{x}} = \frac{0.003412}{\sqrt{20}} = 0.00076.$$

Initially, we will assume that $s_{\bar{x}} = \sigma_{\bar{x}} = 0.00076$ to illustrate how confidence intervals are calculated. We will then relax this assumption and show what changes are required when only $s_{\bar{x}}$ is known.

We shall make two other assumptions at this point. First, we assume that the distribution of sample means can be treated as though it is normal. This assumption is justified by the Central Limit Theorem. We also shall select, arbitrarily, a confidence level of $C = 0.90$. (Confidence levels of 95 and 99 percent are also in common use. Can you imagine telling your boss you are only 50 percent confident? We are 50 percent confident that if you did you would be seeking alternative employment!)

Sufficient data from the RATFINC example is available to obtain a 0.90 confidence interval for μ. The upper limit of this confidence interval is found by computing the value for μ that makes the probability of an $\bar{x}$ as small as 0.4992, equal to $(1 - C)/2 = 0.05$ (see Figure 11-2).

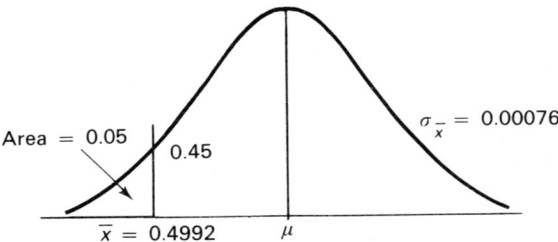

Figure 11-2 Finding the Upper Limit of a Confidence Interval

Using the formula for the standard normal deviate, z, where the subscript on z indicates the tabulated probability:

$$z_{0.45} = \frac{\text{value-mean}}{\text{standard deviation}} = \frac{0.4992 - \mu}{0.00076} = -1.645.$$

The value of $z = -1.645$ is obtained from the normal table to leave 0.05 of the area in the tail of the curve. Solving, we obtain

$$\mu = 0.4992 + 1.645(0.00076) = \bar{x} + z_{0.45}\sigma_{\bar{x}} = 0.5004.$$

This is the upper end of our confidence interval for μ.

Decision Making: Estimation

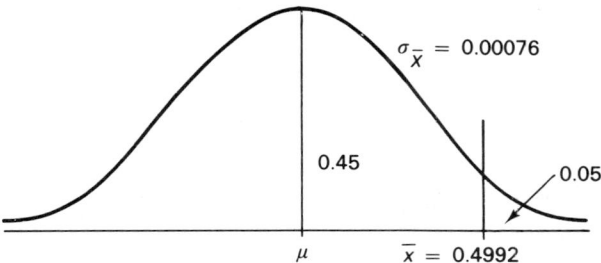

Figure 11-3 Finding the Lower Limit of a Confidence Interval

The lower end of the interval is obtained in a similar fashion. This time, however, we select a value for μ that makes the probability of an $\bar{x}$ as large as 0.4992, equal to 0.05 (see Figure 11-3). Using the formula for z again, we obtain

$$z = \frac{\text{value} - \text{mean}}{\text{standard deviation}} = \frac{0.4992 - \mu}{0.00076} = +1.645$$

and $\mu = 0.4980$. (Since the interval is symmetrical about $\bar{x}$, we could have obtained this value without as much effort.)

We have now found both ends of our 90 percent confidence interval, and the proper confidence interval statement is: The 90 percent confidence interval for the mean is from 0.4980 to 0.5004. Thus,

$$0.4980 \leq \mu \leq 0.5004 \quad \text{is a 90 percent confidence interval for } \mu.$$

We found this interval by taking the $\bar{x}$ and adding and subtracting $z_{C/2}\sigma_{\bar{x}} = z_{0.45}\sigma_{\bar{x}}$ to and from it. Hence, for our 90 percent confidence interval

$$0.4992 - 1.645(0.00076) \leq \mu \leq 0.4992 + 1.645(0.00076).$$

In general terms where σ (hence $\sigma_{\bar{x}}$) is known, normality can be assumed, and a confidence of C is required. The confidence interval is

$$\bar{x} - z_{C/2}\sigma_{\bar{x}} \leq \mu \leq \bar{x} + z_{C/2}\sigma_{\bar{x}}. \tag{11-3}$$

Each possible $\bar{x}$ value leads to its own confidence interval calculated using (11-3). Some of these intervals will include the universe mean and some will not. The above procedure is designed so that 90 percent of the intervals will include μ. Let us see why this is so.

There is a unique value of μ. The distribution of sample means for an n of 20 is illustrated in Figure 11-4. Several possible sample means are also

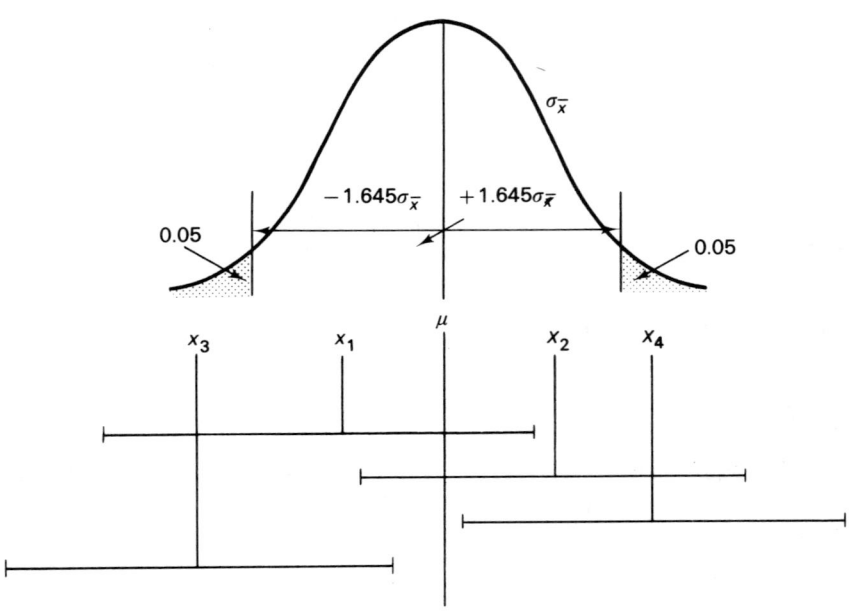

Figure 11-4 Distribution of Sample Means

shown in Figure 11-4. Further, the areas in the tails of the distribution of sample means are found using $\mu \pm z_{C/2}\sigma_{\bar{x}}$ so that $(1 - C)/2$ or 0.05 of the area and hence of the $\bar{x}$ values fall in each tail for our example.

Figure 11-4 illustrates the results of performing this calculation on the $\bar{x}$ values shown in the picture. Each time the $\bar{x}$ falls in the tail ($\bar{x}_3$ and $\bar{x}_4$, for example), the computed confidence interval does not include μ. But only 10 percent of the $\bar{x}$'s fall in the tails. On the other hand, if the $\bar{x}$ value does not fall in the tails (for example, see $\bar{x}_1$ and $\bar{x}_2$ in Figure 11-4), the interval that we calculate will include μ. Since 90 percent of the $\bar{x}$'s are in this range, 90 percent of the intervals we calculate in this way will include the universe parameter, μ. This is the reason the interval we calculate is a 90 percent confidence interval. It may contain μ and it may not, but 90 percent of the intervals calculated in this way will contain μ.

Suppose now that σ (and hence $\sigma_{\bar{x}}$) is unknown. What can be done? All we need to do is use s, our sample estimate, in place of σ ($s_{\bar{x}}$ in place of $\sigma_{\bar{x}}$) and use the t distribution. (The sample size must be large enough, given the distribution of the universe, so that we can assume that $\bar{x}$ is normally distributed.) This gives us formula (11-4), which is analogous to (11-3).

$$\bar{x} - t_{C/2}s_{\bar{x}} \leq \mu \leq \bar{x} + t_{C/2}s_{\bar{x}}, \tag{11-4}$$

DECISION MAKING: ESTIMATION

where $s_{\bar{x}} = s/\sqrt{n}$. For a 90 percent confidence interval and a sample of 20 (19 degrees of freedom), $t_{0.45} = 1.729$. For the RATFINC problem the 90 percent confidence interval will be somewhat larger than when we assumed $\sigma_{\bar{X}} = 0.00076$. The necessity of estimating σ (or $\sigma_{\bar{X}}$) using s (or $s_{\bar{x}}$) introduces more uncertainty into the estimation process. The 90 percent confidence interval is now

$$0.4992 - 1.729(0.00076) \leq \mu \leq 0.4992 + 1.729(0.00076).$$

This, when simplified, gives a confidence interval of from 0.4979 to 0.5005.

The basic ideas are identical in both cases, whether $\sigma_{\bar{X}}$ is known or unknown. The t distribution and equation (11-4) should be used whenever $s_{\bar{x}}$ is available and $\sigma_{\bar{X}}$ is not, and when the distribution of sample means is approximately normal. When $\sigma_{\bar{X}}$ is known, or as an approximation to the t distribution when $n > 30$, the normal distribution and equation (11-3) should be used. The same approach is used below to obtain a confidence interval around an estimate of the universe proportion.

CONFIDENCE INTERVALS FOR p, THE UNIVERSE PROPORTION

In this case we are going to use the normal approximation to the binomial to form a confidence interval. The normal approximation improves as $np(1 - p)$ gets large, and for np and $n(1 - p) \geq 30$ the approximation is acceptable. If the binomial in question can be approximated by a normal, then a confidence interval for the universe proportion, p, can be derived in the same manner as for the sample mean. The sample estimate of p is p' (the number of successes in the sample over the number of trials). The universe standard deviation around p is given by

$$\sigma_{p'} = \sqrt{\frac{p(1-p)}{n}}. \tag{11-5}$$

The estimate of the standard deviation, which we will denote as the sample value, is

$$s_{p'} = \sqrt{\frac{p'(1-p')}{n}}. \tag{11-6}$$

Thus, p' is the sample proportion, the estimate of p and $s_{p'} = \sqrt{p'(1-p')/n}$ is the sample standard deviation of the proportion, the estimate of $\sigma_{p'}$. If n is assumed to be sufficiently large, we can assume that $s_{p'} = \sigma_{p'}$. Under these conditions the confidence interval is

$$p' - z_{C/2} s_{p'} \leq p \leq p' + z_{C/2} s_{p'}, \tag{11-7}$$

where $z_{C/2}$ is obtained from the table of normal probabilities. This calculation does assume, as stated above, that $s_{p'} = \sigma_{p'}$, and that the normal approximation is appropriate. The numbers are easy to generate. For example, suppose that we want a 90 percent confidence interval for the universe proportion when 27 out of 100 successes occurred in the sample:

$$C = 0.90 \text{ (90 percent confidence)}$$
$$p' = 0.27 \text{ (27 out of 100)}$$

so

$$s_{p'} = \sqrt{\frac{0.27(0.73)}{100}} = \sqrt{0.001971} = 0.0444.$$

Then the confidence interval is

$$p' - z_{C/2}s_{p'} \leq p \leq p' + z_{C/2}s_{p'},$$

or

$$0.27 - 1.645(0.0444) \leq p \leq 0.27 + 1.645(0.0444),$$

which gives an interval of from 0.197 to 0.343. The 90 percent confidence interval for the proportion is from 0.197 to 0.343. We have illustrated the procedure even though $np' = 27 < 30$.

Once we find $s_{p'}$, and once we make the necessary assumptions, we proceed exactly as before. If the assumptions cannot be made, other methods based on the basic notions of confidence intervals and the binomial distribution must be used. These methods are not discussed further here.

S.S. A manufacturing firm is investigating the number of labor hours needed to produce a particular item. The item requires a large amount of hand work, and the amount of time is variable, for several reasons. The firm's industrial engineer, Tyman Moshun, measured the labor-hour input for six items. The data are as follows:

12.2, 9.4, 9.1, 11.4, 8.3, 7.2.

Mr. Moshun believes that the distribution of labor hours for individual items is normal, and you are to proceed on that assumption. Obtain 80 percent and 95 percent confidence intervals for the mean labor input. Make any comment about the data that you consider appropriate.

Solution: First, we need estimates of the mean and standard deviation.

$$\bar{x} = \sum_{i=1}^{6} \frac{x_i}{n} = 9.6$$

$$s = \sqrt{\sum_{i=1}^{6} \frac{(x_i - \bar{x})^2}{n - 1}} = \sqrt{\frac{17.74}{5}} = \sqrt{3.548} = 1.884.$$

DECISION MAKING: ESTIMATION

Thus,

$$s_{\bar{x}} = \frac{1.884}{\sqrt{6}} = 0.769.$$

Since $n = 6$, the t distribution with 5 degrees of freedom is appropriate, and the confidence interval is from

$$\bar{x} - t_{C/2}s_{\bar{x}} \quad \text{to} \quad \bar{x} + t_{C/2}s_{\bar{x}}.$$

Using the table of the t distribution, for 5 degrees of freedom, $t_{C/2}$ is 1.476 for $C = 0.80$ and 2.571 for $C = 0.95$. Thus, the two intervals are:

80%: $9.6 - 1.476(0.769)$ to $9.6 + 1.476(0.769)$ or 8.46 to 10.73

95%: $9.6 - 2.571(0.769)$ to $9.6 + 2.571(0.769)$ or 7.62 to 11.58.

Notice that to be 95 percent sure that an interval contains the universe mean, the 95 percent interval must be larger than the 80 percent interval. A comment that can be made about the data is that the number of labor hours seems to be decreasing. Thus, the firm may still be in the learning phase for this product, and the mean may be changing. (We certainly cannot conclude this based solely on the data; it is just something that the firm may wish to investigate before proceeding.) If the mean is changing, any estimate, point or interval, is incorrect.

11-3 Confidence Intervals for the Variance and Standard Deviation*

When we studied confidence intervals for the mean we used the relationship

$$\frac{\bar{x} - \mu}{\sigma_{\bar{x}}} = z_{C/2},$$

when $\sigma_{\bar{x}}$ is known, to obtain an interval estimate of from $\bar{x} - z_{C/2}\sigma_{\bar{x}}$ to $\bar{x} + z_{C/2}\sigma_{\bar{x}}$. When $\sigma_{\bar{x}}$ is unknown, but $\bar{x}$ can still be assumed to be normally distributed, we used the relationship

$$\frac{\bar{x} - \mu}{s_{\bar{x}}} = t_{C/2},$$

to obtain a confidence interval of from $\bar{x} - t_{C/2}s_{\bar{x}}$ to $\bar{x} + t_{C/2}s_{\bar{x}}$.

---

* This section is somewhat more difficult than the other sections. It may be skipped, at this point, without losing continuity or sleep.

We know that if the distribution of sample means is normal, then $(\bar{x} - \mu)/\sigma_{\bar{x}}$ is normally distributed with mean = 0 and standard deviation = 1, so we could look up a value in the table of the normal distribution and use it to construct an interval for μ, around $\bar{x}$. Similarly, if the distribution of sample means is normal, $(\bar{x} - \mu)/s_{\bar{x}}$ is t distributed with $n - 1$ degrees of freedom, and we can use a value from the t distribution table to construct an interval estimate for μ, centered around $\bar{x}$. In both cases, we know the distribution and use the appropriate tables. Also in both cases, we assume that the distribution of sample means is at least approximately normally distributed.

When we discuss confidence intervals for σ_X, we will again need to know what distribution (what table) is appropriate. We will now be assuming that the basic data, the x_i values, as well as the sample means, are normally distributed. If this assumption is invalid, the methods of this section should not be used.

The preceding discussion is both a refresher and a warm-up. It is a new way of looking at our previous study of confidence intervals, stressing the assumption made about the distribution of sample means, and the final distribution used for $(\bar{x} - \mu)/\sigma_{\bar{x}}$ or $(\bar{x} - \mu)/s_{\bar{x}}$ to obtain the interval. It is a warm-up in that we are now going to study a new distribution, the chi-square (χ^2) distribution. This distribution is the appropriate distribution to use in obtaining confidence intervals for the variance or standard deviation.

If we draw a sample of size n from a normally distributed universe and calculate the unbiased estimate of the variance, we obtain

$$s^2 = \sum_{i=1}^{n} \frac{(x_i - \bar{x})^2}{n - 1}.$$

If we divide both sides by σ^2, we have

$$\frac{s^2}{\sigma^2} = \frac{1}{n-1} \sum_{i=1}^{n} \frac{(x_i - \bar{x})^2}{\sigma^2} = \frac{1}{n-1} \sum_{i=1}^{n} \left(\frac{x_i - \bar{x}}{\sigma}\right)^2$$

and multiplying both sides by $n - 1$

$$\frac{(n-1)s^2}{\sigma^2} = \sum_{i=1}^{n} \left(\frac{x_i - \bar{x}}{\sigma}\right)^2.$$

Still assuming that the universe distribution is normal, we know that $(x_i - \bar{x})/\sigma$ is normally distributed. But $[(x_i - \bar{x})/\sigma]^2$ is chi-square (χ^2)-distributed. And, would you believe, the sum of such terms,

$$\sum_{i=1}^{n} \left(\frac{x_i - \bar{x}}{\sigma}\right)^2 = \frac{(n-1)s^2}{\sigma^2}$$

DECISION MAKING: ESTIMATION

is χ^2-distributed with $n - 1$ degrees of freedom. Well it is, and since we will not prove it here, it must be accepted on faith. As before, one degree of freedom is lost when $\bar{x}$ is used as an estimate of μ in computing s^2. The χ^2 distribution is tabulated in Table III, and we will discuss Table III's use later in this section.

New Symbol	Meaning
χ^2	The χ^2 (chi-square) distribution. $$\frac{(n-1)s^2}{\sigma^2} = \sum_{i=1}^{n} \left(\frac{x_i - \bar{x}}{\sigma}\right)^2$$ is χ^2, with $n - 1$ degrees of freedom, if the individual values are normally distributed. (χ is the lowercase Greek letter chi.)

Well, you say, all of that is very nice, but so what? Well, if $(n - 1)s^2/\sigma^2$ is χ^2-distributed with $(n - 1)$ degrees of freedom, we can look up values in the χ^2 table and establish confidence intervals for σ^2, centered around s^2.

That's good news. (Good grief, if that's the good news, what can the bad news be?) The bad news is that χ^2 is not a symmetric distribution, as are the normal and t distributions. In fact, the χ^2 distribution is positively skewed, although the degree of skewness declines for increasing values of n. Several χ^2 distributions, for different degrees of freedom, are shown in Figure 11-5.

Since the χ^2 distribution is not symmetric, the χ^2 table is constructed to give the probability from the point in question out to the extreme right-hand side. That is, it gives the right-hand tail, or $P(\chi^2 \geq K)$ for any value K. Using the χ^2 table to obtain a confidence interval for σ^2, we must find a separate χ^2 value for both tails. For example, for a 90 percent confidence interval we must find the χ^2 value for right-hand tail areas of 0.05 and 0.95. For a 95 percent confidence interval, values are found for 0.025 and for 0.975. For a 90 percent confidence interval, 10 percent is left in the two tails; for a 95 percent confidence interval, 5 percent is left in the two tails. The 10 percent and 5 percent are, respectively, the probability that the interval does not contain the universe value of σ^2.

Once the two χ^2 values are obtained, the confidence interval for σ^2 can be obtained. Since $(n - 1)s^2/\sigma^2$ is χ^2 with $n - 1$ degrees of freedom, then for a 90 percent confidence interval,

$$\chi^2_{0.95} \leq \frac{(n-1)s^2}{\sigma^2} \leq \chi^2_{0.05}.$$

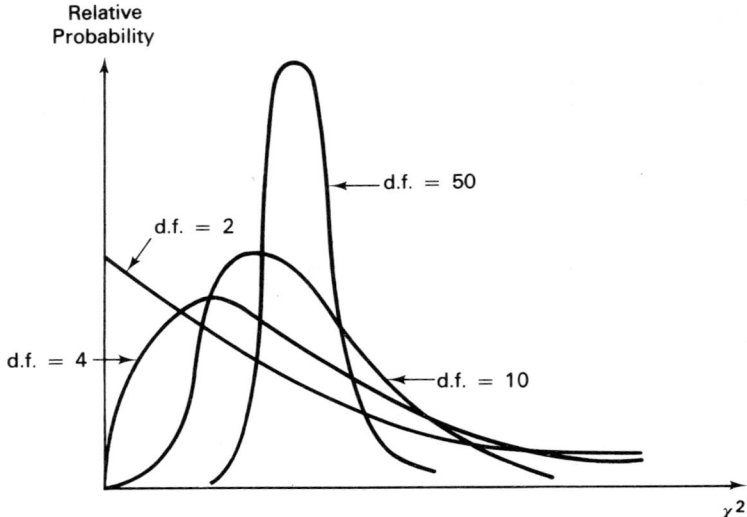

Figure 11-5 χ^2 Distributions

Dividing each term by $(n - 1)s^2$, we obtain

$$\frac{\chi^2_{0.95}}{(n-1)s^2} \leq \frac{1}{\sigma^2} \leq \frac{\chi^2_{0.05}}{(n-1)s^2}.$$

Finally, inverting each term (and reversing the direction of the inequalities) yields

$$\frac{(n-1)s^2}{\chi^2_{0.05}} \leq \sigma^2 \leq \frac{(n-1)s^2}{\chi^2_{0.95}}. \tag{11-8}$$

Several comments are in order here.

1. If following the algebraic manipulations is difficult, do not worry. It is more important to understand what is assumed, how the χ^2 distribution gets into the act, and in general what is happening in the derivation.

2. What is assumed is that the individual values, the x_i values, are approximately normally distributed. The χ^2 distribution gets into the act because

$$\frac{(n-1)s^2}{\sigma^2} = \sum_{i=1}^{n} \left(\frac{x_i - \bar{x}}{\sigma}\right)^2$$

is χ^2-distributed with $n - 1$ degrees of freedom if the individual values are normally distributed.

DECISION MAKING: ESTIMATION

3. The punch line is, a confidence interval with confidence level C for σ^2 is given by

$$\frac{(n-1)s^2}{\chi^2_{(1-C)/2}} \leq \sigma^2 \leq \frac{(n-1)s^2}{\chi^2_{(1+C)/2}}. \tag{11-9}$$

To obtain a confidence interval for σ, take the square root of each term in (11-9). Although this is not a pretty interval, the arithmetic is reasonably easy, as we shall see in the following example.

In the Student Should in the previous section, a manufacturing firm had data on labor input where $n = 6$, $\bar{x} = 9.6$, and $s^2 = 3.548$. For $n = 6$, there are 5 degrees of freedom. We will find a 90 percent confidence interval and a 95 percent confidence interval for σ and σ^2, assuming labor hours for an item are normally distributed.

For a 90 percent confidence interval, using the χ^2 table and 5 degrees of freedom:

$$\chi^2_{0.05} = 11.07 \quad \text{and} \quad \chi^2_{0.95} = 1.145.$$

Then the interval is

$$\frac{(n-1)s^2}{\chi^2_{0.05}} \leq \sigma^2 \leq \frac{(n-1)s^2}{\chi^2_{0.95}}$$

or

$$\frac{5(3.548)}{11.07} \leq \sigma^2 \leq \frac{5(3.548)}{1.145},$$

which is

$$1.603 \leq \sigma^2 \leq 15.493.$$

For a 95 percent confidence interval:

$$\chi^2_{0.025} = 12.83 \quad \text{and} \quad \chi^2_{0.975} = 0.831.$$

The interval is

$$\frac{5(3.548)}{12.83} \leq \sigma^2 \leq \frac{5(3.548)}{0.831},$$

which is

$$1.383 \leq \sigma^2 \leq 21.348.$$

If we want, for example, a 95 percent confidence interval for σ, we take the square root of both values. Thus,

$$\sqrt{1.383} \leq \sigma \leq \sqrt{21.348} \quad \text{or} \quad 1.18 \leq \sigma \leq 4.62$$

is a 95 percent confidence interval for σ.

If the confidence intervals do not look quite right, it is probably because s^2, the estimate of σ^2, is not in the middle of the interval. This is always the case with confidence intervals for σ^2. It happens this way because χ^2 is not a symmetric distribution.

There are at least four reasons for having a section on this difficult material. First, confidence intervals for the variance are occasionally valuable; for example, they can be useful in controlling product quality. Second, this section can help our understanding of the different distributions we are using, what assumptions are being made, and why different distributions are necessary. Third, the χ^2 distribution will be used again, later in this book, and this is as good a place to introduce it as any. Finally, we will need some of the material, the easy part, to perform the following exercise.

S.S. Using $n = 11$ and $s^2 = 4$, find an 80 percent confidence interval for σ, assuming that the universe of individual values is normally distributed.

Solution: First, we require, for 10 degrees of freedom:

$$\chi^2_{0.10} = 15.99 \quad \text{and} \quad \chi^2_{0.90} = 4.87.$$

Then, an 80 percent confidence interval for σ^2 is

$$\frac{(n-1)s^2}{\chi^2_{0.10}} \leq \sigma^2 \leq \frac{(n-1)s^2}{\chi^2_{0.90}}$$

or

$$\frac{10(4)}{15.99} \leq \sigma^2 \leq \frac{10(4)}{4.87},$$

which is

$$2.50 \leq \sigma^2 \leq 8.21.$$

To obtain an 80 percent interval for σ:

$$\sqrt{2.50} \leq \sigma \leq \sqrt{8.21},$$

which is

$$1.58 \leq \sigma \leq 2.87.$$

11-4 Determining the Sample Size

Confidence intervals are easy to calculate once we have the data, but what do we do if the data have not yet been taken? How do we determine the sample size that must be taken in order to obtain an estimate with a certain specified precision? This section investigates that issue for confidence intervals around the mean.

Before a sample size can be chosen for a particular estimation problem, the manager (or researcher) must decide how good an estimate is desired.

DECISION MAKING: ESTIMATION

That is easy, you say; we want a perfect estimate. Well, we cannot have it. Unless one accurately examines the entire universe, the estimate will not be perfect, and even if that is possible, the cost may be prohibitive. What we can obtain is a confidence interval. Suppose we desire an estimate within a specified number of units, L, from the universe value. The confidence interval, with level of confidence, C, is $\bar{x} \pm L$. This section discusses how to find the minimum sample size that will give us the interval with the desired level of accuracy. To obtain that sample size, the standard deviation of the universe must be assumed known. This is an important assumption. (If we had more precise tables of the t distribution, we would not need this assumption.) There are two ways in which this assumption can be made. First, the universe in question may be similar in variability to a previously studied universe, so that we may assume that the previous σ value holds in the new situation. This first case might hold, for example, when studying a diameter of a machined part. The settings may be incorrect, changing the mean of the process, but the variability depends on the machine accuracy, not on the settings, so σ does not change. Second, the standard deviation may be estimated on the basis of a preliminary sample from the new population, and we assume that $s = \sigma$. Then we want to see how much larger a sample should be taken to obtain the specified confidence statement.

To find the required sample size, n, recall how a value such as L, where $\bar{x} - L$ to $\bar{x} + L$ is the confidence interval, is found. (We assume that σ is known and that the sample means are normally distributed.)

$$L = z_{C/2}\sigma_{\bar{x}}, \quad \text{where } \sigma_{\bar{x}} = \frac{\sigma}{\sqrt{n}},$$

so

$$L = z_{C/2}\frac{\sigma}{\sqrt{n}}. \tag{11-10}$$

When C is specified, $z_{C/2}$ is obtained from the normal table. The value of L is also specified in advance, so equation (11-10) reduces to one equation in one unknown. (Our eighth-grade math teacher would expect us to be able to solve one equation in one unknown.)

If

$$L = z_{C/2}\frac{\sigma}{\sqrt{n}}$$

$$\sqrt{n} = \frac{z_{C/2}\sigma}{L}$$

$$n = \frac{(z_{C/2})^2\sigma^2}{L^2}. \tag{11-11}$$

If σ is estimated based on s from a preliminary sample, that value of s will be used for σ in (11-11).

Given a specified value for L, the acceptable distance from $\bar{x}$, and a specified value for C, the probability that $\bar{x}$ is within L of μ (chance of error is $1 - C$), a sample of size n,

$$n = \frac{(z_{C/2})^2 \sigma^2}{L^2}$$

is necessary. The value of n must be rounded up to the next integer if it is between two integers. The value of $z_{C/2}$ is found from the normal table. We assume that σ is known and that the sample means are normally distributed.

For example, suppose an advertising manager knows that the standard deviation of his daily sales during an advertising campaign is $\sigma = 25$. He wants to collect sales data until he can be 99 percent sure that his estimate is within 1 unit of the true mean of the daily sales during the new campaign. He wants to know how many days of sales data he must collect. $C = 0.99$, so $z_{C/2} = z_{0.495} = 2.57$; $L = 1$ and $\sigma = 25$. Thus,

$$n = \frac{(z_{C/2})^2 \sigma^2}{L^2} = \frac{(2.57)^2 (25)^2}{1^2} = 4{,}128.1,$$

which is rounded up to 4,129.

Seeing that over 4,000 days are necessary, and knowing that he cannot collect that many data values, the manager wonders how much data he would need to have $C = 0.90$ and $L = 5$. (These values, it turns out after some discussion, are actually sufficient for his purpose.)

Then
$$C = 0.90, \qquad z_{C/2} = z_{0.45} = 1.645.$$

$$n = \frac{(z_{C/2})^2 \sigma^2}{L^2} = \frac{(1.645)^2 (25)^2}{(5)^2} = 67.6 \text{ or } 68.$$

This sample size is sufficiently large for us to assume that the sample means are normally distributed, relying on the Central Limit Theorem.

Suppose, however, that this is still too large, according to the manager. Suppose he asks for the result corresponding to an L of 10. Then we obtain $n = 17$, which the advertising manager says is feasible. With $n = 17$, if the distribution of individual days' sales is unimodal and nearly symmetric, we can assume that the distribution of sample means is close to normal. If $L = 10$ and $C = 0.90$ are satisfactory to the manager, the sample-size problem is resolved.

DECISION MAKING: ESTIMATION

In practice, considerations such as budgets or lack of time often determine the maximum sample size. The methods of this section can be used iteratively to find what combination of confidence level and interval size is desirable, given the data we can afford to obtain. Once $n = 4129$ is suggested, however, the manager might as well take the data that he can reasonably collect and then make his confidence-interval statements. The methods employed here come into play when a feasible sample will satisfy the specifications. Then some potential sampling cost can be avoided by a careful determination of the necessary sample size.

S.S. A manufacturing firm has a boring machine that places holes in metal, with a standard deviation of 0.002 inch. However, the mean diameter tends to drift out of specification. To check the mean diameter, they sample occasionally. (If the $\bar{x}$ value is far enough away from the diameter's specified value, they will stop the process and reset the machine.) To guarantee the quality level desired, they want to be 90 percent sure of obtaining $\bar{x}$ values within 0.0005 inch of the true mean. How large a sample do they have to take to achieve the indicated confidence?

Solution: The values pertinent to the size of n are $\sigma = 0.002$, $L = 0.0005$, and $C = 0.90$. $C = 0.90$ implies that $z_{C/2} = 1.645$. Then with $L = 0.0005$ and $\sigma = 0.002$ we obtain

$$n = \frac{(z_{C/2})^2 \sigma^2}{L^2} = \frac{(1.645)^2 (0.002)^2}{(0.0005)^2} = 43.3.$$

They must take a sample of size 44 to achieve the desired result. This sample size is sufficiently large to justify the normality assumption.

11-5 A Warning

In making and using estimates, a manager must keep several things in mind. First, using any statistic as an estimate (such as using the sample mean as an estimate of the universe mean) implies that all the data values are drawn from the single universe whose mean is to be estimated. Otherwise, proper estimates cannot be made. An example where the mean changes through time is the labor hours required per unit, when learning is present. Second, any estimate, whether a point estimate or an interval estimate, requires some assumptions and possesses certain properties. For example, an unbiased estimator of σ^2 is formed by dividing $\sum_{i=1}^{n}(x_i - \bar{x})^2$ by $n - 1$ instead of by n. For another example, the t distribution is used to obtain an interval estimate for μ if σ is unknown. However, we still, in that case, must assume that the distribution of sample means is normally distributed.

The point is that estimation is fraught with pitfalls. The manager may be looking at irrelevant data, or the data may be relevant but the statistical assumptions are not met. If all these hazards are circumvented, the manager must still remember that estimates are made on a probabilistic basis and can be incorrect. In basing any decisions on an estimate, the chance of an incorrect estimate must always be considered, and the cost if it is incorrect should be incorporated in the decision-making process where feasible.

In spite of all of the above qualifications, however, estimation is a common use of statistics, and it has proved valuable in many situations. Being precise about how uncertain we are (such as "we are 90 percent sure that the universe mean is between 8 and 12") can help managers make better decisions. We will return throughout the remainder of this book to the notions introduced in this chapter.

TECHNICAL NOTE 11-1: DERIVATION OF AN UNBIASED ESTIMATOR OF σ^2

This technical note is an exercise in algebra and the use of expectations. As such it will be useful to work through. We will start with the

$$s^2 = \frac{1}{n} \sum_{i=1}^{n} (x_i - \bar{x})^2$$

formula and demonstrate that a correction factor is necessary to make it unbiased. We will use the following facts in the derivation. To simplify the notation, $\sum_{i=1}^{n} x_i$ will be written as $\sum x$.

1. $S^2 = \sum (X - \bar{X})^2 / n$, where s is a random variable calculated from $\bar{X}$ and n.
2. $\sigma_{\bar{X}}^2 = \sigma^2 / n$.
3. $\sigma^2 = E(X^2) - [E(X)]^2 = E(X^2) - \mu^2$.
4. $\sigma_{\bar{X}}^2 = E(\bar{X}^2) - \mu^2$.

First, taking expectations,

$$E(S^2) = E\left[\frac{\sum (X - \bar{X})^2}{n}\right] = E\left(\frac{\sum X^2}{n} - \bar{X}^2\right) = \frac{\sum E(X^2)}{n} - E(\bar{X}^2), \quad (11\text{-}12)$$

since the expectation of a sum equals the sum of the expectations.

DECISION MAKING: ESTIMATION

Now, substituting into (11-12) using fact 3 to obtain $E(X^2)$ and fact 4 to obtain $E(\bar{X}^2)$,

$$E(S^2) = \frac{\sum \sigma^2 + \mu^2}{n} - (\sigma_{\bar{X}}^2 + \mu^2) = \frac{n\sigma^2 + n\mu^2}{n} - \sigma_{\bar{X}}^2 - \mu^2 = \sigma^2 - \sigma_{\bar{X}}^2.$$

(11-13)

Using fact 2 we can write (11-13) as

$$E(S^2) = \sigma^2 - \frac{\sigma^2}{n} = \sigma^2 \frac{n-1}{n}.$$

Thus, if we want $E(S^2) = \sigma^2$, we must multiply the original S^2 formula by $n/(n-1)$. That is,

$$\frac{n}{n-1} \frac{\sum (X - \bar{X})^2}{n} = \frac{\sum (X - \bar{X})^2}{n-1}$$

is an unbiased estimator of σ^2, since

$$E\left[\frac{\sum (X - \bar{X})^2}{n-1}\right] = \frac{n}{n-1} \frac{n-1}{n} \sigma^2 = \sigma^2.$$

New Symbols

Symbol	Meaning
C	The level of confidence for a confidence interval. The probability that the confidence interval contains the universe parameter is C. Statements such as a "95 percent confidence interval" or a "0.95 confidence interval" are used interchangeably.
χ^2	The χ^2 (chi-square) distribution. $$\frac{(n-1)s^2}{\sigma^2} = \sum_{i=1}^{n} \left(\frac{x_i - \bar{x}}{\sigma}\right)^2$$ is χ^2 distributed with $n-1$ degrees of freedom. χ_C^2, for example, means the reading from the χ^2 table, for the appropriate degrees of freedom, that has C of the distribution above that value. C may be 0.95, 0.05, or any value we seek (between 0 and 1).

PROBLEMS

11-1. Briefly define a statistic. What statistic is used to estimate the universe mean?

11-2. The sample standard deviation, s, is used in estimating both μ, the universe mean, and σ, the universe standard deviation. In what kind of estimate is it used in both cases? How is it used in estimating μ?

11-3. Briefly define unbiased, efficient, consistent, and sufficient as they apply to estimators. One at a time, indicate why an estimator should have these characteristics.

11-4.* There are two formulas for s^2, the sample variance, given in Section 11-1. Which one (or ones) is unbiased? Which one (or ones) is consistent? If $n = 100$, how large a difference will there be?

11-5.* A production engineer is interested in the average length of a batch of 1,000,000 10-centimeter nails. If the engineer says that the estimate is to be exact, how large a sample must be taken? Suppose that $\sigma = 0.1$ centimeter. How large does the sample have to be to get a confidence interval of $\bar{x} \pm 2\sigma_{\bar{x}}$ to be only 0.001 centimeter wide?

11-6. A consumer testing organization examining the gas mileage of a particular car found that $\bar{x} = 16.8$ and $s_{\bar{x}} = 0.45$ for $n = 9$. They also know, using the t table, that:

$P(\bar{x} \leq 16.8 \mid \mu = 17.64, s_{\bar{x}} = 0.45, \text{d.f.} = 8) = 0.05$

$P(\bar{x} \geq 16.8 \mid \mu = 15.96, s_{\bar{x}} = 0.45, \text{d.f.} = 8) = 0.05$.

(a) Use these values to make a confidence-interval statement.
(b) If the organization (incorrectly) assumed that $s_{\bar{x}} = \sigma_X$, how different would the confidence interval be?

11-7.* A quality control manager in the Measure-Right Ruler Company is examining her company's metric rulers. Each ruler is supposed to be 100 centimeters long, and each ruler must be within 0.2 centimeter of that figure. The manager knows from past experience that $\sigma = 0.06$ for the operation and that the universe of lengths is normally distributed. She wants to examine the output to be sure that $\mu = 100.0$. Her data, for a sample of 100 units, give $\bar{x} = 99.90$.
(a) Construct a 95 percent confidence interval for the mean.
(b) If $\mu = 99.90$, what is the probability that an individual ruler will be outside the allowable limits?
(c) Should the manager stop the process and reset it?

11-8. A commercial bank has just started an advertising campaign designed to increase its loan activity. The managers know that the average

daily initiation of loans before the campaign was exactly $2.3 million. During the 10 days since the campaign began, they have obtained average daily loan initiations of $2.36 million, with a standard deviation of $s_{\bar{x}} = 0.03$. (Assume the days of the week are the same in terms of the loan volume transacted.)
(a) Obtain a 95 percent confidence interval for the mean daily loan initiation after the campaign.
(b) Obtain a 95 percent confidence interval assuming (incorrectly) that $s_{\bar{x}} = \sigma_{\bar{x}}$.
(c) Do you think that the mean has increased?

11-9.* The safety division of a large paint factory is interested in the proportion of employees who smoke. A random sample of 100 is selected, and 80 reported that they smoke. (Use the normal approximation.)
(a) What is the point estimate of the true proportion of smokers in the factory?
(b) Set up 95 percent confidence limits for the estimate of the true proportion. Do the same for 90 percent limits and 50 percent limits. Which of these is the most useful and why?

11-10. The Wojax Corporation produces a number of small electrically powered home appliances and is considering marketing a new model of its popular electric toothbrush. These toothbrushes are powered by a rechargeable battery in the handle, and a long-lived battery is an essential characteristic for a large sales volume. The battery, obtained from an outside supplier and used in the older model, is considered a good one, and it will be used in the new model. Tests on this battery have shown that a random sample of 25 batteries drawn from a large shipment has an average (mean) life, under usual operating and recharging conditions, of 400 hours, with a sample variance of 2,500 hours [computed using $\sum (x - \bar{x})^2/n$]. If the sample is drawn from a shipment of 25,000 batteries, determine an unbiased point estimate of the true mean life in the shipment, and of the shipment variance, and a 95 percent confidence interval estimate for the mean.

11-11. The weight of coins is often important in establishing both the authenticity and the value. If 16 coin dealers weighed a Phoenician tetradrachm and obtained a mean weight of 14.51 grams and a standard deviation of 0.08 gram, construct a 99 percent confidence interval for the true weight of this coin.

11-12.* The training department of a large company would like to know how many hours they can expect a trainee to take to complete a self-study course in office procedures. If 100 randomly selected trainees took an average of 8.2 hours with a standard deviation of 0.8 hour, what can we say with a probability of 0.95 about the average time trainees should expect to spend on the course?

11-13. From 100 observations on the output from the manufacture of steel plates, the mean compressive strength was 62,516 pounds per square inch (psi) and a 99 percent confidence interval for the population mean was found to be 60,304 to 64,728 pounds. Find the standard deviation of compressive strength of the output.

11-14. A quality control manager must be fairly certain that the strength of a certain type of hemlock beam is able to withstand pressure of 2,000 psi before splitting. From each shipment received by the company, the manager draws a sample and rejects the entire shipment if the sample mean breaking strength is less than 2,000. In setting a sampling policy, the inspector decided to draw samples large enough so that the probability of having a sample mean differ by more than 50 psi from the population mean would be no greater than 0.05. The standard deviation of beam strength based upon prior samples is 100 psi.
(a) How large should the manager's samples be if each shipment contains thousands of beams?
(b) From a sample of 500 beams, what is the probability the shipment is rejected if the true mean beam strength for the shipment is 2,100 psi?

11-15. Suppose that the management of a restaurant chain uses a mean of $\bar{x} = 19.6$ minutes as an estimate of the true average time it takes one of their cooks to prepare a food order. Making use of the fact that the sample standard deviation of the values in this sample of size $n = 225$ is 5.5 minutes, what can we assert with a probability of 0.99 about the possible size of their error?

11-16. A department store uses a mean of $\bar{x} = 6.20$ minutes to estimate the true mean time required to take telephone orders. Given the fact that 25 observations were used to obtain this mean and that the sample standard deviation was 2.3 minutes, with a probability of 0.95, what can we assert about the possible size of their error?

11-17. A card shop wishes to estimate the amount customers spend on the average. A random sample of 49 persons is selected and yields a sample average of $3.15 and a standard deviation of $1.30. If it is known that the standard deviation of the amount spent is actually $1.05 per customer, what interval estimate of the average could be made with a confidence level of 95 percent?

DECISION MAKING: ESTIMATION   **361**

11-18.* Because of increasing cash-flow problems, the management of Credit Manufacturing Company believes that the best way to head off insolvency problems is to issue bonds with detachable warrants. In order to sample stockholder opinion concerning the proposed issue, management randomly selected 100 of the company's 10,000 stockholders and asked that they complete a questionnaire. Of the 100, 70 were in favor of the issue and 30 were opposed. Find a 95 percent confidence interval for the actual proportion of shareholders favoring the new issue.

11-19.* A credit card company wants to determine the mean income of its cardholders. A random sample of 225 cardholders was drawn, and the sample average income was $11,556 with a sample standard deviation of $6,452.
   (a) Construct a 99 percent confidence interval for the mean income.
   (b) Management decided that the confidence interval in part (a) was too large to be useful. In particular, the managers want to estimate the mean income within $500 on either side with a confidence level of 99 percent. How large a sample should be selected?

11-20. The research department of a pharmaceutical company has developed a new chemical compound which it hopes will increase the yield in the manufacture of an important drug. The present manufacturing process for the drug is carried out by mixing a number of chemicals under strict temperature and pressure specifications. From past trials conducted by varying experimental conditions, a set of formulas were developed that approximate the yield obtainable by altering the amounts of chemicals, temperature, and pressure; and the recipe in use at present is thought to yield the maximum amount of product. The research department wants to determine the mean yield of the drug that will result from adding the new compound to the manufacturing process.

In a series of trials, the new compound was added at random times to each of 25 batches of the drug. The yield from the trials was recorded as a percentage of the yield using the new compound to the yield that would have resulted from the manufacturing process without the new compound. The mean yield ratio from the 25 trials was 115.6 percent, which means that the sample produced an average of 15.6 percent more of the drug than the old manufacturing process. The sample standard deviation of this yield ratio was 30.2. Compute a 95 percent confidence interval for the true yield ratio using the new compound. Does the evidence suggest that the new compound is effective? Why or why not?

11-21. Two machines in a manufacturing plant are used to produce the same item, which is utilized in the assembly of specialized tools. The item has a protruding dowel 16 inches in length. Any deviation from this 16-inch length affects the precision of the tools. Random samples of size 25 were drawn from each machine during a particular day. The mean length of the dowel from the first machine was 16.10 inches with a sample standard deviation of 0.2 inch, and the mean length of the dowel from the second machine was 15.94 inches with a sample standard deviation of 0.3 inch. Construct 95 percent confidence intervals for the mean length of the population of dowels produced from each machine, and interpret each of these confidence intervals.

11-22.* The production manager of a synthetics plant wants to determine the mean breaking strength of a new type of nylon cord. In a random sample of 25 cords the sample mean was 10.2 pounds and the sample standard deviation was 4.0 pounds. Find a 95 percent confidence interval for the mean breaking strength of the new cord.

11-23. To assist in the design of an elevator system for a new building, the architects for the building hired observers to view the operation of elevators in another structure of similar design. From observations during 25 business days, observers noted that the average number of first-floor stops per day was 160 with a standard deviation of 15.3. Find a 95 percent confidence interval for the population mean.

11-24. The Perish Publishing Company desires to estimate the proportion of teachers among the members of the National Society for the Advancement of Mathematics. The director of research wishes the estimate to be within 2 percent of the actual proportion of teachers with a reliability of 95 percent. How large a sample must be taken if there are 10,000 members in the Society? [*Hint:* Assume that the normal probability law with mean p and variance $p(1 - p)/n$ may be used to approximate the binomial. Now the necessary sample size varies directly with the unknown proportion of teachers. Thus, to assure a 95 percent confidence interval, the maximum possible value of $p(1 - p)$ can be used. This occurs at $p = \frac{1}{2}$.]

11-25. Playtime Toy Company plans to introduce a new toy to the market. The company distributes its existing products through a large number of retail stores. In order to assess the potential success of the new toy, Playtime plans to introduce the toy into a random sample of retail stores for 1 month. Because sampling is expensive, the marketing managers of Playtime want to select no more stores than are necessary to obtain an accurate estimate. The managers estimate

DECISION MAKING: ESTIMATION

that the initiation of the sample design would cost $1,000 and that the cost of sampling would increase from that amount by $500 for every store included in the sample. From past studies with similar products, the managers estimate the variance of monthly sales for this product to be $10,000 in each store. If the budget limitation for sampling is set at $13,500, what is the greatest confidence level that could be achieved for a symmetric confidence interval, $78.40 in width, for the mean annual sales per store?

11-26.* A credit card company is interested in the difference between the proportion of defaults from customers in the East and customers in the West. From two random samples of the firm's customer accounts, the following results were obtained:

	East	West
Proportion of defaults	0.10	0.20
Sample size	100	200

Find a 95 percent confidence interval for the difference between the proportion of defaults in the West and the proportion of defaults in the East. Use a normal approximation and assume that the event that an Eastern customer defaults is independent of the event that a customer in the West defaults. (*Hint:* The difference between two normally distributed random variables is, again, a normally distributed random variable.)

11-27. When are the t distribution and χ^2 distribution used in forming confidence intervals?

11-28. A sales manager is interested in the difference between technically trained salesmen and nontechnical salesmen. The four technical salesmen sold 24, 29, 32, and 41 units last month. The four nontechnical salesmen sold 22, 29, 30, and 41 units:

Technical	Nontechnical	Difference
24	22	+2
29	29	0
32	30	+2
41	41	0
126	122	+4

(a) Construct an 80 percent confidence interval using each of the three columns of data. Use the unbiased estimate of the variance. Assume that these observations are drawn from normal universes.

(b) What does the confidence interval constructed using the third column mean?

11-29. In Problem 11-28, suppose the sales manager were to report the data on sales as follows:

Technical	Nontechnical	Difference
24	41	−17
29	30	−1
32	29	+3
41	22	+19

(a) Construct an 80 percent confidence interval on the third column.
(b) What is wrong with this approach?

11-30.* A candy manufacturer is interested in the variability of their process that produces 1-ounce candy bars. (They also control the average, of course, but if variability is high they have to set the process to produce slightly heavy candy bars.) On a recent sample of 25 candy bars, they obtained $\bar{x} = 1.02$ and $s = 0.016$. Form a 90 percent confidence for both μ and σ^2.

11-31. A salesman for a manufacturer of metal parts claims that his company can provide a particular component whose diameter measurement has a variance of only 0.000081 centimeter. From a large shipment of the component, one purchaser drew a random sample of 25 and found the sample standard deviation to be 0.0130. (The sample variance equals 0.000169.) Assuming that the sample was drawn from a normally distributed population, can the salesman's claim be supported? (Find a 95 percent confidence interval on σ^2 and see if the claimed value is included.)

11-32. A canning company must control the mean amount of fruit packed in its cans as well as the variability for each can. To determine how consistently the process packs 16-ounce cans of cling peaches, the plant manager selected a random sample of 30 cans from a day's production volume and measured the contents of each can. The sample results are as follows:

Can No.	$X \equiv$ Contents (ounces)	X^2	Can No.	$X \equiv$ Contents (ounces)	X^2
1	16.1	259.21	16	15.7	246.49
2	16.2	262.44	17	15.8	249.64
3	16.0	256.00	18	15.8	249.64
4	15.5	240.25	19	15.8	249.64
5	16.3	265.69	20	15.9	252.81
6	15.9	252.81	21	15.7	246.49
7	15.8	249.64	22	16.0	256.00
8	15.2	231.04	23	16.1	259.21
9	15.9	252.81	24	16.0	256.00
10	15.9	252.81	25	15.8	249.64
11	16.2	262.44	26	15.9	252.81
12	16.0	256.00	27	15.8	249.64
13	15.8	249.64	28	15.8	249.64
14	16.1	259.21	29	16.1	259.21
15	16.1	259.21	30	16.3	265.69
			Total	477.5	7,601.75

(a) Find a 95 percent confidence interval for the mean number of ounces filled in the production process. A computationally efficient formula for determining s^2 is

$$s^2 = \frac{\sum_{i=1}^{n}(x_i^2) - n\bar{x}^2}{n-1}.$$

(b) Find a 95 percent confidence interval for the process variance.

Chapter **12**

Decision Making: Tests of Hypotheses

By now you may have an hypothesis that statistics books are longer than any other type of book. (This is not true; dictionaries, for example, are much longer, on the average.) If we specifically wanted to see if statistics books were longer than, say, history books, we could take a sample of both kinds of books and test the hypothesis that statistics books are longer on the average than history books. Along the way, we would determine if we believe the hypothesis to be true, and how sure we are of the result.

Hypothesis testing is, as the preceding discussion implies, a process composed of three parts: framing a statement of belief (or nonbelief), collecting appropriate data, and performing some statistical analysis of the data which leads to a conclusion. The statement of belief is called an *hypothesis*. The decision maker may not believe it at all, and may want to prove it false. However, in hypothesis testing it is necessary that some specific hypothesis be tested and accepted or rejected. In the above case, the hypothesis is that statistics books are longer on the average than history books. In the data-collection phase we would obtain the average length of several books of each kind. Then the statistical analysis will either support the hypothesis or cast doubt on it. If enough doubt is cast, the decision maker will be led to reject the hypothesis and accept the implied alternative that statistics books are not longer. How the hypothesis to be tested is chosen is discussed in Section 12-1.

Another example of a hypothesis is: Does the red box of Monstrous Yucchy Flakes sell more units per day on the average, in a specified region, than the green box? Suppose the data-collection phase shows that the average red-box sales is 12.4 cases per day, while average green-box sales is 11.9. (The data here are incomplete in that standard deviations and sample sizes are not given.) The test procedure is a way of determining whether the red-box sample average of 12.4 is enough larger than the green-box sample average of 11.9 to convince us that the universe red-box average is larger than the universe green-box average. Our answer will be a probabilistic one. For example, the data may indicate that a difference in sales averages as large or larger than the one we observed (red-box average greater than green-box average by 0.5) would occur only 1 (or 5 or 10 or some other) percent of the time, if the universe average of red-box sales were actually less than or equal to the green-box universe average. If this probability is small enough, we can accept the hypothesis that average sales for the red box is larger. "Small enough" depends on how confident the manager insists on being. Typically, probability levels of 0.01, 0.05, or 0.10 are used.

If the probability of the result is not "small enough" to convince us that the red-box universe average exceeds the green-box universe average, we can either take more data (if that is possible within time and money constraints), or we can remain unconvinced. In the latter case, we will go ahead with whatever decision we would have made without the data.

We should note that both the hypothesis and the probability level to be used should be established before the sample data are observed. Otherwise, the hypothesis and the test may not be independent of the sample values. While it is reasonable to use data to formulate a hypothesis, the hypothesis must be tested using new data, and not the observations that led to the formation of the hypothesis.

In the example above, the stated hypothesis was that μ_R (mean sales of the red boxes) is greater than μ_G (mean sales of green boxes). That is, $\mu_R > \mu_G$ is one hypothesis. Alternatively, we could state a hypothesis as $\mu_R \leq \mu_G$. These two hypotheses cover all possibilities for μ_R and μ_G. That is, one of these two hypotheses must be true. In this chapter we will deal only with cases where two hypotheses cover all possibilities. One of the hypotheses will always be designated as the one to accept if the data are wishy-washy or not convincing. How this designated hypothesis is chosen is discussed in the next section. Hypothesis testing also requires that a significance level, which is just a probability, be chosen. The significance level is the 1, 5, or 10 percent probability mentioned above that must be "small enough."

The hypothesis $\mu_R > \mu_G$ is an hypothesis involving two means. An easier problem to deal with involves only one mean. For example, suppose we wish to learn if packaging Monstrous Yucchy Flakes in colored boxes has had some overall effect on sales. Before introducing the new colorful packaging, the product was sold in plain white boxes. The average daily sales of the white box in the region was 22.8 cases. The firm that makes Monstrous Yucchy Flakes is interested in the total packaging effect. That is, they want to know: Is $\bar{x}_R + \bar{x}_G = 24.3$ enough greater than 22.8, the old mean rate, to convince the firm that the new packaging's overall effect is positive? (They had several years of data for the previous packaging, so 22.8 is assumed to be the universe mean if the new packaging has had no effect.) Thus, the hypothesis that the new mean rate exceeds 22.8 involves only one mean. The data are certainly in the direction needed to support the hypothesis since 24.3 exceeds 22.8, but the statistical question relates to how likely an $\bar{x}$ value as high or higher than 24.3 is if the mean is really 22.8. That is, is 24.3 enough larger than 22.8 that there is only a 5 percent (or 1 percent or some other level) chance of an $\bar{x}$ value as high or higher than 24.3 if the mean is 22.8? This chapter will show how to answer questions of this sort.

12-1 Introduction to Hypothesis Testing

Most of the data for the Monstrous Yucchy Flakes example are introduced above. However, the sample sizes and sample standard deviations were not given. These values are shown in Table 12-1. The entire set of data is not given here because of the amount of data involved; the data set would consist of numbers such as 12.15 (cases of red boxes sold on the first day), meaning that 12 cases of 20 plus 3 boxes were sold.

Table 12-1 Average Daily Sales Figures for Monstrous Yucchy Flakes

---

Old average = 22.8 (universe mean before the change to colored boxes)
New data:
    Overall sales; statistics[a] computed based on 25 data values on total sales of both box colors:

$$\bar{x}_n = 24.3, \quad s = 5.0.$$

    Broken down by color; statistics[a] computed based on 25 data points for each box color:

$$\bar{x}_R = 12.4, \quad s_R = 4.0$$
$$\bar{x}_G = 11.9, \quad s_G = 3.7.$$

---

[a] The unbiased estimator of s is used in all three cases.

The basic question relates to a comparison of daily sales with the new packaging to daily sales before the change. The hypothesis is: $\mu_n > 22.8$.

Decision Making: Tests of Hypotheses

Before testing that hypothesis, we pause to give several definitions, including the definition of an hypothesis.

> An *hypothesis* is a value or set of values for an unknown parameter such as the mean or standard deviation.
> A *simple hypothesis* consists of one possible parameter value, such as: $\mu = 10$.
> A *composite hypothesis* consists of more than one possible parameter value, such as: $\mu \geq 10$. (This includes all values that equal or exceed 10, an infinite number of possible values.)

Examples of different kinds of hypotheses are given below, where H_0 and H_1 are used to denote the two hypotheses in each case.

1. H_0: The percentage of patients cured by the new treatment is the same as it was under the old treatment, where it was 45 percent.
   H_1: The cure rate is different from 45 percent.
2. H_0: The average demand is below the breakeven value of 1,000 units.
   H_1: The average demand is greater than or equal to the breakeven value of 1,000 units.
3. H_0: The average diameter of a machined part is between 0.499 and 0.501 inch.
   H_1: The average diameter of the machined part is not within that range.

In case 1, H_0 is a simple hypothesis. The remaining hypotheses are composite hypotheses, in that many possible parameter values are included. For example, in case 3, H_0 contains all possible values of the mean between 0.499 and 0.501. An alternative way to state that hypothesis is: H_0: $0.499 \leq \mu \leq 0.501$. Two more examples, involving simple hypotheses, are as follows:

4. H_0: $\mu = 0$ (a simple hypothesis).
   H_1: $\mu \neq 0$ (a composite hypothesis).
5. H_0: $\mu = 1$ (a simple hypothesis).
   H_1: $\mu = 2$ (a simple hypothesis).

In case 5, no other values for μ are possible.

In each case, we denote the two hypotheses by H_0 and H_1. The hypothesis denoted by H_0 will always be the hypothesis that we choose to accept unless the data strongly suggest that it is false. H_1 will be the hypothesis we accept only if the data strongly imply that H_0 is false, and, thus, H_1 is true. Never missing an opportunity to come up with new words, statisticians have called these two hypotheses the *null hypothesis*, H_0, and the *alternate hypothesis*, H_1. In this chapter we will deal only with situations involving two hypotheses.

New Symbol	Meaning
H_0	The null hypothesis. The hypothesis we will accept unless strongly convinced otherwise.
H_1	The alternate hypothesis.

At this point you may wonder why this chapter is titled "Decision Making: Tests of Hypotheses." Where does the "Decision Making" come into the picture? Based on the data at hand, we can either accept H_0 as the correct hypothesis or accept H_1 as the correct hypothesis. That is the only decision to be made. Based on that decision, certain other actions might be taken. For example, if we accept the hypothesis that sales have not changed, we may return to the old packaging for Monstrous Yucchy Flakes. However, hypothesis testing is concerned only with which hypothesis to accept, not what happens after that point. When we design a hypothesis test, we will also define the possible data sets that will lead us to (make the decision to) accept H_0 and the possible data sets that lead us to (make the decision to) accept H_1.

This means that we must make two choices. First, we must determine which hypothesis is to be the null hypothesis. Second, we must determine the set of sample values (typically for some statistic) that lead to the acceptance of the null hypothesis. Both of these decisions are related to the kinds of errors we might make.

12-2 Selecting the Null Hypothesis

The choice of a null hypothesis is based on the relative seriousness of the different errors that we can make. Errors mean incorrect decisions. There are two kinds of incorrect decisions that we can make in hypothesis testing. We can incorrectly choose H_1. That is, we can accept H_1 when H_0 is true. Alternatively, we can accept H_0 when H_1 is true. Some books refer to this latter error as "not accepting H_1 when it is true," arguing that although the data are not what we would expect under H_0, they are not sufficiently contradictory to allow us to reject H_0. Although it is true that the data often do not strongly support H_0, this is merely a semantic difference; we will use "accept H_0" as the phrase of choice. The important question in both situations is what real decisions are to be made based on the data. Figure 12-1 summarizes the correct and incorrect decisions that can be made.

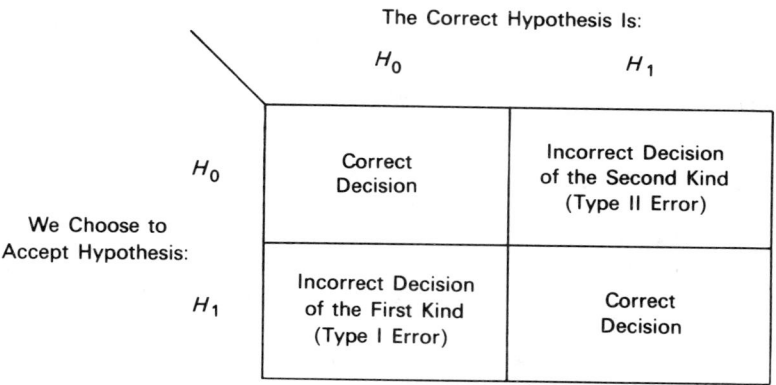

Figure 12-1 Decisions in Hypothesis Testing

The two correct decisions are accepting H_0 when it is true and accepting H_1 when it is true. The two incorrect decisions are accepting H_1 when H_0 is the true hypothesis, and accepting H_0 when H_1 is the true hypothesis. These errors are called "incorrect decision of the first kind" and "incorrect decision of the second kind," respectively. Since these phrases are rather lengthy, the shorter phrases Type I error and Type II error are typically used.

A *Type I error* in hypothesis testing is the error of accepting H_1 when H_0 is true (incorrectly rejecting H_0).

A *Type II error* in hypothesis testing is the error of accepting H_0 when H_1 is true (incorrectly accepting H_0).

Examples of Type I and Type II errors can be given using the following hypotheses.

H_0: The average demand is less than or equal to the break-even value of 1,000 units.

H_1: The average demand is greater than the break-even level of 1,000 units.

A Type I error is believing, on the basis of the sample, that average demand is above the break-even level (believing that H_1 is true) when, in fact, it is not above the break-even level (and H_0 is true). A Type II error is believing that average demand is below the break-even level (believing that H_0 is true) when in fact it is above the break-even level (and H_1 is true). As a mnemonic device, remember that the Type I error is the error of incorrectly accepting H_1. This error is the one that will be set at a low value, such as

0.01 or 0.05. (Nearly everyone has trouble remembering which error is which. If you have trouble, you can always do what the authors do; look it up in the book!)

When making the decision regarding which hypothesis to accept, we do not, of course, know which hypothesis is correct. But the choice of the null hypothesis is based on which of the possible errors is more costly. We select the null hypothesis so as to avoid the more costly error. Suppose, for example, that an ultimate decision to build a new plant or not is to be based on whether the manager believes that the average demand will exceed the breakeven level or not. Two types of potential costs are present:

1. The out-of-pocket cost if the plant is built but demand turns out to be below break-even.

2. The lost profits if the plant is not built and demand would have exceeded break-even.

It is typical for a manager to act as if cost 1 is the more costly error, implying that out-of-pocket losses are more important than potential profits foregone. (It is also true that a new plant with inadequate demand is painfully obvious. If the plant is not built, the manager may never know what was missed, but no one else will know either.)

If the costs of building in the face of inadequate demand are viewed as more important than potential profits foregone, the manager will attempt to avoid building unless quite sure demand is above the break-even level. Hence,

H_0: Average demand is less than or equal to the break-even level.

H_1: Average demand is greater than the break-even level.

The manager would only accept H_1, and proceed to build the plant, if the data strongly suggest that H_1 is true. In some situations, however, the losses from potential profits foregone may be deemed the more important cost to avoid. If so, the hypotheses would be reversed.

Since this is not an easy concept, let us try one more example. Suppose the federal government is considering giving approval for a pharmaceutical company to release a new drug to the public. From the government's position, the "costs" are:

1. The "costs" due to potential harmful side effects if the drug is not sufficiently safe.

2. The "costs" incurred by not helping sick people if the drug is sufficiently safe.

DECISION MAKING: TESTS OF HYPOTHESES

The governmental agencies charged with this decision have traditionally considered the costs in case 1 more severe (in part because of their concern that the manufacturer may give them inadequate attention). If so, the government would select:

H_0: The drug is not sufficiently safe.

H_1: The drug is sufficiently safe.

By these choices the government avoids releasing the new drug unless their data strongly suggest that the harmful side effects are at most minor, and often they are avoidable entirely through proper use or prescription.

Now let us return to the case of Monstrous Yucchy Flakes and choose a null hypothesis. The two hypotheses relate to the overall effect of the new colorful packaging. They are: (1) the new packaging has led to an increase in sales, and (2) the new packaging has not led to an increase in sales. Which of these hypotheses should be denoted as the null hypothesis? It depends on the possible decisions under consideration. Suppose the decision involved is whether to continue the new packaging or return to the old packaging. The important costs in both cases include the opportunity loss of foregone profits and the cost of packaging. Suppose that it is cheaper to package the product using the old, plain packaging, and the manager wants to maintain the new, more expensive, packaging only if it clearly has increased demand. Then the hypotheses are:

H_0: $\mu_n \leq 22.8$ (there has been no increase in average sales)

H_1: $\mu_n > 22.8$ (there has been an increase in average sales).

However, if packaging is an insignificant portion of the total cost, the manager may want to use the new packaging if there is any chance that it has increased sales. Or, alternatively, the manager does not want to bother to change back unless convinced it is necessary. In this case:

H_0: $\mu_n \geq 22.8$

H_1: $\mu_n < 22.8$.

But what do we do if one type of cost is about the same as the other? The answer is that instead of simply having two decisions, accept H_0 or H_1, there is a third possibility. Namely, take more data until we are convinced that either H_0 or H_1 is true. The sample data will be consistent with one of the two alternatives. The problem can then be set up to test whether the alternative with which the sample is *not* consistent can be rejected. For example an $\bar{x}$

of 24.3 is consistent with $\mu_n > 22.8$. Hence $\mu_n \leq 22.8$ would be selected as the null hypothesis. If the null hypothesis can not be rejected, additional data is required. This approach is a viable alternative, and we will discuss it further in Section 12-5. For now, we assume that the standard formulation holds, in which we will accept a prespecified H_0 unless the data strongly indicate that it is false.

Having discussed the choice of the null hypothesis, we must deal with the second choice mentioned previously. We must decide what sample values will cause us to accept H_0. Earlier, we said that a hypothesis test would give all possible data sets for which H_0 would be accepted and all possible data sets for which H_1 would be accepted. That is, for any set of sample data obtained, a decision must be implied by that data. When the hypotheses relate to the universe mean, as in the Monstrous Yucchy Flakes example, we must have a specified hypothesis to accept for each possible value of the sample mean. This specification is called a *decision rule*.

> A *decision rule* gives a unique decision for each possible sample that might be observed. Typically, the rule is based on the sample statistic related to the universe parameter of interest.

This sounds more complex than it is. As an example, we might have the following decision rule for the Monstrous Yucchy Flakes problem. Where the hypotheses are:

H_0: $\mu_n \leq 22.8$ (demand has not increased)

H_1: $\mu_n > 22.8$ (demand has increased)

Accept H_1 if $\bar{x} \geq 24.0$; otherwise, accept H_0.

For every possible sample value (for every $\bar{x}$), a decision is implied. In hypothesis testing, a decision rule can be stated by giving the values of the statistic for which H_1 will be accepted. It is implied that all other values lead to the acceptance of H_0. The set of values of the statistic for which H_1, the alternate hypothesis, is accepted, is called the *critical region*. Giving a critical region is a complete statement of how an hypothesis will be tested. Observing the sample, computing values for the statistic of interest, and seeing which hypothesis is actually accepted is an application of the hypothesis test. In the case above, H_1 is accepted if $\bar{x} \geq 24.0$, and, since $\bar{x} = 24.3$ occurred, H_1 would be accepted. The critical region is denoted by the value(s) of the sample statistic which separate the region where H_0 is accepted from the region where H_1 is accepted. Statisticians define such numbers to be *critical values*. If the hypothesis concerns the universe mean, the critical value(s) is a value of the sample mean and is denoted by $\bar{x}_c$. In the Monstrous Yucchy Flakes case, $\bar{x}_c = 24.0$.

Decision Making: Tests of Hypotheses

New Symbol	Meaning
$\bar{x}_C$	The *critical value* for an hypothesis test concerning the mean. It bounds the critical region, which contains sample mean values that lead to the acceptance of H_1. The critical region is $\bar{x} \geq \bar{x}_C$, $\bar{x} \leq \bar{x}_C$ or $\bar{x}$ outside the interval defined by two such critical values.

At this point, 24.0 was just "pulled out of the air." We do not know what error probabilities are implied by the choice of 24.0 as the critical value. That discussion is left to the next section. Before turning to that discussion, a few more examples of possible critical regions, for different hypotheses, will be given. They will be given without justification of the particular numbers chosen, because we are only concerned with the general idea at this time.

H_0: $0.499 \leq \mu \leq 0.501$

H_1: $\mu > 0.501$ or $\mu < 0.499$ (that is, μ is not in the H_0 range).

Since H_0 is what we will believe unless the data strongly indicate that is is false, if $\bar{x}$ is in the H_0 range or "close" to it, we will accept H_0. Hence, $\bar{x}$ values of $0.498 < \bar{x} < 0.502$ might cause us to accept H_0. The critical region, where H_1 is accepted, is then defined by two critical numbers, $\bar{x}_{C,\text{lower}} = 0.498$ and $\bar{x}_{C,\text{upper}} = 0.502$. H_1 would be accepted, then, for $\bar{x} \geq 0.502$ or $\bar{x} \leq 0.498$.

In general, all values of the statistic ($\bar{x}$ here) that are in H_0 (here 0.499 to 0.501) lead to acceptance of H_0, and some other values of the statistic also do. The critical region, where H_1 is accepted, will be smaller than the H_1 region itself. Thus, even though $\bar{x} = 0.4985$ is within the H_1 range, it still leads to the acceptance of H_0, using our arbitrarily selected critical region. The critical region here contains values of $\bar{x}$ that are either above a certain level (0.502 here) or below a certain level (0.498). This is the type of critical region that always occurs when the null hypothesis is that the mean is within a certain interval or equal to a given value. This type of null hypothesis and the related critical region are shown in Figure 12-2, along with the other basic forms that we have discussed. Figure 12-2 has two lines, one showing values of μ and the other showing sample values of $\bar{x}$, for each situation.

Figure 12-2a and b represent the discussion of the preceding paragraph. Figure 12-2c represents the case of Monstrous Yucchy Flakes, where H_0 is $\mu \leq 22.8$. Figure 12-2d represents the case where H_0 is that $\mu \geq 22.8$. In all

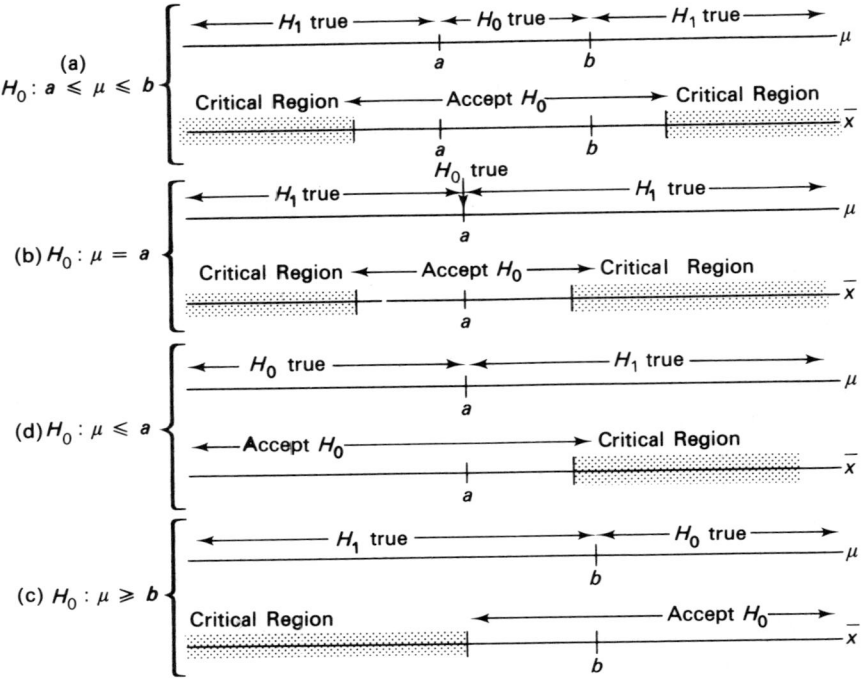

Figure 12-2 Null Hypotheses and Critical Regions

cases, the critical region lies entirely outside the null-hypothesis region. The statistical properties of the critical regions (decision rules) shown in Figure 12-2 will be studied in the next section.

S.S. An advertising manager is interested in the effect that a new advertising campaign is having. The average sales before the new theme was introduced were 3,000 units per month. Average daily sales during the 3 months since the new campaign began have been 3,124. You are to:

(a) Formulate a null hypothesis and an alternate hypothesis. Explain your choice in terms of the cost of a wrong decision.

(b) Using your answer to part (a), indicate what a Type I error is; what incorrect advertising decision might it cause?

(c) Show a possible critical region for your hypotheses and give the critical value. You need not support your choice statistically; just give a region of the proper form.

(d) Are there any potential problems for the manager to contemplate?

Decision Making: Tests of Hypotheses

Solution:

(a) The choice is unclear. If the new campaign is costly, the manager may elect to return to the old theme unless the new one is proved superior. Alternatively, the cost justification might be that the firm was operating profitably before, and we dare not risk getting worse. Then

$H_0: \mu_n \leq 3{,}000$

$H_1: \mu_n > 3{,}000.$

There are also several possible justifications for setting the hypotheses the other way. The answers for parts (b), (c), and (d) assume that the hypotheses are as given here.

(b) A Type I error is incorrectly believing the new campaign is superior when in fact it is not. The potential decision error would be to continue the new theme when the firm should return to the old one.

(c) One critical region would include all sample means greater than or equal to 3,100. That is, $\bar{x}$ must exceed 3,000 by several units (100 here) before we are convinced that H_1 is true. If a sample mean greater than or equal to 3,100 occurs, H_1 is accepted.

(d) There are several things to occupy the manager's mind. First, is 3,000 the universe value of the old mean, or just an estimate? Second, and more important, a new advertising campaign requires time to "settle down"; the first few monthly data points may not be representative of the long-run average for the campaign.

12-3 Computing $\bar{x}_c$ Values: An Example

In Section 12-2, we learned how to select the null hypothesis. We also learned that of the two types of possible errors, we are more concerned about keeping the probability of a Type I error small. A manager would like to know, in an hypothesis testing situation, how likely an error of each type is. A Type I error, accepting H_1 when it is false, is deemed more costly than a Type II error. In fact, the null hypothesis was selected as the hypothesis we will accept unless the data strongly indicate that it is false. We wish to make the probability of a Type I error small. A common choice is to set this probability (somewhat arbitrarily) at 0.05 or 0.01. This probability of a Type I error, which is selected by the decision maker, is called the *level of significance*, or *significance level* of the hypothesis test.

> The *level of significance* is the largest probability allowed of accepting H_1 when H_0 is true, in an hypothesis test. Significance is

frequently referred to in phrases such as: "significant at the 5 percent level" or "significant at the 0.01 level." We will denote the level of significance by α.

New Symbol	Meaning
α	The level of significance. The largest probability allowed of a type I error.

Suppose that the manager selects a level of significance (α value) of 0.05 in the Yucchy Flakes problem. How can this value be used to establish the critical region? First, we note that this is by definition the largest chance the manager is willing to take of accepting H_1 when, in fact, H_0 is true. The largest chance occurs in this case when μ is at the highest possible value consistent with H_0, namely $\mu = 22.8$. Figure 12-3 shows the sampling

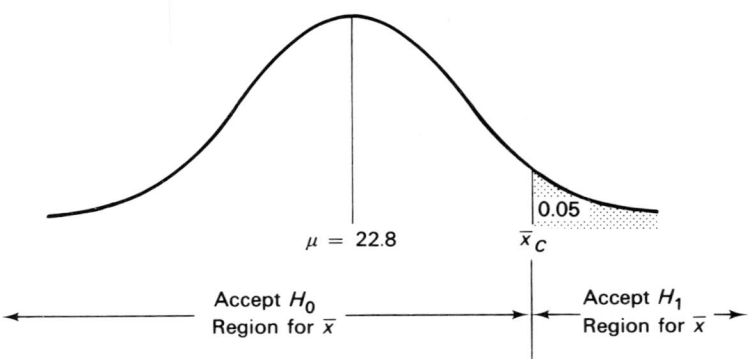

Figure 12-3 Determination of the Critical Region: Monstrous Yucchy Flakes Problem

distribution of $\bar{x}$. Given $\mu = 22.8$, the $\bar{x}$ value that leaves 5 percent of the area in the right tail is the critical value, since that $\bar{x}_c$ value allows at most a 5 percent chance of a Type I error. The chance of error is 5 percent if $\mu = 22.8$; it is less than 5 percent if $\mu < 22.8$.

The data values for Monstrous Yucchy Flakes are $\bar{x} = 24.3$, $s = 5.0$, and $n = 25$, so $s_{\bar{x}} = 5.0/\sqrt{25} = 1.0$. We will assume that the sample means are normally distributed, and, since $s_{\bar{x}}$ is an estimate of $\sigma_{\bar{x}}$, $(\bar{x} - \mu)/s_{\bar{x}}$ is t-distributed. As shown in Figure 12-3, we want the $\bar{x}_c$ that leaves 5 percent in the tail. Thus,

$$\frac{\bar{x}_c - 22.8}{1.0} = t_{0.45} = 1.711.$$

Decision Making: Tests of Hypotheses

Solving, we obtain

$$\bar{x}_C = 22.8 + 1.711(1.0) = 22.8 + t_{0.45} s_{\bar{x}} = 24.511.$$

The decision rule is

$$\text{Accept } H_1 \text{ if } \bar{x} \geq \bar{x}_C$$

$$\text{Accept } H_0 \text{ if } \bar{x} < \bar{x}_C.$$

Since $\bar{x} = 24.3$ was the sample result, H_0 is accepted; the manager is not 95 percent sure that sales have increased with the new packaging.

If the manager had set $\alpha = 0.01$, the $\bar{x}_C$ value would be computed using

$$\frac{\bar{x}_C - 22.8}{1.0} = t_{0.49} = 2.492$$

and

$$\bar{x}_C = 22.8 + 2.492(1.0) = 25.292.$$

H_0 would still be selected since $\bar{x} = 24.3 < \bar{x}_C = 25.292$. If, however, $\alpha = 0.10$ were chosen, then

$$\frac{\bar{x}_C - 22.8}{1.0} = t_{0.40} = 1.318$$

and

$$\bar{x}_C = 22.8 + 1.318(1.0) = 24.118.$$

The manager can be 90 percent (significance level, $\alpha = 0.10$) sure that the new packaging has increased sales, even though he cannot be 95 percent ($\alpha = 0.05$) or 99 percent ($\alpha = 0.01$) sure. In practice it is difficult to choose an α level. We can say that the more serious a Type I error is (as compared to a Type II error), the lower the significance level (α) should be. The decision should be made based on information about the costs and probabilities of incorrect decisions. The probability of incorrect decision is discussed in the next section. Methods that explicitly introduce costs into the analysis are introduced in Chapter 13. Several other hypothesis-testing situations are discussed in Section 12-5, and one more example is given in the Student Should below.

S.S. A boring machine is considered to be running acceptably if it is boring holes with a mean between 0.50 and 0.51 inch. From past experience, we know that $\sigma = 0.02$. Since stopping the process is more expensive than producing some defective parts that will have to be scrapped, the firm assumes that the machine is working properly unless convinced otherwise. We are to use $\alpha = 0.05$ and a sample of size 100.

Figure 12-2a shows that there are two critical numbers in this situation. This means that there are also two directions in which errors can be made,

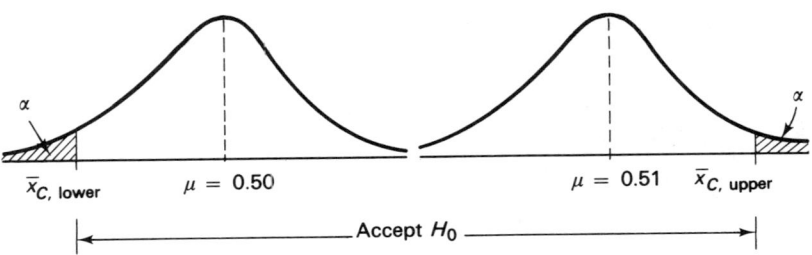

Figure 12-4 Determining the Critical Region in the Boring Machine Problem

and the error is largest when $\mu = 0.50$ *or* when $\mu = 0.51$. A picture of the distribution of sample means for both cases is shown in Figure 12-4, and the critical values are shown. For a given μ, only one of the two critical values is likely to come into play, and, hence, all of the 0.05 probability may be placed in the tail. (For example, if $\mu = 0.50$, there is essentially no area under the curve above $\bar{x}_{C,\,upper}$. Hence the way to reject the null hypothesis is to obtain an $x < \bar{x}_{C,\,lower}$. We want the maximum probability of rejecting H_0 to be 0.05 if $\mu = 0.50$, and we may place the entire 0.05 in the lower tail when calculating $\bar{x}_{C,\,lower}$.)

State the null and alternate hypotheses and find the two critical values. If $\bar{x} = 0.511$ is the sample mean, which hypothesis is accepted?

Solution:

$$H_0: 0.50 \le \mu \le 0.51$$

$$H_1: \mu > 0.51 \text{ or } \mu < 0.50$$

$$\sigma_{\bar{x}} = \frac{\sigma}{\sqrt{n}} = \frac{0.02}{\sqrt{100}} = 0.002.$$

The normal distribution can be used as the distribution of the sample means even if the universe is not normal, because of the Central Limit Theorem and since $\sigma_{\bar{x}}$ is known. (If $\sigma_{\bar{x}}$ were not known, with $n = 100$ we could still use the normal to approximate the t distribution.)

$$-z_{0.45} = \frac{\bar{x}_{C,\text{lower}} - 0.50}{\sigma_{\bar{x}}} \quad \text{and} \quad z_{0.45} = \frac{\bar{x}_{C,\text{upper}} - 0.51}{\sigma_{\bar{x}}}$$

Thus,

$$\bar{x}_{C,\text{lower}} = 0.50 - 1.645(0.002) = 0.497$$

and

$$\bar{x}_{C,\text{upper}} = 0.51 + 1.645(0.002) = 0.513.$$

We will accept H_0 if: $0.497 < \bar{x} < 0.513$, and we will accept H_1 if: $\bar{x} \le 0.497$ or $\bar{x} \ge 0.514$. Since $\bar{x} = 0.513$, H_0 is accepted.

12-4 Probability of Making Incorrect Decisions

In solving for the critical region in the Monstrous Yucchy Flakes problem, we used the probability of 0.05 of making an incorrect decision. The incorrect decision was to accept H_1 when $\mu = 22.8$ (and thus H_0) was true. But no

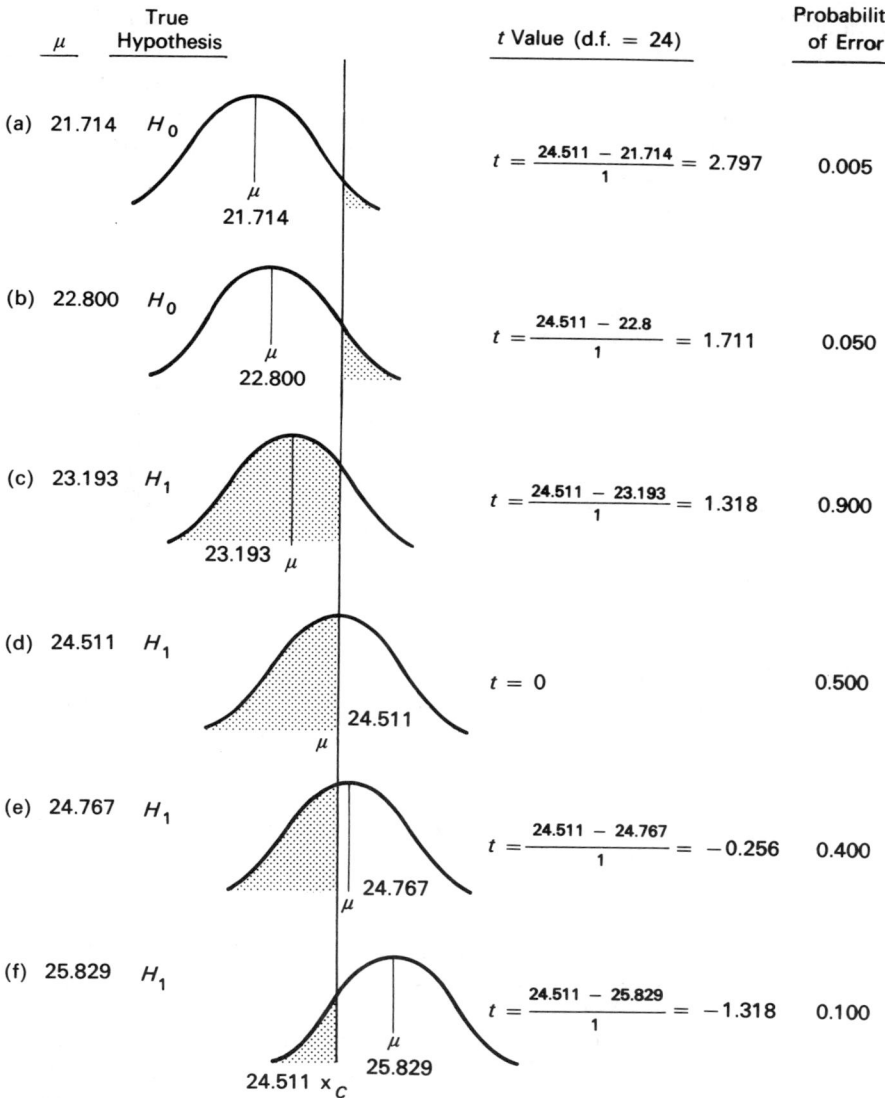

Figure 12-5 Calculation of Error Probabilities: $\bar{x}_C = 24.511$

matter what the value of μ, it is always possible to make a mistake. For example, if $\mu = 22.9$, then H_1 (H_1 is $\mu > 22.8$) is correct, and it would be a mistake to accept H_0. If $\mu = 21.4$, H_0 is true, and it would be a mistake to accept H_1. We will calculate the probability of some of these mistakes, using the previously calculated value of $\bar{x}_C = 24.511$. This is done in Figure 12-5 for six different values of μ. (In every case, we are finding the error probability for an assumed value of the mean. The six chosen values correspond to values for the t distribution that are found in Table II.) The probability of a wrong decision for each value of μ is indicated by the shaded area, which is the area of $\bar{x}$ values where the wrong hypothesis will be accepted. For example, when $\mu = 22$, we incorrectly accept H_1 if $\bar{x} \geq \bar{x}_C = 24.511$. If $\mu = 23.193$, we incorrectly accept H_0 if $\bar{x} < \bar{x}_C = 24.511$. The errors shown in Figure 12-5 are summarized in Table 12-2.

Table 12-2 Errors and their Probabilities

If μ Is:	The Correct Hypothesis Is:	The Error Is:	The Probability of Error Is:
21.714	H_0	Type I	0.005
22.800	H_0	Type I	0.050
23.193	H_1	Type II	0.900
24.511	H_1	Type II	0.500
24.767	H_1	Type II	0.400
25.829	H_1	Type II	0.100

One other important error probability, not shown in Table 12-2, can be obtained without further calculations. If the probability of a Type I error is 0.05 for $\mu = 22.8$, then the probability of a Type II error at 22.80001 (just above 22.8) is 0.95. This is so since the probability of accepting H_0 is still 0.95, but now that is the wrong thing to do. That is, the largest probability of a Type II error is always $1 - \alpha$.

The above data can be used to plot an *error characteristic curve* for the present situation (see Figure 12-6).

> The *error characteristic curve* pictorially shows the probability of error, in an hypothesis-testing situation involving an unknown mean, for all possible values of the unknown mean.

The error characteristic curve depicts the probability of incorrectly accepting H_0 to the right of 22.8, and 1 minus that probability (which is the probability of incorrectly accepting H_1) to the left of 22.8. If, in another example, H_0 was of the form $\mu \geq a$ (rather than the $\mu \leq 22.8$ here), the

DECISION MAKING: TESTS OF HYPOTHESES

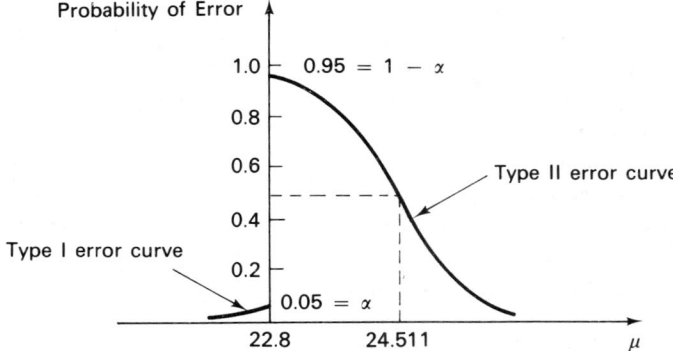

Figure 12-6 Error Characteristic Curve for the Monstrous Yucchy Flakes Problem

error characteristic curve is shaped like the mirror image of Figure 12-6 (see Figure 12-7a). If H_0 is of the form $\mu = a$, the error characteristic curve is shaped like the curve in 12-7b. In Figure 12-7b, only one point exists depicting the Type I error, since H_0 is true only at one point. At $\mu = a$, the probability of a Type I error is α. At all other points, the error curve depicts a Type II error.

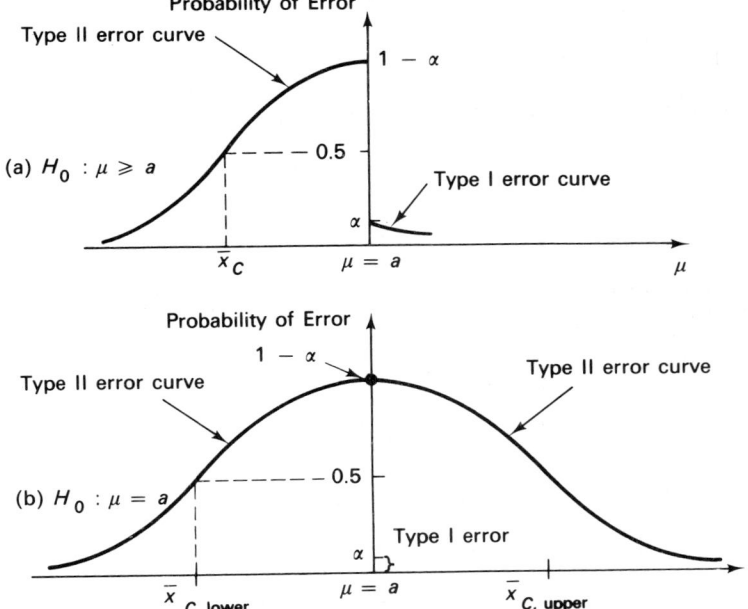

Figure 12-7 Error Characteristic Curves for Two Types of Null Hypothesis

It would be no fun, but we could compute as many points on an error characteristic curve as we wish. However, since the 0.5, α, and $1 - \alpha$ probabilities occur at known values of μ, and since the general shape of an error characteristic curve is known, two or three other points are sufficient to accurately sketch the curve.

Sometimes error probabilities are shown by plotting the probability of rejecting H_0 for all values of μ. This curve is called the *power curve*. It gives the same information as the error characteristic curve, but in different form. In the Yucchy Flakes problem the power curve plots the Type I error for $\mu \leq 22.8$ and one minus the Type II error for $\mu > 22.8$.

> The *power* of an hypothesis test refers to the ability of the test to reject the null hypothesis when it is false. Given two possible hypothesis tests with the same significance level α (they may have different critical values), the one that has a lower curve for Type II error is the more powerful test.

For the same significance level, α, a more powerful test will typically have a lower curve for both types of error. As an example, a more powerful test for the Monstrous Yucchy Flakes problem would use $n = 100$ and $\bar{x}_c = 23.623$. Figure 12-8 illustrates this case; the dashed line shows the error curve for the more powerful test. (Nonbelievers may make the calculations to verify that a more powerful test is obtained with $n = 100$ and $\bar{x}_c = 23.623$.)

S.S. A quality control manager is interested in keeping the weight of a particular part to less than 2.0 pounds. She wants to inspect a few items occasionally to see if the items are too heavy. If they are too heavy, the

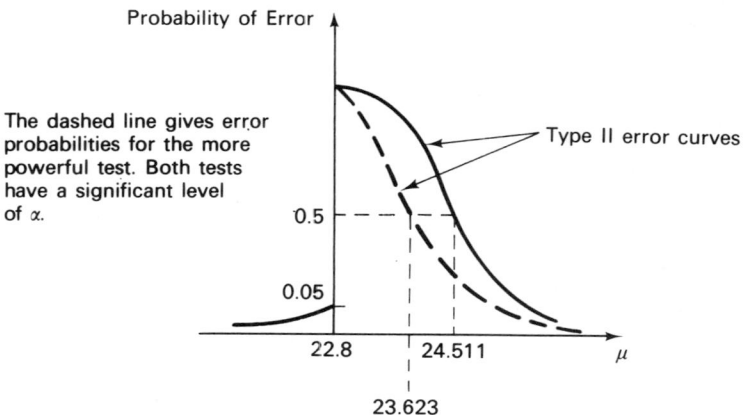

Figure 12-8 Illustration of a More Powerful Test

DECISION MAKING: TESTS OF HYPOTHESES

firm must perform major maintenance on the machine. She has used the following hypotheses:

$$H_0: \mu \leq 2.0$$
$$H_1: \mu > 2.0.$$

She knows that $\sigma = 0.04$ and $n = 16$, so $\sigma_{\bar{x}} = 0.01$.

(a) Using $\bar{x}_c = 2.01$, find a few error probabilities and sketch the error characteristic curve. Assume that the individual weights are normally distributed.

(b) Find an $\bar{x}_c$ that gives a significance level of 0.05.

Solution:

(a)

μ	Error of Type	Probability of Error
1.99	I	0.0228
2.00	I	0.1587
2.005	II	0.6915
2.01	II	0.5000
2.015	II	0.3085
2.02	II	0.1587

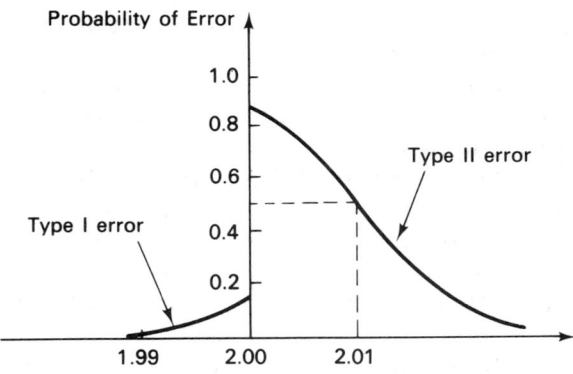

(b) $\bar{x}_c = 2.00 + z_{0.45}\sigma_{\bar{x}} = 2.01645.$

12-5 Hypothesis Tests

In the previous sections, the basic structure of hypothesis testing has been developed. In this chapter, the design of several types of hypothesis tests is discussed. In this section, we will deal with tests of the following hypotheses, involving means, μ, and proportions, p.

1. $H_0: \mu \leq a$ $H_0: p \leq a$
    or
    $H_1: \mu > a$ $H_1: p > a$
2. $H_0: \mu \geq a$ $H_0: p \geq a$
    or
    $H_1: \mu < a$ $H_1: p < a$
3. $H_0: \mu = a$ $H_0: p = a$
    or
    $H_1: \mu \neq a$ $H_1: p \neq a$.

(In all of the first three cases, a is some constant.)

4. $H_0: \mu_1 \leq \mu_2$ $H_0: p_1 \leq p_2$
    or
    $H_1: \mu_1 > \mu_2$ $H_1: p_1 > p_2$
5. $H_0: \mu_1 \geq \mu_2$ $H_0: p_1 \geq p_2$
    or
    $H_1: \mu_1 < \mu_2$ $H_1: p_1 < p_2$
6. $H_0: \mu_1 = \mu_2$ $H_0: p_1 = p_2$
    or
    $H_1: \mu_1 \neq \mu_2$ $H_1: p_1 \neq p_2$.

Hypotheses involving proportions are dealt with in the same manner as hypotheses involving means when σ is known. The normal approximation to the binomial must be appropriate if we are to proceed with tests about p. The first three hypotheses above deal with only one parameter, while the last three deal with two parameters. In these latter situations there will be two samples, one drawn from each of the populations being compared. (Examples of two such sample statistics are the $\bar{x}_R$ and $\bar{x}_G$ values in the Monstrous Yucchy Flakes problem, representing average daily sales of red and green boxes.) An important assumption that is necessary in a two-parameter hypothesis test is that the two populations are independent. The test we make will be based on that assumption.

In the first three cases, the statistic to be used in the test is the sample mean or sample proportion, and the critical region will be found using the methods discussed in Section 12-3. In the second three cases, the statistics used for the hypothesis test will be $\bar{x}_1 - \bar{x}_2$ or $p'_1 - p'_2$, the differences between the two sample statistics. The null hypotheses in 4, 5, and 6 then are written $\mu_1 - \mu_2 \leq 0$, $\mu_1 - \mu_2 \geq 0$, and $\mu_1 - \mu_2 = 0$, respectively, for means; $p_1 - p_2$ is substituted for $\mu_1 - \mu_2$ for proportions. Once the proper assumptions and preparatory calculations are made, these cases will be treated in the same manner as the first three cases.

Decision Making: Tests of Hypotheses

Under each category below, we will study the case where σ is known and the case where σ is unknown. In the σ-unknown case, when the sample size is small but the sample means are assumed to be normally distributed, a t value will be used instead of a z value. In all cases, we will assume that the sample means are normally distributed. If σ is unknown and the sample size exceeds 30, the normal distribution can be used as an approximation to the t distribution. Now, on with the show.

Case 1. $H_0: \mu \leq a$, where a significance level of α is desired.
$H_1: \mu > a$.

The critical region (where H_1 is accepted) is $\bar{x} \geq \bar{x}_C$, so $\bar{x}_C$ is the value we need. When $\sigma_{\bar{X}}$ is known,

$$\bar{x}_C = a + z_{0.5-\alpha}\sigma_{\bar{X}}.$$

This guarantees that the probability of an $\bar{x}$ value higher than $\bar{x}_C$ is α, if the mean is exactly equal to a. Thus, the probability is at most α for any μ in the H_0 range. For example, in the Monstrous Yucchy Flakes example, if we were to assume (inappropriately) that $\sigma = s = 5$, then

$$\bar{x}_C = 22.8 + 1.645(1) = 24.445,$$

using the normal distribution and

$$\sigma_{\bar{X}} = 1.0, \quad \alpha = 0.05, \quad a = 22.8.$$

If $\sigma_{\bar{X}}$ is unknown, as is really the case in the Monstrous Yucchy Flakes problem, then

$$\bar{x}_C = a + t_{0.5-\alpha}s_{\bar{x}}.$$

For example, with $n = 25$, there are 24 degrees of freedom, and $t_{0.5-\alpha} = t_{0.45} = 1.711$. Then, as we found before,

$$\bar{x}_C = 22.8 + 1.711(1) = 24.511.$$

This implies that H_1 is accepted only if $\bar{x} > 24.511$ occurs. In the example $\bar{x} = 24.3$ occurred. Applying the test, we see that H_0 is accepted. Remember that 24.3 does not really validate H_0; in fact it is more consistent with H_1. However, by assuming H_0 to be $\mu \leq 22.8$ we assumed the costs were such that accepting H_0 is the preferred action, unless we are convinced that sales have increased. Rather than stating that H_0 is accepted, we might say that "H_0 cannot be rejected at the 0.05 level." This phrase is more descriptive of what can be said on the basis of the statistic.

If we are dealing with proportions, the equivalent hypotheses are:

$$H_0: p \le a$$
$$H_1: p > a.$$

H_0 will be accepted for all values less than some p'_c, a critical value. The only difference is that we use the normal approximation to the binomial, where

$$\sigma_{p'} = \sqrt{\frac{p(1-p)}{n}},$$

and the estimate of $\sigma_{p'}$, based on the sample proportion, is

$$s_{p'} = \sqrt{\frac{p'(1-p')}{n}}.$$

If the assumption of normality is appropriate, which requires np and $n(1-p)$ of at least 30 (a somewhat less stringent rule of thumb is $np(1-p) \ge 3$), we compute

$$p'_c = a + z_{0.5-\alpha}\sigma_{p'}.$$

New Symbol	Meaning
p'_c	A *critical value*, analogous to $\bar{x}_C$, to be used in hypothesis tests involving proportions.

Since the hypothesis is tested assuming H_0 is true at the $p = a$ value,

$$\sigma_{p'} = \sqrt{\frac{a(1-a)}{n}}$$

is used. In every hypothesis testing situation in this section for the proportion, the technique is identical to the σ known case for the mean, except that $\sigma_{p'}$, computed using

$$\sigma_{p'} = \sqrt{\frac{a(1-a)}{n}},$$

replaces $\sigma_{\bar{x}}$.

Case 2. $H_0: \mu \ge a$ $H_0: p \ge a$
                    or
$H_1: \mu < a$ $H_1: p < a.$

The critical region is of the form $\bar{x} \leq x_C$ or $p' \leq p'_C$. When $\sigma_{\bar{X}}$ or $\sigma_{p'}$ is assumed known:

$$\bar{x}_C = a - z_{0.5-\alpha}\sigma_{\bar{X}} \quad \text{or} \quad p'_C = a - z_{0.5-\alpha}\sigma_{p'}.$$

As in the above case,

$$\sigma_{p'} = \sqrt{\frac{a(1-a)}{n}}$$

is used to evaluate $\sigma_{p'}$. When $\sigma_{\bar{X}}$ is unknown and $s_{\bar{x}}$ is used as an estimate, then for $\bar{x}_C$ only:

$$\bar{x}_C = a - t_{0.5-\alpha}s_{\bar{x}}.$$

These critical regions guarantee that a sample value less than or equal to $\bar{x}_C$ or p'_C occurs with probability α if the universe parameter value (μ or p) is a. Thus, for a value higher than a, the probability of a Type I error is less than α. It is important to understand the logic behind this argument. Basically the argument is that $\bar{x}_C$ or p'_C is chosen to guarantee exactly an α probability of a Type I error when $\mu = a$ or $p = a$. When $\bar{x}_C$ or p'_C is even farther away from μ (as is the case for any $\mu > a$ or $p > a$), the probability of error decreases.

Case 3. $H_0: \mu = a \qquad H_0: p = a$
$\qquad\qquad\qquad$ or
$\qquad H_1: \mu \neq a \qquad H_1: p \neq a.$

The critical region (where H_1 is accepted) includes all values of $\bar{x}$ (or p') that are either below an $\bar{x}_{C,\text{lower}}$ or $p'_{C,\text{lower}}$) or above an $\bar{x}_{C,\text{upper}}$ (or $p'_{C,\text{upper}}$). Figure 12-2b gives a pictorial representation of the critical region, and Figure 12-9 shows the Type I errors. This time, to have a maximum Type I error of α, we must have a maximum Type I error of $\alpha/2$ in each of the two directions. That is, if $\mu = a$ (or $p = a$) there must be an $\alpha/2$ probability of obtaining a sample value below the lower critical value ($\bar{x}_{C,\text{lower}}$ or $p'_{C,\text{lower}}$) and an $\alpha/2$ probability of obtaining a sample value above the upper critical value ($\bar{x}_{C,\text{upper}}$ or $p'_{C,\text{upper}}$). This type of hypothesis test is called a *two-tailed test*, whereas the previous type of hypothesis tests we have examined is called *one-tailed tests*.

Figure 12-9 is drawn for the case involving $\mu = a$. The $p = a$ case is identical if normality is assumed. A two-tailed test applies when we are trying to show that something has changed, but we are not sure in what direction the change has occurred. For example, we could have used this form of hypothesis for the Monstrous Yucchy Flakes example: $H_0: \mu = 22.8$ and $H_1: \mu \neq 22.8$. Using the approach we used earlier implies that only a positive change is of interest. If only one direction of change is of interest to

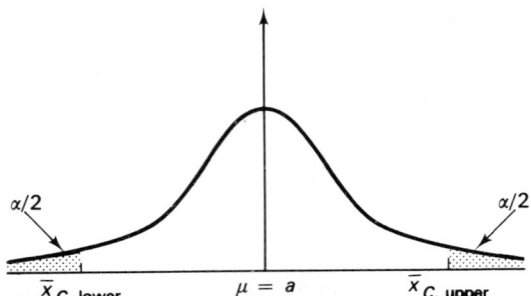

Figure 12-9 Type I Errors for a Two-Tailed Test

us, or if we know what kind of change to expect, a one-tailed test is appropriate. Otherwise, a two-tailed test should be used. The reason for preferring a one-tailed test *if* the direction of change is known is that all the Type I error probability can be placed in one tail. This enables us to accept H_1 more frequently. (Thus, a one-tailed test is more powerful, since it is less likely to incorrectly reject H_1.) For example, if $\alpha = 0.05$ and an $\bar{x}$ value occurs such that a value that high or higher would occur 3 percent of the time if the null hypothesis is true, a one-tailed test would accept H_1, while a two-tailed test would accept H_0.

Moving on to find the critical region:

$$\bar{x}_{C,\text{lower}} = a - z_{0.5-\alpha/2}\sigma_{\bar{x}}$$

and

$$\bar{x}_{C,\text{upper}} = a + z_{0.5-\alpha/2}\sigma_{\bar{x}},$$

if σ is known.

$$\bar{x}_{C,\text{lower}} = a - t_{0.5-\alpha/2}s_{\bar{x}}$$

and

$$\bar{x}_{C,\text{upper}} = a + t_{0.5-\alpha/2}s_{\bar{x}},$$

if σ is unknown.

For proportions:

$$p'_{C,\text{lower}} = a - z_{0.5-\alpha/2}\sigma_{p'}$$

and

$$p'_{C,\text{upper}} = a + z_{0.5-\alpha/2}\sigma_{p'},$$

using the normal approximation, as before with

$$\sigma_{p'} = \sqrt{\frac{a(1-a)}{n}},$$

which is the correct value of $\sigma_{p'}$ under the null hypothesis.

DECISION MAKING: TESTS OF HYPOTHESES

These critical regions guarantee that, for example, an $\bar{x}$ value lower than $\bar{x}_{C,\text{lower}}$ will occur with probability $\alpha/2$ if $\mu = a$. Symbolically,

$$P(\bar{x} \leq \bar{x}_{C,\text{lower}} \mid \mu = a) = \alpha/2.$$

The two tails combine to yield a Type I error equal to $\alpha/2 + \alpha/2 = \alpha$.

For an example, suppose that 10 percent of the movie customers at a local theater bought popcorn before a new automatic buttering machine was installed. During the first week after installation, 285 of the 2,500 customers bought popcorn. The management might use a two-tailed test to determine if a change in the percentage has occurred, since, ahead of time, they might feel they do not know if the change would be beneficial or detrimental. They want a significance level of 0.01.

$$H_0: p = 0.10$$

$$H_1: p \neq 0.10$$

$$p' = \frac{285}{2500} = 0.114.$$

$$\sigma_{p'} = \sqrt{\frac{a(1-a)}{n}} = \sqrt{\frac{(0.1)(0.9)}{2{,}500}} = 0.006.$$

The normal approximation is justified. Now, the critical region can be obtained.

$$p'_{C,\text{lower}} = a - z_{0.5-\alpha/2}\sigma_{p'} = 0.10 - z_{0.495}(0.006)$$

$$= 0.10 - 2.576(0.006) = 0.085$$

$$p'_{C,\text{upper}} = 0.10 + 2.576(0.006) = 0.115.$$

Since p' is not above $p'_{C,\text{upper}}$, and since p' is not below $p'_{C,\text{lower}}$, H_1 is not accepted; we accept H_0. But something is wrong here! A larger percentage of customers bought popcorn, but they still reject H_1. That is true because of the choice of $\alpha = 0.01$. If they had chosen $\alpha = 0.03$ (in which case they still could be 97 percent sure of making the correct decision), H_1 would be accepted. In reporting the results, we should say that H_1 can be accepted at the 0.03 level but not at the 0.01 level. A great danger in hypothesis testing is in setting the α level arbitrarily, not quite making it, and then making what is probably an incorrect decision. Section 12-6 discusses this notion further. We also note that if a one-tailed test had been used, as might be appropriate in this case, we would have accepted the alternate hypothesis.

Case 4. $H_0: \mu_1 \leq \mu_2$ $H_0: p_1 \leq p_2$
            or
$H_1: \mu_1 > \mu_2$ $H_1: p_1 > p_2$.

The problem here involves two $\bar{x}$ or p' values rather than one. Both sample values are subject to error, whereas in comparing $\bar{x}$ to 22.8 as we did above, 22.8 is assumed to be the universe parameter value and, hence, not subject to error. The statistic we will use in the test is $\bar{x}_1 - \bar{x}_2$ ($p'_1 - p'_2$ for problems involving proportions). The hypotheses can be restated to reflect the test statistic as:

$H_0: \mu_1 - \mu_2 \leq 0$ $H_0: p_1 - p_2 \leq 0$
            or
$H_1: \mu_1 - \mu_2 > 0$ $H_1: p_1 - p_2 > 0$.

The only information we need now is the standard deviation of $\bar{x}_1 - \bar{x}_2$ (or $p'_1 - p'_2$). Once this is known, a hypothesis test can be constructed as before. From Section 7-3, we know that

$$\sigma^2_{\bar{X}_1-\bar{X}_2} = \sigma^2_{\bar{X}_1} + \sigma^2_{\bar{X}_2},$$

if X_1 and X_2 are independent. Thus, assuming independence, we can write

$$\sigma_{\bar{X}_1-\bar{X}_2} = \sqrt{\sigma^2_{\bar{X}_1} + \sigma^2_{\bar{X}_2}}. \qquad (12\text{-}1)$$

Similarly,

$$\sigma_{p'_1-p'_2} = \sqrt{\sigma^2_{p'_1} + \sigma^2_{p'_2}} = \sqrt{\frac{p_1(1-p_1)}{n_1} + \frac{p_2(1-p_2)}{n_2}} \qquad (12\text{-}2)$$

Then, since $a = 0$ here,

$$\bar{x}_C = 0 + z_{0.5-\alpha} \sigma_{\bar{X}_1-\bar{X}_2},$$

which is similar to case 1 above, or

$$p'_C = 0 + z_{0.5-\alpha} \sigma_{p'_1-p'_2}.$$

For example, in the Monstrous Yucchy Flakes case, suppose that $\alpha = 0.05$ is chosen and the hypotheses are

$H_0: \mu_R \leq \mu_G$

$H_1: \mu_R > \mu_G$.

(That is, we wish to see if sales of the red box are significantly better.) The data are: $\bar{x}_R = 12.4$, $\bar{x}_G = 11.9$, $s_R = 4.0$, $s_G = 3.7$, and $n = 25$ for both samples. For the moment, assume that the σ values are known; that is, assume $s_R = \sigma_R$ and $s_G = \sigma_G$, so

$$\sigma_{X_R - X_G} = \sqrt{\sigma_{\bar{X}_R}^2 + \sigma_{\bar{X}_G}^2} = \sqrt{\left[\frac{(4.0)^2}{25}\right] + \left[\frac{(3.7)^2}{25}\right]}$$

$$= \sqrt{0.64 + 0.5476} = \sqrt{1.1876} = 1.09.$$

Then

$$\bar{x}_C = 0 + z_{0.45}\sigma_{X_1 - X_2} = 1.645(1.09) = 1.79.$$

The management will only believe that $\mu_R > \mu_G$ if $\bar{x}_R - \bar{x}_G \geq 1.79$. That is not the case ($\bar{x}_R - \bar{x}_G = 0.5$ for our data), so the null hypothesis is accepted.

If σ is not assumed to be known in problems involving means, we may be able to use a matched pairs approach. If the data are taken pairwise (for example, one green-box sales matched with one red-box sales from the same time and place), then the $x_1 - x_2$ values can be obtained one at a time. Those values are then treated as the data ($y = x_1 - x_2$), and we test:

$$H_0: \mu_y \leq 0 \quad (\mu_{x_1} \leq \mu_{x_2})$$
$$H_1: \mu_y > 0 \quad (\mu_{x_1} > \mu_{x_2}).$$

This *matched-pairs* approach is a simple solution if a reasonable means of matching exists. If there are different numbers of readings from the two populations, it cannot be used, and it can be used for the same number of readings only if the data were taken in a pairwise fashion. If they were not so taken, we can obtain an approximate answer using equation (12-1), by substituting the sample variances for $\sigma_{\bar{X}_1}^2$ and $\sigma_{\bar{X}_2}^2$ and using d.f. $= n_1 + n_2 - 2$.

For the situation involving proportions, the test is always made assuming the null hypothesis is true at the point where $p_1 = p_2$. The results of the two samples are treated as though they came from a single population, and an estimate of this single value of p is given by

$$p' = \frac{n_1 p_1' + n_2 p_2'}{n_1 + n_2}. \tag{12-3}$$

Using p' from (12-3) for both p_1 and p_2 in (12-2) gives

$$s_{p_1' - p_2'}^2 = \frac{p'(1-p')}{n_1} + \frac{p'(1-p')}{n_2} = p'(1-p')\left(\frac{1}{n_1} + \frac{1}{n_2}\right). \tag{12-4}$$

Everything required for cases (5) and (6) has been introduced. Independence must still be assumed. Then using either $\sigma_{X_1-X_2}$, $s_{\bar{x}_1-\bar{x}_2}$ or $s_{p_1'-p_2'}$, case (5) is analogous to case (2) and case (6) is analogous to case (3). Thus only the formulas for the critical regions will be given.

Case 5. $H_0: \mu_1 \geq \mu_2 \qquad H_0: p_1 \geq p_2$
or
$H_1: \mu_1 < \mu_2 \qquad H_1: p_1 < p_2$
$\bar{x}_C = 0 - z_{0.5-\alpha}\sigma_{X_1-X_2}$ if σ_{X_1} and σ_{X_2} are known
$\bar{x}_C = 0 - t_{0.5-\alpha}s_{\bar{x}_1-\bar{x}_2}$ if σ_{X_1} and σ_{X_2} are unknown.

The t value is taken from the t table with $n_1 + n_2 - 2$ degrees of freedom. Finally,

$$p_C' = 0 - z_{0.5-\alpha}s_{p_1'-p_2'},$$

since the sample results will always be used in this case. Equations (12-3) and (12-4) are used to compute $s_{p_1'-p_2'}$.
In all cases, H_1 is accepted only if the sample value $(\bar{x}_1 - \bar{x}_2$ or $p_1' - p_2')$ is less than the critical value. We observe that case 5 is redundant, since, if we want to use $\mu_1 \geq \mu_2$ as the null hypothesis, we could study $\mu_2 - \mu_1$ instead of $\mu_1 - \mu_2$. Then case 4 could be used. Both are included here to show the analogy with cases 1 and 2.

Case 6. $H_0: \mu_1 = \mu_2 \qquad H_0: p_1 = p_2$
or
$H_1: \mu_1 \neq \mu_2 \qquad H_1: p_1 \neq p_2$
$\left.\begin{array}{l}\bar{x}_{C,\text{lower}} = 0 - z_{0.5-\alpha/2}\sigma_{X_1-X_2}\\ \bar{x}_{C,\text{upper}} = 0 + z_{0.5-\alpha/2}\sigma_{X_1-X_2}\end{array}\right\}$ if σ_{X_1} and σ_{X_2} are known
$\left.\begin{array}{l}\bar{x}_{C,\text{lower}} = 0 - t_{0.5-\alpha/2}s_{\bar{x}_1-\bar{x}_2}\\ \bar{x}_{C,\text{upper}} = 0 + t_{0.5-\alpha/2}s_{\bar{x}_1-\bar{x}_2}\end{array}\right\}$ if σ_{X_1} and σ_{X_2} are unknown.

Finally,

$$p_{C,\text{lower}}' = 0 - z_{0.5-\alpha/2}s_{p_1'-p_2'}$$

$$p_{C,\text{upper}}' = 0 + z_{0.5-\alpha/2}s_{p_1'-p_2'}.$$

where the sample standard deviation is computed using equations (12-3) and (12-4).
In all cases, H_1 is accepted only if the sample value $(\bar{x}_1 - \bar{x}_2$ or $p_1' - p_2')$ is either below the lower critical value or above the upper critical value.

DECISION MAKING: TESTS OF HYPOTHESES

There are a lot of cases described in this section. Trying to memorize all the critical regions would be futile. We can look them up, or, better still, understand how they are derived. Each $\bar{x}_c$ or p'_c value is found by using a constant value, a, taken from the null-hypothesis statement, and setting $\bar{x}_c$ to allow a Type I error of either α (one-tailed test) or $\alpha/2$ (two-tailed test). This is done by adding or subtracting the appropriate number of standard deviations to or from a. [One way to check your critical value(s) is to see if it lies outside the H_0 range; it should be outside H_0.]

S.S. A book publisher is interested in whether new editions of textbooks sell better than the old edition would. They have some data, on sales before and after a new edition:

Book Number	First Year Sales of the New Edition	Previous Year Sales of the Old Edition	New – Old Difference
1	12,000	8,000	4,000
2	12,000	11,000	1,000
3	2,500	3,000	– 500
4	7,500	4,000	3,500
5	15,000	15,000	0

Assume the universe distribution of sales for both editions to be normally distributed. Data of interest, using unbiased estimators for the standard deviation, are:

$$\bar{X}_{new} = \bar{X}_1 = 9{,}800 \quad \text{and} \quad s_1 = \sqrt{23{,}825{,}000} = 4{,}881$$

$$\bar{X}_{old} = \bar{X}_2 = 8{,}200 \quad \text{and} \quad s_2 = \sqrt{24{,}700{,}000} = 4{,}970$$

$$\bar{X}_{difference} = \bar{X}_D = 1{,}600 \quad \text{and} \quad s_{difference} = s_D = \sqrt{4{,}175{,}000} = 2{,}043.$$

Also:

$$s^2_{\bar{x}_1 - \bar{x}_2} = \frac{23{,}825{,}000}{4} + \frac{24{,}700{,}000}{4}$$

$$= 9{,}705{,}000 \quad \text{and} \quad s_{\bar{x}_1 - \bar{x}_2} = \sqrt{9{,}705{,}000} = 3{,}115;$$

$$s_{\bar{x}_2} = \frac{4{,}970}{\sqrt{5}} = 2{,}223, \quad s_{\bar{x}_1} = \frac{4{,}881}{\sqrt{5}} = 2{,}183, \quad s_{\bar{x}_D} = \frac{2{,}043}{\sqrt{5}} = 914.$$

The company's null hypothesis is that no increase has occurred. This null-hypothesis choice is made because revisions are expensive. We will assume that the firm always uses $\alpha = 0.10$ as the significance level.

(a) Using the difference data only, test the null hypothesis, H_0: $\mu_D \leq 0$, against the alternate hypothesis. This is the method of matched pairs, which can be used in this case since the matching is based on a common book.

(b) Test the hypothesis $H_0: \mu_1 - \mu_2 \leq 0$ against the alternate hypothesis. The difference, namely 1,600, is the statistic to be used in applying both this test and the test in question (a).

(c) The firm feels that books selling 6,500 or more copies break even. Test the hypothesis, $H_0: \mu_1 \leq 6,500$, against the alternate hypothesis.

Solution:

(a)
$$\bar{x}_C = 0 + t_{0.5-\alpha} s_{\bar{x}_D} = 0 + t_{0.4}(914)$$
$$= 1.533(914) = 1,401.$$

(The t-distribution value with 4 degrees of freedom is used.) The alternate hypothesis can be accepted since $\bar{x}_D = 1,600 > \bar{x}_C$.

(b) In this case, the hard work has been done for you.
$$\bar{x}_C = 0 + t_{0.4} s_{\bar{x}_1 - \bar{x}_2} = 1.397(3,115) = 4,352$$
$$(\bar{x}_C = 4,352 > \bar{x}_1 - \bar{x}_2 = 1,600).$$

This time, 8 degrees of freedom, $n_1 + n_2 - 2$, are used. The alternate hypothesis cannot be accepted. You will note that the matched-pairs technique accepted H_1, whereas the other approach did not. That is because the matched-pairs approach eliminates the variability within the "before" or "after" data that is not related to the difference, and centers on the variability within the differences themselves. (It is a more powerful test to use when it is justified by a common matching unit.)

(c)
$$\bar{x}_C = 6,500 + t_{0.4}(s_{\bar{x}_1})$$
$$= 6,500 + 1.533(2,183) = 9,846.$$

We cannot accept the alternate hypothesis since $\bar{x}_1 = 9,800 < 9,846 = \bar{x}_C$. A value of 9,800 certainly looks larger than 6,500, but $\bar{x}_C$ turns out to be higher than 9,800 because of the small sample size. (A larger sample size would reduce $t_{0.4}$ and probably reduce $s_{\bar{x}_1}$. We cannot say what would happen to $\bar{x}_1$.) Rather than accepting H_0 in this case, the firm would be well advised to take more data if cost and time permit.

12-6 A Warning

The first warning related to hypothesis testing is one that we have seen before: be aware of the assumptions. In various parts of this chapter, normality (of the distribution of the sample means or proportions) is assumed, $\sigma_{\bar{x}}$ or $\sigma_{p'}$ is assumed known, and, in cases 4, 5, and 6, $\bar{x}_1$ and $\bar{x}_2$ are assumed to be based on independent observations. The normality assumption holds for the distribution of sample means because of the Central Limit Theorem unless the sample size is small and the basic universe is drastically nonnormal. The assumption of a known standard deviation is only made when the sample size used in estimating the standard deviation is reasonably large. The manager or researcher should always be aware of these assumptions and the possible impact on the conclusions if they are not valid.

Cases 1, 2, and 3 of Section 12-5 use only the normality assumption and occasionally the known standard deviation assumption. For cases 4, 5, and 6 of Section 12-5, the additional assumption of independence between the two target populations must be made. How true the independence assumption is, or how far it may lead us astray, is hard to predict. The amount of potential error varies greatly; the manager should be especially alert to this assumption.

Apart from the assumptions, the major warning regarding hypothesis testing is that the tests must be used intelligently. That sounds harmless enough, but frequently it does not happen. The basic problem is that while the alternate hypothesis may not be accepted at the designated level of significance, the data may still be in the direction of H_1. For example, in the Monstrous Yucchy Flakes problem, $\bar{x}$ might exceed 22.8 but be less than $\bar{x}_C = 24.511$. When this happens, it seems incorrect to accept H_0. Several examples of this type occurred throughout this chapter. H_0 is chosen because a Type I error is deemed more costly than a Type II error, but it may not be much more costly. If the costs are roughly equal, a regular hypothesis testing approach allows too high a probability of a Type II error. (Refer to the figures in Section 12-4.)

The classical methods we have studied in this chapter are not explicit about the costs of the various errors. They merely require the decision maker to judge which type of error is more costly. Sometimes this is all that can be done. However, it makes it difficult to make a decision about how much Type I error to tolerate, knowing that a small chance of error here makes the probability of a Type II error even higher. Hence, significance levels are set arbitrarily, frequently at the 0.05 or 0.01 level. If the error costs can be quantified in a decision problem, more powerful techniques exist. We will study these techniques in Chapter 13.

Another option that is available to management but is ignored in the basic hypothesis-testing techniques is the option of taking more data, until management is convinced one way or the other. In making this decision we should consider the significance level used in the test and its reasonableness, given the particular costs involved. Although this may seem inappropriate, α levels are often chosen arbitrarily. If so, accepting H_0 may be inappropriate. In the case 3 example in Section 12-5, $\alpha = 0.01$ was chosen, and H_1 could not be accepted. If $\alpha = 0.03$ had been chosen, H_1 would be accepted. Thus the evidence strongly favors H_1, even though the 0.01 level is not attained.

In such a situation, management might choose to collect more data rather than accept the H_1 hypothesis which is probably incorrect.

The "more data" option deserves further comment. The idea is that after data are received, there are really three decisions, not just two. They are: accept H_0, accept H_1, and take more data. If the supply of data is large enough, the cost of sampling sufficiently low, and time permits, a manager can continue to take data until either H_0 or H_1 is accepted at the selected significance level. If at some point no further data can be taken, the hypothesis consistent with the sample result is selected.

Hypothesis testing is a well-developed and useful set of methods. We have discussed some of these methods in this chapter, others are discussed later, and still others will not be discussed. It is very useful to know that H_1 can be accepted at the 5 percent or at the 1 percent level. Better decisions can be made using such information. But hypothesis testing is a relatively rigid technique, and management problems are complex and demand flexibility. Proper management decision making must include examining the statistics and the related hypothesis test, as well as being willing to use judgment. Proper judgment can only be exercised when the final decision to be made is kept in mind. For example, the decision in the Monstrous Yucchy Flakes example is choosing what type of packaging to use. Accepting H_1 or H_0 is just a means to the end of deciding on packaging materials. A manager must not lose sight of the real reason for taking the data when constructing and testing an hypothesis. The manager should also consider the Type III and Type IV errors which are not statistical in nature. The Type III error is solving the wrong problem. The Type IV error is solving the right problem too late to be of value to anyone.

New Symbols

Symbol	Meaning
H_0	The null hypothesis. The hypothesis we will accept unless strongly convinced otherwise.
H_1	The alternate hypothesis.
α	The significance level. The maximum probability of a Type I error.
$\bar{x}_C$	A *critical value*, used in defining a critical region in hypothesis testing about means. For example: accept H_1 for $\bar{x} \leq \bar{x}_C$. If the critical region involves two segments, $\bar{x}_{C,\text{upper}}$ and $\bar{x}_{C,\text{lower}}$ are used.
p'_C	A *critical value* to be used in defining a critical region in hypothesis tests about proportions; $p'_{C,\text{upper}}$ and $p'_{C,\text{lower}}$ are used as stated above for $\bar{x}_C$.

DECISION MAKING: TESTS OF HYPOTHESES

Key Formula

Formula	Used to Compute:
$s^2_{p_1'-p_2'} = p'(1-p')\left(\dfrac{1}{n_1} + \dfrac{1}{n_2}\right)$	The variance of the difference between two proportions based on sample data. The formula is used because the null hypothesis is tested at $p_1 - p_2 = 0$ and $p' = \dfrac{p_1' n_1 + p_2' n_2}{n_1 + n_2}.$

The formulas for computing critical values are summarized in Section 12-5.

PROBLEMS

12-1. Define hypothesis. What is the difference between a simple hypothesis and a compound hypothesis?

12-2.* How is the null hypothesis selected? Is the choice of a null hypothesis always clear?

12-3. Define a Type I error and a Type II error. Are these the only things that a manager should worry about in applying statistics to a real problem?

12-4. Murphy's law maintains that "If something can possibly go wrong, it will." How is this law similar to the null hypothesis in an hypothesis test?

12-5.* Is a one-tailed test or a two-tailed test more powerful? Why? When should a two-tailed test be used?

12-6. Use the concept of an error curve to discuss Type I and Type II errors and to discuss the power of a test. Use the following picture to guide the discussion.

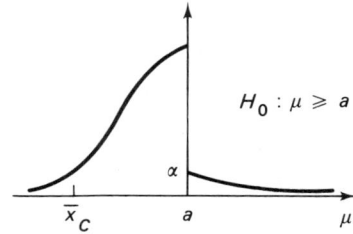

400   CHAPTER 12

12-7. A book company is interested in the sales of its product line in a certain geographic region. They are trying a new system of marketing in the area, dividing sales responsibility by topics rather than by geographic subregion. They want to know if sales have improved under the new system.
(a) What data should they collect, and what should they be wary of with regard to the data?
(b) What are the costs of making wrong decisions? Formulate an alternate and a null hypothesis based on these cost considerations. (Specify the hypotheses even though the choice may not be clear.) Using your null hypothesis, explain what a Type I error would be.
(c) Show how the hypothesis would be tested.

12-8. Following the pattern of problem 12-7, state a problem from the health care area. What data would you collect to analyze the problem? What are the possible costs of a wrong decision? What would the hypotheses be? What decision error would a Type I error cause? What form would the hypothesis test have? (Use an actual problem; a recent newspaper may be used as a source for ideas.)

12-9. Repeat Problem 12-8 for a problem faced by a local, state, or federal government.

12-10. State null and alternate hypotheses in the following situations.
(a) A production manager for a soft drink manufacturer wants to test a consumer group's contention that the plant fills its 12-ounce cans with less than 12 ounces of fluid.
(b) A quality control inspector wants to test whether the average width of a certain part is 2.5 cm.
(c) A manager wants to know if the life of a certain type of transistor is at least 150 hours.

12-11. Consider the following figure for a simple hypothesis test, where we assume there are only two possible values for μ:

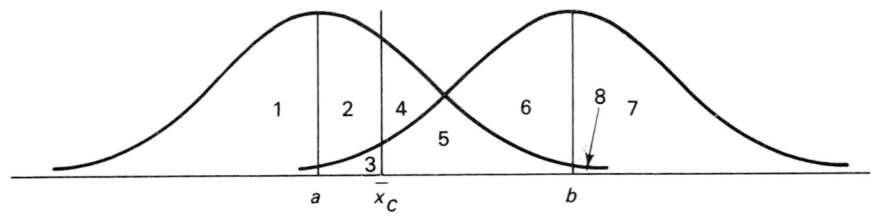

If the manager wishes that the Type I error be less than the Type II error:
(a) State the null hypothesis.
(b) State the decision rule.
(c) What is the probability [which area(s)] that the alternate hypothesis will be accepted when it is false?
(d) What probability is given by the total area in 4, 5, and 8?

12-12.* Suppose that we have: $H_0: \mu \leq 25$, $H_1: \mu > 25$, and $\bar{x}_C = 27.0$. If $\alpha = 0.05$, what is the maximum Type II error and for what value of μ does it occur? If we increase the sample size, maintain $\alpha = 0.05$, and $\bar{x}_C = 26.0$ results, will the new test be more powerful than the first test? What will the maximum Type II error be now?

12-13.* The retention of good customers for a precision tool manufacturer depends critically upon the quality of the final product. In the production of a particular tool, a hole precisely 0.500 centimeter in diameter must be bored in a metal plate. Since quality is so important, sample diameters are frequently inspected to ensure the proper functioning of the manufacturing process. In the most recent sample of 100 plates, the mean diameter was 0.503 centimeter with a standard deviation of 0.010 centimeter.
(a) State an appropriate null hypothesis and an appropriate alternate hypothesis.
(b) Test the hypothesis at the $\alpha = 0.01$ level of significance.

12-14.* A spark-plug manufacturer claims that its new spark plug will last 20,000 miles in city-driving conditions. To test the assertion, the manufacturers obtained data from 25 cars and found $\bar{x} = 20,400$. The hypothesis test used was as follows:

$$\bar{x}_C = 20,200$$
$$H_0: \mu \leq 20,000$$
$$H_1: \mu > 20,000$$

Accept H_1 if $\bar{x} \geq \bar{x}_C = 20,200$.

(a) Find the probabilities of error, using the hypothesis test above, if $\mu = 19,800$; if $\mu = 20,000$; if $\mu = 20,400$. Sketch the error curve using these values. Based on previous data on the variability of spark-plug life, they believe that $\sigma = 600$.
(b) Is there any potential problem (or problems) with the data?

12-15. In Problem 12-14, still assuming that $\sigma = 600$, compute $\bar{x}_C$ when $\alpha = 0.05$; when $\alpha = 0.10$. In both cases, will the Type II error probability when $\mu = 20{,}200$ be higher or lower than when $\bar{x}_C = 20{,}200$?

12-16. In the production of rolls of paper, long rolls are first produced and then cut into the appropriate sizes. The wasted paper depends critically upon the cutting pattern and the reliability of the cutting process. For the cutting pattern currently in use, the total waste on each long roll should be 0.75 inch in length.
   (a) State an appropriate null hypothesis and a two-sided alternate that would be applicable to an examination of the mean waste in the current cutting process.
   (b) Using the $\alpha = 0.05$ level of significance, state an appropriate decision rule to follow if a sample of the waste on 225 rolls yields a standard deviation of 0.25 inch.
   (c) Sketch an error characteristic curve for the hypotheses in parts (a) and (b).
   (d) Use the alternate hypothesis $H_1: \mu > 0.75$ to formulate a decision rule and to sketch an error characteristic curve.

12-17. In each of the following examples, a statistical procedure has been used inappropriately. In each case, specify the inappropriate procedures.
   (a) A television tube manufacturer tested 25 tubes and found the mean life to be 5,000 hours with a standard deviation of 250 hours. If the tubes were operating according to specifications, the mean life would be 5,200 hours. Since the z statistic (normality is assumed),
   $$z = \frac{5{,}000 - 5{,}200}{250/\sqrt{25}},$$
   is less than -1.96, the plant manager concluded at the $\alpha = 0.05$ level that the tubes were operating at a lower level than specifications demanded.
   (b) In a cost accounting system which utilizes standard costs, a manufacturer has set limits within which a deviation from standards may occur and yet not cause concern. Let X be a random variable which represents a deviation from standards. From past experience, the manufacturer knows that if the production process is in control, deviations will occur, but the distribution of deviations may be approximated by a normal distribution with $\mu = 0$ and $\sigma^2 = 10{,}000$. The manufacturer is willing to tolerate a 5 percent chance of not investigating an out-of-control situation and has set tolerance limits such that a deviation will not be investigated as long as
   $$-165 \leq x \leq +165.$$

DECISION MAKING: TESTS OF HYPOTHESES

12-18.* The manufacturer of a particular nonprescription drug claims that it relieves the symptoms associated with hay fever for 80 percent of the population. A test of 20 individuals found that the symptoms were relieved in 17 of the 20 cases.
(a) Using the normal approximation to the binomial, is the manufacturer's claim validated? Use $\alpha = 0.05$.
(b) Is it reasonable to use the normal approximation here?

12-19. A production manager has several machines, each of which bores holes in a particular assembly. One of these machines is set to produce holes averaging 0.75 inch in diameter. A sample of 100 units resulted in $\bar{x} = 0.753$ and $s = 0.02$. She wants to know if the machine is in perfect adjustment. If she cannot know that, she would like to be able to decide if the machine is out of adjustment currently.

12-20. A group of persons concerned with environmental protection believes that rainfall is decreasing in a geographic region due to increasing erosion of the soil in the region. (They claim that the ability of the ground to hold moisture allows the moisture to evaporate and then fall as rain again.) To validate their claim they take total rainfall readings at 50 points in a state. For the last several years, the average rainfall at these locations has been 35 inches. The average rainfall last year was 32.0 with $s_{\bar{x}} = 1.2$.
(a) Is last year's rainfall significantly lower than 35.0 inches, using a standard method of analysis and $\alpha = 0.10$?
(b) They should not proceed using a standard analysis here. Why not?

12-21. In Problems 12-14 and 12-15 we assumed that σ and hence $\sigma_{\bar{x}}$ were known. Suppose, now, the firm computes s from the sample of 25 cars and finds that $s = 600$. Find $\bar{x}_c$ for $\alpha = 0.05$ and $\alpha = 0.10$, using the fact that only s is known. Are these values smaller or larger than their counterparts in Problem 12-15? Why?

12-22.* A company produces small electric drills. A testing agency, trying to determine how long one will last, tested three units from the manufacturer. The figures, in hours, were 3,100, 3,600, and 3,700. They would like to see if average life is greater than 2,500 hours. Test that hypothesis at the 0.01 level and explain the result. State any assumptions that you must make in order to proceed.

12-23.* A grocery store chain is interested in determining if regional differences in buying habits exist, specifically with regard to average dollar value purchased. In particular, they want to know if store 1, in a metropolitan area, has an average sale per customer that is higher than store 2, in a rural area. They have data from the two stores; each set of data represents several thousand customers. They want to know if the average bill at store 1 is significantly larger than the average bill at store 2.

	Store 1	Store 2
$\bar{x}$	45.26	42.71
$s_{\bar{x}}$	0.24	0.18

12-24.* In Problem 12-23, what would change (and perform the changes) if instead of asking if μ_1 is greater than μ_2, the firm posed the question as: Is μ_1 different from μ_2? (The implication would be that they have no prior reason to suspect that the average sale in the metropolitan area is larger than the average sale in the rural area.)

12-25. A chemical company has developed a new fertilizer which it believes will increase the yield of corn. In an experiment, the new fertilizer was applied to 9 one-acre plots and a standard fertilizer was applied to a different set of 9 one-acre plots. The first set of plots yielded a mean of 58.6 bushels per acre with a standard deviation of 11.1 bushels, and the second set of plots yielded a mean of 50.3 bushels per acre with a standard deviation of 9.9 bushels. For each of the following null hypotheses, (1) express H_0 and H_1 symbolically and (2) test the hypothesis at the $\alpha = 0.05$ level of significance (assume normality).
(a) Each of the fertilizers produces the same yield per acre for corn.
(b) The new fertilizer is at least as effective as the old fertilizer.
(c) The old fertilizer is at least as effective as the new fertilizer.

12-26. A political organization wants to know if two counties of a state are significantly different in their opinions about a transportation bond issue. Two hundred individuals were sampled in each region. In the first region, 112 of the individuals indicated support of the bond issue. In the second region, 104 of the individuals indicated support of the bond issue.
(a) Are the two proportions significantly different? Use $\alpha = 0.10$.
(b) Is the average proportion $(112 + 104)/400 = 0.54$ significantly larger than 0.50? In other words, do the counties together favor the bond issue?

12-27.* In order to determine whether the views concerning a pending Congressional bill differ between voters in the North and voters in the South, a pollster randomly selected 400 voters in the North and 225 voters in the South. The following results were noted:

Voters	Number Opposing the Bill	Number Favoring the Bill	Total
North	150	250	400
South	90	135	225

Formulate appropriate hypotheses and perform a test at the $\alpha = 0.05$ level of significance.

12-28. The company in Problem 12-22, having found out that three data points are insufficient to conclude much of anything, have taken more data. They have also taken data on a competitive drill, and they want to know if their drill has an average life different from the competitive model. The data are:

Firm's model: $\bar{x}_1 = 3{,}510,\ s_1 = 140,\ n_1 = 14$
Competitive model: $\bar{x}_2 = 3{,}380,\ s_2 = 80,\ n_2 = 10$.

(a) Is the average life of their model significantly different from the average life of the competitive model? Use $\alpha = 0.05$.
(b) In part (a), a two-tailed test is implied. If they had phrased the question differently, a one-tailed test would be appropriate. Using a one-tailed test, is the average life of their model significantly longer than the average life of the competitive model? Use $\alpha = 0.05$. (Can you answer the question without making any new calculations?)

12-29. A production process is supposed to make steel bars with an average weight between 16.0 and 18.0 pounds. In order to determine whether the process is operating correctly, the plant manager periodically selects 25 bars and finds the average weight. The standard deviation of the production process is known to be 1.5 pounds, and the weights are normally distributed. The manager uses the following decision rule: If the sample average weight is between 15.4 and 18.6 pounds, conclude that the process is operating correctly; otherwise, investigate.

(a) State the null and alternate hypotheses.
(b) What is the greatest probability of a Type I error?
(c) Sketch an error characteristic curve for this problem using the following values for μ:

15.0, 15.4, 15.8, 16.2, 16.6, 17.0, 17.4, 17.8, 18.2, 18.6 and 19.0.

12-30. In a manner similar to that of determining the sample size necessary to obtain a confidence interval of a certain size when σ is known

(Section 11-4), we can determine a sample size necessary to distinguish between two means of specified values. This occurs, for example, if a machine produces parts of 0.500 inch but may jump to where $\mu = 0.502$ inch. This problem is examined in this question and the next.

(a) Suppose we know that $\sigma = 0.01$. How large a sample size is necessary (see Section 11-4) to obtain a 90 percent confidence interval around an $\bar{x}$ value that is less than or equal to 0.002 inch in width. (That is, $\bar{x} \pm 0.001$.)

(b) Does this value for n guarantee:

$$P(\bar{x} \leq 0.501 \mid \mu = 0.502, \sigma = 0.01) \leq 0.05$$
$$P(\bar{x} \geq 0.501 \mid \mu = 0.500, \sigma = 0.01) \leq 0.05?$$

(c) Discuss the meaning of the statements in (b) as they relate to a hypothesis test where H_0 is $\mu = 0.500$ and H_1 is $\mu = 0.502$.

12-31. (This problem should be worked only after solving Problem 12-30.) In Problem 12-30, we attacked a problem of the form:

$$H_0: \mu = 0.500$$
$$H_1: \mu = 0.502.$$

We wanted to be 95 percent sure of both not accepting H_0 when H_1 was true and not accepting H_1 when H_0 is true. A sample size of 271 was found to be sufficient to guarantee those goals. However, we may not want, in many situations, to have the two error probabilities equal. To guarantee a 0.05 error probability within 0.001 inch, 0.001 was set equal to $1.645\sigma_{\bar{x}}$. To guarantee a 10 percent error probability, the interval (0.001, for example) would be set equal to $1.28\sigma_{\bar{x}}$. To achieve a 5 percent limit on the probability of a Type I error, and a 10 percent limit on the probability of a Type II error, the entire interval $(0.502 - 0.500 = 0.002)$ would be set equal to $(1.645 + 1.28)\sigma_{\bar{x}}$. This can be stated in general as follows:

$$\text{Given } a < b: \quad H_0: \mu = a$$
$$H_1: \mu = b$$

(implying that no other values are possible or at least of interest). A sample size necessary to guarantee a limit of α on Type I error (accepting $\mu = b$ when $\mu = a$ is true) and a limit of β on Type II error (accepting $\mu = a$ when $\mu = b$ is true) can be found using

$$(b - a) = (z_\alpha + z_\beta)\sigma_{\bar{x}} = \frac{(z_\alpha + z_\beta)\sigma}{\sqrt{n}}.$$

Decision Making: Tests of Hypotheses

We assume that the sample mean is normally distributed and that σ is known. Then we can solve for n, using

$$n = \left[\frac{(z_\alpha + z_\beta)\sigma}{b - a}\right]^2, \quad \text{rounded up to the nearest integer.}$$

We can find a critical value $\bar{x}_C$ (for $\bar{x}$ values less than $\bar{x}_C$ we accept H_0 and for $\bar{x}$ values equal to or above $\bar{x}_C$ we accept H_1), using

$$\bar{x}_C = a + \frac{z_\alpha \sigma}{\sqrt{n}}, \quad \text{where } n \text{ is found using the above formula.}$$

(a) For the problem used in the discussion ($a = 0.500$ and $b = 0.502$), with $\alpha = 0.05$ and $\beta = 0.10$, find the appropriate value of n.

(b) If we want to attack a problem of the form:

$$H_0: \mu = a$$
$$H_1: \mu = b, \quad \text{where } a > b,$$

what changes are required in the above analysis?

12-32. Suppose that you have just taken the position of assistant inspector for a manufacturing firm. You are told that one machine, from whose output of 100, 16 parts are electronically sampled, is controlled by a computer which directs it to produce parts with either 10 or 9.5 inches between the centers of two holes in each part. The machine is known to produce parts for which this distance has a standard deviation of 0.48 inch in either case, and you are told that if the average distance between centers in the sample from a run is greater than 9.80 inches, the parts all go into a bin marked "10 inches"; otherwise, they go into a bin marked "9.5 inches."

(a) What was the company's implied decision as to the probability that they are willing to accept of placing runs with $\mu = 9.5$ into the bin marked "10 inches"? What probability are they implicitly accepting of putting a run made at $\mu = 10$ into the bin marked "9.5 inches"?

(b) Repeat part (a) using an $\bar{x}_C$ of 9.75.

12-33. A candy maker makes and sells bags of candy which are supposed to weigh at least 16 ounces. The bag-filling machine is getting old, and, as a result, the weight of candy in each bag now varies substantially. The historical distribution of the weight of candy in each bag has a roughly normal shape with a mean that can be set and a standard deviation of 1.45 ounces. The candy maker wants the bags to weigh at least 16 ounces, since if they do not, he will experience a decline in goodwill over time. It is not possible to inspect all 400 bags from each day's run before sales the next day. The manager would rather sell the bags at a substantial markdown if he believes there are very many with a weight of less than 16 ounces. At what level should the process be set so that no more than 1 percent of the bags will contain less than 16 ounces of candy?

Chapter **13**

Statistical Decision Theory

Within the last several decades important advances have been made in adapting statistical methods to decision making under uncertainty. In particular, substantial effort has been devoted to finding better methods of selecting among alternative courses of action involving uncertain payoffs. These methods have been expanded to treat the question of how new information can be incorporated into the decision process and whether such new information should be obtained, knowing its expected cost. This approach has been given the title "statistical decision theory."

In this chapter we will follow a dual approach in describing statistical decision theory. We will first introduce the ideas using as an example the drawing of identically sized chips from a vase or urn. (Every honest-to-goodness probability text has an urn example.) We shall then apply these ideas to a simplified managerial problem.

There are several reasons for using the abstract urn example. The urn model allows us to simplify the problem by focusing on the essential concepts without many of the encumbering computational complexities. Unfortunately, in the process we lose the realism that motivates much of our interest in the subject. Yet, as we have argued elsewhere in this book, any actual problem must be simplified before we can deal with it effectively. Naturally, we hope to retain the essential elements of the problem after simplification. One very important extra dividend from the abstract urn approach should result. When actual managerial problems are reduced to their basic mathematical formulation, they are usually similar to one of a few basic structures. One such structure is our urn problem. With very minor alterations many managerial problems of decision making under uncertainty will be structurally identical to our urn problem or one of its variations. Hence, if we understand the urn problem, we can adapt our understanding to many managerially similar situations.

13-1 An Urn Problem and a Managerial Problem

For our urn problem, suppose that there are 100 identically shaped urns in the laboratory of a slightly mad mathematician whom we are visiting. The urns are opaque and of two kinds only. The urns are indistinguishable except that the first type of urn, call it urn 1, has the symbol θ_1 (theta sub one) on its bottom. The second type of urn, call it urn 2, has the symbol θ_2 (theta sub two) on its bottom. (These symbols are not visible unless we turn the urn over and look at the marking.) Now suppose that one urn is selected completely at random by the mad mathematician and set before us. He now presents us with three choices or actions:

a_1: State that the urn is a θ_1 urn

a_2: State that the urn is a θ_2 urn

a_3: Refuse to become involved in such nonsense.

Before we can make a reasonable selection among these three actions we need to know what is to be gained or lost by our choice. Suppose the mad mathematician states that if we call "θ_1 urn" and are correct, he will pay us $50. If we call "$\theta_2$ urn" and the urn is a θ_2 urn, we win $80. Since he is a rich and generous old professor, we accept his promise as true. But he says, "I must not bear all the risk. If you wish to play, it will cost you $20.00." One other fact is known to us. Namely, we know or are told that 70 (that is, 70 percent) of the urns have a θ_1 on their bottoms and the rest (30 percent) have a θ_2 on their bottoms. In deciding whether or not to accept his offer, we accept the facts to be as stated and we are concerned only with our own monetary gains and losses.

Some students, particularly of management, find it difficult to abide abstract mathematical problems. If you are one of these, try to picture the urns perhaps as spittoons which cost $20 each but have the indicated profitable results on our barbering business based on the differential appeal of the note resounding from a well-aimed shot.

After a detailed examination of this urn problem in the next section, we will study the following simple managerial problem. The Flimsy Fabricating

STATISTICAL DECISION THEORY

Company is considering the introduction of a new and improved mousetrap into its product line. Hap Hazard, president of Flimsy, is aware of the old adage, "Build a better mousetrap and the world will beat a path to your door." Hap's mousetrap emits a cheesy odor, thereby attracting the mouse while saving on both cheese and fingers. The presence of the mouse in the trap is detected by a weight sensor in the floor, causing the trap to shut. The mouse is then conditioned to desire a life in the wide-open spaces. When the conditioning is complete, the trap plays a chorus or two of "Born Free," which announces to the owner that the mouse may be released.

The new trap can be easily marketed in the company's present line. However, new productive machinery will be required. The firm is considering three possible actions:

a_1: Buy a large-capacity machine at a cost of $150,000

a_2: Buy a cheaper, small-capacity machine at a cost of $90,000

a_3: Neither purchase a machine nor market the mousetrap.

The appropriate decision depends on the required level of production, which in turn depends on the level of demand for the new mousetrap. Unfortunately, Hap Hazard does not know exactly what demand level the firm will face. But since either machine will last for 10 years, Hap is interested in what demand may be over this 10-year period. To simplify the problem in his initial analysis, President Hazard has partitioned the possible demand levels into three situations.

θ_1: Demand is initially strong and remains strong for the entire 10 years.

θ_2: Demand is initially strong, but the product fails to produce repeat customers in sufficient quantities. Demand drops to a lower level after the first 2 years.

θ_3: The product never captures the consumers' attention and demand remains at a low level for the entire 10 years.

After several lengthy meetings, the management of the Flimsy Fabricating Firm is able to establish the following additional estimates for the likelihood of the demand conditions and the payoff values to the company under each.

AS TO DEMAND

The managers believe that it is twice as likely that demand will be sustained at a high level for the full 10 years as it is that high initial demand will drop off after 2 years. They also estimate that it is equally likely that demand will drop off after 2 years as it is that demand will be low the entire time.

If we let $p_j \equiv$ the probability of state θ_j, then we know that $p_1 = 2p_2$ and $p_2 = p_3$. Since the probabilities must sum to 1,

$$p_1 + p_2 + p_3 = 1$$

and substituting,

$$2p_2 + p_2 + p_2 = 1.$$

Hence,

$$p_2 = p(\theta_2) = 0.25, \quad p_1 = p(\theta_1) = 0.50, \quad p_3 = p(\theta_3) = 0.25.$$

AS TO THE PAYOFF VALUES

The firm's objective is to maximize its contribution to company fixed costs and profits. Management estimates that:

1. If it buys the large-capacity machine and demand is at a high level, the firm will make $100,000 per year in contribution to fixed costs and profit, over and above the costs of operating and maintaining the machine. However, if demand is low, the company will make only $10,000 per year above operating costs.

2. If the small-capacity machine is obtained, it will provide a contribution of $50,000 per year if demand is high. However, if demand is low, a small machine would still be efficient and yield $20,000 per year over operating costs. (The returns will not be discounted to allow for the time value of money in this example.)

We now have sufficient data to resolve both problems: the urn problem and the managerial problem. Note how much simpler the urn problem was to state (and to grasp).

13-2 Decision-Theory Model

A close examination of either problem situation described in Section 13-1 will suggest that it is composed of the following four elements:

1. Actions open to the decision maker; these will be designated by the symbols a_1, a_2, or in general by a_i.

STATISTICAL DECISION THEORY

2. States or conditions not (completely, at least) under the decision maker's control; these states will be designated by the symbols θ_1, θ_2, and θ_3, or in general by θ_j.

3. Payoff values to the decision maker for each action and under each state. We shall denote these values by the symbols v_{ij}. (Hence v_{23} is the value if the decision maker selects action 2, $i = 2$, and the state which actually occurs is state 3, $j = 3$.)

4. The probabilities of the states; these will be designated by p_1, p_2 or in general by p_j or by $f(\theta_j)$.

These four elements of our decision problem can be illustrated using a decision tree. First, consider the urn problem. Its decision tree is illustrated in Figure 13-1. The square represents a decision to be made and the circles indicate chance events. The payoff value, after the action costs are subtracted, is shown at the far right end of the decision-tree branch.

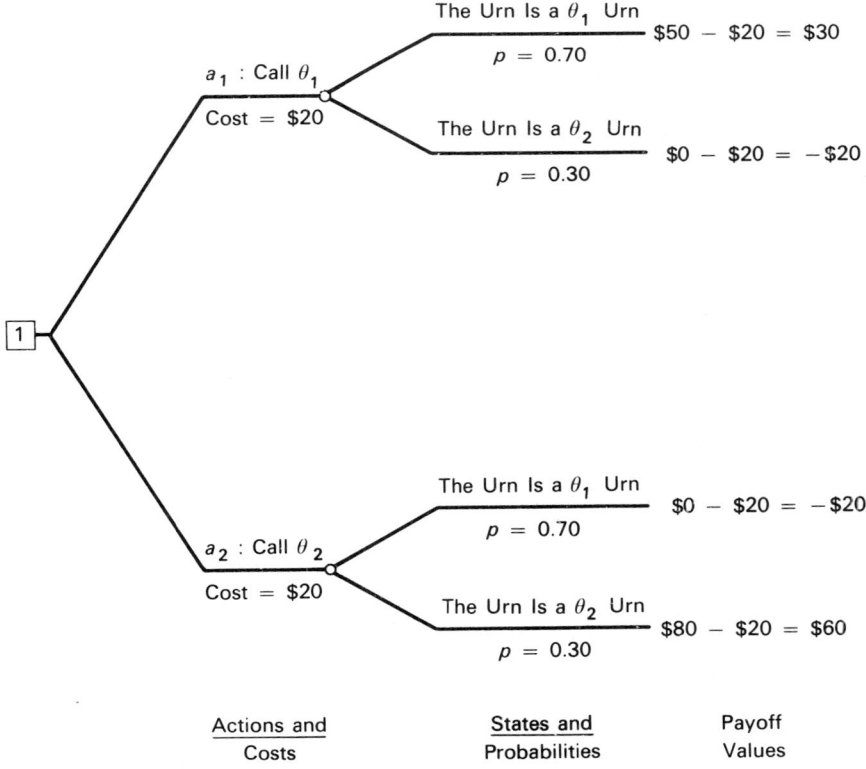

Figure 13-1 Decision Tree: Urn Problem

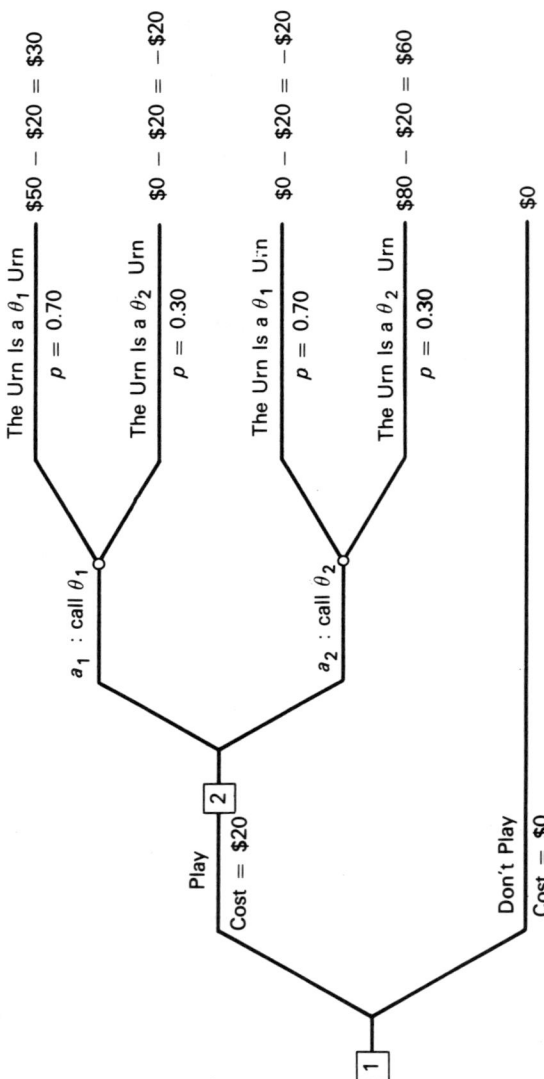

Figure 13-2 Expanded Decision Tree: Urn Problem

But wait! What happened to the third action: "I refuse to become involved in such nonsense"? We know that the payoff value from selecting this action is $0. It involves no change in our wealth and there is no uncertainty in it. Hence, this action needs no evaluation. We will use it for deciding whether or not to play. Graphically, we could say the decision maker really has two choices (made one right after the other), as depicted in Figure 13-2. (The same line could be added to Figure 13-1 as well.)

Sometimes the data from a problem like this one are most easily visualized in tabular rather than decision-tree form. The data from our urn problem are given in Table 13-1. The payoff values are placed in the body of the table. (No notation is made in the table that these are payoff values. It is understood.) The payoff values represent the winnings less the cost. For example, if we select action a_1 ("call θ_1") and the urn is a θ_1 urn, the payoff to us is $50 less the $20 fee to play or $30. The other tabular entries are obtained similarly. We may use either the decision-tree or decision-table method to work a problem, depending on which seems easier for the problem at hand.

Table 13-1 Decision Table: Urn Problem

States and Their Probabilities

Action	$\theta_1 \equiv$ Urn Is a θ_1 Urn, $f(\theta_1) = 0.70$	$\theta_2 \equiv$ Urn Is a θ_2 Urn, $f(\theta_2) = 0.30$
a_1: "call θ_1"	$30	−$20
a_2: "call θ_2"	−$20	$60
a_3: don't play	$0	$0

S.S. Construct either a decision tree or a decision table for the mousetrap problem.

Solution: Using the probabilistic data given earlier we may construct both the decision tree (see Figure 13-3) and the decision table (Table 13-2).

The payoff values for the decision tree in Figure 13-3 are obtained by subtracting the cost of purchasing the machine from the total contribution. Hence the payoff value from buying the large machine and experiencing sustained high demand is 10($100,000) − $150,000 = $850,000. This represents $100,000 for 10 years less the equipment's cost. The rest of the payoff values are as follows:

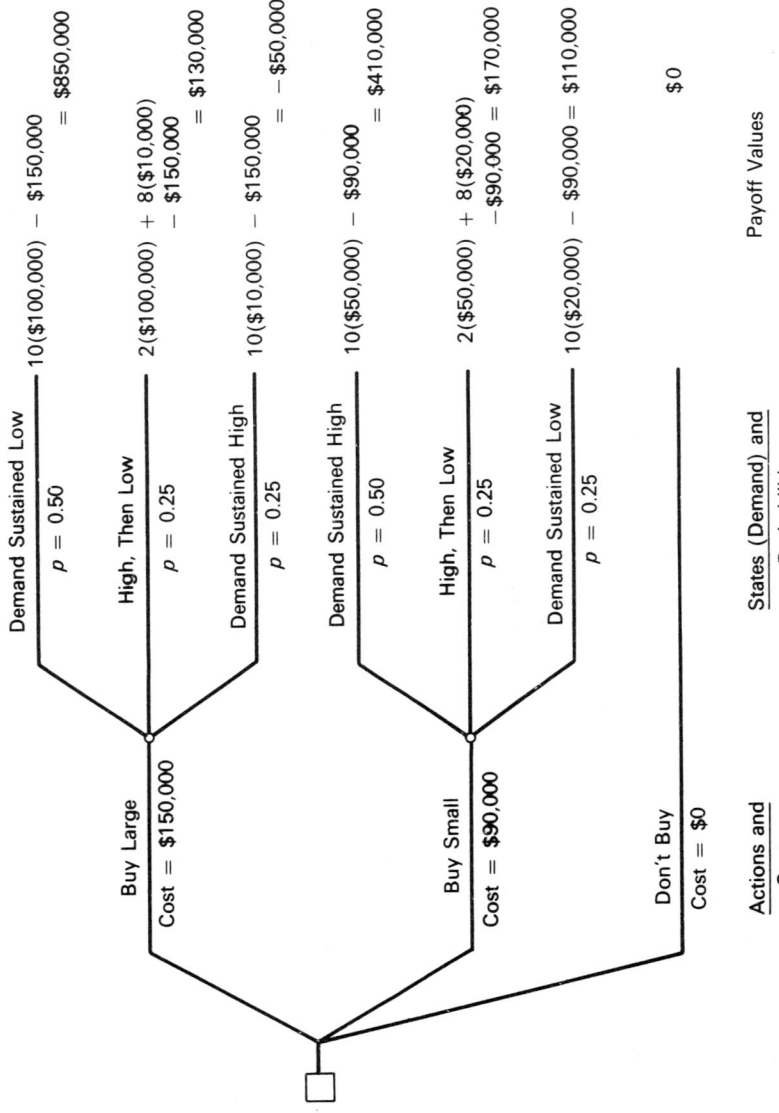

Figure 13-3 Decision Tree: Mousetrap Problem

Statistical Decision Theory

Payoff Value		Years High Demand		Years Low Demand		Machine Cost
$130,000	=	2($100,000)	+	8($10,000)	−	$150,000
−$50,000	=			10($10,000)	−	$150,000
$410,000	=	10($50,000)			−	$90,000
$170,000	=	2($50,000)	+	8($20,000)	−	$90,000
$110,000	=			10($20,000)	−	$90,000

The related decision table is given in Table 13-2.

Table 13-2 Decision Table: Mousetrap Example

States and Their Probabilities

Action	$\theta_1 \equiv$ Sustained High $p_1 \equiv f(\theta_1) = 0.50$	$\theta_2 \equiv$ High, Then Low $p_2 \equiv f(\theta_2) = 0.25$	$\theta_3 \equiv$ Sustained Low $p_3 \equiv f(\theta_3) = 0.25$
a_1: buy large	$850,000	$130,000	−$50,000
a_2: buy small	$410,000	$170,000	$110,000
a_3: don't buy	$0	$0	$0

13-3 Making an Action Choice: The Bayesian Decision Criterion

We have made substantial progress. Our problem has been analyzed, and we are now ready to make a decision. Indeed, perhaps the most important idea to be learned from statistical decision theory is a method of thinking about problems requiring decisions under uncertainty. By attempting to structure a problem in decision-theory form, we often get a better idea of just what the problem is, what the critical variables are, which estimates are most crucial to a choice of action, whether more data are needed, if any reasonable alternatives have been overlooked, and so on.

But let us not rest on our laurels. There are decisions to be made! A *decision* is defined as a choice of one action from those available. There are several different methods for choosing among actions. They are called *decision rules*. Before considering these methods, however, let us consider first the concept of dominance. We do so because dominated actions can be discarded. An action, a_k, is said to *dominate* another action, a_m, if the payoff values for a_k are at least as good as those for a_m under all states and better for at least one state. When this is the case, action a_m need no longer be considered.

For our urn problem no one action dominates any other; no action can be discarded because it is dominated. This is most easily seen in Table 13-1. If we compare action a_1 to a_3, we find a_1 is better under θ_1 and a_3 is better under θ_2. Neither action dominates the other. This is also the case if we compare a_1 with a_2 or a_2 with a_3.

However, in the mousetrap example, action a_2 dominates action a_3. As we can see from Table 13-2, no matter what the state, the payoff value for action a_2 exceeds that for action a_3. Action a_2 dominates action a_3. Hence, we may discard action a_3 from further consideration. This is true even though we have not decided that action a_2 is better than action a_1.

We now turn to a consideration of several different methods of selecting among nondominated actions. These methods are called decision rules. Different decision makers will prefer different decision rules and a single decision maker may even change from one rule to another in different situations. The four decision rules discussed here are the more common, but they do not exhaust all the possibilities. We will argue for using the last of the four rules introduced here, the Bayesian or expected value decision rule.

MAXIMUM PROBABILITY DECISION RULE

This rule is used as follows:

1. Find the state with the largest probability.
2. Select the action that gives the best payoff value for the state indicated in step 1.
3. If there are ties for the best payoff, move to the state with the second largest probability and repeat steps 1 and 2 for the actions tied after step 2.

Using the urn problem, state θ_1 has the largest probability. Hence, we would select action a_1 because it has the largest payoff value for this state.

We observe that this procedure ignores (some would say avoids) worrying about anything more than which state is most likely. Many of our probabilistic data are ignored, and so are the payoffs for the states with smaller probabilities. It is easy to see how this rule could lead to bad action choices. Bad action choices are particularly likely when the state probabilities are nearly equal or when there are large differences among the payoff values across states.

MOST FAVORABLE PAYOFF VALUE DECISION RULE

This rule works as follows:

1. Scan the decision tree or table for the most favorable payoff value.
2. Select the action that may lead to this payoff value.

STATISTICAL DECISION THEORY

3. If there are ties for the largest payoff, move to the next most favorable payoff value for the actions tied in step 2 and repeat steps 1 and 2.

Using this approach on the urn problem, the largest payoff is $60 and, hence, this rule directs the decision maker to select action a_2 (see Table 13-1). This approach ignores even more of the available data than the previous rule, concentrating, as it does, on only the most favorable payoff value.

It is worth noting here that the units in which the payoff values are given in both problems introduced in Section 13-1 are "good things." We want to maximize them. In many problems the units in which the payoff values are given are costs, losses, failures, injuries, or some other undesirable quantity. We want to minimize these. That is why we used the words "most favorable" rather than the terms "largest" or "maximum." In a problem involving costs, this rule would have us find the minimum payoff value.

MINIMAX DECISION RULE

This rule works as follows:

1. For each action, identify the worst payoff value.
2. Find the best of the values selected in step 1.
3. Pick the action associated with the payoff value selected in step 2.
4. If ties result in step 2, repeat steps 1 and 2 for the tied actions using the next worst payoff. (When the payoffs are favorable, for example, profits, this rule is often called the *maximin decision rule*. We will consistently use the terminology minimax decision rule.)

In our urn problem, we identify −$20 as the worst payoff under action a_1. It is also the worst payoff under action a_2. The worst payoff value under action a_3 is $0. The best of these three payoffs is the $0 payoff value for action a_3. Thus, we would select action a_3 and not play. Action a_3 maximizes our minimum gain. We might call action a_3 the "do nothing" action. The minimax rule often leads to a "do nothing" action because most substantive choices involve some chance of loss.

It is often argued in support of this decision rule that the consequences associated with the worst occurrence are often so serious that they must be avoided. But this implies that the numerical payoff values are not adequate indicators of the consequences to the decision maker of that action and state combination. When this is so, we need better means of measuring the consequences of a given outcome; we need better measures of the payoffs and not a different procedure. We will return to this issue in Section 13-7. For now, we assume that the dollar payoff values capture the impact of each outcome on the decision maker.

The minimax decision rule ignores the probabilities of the states. Hence, it can result in an action choice that is dictated by one large potential loss where the likelihood of that loss is extremely low. In fact, given the "do nothing" alternative that is usually available, the minimax rule would reject any alternative that could lead to a loss under any state. As any manager knows, there are few if any actions that do not entail some likelihood of loss. For these reasons, we do not recommend the minimax decision rule. (The rule may be appropriate to anyone wishing to act consistently with Murphy's law. You may recall that Murphy's law states that "If something bad can happen, it will.")

BAYESIAN (OR EXPECTED-VALUE) DECISION RULE

This rule works as follows:

1. Consider the first action. Multiply the payoff values for that action under each state by the probability of that state and sum over all states. This procedure gives the expected payoff value for that action.
2. Repeat step 1 for all actions.
3. Select the action with the most favorable expected value.

Let us try this on the urn problem. The expected value for the first action, $E(a_1)$, is given by

$$E(a_1) = 0.7(\$30) + 0.3(-\$20) = \$15.$$

Similarly, the expected values for actions 2 and 3 are

$$E(a_2) = 0.7(-\$20) + 0.3(\$60) = \$4$$
$$E(a_3) = 0.7(\$0) + 0.3(\$0) = \$0.$$

The Bayesian decision rule directs us to select action a_1 since the expected payoff is highest. The highest value is chosen in this problem since the payoffs are profits to us. If the payoffs had been in cost terms, then the lowest expected payoff would have determined the appropriate action choice.

New Symbol	Meaning
$E(a_i)$	The expected payoff value for action a_i

$$E(a_i) = \sum_{j} v_{ij} p_j, \qquad (13\text{-}1)$$

where the summation is across all states.

Statistical Decision Theory

We prefer the Bayesian decision rule in cases where the payoff values adequately reflect the consequences of each outcome to the decision maker. The rule relies on the use of expected values and uses all the data at hand. It weights each payoff by its likelihood. The method requires the manager to establish the possible states and their probabilities, determine the available actions, and calculate the payoffs for each state-action pair. If this can be done, the Bayesian decision rule provides a rational choice mechanism.

This decision rule has been criticized on several fronts. First, some argue that each decision is unique. Hence, a rule based on averages (expectations are averages) is inappropriate since it is based on the concept of working over the long run and consequently only over a series of identical decision cases. But if we can view each decision as one of many made under uncertainty and under similar, if not identical, circumstances, then it may be reasonable to treat each decision as simply a single trial in a long series of similar trials. After all, each flip of a coin is unique. The last flip is the only last flip, for example. In the final analysis, it depends on whether the decision maker wishes to be guided by the averages.

A second criticism made of the Bayesian decision rule is that some of the consequences may not be adequately reflected by the payoff values. If the decision maker believes that some payoff values are more important than others (for example, one might mean bankruptcy) or that, in addition to the expected payoff for each action, the decision maker would also be influenced by the variance, then the payoffs do not adequately reflect the total impact of the outcomes. In this case, we must find new payoff values that do reflect the consequences or abandon the method. We return to this problem in Section 13-7.

In recommending the expected-value approach, we recognize that we are speaking normatively rather than descriptively: Decisions may not always be made this way today, but we believe they should be. As a final note, we observe that a good decision before the uncertainty is resolved (*ex ante*) may turn out to be a bad decision after all the facts are in (*ex post*). In the urn problem, we may select action a_1 and call θ_1 because the expected payoff is higher, only to find that the urn before us has a θ_2 on its bottom side. We wish now that we had selected action a_2 and called θ_2. But, as we all know, that is hindsight. If the skier had known that he would break his leg, he might have elected to forego his final run of the day.

S.S. Answer the following questions dealing with the mousetrap example:

(a) What action is selected by the maximum probability decision rule?

(b) What action is selected by the most favorable payoff value decision rule?

(c) What action is selected by the minimax decision rule?

(d) What action is selected by the Bayesian decision rule and why?

Solution: As we already know, action a_3 is dominated. Hence, we ignore it here.

(a) The state with the greatest likelihood is state θ_1. Scanning the payoff values for this state, the most favorable is $850,000. Since this is the payoff for action a_1, we elect to buy the large machine.

(b) This same value, $850,000, is also the most favorable payoff, and so the most favorable payoff decision rule also selects action a_1.

(c) The worst payoff under action 1 is −$50,000 if state θ_3 occurs. The worst payoff under action 2 is $110,000, also for state θ_3. The minimax rule selects the action that may lead to the better of these possible bad results. In this case, the rule elects action a_2 since $110,000 is better than −$50,000; we are directed to buy the small machine.

(d) Computing the expected values:

$E(a_1) = 0.50(\$850,000) + 0.25(\$130,000) + 0.25(-\$50,000) = \$445,000$

$E(a_2) = 0.50(\$410,000) + 0.25(\$170,000) + 0.25(\$110,000) = \$275,000$

$E(a_3) = 0$ (dominated; could have been ignored).

Since the payoff values are again "good," they are profits, we select the action with the largest expected value. The Bayesian decision rule directs us to take action a_1 and buy the large machine. (The time value of money has been ignored.)

13-4 Opportunity Loss

Many decision makers think in terms of what possible benefits an action may cause them to forego. That is, they think in terms of the differences between the payoffs for the same state. We can illustrate this concept using the urn problem. Suppose that the state is actually θ_2, although we do not know it yet. Then we would prefer to select action a_2 and call θ_2. We would win $60. If we were to call θ_1 instead, our loss is −$20, a net difference of $80. This $80 represents our *opportunity loss* from choosing action a_1 under state θ_2. If we had selected action a_1, and the state were θ_2, we could have made $80 more. Similarly, if the state were θ_2 but we selected action a_3, that is, if we chose not to play, our return is $0, and the opportunity loss is

Statistical Decision Theory

$60 − $0, or $60. This is the opportunity loss for action a_3 under state θ_2. The opportunity loss under state θ_2 for action a_2 is $60 − $60 = $0. Note that the opportunity loss is conditional on the state.

For a given state the *opportunity loss* for any action is defined to be the absolute value of the difference between the best payoff value under the given state and the payoff for the action being considered. In symbols,

$$L(a_i \mid \theta_j) = |\text{payoff}(a_j^* \mid \theta_j) - \text{payoff}(a_i \mid \theta_j)| = |v_j^* - v_{ij}|, \quad (13\text{-}2)$$

where a_j^* is the best action to take if θ_j is the actual state and v_j^* is the payoff under a_j^* if θ_j occurs. An opportunity loss is also referred to as a regret by some, since such a loss is regretted by the decision maker. The vertical lines around the payoff expressions direct us to take the absolute value, while the vertical line between the action and state symbols stands for the word given.

To find the opportunity losses, proceed as follows:

1. Select a state and find the best payoff across all actions for that state.

2. Take the difference between this best payoff and the payoff for each action under the state selected in step 1. The absolute value of this difference gives the opportunity loss for each action under the state selected.

3. Repeat steps 1 and 2 for all states.

Let us try it for our urn problem. Consider state θ_1. The best payoff is $30 under action a_1. This completes step 1. The opportunity losses under state θ_1 for each action (step 2) are:

$$\begin{aligned}
\text{opportunity loss for action } a_1 \text{ under } \theta_1 &\equiv L(a_1 \mid \theta_1) \\
&= |\$30 - \$30| \\
&= \$0
\end{aligned}$$

$$\begin{aligned}
\text{opportunity loss for action } a_2 \text{ under } \theta_1 &\equiv L(a_2 \mid \theta_1) \\
&= |\$30 - (-\$20)| \\
&= \$50
\end{aligned}$$

$$\begin{aligned}
\text{opportunity loss for action } a_3 \text{ under } \theta_1 &\equiv L(a_3 \mid \theta_1) \\
&= |\$30 - \$0| \\
&= \$30.
\end{aligned}$$

The opportunity losses for states θ_1 and θ_2 are given in Table 13-3. Try to compute the opportunity losses for state θ_2 for yourself. We must keep in mind that the opportunity losses are unfavorable payoffs to the decision maker. The opportunity-loss table can be used to make decisions. In doing so, we imply that the decision maker is as concerned about foregoing possible benefits as with out-of-pocket expenditures of the same magnitude.

Table 13-3 Opportunity-Loss Table: Urn Problem

	States and Their Probabilities	
Action	$\theta_1 \equiv$ Urn Is a θ_1 Urn $f(\theta_1) = 0.70$	$\theta_2 \equiv$ Urn Is a θ_2 Urn $f(\theta_2) = 0.30$
a_1: "call θ_1"	$0	$80
a_2: "call θ_2"	$50	$0
a_3: don't play	$30	$60

All the decision rules we used before can be used here, but we will discuss only the minimax and Bayesian decision rules.

Let us first apply the minimax decision rule to the opportunity-loss table. For action a_1, the maximum loss is $80 and for action a_2 it is $50. The largest loss for action a_3 is $60. The minimum of these losses is $50, and hence a_2 is the minimax action. Applying the minimax decision rule to the opportunity losses leads to a different choice than when this rule was applied to the original payoff values. Applying the minimax decision rule to the opportunity losses will typically lead to a substantive action choice rather than the "do nothing" alternative it favors when used with the original payoffs values. To distinguish the application of the minimax decision rule to opportunity losses from its application to payoffs, the phrase *minimax-loss decision rule* or *minimax-regret decision rule* is used.

A more appealing choice is again to use the Bayesian decision rule. Applying this decision rule to the opportunity losses in Table 13-3 gives

$$\text{expected loss } a_1 \equiv E[L(a_1)] = |0.7(\$0) + 0.3(\$80)|$$
$$= \$24 \quad \text{(minimum expected loss)}$$
$$\text{expected loss } a_2 \equiv E[L(a_2)] = |0.7(\$50) + 0.3(\$0)|$$
$$= \$35$$
$$\text{expected loss } a_3 \equiv E[L(a_3)] = |0.7(\$30) + 0.3(\$60)|$$
$$= \$39.$$

Action a_1 is selected because it has the lowest expected opportunity loss.

New Symbol	Meaning
$E[L(a_i)]$	The expected opportunity loss for the action a_1.

STATISTICAL DECISION THEORY

Two important observations should be made. First, since the entries in Table 13-3 are bad things, the action choice according to the Bayesian decision rule is action a_1; it has the smallest expected opportunity loss. This is the same choice as resulted from applying the Bayesian decision rule to the original payoff values in Table 13-1. This will always be so. Moreover, the marginal advantage of action choice a_1 over a_2 and a_3 is identical to what it was under the original payoff values. For example, consider actions a_1 and a_2:

$$\text{expected payoff value for action } a_1 = \$15$$
$$\underline{\text{expected payoff value for action } a_2 = \$\ 4}$$
$$\text{expected additional payoff for action } a_1 = \underline{\underline{\$11}}$$

$$\text{expected opportunity loss for action } a_2 = \$35$$
$$\underline{\text{expected opportunity loss for action } a_1 = \$24}$$
$$\text{expected lower loss for action } a_1 = \underline{\underline{\$11.}}$$

The advantage of action a_1 over a_2 is $11 in either case. The Bayesian decision rule always leads to the same action choice, whether we are dealing with the original payoff values or with opportunity losses. This is a comforting state of affairs. Another way to say this is that the opportunity losses are implicit in the payoff table. This must be so since the opportunity loss table can be obtained (and it was) from the payoff table.

Opportunity losses are worth studying because in many situations it is easier to work with them. Further, managers often think in these terms. For example, in a number of decision problems there are costs which are constant for all actions for a given state. Such costs can be ignored for that state over all actions. If this is true for all states (that is, given the state, the variable affects all actions identically), then the variable can be ignored. (This is so even if there is a different effect under each state.) What is important is that the effect is the same across all actions. Summarizing, a variable whose effect is the same for all actions over a given state can be ignored for *that* state. If this holds for each state (even though the effect may be different from state to state), the variable may be ignored completely. It will not affect the decision and it will not affect the opportunity losses, which represent differences between decisions. Thus, the use of losses allows us to focus only on those factors that will cause differences in the payoffs between actions.

We also note, unfortunately, that after the decision is made and the outcome, hence the payoff, known, the actual opportunity loss may not be known. This is so because in most real situations, the actual results may differ from those we expected, and we cannot be sure of the payoff that would have resulted had we selected some other action. Thus, we do not know the actual opportunity loss after the fact. This complicates the process of evaluating past decisions. (We shall typically use the term "loss" in place of "opportunity loss" from here on.)

S.S. Develop the opportunity-loss table for the mousetrap example and compare actions a_1 and a_2 using the Bayesian decision rule. Action a_3 is dominated and hence can be excluded from consideration.

Opportunity Losses: Mousetrap Problem

States and Their Probabilities

Action	$\theta_1 \equiv$ Sustained High $p_1 = 0.50$	$\theta_2 \equiv$ High, Then Low $p_2 = 0.25$	$\theta_3 \equiv$ Sustained Low $p_3 = 0.25$
a_1: buy large			
a_2: buy small			

Solution: The opportunity losses for each action are: for a_1: $0, $40,000, $160,000, and for a_2: $440,000, $0, $0. Note that there is always at least one zero in each column. It is the opportunity loss from selecting the best action for the state of that column.

The expected values are:

$$E[L(a_1)] = 0.50(\$0) + 0.25(\$40,000) + 0.25(\$160,000) = \$50,000$$

$$E[L(a_2)] = 0.50(\$440,000) + 0.25(\$0) + 0.25(\$0) = \$220,000.$$

The Bayesian decision criterion selects action a_1 and, again, action a_1 is as superior to a_2 as it was when the analysis was made in terms of the payoff values.

Payoff values:

$$E(a_1) - E(a_2) = \$445,000 - \$275,000 = \$170,000$$

Opportunity losses:

$$E[L(a_2)] - E[L(a_1)] = \$220,000 - \$50,000 = \$170,000.$$

13-5 Multiple Decisions Over Time

Up to this point we have considered problems where a single decision was all that was required. These are called *single-stage decision problems*. Consider now the case where a second decision may be required at a later point. Such problems are called *multistage problems*.

STATISTICAL DECISION THEORY

We may expand our mousetrap example to illustrate this notion. Suppose that at the end of the first 2 years management wishes to consider the purchase of a second small machine. This would make sense only if demand were to be high during the first 2 years. Additionally, it gives greater flexibility to the initial "buy small" alternative. Can we take this contingent decision, which is 2 years away, into account in our immediate purchase decision? (You will be gratified to know that we can, and that is just what this section is all about.)

DECISION TREE FOR MULTISTAGE PROBLEM

We begin by drawing a decision tree in Figure 13-4 for the expanded problem. The squares again represent points where the manager must choose. They are numbered 1 and 2. The circles represent random events occurring with the probabilities indicated on the branches immediately to their right. The new branch on our tree represents the decision to be made after 2 years if initial demand is high. This is the decision to expand or not. It is indicated by the box with a 2 in it. The manager may buy the second small machine (expand) at a cost of $90,000. Even though the two small machines produce as much as one large one, they are not as cheap to buy or to operate. Added personnel and maintenance reduce this contribution (excluding purchase cost) during the last 8 years to $90,000 per year or $10,000 per year, depending on whether the level of demand is high or low during those last 8 years. The contribution figures for the "don't expand" alternative are taken from the original statement of the problem in Section 13-1.

The probability demand will be high in the first 2 years is the sum of the probabilities that it will be sustained high and that demand will be high for 2 years and then low.

$$P(\text{high in first 2 years}) \equiv P(\text{initially high}) \equiv P(\text{IH})$$
$$= P(\text{demand sustained high}) + P(\text{high, then low})$$
$$= \tfrac{1}{2} + \tfrac{1}{4} = \tfrac{3}{4}.$$

The probabilities for the demand levels following the expand decision point are conditional probabilities. They depend, are conditioned, on the fact that at the end of year 2 we know that demand has been high during the first 2 years. That is how we arrive at decision square 2. The conditional probability on the top branch after decision square 2 in Figure 13-4 is

$$P(\text{sustained high} \mid \text{high in first 2 years}) \equiv P(\text{SH} \mid \text{IH}).$$

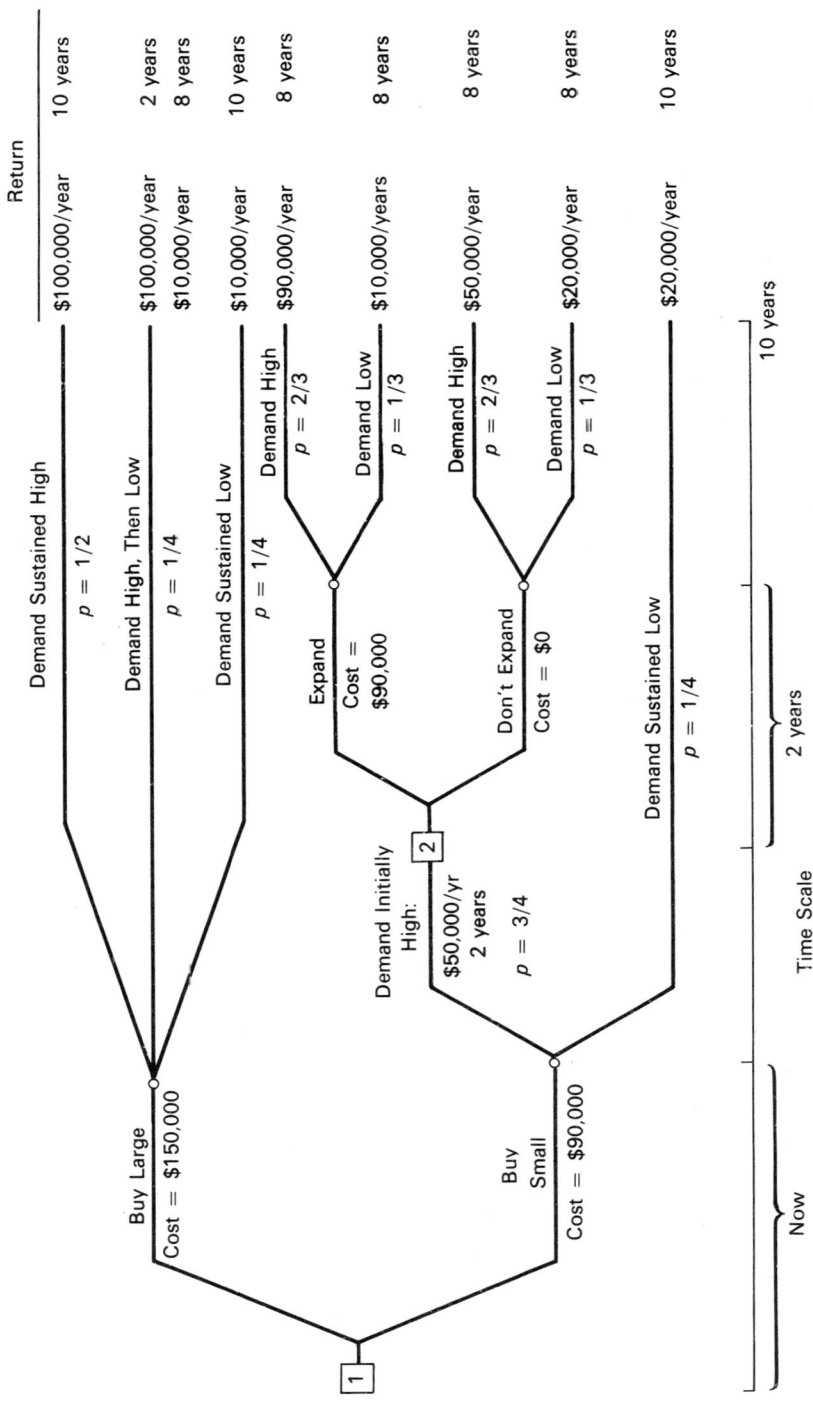

Figure 13-4 Decision Tree: Multistage Mousetrap Problem

STATISTICAL DECISION THEORY

Using the formula for conditional probability, this probability is found to be

$$P(SH \mid IH) = \frac{P(SH \text{ and } IH)}{P(IH)} = \frac{\frac{1}{2}}{\frac{3}{4}} = \frac{2}{3}.$$

(We could also have used Bayes' theorem to find this probability.) The probability that demand will be high in the last 8 years increases if we know it was high in the first 2 years. The first 2 years provide information about the last 8 years. This is the reason we consider the alternative to buy a small machine initially and then expand if demand is high. The probability on the lower branch is $1.00 - \frac{2}{3} = \frac{1}{3}$. All the data we need to make a decision using the Bayesian decision rule are shown in Figure 13-4.

In a real problem, we may consider expanding, given we buy a small machine, any time substantive new information is obtained. Here we simplify the problem by assuming this information will be available at the end of the second year.

ROLL-BACK PROCEDURE

The easiest way to work multistage decision problems is to work them backwards. (Wouldn't you know!) By working backwards we mean that the last

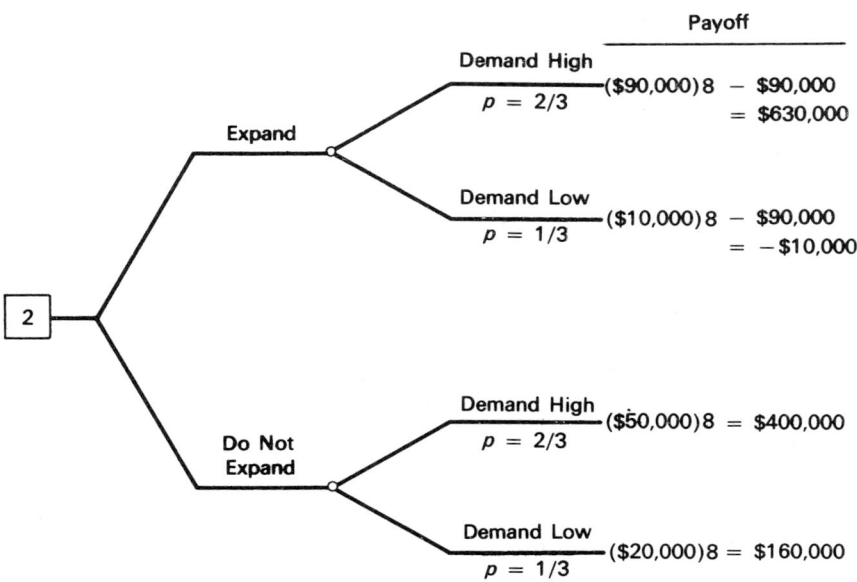

Figure 13-5 Decision Tree: Multistage Mousetrap Problem Evaluation of Decision 2: Expand or Not

decision is considered first. Thus, we turn now to an evaluation of decision 2 in the mousetrap example; to expand or not. This decision is illustrated in Figure 13-5.

The probabilities and payoffs are obtained directly from Figure 13-4. The expectations are as follows:

$$E(\text{expand}) = \tfrac{2}{3}(\$630,000) + \tfrac{1}{3}(-\$10,000) = \$416,667$$
$$E(\text{do not expand}) = \tfrac{2}{3}(\$400,000) + \tfrac{1}{3}(\$160,000) = \$320,000.$$

The more favorable expectation attaches to the decision expand. Hence, if we were to buy the smaller machine initially and demand were to be high in the first 2 years, we should expand. But should we buy the smaller machine initially? Let us see. Either we can place the $416,667 figure at the square numbered 2 in Figure 13-4 and use that figure, or we can draw a new picture. We shall do the latter to help us see the remaining problem to be solved and

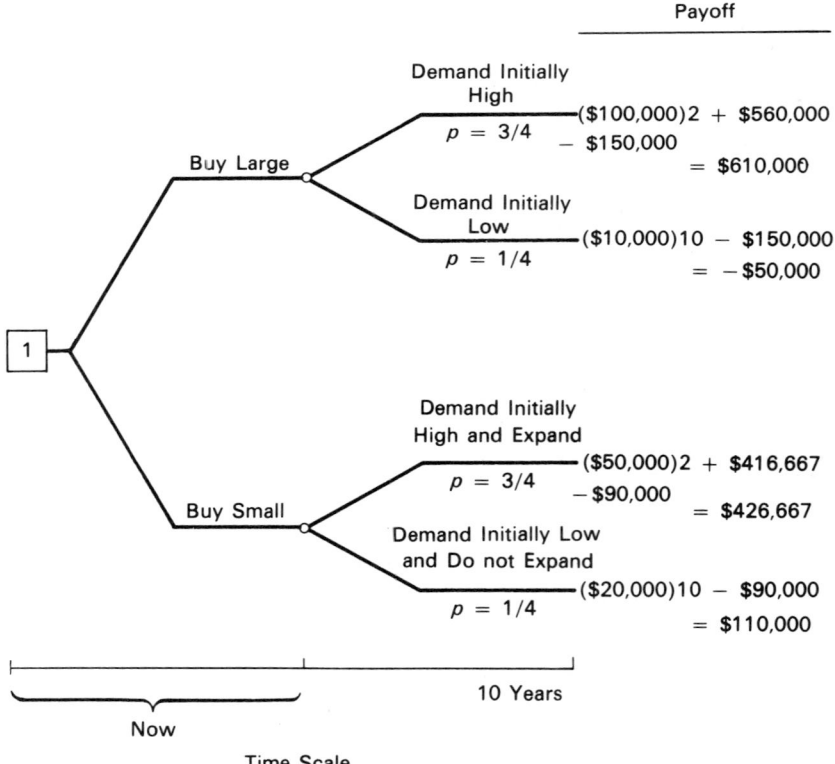

Figure 13-6 Decision Tree: Multistage Mousetrap Problem: Decision 1 (Decision 2 Resolved: Expand)

STATISTICAL DECISION THEORY

the generality of the method. This is done in Figure 13-6. We now evaluate decision 1 knowing that, if we buy the small machine and demand is initially high, our best choice is to expand and this choice has an expectation of $416,667.

Most of the numbers in Figure 13-6 are again obtained directly from Figure 13-4. Some explanatory comments are in order, however. (1) The analysis for decision 1 admits to only two states: demand initially high and demand initially low. (2) Given the decision to buy small, coupled with initially high demand, the payoff for the last 8 years is the value calculated for expansion using Figure 13-5, namely $416,667. (3) Given the decision to buy large coupled with initially high demand, the return for the last 8 years is an expectation based on the state of demand in the last 8 years. As we previously determined, there is a $\frac{2}{3}$ probability that demand will remain high and a $\frac{1}{3}$ probability that demand will drop to low. Hence, the expectation for the last 8 years for the top branch of the decision tree in Figure 13-6 is

$$\tfrac{2}{3}(\$100,000)(8) + \tfrac{1}{3}(\$10,000)(8) = \$560,000.$$

If demand is initially low, it cannot in this problem move to high in the last 8 years.

Using these facts, the payoffs in Figure 13-6 are determined as follows:

Branch Description	First 2 Years	+	Last 8 Years	−	Cost	=	Payoff
1. Buy large—demand initially high	$100,000(2)	+	$560,000	−	$150,000	=	$610,000
2. Buy large—demand initially small	10,000(2)	+	10,000(8)	−	150,000	=	−$50,000
3. Buy small—demand initially large	50,000(2)	+	416,667	−	90,000	=	426,667
4. Buy small—demand initially small	20,000(2)	+	20,000(8)	−	90,000	=	110,000

Using these data, the expectations for the two actions at decision point 1 are:

$$E(\text{buy large}) = \tfrac{3}{4}(\$610,000) + \tfrac{1}{4}(-\$50,000) = \$445,000$$

$$E(\text{buy small}) = \tfrac{3}{4}(\$426,667) + \tfrac{1}{4}(\$110,000) = \$347,500.$$

The new situation leads to the same action choice, namely to buy large initially. A reevaluation is required because of the flexibility introduced into the action choice "buy small." In the multistage problem, we get a chance to respond to the demand level observed in the first 2 years if we elect to buy the small machine initially. This option is valuable and increases the expected value of initially buying a small machine from $275,000 to $347,500, but the increase is not sufficient to alter the decision. The change did not affect the "buy large" alternative and its expectation is unchanged.

The method illustrated for solving multistage problems is not the only possibility, but it, perhaps, best illustrates how an r-decision-stage problem (r equals 2 here) is successively broken down beginning at the end and working backwards. The value from each stage is carried back to solve the next stage. This principle is used often in mathematics to solve dynamic problems.

We observe, in conclusion, that much is left out of the mousetrap problem. For one thing, we have ignored the time pattern of the cash flows. A dollar today is worth more than a dollar 10 years away. We have also ignored taxes and other possible alternatives available to the firm. Also, in solving the multistage version of this problem, we have not used decision tables. A decision table could be used to solve the decision of whether to expand or not and then a second table could be used to solve the initial purchase decision. Let us try this approach for our review problem.

S.S. Solve the multistage mousetrap problem using two decision tables. Solve the "expand decision" portion first. (Action a_3 can again be ignored.) The states used in the first table for decision 2 should reflect only what can occur in the last 8 years.

Decision Table: Expand or Not: Mousetrap Problem

States and Probabilities

Action	$\theta_1 \equiv$ High Demand $f(\theta_1) = \frac{2}{3}$	$\theta_2 \equiv$ Low Demand $f(\theta_2) = \frac{1}{3}$
a_1: expand		
a_2: don't expand		

Decision Table: Immediate Buy Decision: Mousetrap Problem

States and Probabilities

Action	$\theta_1 \equiv$ Initial Demand High $f(\theta_1) = \frac{3}{4}$	$\theta_2 \equiv$ Initial Demand Low $f(\theta_2) = \frac{1}{4}$
a_1: buy large		
a_2: buy small and ―――		

STATISTICAL DECISION THEORY

Solution: The values for the "expand or not" table are as follows:

Payoff Values

Action	High Demand: State $\theta_1: f(\theta_1) = \frac{2}{3}$	Low Demand: State $\theta_2: f(\theta_2) = \frac{1}{3}$
Expand	8($90,000) − $90,000 = $630,000	8($10,000) − $90,000 = −$10,000
Don't expand	8($50,000) = $400,000	8($20,000) = $160,000

$$E(\text{expand}) = \tfrac{2}{3}(\$630,000) + \tfrac{1}{3}(-\$10,000) = \$416,667$$

$$E(\text{don't expand}) = \tfrac{2}{3}(\$400,000) + \tfrac{1}{3}(\$160,000) = \$320,000.$$

The optimal action is to expand with an expected value of $416,667. This is the answer we obtained earlier. For the "immediate buy" table we have the following values (the values for the action "buy large" were computed in Table 13-2):

Payoff Values

Action	Initial Demand High: State $\theta_1: f(\theta_1) = \frac{3}{4}$	Initial Demand Low: State $\theta_2: f(\theta_2) = \frac{1}{4}$
a_1: buy large	$100,000(2) + $560,000 −$150,000 = $610,000	$10,000(2) + $10,000(8) −$150,000 = −$50,000
a_2: buy small and expand if demand high	$50,000(2) + $416,667 −$90,000 = $426,667	$50,000(2) + $20,000(8) −$90,000 = $110,000

$$E(a_1) = \tfrac{3}{4}(\$610,000) + \tfrac{1}{4}(-\$50,000) = \$445,000$$

$$E(a_2) = \tfrac{3}{4}(\$426,667) + \tfrac{1}{4}(\$110,000) = \$347,500.$$

The best action is to buy the large machine. Fortunately, the answer is the same as we obtained before. It is always comforting when this happens.

13-6 Decision To Obtain More Information

Sometimes we must act on our information immediately or at least before it is possible to obtain additional information. In other cases, the option to obtain additional information, usually at some cost, is open to us. In this section we will examine two questions: (1) Is it worthwhile to obtain more information by taking a sample? (2) Once it is obtained, how can it be combined with what we already know?

EXPECTED VALUE OF PERFECT INFORMATION

Suppose that in our urn problem there was a way to know whether the urn was a θ_1 or a θ_2 urn before we elected to play or not. That is, suppose we knew the state before we had to select an action. That would give us the advantage since we could not lose with this information. But we might wonder, first, what we would do with this information and, second, how much we would pay for it?

Consider the first question. If we knew the urn was a θ_1 urn, we would call θ_1 and win a net of $30. Since 70 percent of the urns are θ_1 urns, the urn selected randomly by the mad mathematician will be a θ_1 urn 70 percent of the time. We would win $30 70 percent of the time. If the urn were a θ_2 urn, we would know this, too, and call θ_2, winning a net of $60. Since 30 percent of the urns are θ_2 urns, we would win $60 30 percent of the time. Hence, our expected winnings, if we knew the state before selecting an action, would be

$$\$30(0.7) + \$60(0.3) = \$39.$$

But would we pay $39 for this information? (Try to answer before reading on.) The answer depends on how well we can do without the information. When we analyzed this problem in Section 13-3 we found, using the Bayesian decision rule, that action a_1, with an expected value of $15, was best. Since we can expect to make $15 without the actual state information, and we can expect to make $39 with it, perfect information on the state is worth the difference: $39 − $15 = $24. This is the *expected value of perfect information* (EVPI). It is an expected value since the winnings are weighted by their likelihoods; it is an average.

EVPI ≡ (the expected payoff with certain knowledge of the state) less (the expected payoff using the best action when we do not have certain knowledge of the state).

In Section 13-4 we found that the expected opportunity loss of the optimal action, a_1, was also $24. This is not a coincidence. The expected value of perfect information is always identical to the expected opportunity loss of the optimal action.

$$\text{EVPI} \equiv \text{the expected opportunity loss of the optimal action} = E[L(a_i^*)], \quad (13\text{-}3)$$

where a_i^* is the optimal action.

But why all this fuss over such an impractical idea? No one ever knows exactly what the state will be. Yet, as we have said, there is often the alternative of obtaining sample information rather than making an immediate

Statistical Decision Theory

action choice. Sampling gives us a better knowledge of the state, but this knowledge is not perfect knowledge. Since sampling costs money, how much is a particular sample worth? We cannot give a final answer to that question as yet. But we can say that it is worth no more than the EVPI. The EVPI is the value of certain state knowledge obtained at no cost and sampling can never do this well. Hence, in the present example, any sample costing more than $24 can be rejected as too expensive.

Now let us expand our urn problem somewhat by assuming each opaque urn contains 10 chips, some of which are red and some of which are white. The numbers are:

	Number Red	Number White
Each θ_1 urn	6	4
Each θ_2 urn	8	2

The urn composition is depicted in Figure 13-7.

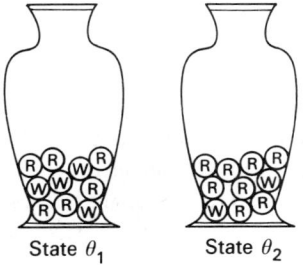

Figure 13-7 Urn Composition

Now suppose that we are offered the opportunity (by the gracious mathematician) to draw one chip from the urn before us and note its color (and do a little arithmetic) before making an action choice. Our benevolent mathematician places a cost of $1 on sampling one chip. Should we accept?

Since $1 is less than the expected value of perfect information, $24, we cannot immediately reject the sampling alternative. But we should not accept it just yet either. What we need to know is how much better we can do with the information the sample could provide. This requires that we learn how to integrate the possible results from sampling with what we already know in order to revise our probabilities on the states. This is the task of the next subsection.

S.S. What is the expected value of perfect information for the mousetrap problem as it was originally stated in Section 13-1. (Use Table 13-2.)

Solution: Using the values in Table 13-2 and the expectation of the best action, a_1:

EVPI = 0.50(\$850,000) + 0.25(\$170,000) + 0.25(\$110,000) − \$445,000
= \$50,000.

Hence any sample costing more than \$50,000 can be immediately rejected.

REVISION OF PROBABILITIES BASED ON SAMPLE INFORMATION

The more alike the two urns are in the composition of the chips inside, the less information a sample of a given size gives us to distinguish between them. In other words, the more similar the composition of the populations to be sampled, the larger the sample required to differentiate among them.

The sampling procedure described represents sampling from a binomial process. The sample is of size 1. The result is either red or white. Suppose that the sample result is red. We know the following probabilities:

probability of a θ_1 urn before sampling = 0.7 = $f_0(\theta_1)$
probability of a θ_2 urn before sampling = 0.3 = $f_0(\theta_2)$.

These are the probabilities of the states prior to sampling. They are called, naturally enough, the *prior* probabilities of the states or simply *prior probabilties*. The zero subscript on the letter *f* is to indicate that the probabilities are prior probabilities. Given the characteristics of the two types of urns, as illustrated in Figure 13-7, we also know:

probability of a red given urn θ_1 = 0.6 = $f(r \mid \theta_1)$
probability of a red given urn θ_2 = 0.8 = $f(r \mid \theta_2)$.

The question remains of how we can put this information together to revise our prior probability estimates on states θ_1 and θ_2; that is, how can we revise $f_0(\theta_1)$ and $f_0(\theta_2)$ based on the sample result? We want to obtain $f(\theta_1 \mid r)$, the probability of urn 1 given a red chip is drawn, and $f(\theta_2 \mid r)$, the probability of urn 2 given a red chip is drawn. These are the revised state probabilities based upon a sample of one chip which turns out to be red. To accomplish this revision, we can use Bayes' Theorem, which was introduced in Section 6-4 as formula (6-18). We note that a red chip is more likely to be drawn if the urn before us is a θ_2 urn. Hence, a red chip should cause us to revise $f_0(\theta_2)$ upward and $f_0(\theta_1)$ downward.

Statistical Decision Theory

Using Bayes' Theorem:

$$f(\theta_1 \mid r) = \frac{f(r \mid \theta_1)f_0(\theta_1)}{f(r \mid \theta_1)f_0(\theta_1) + f(r \mid \theta_2)f_0(\theta_2)}$$

$$= \frac{0.6(0.7)}{0.6(0.7) + 0.8(0.3)}$$

$$= \frac{0.42}{0.66}$$

$$\approx 0.6364.$$

This is our revised probability, after the sample, of state θ_1. To emphasize this point, we write

$$f_1(\theta_1) = f(\theta_1 \mid r) \approx 0.6364.$$

The subscript 1 will indicate a revised probability. The revision process has resulted in $f_1(\theta_1)$ being less than $f_0(\theta_1)$, as expected.

Similarly, $f_1(\theta_2)$ can be calculated to be approximately 0.3636. Or it can be obtained from the knowledge that the sum of the two state probabilities must be equal to 1, since they are the only two possible states.

$$f_1(\theta_1) + f_1(\theta_2) = 1.$$

New Symbol	Meaning
$f_0(\theta_j)$	The prior probability of state θ_j.
$f_1(\theta_j) = f(\theta_j \mid \text{sample result})$	The revised probability of state θ_j based on the sample result and the prior probability.

If the sample result were white instead of red:

$$f_1(\theta_1) = f(\theta_1 \mid w) = \frac{f(w \mid \theta_1)f_0(\theta_1)}{f(w \mid \theta_1)f_0(\theta_1) + f(w \mid \theta_2)f_0(\theta_2)}$$

$$= \frac{0.4(0.7)}{0.4(0.7) + 0.2(0.3)}$$

$$= \frac{0.28}{0.34}$$

$$\approx 0.8235.$$

Hence, $f_1(\theta_2) = f(\theta_2 \mid w) = 0.1765$ if a white chip is drawn. The revised state probability depends on the sample result. In the case of drawing a white chip, the revised probability on a θ_1 urn increased, since urn θ_1 contains the higher proportion of white chips.

Sometimes the calculation of the revised state probabilities is seen more easily in tabular form. Table 13-4 summarizes the calculations that we have just made for the case where the sample result is red. A second table would be needed for the sample result white.

Table 13-4 Calculation of Revised State Probabilities: Sample Result Red

(1)	(2)	(3)	(4)	(5)
State θ_j	Prior State Probability $f_0(\theta_j)$	Probability of the Specific Sample Obtained Given State θ_j $f(r \mid \theta_j)$	Joint Probability (Col. 3) × (Col. 2) $f(r \mid \theta_j)f_0(\theta_j)$ or $f(r \cap \theta_j)$	Revised State Probability (Col. 4) ÷ (Sum Col. 4) $f_1(\theta_j)$
θ_1	0.70	0.60	$(0.7)(0.6) = 0.42$	$\dfrac{0.42}{0.66} \approx 0.6364$
θ_2	0.30	0.80	$(0.3)(0.8) = 0.24$	$\dfrac{0.24}{0.66} \approx 0.3636$
Sum	1.00		0.66	1.00

Using the revised state probabilities, we can compute the expected value of each action under each sample result.

If sample result is red:

$E(a_1) = 0.6364(\$30) + 0.3636(-\$20) = \$11.92$ best choice

$E(a_2) = 0.6364(-\$20) + 0.3636(\$60) = \$9.09.$

If sample result is white:

$E(a_1) = 0.8235(\$30) + 0.1765(-\$20) = \$21.27$ best choice

$E(a_2) = 0.8235(-\$20) + 0.1765(\$60) = -\$5.88.$

Actually, we did not need to evaluate the actions based on the sample result white. Since the best action with no sample is a_1, "call θ_1," and we know without any calculations that the result white chip increases our belief that

STATISTICAL DECISION THEORY

θ_1 is the actual state, we know that our action choice will not change if a white chip is drawn. This is not so for a red chip. A red chip is more likely if the urn is a θ_2 urn. Hence, if a red chip is drawn, the likelihood (represented by the revised probability on state θ_2) that we have a θ_2 urn before us rises. In the present example, one red chip does not raise this probability sufficiently to alter our action choice. In the present problem, we still select action a_1 regardless of the color of the chip selected for a sample of size 1.

Since a sample of size 1 cannot change our action choice no matter whether it is red or white, it is not worth taking. A sample that will not alter our action choice is worth nothing to us. Hence, we would not take it at any price.

Suppose, however, that we are offered a sample of size 2 for $2. We shall assume that a chip is drawn, its color noted, and that it is then replaced with thorough mixing before the second chip is drawn. This assures that both chips are drawn from the same composition of chips. The results and revised probabilities are calculated, using the binomial, as follows.

If the result is two red:

$$f(2 \text{ red} \mid \theta_1) = b(2; 2, 0.6) = (0.6)^2 = 0.36$$

$$f(2 \text{ red} \mid \theta_2) = b(2; 2, 0.8) = (0.8)^2 = 0.64$$

$$f_1(\theta_1) = \frac{0.36(0.7)}{0.36(0.7) + 0.64(0.3)} = \frac{0.252}{0.444} \approx 0.5676$$

$$f_1(\theta_2) \approx 1 - 0.5676 = 0.4324$$

$$E(a_1) = 0.5676(\$30) + 0.4324(-\$20) = \$8.38$$

$$E(a_2) = 0.5676(-\$20) + 0.4324(\$60) = \$14.59. \quad \text{best choice}$$

If the result is one red and one white:

$$f(1 \text{ red, } 1 \text{ white} \mid \theta_1) = b(1; 2, 0.6) = 2(0.6)(0.4) = 0.48$$

$$f(1 \text{ red, } 1 \text{ white} \mid \theta_2) = b(1; 2, 0.8) = 2(0.8)(0.2) = 0.32$$

$$f_1(\theta_1) = \frac{0.48(0.7)}{0.48(0.7) + 0.32(0.3)} \approx 0.7778$$

$$f_1(\theta_2) \approx 1 - 0.7778 = 0.2222$$

$$E(a_1) = 0.7778(\$30) + 0.2222(-\$20) = \$18.89 \quad \text{best choice}$$

$$E(a_2) = 0.7778(-\$20) + 0.2222(\$60) = -\$2.23.$$

If the result is two white:

$$f(2 \text{ white} \mid \theta_1) = (0.4)^2 = 0.16$$

$$f(2 \text{ white} \mid \theta_2) = (0.2)^2 = 0.04$$

$$f_1(\theta_1) = \frac{0.16(0.7)}{0.16(0.7) + 0.04(0.3)} \approx 0.9032$$

$$f_1(\theta_2) \approx 1 - 0.9032 = 0.0968$$

$$E(a_1) = 0.9032(\$30) + 0.0968(-\$20) = \$25.16 \qquad \text{best choice}$$

$$E(a_2) = 0.9032(-\$20) + 0.0968(\$60) = -\$12.25.$$

A sample of 2 has the power to alter our decision, but only if two reds are drawn. Only then will we change our action choice from a_1 to a_2. If one or no reds are drawn, the best action is still a_1, and we save nothing. If two reds are drawn, the advantage of altering our action choice to action a_1 is $\$14.59 - \$8.38 = \$6.21$. This will occur as often as we draw two reds, and the probability of drawing two reds is given by the denominator of Bayes' theorem to be 0.444. (If you do not recall this fact, reread Section 6-4 to see that the denominator in Bayes' theorem is the probability of the sample result.) Hence, the value of the sample is $0.444(\$6.21) = \$2.76$. We should be willing to pay $2.00 for the sample, or indeed any amount up to $2.76. Note that we ignored the results for two whites and for one white and one red in this calculation, since these results cannot alter the best decision without sampling. In these 2 cases the value of the sample is zero.

An alternative but more complex calculation of this result is provided in Table 13-5. It requires us to calculate the expectation for each action. Using the first and easier method, we can see after making the calculations for one white in two that neither it nor the result two whites will change our original action choice. Two whites makes urn θ_1 even more likely, since there is a larger proportion of white chips in the θ_1-type urns. Hence, if one red in two will not alter our choice of action a_1, certainly zero reds will not, and we need not calculate the revised probabilities for two whites. Further, we can ignore the expectations for one and for two whites in calculating the expected value of the sample since neither alters the decision. (Actually, knowing the results for a sample of size 1, we could have predicted that if one red in one draw is not sufficient to change our action choice from a_1, then one red and a white offering conflicting information will also not change our action choice.)

Table 13-5 Expected Value of a Sample of Size 2

(1) Sample Result	(2) Best Action	(3) Expected Value of Best Action	(4) Probability of Sample*	(5) (Col. 3) × (Col. 4)
2 red	a_2: call θ_2	$14.59	0.444	$6.478
1 red, 1 white	a_1: call θ_1	18.89	0.432	8.160
2 white	a_1: call θ_1	25.16	0.124	3.120

Expectation with a sample of size 2 (sum column 5)	$17.76
Expectation with no sample of taking best action: a_1	15.00
Expected value of a sample of size 2 (Difference)	$2.76

*f(sample and θ_1) + f(sample and θ_2) = f(sample $\cap$ θ_1) + f(sample $\cap$ θ_2) = f(sample $|$ θ_1)$f(\theta_1)$ + f(sample $|$ θ_2)$f(\theta_2)$. This calculation gives the denominator of Bayes' theorem and each entry in this column.

The sample of size 2 has an expected value of $2.76 and hence is worth the $2 cost. We should not pay more than $2.76 for it under any conditions. By similar analysis we could, given the cost of sampling, determine the best size sample to take. This is done by finding that sample for which the difference between its expected value and its cost is a maximum. This we leave as an exercise.

In general, it is difficult to estimate the amount of information in a sample using our intuition. Suppose, in the example we have been considering, that the binomial process yields a sample of 8 reds in 10 chips, what probability, estimated without calculations would we place on the state θ_2 (that the urn is a θ_2 urn)? Remember that the prior probability is 0.30. Try to estimate it intuitively. The correct answer is about 52 percent. If you had difficulty with this estimation, it suggests the value of making the calculation using Bayes' theorem.

S.S. Consider again the urn problem for a sample of size 1 costing $1, if the composition of chips in the two urns is changed to:

Urn θ_1: 6 red and 4 white

Urn θ_2: 1 red and 9 white.

The prior probabilities and payoffs are left unchanged.

(a) Can a sample of one change the best action?

(b) Is the sample worth its cost?

Solution:

(a) The best action without sampling is a_1, call θ_1. Only a white chip can change our action; a red will not. So we need only evaluate the result "white chip." The probabilities are

$$f(w \mid \theta_1) = 0.4 \quad \text{and} \quad f(w \mid \theta_2) = 0.9.$$

Thus,

$$f(\theta_1 \mid w) = f_1(\theta_1) = \frac{0.4(0.7)}{0.4(0.7) + 0.9(0.3)} = \frac{0.28}{0.55} \approx 0.5091$$

and

$$f_1(\theta_2) \approx 1 - 0.5091 = 0.4909.$$

The expected payoffs are

$$E(a_1) = 0.5091(\$30) + 0.4909(-\$20) = \$5.45$$

$$E(a_2) = 0.5091(-\$20) + 0.4909(\$60) = \$19.27.$$

The larger payoff is $19.27 and the best choice is now a_2. The sample can change our action, but only if a white chip is drawn.

(b) The value of the sample in this case is

$$(\$19.27 - \$5.45)(0.55) = \$7.60.$$

This value is higher than before because a white chip is a stronger indicator of a θ_2 urn in this situation.

Sampling as we described it for our urn problem was from a binomial process. If we do not replace the chip before a second is drawn, sampling is hypergeometric and the conditional probabilities of the sample results change (see Problem 13-20). Sometimes we sample from a normal universe and again the use of Bayes' theorem is not the same. In Technical Note 13-3, we will describe a problem involving sampling from a normal process. But we will not digress to consider all forms of sampling. The principle is the same in each case, but the calculations can, unfortunately, become messy.

13-7 Evaluating Payoff Values

It is not unusual in situations involving monetary payoff values for the monetary consequences to fail to reflect the full impact of the outcome on the decision maker. In still other problems the payoffs may not even involve money.

For example, a large monetary loss that may result in bankruptcy is likely to have a greater impact on the decision maker than the monetary value alone

Statistical Decision Theory

would indicate. In other cases, factors difficult to quantify may be involved: for example, the effects of an action on morale, safety, or employee turnover, all of which are difficult and perhaps impossible to quantify. In these situations, the decision maker may elect to modify the monetary payoffs or ignore them altogether. Other decisions may be made: to increase market penetration while experiencing a short-run loss in profits, avoid governmental intervention, improve the environment, or merely to survive. All of these decisions have a long-run monetary impact on the organization, but the costs or benefits may not be easily estimated.

A true story may illustrate the inadequacy of monetary payoffs in certain situations. Once upon a time there was a young woman who worked as chief teller in her father's bank, located in a small Midwestern town. She saw many worthwhile projects in her town left undone because of a lack of funding. She also saw all that money lying idle in daddy's bank. Well, lo and behold, many of those projects later found financing from an unknown benefactor.

All seemed well until one day, notice of the impending arrival of the state bank examiners was received. By this time our heroine was into the bank for several hundred thousand. She could await the examiners and certain discovery or she could try to replace the money. Her choice was to "borrow" (for she always intended to repay the money) another $50,000 and head for the horse races. Now, as we all know, her monetary expectation at the races was negative. But would she go to jail longer for another $50,000? And this was the only way she saw to replace the "borrowed" funds. Certainly the monetary values and expectations were not adequate indicators of the consequences in this case, and they were not used in her action choice.

In this section, we will discuss several means by which nonmonetary factors can be incorporated into a decision-theory approach.

MODIFICATION OF THE MONETARY VALUES

In some problems, the nonmonetary factors may favor one action. If that action already has the best expectation, it is the best decision. If the action favored by the nonmonetary factors does not have the best monetary expectation, we may ask whether the net nonquantifiable factors are worth the difference. Thus, suppose the expected profit from action a_1 is $100, while the expected profit from action a_2 is only $50. Suppose, further, that action a_2 has safety features for the employees not possessed by action a_1. If this is the only difference, we might select between the two actions by determining whether the safety features attached to action a_2 are worth more or less than the $50 difference in expected values.

In many problems, there will be some nonmonetary factors that favor one alternative and some that favor another. Furthermore, the differences may not be the same across all states. One possible approach in this situation is to try to find the monetary value of each of these factors under each state. But this is not an easy task and it may be impossible. A decision maker may be satisfied in such situations to locate an acceptable action considering the major factors rather than attempt to locate the best decision. Further analysis and evaluation may not be worth the cost. Indeed, this suggests that a wise decision maker will leave out items that are difficult to quantify unless they are likely to be of major importance. The discussion also suggests that the potential benefits from an improved action choice should be balanced with the cost of the additional analysis.

It is often possible to use the concept of insurance to evaluate outcomes where the consequences to the decision maker are not adequately reflected by the monetary values. For example, suppose that for some action choice two possible events may occur. The first is a profit of, say, $50,000 with a probability of 0.95 and the second is a fire or other catastrophic event with probability 0.05, which would result in the loss of substantial business and possible bankruptcy. A manager might wish that insurance could be purchased that would shift this risk to someone else, namely the insurance company. Suppose the manager would pay $10,000 but no more for this insurance; then the value of the action choice described above is the expected monetary payoff less the amount the manager would pay for the insurance. This value is 0.95($50,000) − $10,000 = $37,500. The implied impact of the fire to the manager, call it x, is given by

$$x(0.05) = \$10,000$$
$$x = \$200,000.$$

The manager can use this figure to evaluate the decision. If the fire is less significant than this, then $10,000 is too much to pay for the insurance.

$$-\$200,000(0.05) + \$50,000(0.95) = \$37,500.$$

But of greater importance, perhaps, the manager can use this approach even when insurance is not actually available. By asking how much such insurance would be worth, if it were available, similar consequences can be evaluated. If the manager would not purchase insurance to avoid some outcome, then the monetary value itself may be used.

Since insurance companies wish to make a profit through the service they provide, they will charge a premium above the pure monetary expectation. Where the monetary values adequately reflect the consequences, the decision maker will not be willing to pay this premium. Only if the potential event leads to a loss with an impact greater than its monetary value will the decision

STATISTICAL DECISION THEORY   445

maker consider paying the extra amount charged by the insurance company. That is why most organizations do not insure "small" potential losses. In these cases the monetary consequences adequately incorporate the impact on the organization, and it will "self-insure"; that is, it will not purchase insurance.

S.S. A firm is considering purchase of a chemical from an alternative source for $35,000 to use as the major ingredient in its final product. It is considering this alternative because its new plant has not yet been debugged. However, the firm is worried that this information may be learned by its major customer, who might then shift this very profitable business elsewhere. The decision problem is diagrammed in Figure 13-8. Also, the manager would pay $20,000 for insurance against the information leak, but unfortunately such insurance is not available. The firm's internal processing cost for the same chemical is $30,000 but there is one chance in three that the process will fail.

(a) What value is the manager implicitly placing on the consequences of the information leak?

(b) What decision should be made?

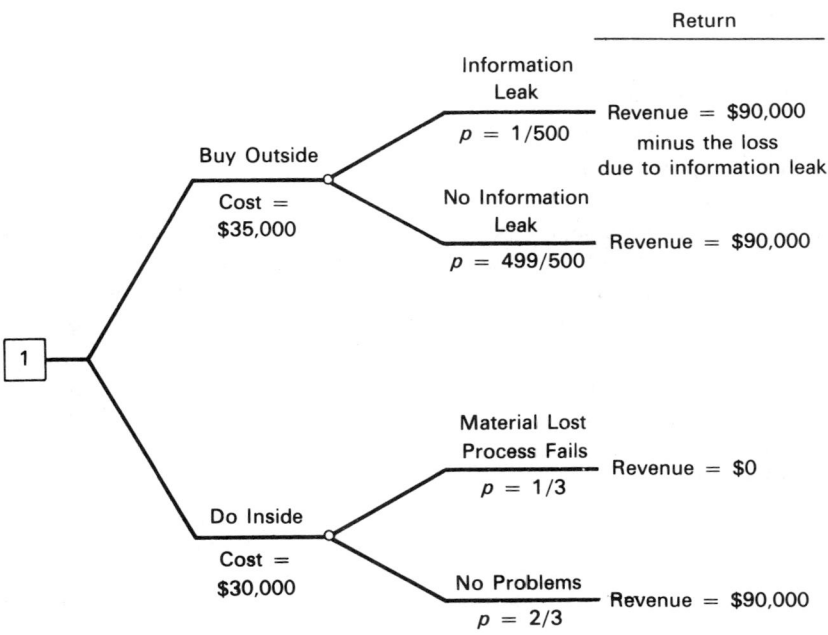

Figure 13-8 Decision Tree: Information-Leak Problem

Solution: The decision maker has elected to pay $20,000 to assure revenues of $90,000 without an information leak. Thus, the cost of taking the buy action is equivalent to the $35,000 material cost plus the $20,000 the decision maker is willing to give up to assure the secret is kept. The expected value of this alternative is $90,000 − $35,000 − $20,000 = $35,000. The expected value of "do it inside" is

$$\tfrac{1}{3}(-\$30,000) + \tfrac{2}{3}(\$90,000 - \$30,000) = \$30,000.$$

The best action is purchase the chemical outside and assure delivery. The value implicitly placed on the information leak is

$$x(\tfrac{1}{500}) = \$20,000$$

$$x = \$10,000,000.$$

The secret is apparently very important to the manufacturer. However, it is not sufficient to cause the firm to produce inside. The expected value of buying can also be determined by

$$E(\text{buy outside}) = \tfrac{1}{500}[-\$10,000,000 + (\$90,000 - \$35,000)]$$

$$+ \tfrac{499}{500}[\$90,000 - \$35,000]$$

$$= \$35,000.$$

(If the revenues from delivery in this problem had been $60,000 rather than $90,000, the best decision would have been to make inside.)

NONMONETARY PAYOFFS: UTILITIES

One alternative to monetary payoffs is to use some other measure. The literature on decision theory suggests measuring the impact of the various outcomes using utilities. The utility of a payoff is a measure of its importance to the decision maker. The theoretical development of utility measures in decision theory is quite far along, but the method is still not operational at this time. For this reason and for space considerations, we elect not to fully investigate this alternative. Following the next paragraph, we will discuss one method whose rationale relies on utility theory both because it is relatively easy to understand and because it covers a set of interesting problem situations.

Another alternative that can be used in problems involving nonmonetary payoffs is to find some problem-specific numerical variable that adequately reflects the decision maker's objective. For example, market share might be an appropriate measure to use in some business problems. A health care institution might adopt patient days as a payoff measure; a crime program might be interested in the drop in serious crimes; a local flood-relief program

STATISTICAL DECISION THEORY

might focus on the number of people served or resettled. Nonmonetary measures are especially necessary in not-for-profit organizations when the administrator wishes to measure benefits instead of costs. Alternatively, if the administrator wishes to achieve a given result at minimum cost, a cost-per-unit-benefit ratio might be used as the payoff measure.

One method that can be used relies on a procedure that works as follows:

1. In terms of the relevant payoffs, define two outcomes. The first must be at least as bad as the worst outcome possible and the second at least as good as the best outcome possible. Call these O_w and O_b, respectively.

2. Given a specific outcome, call it O_s, find the probability of O_b in a gamble involving only O_w and O_b that makes the decision maker indifferent between O_s for certain and the gamble involving O_w and O_b. [That is, find a probability p such that p(payoff for O_b) + $(1 - p)$(payoff O_w) = payoff for O_s.]

3. Do this for all outcomes and use the probabilities found in step 2 as the payoff values.

4. Use the Bayesian decision rule to find the best action using these (probabilistic) payoff values to select an action.

Consider the following example involving a market-expansion problem with the data given in Table 13-6. The actions represent marketing strategies.

Table 13-6 Payoffs: Market-Expansion Problem

States and Their Probabilities

Action	$\theta_1: f(\theta_1) = 0.5$	$\theta_2: f(\theta_2) = 0.2$	$\theta_3: f(\theta_3) = 0.3$
a_1	Weak penetration over wide area	Moderate penetration over small area	Strong penetration over moderate area
a_2	Moderate penetration over wide area	Strong penetration over small area	Moderate penetration over small area

Neither action dominates and this gives us a chance to try the suggested method. First, we define O_w and O_b:

$$O_w \equiv \text{weak penetration over a small area}$$
$$O_b \equiv \text{strong penetration over a wide area.}$$

These outcomes are at least as bad and as good, respectively, as any in the problem. This completes step 1.

Consider the outcome "strong penetration over a moderate area," action a_1 and state θ_3. We now ask the manager to assume that this outcome is assured. What probability of winning O_b in a gamble involving only O_b and O_w would leave the manager indifferent between the gamble and the assured outcome of a strong penetration over a moderate area? In other words, what probability of attaining a strong penetration over a wider area, versus a weak penetration over a small area, would cause this option to be equivalent to the assured position?

Suppose that the manager has a preference for a strong penetration since it is more difficult for a competitor to counter. After considerable agonizing, it is established that only if the probability of obtaining O_b is 0.98 (thus, the probability of O_w is 0.02) would the gamble involving O_b and O_w in place of the assured outcome of a strong penetration over a moderate area be accepted. At a lower probability of winning O_b, the manager would prefer the assured position. We may use 0.98 as the payoff value for action a_1 under state θ_3. The other "payoff values" are determined in a similar fashion. Suppose that after substantial questioning the "payoff values" are as indicated in Table 13-7. Then computing the expected values gives

Table 13-7 Payoff Values: Market-Penetration Problem

Action	$\theta_1: f(\theta_1) = 0.5$	$\theta_2: f(\theta_2) = 0.2$	$\theta_3: f(\theta_3) = 0.3$
a_1	0.05	0.20	0.98
a_2	0.30	0.72	0.20

States and Their Probabilities

$$E(a_1) = 0.5(0.05) + 0.2(0.20) + 0.3(0.98) = 0.359$$

$$E(a_2) = 0.5(0.30) + 0.2(0.72) + 0.3(0.20) = 0.354.$$

In this case action a_1 is preferred using the Bayesian decision rule. The technique is, as you might guess, not an easy one to use. In the first place, managers are not used to thinking in these terms. Second, the numbers obtained are fuzzy and there is no guarantee, even if the initial process were repeatable, that it would produce exactly the same payoff values.

It is also difficult to explain to managers just why this approach works. In

Statistical Decision Theory

fact, we have not given any rationale as to why the procedure should work or even why it is logical. The logic of the approach relies on utility theory and requires the (reasonable) assumptions of that theory. Since it is relevant to several common problem situations, we have elected to describe the approach, although a complete description of utility theory would take us too far afield.

S.S. Try this on yourself. Suppose that you own a lottery ticket which has survived several draws, until at this time it is one of 10 left in the running for the first prize of $1,000,000. (You should be so lucky!) One ticket will be drawn at random and win this prize. The other nine will each win $10,000. Suppose now that someone is willing to buy your ticket. At what price (what certain amount) would you be indifferent between (1) selling your ticket and (2) waiting for the draw? At any lower price you would prefer to hold the ticket and take your chances. At any price above the one you set, you prefer the certain cash obtained from the sale. (There is no single best answer for everyone.) What does this imply about the utility for money here? Letting $O_w = \$10,000$ and $O_b = \$1,000,000$, what is your utility for the price you selected?

Solution: As we indicated, there is no correct answer. Different people will evaluate this gamble differently. This is because the monetary amounts are sufficiently large that the consequences they imply are neither reflected adequately by the amounts involved nor are they constant from person to person. The expected monetary value of the ticket is $0.1(\$1,000,000) + 0.9(\$10,000) = \$109,000$. But most, if not all, of us would be happy to find a buyer who would pay us substantially less than that (but more than $10,000, of course). If you set a price less than $109,000, this implies that your utility for money decreases as you get more of it. You would be called a risk avoider in utility theory terms. In any case, the utility, p, of the selling price you chose is given by setting the selling price you chose equal to $p(1,000,000) + (1-p) \times (10,000)$. The amount you would have sold the ticket for is called the certain monetary equivalent for the gamble to which it relates. It can be used in some situations to evaluate forks on a decision tree that cannot be evaluated using the expected-value approach. However, when the probability of one outcome becomes quite small or when there are numerous branches at some node in a decision tree, this simple technique is more difficult to apply. Nevertheless, the lottery-ticket idea can be extended to such complex situations, and it gives us another way to deal with problems where the consequences are not adequately reflected by the monetary amounts. [The interested reader is referred to Chapter 4 of Howard Raiffa's easy-to-read book *Decision Analysis* (Reading, Mass.: Addison-Wesley Publishing Company, Inc., 1968) for a more complete discussion and extensions.]

13-8 A Warning

The material we have studied here is not easy. (You were aware of that, you say?) Yet even so, we may have simplified too much in our attempt to get at basic ideas. By working only with problems that have already been defined and put in decision form ready for solution, a sticky wicket or two is avoided. But we should remember that this is one of the most difficult and important parts of problem solving.

Further in the problems we have examined, only one uncertain variable such as demand has been present. In more realistic problems, factors including costs, market size, equipment life, service, rate of sales growth, and the impact on other products are all relevant. Moreover, these factors may not be independent. The number of actions and states can stagger the imagination. Some actions may be dropped due to legal limitations such as are involved in the Pure Food and Drug Act. Others may be discarded because of company policies, social mores, or the dictates of public relations. Hence, a company may immediately discard an action that involves dumping industrial waste materials into a local river. Still other actions may be rejected due to convention or company practice, or even a reluctance to disturb the status quo. Yet, even after these reductions, the action set is likely to remain large.

The complexity of most real problems deters most managers from attempting to find the optimal action. Instead, the choices which come quickly to mind are evaluated first, and the best of these is selected. Time and funds permitting, the manager may elect to accept the action deemed best at the moment or to search for another action which is then evaluated and compared with the best previous one. This is itself a sequential decision problem. Actions are located one at a time, evaluated, and a decision is made to (1) take one of the actions available, (2) sample or test to obtain a better basis for choosing from among the present alternatives, or (3) continue searching for better alternatives. Typically the decision maker stops this process when an action that is good enough is found (this behavior is called *satisficing*) or when time or cost limitations are binding.

One other difficulty in our approach, which we have ignored, deserves mention. The payoffs in a decision problem represent the impact of various actions and states occurring together. Determining this impact is difficult to do in the case of a single decision maker. It is much more difficult to do when a group is involved. In group decision making, the group making the decision may be difficult to identify. Conflicting goals and differences in goal importance to the individuals, as well as the relative importance of an individual's views in a group, are not easily established. This makes it harder to be sure that the problem formulation defines the effect of the choices on the organization.

STATISTICAL DECISION THEORY

TECHNICAL NOTE 13-1: ESTABLISHING PRIOR STATE PROBABILITIES

Several times we have referred to the prior probabilities over the states. But just how are these established? They may, on the one hand, represent historical or experimental relative frequencies. On the other hand, they may be our subjective estimates of the probabilities of the various possible states based on familiarity and experience: our betting odds. But more likely they will represent a blending of both. In this note, we will be concerned primarily with developing priors based on historical information. After this task is done, the decision maker should (1) check the implications of the probability estimates, and (2) determine what alterations are required because the situation has changed from the one that produced the historical data: for example, a new machine, a new employee, a new competitor, or altered economic conditions could require modification of the historical data.

WHEN SUBSTANTIAL HISTORICAL DATA ARE AVAILABLE

Suppose that a retailer has information on weekly demand which is considered relevant to estimating future demand (Table 13-1A). This information is

Table 13-1A Weekly Demand

Quantity	Frequency	Relative Frequency
8	3	0.02
9	15	0.10
10	9	0.06
11	30	0.20
12	60	0.40
13	24	0.16
14	9	0.06
	150	1.00

reproduced in Figure 13-1A. Based on experience, the manager may believe that a smooth curve rising to a single peak and gradually tapering off on either side is most reasonable. The manager believes the unevenness in the historical data is not indicative of the underlying relationship, which should be smooth. The estimated demand probabilities are read from the smooth curve on the graph and then adjusted to sum to unity. This is done in Table 13-1B. Finally, the probabilities are checked for reasonableness by

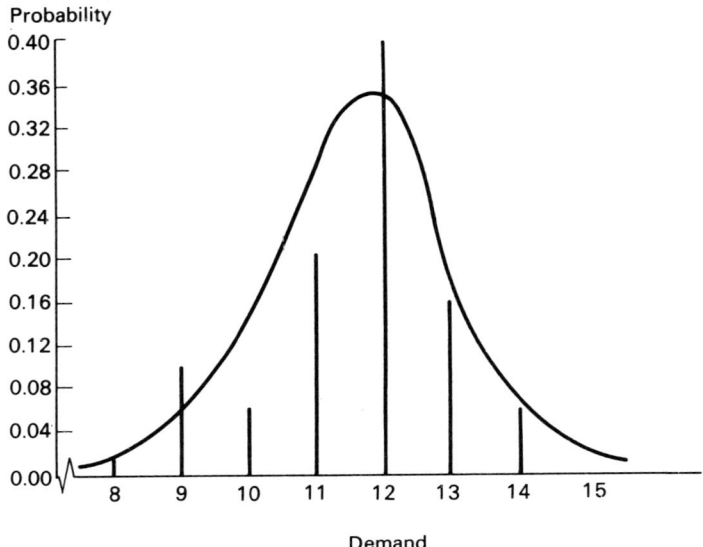

Figure 13-1A Plot of Relative Demand Frequency

seeing if combinations of probabilities still agree with the decision maker's beliefs. For example, a demand for 11 or 12 units should be more likely than all other levels combined, if the final estimates shown in Table 13-1B are to be useful and the procedure successful.

<table>
<tr><td colspan="3">Table 13-1B Adjusted Demand</td></tr>
<tr><td>(1)</td><td>(2)</td><td>(3)</td></tr>
<tr><td>Demand</td><td>Curve Ordinate
from Figure 13-1A</td><td>Adjusted Probability
Col. (2) ÷ Sum Col. (2)</td></tr>
<tr><td>8</td><td>0.02</td><td>0.02</td></tr>
<tr><td>9</td><td>0.06</td><td>0.06</td></tr>
<tr><td>10</td><td>0.13</td><td>0.12</td></tr>
<tr><td>11</td><td>0.26</td><td>0.24</td></tr>
<tr><td>12</td><td>0.34</td><td>0.31</td></tr>
<tr><td>13</td><td>0.20</td><td>0.18</td></tr>
<tr><td>14</td><td>0.06</td><td>0.06</td></tr>
<tr><td>15</td><td>0.01</td><td>0.01</td></tr>
<tr><td></td><td>1.08</td><td>1.00</td></tr>
</table>

WHEN SPARSE HISTORICAL DATA ARE AVAILABLE

Suppose the manager had only nine observations on demand (4, 5, 7, 8, 9, 10, 11, 12, and 14). In this case, it may not seem that much can be done. Certainly a frequency distribution would not be useful. However, it can be shown that the rth smallest observation in a set of n ordered observations is an estimate of the $r/(n + 1)$ fractile. [See W. Feller, *An Introduction to Probability Theory and Its Applications*, Vol. 1, 2nd ed. (New York: John Wiley & Sons, Inc., 1968, pp. 211–212).] Hence, four is a reasonable estimate of the first decile, five is a reasonable estimate of the second decile, and so on. The cumulative frequency implied by these data is graphed in Figure 13-1B.

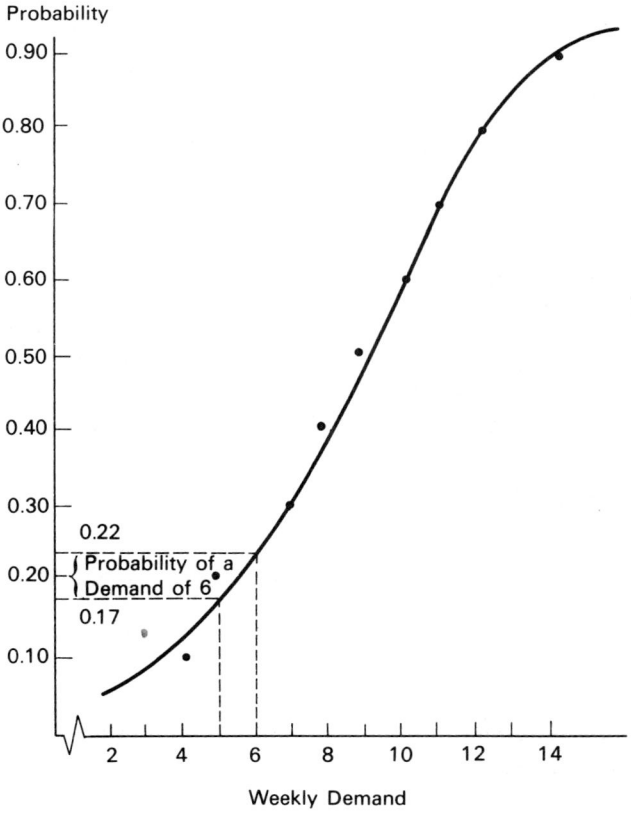

Figure 13-1B Sparse Demand Data: Cumulative Probability

A smooth curve is again fitted to remove irregularities, if this seems appropriate, and any desired cumulative probability is read from the graph. The probability of a single demand value can be found by subtracting successive values obtained from the graph. For example, finding the probability of a demand for six units is illustrated on the graph. This probability is approximately 0.05. Even such crude approximations can be helpful to the manager in assigning probabilities to states.

TECHNICAL NOTE 13-2: FRACTILE RULE

In some problems, such as ones dealing with demand, there is a different and unique best action for every state. If the state of demand is 1, then "stock 1" is the best action. If demand is 2, then we should stock 2; and so on. Such problems are called *equivalent state-action problems* (ESAP).

DISCRETE STATE VARIABLE

Certain ESAP problems have opportunity losses that change linearly as the state changes. Suppose, for example, that an item yields a contribution margin of $2 per unit sold. If an item is unavailable for sale, the margin is lost. If an item is stocked but not sold, the carrying costs, of say $1 per unit, are lost. If these amounts are constant per unit, the opportunity losses for any action graph as two straight lines with constant (but different) slopes. (That is, the unit contribution margin or inventory cost must be the same for every unit.) This is illustrated in Figure 13-2A for the action stock six with a contribution margin of $2 and carrying costs of $1.

Although such problems can be solved using the tabular methods we already know, a special and simpler technique will also work. This procedure directs us to perform the following steps.

1. Define the opportunity loss of underestimating the state by one unit (here, underestimating demand and being short, or under, by one unit): b_u.

2. Define the opportunity loss of overestimating the state by one unit (here, overestimating demand and having an excess of one unit): b_o.

3. Form the ratio $b_u/(b_o + b_u)$.

4. Cumulate the state probabilities, and find the state for which the accumulated probability first exceeds the ratio found in step 3.

5. Select the best action for this state. (If the cumulative probability just equals the ratio for some state, we are indifferent to the best action for this state or the one just before it.)

STATISTICAL DECISION THEORY

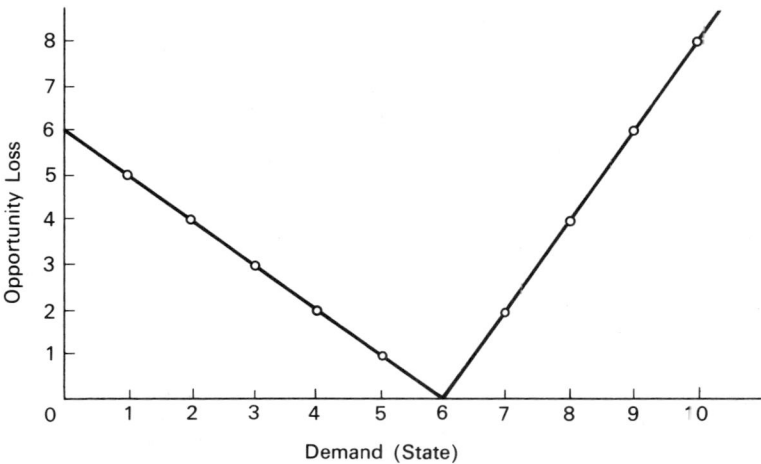

Figure 13-2A Opportunity Losses: ESAP Demand Problem: Action Stock Six

For our demand problem $b_u = \$2$, $b_o = \$1$, and $b_u/(b_o + b_u) = \frac{2}{3} = 0.67$. Suppose our demand probabilities are those given in the last column of Table 13-1B. Cumulating probabilities, we find that $p(d \leq 11) = 0.44$ and $p(d \leq 12) = 0.75$. Hence the state satisfying step 4 is demand equals 12. Selecting the best action for this state directs us to stock twelve units.

This problem is a common type of problem, and it is, therefore, useful, to have a simple solution procedure for it. The five steps above give the Bayesian decision solution much more easily and quickly than the tabular approach used in the chapter. Due also to the common nature of this problem, it is useful to describe it by its characteristics. Problems lending themselves to this solution technique have a different best action for each state (equivalent state-action problems, ESAP) and are found with loss functions that are linear in at most two pieces (with piecewise linear loss, WPLL). Hence we will call them ESAPWPLL (E-SAP-WA-PULL) problems!

The five-step method saves calculating all the expectations. Only two opportunity losses are required, the opportunity losses from overestimating and from underestimating the state by one unit.

CONTINUOUS STATE VARIABLE

If the state variable is continuous, a payoff or loss table cannot be used. However, the approach developed in this technical note still works. Suppose, merely for illustration, that $b_u = 1$, $b_o = 3$, $b_u/(b_o + b_u) = 0.25$, and that the relevant probability density function over the continuous state variable, θ, is

$$f(\theta) = \begin{cases} \frac{1}{8}\theta & 0 \leq \theta \leq 4 \\ 0 & \text{elsewhere.} \end{cases}$$

Then we wish to find that value of θ, call it θ_*, for which

$$\int_0^{\theta_*} \tfrac{1}{8}\theta \, d\theta = \tfrac{1}{4}.$$

Solving, $\theta_*^2/16 = \tfrac{1}{4}$ and $\theta_* = 2$. We are therefore led to pick the best action for the state $\theta = 2$.

We have not proved that our five-step solution technique works for ESAPWPLL problems. We have only illustrated it. A proof can be found in T. Dyckman, S. Smidt, and A. McAdams, *Managerial Decision Making Under Uncertainty* (New York: Macmillan Publishing Co., Inc., 1969).

TECHNICAL NOTE 13-3: MEAN RULE

Some problems involve a very large (perhaps even an infinite) number of states but fewer actions. This is the case when, for example, the values of the state variable are given by a continuous probability distribution, but when only two actions are being considered, say whether or not to build a new facility. In such problems one action will be best over several states while the other action is better over the other states. Hence, we have no equivalence among the states and actions (NESAP ≡ nonequivalent state-action problems).

Suppose, further, that the payoffs (not the opportunity losses) are linear functions of the state variable. Under these conditions it can be shown that the Bayesian decision criterion reduces to two steps:

1. Find the mean value of the probability distribution of the state variable.
2. Pick the best action for that value.

Suppose, for example, that the payoff values for our new facility problem depend again on the demand for the final service as follows:

a_1: Build facility: $-\$5{,}000{,}000 + \$4(25{,}000)\mu$

a_2: Don't build: $\$0$,

where $\$5,000,000$ is the building cost and $\$4$ is the margin for each unit of the service demanded. Assume that the company has 25,000 potential customers and μ represents average annual demand per customer. Suppose, further, that μ is assumed by the manager to follow a normal probability distribution with a mean of 55 and variance of 9.

Using the two linear payoff functions, the organization's break-even average customer demand level, μ_b, is found by setting

$$-\$5{,}000{,}000 + \$4(25{,}000)\mu_b = 0.$$

Hence, $\mu_b = 50$. If $\mu > 50$, the facility should be built. If $\mu < 50$, the organization loses money. Using the two-step decision rule the manager has told us that $E(\mu) = 55$ and, since this exceeds 50, we should build. Furthermore the expected payoff from this action choice is $-\$5{,}000{,}000 + \$4(25{,}000)(55) = \$500{,}000$.

This procedure can be expanded to any finite number of actions, not just two. Since it is so useful, we shall give it a name, too, based on its characteristics of (1) nonequivalent state-action problem (NESAP) and (2) with linear payoff functions (WLPF) or NESAPWLPF (KNEE-SAP-WOOLP). (Try and pronounce that one!)

The rule is easily proved and we will do so only to show you that such things are possible. Consider two linear payoff functions, call them g_1 and g_2:

$$g_1 = C_1 + b_1\mu$$
$$g_2 = C_2 + b_2\mu,$$

where μ is the state variable (we have used θ as the state variable previously to illustrate how versatile we are), $C_1 < C_2$ and $b_1 > b_2$. (If $C_1 > C_2$ and $b_1 > b_2$, then action g_1 would dominate g_2.)

The break-even value is found by equating the two payoffs $C_1 + b_1\mu_1 = C_2 + b_2\mu_2$, which yields

$$\mu_b = \frac{C_2 - C_1}{b_1 - b_2}.$$

As long as $E(\mu) > \mu_b = (C_2 - C_1)/(b_1 - b_2)$, the expected payoff for action g_1 is higher than for action g_2. This is demonstrated next. Suppose that

$$E(\mu) > \frac{C_2 - C_1}{b_1 - b_2}.$$

Then since $b_1 - b_2 > 0$,

$$(b_1 - b_2)E(\mu) > C_2 - C_1$$

or

$$b_1 E(\mu) + C_1 > b_2 E(\mu) + C_2,$$

which implies that
$$E(b_1\mu + C_1) > E(b_2\mu + C_2),$$
and therefore
$$E(g_1) > E(g_2).$$

The expected payoff value of action 1 exceeds that for action 2.

In the problem just worked, the state variable was continuous and hence the tabular method introduced in the chapter could not be used. If the state variable were discrete for a NESAPWPLF problem, we could use either the tabular method or the mean rule that we have just developed. For a simple example consider the payoff table given in Table 13-3A.

Table 13-3A Payoffs (Profits)

States and Their Probabilities

Action	$\theta_1 = 1$ $p(\theta_1) = 0.2$	$\theta_2 = 2$ $p(\theta_2) = 0.5$	$\theta_3 = 3$ $p(\theta_3) = 0.2$	$\theta_4 = 4$ $p(\theta_4) = 0.1$
a_1	10	20	30	40
a_2	40	30	20	10
a_3	27	27	27	27

Using the tabular approach, we obtain for the expectations:
$$E(a_1) = 22, \quad E(a_2) = 28, \quad E(a_3) = 27.$$

Hence, using the Bayesian decision rule, the best action is a_2.

Using the mean rule, we first calculate
$$E(\theta) = 0.2(1) + 0.5(2) + 0.2(3) + 0.1(4) = 2.2.$$

Assuming linearity of the payoff functions (required for a NESAPWLPF) the value of the actions at $\theta = 2.2$ would be
$$a_1 = 22, \quad a_2 = 28, \quad a_3 = 27.$$

These numbers are found using linear interpolation between the tabulated values.

For example, consider a_2. Here $\theta = 2.2$ is two tenths of the way from θ_2 to θ_3. Hence, the payoff should be (given linearity) two tenths of the way from the payoff for θ_2 (which is 30 for a_2) to the payoff for θ_3 (which is

20 for a_2). Thus, $30 - 0.2(30 - 20) = 28$. Again the best action is a_2, which was the expected value that we found using the tabular approach.

The discussion to this point has assumed nothing about the form of the probability function over the states. The method is very general. But suppose that a sample is taken and the data used to revise the state probabilities. If discrete states are involved, the methods of Section 13-6 apply. Unfortunately, these methods cannot easily be applied if the state variable is continuous. We need some new procedures to deal with the continuous-state-variable case. First, we note that the decision in NESAPWLPF problems depends only on the mean and, hence, after revision on the revised mean, $E_1(\mu)$. It is not easy to calculate $E_1(\mu)$ in most cases. However, if the probability distribution over the states is normal (or near normal), then the revised mean is given by

$$E_1(\mu) = \frac{E_0(\mu)[1/\sigma_0^2(\mu)] + \bar{x}(1/\sigma_{\bar{x}}^2)}{[1/\sigma_0^2(\mu)] + (1/\sigma_{\bar{x}}^2)}, \qquad (13\text{-}3.1)$$

where

$E_1(\mu) \equiv$ revised expectation of μ
$E_0(\mu) \equiv$ prior expectation of μ
$\sigma_0^2(\mu) \equiv$ prior variance of μ
$\bar{x} \equiv$ sample mean
$\sigma_{\bar{x}}^2 \equiv$ variance of the sample mean.

If $\sigma_{\bar{x}}^2$ is unknown, $s_{\bar{x}}^2 = s^2/n$ may be used.

Formula (13-3.1) looks worse than it is. (Don't they all?) In fact, it is a weighted mean where each of the two estimates of the mean, $E_0(\mu)$ and $\bar{x}$, is weighted to obtain the revised expectation. The weights are the reciprocals of the variances of each estimate. This is reasonable, since the smaller the variance is, the less the variability and the more information there is in the estimate. In the formula, the smaller the variance, the larger the weight because the reciprocal is involved. As the sample size increases, more information is obtained. This is reflected in the fact that $\sigma_{\bar{x}}^2 = \sigma^2/n$ declines and $1/\sigma_{\bar{x}}^2$ increases.

Assume that a sample of 100 yields an $\bar{x}$ of 48.5 and a sample variance, s^2, of 300; hence $s_{\bar{x}}^2 = \frac{300}{100} = 3$. Then, since we were given that $\sigma_0^2(\mu) = 9$,

$$E_1(\mu) = \frac{55(\frac{1}{9}) + 48.5(\frac{1}{3})}{\frac{1}{9} + \frac{1}{3}} = 50.1.$$

Using the revised mean, the expected value still exceeds the break-even value, so we build.

An important question concerns how a manager might determine the mean and variance of a probability distribution which the manager believes can reasonably be approximated by a normal probability law. This can be done if the manager can answer two questions concerning the state variable, call it μ again. They are:

1. What value is as likely to be exceeded as not reached by the variable in question? Call it $E_0(\mu)$.
2. Select a range, symmetric about $E_0(\mu)$, so that it is just as likely that μ will be within as outside this range.

The answer to question 1 gives us the median, which is also the mean for any symmetrical probability law. It is easier for managers to think about medians than directly about means. The answer to the second question gives a ± 0.67 standard deviation interval around the mean, since such an interval contains 50 percent of the area.

For our problem, the manager's responses would have been:

To question 1: 55.
To question 2: 53 to 57.

Thus, we establish from answer 1 that $E_0(\mu) = 55$ and from the answer to question 2 that

$$z_{0.25} = +0.67 = \frac{\mu - E_0(\mu)}{\sigma_0(\mu)} = \frac{57 - 55}{\sigma_0(\mu)},$$

so that $\sigma_0(\mu) = 3$ and $\sigma_0^2(\mu) = 9$.

A final question concerns whether a second sample should be taken given that we now have $E_1(\mu) = 50.1$. To answer this question we also need to determine $\sigma_1(\mu)$. Given normality of the distribution over μ,

$$\frac{1}{\sigma_1^2(\mu)} = \frac{1}{\sigma_0^2(\mu)} + \frac{1}{\sigma_{\bar{X}}^2}. \tag{13-3.2}$$

For our sample, where $\sigma_{\bar{X}}^2 = s_{\bar{X}}^2 = 3$,

$$\frac{1}{\sigma_1^2(\mu)} = \tfrac{1}{9} + \tfrac{1}{3} = \tfrac{4}{9}$$

and

$$\sigma_1(\mu) = \tfrac{3}{2}.$$

Given these two values and normality, there is a probability of $p(z < [(50 - 50.1)/\tfrac{3}{2}]$, or about 0.47, that the actual mean may be below the break-even value. In this case, we would likely recommend that a decision

Statistical Decision Theory

be delayed, if possible, until more data are obtained. (Note the similarity of the solution to this question with hypothesis testing. The technique is the same as testing the null hypothesis that $\mu \geq 50.1$ given an $\bar{x} = 50$.)

If a second sample is taken, then in the revision process all the data prior to the second sample determines $E_0(\mu)$ and $\sigma_0^2(\mu)$. We have, therefore, that $E_0(\mu) = 50.1$ and $\sigma_0^2(\mu) = \frac{9}{4}$.

New Symbols

Symbol	Meaning
a_i	Action i.
θ_j, μ_j, etc.	State j.
$E[L(a_i)]$	Expected opportunity loss from taking action a_i.
$f_0(\theta_j)$	The prior probability of state θ_j.
$f_1(\theta_j) = f(\theta_j \mid \text{sample result})$	The revised probability of state θ_j given the sample result.
ESAPWPLL	Equivalent state-action problem with piecewise linear losses.
NESAPWLPF	Nonequivalent state-action problem with linear payoff functions.

Key Formulas

Formula	Used to Compute:
$E(a_i) = \sum v_{ij} p_j$	The expected payoff from action a_i, where v_{ij} is the payoff for action i and state j.
$L(a_i \mid \theta_j) = \lvert v(a_j^* \mid \theta_j) - v(a_i \mid \theta_j)\rvert$ $= \lvert v_j^* - v_{ij}\rvert$	The opportunity loss from selecting action a_i given state θ_j.
$\text{EVPI} = E[L(a_i^*)]$	The expected value of perfect information.
$\dfrac{b_u}{b_o + b_u}$	The critical ratio for an ESAPWPLL problem
$E_1(\mu) = \dfrac{E_0(\mu)[1/\sigma_0^2(\mu)] + \bar{x}(1/\sigma_{\bar{X}}^2)}{[1/\sigma_0^2(\mu)] + (1/\sigma_{\bar{X}}^2)}$	The revised expectation of the state variable for a normal prior and normal sampling.
$\dfrac{1}{\sigma_1^2(\mu)} = \dfrac{1}{\sigma_0^2(\mu)} + \dfrac{1}{\sigma_{\bar{X}}^2}$	Reciprocal of the revised variance for a normal prior and normal sampling.

PROBLEMS

13-1.* A retailer has shelf space for four units of highly perishable and identical items which are destroyed at the end of the day if they are not sold. The unit cost of the items is $2.00, and the selling price is $4.00. This yields a contribution of $2.00 per item sold. Set up a payoff table and an opportunity-loss table.

(a) If nothing is known about the possible level of demand for the item, how many items should the retailer stock to maximize his minimum possible payoff? Minimize the maximum possible opportunity loss?

(b) Suppose that the probabilities of demands of 0, 1, 2, 3, and 4 are 0.10, 0.30, 0.40, 0.10, and 0.10, respectively. What action would you advise using the Bayesian decision rule?

(c) In part (b), does it matter whether the payoff table or the opportunity loss table is used?

13-2. A manager receives a shipment of parts and knows that the proportion of defective items in the shipment is either 10 or 20 percent, depending on the nature of the process used to manufacture the parts. Suppose the manager, based on past history, assigns an 0.6 probability to the possibility that the shipment's parts are 10 percent defective. Suppose further that the manager has only two actions open: either accept or reject the shipment. The opportunity losses are as given in the accompanying table.

Loss Table

	States	
Action	$\theta_1: 10\%$	$\theta_2: 20\%$
a_1: accept	$0	$20,000
a_2: reject	$16,000	$0

(a) What action should be taken using the Bayesian decision rule?
(b) Construct the payoff table from which this loss table resulted.
(c) Can you think of a third action alternative?

13-3. A wildcatter must decide whether or not to drill for oil at a specific location. Assume that the payoffs are based on the following monetary values:

Value of a producing well: $900,000
Cost of drilling: $300,000
Value of a dry well: $0

(a) Set up the payoff and loss table for this problem.
(b) Using the Bayesian decision rule, which action should be taken if the prior probability of oil is 0.45?
(c) Suppose that seismograph recordings can be taken before making a decision to drill or not. What is the expected value of perfect information? (If the recordings cost $25,000 each, is a sample of size 2 worth taking?) Assume that the results are independent. (The probability that the recording indicates oil when, in fact, there is oil is 0.7. The probability that the recording indicates no oil if, in fact, there is no oil is 0.6. The probabilities are based on prior drillings.)

13-4.* In Problem 13-3, find the optimal sample size if sampling costs $10,000 per item.

13-5.* A contractor has to choose between two jobs that require the same time and resources. The first promises a profit of $80,000 with a probability of $\frac{3}{4}$ or a loss of $20,000 (due to strikes and other delays) with a probability of $\frac{1}{4}$. The second promises a profit of $120,000 with a probability of $\frac{1}{2}$ or a loss of $30,000 (due to the effects of weather on completion time) with a probability of $\frac{1}{2}$. The losses are negative profits, not opportunity losses. Which job should the contractor select using the Bayesian decision rule?

13-6.* A company has $100,000 available to invest in a new plant. If business conditions remain unchanged, the investment will return 15 percent per year, but, if there is a mild recession, it will return only 3 percent per year. Alternatively, the money can be invested in government bonds for a sure return of 5 percent. What probability must management assign to the condition "recession" before the investment in government bonds is the better choice using the Bayesian decision rule?

13-7. Given the following payoff table, construct an opportunity-loss table:

Payoff Table

Action	θ_1	θ_2	θ_3	θ_4
a_1	10	12	16	2
a_2	14	9	12	10
a_3	7	15	6	9

State

13-8.* A producer of fountain pens has decided to produce a new design specifically for the military. The president thinks that there is a 50–50 chance that the average variable cost of production will be above or below $4.00 per pen. She also feels it is equally likely that this cost will be in the range $3.50 to $4.50 or outside this range. Suppose, further, that she is willing to accept a normal curve as suitably approximating her subjective judgments over the range of possible costs.
  (a) What is the standard deviation of this distribution?
  (b) What is the probability that the average variable cost will be between $4.20 and $4.30?
  (c) If an average variable cost of $3.25 is required to break even, given a fixed contract price, what is the chance that the manufacturer will fail to break even?

13-9. Suppose that a decision maker claimed no knowledge in a given situation of what probabilities were relevant to the possible states. Although he could tell you his action choices and the possible states, he is unwilling to engage in any discussion concerning the state likelihoods. What, if anything, might be done to assist in the decision process?

13-10.* Consider the cost payoff table shown. Are any actions dominated? If so, which ones?

	State		
Action	θ_1	θ_2	θ_3
a_1	20	15	10
a_2	25	10	15
a_3	20	5	20
a_4	15	20	10
a_5	25	5	10
a_6	10	15	20

13-11.* Josef Tomaso's utility for money (not for the change in his amount of money) is given by $U = 10x - x^2$ for $0 \leq x \leq 3$, where x is in thousands of dollars. He now has $1,000.
  (a) Suppose he can invest his $1,000 with the following expectations (gains include return of the $1,000):

  Gain $0 with probability 0.3 (no change in his monetary position)
  Gain $1,000 with probability 0.3
  Gain $2,000 with probability 0.2
  Lose $1,000 with probability 0.2 (loses his investment).

Statistical Decision Theory

Should he invest?

(b) If he is offered the opportunity to gain $1,000 or nothing, and he believes that each outcome is equally likely, what should he be willing to pay for the opportunity? The payment is not recoverable in either case.

13-12. The Beer-Smud Sporting Goods Company is trying to decide whether or not to carry a new line of lightweight boxing gloves. The decision will rest upon profit considerations only. They estimate that each pair can be sold for $5.00 while purchase and handling costs amount to $2.00 a pair. The floor space to be used has no alternative use. Depreciation on the floor space is estimated at $20 per day. Assume that demand is approximated by the accompanying table. What is the optimal level of gloves to stock? If the space could also be used to stock ski gloves, how could this factor be included in the analysis? Gloves not sold are disposed of at a net return of $0.

Units Demanded	Number of Days	Relative Frequency	Cumulative Frequency
0	0	0	0
1	1	$\frac{1}{60}$	$\frac{1}{60}$
2	2	$\frac{2}{60}$	$\frac{3}{60}$
3	3	$\frac{3}{60}$	$\frac{6}{60}$
4	5	$\frac{5}{60}$	$\frac{11}{60}$
5	8	$\frac{8}{60}$	$\frac{19}{60}$
6	11	$\frac{11}{60}$	$\frac{30}{60}$
7	10	$\frac{10}{60}$	$\frac{40}{60}$
8	7	$\frac{7}{60}$	$\frac{47}{60}$
9	5	$\frac{5}{60}$	$\frac{52}{60}$
10	4	$\frac{4}{60}$	$\frac{56}{60}$
11	3	$\frac{3}{60}$	$\frac{59}{60}$
12	1	$\frac{1}{60}$	$\frac{60}{60}$
Total	60	1	

13-13.* Requires Technical Note 13-3. The Craig Machine and Foundry Company is considering the purchase of a sharpening machine which will obviate a good deal of expensive labor hours. The machine costs $11,400 and has a life of 1 year. It can be sold for $1,000 net of disposal costs at that time. The production vice president estimates that each labor hour saved is worth $5.00 (the incremental labor cost per hour) and that he expects to save 2,000 labor hours during the year if the machine is purchased. After some questioning the vice president modified his initial statement to say that he is 50 percent sure the labor hours saved will exceed 2,000,

but there is a 25 percent chance the hours saved will be less than 1,800. He further states that a normal probability distribution adequately reflects the shape of the hours-saved distribution.
(a) Accepting the data, should the machine be purchased?
(b) If a sample of historical data on 10 similar saving situations yielded a mean yearly saving of 2,050 hours with a variance of 100,000, should the machine be purchased?
(c) Suppose the sample in (b) yielded a mean of 2,100 hours with the same variance. Should the machine be purchsed?

13-14. The president of Aviation Electronics is considering how to approach a proposal for building a vertical-takeoff-and-landing plane for use by the military special forces and navy combat ships. If the company submits a proposal, it may obtain the contract or lose it to a competitor. If it wins the contract, it must then elect whether to build the plane in a new facility from the start or use its present plant. The new plant would be more costly in the construction phase but less costly to operate than using existing facilities if the final prototype is accepted and a long-run production contract obtained. Of course it is possible to build a new plant at this time, but it would create delays in meeting orders and a subsequent loss of revenue. The design engineers are at the moment considering three possible designs, only one of which can be submitted. Diagram the decision problem. Denote the final return at the end of each branch of the decision-tree diagram by v_i. Denote those portions of the diagram that are management decisions and those that are chance events. (Assume that the final dollar value of the contract order, if development is successful, is fixed and known. That is, it is a certain number rather than an unknown as in most real cases.) Are the v_i's uncertain? Why or why not? Indicate for each stage how many different costs or probabilities are involved.

13-15. A government official wished to decide the best way to control crop damage from the gypsy moth. Three methods for attacking the pest are: (1) spray with DDT; (2) use a scent to lure and trap males, so that those that remain must compete for mating with a much larger number of males that have been sterilized in a laboratory and then released; and (3) spray with a juvenile hormone, which prevents the larvae from developing into adults. The net improvement (cost of crops saved less spraying cost) in current and future crop losses using DDT is zero at any stage of the program, for it is assumed that DDT will never completely eradicate the moth. The scent-lure program has a probability of 0.5 of leaving a low number of native males, and an 0.5 chance of a high number. Once the scent lure results are known, a later choice must then be made either to switch back to DDT or to release sterile males. The cost of the scent lures is $5 million and the cost for sterilizing and releasing the males is an additional $5 million. But if this two-phase program is successful, that is, a low survival rate for fertile males, the present and future crop savings will be $30 million, $20 million over cost. If scent lures leave the remaining native male population small, there is a 90 percent chance of success using sterile males; otherwise there is only a 10 percent chance of success with sterile males. A failure results in zero crop savings. The juvenile hormone must be synthesized at a cost of $6 million. There is only a 0.20 probability that the resulting product will work. If it does, the crop savings would be $50 million, as the gypsy moth would become extinct. If not, crop savings would be zero. Construct a decision-tree diagram for the official's decision. Using crop savings minus cost as the payoff measure, determine the maximum expected payoff course of action. Is anything left out of the analysis?

13-16.* Consider the following opportunity-loss table for stocking an item. The item sells for $10 and costs $8. Unsold items can be returned weekly for a credit of $7. Suppose the manager could stock almost six items, that she has no ideas concerning demand, and hence elects the action that minimizes the maximum opportunity loss. What action would she take? Now assume that she can stock seven items, what action would she take? Why do the answers here raise problems for the minimax strategy?

Stock Action	Demand							
	0	1	2	3	4	5	6	7 or More
0	0	2	4	6	8	10	12	14
1	1	0	2	4	6	8	10	12
2	2	1	0	2	4	6	8	10
3	3	2	1	0	2	4	6	8
4	4	3	2	1	0	2	4	6
5	5	4	3	2	1	0	2	4
6	6	5	4	3	2	1	0	2
7	7	6	5	4	3	2	1	0

13-17. Solve Problem 13-1 using the fractile rule given in Technical Note 13-2.

13-18. Indicate the major difference(s) between classical hypothesis testing and decision theory.

13-19. Fill out the following table using Y ≡ yes and N ≡ no. What limitations are there to Technical Notes 13-2 and 13-3?

	Can Be Used for:	
Methods	Discrete States	Continuous States
Decision-table (or tree) methods: Chapter 13		
Technical Note 13-2		
Technical Note 13-3		

13-20. Solve the urn problem in Section 13-6 for a sample of size 2 without replacement.

13-21. Using an area in which you have a personal interest, try to think of a simple situation involving two states and two actions where the payoffs are nonmonetary. Construct a payoff matrix in words and think through a solution using the approach described in the subsection of Section 13-7 titled "Nonmonetary Payoffs." For example, a simple economic case would be whether to pass a tax cut if a recession seemed likely. The payoff matrix might look like the one below. (*Note:* The problem is oversimplified. It is useful to see where.)

	State of Economy	
Action	θ_1: Recession	θ_2: No Recession
a_1: pass tax cut	Increased spending More employment Consumer confidence up Some inflation	Greatly increased spending More employment Little change in consumer confidence Substantial inflation
a_2: no tax cut	Decreased spending Substantial unemployment Consumer confidence down No inflation	Little effect on spending Little effect on unemployment Consumer confidence up Some inflation

13-22. An organization is considering establishing a line of credit with a local bank. The cost of a line of credit, which permits the organization to borrow on demand, is 1 percent of the line whether it is used or not. If the organization borrows, it pays an additional 9 percent

STATISTICAL DECISION THEORY

on the amount borrowed. Funds needed above the line of credit established must be obtained at a higher cost. These additional funds cost the organization 12 percent.

(a) If the bank makes lines of credit available in 100,000 dollar units and if the organization believes its needs are expressed by the following probability mass function, find the optimal line of credit.

$$f(0) = 0.1, \quad f(100{,}000) = 0.3, \quad f(200{,}000) = 0.4,$$
$$f(300{,}000) = 0.1, \quad f(400{,}000) = 0.1.$$

(b) Requires Technical Note 13-3. If the bank allows any dollar amount as a line of credit and the organization believes that the following probability density function reflects the likelihood of its needs, find the optimal line of credit.

$$f(\theta) = \begin{cases} \frac{1}{2} - \frac{1}{8}\theta & 0 \le \theta \le 4 \\ 0 & \text{otherwise,} \end{cases}$$

where θ is in units of 100,000.

13-23. For the urn example in Section 13-6, evaluate a sample of size 3. The cost of each item sampled is $1. Assume a binomial process at each stage. That is, assume that a chip, once drawn, is replaced and mixed thoroughly before another is drawn. Does a sample if size 3 yield a higher expectation than a sample of size 2? How would we find the optimal sample size in this problem?

13-24. Requires Technical Note 13-2. The Navy is considering the question of how many spare aircraft engines it should carry aboard its large aircraft carriers while they are on yearly deployment.

(a) Would you expect this to be a problem that fits the fractile rule? Why?

(b) What major problems do you see in solving the problem?

(c) Would you expect the over- and underestimation cost to be equal? Explain.

(d) If yearly usage on similar ships is given by the following data, estimate a relevant probability mass function. Do you see problems with the data? What type of probability mass function studied earlier in this book might fit the data reasonably well? Use this type of pmf to estimate the probabilities based on the data.

Number of Engines Needed	Number of Ship-Years	Number of Engines Needed	Number of Ship-Years
0	0	8	14
1	2	9	6
2	5	10	5
3	10	11	3
4	15	12	2
5	16	13	1
6	18	14	0
7	12	15	1
			110

13-25.* Consider the following game. Suppose that an urn contains five chips. Two chips have the number 12 on them, one has the number 11, one the number 9, and one the number 6. A chip is drawn from the urn at random and the number on it is placed in the expression $1|X - N|$ in place of N. Now suppose that you can pick any real number X to place in the expression. Suppose you must pay (that is, you lose) the value obtained in dollars. Thus, if you selected $X = 7$ and a chip with an 11 is drawn, you must pay $1|7 - 11| = 1|-4| = 1(4) = \$4$. What choice of X (before the chip is drawn) minimizes your expected payment? What is the expected value of perfect information on the chip to be drawn?

13-26. Requires Technical Note 13-3. Suppose in a decision problem that there are two actions, each with a linear payoff function over a continuous state variable. Suppose further that the prior probability function over the state variable is normal with a mean of 50 and a variance of 100 and the break-even state value is 40. Thus, if we knew that the state were below 40, one action would be best, while if the state were above 40, the other action would be best.
(a) If sampling is relatively inexpensive and the potential losses from taking the wrong action are large, would another sample be reasonable? Why?
(b) Suppose that a sample of 400 is taken and it yields a mean of 35 and a standard deviation of 100. What is the revised mean?

13-27. Requires the Technical Notes and much patience. The Ohio Coal Company is considering the purchase of a service (breakdown) contract on its new coal auger. The contract is available from the manufacturer who sold the auger to Ohio Coal. The contract requires a fixed yearly fee of $1,000 and $10 per man-repair hour. On the other hand, if the firm does its own repair, the cost is $15 per man-repair hour. Past data, considered relevant to the present

Statistical Decision Theory

problem, suggest that the yearly probability distribution is rectangular over its entire range of zero to 600 hours.
(a) Is this problem workable
  (1) Using the fractile rule?
  (2) Using the mean rule?
  (3) Using the methods within Chapter 13 but not the Technical Notes?
  (4) Using the methods of Chapter 12?
  (5) None of the above?
(b) Should we ignore machine downtime? Why?
(c) Under what condition(s) is it proper to omit the cost of repair parts required? (Assume that they would be extra and are not included in the contract.)
(d) Would acceptance of the outside contract increase or decrease preventive maintenance expenditures by Ohio Coal? Why? Suggest a simple but appropriate method of including this extension in the present problem.
(e) Develop an expression for the appropriate probability function and show that it is a proper probability function.

For parts (f), (g), and (h) use the following probability function:

$$f(x) = \begin{cases} \frac{1}{3} - \frac{1}{18}x & 0 \le x \le 6 \\ 0 & \text{otherwise} \end{cases}$$

(x is measured in hundreds).
(f) Should the firm purchase the service contract? Why?
(g) Suppose that another service firm offers to do the same job for a fixed yearly fee of \$3,200 regardless of the hours required. Given the original statement of the problem, should this new offer be accepted?
(h) Suppose a contract were available whereby Ohio Coal could buy a given number of repair hours at \$4 per hour whether used or not. Any repair hours actually used under the contract would cost an additional \$6 per hour or a total of \$10. Extra hours needed beyond those contracted for would cost \$15. How many hours should be purchased in a yearly contract if the equipment had to be repaired? [Ignore parts (a)–(g).]

Chapter 14

Analysis of Variance

The Knudsen Boltz Company is a manufacturing firm that produces small metal parts for sale to both industrial and retail customers. They produce most of these items at five different plant locations, and each location uses a slightly different production layout. In addition, production control is handled differently at each of the locations, and each location obtains its raw materials from different suppliers at different costs.

Mr. Boltz, who is the chairman of the board and chief executive officer, is interested in investigating the relative efficiencies of the five locations. To do so, he has directed Bob Cratchet, his chief accountant, to obtain some average cost-of-production data. Mr. Cratchet observed that the firm makes several items that are produced at all five plant locations. He decided to choose one of the items common to all locations in the study. Mr. Cratchet used the average cost figures that were computed by the accounting staff in a cost study performed on four batches of this product at each of the plants, collected between 6 and 10 months ago. The average unit costs in dollars are shown in Table 14-1.

Mr. Boltz examined the data and decided that Plant 4, which is the oldest of the five plants, was the best plant (in terms of average unit cost), and that Plant 5, which is the newest plant, was the worst plant (in terms of average unit cost). Mr. Cratchet, being surprised at this result, suggested to Mr. Boltz that the results should be tested for significance, using the methods of Chapter 12. Toward that end, Mr. Cratchet tried to specify a null and an alternate hypotheses. (He should have written the hypotheses before the data were examined, but at this point he feels that he must go ahead with some analysis.)

Analysis of Variance

Table 14-1 Average Unit Cost Figures (dollars) for Four Batches of an Item at Five Locations of the Knudsen Boltz Company

Plant Location Number	Batch Number				Mean
	1	2	3	4	
1	0.41	0.41	0.42	0.45	$\bar{x}_1 = 0.4225$
2	0.39	0.41	0.41	0.42	$\bar{x}_2 = 0.4075$
3	0.42	0.40	0.39	0.41	$\bar{x}_3 = 0.4050$
4	0.35	0.39	0.37	0.39	$\bar{x}_4 = 0.3750$
5	0.42	0.44	0.46	0.51	$\bar{x}_5 = 0.4575$
					(grand mean = 0.4135)

Chapter 12 deals with either one mean or two means at a time. Thus, if Mr. Cratchet is to compare the five plant location cost averages, he might examine all possible pairs. If we denote the five averages by μ_1, μ_2, μ_3, μ_4, and μ_5, then he could test:

$$H_0: \mu_1 = \mu_2;\ H_0: \mu_1 = \mu_3;\ \ldots;\ H_0: \mu_4 = \mu_5$$
$$H_1: \mu_1 \neq \mu_2;\ H_1: \mu_1 \neq \mu_3;\ \ldots;\ H_1: \mu_4 \neq \mu_5.$$

There would be 10 possible pairs of means to test (computed using $_5C_2$, the number of combinations of two items chosen from a set of five items). Given 10 possible null hypotheses, you would think the firm might find at least one case in which H_0 could be rejected! In fact, if $\alpha = 0.10$, then we expect 1 out of 10 of the null hypotheses to be rejected even if all 10 of the null hypotheses are true. That is, if we subject a true null hypothesis to enough tests, eventually it will be incorrectly rejected. For this reason, specific hypotheses (as few as possible) should be specified before the data are obtained, and the hypotheses should be chosen based on a theory or belief. Otherwise, if one of several null hypotheses is rejected, it is impossible to determine how likely it is that the result was simply due to chance. For example, if Mr. Cratchet were to test $H_0: \mu_4 = \mu_5$, and find that he is able to reject H_0, he could not really be convinced of the significance of his conclusion. (In addition to this problem, it is tedious to test 10 pairs of hypotheses.)

Mr. Cratchet and Mr. Boltz decided to think back, to before the collection of the data, to what their specific hypotheses would have been at that time. Mr. Boltz feels that he did not know, ahead of time, which plant he would have suspected of having the lowest costs. In fact, he did not think of hypotheses involving two plants at a time. His null hypothesis would have been that all five plants had the same costs, and the alternate, which he would accept only if convinced, would have been that there are real differences among the plants. (If he finds that differences do exist, he will begin other investigations to isolate differences and determine the reason for their existence.) That is, Mr. Boltz would specify

$H_0: \mu_1 = \mu_2 = \mu_3 = \mu_4 = \mu_5$ (all five means are equal)
H_1: all five means are not equal.

Such a null hypothesis must be tested in one test. If, as mentioned above, all possible pairs are tested, we may find some spurious significant results. Fortunately for Mr. Cratchet and Mr. Boltz, this chapter provides a technique for testing this important kind of hypothesis. The technique is called *analysis of variance*, in spite of the fact that it is a test regarding means. The reason for the title is discussed in the next section.

In general, analysis of variance is used when the same data (such as costs, above) are collected under several different conditions (such as at each of five locations), and the null hypothesis is that the factor (or independent variable) in question (location) has no effect on the dependent variable (cost). There may be only one independent variable (such as location, above), or there may be more than one, studied simultaneously. Section 14-1 covers analysis of variance for one independent variable, continuing the Knudsen Boltz example, and Section 14-2 discusses analysis-of-variance techniques when two or more independent variables (factors) are involved.

There is one other characteristic of analysis-of-variance problems worth mentioning. The independent variable is measured on at most an ordinal scale, while the dependent variable is measured on at least an interval scale. In the Knudsen Boltz example, location is measured on a nominal scale and cost is measured on a ratio scale.

14-1 One-Factor Analysis of Variance

In general, one-factor analysis of variance involves taking data under several conditions for (or levels of) that factor. The null hypothesis is that the factor has no effect, which implies that the several samples are each drawn from a single population. If there are L levels of the factor ($L = 5$ in the Boltz example), then H_0 is

$H_0: \mu_1 = \mu_2 = \cdots = \mu_L$
H_1: not all the means are equal (the factor does have some effect).

New Symbol	Meaning
L	The number of levels of the factor in a one-factor analysis of variance.

ANALYSIS OF VARIANCE

If the alternate hypothesis is true, there will (probably) be differences among the $\bar{x}$ values collected at each level of the factor under consideration. We can measure the degree to which the sample $\bar{x}$ values are different between factor levels by computing $s_{\bar{x}}^2$, the sample variance of the $\bar{x}$ values. We compute this $s_{\bar{x}}^2$ using only the $\bar{x}$ values for the L levels of the factor.

$$s_{\text{between}}^2 = s_B^2 = s_{\bar{x}}^2 = \sum_{i=1}^{L} \frac{(\bar{x}_i - \bar{\bar{x}})^2}{L-1}, \qquad (14\text{-}1)$$

where

$$\bar{\bar{x}} \equiv \text{the grand mean of all the data values, } \sum_{i=1}^{L} \frac{\bar{x}_i}{L}.$$

$s_{\bar{x}}^2$ can be thought of as a measure of the variation *between* different factor-level means. If $s_{\bar{x}}^2$ is large enough, we will be led to believe that real differences between the means exist; then we will accept the alternate hypothesis. However, $s_{\bar{x}}^2$ must be compared to some measure of the variability of the data. It is only large or small in comparison with some such measure. The measure that we use is a measure of the within-level variance. One indication of a factor-level effect is if the variation between factor levels exceeds the variation within factor levels. It is this relationship that forms the basis for the hypothesis test.

We can compute a measure of the variation *within* factor levels by computing the sample variance for each factor level and averaging them. In the Knudsen Boltz example, that would involve computing a sample variance for each plant and averaging the five values obtained. In general,

$$s_{\text{within}}^2 = s_W^2 = \frac{1}{L} \sum_{i=1}^{L} s_i^2 = \frac{1}{L} \sum_{i=1}^{L} \left[\sum_{j=1}^{n} \frac{(x_{ij} - \bar{x}_i)^2}{n-1} \right], \qquad (14\text{-}2)$$

where

$x_{ij} \equiv j$th data value taken at factor level i

$n \equiv$ number of data values taken at each factor level.

New Symbol	Meaning
s_W^2	The average variance within factor levels. It is computed using equation (14-2).
s_B^2	The variance of the $\bar{x}$'s between factor levels. It is computed using equation (14-1).

We are assuming here that the same number of data values are obtained for each factor level. In our example, $n = 4$. It will usually be the case that the same number of data values are obtained for each level of the factor. (If not, modifications of the formulas, not shown here, can be given.)

Assuming that the universe (or within-factor-level) variance, σ^2, is the same for all levels of the factor, s_W^2 is an estimate of σ^2. Further, $s_{\bar{x}}^2$ is an estimate of $\sigma_{\bar{X}}^2 = \sigma^2/n$, if all the means are the same. Hence, $ns_{\bar{x}}^2$ is also an estimate of σ^2, if the null hypothesis is true. If the means are not equal (that is, if the null hypothesis is false), then $s_{\bar{x}}^2$ will exceed $\sigma_{\bar{X}}^2$ on the average, since some of the $s_{\bar{x}}^2$ value will be caused by the differences in the means.

We can now form the *variance ratio*:

$$F = \frac{ns_B^2}{s_W^2} = \frac{ns_{\bar{x}}^2}{s_W^2}. \tag{14-3}$$

If the null hypothesis is true, both parts of the ratio are estimates of σ^2 and the ratio should be approximately equal to 1. If the null hypothesis is false, then the numerator will exceed the denominator, and the ratio will be larger than one. If the ratio is large enough, that is, if it exceeds a critical value, we will reject the null hypothesis and accept the alternate hypothesis.

Finally, Mr. Boltz can be told why the chapter is called analysis of variance, even though it discusses a hypothesis test about means. The variances are used to make the test. We obtain two estimates of the variance σ^2, $ns_{\bar{x}}^2$ based on the variability *between* factor level means, and s_W^2 based on the variability within factor levels. If the null hypothesis is true, the two variance measures are expected to be equal, and their ratio should be close to 1. (Because of random variation, the ratio may be either higher or lower than 1.) If the alternate hypothesis is true, the ratio will (probably) exceed 1. In any case, only values of the ratio exceeding 1 cause us to question the null hypothesis.

F DISTRIBUTION

If the data taken for factor level i are drawn from a normally distributed universe with mean $= \mu_i$ and variance $= \sigma^2$, which is the same at all factor levels, the variance ratio given by (14-3) is distributed according to the F distribution. A small portion of the F distribution is tabulated in Table 14-2. More extensive tables are provided at the end of the book.

Like the t distribution and the χ^2 distribution, the F distribution is characterized by degrees of freedom. Unlike the t distribution and the χ^2 distribution, the F distribution has two different degrees of freedom, one for the numerator of the ratio and one for the denominator. The numerator uses L data points

Analysis of Variance

Table 14-2 Portion of the F Table

Degrees of Freedom for the Denominator	α	Degrees of Freedom for the Numerator			
		3	4	5	6
14	0.05	3.34	3.11	2.96	2.85
	0.01	5.56	5.04	4.69	4.46
15	0.05	3.29	3.06	2.90	2.79
	0.01	5.42	4.89	4.56	4.32
16	0.05	3.24	3.01	2.85	2.74
	0.01	5.29	4.77	4.44	4.20

(the number of $\bar{x}_i$ values), and one degree of freedom was lost in calculating $\bar{x}$. Thus, the numerator has $L - 1$, the number of levels of the factor minus one, degrees of freedom. The denominator uses nL data values (all the individual data points), and loses L degrees of freedom due to calculating one mean for each factor level. Thus, the denominator has $nL - L = L(n - 1)$ degrees of freedom.

Thus, for a one-factor analysis of variance, if the assumptions are met the variance ratio follows an F distribution with $(L - 1, L(n - 1))$ degrees of freedom, where the numerator's degrees of freedom are given first. The assumptions are:

1. Each factor level's data is provided by a random sample drawn from a normally distributed universe.

2. The variance is the same for each level of the factor.

Slight deviations from normality do not cause serious difficulties, so that, unless the basic distribution is known to be strongly nonnormal, it is fairly safe to proceed with an F test.

The F test consists of computing the variance ratio, then looking up the critical F value, denoted by F_α, where the α level is the significance level set

New Symbol	Meaning
F	The variance ratio, used to test hypotheses.
F_α	The F value, from the table, that assures a maximum probability of a Type I error equal to α. F_α is used as the critical value.

by management. If the computed variance ratio exceeds the critical value, the alternate hypothesis is accepted; otherwise, the null hypothesis is accepted.

As an example, we return to the Knudsen Boltz Company data from Table 14-1. We can illustrate the use of equations (14-1), (14-2) and (14-3) and the F table, while simultaneously helping Messrs. Cratchet and Boltz to test their hypothesis. Before proceeding, we should note that while differences in the mean may exist, we assume that σ^2 is the same at all plants. We also assume that the universes are roughly normally distributed.

Table 14-1 and the hypotheses are repeated below for convenience.

Table 14-1 Average Unit Cost Figures (dollars) for Four Batches of an Item at Five Locations for the Knudsen Boltz Company

Plant Location Number	Batch Number				Mean
	1	2	3	4	
1	0.41	0.41	0.42	0.45	$\bar{x}_1 = 0.4225$
2	0.39	0.41	0.41	0.42	$\bar{x}_2 = 0.4075$
3	0.42	0.40	0.39	0.41	$\bar{x}_3 = 0.4050$
4	0.35	0.39	0.37	0.39	$\bar{x}_4 = 0.3750$
5	0.42	0.44	0.46	0.51	$\bar{x}_5 = 0.4575$
					(grand mean = 0.4135)

H_0: $\mu_1 = \mu_2 = \mu_3 = \mu_4 = \mu_5$ (location does not affect cost)

H_1: the means are not all equal (there are differences in average unit costs among the locations).

Using the notation introduced above,

$$L = 5, \quad \bar{x} = 0.4135, \quad n = 4$$

degrees of freedom for the numerator $= L - 1 = 4$

degrees of freedom for the denominator $= L(n - 1) = 15$.

Mr. Boltz elects to use $\alpha = 0.01$. He wants to be strongly convinced of finding a cost difference between plants before he begins the more expensive cost analysis.

Analysis of Variance

Using equation (14-1),

$$s_B^2 = s_{\bar{x}}^2 = \sum_{i=1}^{5} \frac{(\bar{x}_i - \bar{\bar{x}})^2}{L - 1}$$

$$= \frac{[(0.4225 - 0.4135)^2 + (0.4075 - 0.4135)^2 + (0.4050 - 0.4135)^2 + (0.3750 - 0.4135)^2 + (0.4575 - 0.4135)^2]}{5 - 1}$$

$$= \frac{0.000081 + 0.000036 + 0.0000722 + 0.0014822 + 0.001936}{4}$$

$$= \frac{0.0036074}{4} = 0.0009018.$$

Using equation (14-2),

$$s_W^2 = \tfrac{1}{5}(s_1^2 + s_2^2 + s_3^2 + s_4^2 + s_5^2),$$

where

$$s_1^2 = \frac{[(0.41 - 0.4225)^2 + (0.41 - 0.4225)^2 + (0.42 - 0.4225)^2 + (0.45 - 0.4225)^2]}{3}$$

$$= \frac{0.00015625 + 0.00015625 + 0.00000625 + 0.00075625}{3}$$

$$= \frac{0.0010750}{3} = 0.0003583$$

$$s_2^2 = \frac{[(0.39 - 0.4075)^2 + (0.41 - 0.4075)^2 + (0.41 - 0.4075)^2 + (0.42 - 0.4075)^2]}{3}$$

$$= \frac{0.000475}{3} = 0.0001583$$

$$s_3^2 = \frac{0.000500}{3} = 0.0001667$$

$$s_4^2 = \frac{0.001100}{3} = 0.0003667$$

$$s_5^2 = \frac{0.004475}{3} = 0.0014917$$

$$s_W^2 = \tfrac{1}{5}(0.0003583 + 0.0001583 + 0.0001667 + 0.0003667 + 0.0014917)$$
$$= \tfrac{1}{5}(0.0025417) = 0.00050834.$$

Finally, after much arithmetic, Mr. Boltz can use equation (14-3):

$$F = \frac{ns_{\bar{x}}^2}{s_W^2} = \frac{4(0.0009018)}{0.00050834} = 7.097.$$

The degrees of freedom are 4 (numerator) and 15 (denominator). Table 14-2 for the F distribution gives entries for $\alpha = 0.05$ and $\alpha = 0.01$. For the degrees of freedom (4, 15), the table entries are shown in a portion of the F table. The circled values, $F_{0.05}$ and $F_{0.01}$, are the ones we are after. Our computed F statistic exceeds the $\alpha = 0.01$ value of 4.89, so Mr. Boltz can accept the alternate hypothesis. He and Mr. Cratchet should study the cost differentials in more depth to ascertain the reasons for the differences. That is,

$$7.097 > 4.89 \text{ (the critical value for } \alpha = 0.01),$$

so we can accept H_1 (and reject H_0).

Mr. Cratchet is prepared to begin the cost analysis to discover the source of the differences. However, before expending much effort on the investigation, he should reexamine the original data values, which are given in Table 14-1.

It may be surprising that the newest plant had the highest cost and that the oldest plant had the lowest cost. These two values (0.3750 and 0.4575) account for much of the variation *between* levels of the location factor. But we must remember that accounting numbers have been used, and they may include some fixed costs. (It is usually more meaningful to consider only variable costs when considering production cost differentials, since fixed costs are committed and cannot be affected by production-level decisions.) In particular, depreciation expense for an old plant may be near zero, while depreciation cost for a new plant may be very high, and neither is affected by present production decisions. It is possible, therefore, that the new plant has a lower variable cost of production than the old plant, and it is the variable cost that is relevant to production decisions. You may study these concepts in other courses (such as accounting). The thing to remember is to always examine the original data for possible defects. (In our example, it would have been appropriate if we had done this before doing the analysis.)

Mr. Boltz believes he has spotted another problem. In general, he believes, the costs for the later batches tend to be higher than the costs for the earlier batches (see Table 14-1). There are several possible explanations for this variation, but Mr. Boltz does seem to be correct that it is present. If there is some other systematic variation (other than that caused by location), we should try to account for it. Depending on the nature of the variation, there

ANALYSIS OF VARIANCE   481

are many methods that can be used. One of these, multiple factor analysis of variance, is discussed in the next section. [Other methods are discussed in books on experimental design, such as: W. G. Cochran and G. M. Cox, *Experimental Designs* (New York: John Wiley & Sons, Inc., 2nd ed., 1957) and B. J. Winer, *Statistical Principles in Experimental Design* (New York: McGraw-Hill Book Company, 1963).]

Before proceeding to a Student Should, one additional topic is discussed. The preceding discussion of one-factor analysis of variance uses formulas that help us understand the logic of the hypothesis test involved. When the actual computations are to be made, one of two things should happen. First, a canned computer routine can be used to perform the calculations and provide the F ratio (the variance ratio) to compare against F_α (the critical value). If we left our computer at home, we may have to perform the calculations by hand. In this case, better formulas can be found which ease the computational burden. These formulas are given next. They can be derived from equations (14-1), (14-2), and (14-3), but we will simply give the results.

COMPUTATIONAL FORMULAS

Total of the data values from factor level i: $\quad T_i = \sum_{j=1}^{n} x_{ij} = n\bar{x}_i \quad$ (14-4)

Total of all data values: $\quad T = \sum_{i=1}^{L} T_i = \sum_{i=1}^{L} \sum_{j=1}^{n} x_{ij}.$

(14-5)

Then

$SS_B \equiv$ sum of squares between factor levels: $\quad \sum_{i=1}^{L} \left[\dfrac{T_i^2}{n} \right] - \dfrac{T^2}{Ln} \quad$ (14-6)

$SS_W \equiv$ sum of squares within factor levels: $\quad \sum_{i=1}^{L} \sum_{j=1}^{n} x_{ij}^2 - \sum_{i=1}^{L} \dfrac{T_i^2}{n}$

(14-7)

$SS_T \equiv$ total sum of squares: $\quad\quad SS_B + SS_W. \quad$ (14-8)

[Formula (14-8) is merely a statement that the total sum of squares is made up of the two components. We will not need the total sum of squares in the F test.]

Finally, the variance ratio is formed using the mean sum of squares between factors [$MS_B = SS_B$ divided by the degrees of freedom $= SS_B/$

($L - 1$) and the mean sum of squares within factors $MS_W = SS_W$ divided by the degrees of freedom $= SS_W/L(n - 1)$].

$$F = \frac{MS_B}{MS_W} = \frac{SS_B/(L - 1)}{SS_W/L(n - 1)}. \quad (14\text{-}9)$$

New Symbols	Meaning
$T_i, T, SS_B, SS_W, SS_T, MS_B, MS_W$	Values used in analysis of variance. They are computed using equations (14-4) through (14-9).

Believe it or not, the above formulas will be easier to calculate than (14-1), (14-2), and (14-3). The results are usually reported in an *analysis-of-variance table*. Table 14-3 shows the form for a one-way analysis-of-variance table.

Table 14-3 One-Way Analysis-of-Variance Table

Source of Variation	Sum of Squares	Degrees of Freedom	Mean Sum of Squares
Between-factor levels	SS_B	$L - 1$	$MS_B = \dfrac{SS_B}{L - 1}$
Within-factor levels	SS_W	$L(n - 1)$	$MS_W = \dfrac{SS_W}{L(n - 1)}$
Total	SS_T	$Ln - 1$	

The variance ratio (F ratio) is computed as the ratio of the two numbers in the far-right-hand column of Table 14-3.

Let us try the computational technique and tabular representation on the Knudsen Boltz data, given in Table 14-1.

Using equation (14-4),

$$T_1 = 0.41 + 0.41 + 0.42 + 0.45 = 1.69$$

$$T_2 = 1.63, \quad T_3 = 1.62, \quad T_4 = 1.50, \quad T_5 = 1.83.$$

Using equation (14-5),

$$T = 8.27 = T_1 + T_2 + T_3 + T_4 + T_5.$$

Using equation (14-6),

$$SS_B = \left[\frac{(1.69)^2}{4} + \frac{(1.63)^2}{4} + \frac{(1.62)^2}{4} + \frac{(1.50)^2}{4} + \frac{(1.83)^2}{4}\right] - \frac{(8.27)^2}{5(4)}$$

$$= [0.714025 + 0.664225 + 0.65610 + 0.56250 + 0.837225] - 3.419645$$

$$= 3.434075 - 3.419645 = 0.014430.$$

Using equation (14-7),

$$SS_W = [(0.41)^2 + (0.41)^2 + (0.42)^2 + (0.45)^2 + (0.39)^2 + (0.41)^2$$
$$+ (0.41)^2 + (0.42)^2 + (0.42)^2 + (0.40)^2 + (0.39)^2$$
$$+ (0.41)^2 + (0.35)^2 + (0.39)^2 + (0.37)^2 + (0.39)^2$$
$$+ (0.42)^2 + (0.44)^2 + (0.46)^2 + (0.51)^2]$$
$$- \left[\frac{(1.69)^2}{4} + \frac{(1.63)^2}{4} + \frac{(1.62)^2}{4} + \frac{(1.50)^2}{4} + \frac{(1.83)^2}{4}\right]$$

$$= 3.4417 - 3.434075 = 0.007625.$$

(Notice that 3.434075 was calculated as part of SS_B above, so that some effort can be saved. Satchel Paige, the immortal baseball pitcher, in arguing against exercise, used to have as one of his rules "Don't hurt the body." While doing excess calculations may not hurt the brain, we believe that we should reserve our intellectual exercise to understanding the material and doing only necessary calculations.)

Finally, using equation (14-9),

$$F = \frac{MS_B}{MS_W} = \frac{SS_B/(L-1)}{SS_W/L(n-1)} = \frac{0.014430/4}{0.007625/15} = 7.097.$$

The answer is the same as before, which is comforting. But a warning is appropriate here. The first time through the calculation, we rounded the numbers to the nearest thousandth, and the F ratio turned out to be 6.8. Small errors can become magnified through the calculational process. (For example, consider $3.434 - 3.420 = 0.014$, which is the SS_B calculation to the nearest thousandth; but 0.014 is over 3 percent less than 0.014430, and that error alone would account for a difference of over 0.2 in the F ratio.) We are not necessarily protected from errors of this sort when using a computer routine, either. However, if the F ratio is over 7 and the critical value is less than 5, we can relax. The rounding error will not make that much difference.

As a final step, Table 14-4 displays the above data in an analysis-of-variance table. The F ratio (variance ratio) is the ratio of the two numbers in the last column; $F = 0.0036075/0.0005083 = 7.097$.

Table 14-4 Analysis-of-Variance Table for the Knudsen Boltz Company Example

Source of Variation	Sum of Squares	Degrees of Freedom	Mean Sum of Squares
Between-factor levels	0.014430	4	0.0036075
Within-factor levels	0.007625	15	0.0005083
Total	0.022055	19	

S.S. Marge Snedden, president of an advertising agency, is interested in studying the effect of family income on the amount of yearly clothing expenditures by an individual. The reason for the study is that the agency would like to know where to advertise, and they know that some magazines draw more high income readers than others. However, advertisements in these magazines are also more expensive, so they would like to see if the extra expenditure is worthwhile.

The null hypothesis is that income has no effect on annual clothing expenditures. Families were classified as low, medium, or high income. Four individuals in each level were studied. (In a real study, more data would be obtained, and a computer would be used for the calculations. Here we want the computations to be short enough to allow hand computation.) The data are shown in Table 14-5.

Table 14-5 Annual Clothing Expenditures for 12 Individuals (dollars)

Family Income	Individual				Mean
	1	2	3	4	
Low	140	130	100	110	120
Medium	220	270	230	180	225
High	300	370	500	450	405

(a) Write the null hypothesis and alternate hypothesis symbolically.

(b) Perform a hypothesis test using $\alpha = 0.01$ and equations (14-1), (14-2), and (14-3).

(c) Verify the result by performing the test using formulas (14-4) to (14-9). Display the results in an analysis-of-variance table, and interpret the result of your test in a sentence.

ANALYSIS OF VARIANCE

(d) Income is not the only factor that might help explain clothing expenditures. List at least two other such factors.

Solution:

(a) $H_0: \mu_1 = \mu_2 = \mu_3$
H_1: not all the means are equal; μ_1 is the mean clothing expenditure for the low-income group, μ_2 is the mean for the medium-income group, and μ_3 is the mean for the high-income group.

(b) $\bar{x} = 250$, so

$$s_{\bar{x}}^2 = \sum_{i=1}^{3} \frac{(\bar{x}_i - \bar{x})^2}{2} = \frac{(120 - 250)^2 + (225 - 250)^2 + (405 - 250)^2}{2}$$

$$= \frac{(16{,}900 + 625 + 24{,}025)}{2} = 20{,}775$$

$s_W^2 = \tfrac{1}{3}(s_1^2 + s_2^2 + s_3^2)$

$$s_1^2 = \frac{(140 - 120)^2 + (130 - 120)^2 + (100 - 120)^2 + (110 - 120)^2}{3}$$

$= 333.33$

$s_2^2 = 1{,}366.67$

$s_3^2 = 7{,}766.67$

Thus

$$s_W^2 = \tfrac{1}{3}(333.33 + 1{,}366.67 + 7{,}766.67) = 3{,}155.56$$

and

$$F = \frac{n s_{\bar{x}}^2}{s_W^2} = \frac{4(20{,}775)}{3{,}155.56} = 26.33.$$

The F table, with $L - 1 = 2$ degrees of freedom in the numerator and $L(n - 1) = 9$ degrees of freedom in the denominator, has $F_{0.01} = 8.02$. Since $26.33 > F_{0.01}$, we can accept H_1.

(c) $T_1 = 480$, $T_2 = 900$, $T_3 = 1620$, and $T = 3{,}000$.

$SS_B = (480^2/4 + 900^2/4 + 1{,}620^2/4) - 3{,}000^2/12$

$= (57{,}600 + 202{,}500 + 656{,}100) - 750{,}000 = 166{,}200$

$SS_W = (140^2 + 130^2 + 100^2 + 110^2 + 220^2 + 270^2 + 230^2 + 180^2$
$\qquad\qquad + 300^2 + 370^2 + 500^2 + 450^2)$
$\qquad - 916{,}200 = 28{,}400$

$$F = \frac{MS_B}{MS_W} = \frac{SS_B/(L-1)}{SS_W/L(n-1)} = \frac{166{,}200/2}{28{,}400/9} = 26.33.$$

The answer is the same under both methods. In both cases we accept H_1. In words, this means that we have shown that family income has a significant effect on an individual's clothing expenditures. An analysis-of-variance table is given in Table 14-6. As before, the F ratio is the ratio of the two numbers in the last column.

Table 14-6 Analysis-of-Variance Table for the Clothing-Expenditure Problem

Source	Sum of Squares	Degrees of Freedom	Mean Sum of Squares
Between-factor levels	166,200	2	83,100.0
Within-factor levels	28,400	9	3,155.56
Total	194,600	11	

(d) Examples of some other factors that might affect clothing expenditures are: type of work, sex, age, marital status, and geographic region.

14-2 Two-Factor Analysis of Variance*

In the Student Should at the end of Section 14-1, we listed several other factors that might help explain clothing expenditures. We can perform a multiple-factor analysis of variance, including more than one factor, to help us explain more of the variability in the data. In this section, we will discuss multiple-factor analysis of variance and give the computational forms for two-factor analysis of variance.

Mr. Boltz will be pleased to know that more than one factor can be investigated at a time, since he believes that his data can be explained more fully using two factors. In particular, he is concerned about the differential cost of items produced on the second shift. Suppose he now tells us that columns 1 and 2 of Table 14-1 represent batches that were produced on the day shift (7:00 A.M. to 3:00 P.M.), and columns 3 and 4 represent items that were produced on the second (or "swing") shift (3:00 P.M. to 11:00 P.M.). They do not currently operate a third (or "graveyard") shift. He believes the second-shift costs may be different due to a shift premium (extra pay for working on the second shift) and, more importantly, a different level of productivity. Table 14-7 shows the Table 14-1 data, recast to show which costs came from which shift.

---

* This section is difficult and tedious. The main ideas in analysis of variance can be obtained from Section 14-1.

Analysis of Variance

The new form for the Knudsen Boltz data has 10 "cells," with $n = 2$ data values in each cell. Each data value may now be written as x_{ijk}, where i is the number of the plant, j is the number of the shift, and k is the number of the data value in the cell. In general:

$i = 1, \ldots, L_1$ ($L_1 \equiv$ the number of levels of the first factor)

$j = 1, \ldots, L_2$ ($L_2 \equiv$ the number of levels of the second factor)

$k = 1, \ldots, n$ ($n \equiv$ the number of data values in each cell).

In our example $L_1 = 5$, $L_2 = 2$, and $n = 2$.

Table 14-7 Average Unit Cost Figures (dollars) for the Knudsen Boltz Company Data

Plant Location Number	Shift Number			
	1: (7 A.M.–3 P.M.)	2: (3 P.M.–11 P.M.)		
1	0.41	0.42		
	0.41	0.45	$\bar{x}_1.. = 0.4225$	
2	0.39	0.41		
	0.41	0.42	$\bar{x}_2.. = 0.4075$	
3	0.42	0.39		
	0.40	0.41	$\bar{x}_3.. = 0.4050$	
4	0.35	0.37		
	0.39	0.39	$\bar{x}_4.. = 0.3750$	
5	0.42	0.46		
	0.44	0.51	$\bar{x}_5.. = 0.4575$	
	$\bar{x}_{.1.} = 0.404$	$\bar{x}_{.2.} = 0.423$	$\bar{x} = 0.4135$	

$\bar{x}_{.j.}$ denotes the mean of all the data values at the jth level of the second factor, and $\bar{x}_{i..}$ denotes the mean of all the data values of the ith level of the first factor. For example:

$$\bar{x}_{1..} \equiv \text{mean of the plant 1 values} = \frac{0.41 + 0.41 + 0.42 + 0.45}{4}$$

$$= 0.4225$$

$$\bar{x}_{.2.} \equiv \text{mean of the second shift values} = \frac{0.42 + 0.45 + \cdots + 0.51}{10}$$

$$= 0.423.$$

In addition, we will define $\bar{x}_{ij.} \equiv$ the mean of the values within the cell having the ith level of factor 1 and the jth level of factor 2. These values, each an average of two numbers, are 0.41, 0.40, 0.41, 0.37, 0.43, 0.435, 0.415, 0.40. 0.38, and 0.485, reading down column 1 first, and then down column 2, The new symbols introduced thus far are summarized below.

New Symbol	Meaning
L_1	The number of levels of factor 1.
L_2	The number of levels of factor 2.
$\bar{x}_{i..}$	The mean of the data values at the ith level of factor 1.
$\bar{x}_{.j.}$	The mean of the data values at the jth level of factor 2.
x_{ijk}	The kth data value observed at the ith level of factor 1 and the jth level of factor 2.
$\bar{x}_{ij.}$	The mean of the data values within the cell having the ith level of factor 1 and the jth level of factor 2.

When testing significance in a two-factor analysis-of-variance problem, we will be able to decide if factor 1 has any effect and if factor 2 has any effect. In addition, we will be able to see if the two factors produce any combined effect over and above their individual effects. That is, we can see if there is any *interaction effect* caused by a combination of the two factors. The individual effects are called *main effects*, to distinguish them from the *interaction effect*.

As an example of an interaction effect, it might be true that the second shift is usually as productive as the first shift, but at plant 5 poor management control of the second shift has caused higher costs. That is, the combination of plant 5 and the second shift causes higher costs, over and above the main effects of plant location and shift number. When interaction effects exist, the effect of one factor is conditional on the level of another factor. In our two-way analysis of variance, we will test both main effects and the interaction effects for Mr. Boltz.

Before discussing the computations required, we will reiterate the assumptions of analysis of variance and make a comment about Mr. Boltz's hypotheses. The assumptions are:

1. The data in each cell (for each combination of factors) are drawn randomly from a normally distributed universe.
2. The variance is the same for each of the universes.

As before, slight deviations from normality will not cause us any difficulties.

ANALYSIS OF VARIANCE

The comment about Mr. Boltz's hypotheses relates to the second factor. He stated that he believes the second shift produces items at a different cost than the first. Thus, there are only two means relevant to testing this assertion, and the hypotheses related to this factor are well defined. This main effect could be tested using a t test, such as we studied in Chapter 12, since only two means are to be tested. However, he could not investigate the main effect of plant location or the interaction effect unless he proceeds with the two-factor analysis of variance. It is always wise to ask if a simpler test will do, but here we and Mr. Boltz choose to use a two-way analysis of variance because it will produce the added information that he desires.

The computations for a two-way analysis of variance again involve measures of variation. We will deal directly with the sum of squares and mean sum of squares here. The hypotheses to be tested can be stated in words.

1. H_0: Factor 1 (plant location) has no effect.
   H_1: Factor 1 has an effect.
2. H_0: Factor 2 (shift) has no effect.
   H_1: Factor two has an effect.
3. H_0: There is no interaction effect.
   H_1: There is an interaction effect.

To test these hypotheses, we can use the following calculations.

Sum of squares between levels of factor 1

$$\equiv SS_B(1) = L_2 n \sum_{i=1}^{L_1} (\bar{x}_{i..} - \bar{x})^2 \quad (14\text{-}10)$$

Sum of squares between levels of factor 2

$$\equiv SS_B(2) = L_1 n \sum_{j=1}^{L_2} (\bar{x}_{.j.} - \bar{x})^2 \quad (14\text{-}11)$$

Sum of squares within cells $\equiv SS_W = \sum_{i=1}^{L_1} \sum_{j=1}^{L_2} \sum_{k=1}^{n} (x_{ijk} - \bar{x}_{ij.})^2 \quad (14\text{-}12)$

Sum of squares of the effect caused by interactions

$$\equiv SS_I = n \sum_{i=1}^{L_1} \sum_{j=1}^{L_2} (\bar{x}_{ij.} - \bar{x}_{i..} - \bar{x}_{.j.} + \bar{x})^2. \quad (14\text{-}13)$$

New Symbol	Meaning
$SS_B(1), SS_B(2), SS_W, SS_I$	Values used in a two-factor analysis of variance. They are computed using equations (14-10) to (14-13).

Equation (14-10) measures the variation that can be explained by factor 1; thus (14-10) measures the squared deviation between the means at the different levels of factor 1. The $L_2 n$ term is present because each $\bar{x}_{i..}$ is computed using $L_2 n$ data values. Equation (14-11) accomplishes the same calculation for factor 2, and $L_1 n$ is present because each $\bar{x}_{.j.}$ is computed using $L_1 n$ data values. The "between" calculations can be tied to Table 14-7 by noting that equation (14-10) requires the means computed at each level of factor 1, then computing the squared deviation of those means (shown in the right-hand margin of Table 14-7) from $\bar{x}$. Equation (14-11) computes means for the levels of factor 2, then computes the squared deviation of those means (shown along the bottom margin of Table 14-7) from $\bar{x}$.

Equation (14-13) computes the portion of the mean in a cell that cannot be explained by the two factors. In particular, a difference of $(\bar{x}_{ij.} - \bar{x})$ exists between the cell mean and the overall mean. But $(\bar{x}_{i..} - \bar{x})$ of that is explained by factor 1 and $(\bar{x}_{.j.} - \bar{x})$ of that is explained by factor 2. The entire difference must have these two factor-explained differences subtracted, to leave only the difference explained strictly by the combination. Thus, $(\bar{x}_{ij.} - \bar{x}) - (\bar{x}_{i..} - \bar{x}) - (\bar{x}_{.j.} - \bar{x}) = (\bar{x}_{ij.} - \bar{x}_{i..} - \bar{x}_{.j.} + \bar{x})$ is the term to be squared and summed. Each one of these squared terms is based on n data values, the number within a given cell.

Finally, equation (14-12) gives us a basis for comparison for the sums of squares. Equation (14-12) looks within a cell to see how much inherent variability the data have. Variability within a cell cannot be explained by any factors or interaction. If that variability is large (relatively), then the variability between levels of a factor may be the result of chance variation. Our F ratios, which are used to test the hypotheses stated above, use this variation within cells in the denominator. If the variation explained by a factor is enough larger than this variation within the cells, we will reject H_0 and accept H_1. As before, given the assumptions made, the variance ratio will be distributed according to the F distribution.

The three hypotheses and their associated ratio are given below.

1. H_0: Factor 1 has no effect.
   H_1: Factor 1 has an effect.

$$F = \frac{\text{mean sum of squares between levels of factor 1}}{\text{mean sum of squares within cells}} \equiv \frac{MS_B(1)}{MS_W}$$

$$= \frac{SS_B(1)/(L_1 - 1)}{SS_W/L_1 L_2(n - 1)}. \tag{14-14}$$

The degrees of freedom associated with $SS_B(1)$ is $L_1 - 1$ since there are L_1 levels, and we lose one degree of freedom because we use $\bar{x}$. The degrees

ANALYSIS OF VARIANCE

of freedom associated with SS_W is $L_1L_2n - L_1L_2 = L_1L_2(n-1)$ because there are L_1L_2n data values, and we lose L_1L_2 degrees of freedom because we use the mean for each of the L_1L_2 cells. Thus, we use $L_1 - 1$ degrees of freedom for the numerator and $L_1L_2(n-1)$ degrees of freedom for the denominator when using the F table.

2. H_0: Factor 2 has no effect.
   H_1: Factor 2 has an effect.

$$F = \frac{\text{mean sum of squares between levels of factor 2}}{\text{mean sum of squares within cells}} \equiv \frac{MS_B(2)}{MS_W}$$

$$= \frac{SS_B(2)/(L_2 - 1)}{SS_W/L_1L_2(n-1)}. \qquad (14\text{-}15)$$

The degrees of freedom associated with $SS_B(2)$ is $L_2 - 1$, the number of levels of factor 2 minus one. Thus the critical value will be found using $L_2 - 1$ degrees of freedom for the numerator and $L_1L_2(n-1)$ degrees of freedom for the denominator.

3. H_0: There is no interaction effect.
   H_1: There is an interaction effect.

$$F = \frac{\text{mean sum of squares, interactions}}{\text{mean sum of squares within cells}} \equiv \frac{MS_I}{MS_W} = \frac{SS_I/(L_1-1)(L_2-1)}{SS_W/L_1L_2(n-1)}. \qquad (14\text{-}16)$$

The degrees of freedom associated with SS_I is $(L_1 - 1)(L_2 - 1) = L_1L_2 - L_1 - L_2 + 1$. This is so first because there are L_1L_2 cell means used; then $L_1 + L_2 - 1$ degrees of freedom are lost because we use the means for each level of each factor in the calculation, but both sets of means must average to $\bar{x}$. Therefore, only $L_1 + L_2 - 1$ of these means are really needed; the last one could be calculated from the first $L_1 + L_2 - 1$ means. Thus, there are $L_1L_2 - (L_1 + L_2 - 1) = (L_1 - 1)(L_2 - 1)$ degrees of freedom associated with SS_I. (This reasoning is harder to follow than the previous degrees-of-freedom arguments. Another way to accept the result is to see that it makes the degrees of freedom total $L_1L_2n - 1$, which is the number of degrees of freedom associated with the entire data set.)

An analysis-of-variance table for a two-way analysis-of-variance problem is given in Table 14-8, followed by a summary of the new symbols used. The three F ratios can be formed using the analysis-of-variance table. They are the first, second, and third values in the last column, respectively, divided by the fourth value in the last column.

Table 14-8 Two-Way Analysis-of-Variance Table

Source of Variation	Sum of Squares	Degrees of Freedom	Mean Sum of Squares
Between factor 1 levels	$SS_B(1)$	$L_1 - 1$	$\dfrac{SS_B(1)}{(L_1 - 1)}$
Between factor 2 levels	$SS_B(2)$	$L_2 - 1$	$\dfrac{SS_B(2)}{(L_2 - 1)}$
Interaction effects	SS_I	$(L_1 - 1)(L_2 - 1)$	$\dfrac{SS_I}{(L_1 - 1)(L_2 - 1)}$
Within cells	SS_W	$L_1 L_2 (n - 1)$	$\dfrac{SS_W}{L_1 L_2 (n - 1)}$
Total	SS_T	$L_1 L_2 n - 1$	

COMPUTATIONAL FORMULAS

As in Section 14-1, there are formulas for two-way analysis of variance that are easier to use in hand computation. These are given below, without discussion, for the times you have to do a two-way analysis of variance by hand.

$$T \equiv \sum_{i=1}^{L_1} \sum_{j=1}^{L_2} \sum_{k=1}^{n} x_{ijk} = \text{overall total of the data values} \qquad (14\text{-}17)$$

$$T_{i..} \equiv \sum_{j=1}^{L_2} \sum_{k=1}^{n} x_{ijk} = \text{total of the data values for the } i\text{th level of factor 1}$$
$$(14\text{-}18)$$

$$T_{.j.} \equiv \sum_{i=1}^{L_1} \sum_{k=1}^{n} x_{ijk} = \text{total of the data values for the } j\text{th level of factor 2}$$
$$(14\text{-}19)$$

$$T_{ij.} \equiv \sum_{k=1}^{n} x_{ijk} = \text{total of the data values within each cell.} \qquad (14\text{-}20)$$

New Symbol	Meaning
$T, T_{i..}, T_{.j.}, T_{ij.}$	Values used in two-factor analyses of variance. They are computed using (14-17) to (14-20).

Analysis of Variance

Then

$$SS_B(1) = \sum_{i=1}^{L_1} \frac{(T_i..)^2}{L_2 n} - \frac{T^2}{L_1 L_2 n} \qquad (14\text{-}21)$$

$$SS_B(2) = \sum_{j=1}^{L_2} \frac{(T._j.)^2}{L_1 n} - \frac{T^2}{L_1 L_2 n} \qquad (14\text{-}22)$$

$$SS_W = \sum_{i=1}^{L_1} \sum_{j=1}^{L_2} \sum_{k=1}^{n} (x_{ijk})^2 - \sum_{i=1}^{L_1} \sum_{j=1}^{L_2} \frac{(T_{ij}.)^2}{n} \qquad (14\text{-}23)$$

$$SS_T = \sum_{i=1}^{L_1} \sum_{j=1}^{L_2} \sum_{k=1}^{n} (x_{ijk})^2 - \frac{T^2}{L_1 L_2 n} \qquad (14\text{-}24)$$

$$SS_I = SS_T - SS_B(1) - SS_B(2) - SS_W. \qquad (14\text{-}25)$$

How on earth, you may ask, can formulas as ugly as (14-17) to (14-25) be considered the "easy" way to make two-way analysis of variance? Two comments are in order. First, we never said they were easy, just easier than equations (14-10) to (14-13). Equations (14-10) to (14-13) are used only to illustrate what variations are being measured. Second, the easiest way to make the calculations is not to use (14-10) to (14-13) or (14-17) to (14-25) but to use a computer. We recommend that approach whenever it is available. Once the sums of squares are calculated, using whichever set of formulas we choose, the F ratios are calculated using equations (14-14), (14-15), and (14-16).

Now, since Mr. Boltz has waited long enough for an answer, we will use his data as an example of both methods of calculation. Using his data, as reported in Table 14-7, Mr. Boltz wants to study the effect of plant location, shift number, and any potential interaction of the two factors. Intuitively, examining the differences among the factor means in Table 14-7, it appears that both main effects are present. It also appears that some interaction effects may exist. For example, the second shift at plant 5 seems to do worse than the main effects can explain, and the second shift at plant 4 does better. It is difficult (if not impossible) to predict the degree of significance of any factor, but we can get a feel for what analysis of variance is doing if we try to predict what the results will be.

First, we will use equations (14-10) to (14-13), and the data in Table 14-7.

$$SS_B(1) = L_2 n \sum_{i=1}^{L_1} (\bar{x}_i.. - \bar{x})^2$$

$$= 2(2)[(0.4225 - 0.4135)^2 + \cdots + (0.4575 - 0.4135)^2]$$

$$= 4(0.0036074) = 0.0144296$$

$$SS_B(2) = L_1 n \sum_{i=1}^{L_2} (\bar{x}_{\cdot j} - \bar{x})^2$$

$$= 10[(0.404 - 0.4135)^2 + (0.423 - 0.4135)^2] = 0.001805$$

$$SS_W = \sum_{i=1}^{L_1} \sum_{j=1}^{L_2} \sum_{k=1}^{n} (x_{ijk} - \bar{x}_{ij \cdot})^2$$

$$= [(0.41 - 0.41)^2 + (0.41 - 0.41)^2 + (0.42 - 0.435)^2$$
$$+ (0.45 - 0.435)^2 + (0.39 - 0.40)^2 + (0.41 - 0.40)^2 + \cdots$$
$$+ (0.42 - 0.43)^2 + (0.44 - 0.43)^2 + (0.46 - 0.485)^2$$
$$+ (0.51 - 0.485)^2] = 0.00355.$$

The second number in each calculation is the mean of the two numbers within the cell. The means in the first row (plant location 1) are 0.410 and 0.435.

$$SS_I = n \sum_{i=1}^{L_1} \sum_{j=1}^{L_2} (\bar{x}_{ij \cdot} - \bar{x}_{i \cdot \cdot} - \bar{x}_{\cdot j \cdot} + \bar{x})^2$$

$$= 2[(0.41 - 0.4225 - 0.404 + 0.4135)^2$$
$$+ (0.435 - 0.4225 - 0.423 + 0.4135)^2$$
$$+ (0.40 - 0.4075 - 0.404 + 0.4135)^2$$
$$+ (0.415 - 0.4075 - 0.423 + 0.4135)^2 + \cdots$$
$$+ (0.485 - 0.4575 - 0.423 + 0.4135)^2]$$

$$= 2(0.000009 + 0.000009 + 0.000004 + 0.000004$$
$$+ 0.00021025 + 0.00021025 + 0.00002025 + 0.00002025$$
$$+ 0.000324 + 0.000324) = 0.00227.$$

The three F ratios are calculated using equations (14-14), (14-15), and (14-16).

Factor 1: $F = \dfrac{MS_B(1)}{MS_W} = \dfrac{SS_B(1)/(L_1 - 1)}{SS_W/L_1 L_2(n - 1)} = \dfrac{0.0144296/4}{0.00355/10} = 10.16.$

The critical value is taken from Table VI in the Appendix using 4 (numerator) and 10 (denominator) degrees of freedom.

$F_{0.01} = 5.99$, so we accept H_1; factor 1 (plant location) does have an effect on costs.

Analysis of Variance

Factor 2: $F = \dfrac{MS_B(2)}{MS_W} = \dfrac{SS_B(2)/(L_2 - 1)}{SS_W/L_1L_2(n-1)} = \dfrac{0.001805/1}{0.00355/10} = 5.08.$

For 1 and 10 degrees of freedom, $F_{0.05} = 4.96$. Thus, we can accept H_1 at the $\alpha = 0.05$ level. At the $\alpha = 0.05$ level, we can accept the conclusion that the shift has an effect on costs.

Interaction: $F = \dfrac{MS_I}{MS_W} = \dfrac{SS_I/(L_1-1)(L_2-1)}{SS_W/L_1L_2(n-1)} = \dfrac{0.00227/4}{0.00355/10} = 1.60.$

For 4 and 10 degrees of freedom, $F_{0.01} = 5.99$ and $F_{0.05} = 3.48$. We must accept the null hypothesis regarding the interaction effect. The data do not show that an interaction effect exists over and above the main effects. This is in spite of the fact that we initially thought the data would support the effect. We should remember, of course, that we have not shown that there is no interaction effect.

Next we will use the "computational formulas," equations (14-17) to (14-25), to obtain a solution.

$T_1.. = 1.69, \quad T_2.. = 1.63, \quad T_3.. = 1.62, \quad T_4.. = 1.50, \quad T_5.. = 1.83$

$T_{.1.} = 4.04, \quad T_{.2.} = 4.23, \quad T = 8.27.$

Also,

$T_{11.} = 0.82, \quad T_{12.} = 0.87, \quad T_{21.} = 0.80, \quad T_{22.} = 0.83,$

$T_{31.} = 0.82, \quad T_{32.} = 0.80, \quad T_{41.} = 0.74, \quad T_{42.} = 0.76,$

$T_{51.} = 0.86, \quad T_{52.} = 0.97.$

From Section 14.1 we know the sum of squares of all the individual data values. That is,

$$\sum_{i=1}^{L_1} \sum_{j=1}^{L_2} \sum_{k=1}^{n} (x_{ijk})^2 = 3.4417.$$

Now

$$SS_B(1) = \sum_{i=1}^{L_1} \dfrac{(T_i..)^2}{L_2 n} - \dfrac{T^2}{L_1 L_2 n}$$

$$= \left[\dfrac{(1.69)^2}{4} + \dfrac{(1.63)^2}{4} + \dfrac{(1.62)^2}{4} + \dfrac{(1.50)^2}{4} + \dfrac{(1.83)^2}{4}\right] - \dfrac{(8.27)^2}{2(5)2}$$

$$= 3.434075 - 3.419645 = 0.014430.$$

$$SS_B(2) = \sum_{j=1}^{L_2} \frac{(T._{j}.)^2}{L_1 n} - \frac{T^2}{L_1 L_2 n}$$

$$= \frac{(4.04)^2}{5(2)} + \frac{(4.23)^2}{5(2)} - 3.419645$$

$$= 3.42145 - 3.419645 = 0.001805.$$

$$SS_W = \sum_{i=1}^{L_1} \sum_{j=1}^{L_2} \sum_{k=1}^{n} (x_{ijk})^2 - \sum_{i=1}^{L_1} \sum_{j=1}^{L_2} \frac{(T_{ij}.)^2}{n}$$

$$= 3.4417 - \left[\frac{(0.82)^2}{2} + \frac{(0.87)^2}{2} + \frac{(0.80)^2}{2} + \cdots + \frac{(0.97)^2}{2}\right]$$

$$= 3.4417 - 3.43815 = 0.00355.$$

$$SS_T = \sum_{i=1}^{L_1} \sum_{j=1}^{L_2} \sum_{k=1}^{n} (x_{ijk})^2 - \frac{T^2}{L_1 L_2 n}$$

$$= 3.4417 - 3.419645 = 0.022055.$$

$$SS_I = SS_T - SS_B(1) - SS_B(2) - SS_W$$

$$= 0.022055 - 0.014430 - 0.001805 - 0.00355$$

$$= 0.00227.$$

These results are shown in an analysis-of-variance table in Table 14-9.

Table 14-9 Two-Way Analysis-of-Variance Table for the Knudsen Boltz Example

Source of Variation	Sum of Squares	Degrees of Freedom	Mean Sum of Squares
Between factor 1 levels	0.014430	4	0.0036075
Between factor 2 levels	0.001805	1	0.001805
Interaction effects	0.00227	4	0.0005675
Within cells	0.00355	10	0.000355
Total	0.022055	19	

All values are the same as with the previous formulas [except that $SS_B(1)$ differs in the sixth decimal place due to rounding]. They always will be the same, of course, and there is no need to do both. The F ratios will be the same as before, and we conclude that both main effects are significant at the 0.05 level. The interaction effect is not shown to be significant by the test.

Based on these results, Mr. Knudsen might impose more controls on

ANALYSIS OF VARIANCE

second shifts. He might begin an investigation to find the reason for plant differences in costs. (Remember that some of the difference may be caused by the way the accounting numbers are derived rather than by real differences in costs.) Mr. Boltz is not finished when the analysis of variance is completed. He still must choose a course of action, based on the results, and implement his plan.

It is easy to wonder, in Mr. Boltz's case or any other situation if as much variability as possible has been explained. We might conjecture that additional factors should be included. Analysis of variance can include more than two factors, but we must be careful in proceeding to do so. At a certain point, extra explanatory factors simply confuse the issue in a managerial setting, making it difficult to proceed. Moreover, the interaction effects become difficult to explain, and occasionally the interaction effects are ignored for this reason. Also, in even a three-way analysis of variance, there are four interaction effects: factor 1 with factor 2, 1 with 3, 2 with 3, and 1, 2 and 3 together. Explaining the fact that 1 with 2 (say plant number and shift number) and 1, 2 and 3 (say factor 3 is average age of the worker) are significant while 2 with 3 and 1 with 3 are not significant requires a little imagination. Using too many explanatory variables also reduces the degrees of freedom available for significance testing.

We will not give formulas for three-or-more-factor analysis of variance. Computer programs exist that will do the calculations, and the reasoning follows the same pattern as above for one- or two-factor analysis of variance.

S.S. The Student Should at the end of Section 14-1 discusses an advertising agency study relating individuals' annual clothing expenditures to annual family income. The data are shown in Table 14-5. Ms. Snedden now tells us that she is concerned about advertising products that are sold to young men, largely in the 18-to-30 age range. The data were taken from this group. Further, the data in columns 1 and 2 were obtained from college men; while columns 3 and 4 were taken from young professional men. Ms. Snedden feels that the student–nonstudent factor may be a significant factor in explaining the data. She wants us to perform the necessary analysis. In particular:

(a) Recast Table 14-5 to show the breakdown into student (level 1) and nonstudent (level 2), in the same manner in which Table 14-7 recasts Table 14-1.

(b) Perform an analysis of variance and test both main effects and the interaction effect using $\alpha = 0.05$. [The hypotheses are those stated in conjunction with (14-14), (14-15), and (14-16). You need not repeat them here.] Make the calculations two ways, first using (14-10) to (14-13), then using (14-17) to (14-25).

(c) Briefly interpret the results and discuss any potential problems in the data.

498 CHAPTER 14

Solution:

(a)

Annual Clothing Expenditures for 12 Individuals (dollars)

	Job Level		
Family Income	1 (*Student*)	2 (*Nonstudent*)	
Low	140	100	$\bar{x}_{1..} = 120$
	130	110	
Medium	220	230	$\bar{x}_{2..} = 225$
	270	180	
High	300	500	$\bar{x}_{3..} = 405$
	370	450	
	$\bar{x}_{.1.} = 238.33$	$\bar{x}_{.2.} = 261.67$	$\bar{x} = 250$

(b) We will use equations (14-10) to (14-13) first. We need

$$\bar{x}_{11.} = 135, \quad \bar{x}_{12.} = 105, \quad \bar{x}_{21.} = 245,$$
$$\bar{x}_{22.} = 205, \quad \bar{x}_{31.} = 335, \quad \bar{x}_{32.} = 475.$$

Also, $L_1 = 3$, $L_2 = 2$, and $n = 2$. Then

$$SS_B(1) = L_2 n \sum_{i=1}^{L_1} (\bar{x}_{i..} - \bar{x})^2$$
$$= 4[(120 - 250)^2 + (225 - 250)^2 + (405 - 250)^2]$$
$$= 4(41,550) = 166,200$$

$$SS_B(2) = L_1 n \sum_{j=1}^{L_2} (\bar{x}_{.j.} - \bar{x})^2$$
$$= 6[(238.33 - 250)^2 + (261.67 - 250)^2] = 1,633.33$$

$$SS_W = \sum_{i=1}^{L_1} \sum_{j=1}^{L_2} \sum_{k=1}^{n} (x_{ijk} - \bar{x}_{ij.})^2$$
$$= [(140 - 135)^2 + \cdots + (450 - 475)^2] = 6,300$$

$$SS_I = n \sum_{i=1}^{L_1} \sum_{j=1}^{L_2} (\bar{x}_{ij.} - \bar{x}_{i..} - \bar{x}_{.j.} + \bar{x})^2$$
$$= 2[(135 - 120 - 238.33 + 250)^2 + \cdots$$
$$+ (475 - 405 - 261.67 + 250)^2]$$
$$= 2[(711.11) + (711.11) + (1,002.78) + (1,002.78)$$
$$+ (3,402.78) + (3,402.78)]$$
$$= 20,466.67.$$

ANALYSIS OF VARIANCE

We can now form the three F ratios.

Factor 1 (family income):
$$F = \frac{MS_B(1)}{MS_W} = \frac{SS_B(1)/(L_1 - 1)}{SS_W/L_1L_2(n-1)} = \frac{166{,}200/2}{6{,}300/6} = 79.14.$$

$F_{0.05} = 5.14$ for 2 degrees of freedom in the numerator and 6 degrees of freedom in the denominator. We accept H_1; factor 1 has some effect.

Factor 2 (student–nonstudent):
$$F = \frac{MS_B(2)}{MS_W} = \frac{SS_B(2)/(L_2 - 1)}{SS_W/L_1L_2(n-1)} = \frac{1{,}633.33/1}{6{,}300/6} = 1.56.$$

$F_{0.05} = 5.99$ for 1 degree of freedom in the numerator and 6 degrees of freedom in the denominator. We accept H_0; at the 0.05 level we cannot show that factor 2 has an effect.

Interaction:
$$F = \frac{MS_I}{MS_W} = \frac{SS_I/(L_1 - 1)(L_2 - 1)}{SS_W/L_1L_2(n-1)} = \frac{20{,}466.67/2}{6{,}300/6} = 9.75$$

$F_{0.05} = 5.14$ for 2 and 6 degrees of freedom. We accept H_1; with $\alpha = 0.05$, we can show that the interaction effect is significant. Next we will use equations (14-17) to (14-25):

$T_{11\cdot} = 270,\quad T_{12\cdot} = 210,\quad T_{21\cdot} = 490,\quad T_{22\cdot} = 410,$

$T_{31\cdot} = 670,\quad T_{32\cdot} = 950,\quad T_{1\cdot\cdot} = 480,\quad T_{2\cdot\cdot} = 900,$

$T_{3\cdot\cdot} = 1{,}620,\quad T_{\cdot 1\cdot} = 1{,}430,\quad T_{\cdot 2\cdot} = 1{,}570,\quad T = 3{,}000.$

Also,
$$\sum_{i=1}^{L_1}\sum_{j=1}^{L_2}\sum_{k=1}^{n} (x_{ijk})^2 = 944{,}600.$$

Now
$$SS_B(1) = \sum_{i=1}^{L_1} \frac{(T_{i\cdot\cdot})^2}{L_2 n} - \frac{T^2}{L_1 L_2 n}$$
$$= \left[\frac{(480)^2}{4} + \frac{(900)^2}{4} + \frac{(1{,}620)^2}{4}\right] - \frac{(3{,}000)^2}{12}$$
$$= 916{,}200 - 750{,}000 = 166{,}200.$$

$$SS_B(2) = \sum_{j=1}^{L_2} \frac{(T_{\cdot j\cdot})^2}{L_1 n} - \frac{T^2}{L_1 L_2 n}$$
$$= \left[\frac{(1{,}430)^2}{6} + \frac{(1{,}570)^2}{6}\right] - 750{,}000 = 1{,}633.33$$

$$SS_W = \sum_{i=1}^{L_1}\sum_{j=1}^{L_2}\sum_{k=1}^{n}(x_{ijk})^2 - \sum_{i=1}^{L_1}\sum_{j=1}^{L_2}\frac{(T_{ij\cdot})^2}{n}$$

$$= 944{,}600$$

$$- \left[\frac{(270)^2}{2} + \frac{(210)^2}{2} + \frac{(490)^2}{2} + \frac{(410)^2}{2} + \frac{(670)^2}{2} + \frac{(950)^2}{2}\right]$$

$$= 944{,}600 - 938{,}300 = 6{,}300.$$

$$SS_T = \sum_{i=1}^{L_1}\sum_{j=1}^{L_2}\sum_{k=1}^{n}(x_{ijk})^2 - \frac{T^2}{L_1 L_2 n}$$

$$= 944{,}600 - 750{,}000 = 194{,}600$$

$$SS_I = SS_T - SS_B(1) - SS_B(2) - SS_W$$

$$= 194{,}600 - 166{,}200 - 1{,}633.33 - 6{,}300 = 20{,}466.67.$$

The values are the same as before, so our conclusions regarding the hypotheses will be the same.

(c) The results show that the main effect for factor 1 is significant; income affects clothing expenditures. The main effect for factor 2 is not significant. The difference in the means (261.67 − 238.33) is not large enough to convince us to accept H_1, especially in light of the small sample size. (More data might find this second factor to be significant.) The interaction effect is significant, indicating that some income–job level pairs have significant effects over and above the main effects. For example, nonstudents from high income families spend more than can be explained by the main effects.

14-3 A Warning

Analysis of variance is used to see if a variable or variables (called factors) affect the variable under investigation (such as cost per unit and annual clothing expenditures in the examples used in this chapter). We have covered how to relate either one or two explanatory variables to the variable under investigation, if certain assumptions are met. As in previous chapters, the main warning regarding the use of these techniques is that we should be aware of all the assumptions we have made to justify the methodology.

We have assumed, in both Sections 14-1 and 14-2, that each cell (the data from each factor level in Section 14-1 or combination of factor levels in Section 14-2) is randomly drawn from a normally distributed universe, and that each of these universes has the same variance. (The null hypothesis states that some of the means are the same, but we do not assume this is so; we test it.) In addition, we have limited our discussion in two ways:

Analysis of Variance

1. The number of data values obtained at each factor level or combination of factor levels is assumed to be the same in the formulas given here.

2. The factor levels studied were assumed to contain all factor levels of interest.

A discussion of these assumptions and limitations will lead us into a few of the topics we have not covered. Failure of the normality assumption is not critical unless the universe is strongly nonnormal. Thus, we can usually proceed to use an F test, which is based on the normality assumption. If the universe is strongly nonnormal, there are other tests that can be applied. One such test, the Kruskal–Wallis test, reduces the data to a rank ordering, then applies a statistical test. This test and related tests are discussed in the final chapter of this book.

The use of an equal number of data values at each factor level or combination of factor levels is unnecessary. However, the formulas are more complicated in this case, and, since there usually is an equal number at each level in a designed experiment, we chose to use the equal-number-of-values form. The formulas for an unequal number are closely analogous to those given here. They can be found in books such as those by Cochran and Cox and by Winer, mentioned in Section 14-1.

Let us turn to point 2 above. There are only five plants in the Knudsen Boltz example. We use all possible levels of this factor, and we seek to learn something only about the five plants. As with all such statements, we statisticians have a term to confuse the issue. The case where all possible levels of a factor are involved is called a *fixed-effects model*. There are analysis-of-variance formulas for the other case, where not all levels are included, and this situation is called a *random-effects model*. The random-effects model is not included here because the types of questions for which it is useful are more often dealt with using regression techniques, and regression is studied in Chapter 15.

The point is that we have discussed the most commonly used analysis-of-variance method, and the logic carries over to other models. However, there are many other techniques that fit other situations. We should be careful to choose the technique whose assumptions fit the situation at hand, and, hopefully, a computer will be available to make the calculations.

A second warning regarding analysis of variance involves the place of these techniques in a managerial decision process. In our example, we saw that more investigation was needed before any actions could be taken. One reason is that we only show that a factor does affect the variable of interest. We do not show how or to what degree the effect occurs. There are methods for predicting the differences for different levels of the factor or factors, and for deriving simultaneous confidence intervals regarding these differences. However, even with these confidence interval statements, the appropriate actions to take are unclear because of the complexity of the statements. The analysis-of-variance technique can best be thought of as a preliminary step, used to indicate the areas to be investigated further.

Another characteristic of analysis of variance is that it uses what can be called an *experimental design*. That is, an investigator decides to collect data (on costs, for example) at each factor level or combination of factor levels. The factors are selected and the data are collected. Then we investigate the relationship, given a complete set of data, as designed. There are entire books written on how to design experiments (and analyze the results), and analysis-of-variance techniques constitute a significant portion of them. Basically, the experimental design chooses which universes we will draw samples from. Each combination of factors is assumed to produce a unique universe, although we assume that each universe has equal variance. Thus, we are comparing universes in an experimental design (and analysis of variance) approach. This is different from the estimation and most of the hypothesis-testing material, where we sought the use of a sample from one universe to learn something about that universe.

In many managerial settings, we cannot have the luxury of designing an experiment. The data may be in hand, or we may be unable to obtain data from all combinations of factors. We frequently must "muddle through" without having complete experimental design data. This makes it imperative that we understand what analysis of variance assumes and what it requires, so that it is used only when it is appropriate. Then it must take its proper place (along with other statistical tools) as one input to the managerial decision process. In some situations where analysis of variance is inappropriate, regression analysis can be used. Regression analysis is the subject of Chapter 15.

New Symbols

Symbol	Meaning
F	The variance ratio used in testing the effects of factors and interactions. The F distribution is used to find critical values.
F_α	The critical value used in hypothesis tests involving the F distribution.
T_i	The total of data values from factor level i in a one-factor analysis of variance. $T_i = n\bar{x}_i$, where $\bar{x}_i$ is the mean of the values at factor level i.
T	The total of all data values. In a one-factor analysis of variance, it is computed as $$T = \sum_{i=1}^{L} \sum_{j=1}^{n} x_{ij}.$$

	In a two-way analysis of variance: $$T = \sum_{i=1}^{L_1} \sum_{j=1}^{L_2} \sum_{k=1}^{n} x_{ijk}.$$
MS_B, MS_W	The mean sum of squares between and within factors. They are equal to the sum of squares divided by the degrees of freedom. $MS_B = SS_B/(L-1)$ and $MS_W = SS_W/L(n-1)$. The ratio of MS_B/MS_W gives the F-ratio value to be used in testing the hypothesis.
L	The number of levels of the factor in a one-way analysis of variance.
L_1	The number of levels of factor 1 in a two-way analysis of variance.
L_2	The number of levels of factor 2 in a two-way analysis of variance.
$\bar{x}_{i..}$	The mean of the data values at the ith level of factor 1 in a two-way analysis of variance.
$\bar{x}_{.j.}$	The mean of the data values at the jth level of factor 2 in a two-way analysis of variance.
$\bar{x}_{ij.}$	The mean of the data values within the cell having the ith level of factor 1 and the jth level of factor 2 in a two-way analysis of variance.
$T_{i..}$	The total of the data values for the ith level of factor 1 in a two-way analysis of variance: $$T_{i..} = \sum_{j=1}^{L_2} \sum_{k=1}^{n} x_{ijk}.$$
$T_{.j.}$	The total of the data values for the jth level of factor 2 in a two-way analysis of variance: $$T_{.j.} = \sum_{i=1}^{L_1} \sum_{k=1}^{n} x_{ijk}.$$
$T_{ij.}$	The total of the data values within the ijth cell (ith level of factor 1, jth level of factor 2): $$T_{ij.} = \sum_{k=1}^{n} x_{ijk}.$$
$MS_B(1)$, $MS_B(2)$, MS_I, MS_W	The mean sum of squares between factor 1, between factor 2, caused by interactions, and within factor levels, respectively. As in one-factor analysis of variance, these values equal the sums of squares divided by degrees of freedom, and F ratios are formed by dividing either $MS_B(1)$, $MS_B(2)$, or MS_I by MS_W. The degrees of freedom for the four mean sums of squares are $L_1 - 1$, $L_2 - 1$, $(L_1 - 1)(L_2 - 2)$, and $L_1 L_2 (n - 1)$, respectively.

Some symbols are defined by the following formulas.

Key Formulas

Formula	Used to Compute
$s_{\bar{x}}^2 = \sum_{i=1}^{L} \dfrac{(\bar{x}_i - \bar{x})^2}{L-1}$	The variance between means taken at the various factor levels.
$s_W^2 = \dfrac{1}{L} \sum_{i=1}^{L} \left[\sum_{j=1}^{n} \dfrac{(x_{ij} - \bar{x}_i)^2}{n-1} \right]$	The variance of the data values taken within the factor levels.
$SS_B = \sum_{i=1}^{L} \dfrac{T_i^2}{n} - \dfrac{T^2}{n}$	The sum of squares between factor levels. T_i and T are defined above.
$SS_W = \sum_{i=1}^{L} \sum_{j=1}^{n} x_{ij}^2 - \sum_{i=1}^{L} \dfrac{T_i^2}{n}$	The sum of squares within factor levels. T_i is defined above.

(See Table 14-2 for a general analysis-of-variance table with these parameters.)

Formula	Used to Compute
$SS_B(1) = L_2 n \sum_{i=1}^{L_1} (\bar{x}_{i..} - \bar{x})^2$	Sum of squares between levels of factor 1, in a two-factor analysis of variance.
$SS_B(2) = L_1 n \sum_{j=1}^{L_2} (\bar{x}_{.j.} - \bar{x})^2$	Sum of squares between levels of factor 2.
$SS_W = \sum_{i=1}^{L_1} \sum_{j=1}^{L_2} \sum_{k=1}^{n} (x_{ijk} - \bar{x}_{ij.})^2$	Sum of squares within cells.
$SS_I = n \sum_{i=1}^{L_1} \sum_{j=1}^{L_2} (\bar{x}_{ij.} - \bar{x}_{i..} - \bar{x}_{.j.} + \bar{x})^2$	Sum of squares caused by interactions.
$SS_B(1) = \sum_{i=1}^{L_1} \dfrac{(T_{i..})^2}{L_2 n} - \dfrac{T^2}{L_1 L_2 n}$	$SS_B(1)$; this formula and the next four are more easily computable formulas. $T_{i..}$ and T are defined above.
$SS_B(2) = \sum_{j=1}^{L_2} \dfrac{(T_{.j.})^2}{L_1 n} - \dfrac{T^2}{L_1 L_2 n}$	$SS_B(2)$. $T_{.j.}$ and T are defined above.
$SS_W = \sum_{i=1}^{L_1} \sum_{j=1}^{L_2} \sum_{k=1}^{n} (x_{ijk})^2 - \sum_{i=1}^{L_1} \sum_{j=1}^{L_2} \dfrac{(T_{ij.})^2}{n}$	SS_W. $T_{ij.}$ is defined above.
$SS_T = \sum_{i=1}^{L_1} \sum_{j=1}^{L_2} \sum_{k=1}^{n} (x_{ijk})^2 - \dfrac{T^2}{L_1 L_2 n}$	SS_T. T is defined above.
$SS_I = SS_T - SS_B(1) - SS_B(2) - SS_W$	SS_I.

(See Table 14-8 for a two-factor analysis-of-variance table for these parameters.)

ANALYSIS OF VARIANCE

PROBLEMS

14-1. When are methods of analysis of variance needed? Why cannot we just use the methods of Chapter 12 in every situation?

14-2. What advantages do the methods presented here have over the methods of Chapter 13, and vice versa?

14-3.* How is an F ratio formed (in general), and what does it assume about the basic data?

14-4. Throughout the next 26 problems, significance levels of 0.05 and 0.01 will be used. Why is that so? Is 0.05 or 0.01 always appropriate for every decision situation? Why or why not? Suppose that a factor is found to be just barely not significant. What should the decision maker do?

14-5.* If an F ratio with 4 degrees of freedom for the numerator and 15 degrees of freedom for the denominator is found to be 3.64, is that value significantly different from zero at the 0.05 level? At the 0.01 level?

14-6.* A distribution center for a food processing company employs three salespersons. The dollar amounts of sales for each of the three salespersons for 3 months are as follows:

Smith	Jones	Doe
14,150	15,600	14,800
14,500	15,125	14,560
15,675	14,900	14,200

Prepare an analysis-of-variance table and test at the 0.05 level of significance the hypothesis that the mean monthly sales do not differ among salespersons. [In this case, you may find it easier to use equations (14-1) and (14-2).]

14-7. The accuracy of a set of precision scales may depend upon the manufacturer of the scale as well as upon the technician who operates the scales. From an experiment in which 4 technicians and 5 different brands of scales were used, the following analysis-of-variance table was prepared:

Two-Way Analysis-of-Variance Table

Source of Variation	Sum of Squares	Degrees of Freedom	Mean Sum of Squares
Between brands	212.6	(3) _____	(7) _____
Between technicians	182.3	(4) _____	(8) _____
Interactions	(1) _____	(5) _____	1.675
Within cells	(2) _____	(6) _____	(9) _____
Total	428.5		

(a) Complete the table by filling in the blanks numbered (1) through (9).
(b) Test at the 0.05 level of significance the hypothesis that brands of scales have no effect upon precision.

14-8. An electronics firm is considering four different materials as conductors in a new type of miniature circuit. The manager believed that five observations per material would be sufficient to decide whether the type of material would significantly affect the performance of the circuit. The engineer tested and rated the materials for conductivity, and the results of the tests were as follows:

Material

1	2	3	4
40	68	66	61
55	42	48	57
56	63	53	52
61	61	56	54
59	55	56	59

(a) Formulate an appropriate hypothesis to be tested.
(b) Using an analysis-of-variance table, test the hypothesis at the 0.05 level of significance.

14-9. Divided Tool Company manufactures metal rods. In order to assure that production will not halt during breakdowns, the company has three machines which produce the rod. Since all rods must be as nearly equal in length as possible, the plant manager sampled output from each of the three machines to determine if the machines were actually producing rods of the same length. The sample results are as follows:

Length of Rod (inches)

Machine		
1	2	3
11.5	12.0	12.2
11.9	12.1	12.0
11.8	12.4	11.9
12.0	12.1	12.0
12.1	12.0	11.8
12.0	12.2	12.3
11.7	12.3	12.1
12.5	12.1	11.9

Prepare an analysis-of-variance table and test at the $\alpha = 0.01$ level of significance the hypothesis that all machines produce rods of the same length. (To simplify calculations, you may want to use the fact that the answer will remain unchanged if a constant is subtracted from each observation.)

14-10.* Growmuch Fertilizer Company has developed two new types of fertilizer which it hopes to market in the next year. To assess the effectiveness of the two fertilizers, an experiment was conducted in which each of the new fertilizers, in addition to a fertilizer already popular with farmers, was applied to 7 different plots (21 total) of corn. The yield of each plot in terms of bushels per acre is as follows:

New Fertilizer 1	New Fertilizer 2	Old Fertilizer
53.0	58.6	52.6
54.3	57.2	51.5
54.2	51.6	52.0
55.6	53.4	52.6
52.1	55.1	50.5
53.0	50.6	53.8
53.5	52.1	54.9

Prepare an analysis-of-variance table and determine at the 0.05 level of significance whether fertilizer affects yield.

14-11. Plastic of Paris Company manufactures several types of synthetic threads which it sells for uses ranging from fishing line to clothing material. Uniformity of threads is extremely important to customers, but the company sometimes experiences shortages of certain raw materials. Substitutes are often used until the desired ingredient is again available. One customer has recently been complaining that threads sent to him frequently break too easily. As a result of this persistent complaint, the company decided to undertake an experiment to test the tensile strength of threads manufactured from four different types of materials. The results of the test were as follows:

Tensile Strength
(pounds per square inch)

Material

1	2	3	4
401	381	400	390
389	379	392	372
392	376	391	386
399	375	396	400
400	389	389	399

Prepare an analysis-of-variance table and test at the $\alpha = 0.05$ level of significance the hypothesis that tensile strength does not differ among material types. Does the answer change if $\alpha = 0.01$ is chosen?

14-12. A company that produces chemical compounds used to retard corrosion in metals is testing five ingredients to see which additive is most effective in controlling rust. Using each additive, a batch of compound was produced, and each of the batches was applied to four pieces of metal. The metal was then subjected to a corrosion-inducing environment. The effectiveness of a batch is measured as the number of pits per square inch formed on the metal plates. From the results shown, prepare an analysis-of-variance table and test at the $\alpha = 0.05$ level of significance whether different ingredients result in differences in effectiveness. Does the answer change if $\alpha = 0.01$ is chosen?

Ingredient

1	2	3	4	5
5.6	6.2	6.1	5.8	6.3
5.2	5.5	6.0	5.7	6.2
5.7	5.1	6.2	5.6	6.2
5.7	5.9	6.0	5.6	6.2

ANALYSIS OF VARIANCE

14-13.* A large distributor of grocery products has experienced a high level of breakage in the transporting of eggs. The company has three types of packaging under consideration, including the type already in use. Five shipments of equal size were prepared using each type of packaging, and the number of eggs broken in each shipment is as follows:

Type of Packaging

1	2	3
102	156	112
110	152	113
122	155	86
103	138	105
99	100	119

Prepare an analysis-of-variance table and test at the $\alpha = 0.01$ level of significance the hypothesis that packaging has no effect upon breakage. Does the answer change if $\alpha = 0.05$ is chosen?

14-14. A company has three suppliers, which provide it with the same product. In order to achieve better control over inventory, the company is attempting to cut down on lead time, the time between when an order is placed and when the product is received. If the average lead time for all three suppliers is about the same, the company will assume that little can be done to decrease lead time. If the company finds that the lead time differs significantly among suppliers, then the company will investigate the differences and attempt to pressure the slower suppliers into speeding up the delivery process. From each supplier, a sample of six delivery receipts were chosen and the dates were compared to the purchase order dates:

Supplier

	1		2		3	
	Delivery Date	Order Date	Delivery Date	Order Date	Delivery Date	Order Date
1	Jan. 20	Jan. 4	Jan. 31	Jan. 6	Jan. 15	Jan. 5
2	Feb. 2	Jan. 19	Feb. 28	Feb. 4	Feb. 11	Feb. 2
3	Mar. 22	Mar. 6	Feb. 28	Feb. 6	Feb. 28	Feb. 13
4	Apr. 1	Mar. 22	Apr. 1	Mar. 4	Apr. 15	Apr. 2
5	Apr. 30	Apr. 9	Apr. 1	Mar. 16	Apr. 15	Apr. 5
6	Apr. 30	Apr. 15	Apr. 30	Mar. 30	May 20	May 6

Using the $\alpha = 0.01$ level of significance, test the hypothesis that lead time does not differ among suppliers. Assume normality for lead time even though it is measured in days.

14-15. A manufacturer of lightbulbs is conducting an experiment in order to determine if four different materials used for filaments will cause a significant difference in the lifetimes of the bulbs. The results of the experiment using 20 bulbs are as follows:

Bulb Life (hours)

Filament Material

w	x	y	z
20,000	25,000	19,000	20,000
21,000	24,000	20,000	21,000
21,000	23,000	18,000	20,000
22,000	23,000	17,000	20,000
20,000	22,000	20,000	21,000

(a) Test at the $\alpha = 0.05$ level of significance the hypothesis that filament material has no effect on bulb life. To simplify your calculations, use the fact that your answer will remain unchanged if you subtract the same constant value from each observation in the above table.
(b) Would the sums of squares change if the analysis were instead performed after converting the data from "hours" into "thousands of hours"? Would the results of the hypothesis test change?

14-16.* A plant foreman suspects that the productivity of workers depends upon the time of day. If he finds that productivity does indeed vary by time of day, he will alter the work-break schedule in an attempt to increase productivity. For a period of one week, he notes the output of the factory workers at different times of day.

Output (units produced per hour)

Time of Day

8–10 a.m.	10–12 a.m.	1–3 p.m.
51	48	49
49	46	50
52	47	48
50	42	47
45	45	49

Analysis of Variance

Prepare an analysis-of-variance table and test at the 0.05 level of significance the hypothesis that output does not vary by time of day.

14-17. Four different drugs are being studied by a pharmaceutical company. The drugs are thought to lower blood platelet levels in patients. In an experiment, the four drugs are administered to patients in a uniform dosage. The resulting platelet counts for the patients are as follows:

	Drug		
1	2	3	4
250	325	159	257
251	276	182	243
259	301	203	252
306	287	191	289
275	295	196	207

Do the platelet counts differ significantly among groups who were administered different drugs?

14-18. A psychologist is conducting an experiment designed to test the effects of coffee with different levels of caffein on the typing abilities of secretaries. A group of 15 secretaries volunteered for the experiment. The secretaries were randomly assigned to three different groups, and each group was tested with a different coffee. The secretaries were given two cups of coffee and then asked to type for five minutes. The effectiveness of the caffein levels was measured by the mean number of mistakes per page for each secretary, with the following results:

	Caffein Level of Coffee	
High	Moderate	97% Caffein-free
1.3	1.8	3.1
1.6	1.7	2.5
2.0	1.1	0.9
1.7	2.3	1.6
1.5	1.5	1.4

(a) Prepare an analysis-of-variance table and test at the $\alpha = 0.01$ level of significance the hypothesis that caffein has no effect upon typing ability.

(b) Criticize this experiment and explain how it might better have been conducted.

14-19.* In the manufacture of a special tool, 5 workers are involved with the operation of three machines. A major task of the plant manager is to schedule the workers in such a way that lost work time is kept at a minimum. If the manager knew that the workers were equally efficient on all machines, the task would not be easy, but at least it would be simplified. As a preliminary step to determining an optimal schedule, the plant manager decided to observe each worker five times on each machine and measure the operating time. Some results from the manager's observations appear below.

$$\sum_{i=1}^{5} \frac{(T_{i..})^2}{15} = 150.62, \qquad \sum_{j=1}^{3} \frac{(T_{.j.})^2}{25} = 183.49$$

$$\sum_{i=1}^{5} \sum_{j=1}^{3} \sum_{k=1}^{5} (x_{ijk})^2 = 594.54, \qquad SS_I = 115.84, \qquad SS_W = 216.86$$

Using these data, prepare a two-way analysis-of-variance table and test at the $\alpha = 0.05$ level of significance the hypothesis that operating time does not vary among workers.

14-20. The yield from a certain chemical process may depend on the temperature and pressure applied during manufacturing. In a series of observations during the manufacturing process, the following results were noted:

Temperature (° Celsius)	Pressure (psi)	Two Yields (cc)	
70	50.0	12.1	12.0
	50.1	12.2	12.2
71	50.0	12.4	12.2
	50.1	12.5	12.4
72	50.0	12.5	12.4
	50.1	12.3	12.6

Prepare an analysis-of-variance table and test at the $\alpha = 0.05$ level of significance whether pressure effects yield at these factor levels. Also, test whether temperature affects yield at these levels.

14-21. In an experiment in which the yield of a certain chemical was observed for five different temperatures and two different catalysts, the results appear graphically as follows:

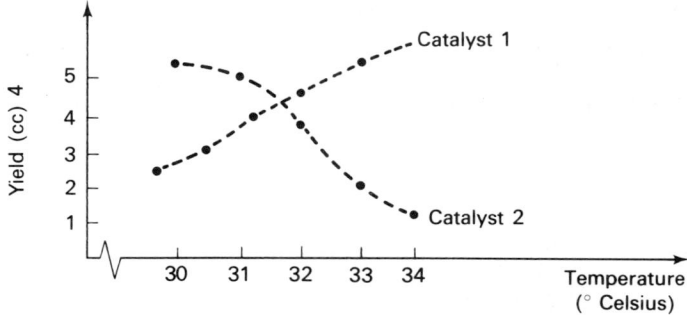

(a) From the graph, does it appear that significant interactions exist between temperature and catalysts? How do you know?
(b) Draw graphs illustrating results in which no interactions exist between temperature and catalysts.

14-22.* The Chamber of Commerce in Central City conducted a small-scale survey of businesses in Central City and two other cities in the state to determine how the mean wage rates differ. From a list compiled in previous studies, the Chamber of Commerce selected two establishments in each of three industries in each city. The results of the survey were as follows:

Mean Hourly Wage Rate

	Retailing	*Manufacturing*	*Banking*
Central City	$3.10	$4.51	$3.97
	3.50	4.29	3.52
City 1	3.89	4.58	3.25
	4.01	4.62	3.28
City 2	3.83	4.57	4.05
	4.00	5.05	4.00

Using an analysis-of-variance table, answer the following questions at the $\alpha = 0.05$ level of significance:
(a) Does the mean hourly wage rate differ significantly among the three cities?
(b) Does the mean hourly wage rate differ significantly among industries?
(c) Is there a significant interaction effect between cities and industries?

14-23. The output from a certain manufacturing process may depend not only upon the type of raw material used but also upon the processing method. The results of an experiment in which the output was noted for different raw materials and processing were as follows:

Output (pounds)

Raw Material	Processing Method			
	1	2	3	4
A	50.6	55.2	62.4	63.7
B	46.7	48.0	65.8	61.5
C	48.9	51.7	66.5	60.8

(a) Criticize this experiment in terms of the usefulness of the results.
(b) Could the problem noted in part (a) be overcome if the experimenter knew ahead of time that no interactions existed between raw materials and processing method?

14-24. In the manufacture of a ceramic type of material, the brittleness of the product depends not only upon the amount of time the material is baked but also upon the temperature of the oven. One company, however, believes that within temperature ranges of 400 to 450°F and baking times of 2 to 3 hours, the brittleness of the product is essentially uniform, thus eliminating the need for close control. To support its claim, the company undertook an experiment whereby different batches of the material were subjected to 400° or 450° temperatures and 2- or 3-hour baking. Brittleness of the product was measured in terms of pounds per square inch necessary to shatter the material, and the results were as follows:

Brittleness (pounds per square inch)

Baking Time (hours)	Temperature (°Celsius)	
	200	250
2	15.1	15.9
	15.6	15.6
	14.2	15.1
	14.3	14.5
3	16.1	16.4
	15.8	16.0
	15.9	15.8
	15.3	16.0

Analysis of Variance

Use an analysis-of-variance table to answer the following questions at an $\alpha = 0.05$ level of significance:
(a) Does temperature significantly affect brittleness?
(b) Does baking time significantly affect brittleness?
(c) Does the temperature effect depend significantly upon baking time?
(d) Could you conceive of an experiment in which the temperature effect would be found insignificant and yet the interaction effect between temperature and baking time would be found significant? If so, how would you interpret such a result?

14-25.* A producer of dog foods maintains an extensive laboratory for purposes of controlling nutritional content and taste appeal for all its products. As part of its laboratory, the company owns an extensive stock of dogs, ranging from mongrels to pure breeds. The company recently developed a new dog food which includes cheese flavoring, and it wants to test the appeal of this food against two other popular brands. From past experience, the company knows that different breeds of dogs crave different flavors, and only through careful choice of ingredients can a dog food be made appealing to a large number of breeds. As part of the testing process for this new food, the company chose 6 pomeranians, 6 poodles, 6 German shepherds, and 6 beagles. The company decided to measure goodness by the weight in ounces of food remaining after the dogs had eaten. Each bowl contained 20 ounces of food, which was more than any of the dogs usually ate at a normal meal. The results of the experiment, 2 days of average leftovers in each class of dog for each dog food, were as follows:

Weight of Food Remaining (ounces)

	Food		
	Cheese	*Brand 1*	*Brand 2*
Pomeranian	12.1	13.2	13.3
	12.9	11.6	13.7
Poodle	11.6	9.8	11.7
	10.3	12.2	11.9
German shepherd	1.5	3.0	4.7
	2.6	2.5	3.1
Beagle	8.2	9.1	9.5
	6.7	7.3	8.9

Construct an analysis-of-variance table and undertake appropriate 5 percent tests to answer the following questions.

(a) Do significant differences exist in the desirability of the different dog foods?
(b) Does the desirability of the different foods depend significantly on the breed of dog? (That is, is there a significant interaction between the two factors?)

14-26. A wood products company produces laminated beams from several varieties of lumber and two types of glue. Researchers for the company want to determine whether different combinations of wood variety and glue can significantly affect the quality of the beams. In an experiment in which a number of beams were constructed, the following results for breaking strength of the wood appear (the pressure readings are in pounds per square inch applied to the 1 square foot in the center of a standard 20-foot-long beam):

	Variety of Wood			
Glue Type	1	2	3	4
A	246	258	227	261
	250	261	228	256
	261	259	215	239
B	225	321	240	242
	227	326	249	243
	238	337	231	215

Using an analysis-of-variance table, answer the following questions at the $\alpha = 0.05$ significance level.
(a) Are there significant breaking-strength differences among varieties of wood?
(b) Are there significant breaking-strength differences between types of glue?
(c) Are there significant interactions?
(d) What wood–glue combination would you use for greatest strength?

14-27.* A furniture manufacturer wants to increase the efficiency of factory operations. One important aspect of the manufacturing operation is the drilling department. The foreman of the department can regulate the speed of work by increasing or decreasing the speed of the drills. But by increasing the speed of drilling, she may also significantly shorten the lives of the tools. To investigate cost–efficiency trade-offs the foreman decided to subject certain drills to different intensities and measure the lives of the drills. Since the firm uses three different brands of drills in the department, the foreman decided to test all three different levels of work intensity. Each drill would be used at the prescribed intensity until it was no longer in satisfactory working condition. The results of the experiment are given below.

Useful Life (in hundreds of hours)

Operating Intensity	Brand X	Brand Y	Brand Z
High	10.6	13.5	14.1
	11.2	12.6	13.5
	8.5	11.9	13.6
Medium	12.1	12.7	12.9
	11.6	12.2	10.3
	10.9	11.6	13.8
Low	14.1	15.0	13.8
	13.5	12.7	15.6
	13.6	13.2	15.1

Using an analysis-of-variance table, test the following hypotheses at the $\alpha = 0.01$ level of significance:

(a) There is no significant difference in the useful lives among different brands of drills.

(b) There is no significant difference in the useful lives of drills subjected to different operating intensities.

(c) There is no significant interaction between operating intensity and brand.

14-28. A tire manufacturer has several different cord materials available for use in making a certain type of truck tire. The choice of material in the past has been based solely upon cost considerations, since it was not known if the type of material used would significantly affect the life of the tire. Recent studies conducted by the firm's engineering department suggest that the type of material may have important influences on the wearability of the tire. In order to gain more evidence concerning this possibility, the plant manager decided to conduct an experiment designed to test tires made with each of the four cord materials, with the following results:

Useful Life of Tires (thousands of miles)

Material

1	2	3	4
12.2	11.6	13.1	12.8
12.1	11.8	12.6	12.0
12.6	12.3	12.3	12.7
12.9	10.9	12.5	13.1

(a) Construct an analysis-of-variance table and test at the $\alpha = 0.05$ level of significance the hypothesis that type of material has no effect on durability.

(b) Would your decision regarding whether or not to reject the hypothesis in part (a) change if you were told that the first two rows of the table refer to results from one truck and the second two rows refer to results from a second truck?

14-29. The output from a chemical process is thought to behave according to the following model:

$$Y_{ij} = \mu + \alpha_i + e_{ij},$$

where

$Y_{ij} \equiv$ output from the jth observation from the ith catalyst

$\mu \equiv$ overall mean output

$\alpha_i \equiv$ amount of output above the overall mean resulting from the use of catalyst i

$e_{ij} \equiv$ random error factor.

In this chemical process, each observation, Y_{ij}, is known to be drawn from a normal distribution with standard deviation σ. Each observation on Y_{ij} has an expected value given by $E(Y_{ij}) = \mu + \alpha_i$. The results of an experiment from this chemical process were as follows:

Output (cubic centimeters)

Catalyst

$i = 1$	$i = 2$	$i = 3$
$Y_{11} = 5.2$	$Y_{21} = 5.7$	$Y_{31} = 4.9$
$Y_{12} = 5.0$	$Y_{22} = 5.8$	$Y_{32} = 4.8$
$Y_{13} = 5.1$	$Y_{23} = 5.7$	$Y_{33} = 4.8$
$Y_{14} = 5.1$	$Y_{24} = 5.6$	$Y_{34} = 5.0$
$Y_{15} = 5.3$	$Y_{25} = 5.6$	$Y_{35} = 5.1$

(a) Prepare an analysis-of-variance table and test at the $\alpha = 0.05$ level of significance the hypothesis that yield from the chemical process does not differ according to the catalyst used.

(b) Show that $\bar{Y}_1 - \bar{Y}_2$ is an unbiased estimator of $\alpha_1 - \alpha_2$. α_i is called the effect of catalyst i.

(c) Find a 95 percent confidence interval for $\alpha_1 - \alpha_2$. Interpret this interval in terms of effects (*Hint:* $SS_W/[L(n-1)]$ is an unbiased estimator for σ, and the t distribution with $L(n-1)$ degrees of freedom is appropriate).

14-30. A detergent manufacturer has commissioned an independent testing laboratory to conduct experiments designed to detect which of a number of popular detergents perform best. As a prelude to the study, laboratory researchers selected five national brands to determine whether the detergents are significantly different in their abilities to remove stains. Five identical washing machines, one for each brand of detergent, were used for the study, and several batches of soiled cloth were prepared. Lengths of cloth were randomly assigned to the machines with each machine receiving 576 square inches of soiled cloth. After running the cloth through the machines, the researchers measured as precisely as possible the amount of stain remaining on the cloth. The experiment was repeated three times with the five detergents assigned randomly to the machines for each replication. The results of the final measurements were as follows:

Amount of Stain Remaining (square inches)

	Brand of Detergent				
Replication	1	2	3	4	5
1	25.2	16.9	31.6	14.1	18.2
2	31.0	22.6	28.9	26.3	26.4
3	26.3	27.0	28.8	21.5	21.5

(a) Construct an analysis-of-variance table and test the homogeneity hypothesis (no brand effect) at the $\alpha = 0.05$ level of significance.

(b) In a subsequent meeting with the manufacturer, questions were raised as to whether the type of machine is important in the performance of the detergents. As a result of the meeting, researchers agreed to conduct a test in which the effect of different machines would be explicitly considered. In the second experiment five different machines were used, and each of the five detergents was tested on each type of machine:

	Machine				
	A	B	C	D	F
Detergent assigned to machine at each of five trials	1	2	3	4	5
	2	3	4	5	1
	3	4	5	1	2
	4	5	1	2	3
	5	1	2	3	4

This experiment design is a type of *Latin square design* in which each detergent is paired with each machine once and only once. The results of the measurement of stain appear as follows:

Amount of Stain Remaining (square inches)

Detergent	Machine				
	A	B	C	D	E
1	25.6	14.3	18.6	19.1	22.3
2	18.1	16.0	21.5	19.8	20.3
3	17.6	31.0	22.7	16.0	14.9
4	20.8	16.4	17.9	18.0	15.7
5	18.5	20.0	22.6	19.5	19.8

Criticize this experiment. Is the design appropriate if interactions exist?

Chapter **15**

Regression Analysis

In Chapters 1 through 13, we were concerned with variables on a one-at-a-time basis. The analysis-of-variance techniques of Chapter 14 provide methods of relating two (or more) variables when one is measured on at least an interval scale and the second is measured at most on an ordinal scale. Regression and correlation analysis, the subjects of this and the next chapter, give us methods for studying the relationships among two (or more) variables, both measured on at least an interval scale. In Chapter 18, we will learn how to treat cases where two or more variables are all measured on at most an ordinal scale.

15-1 Objectives of Regression Analysis

Regression analysis has three basic purposes. First, and most important, the manager may wish to predict the values of one variable using the values of some other variable. This is useful if the variable of interest cannot be observed in time to make a decision, but the value of a related variable can be determined. To use this fact, the manager must establish a relationship between the two variables. This chapter will describe how relationships between variables such as income and spending, costs and activity, and sales and advertising, to mention a few examples, can be established. We will also find methods by which we can obtain a confidence interval for our prediction.

Our attention will be restricted primarily to relationships involving two variables. The variable to be predicted is called the *dependent variable*. It is

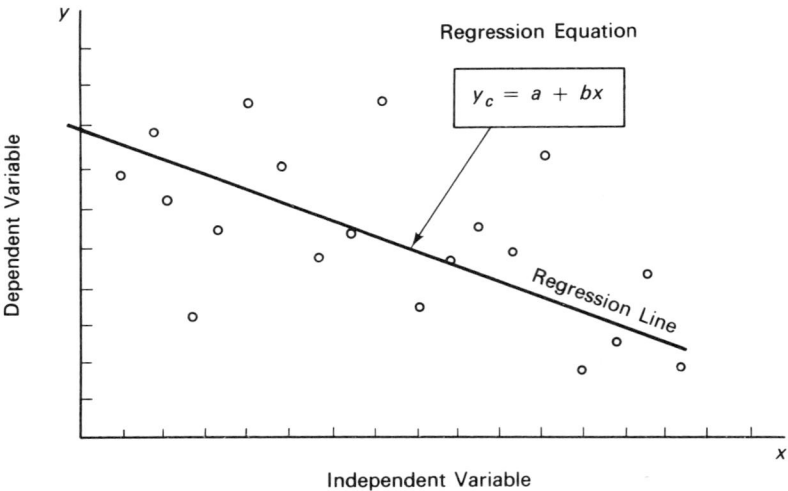

Figure 15-1 Hypothetical Relationship Between Two Variables

found on the left-hand side of the predictive equation, called the *regression equation*, and on the vertical axis of a graph relating the two variables. The variable y is the dependent variable in the hypothetical example illustrated in Figure 15-1. The other variable, whose values will be used to make the prediction, is called the *independent variable*. It is the variable on the right-hand side of the regression equation and on the horizontal axis in Figure 15-1, and this variable is denoted by the letter x.

The relationship depicted in Figure 15-1 is linear, and we shall assume linearity in all the relationships described in this chapter. Although linearity may seem to be a seriously limiting assumption at first, we shall find later that this is not the case. Assuming that linearity is appropriate in describing the relationship between two variables, say spending and income, the regression equation may be written

$$y_c = a + bx, \qquad (15\text{-}1)$$

where

$y_c \equiv$ calculated value of y, spending, given x, income, and the regression equation (it is the value of y read from the line in Figure 15-1 for a given value of x)

$x \equiv$ value of the independent variable, income

$b \equiv$ slope of the regression equation

$a \equiv y$ intercept, the value of y when $x = 0$.

We use the symbol y_c (meaning the calculated value of y) to stand for the points on the line. The symbol y alone stands for the y coordinate of each data point used to establish the relationship. Nineteen of these data points are indicated in Figure 15-1.

Equation (15-1) allows the manager to estimate the value of the dependent variable given a value of the independent variable provided the constants a and b are known. Thus, if a and b are found, by methods to be described in Section 15-3, to be 5 and 10, respectively, and if $x = 2$, then $y_c = 5 + 10(2) = 25$.

The constant a gives the value of y_c when x is set equal to zero. It is called the *y intercept*. In many problem situations, the *y*-intercept value has no reasonable managerial interpretation, but only establishes the height of the line so the line may be used to predict y over the range of values for which the equation is valid. For example, if Figure 15-2 were to describe the

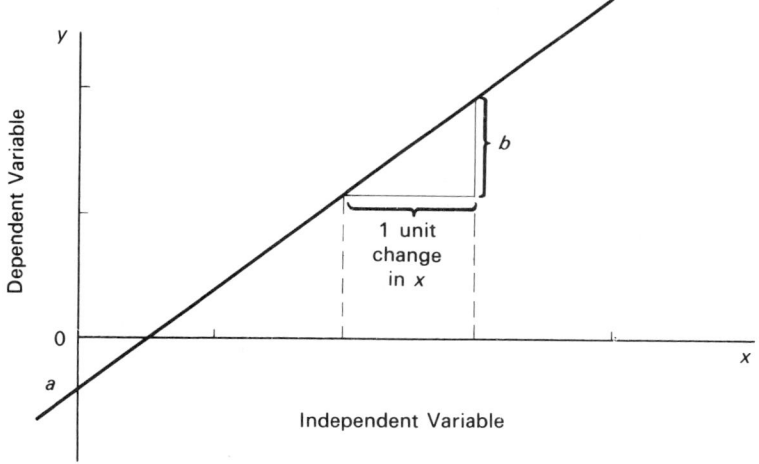

Figure 15-2 Hypothetical Regression Line

relationship of total dollar purchases to income, the value of a is negative. Yet a negative level of purchases is impossible. Indeed, purchases (using savings) would most likely be positive even at a zero level of income. The line is useful for predicting y only for positive x values.

This raises an important issue to which we will return again later. A regression equation should only be accepted over the range of data used to establish it. If income data are available only in the range from $500 per month to $10,000 per month, the resulting regression equation should not be used (or it should be used with great caution) to estimate values of the dependent variable, y, for values of independent variable outside this range. Relationships can and often do change drastically beyond the range of x-values used to establish a given equation.

Figure 15-2 illustrates the value of a. It also illustrates the value of b, the slope of the regression equation. The value of b gives the change in y for a unit change in x. The value of b depends on the units used to measure x. For example, if x is in dollars, then b is the change in y for a \$1 change in x; for example, b might equal \$0.50. Now, if the units of x are changed by a factor of 100 so that x is measured in hundreds of dollars, then b also changes by a factor of 100 to \$50.

The slope is a measure of the importance of a unit change in the independent variable on the dependent variable. The larger the value of b (holding the scale constant), the larger the impact of a change in x on y.

A second purpose of regression analysis is to establish the relationship between the variables for explanatory purposes. In this case, the manager is interested in the relationship itself and, in particular, what the values of the constants a and b are. The manager may not be interested in predicting y but rather in the importance of the variable x. This would be the case if x were under the manager's control, for example if the dependent variable were sales, which is a function of advertising. Advertising is a variable subject to management control. If we have previously established the regression equation using the methods of Section 15-3, then this information is already available. The manager may wish to know how much of the variability in, for example, sales are explained by advertising. In other words, the manager may be interested in how closely they move together, that is, how closely they are correlated. We defer a general consideration of correlation until Chapter 16. But even here we can glimpse policy considerations. If advertising has little effect on sales, promotional effort may be devoted to new programs, promotional activities, or dropped altogether in favor of increased emphasis on quality. The value of the slope coefficient, b, provides some information on such relationships.

A third purpose of regression analysis is to test whether the relationship implied by the equation, in the form of equation (15-1), is statistically significant. Could it have easily occurred by chance? This is essentially a test of the hypothesis that the slope coefficient, b, is significantly different from zero. If so, it means that we are prepared to assume that the relationship found is not simply a random one.

Such questions arise when, for example, an advertising agency claims that its efforts will have a positive effect on sales or an economist suggests that some governmental policy (such as increased federal spending) will have a favorable impact on economic activity (as measured, say, by GNP). Such statements are, in effect, hypotheses concerning the regression relation in general and in particular about the slope coefficient b. As such, these hypo-

REGRESSION ANALYSIS

theses can never be proved, but the statistical hypothesis tests provided in regression analysis allow us to place varying degrees of confidence in such claims.

S.S. The equation $y = k/x$ is sometimes used in economics to express the relationship between the demand for a commodity and its price.

(a) What are the dependent and independent variables in words and symbols?

(b) If a new variable $z = 1/x$ is defined, and if the equation is rewritten in terms of z, is the relationship linear?

(c) What sign would you expect k to have?

Solution:

(a) The dependent variable is demand, y, and the independent variable is price, x.

(b) If $z = 1/x$ is used, the relationship is linear, with an intercept term of zero. The value k is the slope and gives the unit change in demand for a unit change in z.

(c) The sign of k should be positive; otherwise, demand increases as price increases.

15-2 Regression Population

To show how an estimated regression equation is established, we will assume that the entire population is available to us. We shall first examine the relationship between two variables in the entire population. This will allow us to point out the assumptions underlying each step to be made in a regression analysis. We will then show how a sample is drawn from this population and how the observed values of the variables are used to estimate the relationship existing in the population.

Now, in fact, we would seldom if ever have a complete population available. If we did, we would not need the statistical methods of regression analysis, which apply to samples. The assumed population is only for pedagogical purposes and nothing more.

Sometimes it might seem that an entire population is present, such as when the complete sales and advertising records for a 5-year-old firm are available. But these values merely represent a sample from the possible values that might have been observed if other factors (such as personal income or investment) had been different.

Let us consider an example. The Dundee Uranium Dilatometer, or DUD as it is known in the trade, is an apparatus for measuring the expansion of various substances. The ultimate life of each dilatometer is a function of the processing time taken in assembling the DUD and setting its molten core. Suppose that we are interested in this relationship between processing time and ultimate life. The dependent variable, y, is the ultimate life. The independent variable, x, is processing time. Suppose that the entire population of dilatometers we are interested in is the set of 54 for which the data are given in Table 15-1. (We reiterate that this is a contrived situation. We would not normally have an entire population. In this problem, we would in actuality be more interested in what the data portend for the dilatometers for which we do not have information, that is, those to be made in the future and hence not included in the sample.)

The ultimate life values for each of the five separate processing times listed in Table 15-1 represent subpopulations. We shall assume that processing time, the independent variable, is measurable without error or, alternatively, that any measurement error is negligible. One way that this can occur is if x is predetermined or preset. This means that if we wish to draw a sample of, say, five from the population, we may specify in advance the processing time for each sampled item.

Table 15-1 Ultimate Life (weeks) by Processing Time

	Processing Time (hours)				
	4.0	5.0	6.0	7.0	8.0
	185	190	205	255	260
	195	195	250	265	295
	195	200	260	270	305
	205	200	260	270	305
	205	250	260	290	310
	205	250	260	290	310
	210	250	275	290	310
	245	260	275	290	315
	263	260	280	300	315
		270	295	300	335
		282		310	335
				314	349
Total	1,908	2,607	2,620	3,444	3,744
Average	212	237	262	287	312

Before drawing a sample, let us plot the population values from Table 15-1 and the averages of each subpopulation given at the bottom of the table. This is done in Figure 15-3.

Regression Analysis

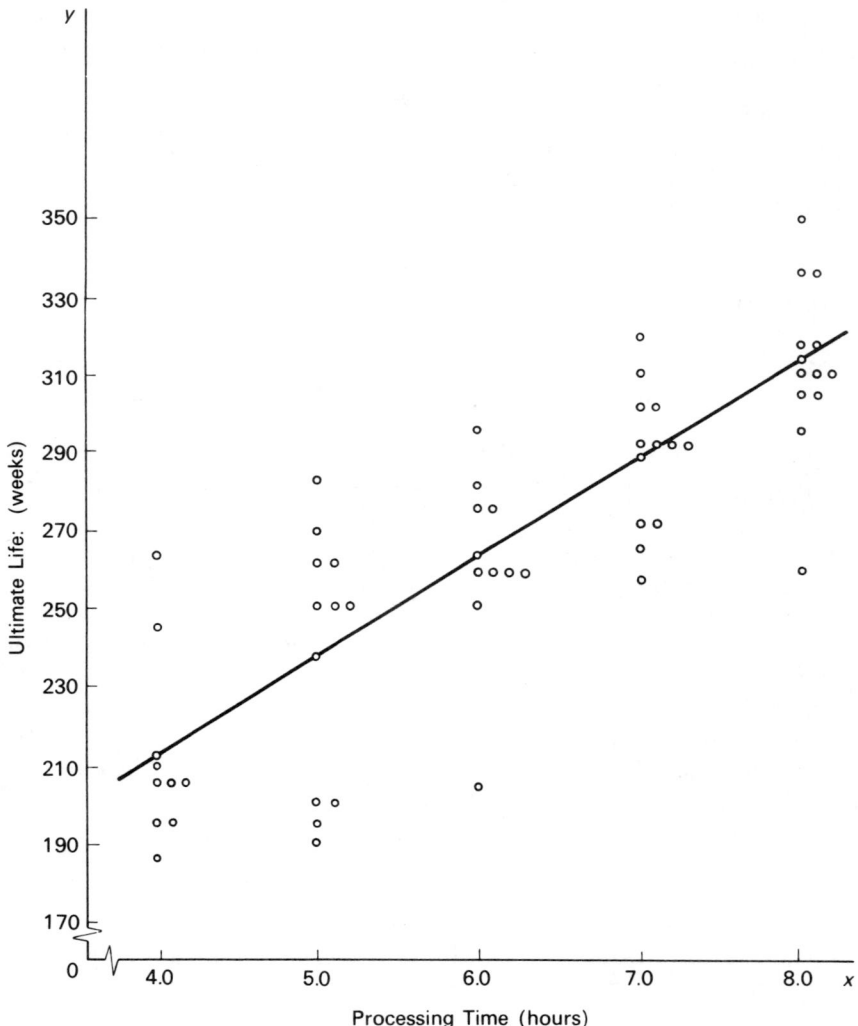

Figure 15-3 Population: Processing Time and Ultimate Life

The values actually observed for y are plotted above their x value and next to one another in a line where there is more than one identical observation. The larger dots, which represent the averages given at the bottom of Table 15-1, fall on a straight line. This line has a slope of 25 and a y intercept of 112. (The slope value can be estimated by inspection. The intercept is found by subtracting four times the slope value from 212, the value of y at

$x = 4$, to obtain the value at $x = 0$.) It is this line that gives the average value of y for each value of x in the population. It is this line that we are attempting to estimate when we sample.

Not all the points fall exactly on the line. Indeed, none of them does. This is because other factors, including measurement error in y, are present. We assume that the line of relationship for the population is linear and can be written in general form as

$$y_c = \alpha + \beta x \tag{15-2}$$

or, specifically, for our example as

$$y_c = 112 + 25x, \tag{15-3}$$

where

$\alpha \equiv y$ intercept for the population

$\beta \equiv$ slope (change in y for a unit change in x) for the population.

Equation (15-2) uses the Greek letters α and β for the intercept and slope, respectively. That is because they are the values in the population; they are the parameters of the relationship. Equation (15-1) used the letters a and b for the slope and intercept of the regression equation. This is because these values are in practice estimated from sample data. Thus, a and b are the sample estimates of α and β.

Suppose now that we only have a sample of five observations from this population. Since we want the relationship to cover as wide a range of predictive values of x as is possible, we select one observation randomly from the process at each of the five levels of x. Suppose we obtain

Sample Point

	1	2	3	4	5
Value of x	4.0	5.0	6.0	7.0	8.0
Value of y	205	250	260	300	315

The question is what these five values of x and y can tell us about the underlying relationship. We shall learn what they can tell us in Section 15-3.

S.S. For the data in Table 15-1 and Figure 15-3, which of the following mathematical expressions are correct?

(a) $\alpha + \beta x = \bar{y}$ given x, where $\bar{y}$ is $E(Y \mid x)$ for this example.

(b) $237 = y_c$ when $x = 5.0$

REGRESSION ANALYSIS   529

(c) $\bar{x} = 6.0$
(d) $\bar{y} = 262$
(e) $P(y = 205) = \frac{4}{54}$
(f) $P(y = 205 \mid x = 4.0) = \frac{1}{3}$

Solution: Items (a), (b), (e), and (f) are correct. The average value of x is given by

$$\bar{x} = \frac{9(4) + 11(5) + 10(6) + 12(7) + 12(8)}{54} \approx 6.13.$$

The average value of y can be found as

$$\bar{y} = \frac{1{,}908 + 2{,}607 + 2{,}620 + 3{,}444 + 3{,}744}{54} \approx 265.24.$$

15-3 Procedures for Estimating the Population Regression Line

Perhaps the first alternative that suggests itself is simply to plot the sample values and draw in a line of relation using a straightedge. This is a perfectly reasonable procedure, and it is illustrated in Figure 15-4. When the plotted points fall closely along a linear path, this approach is relatively accurate as well as easy. But when the points are widely diverse, it is not so simple to do.

Using the graph in Figure 15-4, the slope may be estimated as

$$(322 - 210) \div (8.0 - 4.0) = 28.0,$$

and the intercept is then $210 - 4(28.0) = 98$. Hence, the equation of the line we have estimated using a straightedge is

$$y_c = 98 + 28x. \tag{15-4}$$

This is not the same line as the population line, nor should we expect it to be, since it is based on only a sample of five observations.

The "ad hoc" nature of this line-drawing procedure and its lack of certain agreeable statistical properties has prompted statisticians to look for other ways by which lines might be fitted to the sample data. One way that we could fit a line to the data would be to find a line that would make the absolute vertical deviations of the points from the fitted line as small in total as possible. There are five of these vertical distances for our sample of size 5. They are indicated by the vertical lines from the data points to the fitted line

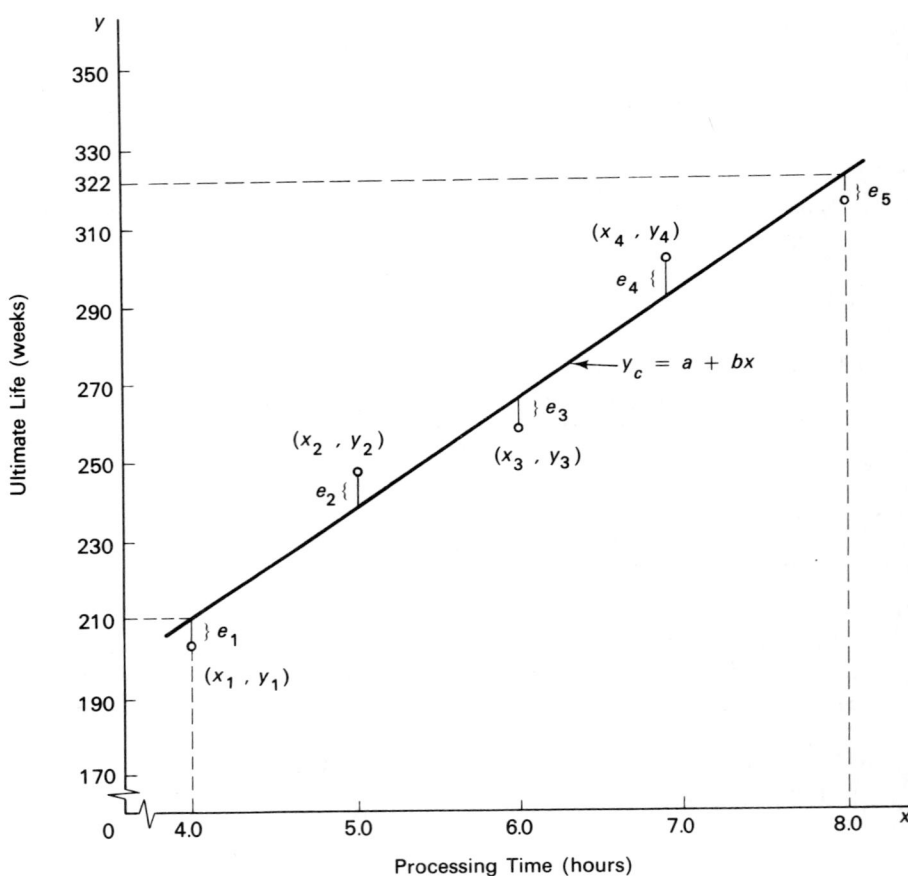

Figure 15-4 Estimated Regression Line Fitted to Five Sample Points Using a Straightedge

in Figure 15-4. These vertical distances are called the *errors* or *residuals*. We may use the symbol e_i, $i = 1, 2, 3, 4$, and 5 to stand for each of these five errors.

The most common method of fitting a regression line to a set of sample data, however, does not minimize the sum of the absolute errors as suggested in the previous paragraph. Rather, it minimizes the total sum of the squares of these errors. In other words, a sample regression line is selected so that the total of the squared deviations of the points (taken in the vertical direction) is as small as possible. There are several reasons for using this approach to fiitting a regression line. A complete discussion is beyond the purpose of this book. However, one reason is simply that it is relatively easy to mathematically fit a line that minimizes squared deviations.

If that does not sufficiently confuse you, let us look at the mathematics. Each sample point, call it point i, has a y coordinate, y_i. The regression line gives a calculated value of y (different from y_i), which we have denoted as y_c and which equals $a + bx_i$. Of course, at the moment, we do not know the constants a and b, but in general form, we can write the error, e_i, for any point as

$$e_i = y_i - y_c = y_i - (a + bx_i), \qquad (15\text{-}5)$$

where each e_i is based on a particular value of x. Now what we want to do is to make the sum of the e_i^2 values as small as possible; that is, we want to minimize them. Hence, we want to

$$\text{minimize} \sum_i e_i^2 = \text{minimize} \sum_i [y_i - (a + bx_i)]^2. \qquad (15\text{-}6)$$

So where does all this gobbledygook get us? Well, not too far, we must admit. The solution to equation (15-6) involves the calculus, and we have therefore placed it in Technical Note 15-1. What we can tell you is that the solution to equation (15-6) yields two new equations (just like rabbits) whose simultaneous solution will give us the values of a and b that we seek. These two equations are called the normal equations in statistics, for reasons that we shall not attempt to justify. The normal equations are given by (15-7), where $n \equiv$ the number of observations, 5 in our example.

The Normal Equations

1. $\sum_i y_i = an + b \sum_i x_i.$ \hfill (15-7)

2. $\sum_i x_i y_i = a \sum_i x_i + b \sum_i x_i^2.$

These two equations can be solved simultaneously for a and b when the x and y values are known.

In Table 15-2, we provide the calculations necessary to solve these equations for our sample of size 5. For the hand calculations used here, but rarely used in practice when computers are available, columns are formed for x, y, xy, and x^2 values.

Substituting the values from Table 15-2 into the normal equations (15-7) yields:

1. $1,330 = 5a + 30b$
2. $8,250 = 30a + 190b.$

Table 15-2 Computations for the Normal Equations: Sample of Size 5 = n

Sample Point	x_i	y_i	$x_i y_i$	x_i^2
1	4.0	205	820	16
2	5.0	250	1,250	25
3	6.0	260	1,560	36
4	7.0	300	2,100	49
5	8.0	315	2,520	64
Total	30.0	1,330	8,250	190
Symbol	$\sum x_i$	$\sum y_i$	$\sum x_i y_i$	$\sum x_i^2$

We must now solve these two equations simultaneously for a and b. One means of doing so that will always work is to first solve equation 1 for a. We obtain

$$a = 266 - 6b.$$

We then substitute this value for a into equation 2 and we get

$$8{,}250 = 30(266 - 6b) + 190b.$$

Solving for b gives $b = 27$, and substituting this value into equation 1 gives $a = 104$. For our particular sample of size 5, we can now write the estimated sample regression line. It is

$$y_c = 104 + 27x. \qquad (15\text{-}8)$$

The equation is different from the one we obtained using a straightedge, equation (15-4). It is also different from the population equation given by equation (15-3). This is because equation (15-8) is based on a sample of only 5 of the 54 observations used to establish equation (15-3). In fact, equation (15-8) represents only one of several thousand equations that we might have obtained from choosing a sample of size 5, where one observation is taken at each of the five values of x.

S.S. Suppose that a sample yields the values

	x_i	y_i
	1	1
	2	11
	3	9
Total	6	21

Compute the regression equation and estimate y for $x = 2$.

REGRESSION ANALYSIS

Solution: Since $\sum x_i y_i = 50$ and $\sum x_i^2 = 14$, the normal equations are

$$21 = 3a + 6b$$
$$50 = 6a + 14b.$$

Solving, $y_c = -1 + 4x$ and $y = 7$ for $x = 2$.

15-4 Using the Regression Equation To Predict

Two assumptions are necessary before we can use equation (15-8) to predict values of y for given values of x. These two assumptions are:

1. The sample values were selected independently.
2. The line of relationship in the population as specified by the regression equation is linear.

If these two assumptions are satisfied, then we may use equation (15-8) to estimate the value of y for a given x. In fact, the equation gives us the estimated average value of y for the given x. Recall that our sample line is an estimate of the population relationship, and the population equation goes through the average value of y for each x. But this is exactly what we want. The estimated average value of y for a given x, the expected value of y for the given value of x, is our best estimate of the individual predicted value of y, and it is our best estimate of the average value of y for the given value of x. Thus, if x is 5, equation (15-8) gives:

1. 239 as the estimated life in weeks for *an item* with a 5-hour processing time.
2. 239 as the estimated *average* life in weeks for *items* with 5 hours of processing time.

EXTRAPOLATION

In using the estimated regression equation for prediction, we must be careful in extrapolating it beyond the data set on which it is based. Equation (15-8), for example, suggests that if the processing time were zero hours, the item will have a life of 104 weeks. In fact, it would be likely that the item would have a zero life. This emphasizes the fact that the relationship may be quite different beyond the available data range. This does not mean that we should not extrapolate (frequently there is no alternative), but rather that we should be careful and realize the increasing chance of error when doing so.

We may wonder why, if the real relationship is linear, 104 is not the life of an item with zero processing time. The reason is that the linearity assumption may be a reasonable approximation only over the actual data range. The population relationship may not be linear, but this does not unduly concern us. Reality is only approximated by our statistical work. The important consideration is that the linear approximation is close enough to provide useful results over the range where it is assumed. This is a managerial determination.

Extrapolation is a common result of many regression problems. Management must predict and this often takes the analysis beyond the data set. The assumption that the relationship estimated on the available data is reasonable in the prediction range must be accepted. A certain amount of caution here is a virtue, yet few great undertakings in this world would ever get off the ground without faith in things not known. The discussion brings to mind the story of the sheep herder standing on the hillside watching his flock. "See," he said to an old friend, "those sheep have been sheared." The old friend peered thoughtfully at the flock for a moment, then said, "Sheared on this side anyway."

INTERPOLATION

Although using the relationship within the data range is much safer, it too can lead to trouble. In our example, before predicting ultimate life on the basis of a processing time of, say, $5\frac{1}{2}$ hours, we should ask whether there is any reason to expect discontinuities in the linear functional relationship between the hourly values in which x is measured. If so, and again this is a managerial decision, care should be exercised in using the regression equation to predict.

OTHER CONSIDERATIONS IN FORECASTING

Regression analysis requires a good deal of data and hence is an expensive process. It should be used only in problem situations which warrant the cost. Also, these data are often of a time-series nature. When this is so, conditions may have changed sufficiently that the older data are no longer relevant. The manager should be careful at all times that important changes in the process such as might be caused by a new machine, new personnel, new materials, or new regulations have not occurred which render the available data obsolete.

One other factor is worth mentioning here. If one variable is to be used to predict a second variable, it must be possible to obtain reasonably good information on the predictor variable. For example, suppose that we wished to predict sales of a particular car model, and we could use either (1) the prices of the model or (2) the effect of credit terms, including downpayment percentage and contract length. In this case, it may be easier to obtain data on the price of the model than on expected credit terms. A variable that is itself difficult to estimate is of limited value in prediction.

On the other hand, developing a regression relationship based on such variables may still be useful in explaining or understanding those economic variables which impact on the dependent variable. Hence, we may learn how credit terms, disposable or discretionary income, the price of gasoline, and other factors relate to or influence new car sales without being concerned with prediction. Understanding economic, social, and technical relationships can be the first step to better managerial decision making.

REGRESSION AND CAUSATION

A relationship between two variables, as established by mathematical methods, does not imply in and of itself a cause-and-effect relationship. The cause-and-effect relationship must be established on the basis of the manager's knowledge of the underlying economic or sociological aspects of the situation.

For example, statistics on sex and level of schooling completed will show a relationship between sex and the years of schooling completed. If one wishes to interpret this relation as a cause-and-effect relationship, it is clear which way the causation must go. The sex of the student is the causal variable, not the other way around. However, there is also a relationship between family income level and schooling. If we choose to read a cause and effect into this relationship, it is not clear which direction the causation runs. The income level facilitates schooling, but schooling also leads to higher income levels.

In other cases the relationship may be misinterpreted. For example, one educator found a positive relationship among elementary school children between foot size and the quality of handwriting. Now few would claim that bigger feet contribute to one's penmanship. On the other hand, foot size is related to age and so, probably, is handwriting skill. Many relationships such as this are not spurious but rather merely misinterpreted or misspecified.

Finding relationships between variables is a useful exercise early in an investigation. It can help in the understanding of a problem. But in the end, if the relationships uncovered cannot be explained in causal terms, our knowledge is of a meager sort. Much is left unanswered, and few managers would feel comfortable basing requests for large expenditure programs on statistical relationships they cannot explain. Managers will, fortunately, look to relationships justified by their experience and not simply ones found empirically.

S.S. Can you explain the following statistically verified relationships?

(a) In northwestern Europe the number of storks' nests is related positively to population in a geographic area.

(b) In the eighteenth century the number of ministers and the number of illegitimate births were positively related in England.

Solution:

(a) Very few of us would invoke the stork-and-baby explanation to explain the first statement. More likely, increased population provides increased buildings and other preferred nesting places for the stork population.

(b) Both may reflect the social conditions of the times. The very cause (or causes) of the illegitimate births may have suggested to young Britons the advantages of saving souls. (Fortunately for us, the Reverend Bayes lived earlier and could apply his efforts to the solution of mathematical puzzles.)

15-5 Placing Confidence Intervals Around Predictions

It is not unusual for a manager to want to place some confidence statement on a prediction. Thus, if the manager predicts that the ultimate life for a processing time of 5 hours is 239 hours, the manager may wish to say instead that the life will be between some range of values with 95 percent confidence.

To make such a statement, we can use the standard deviation just as was done in the material on estimation in Chapter 11. Typically, we will not know the actual value, denoted by σ, and we must turn to a sample estimate. In regression analysis the sample estimate of σ is called the *standard error of the estimate*, and it is denoted by s_e for the sample, or by σ_e for the parameter value. The sample standard error of the estimate is defined by equation (15-9a)

$$s_e = \sqrt{\frac{\sum_i (y_i - y_c)^2}{n - 2}}. \qquad (15\text{-}9a)$$

The standard error is the square root of the average squared error and, hence, is analogous to the standard deviation. It measures the scatter of the points about the fitted regression line. The denominator, $n - 2$, reflects the fact that two degrees of freedom are lost in its calculation, one each for using the basic data to compute a and b. These two constants are needed to compute y_c and hence to compute s_e.

In Table 15-3, we establish the values necessary to compute s_e for our sample of 5 observations.

REGRESSION ANALYSIS

Table 15-3 Calculation of the Standard Error for Our Sample of 5

Sample Point	x_i	y_i	$y_c = 104 + 27x_i$	$y_i - y_c$	$(y_i - y_c)^2$
1	4.0	205	212	−7	49
2	5.0	250	239	11	121
3	6.0	260	266	−6	36
4	7.0	300	293	7	49
5	8.0	315	320	−5	25
Total					280

Formula (15-9a), which defines the standard error, is cumbersome to use for calculations. Hence, we typically use an equivalent form given by

$$s_e = \sqrt{\frac{\sum_i y_i^2 - a \sum_i y_i - b \sum_i x_i y_i}{n - 2}}. \qquad (15\text{-}9b)$$

However, using Table 15-3 and equation (15-9a) this one time,

$$s_e = \sqrt{\frac{\sum_i (y_i - y_c)^2}{n - 2}} = \sqrt{\frac{280}{3}} = \sqrt{93.33} = 9.66.$$

Using this value, we may establish confidence intervals around any prediction as long as two additional assumptions are met. The additional assumptions required are:

3. That the conditional distributions of y, for any given x, around the expected (average) y for that x are normal.

4. Each of these distributions has the same standard deviation.

These two assumptions are illustrated in Figure 15-5, and they are in addition to the assumptions of randomness and linearity made at the beginning of Section 15-4. In words, each of the four normal distributions drawn in Figure 15-5 is identical to the others in terms of its standard deviation. Each has its own individual mean centered on the population regression line.

We could not leave this section without mentioning that the requirement of constant variance is called homoscedasticity by statisticians. We thought you could at least impress your friends with that one. (Non-constant variance in y as x changes is called heteroscedasticity, or could you have guessed?)

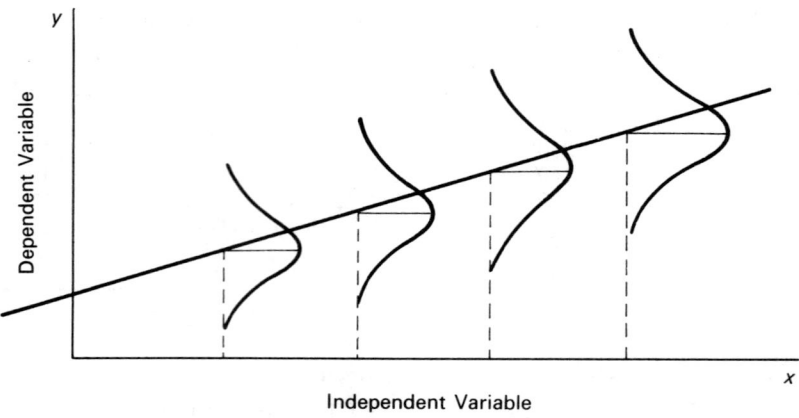

Figure 15-5 Normality and Constant Variance

CONFIDENCE INTERVAL FOR $\bar{y}$,
THE AVERAGE VALUE OF y FOR A GIVEN x

We know that the estimated average value of y for an x of 5 hours is 239. This is given by the value of y_c in the sample regression line for $x = 5$. But how can we place, say, a 95 percent confidence interval around this estimate? That is, how can we find a 95 percent confidence interval for the mean value of y. We begin by observing that the sample regression line may be in error either in the height at which the line is placed in relation to the x axis, or in the slope of the line as measured by the slope coefficient, b. After all, the line is based on a sample. In the universe, the error in the height of the line is measured by the variance of the height of the line at the mean value of x, namely at $\bar{x}$. This variance is given by

$$\sigma_h^2 = \frac{\sigma_e^2}{n}. \tag{15-10}$$

(The concept is similar to the variance in the sample means.) The universe variance of the slope coefficient of the regression equation is given by

$$\sigma_b^2 = \frac{\sigma_e^2}{(n-1)\sigma^2}, \tag{15-11}$$

where σ^2 is the variance of the independent variable x. Typically, these values are unknown and must be estimated using their sample counterparts. Thus,

$$s_h^2 = \frac{s_e^2}{n} \quad \text{at } x = \bar{x} \tag{15-12}$$

Regression Analysis

and

$$s_b^2 = \frac{s_e^2}{(n-1)s^2} = \frac{ns_e^2}{n(\sum_i x_i^2) - (\sum_i x_i)^2}. \tag{15-13}$$

For our problem s_h^2 is approximately 18.67 and s_b^2 is approximately 9.33.

Now, in specifying a confidence limit, we find that our confidence decreases as we near the boundaries of our data; that is, more uncertainty enters as we approach the highest or lowest values of x in our sample. This is because the error in estimating the expected value of y is the error in the regression line. This error is from two sources: the error in the height, and the error in the slope. The error in the slope becomes more significant as we depart from the center of the data as represented by $\bar{x}$ since the line deviates further from the universe relationship. This is reflected in formula (15-14).

Variance of $\bar{y}$, the *average* value of y, at x equal to a specific value, call it x_0:

$$s_{\bar{y}}^2 = \frac{s_e^2}{n} + (x_0 - \bar{x})^2 s_b^2. \tag{15-14}$$

For our sample, and for a processing time of 5 hours, for example:

$$s_{\bar{y}}^2 = 18.67 + (5 - 6)^2\, 9.33 = 28 \quad \text{and} \quad s_{\bar{y}} = 5.29.$$

Thus, to find a 95 percent confidence interval for the population mean, given an average of 239 for a processing time of 5 hours, we write $239 \pm t_{0.475}(5.29)$. The 95 percent confidence interval for $\bar{y}$ at $x = 5$ is $239 \pm t_{0.475}(5.29)$. Since t has $n - 2$ degrees of freedom, the confidence interval is $239 \pm 3.182(5.29)$, or from about 222 to 256. Intervals computed by this procedure will contain the (conditional) universe mean 95 percent of the time.

This is the 95 percent confidence interval for the *average* ultimate life for a processing time of 5 hours. The t distribution is appropriate because the standard deviation is based on a sample. (Note that the assumption of normality for the underlying distribution of y values for each value of x is necessary before the t distribution is appropriate for this test.)

CONFIDENCE LEVEL FOR $\hat{y}$, THE INDIVIDUAL VALUE OF y FOR A GIVEN x

We may instead desire the confidence interval for the life of an individual unit with a processing time of 5 hours. This is useful, for example, when we desire a confidence interval for next year's sales or the next individual value of any dependent variable. Once more the best single estimate of this value,

$\hat{y}$, is the value of y_c calculated from the regression equation. At $x = 5$, this is again 239. However, the variability of the predicted individual value, $\hat{y}$, is greater than for the average, $\bar{y}$, because the individual points vary more than the averages. There is variability around the line as well as variability in the line. (In other words, there is variability around the average as well as in the average itself.) To obtain an estimate of this variability, we need merely add the variability of the individual points around the line to the variability of the average. But we already have a measure of the variability around the line, s_e^2.

Variance of $\hat{y}$, an individual value of y, at x equal to a particular value, call it x_0:

$$s_{\hat{y}}^2 = s_e^2 + \frac{s_e^2}{n} + (x_0 - \bar{x})^2 s_b^2 = s_e^2 + s_{\bar{y}}^2 \quad (15\text{-}15)$$

For our sample, and again at $x = 5$, $s_{\hat{y}}^2 = 93.33 + 28.00 = 121.33$: thus $s_{\hat{y}} \approx 11$. Note that the variability around the line dominates this variance measure. This is generally the case, and for most purposes it would be sufficient to simply let $s_{\hat{y}}^2 = s_e^2$ as an approximation.

Using the t value for $n - 2$ degrees of freedom and a 95 percent confidence level, we obtain the 95 percent confidence interval for an individual value of y at $x = 5$ to be

$$239 \pm 3.182(11)$$

or from about 205 to 273. This tells us that when $x = 5$, 95 percent of the individual lives are contained between 205 and 273.

S.S. A major hospital wishes to have sufficient blood on hand so that there will be less than one chance in a thousand of not having sufficient whole blood immediately available for all needs in any given month. It has developed a regression equation of monthly usage versus time.

(a) What problems do you see in using this equation to predict future usage?

(b) Assuming that we have a valid predictive equation, the 0.001 probability should be calculated for (the average monthly usage, an individual month's usage) and (be placed in the upper tail only, divided equally in both tails). Choose one phrase in each set of parentheses.

Solution:

(a) There are many possible problems. Perhaps those most specific to this example include: different months may require different predictions (a seasonal pattern), the future needs of the hospital may be different from past needs due to changing or changed conditions, and no allowance is made for

REGRESSION ANALYSIS   541

blood-type or cross-matching problems. You may well have thought of others.

(b) The hospital wishes to assure that it has sufficient supplies for each month, hence, it is concerned with an individual month. It is also worried only about not having enough, so the 0.001 value should be placed entirely in the upper tail. Assuming sufficient data are present so that the normal distribution can be used to approximate the t distribution, the amount of blood required for a given month would be calculated as

$$\hat{y} + 3.09 s_{\hat{y}},$$

where both $\hat{y} = a + bx_0$ and $s_{\hat{y}}$ are based on the particular month to be predicted, namely x_0. If we had been estimating the same safety factor on the average for that month, the appropriate calculation is given by $\bar{y} + 3.09 s_{\bar{y}}$, where $\bar{y} = a + bx_0$ and x_0 are for that particular month. Here $s_{\bar{y}} < s_{\hat{y}}$ and the safety level would be set lower.

15-6 Testing Hypothesis

We are frequently interested in whether a relationship really exists or whether it is simply a chance result. If there is no relationship between x and y, then a change in x should have no effect on y whatsoever. In other words, the slope of the population regression line is zero; $\beta = 0$. But a sample will almost inevitably yield $b \neq 0$. The question, then, is whether b is sufficiently different from zero that we can reject the null hypothesis $\beta = 0$. Hence, we wish to test the null hypothesis

$H_0: \beta = 0$ against the alternate hypothesis

$H_1: \beta \neq 0.$

In some cases, including the ultimate life example that we are considering in this chapter, it may be that the relationship portends a $\beta > 0$; processing time should increase the ultimate life. (In other problems, the relationship expected could be, alternatively, $\beta < 0$.) When either of these situations holds, we can make a one-tailed hypothesis test. The form of the one-tailed test for the processing-time example is:

$H_0: \beta \leq 0$ against the alternate hypothesis

$H_1: \beta > 0.$

The advantage of a one-tailed test is that it is more powerful; that is, the null hypothesis can be rejected at the same significance (α) level with a smaller sample size. But the directional relationship must be posited by the manager before the data are gathered.

For our problem, this test is made assuming the population of b values is normal, with $E(\beta) = 0$ and using s_b as the standard deviation of the b values. Since we are using s_b (rather than σ_b) and the sample is small, a t test is used where the appropriate t distribution again has $n - 2 = 3$ degrees of freedom. The appropriate one-tailed test with $H_0: \beta \leq 0$ versus $H_1: \beta > 0$ is illustrated in Figure 15-6 for our problem. An 0.01 significance level is

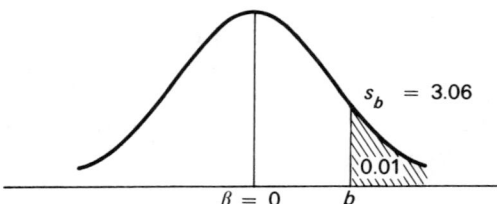

Figure 15-6 t Test on the Slope Value

arbitrarily assumed, yielding a t value of 4.54. The slope value that leaves 0.01 of the area in the tail is given by

$$4.54 = \frac{b - 0}{3.06}$$

or $b = 13.9$. We could call this the critical value for b, or b_c. Since our value of $b = 27$ is well above 13.9, we have a significant relationship well beyond the 0.01 significance level.

Tests could also be made on the intercept value, but such tests are seldom of interest. We elect not to discuss them.

Using the data we have generated in this section, we could also place a confidence level around the slope value obtained for our sample. Again using the t distribution, we obtain for the 95 percent confidence interval,

$$b \pm t_{0.475} s_b$$
$$27 \pm 3.182 s_b$$

since $t_{0.475} = 3.182$ for 3 degrees of freedom. Thus, the confidence interval is

$$27 \pm 3.182(3.06)$$

or approximately 17.3 to 36.7. Hence, we can say with 95 percent confidence that an increase of one more hour of processing time has the effect of in-

Regression Analysis

creasing the ultimate average life somewhere between 17.3 to 36.7 weeks on the average.

S.S. The following relationship was found between rail transportation time (y) in days and railroad distance in miles (x) based on 1,000 shipments:
$$y_c = 4.02 + 0.009x, \qquad s_e = 2.02$$
If $s_b = 0.0036$, could this relationship be due to chance at the 0.01 significance level?

Solution: The solution depends on whether a two-tailed or one-tailed test is used, given that α is already set at 0.01. Since we believe that rail transportation time should be directly related to rail distance, we test $H_0: \beta \leq 0$ versus $H_1: \beta > 0$. Because the sample is large, we may use the normal curve to find the critical value of b.

$$b_c = 0 + z_{0.49}s_b = 0 + 2.33(0.0036) = 0.008.$$

Since $0.008 < 0.009$, we reject H_0. If a two-tailed test had been used, $z_{0.495} = 2.58$ and $0 + 2.58(0.0036) > 0.009$. The null hypothesis is accepted. This illustrates both the value of knowing enough about the managerial situation to predict the direction of the relationship and the arbitrariness of selecting significance levels.

15-7 Cross-Section and Time-Series Data in Regression Analysis

There are two types of sample data typically used in regression analysis: cross-section data and time-series data. Cross-section data consist of a set of observations at a given moment of time—for example, the consumption expenditures of households in the United States in 1976, the amount of apricots produced in the year 1977 by state, sales of oogah horns by type of retail outlet, and so on. In each of these examples, an individual data point is distinguished from the others by the fact that it relates to a different place. Each occurs at the same point in time.

Time-series data, on the other hand, consists of a set of ordered observations on a variable at regular intervals over time. In a time series, each successive observation is separated from the others by its occurrence at a different time rather than by its occurrence at a different place. A sample comprising the total output of apricots in Pawtucket by year, from 1950 to 1977, is a set of observations ordered in time and, hence, it constitutes time-series data. Similarly, the sales of a given item by month or week is a time series as is the number of tourists going to Europe from the United States, by year from 1950 to 1977. The number of tourists coming from each state in 1975, however, would constitute cross-section data.

Random sampling, as we know, is a procedure for selecting a sample in such a way that every item in the population has the same chance of being included in the sample at each trial. Furthermore, if a sample of a fixed size, say n, is drawn, then each possible sample of size n is selected from the population in such a way that it has the same probability of being drawn as any other possible sample of size n from the same population. Thus, in selecting a random sample of size 5 from a lot consisting of 1,000 manufactured parts, every possible sample of size 5 has the same chance of being selected from this lot. In practice, we use some chance device, such as a table of random numbers, to be sure that the selection process yields a random sample.

If, however, we consider a time series consisting, say, of the average yearly prices of apricots from 1950 to 1977 in consecutive order, the prices of any two consecutive years have been selected in order of time sequence, and, hence, the data do not constitute a random sample. If the twenty-eight observations on price from 1950 to 1977 are to be treated as a random sample, it is necessary to assume that the price of apricots in one year is independent of the price of apricots in the following year, and that these prices have been selected at random from a "random economy."

However, this is not normally the case in real-world problems. In fact, the prices in successive years are closely related; prices in preceding years can be of great help in predicting prices in the following years. Furthermore, when we draw a random sample of size n, say 1,000 families, from all the families in the United States, we are drawing the sample from the same population. When we have a time series, there is no guarantee that each item is drawn from the same population, since the characteristics of the population may have changed as time passes.

The advantage of using (or assuming) a random sample is that we can use statistical methods to assess the reliability of the estimated results. Statistical methods are based on probability theory, but this theory applies only if the data used are obtained through random sampling. In the case of a time series, however, the data do not arise from random sampling. We cannot assess the reliability of the estimates. For instance, suppose that we estimate a regression equation involving the sales of some consumer durable (as the dependent variable) and U.S. personal income (as the independent variable). Suppose, further, that the data are the yearly figures between 1950 and 1977. Sales in 1 year will affect sales in subsequent years as a result of the durable nature of the item in question. Hence, the assumption of independence in the observations is violated.

Because of the nature of a time series, businessmen and many economists

REGRESSION ANALYSIS

prefer to use cross-sectional data. If cross-sectional data are used, we can accept the assumptions underlying random sampling and, hence, we can make probabilistic statements relating to our estimated results. However, frequently the problem or data base requires the use of time-series data. When this is so, care must be exercised in drawing conclusions based on statistical analysis.

S.S. Indicate whether the data below are cross-sectional or time series.
  (a) Advertising expenditures by firm X, 1960–1977
  (b) Advertising expenditures in 1977 by automobile firms
  (c) Sales of toasters by retail outlet
  (d) Major crimes by city
  (e) Accidents by county
  (f) Costs of X-ray treatment over the last 10 years in hospital Y

Solution: Items (a) and (f) are time series. The rest give cross-sectional data.

15-8 Limitation of Linearity

It has been assumed in this chapter, as we have noted several times, that the regression line is linear. This might appear to seriously limit the techniques discussed since many relationships are not linear. Several comments are in order. First, even when theoretical considerations imply a nonlinear relationship, a linear approximation may still be appropriate (that is, it may be reasonably close) over the range of x and y values observed.

Second, data can often be transformed to make them linear. For example, if the relationship is of the form $y = ax^b$, taking logarithms of both sides gives a linear expression in the logs of the original variables: $\log y = \log a + b \log x$. Many other transformations are also possible, but we will be satisfied with just the idea of transforming data here. Finally, in some common cases, the techniques have been worked out for fitting regression lines to other than linear data. We shall not discuss these more complex curve-fitting techniques.

S.S. What could be done to make the following relationships (implied by the economic theory of the situation) linear in the independent variable?
  (a) $y = a + bx^2$
  (b) $y = ab^x$

Solution:

(a) Define a new variable $z \equiv x^2$, and we have $y = a + bz$.

(b) Take logs of both sides, which gives $\log y = \log a + x \log b$. This equation is linear for the log of y in the form $\log y = a' + b'x$, where $a' = \log a$ and $b' = \log b$. To complete the regression, we would compute $\log y$ for all values of y and then use the normal equations to find the regression equation for the variables x and $\log y$. Once estimates of a' and b' are obtained, a and b are found by solving $a = 10^{a'}$ and $b = 10^{b'}$ or by taking antilogs.

15-9 Multiple Relationships

Relationships may often appear which have apparently no reasonable cause-and-effect relationship. But this is often because the investigator has not searched far enough. For example, a plot of applications against tuition at a major private graduate school of business shows a strong and significantly positive relationship. This implies at first blush that as tuition is increased applications rise. If true, this school should raise tuition further. But common sense tells us that this cannot be the case; tuition should not be raised faster than it is by other, similar schools. What is wrong here?

Some might say this is a spurious relationship. We prefer to argue that it is simply a misinterpreted relationship. The simplest explanation is that both variables have increased over time in response to several economic variables whose changes are, in turn, captured by the time variable.

Another way of addressing this question is to say that in most situations more than one independent variable is involved. Applications to graduate business schools are influenced by opportunities in the job market, family income levels, and the number of college graduates, as well as tuition. Automobile sales are affected by credit terms, the stock of used cars, and used car prices, as well as by new car prices. If we may assume that each of these variables is linearly related (possibly after a transformation is applied to the total relationship) to the single dependent variable, we might write the regression equation in the form

$$y = b_0 + b_1 x_1 + b_2 x_2 + \cdots + b_r x_r. \tag{15-16}$$

This equation is used, much as before, to predict, to form confidence intervals based primarily on s_e, and to make tests of hypotheses on the slope coefficients each based on its own s_{b_i}. There is a great deal more calculational effort required, but it is straightforward (and is always done these days by a computer). This extended analysis is called *multiple regression analysis*.

REGRESSION ANALYSIS

In multiple regression analysis, it is more difficult to know when we are extrapolating due to the combination of variables involved. And for prediction purposes, it is now necessary to establish the values of all r independent variables before the value of the dependent variable can be estimated. Moreover, it is not unusual for some of the independent variables to move together (we say they are correlated). When this is true, problems arise in interpreting the slope coefficients (which are then biased), constructing confidence intervals around y_c, and conducting significance tests. Fortunately, however, correlated independent variables do not bias the prediction itself.

Perhaps the most important managerial task is in selecting an appropriate set of independent variables. This is where the understanding that is often uniquely the manager's comes into play. If the manager does a good job of selecting the appropriate set of variables and obtaining accurate data, and if the manager is careful to check for the logic of the net effect of the variables that appear in the regression equation, then the computer can do the number-crunching task.

In selecting variables, the manager should pick those (1) that have some causal relationship to the dependent variable and (2) that are easier to estimate than estimating the dependent variable directly. Of course, there should be data available on the variables selected and such data should be reasonably accurate, if the variable is to be useful in prediction.

Finally, the manager should carefully examine the regression equation and the associated statistics. A large standard error implies low confidence in predictions and may suggest that important independent (explanatory) variables have been omitted. An unreasonable sign for the coefficient of an important variable may suggest problems with the data or, again, the omission of an important variable from the data set. Thus, if we found that in predicting the sales of a particular model car, the price of the model, as an independent variable, had a negative coefficient, we should suspect that something is wrong with the analysis. In this regard, it is, however, necessary to remember that each coefficient, b_i, in a multiple regression equation gives the *net* effect of that variable with all other independent variables in the model held constant.

S.S. Consider the number of job offers a student may receive upon graduation. List at least five variables that might have an effect on this number and in which direction this effect would be expected to operate (that is, the sign of its regression coefficient). Pick variables for which values could be estimated.

Solution: Some factors and the expected sign of their regression-slope coefficient are: years of work experience (+), rank in class (probably +), economic prosperity as measured perhaps by the recent change in GNP (+), number of graduates in the area (−), and grade-point average (+) (although this variable is similar to rank in class). Some other variables that might be useful, assuming that an adequate measuring scale could be devised, are: preparation in terms of course hours for job sought, maturity, quality of references, willingness to relocate, appearance, leadership qualities, expression, moral character, quality of school and degree program, and so on.

15-10 A Warning

Some relationships involving one or more independent variables can be classified as spurious. We will use this term to refer to those relationships created by either improper methods of selecting the data or by the arithmetic operations performed on the data. An example of both is given next.

In a plant, the manager had an employee survey the attitudes of the plant's work force toward a union shop. The relationship was quite different than when an independent investigator repeated the study. In this case, a spurious rather than the true relationship was determined, because of the biased responses caused by the choice of interviewer.

For a mathematical example, suppose that the purchase of some consumer durable is related to total personal income in a geographic area and that the figures (in thousands of dollars) are as follows:

	\multicolumn{5}{c}{Period}				
	1	2	3	4	5
y, purchases	30	30	30	30	30
x, income	20	50	80	110	140

Now this set of figures shows no relationship between y and x. The dependent variable is constant regardless of the value of x. (For example, the product might be sailboats.) But suppose that we elect to put both series on a per capita basis by dividing by population, z, which takes on the values 2, 3, 4, 5, and 6 successively over the five time periods. We then obtain the following data:

	Period				
	1	2	3	4	5
y/z, purchases per capita	15	10	7.5	6.0	5.0
x/z, income per capita	10	17	20	22	23.5

A strong negative relationship appears between the variables y/z and x/z. A spurious correlation has been introduced. Dividing the constant purchasing series by an increasing population series creates a negative trend over time. This declining series is then regressed against an increasing income per capita series; income is increasing faster than population in the geographical area under consideration. The result is different when variables are considered on a per capita basis. Some relationships are best studied on a per capita basis, for example, expenditures on food. This one is not.

The manager should determine in advance what the appropriate form of the variable should be. Sometimes ratios are more relevant. For example, in studying purchases, constant dollars (that is, dollars deflated to represent purchasing power) may be more meaningful than actual dollars. Here, as is true in so many other places in this book, there is no substitute for a thorough understanding of a problem by the decision maker.

TECHNICAL NOTE 15-1: DERIVATION OF THE NORMAL EQUATIONS

The normal equations arise from selecting a and b so as to minimize equation (15-6), where the summation is taken over all sample points.

$$\text{Minimize} \sum_i [y_i - (a + bx_i)]^2. \qquad (15\text{-}6)$$

Taking partial derivatives with respect first to a and then to b yields

$$\frac{\partial}{\partial a}\left[\sum_i (y_i - (a + bx_i))^2\right] = -2 \sum_i (y_i - a - bx_i) \qquad (15\text{-}6a)$$

and

$$\frac{\partial}{\partial b}\left[\sum_i (y_i - (a + bx_i))^2\right] = -2 \sum_i (y_i - a - bx_i)x_i. \qquad (15\text{-}6b)$$

Setting these two equations equal to zero and solving:

$$\begin{array}{ll} (15\text{-}6a) & (15\text{-}6b) \\ -2 \sum_i (y_i - a - bx_i) = 0 & -2 \sum_i (y_i - a - bx_i)x_i = 0 \\ \sum_i y_i - \sum_i a - \sum_i bx_i = 0 & \sum_i y_i x_i - \sum_i ax_i - \sum_i bx_i^2 = 0 \\ (1) \sum_i y_i = na + b \sum_i x_i. & (2) \sum_i y_i x_i = a \sum_i x_i + b \sum_i x_i^2. \end{array}$$

Equations 1 and 2 are the normal equations (15-7).

New Symbols

Symbol	Meaning
a	The y intercept calculated using the sample values.
b	The slope coefficient calculated using the sample values.
e_i	Error measured by the vertical distance between the y value for the ith sample point and the value given for that point using the regression equation.
β	The slope of the regression line for the universe.
α	The intercept of the regression line for the universe.
σ_e	The universe standard error of estimate. A measure of the variability around the regression line.
σ_h^2	The universe variance in the height of the regression equation.
σ_b^2	The universe variance of the slope term in the regression equation.

Key Formulas

Formula	Used to Compute:
$y_c = a + bx$	The regression equation for a sample.
$y_c = \alpha + \beta x$	The regression equation for the universe.
$\sum_i y_i = an + b \sum_i x_i$	
$\sum_i x_i y_i = a \sum_i x_i + b \sum_i x_i^2$	The normal equations.
$s_e = \sqrt{\dfrac{\sum_i y_i^2 - a \sum_i y_i - b \sum_i x_i y_i}{n - 2}}$	Standard error of estimate (calculational formula).
$s_h^2 = \dfrac{s_e^2}{n}$	Variance of the height of the regression equation at $x = \bar{x}$.
$s_b^2 = \dfrac{n s_e^2}{n \sum_i x_i^2 - (\sum_i x_i)^2}$	Variance of the slope term.
$s_{\bar{y}}^2 = \dfrac{s_e^2}{n} + (x_0 - \bar{x})^2 s_b^2$	Variance of $\bar{y}$ at $x = x_0$.
$s_{\hat{y}}^2 = s_e^2 + s_{\bar{y}}^2$	Variance of $\hat{y}$ at $x = x_0$.

REGRESSION ANALYSIS

PROBLEMS

15-1.* For the chapter example involving processing time and ultimate life of dilatometers, test the null hypothesis that $\alpha = 0$. Use a 0.05 significance level.

15-2. Suppose that you were attempting to predict refrigerator sales in dollars for the entire industry next year, year t, using regression analysis. Check the variables listed below which would be good predictors and indicate why they would be good predictors.
(a) Industry advertising expenditures, year t.
(b) Population, January 1 year t.
(c) Level of personal income, year $t - 1$.
(d) Change in personal income, year $t - 1$ to t.
(e) Level of consumer confidence, January 1 year t.
(f) Price of refrigerators, year t.
(g) Sales of automobiles, year t.
(h) Sales of cabbage, year t.

15-3.* (a) Determine the regression line for the following data:

Units of Output	Dollars of Cost
m	n
1	4
2	7
3	10

(b) Explain the slope and intercept values in terms of the variables m and n.

15-4. The equation $Y = ab^t$ can be used to express the way in which a company's production (or sales) grow with time, t.
(a) Give the dependent and independent variables.
(b) Is this a linear functional relationship?
(c) Can you make it into a linear relation? If so, what is the slope and what is the intercept? What signs do you expect a and b to have?
(d) Suppose that for a firm $Y = 22(1.05)^t$, where $Y \equiv$ sales in millions of units and $t \equiv 0$ in 1960. Predict sales for 1985 and indicate the reasonableness of the value of a. Are there problems in the prediction?

15-5.* The personnel director for a large organization has estimated the following relationship between an individual's score on a preliminary achievement test and work efficiency rating based on data for 30 individuals.

$$y_c = 16.3 + 1.8x \quad (s_b = 0.6)$$

where

$$y \equiv \text{job efficiency rating}$$
$$x \equiv \text{score on preliminary achievement test}$$
$$s_e^2 = 0.75.$$

(a) Can we say (at the 0.05 probability level) that a linear relationship exists between x and y?

(b) Assuming that a relationship does exist between x and y and that the relationship was estimated from x values ranging between 40 and 90 with an average of 70, find a 95 percent confidence interval for the efficiency rating of an employee who scored 55 on the preliminary exam.

15-6.* Suppose that the manager of a winery is interested in forecasting the sales of wine in 10 regions. In making his forecast, he assumes that demand for his product depends on advertising expenditures. As a forecasting device, he decides to use simple regression analysis. Given the following data, perform the analyses required in (a) through (f).

(a) Determine the estimated equation for the relationship. Assume a linear function $y = a + bx$.

(b) Is the sign of the slope value as expected?

(c) Find the standard of error of the estimate.

(d) Find the standard deviation of the regression coefficient b.

(e) Construct a 95 percent confidence interval estimate of the regression coefficient b.

(f) Conduct a test to determine whether the regression coefficient b is significant. Use $\alpha = 0.05$.

1974 Sales Data for 10 Selected Regions

Regions	Sales (thousands of units)	Advertising (thousands of dollars)
A	2	$1
B	2	0
C	2	2
D	3	0
E	3	1
F	4	3
G	5	4
H	5	2
I	6	4
J	6	3

REGRESSION ANALYSIS

15-7. It is believed by many economists that the Hawaiian economy depends significantly upon military expenditures. To see whether or not this is true, the relationship between total employment and military expenditures may be estimated. Given the following data, perform the operations required in parts (a) through (f) of Problem 15-6. Do you believe the assertion given at the start of this problem based on the data?

Year	Total Employment (thousands)	Federal Expenditure (millions of dollars)
1950	173	$151
1951	198	202
1952	204	239
1953	206	242
1954	202	232
1955	218	253
1956	223	279
1957	228	292

15-8. Suppose a marketing manager maintains that demand for her product depends on the disposable incomes of each region. As a forecasting device, she will use a simple regression analysis ($y = a + bx$), where the independent variable is disposable income. Given the following data, perform the operations required in parts (a) through (f) of Problem 15-6. Do you accept the marketing manager's claim?

1964 Sales Data for 10 Selected Regions

Region	Sales (thousands of units)	Disposable Income (millions of dollars)
A	1	$10
B	1	20
C	2	20
D	2	30
E	2	20
F	3	40
G	4	30
H	4	50
I	5	60
J	6	50

15-9. The following are total investments of the Federal Reserve Member Banks from 1949 to 1958, where t represents the years starting with $1949 = 1$.

Year	Investments, Y (billions of dollars)	Year, t
1949	$4.3	1
1950	4.0	2
1951	3.9	3
1952	4.0	4
1953	4.0	5
1954	4.5	6
1955	3.8	7
1956	3.4	8
1957	3.4	9
1958	4.1	10

(a) Find the least-squares trend line.
(b) Find the standard error of estimate.
(c) Find the estimated investment for year 11.
(d) Find the standard deviation of the regression coefficient b.
(e) Conduct a test to determine whether or not the slope coefficient differs significantly from zero.
(f) Do you see any problems in using regression analysis on these data? If so, what are they?
(g) What assumptions, if any, are implicit to your estimate in part (c)?

15-10. In an attempt to measure the relation of the quantity of steel sold to steel prices, and more specifically to measure the elasticity of demand, defined as the ratio of a small percentage change in quantity to the corresponding small percentage change in price, five different relationships were fitted to the data. These involved different variables and different shapes, but all seemed reasonable. The resulting elasticities ranged from +0.52 (indicating that a given percentage increase in price will be accompanied by a percentage *increase* in quantity sold that is half as great) to −0.88 (indicating that a given percentage increase in price will be accompanied by a percentage *decrease* in quantity sold that is seven-eighths as great). What can we learn from this example?

15-11. A mechanized truck farm is experimenting with different levels of a new fertilizer in hopes of finding the relationship between the amount used and yield. They have gathered the following data involving eight nearly identical tracts of land.

Required:
(a) Determine the linear regression equation and the standard error of estimate.

Tract	Fertilizer Used (tons)	Crop Yield per Acre (hundreds of bushels)
1	10	6
2	20	7
3	25	9
4	30	13
5	30	10
6	35	9
7	42	14
8	48	12

    (b) Estimate the crop yield for an application of 50 tons of fertilizer. Does this estimate require any assumptions?
    (c) Does the assumption of linearity in part (a) seem reasonable? Explain.
    (d) Why is it important to have nearly identical plots?
    (e) What other variables might one wish to examine simultaneously with fertilizer?

15-12.* In Problem 15-11 perform the following calculations.
    (a) Test the hypothesis that the slope of the actual relationship is zero. Use a significance test at the 5 percent level and a two-tailed test.
    (b) Would a one-tailed test have been more appropriate? What would the null hypothesis be? Given the result in part (a), can a conclusion be drawn without further calculations?
    (c) Estimate the yield for 1 acre for a fertilizer application of 20 tons and place a 95 percent confidence limit on this estimate.

15-13. The U.S. Forest Service has been studying the relationship between facilities provided and area burned in forest fires. There are many facilities involved, including assistants hired during the fire season, fire control apparatus, fire crew members, and so on. To make the study manageable, all optional services provided have been reduced to a cost per year and combined. The data on 10 counties are:
    Operating Costs of Optional Service (thousands of dollars):

    2.8   1.8   0.8   2.0   2.4   2.8   0.8   1.4   2.2   1.2

    Protected Area Burned (thousands of acres):

    0.09   0.24   0.50   0.35   0.11   0.21   0.40   0.32   0.22   0.43

    (a) Derive a regression equation. Is the sign of the slope value as expected?

(b) Predict the average acreage expected to be lost if optional services costing $2,500 are obtained.
(c) Is the relationship significant? How do you know? (Use an 0.05 significance level.)
(d) Give a 95 percent confidence level for the average protected area burned estimated in (b).
(e) Should the ranger's salary be included in the costs?
(f) What problems do you see in using the regression equation?

15-14. The following table shows the number of weeks that nine waitresses have been employed in a certain restaurant, and the number of customers each can serve on the average in 1 hour.

Weeks Employed	Customers Served per Hour	Weeks Employed	Customers Served per Hour
4	13.3	10	26.6
7	18.5	1	8.3
2	10.1	9	21.1
14	32.4	5	14.6
		12	28.2

(a) Fit a least-squares line to these data, and plot the line together with the actual data on one diagram.
(b) If a waitress has been employed in this restaurant for 8 weeks, how many customers "should" she be able to serve in 1 hour?
(c) Would it be reasonable to use the equation obtained in part (a) to predict how many customers a waitress should be able to serve if she has been employed in the restaurant for 5 years? Explain your answer.

15-15. A company controller is studying the relationship between her company's annual expenditures for research and development and the subsequent year's sales revenue. Let S_t be the company's sales revenue in year t and let R_t be the company's research and development expenditure for year t.
(a) Write an equation which might describe the relationship between the two variables. Assume linearity and include an error term.
(b) Do you believe the model in part (a) is complete? Why or why not?
(c) If you were to estimate this regression equation from a set of data, what result might cause you to question the assumption of constant variance? What effect would such a result have on the use of the regression equation?

Regression Analysis

15-16.* The following table shows 15 weeks' sales of a department store in Boston, Massachusetts, and of its suburban branch store:

Sales of Boston Store (thousands)	Sales of Suburban Store (thousands)	Sales of Boston Store (thousands)	Sales of Suburban Store (thousands)
$64	$32	$63	$30
60	22	60	18
71	51	71	50
71	47	64	32
66	25	69	40
63	24	64	38
66	35	68	45
70	51		

(a) Find the equation of the least-squares line which will enable us to predict the sales of the suburban store using the sales of the Boston store.

(b) Predict the sales of the suburban store for a week in which the sales of the Boston store are $67,000.

(c) Do you think it would be better to use sales of the Boston store in period t to predict sales for the suburban store in period $t + 1$?

15-17.* In the manufacture of a product, the relationship between the yield of the product and the amount of a yield-increasing chemical added to the production process was estimated to be

$$y = 20.000 + 0.463x,$$

where

$y \equiv$ yield, in cubic centimeters of the product

$x \equiv$ amount, in cubic centimeters of chemical added to the process.

(a) The product is sold for $3.16 per cubic centimeter, and the yield-increasing chemical is purchased for $1.75 per cubic centimeter. How much of the chemical should be purchased and added to the process if the company wants to maximize its contribution to fixed costs and profits?

(b) What is the highest price at which the company would be willing to purchase the chemical?

15-18.* An agricultural expert has estimated the relationship between corn yield and the amount of fertilizer applied to plots of land in a particular area. The following data represent results from a series of experiments.

$x \equiv$ amount of fertilizer applied, in tons

$y \equiv$ yield per acre, in hundreds of bushels

$$n = 20, \qquad \Sigma y = 95,$$
$$\Sigma x = 280, \qquad \Sigma y^2 = 867.35,$$
$$\Sigma x^2 = 5{,}116, \qquad \Sigma xy = 1{,}929.91.$$

(a) Estimate the regression line.
(b) Is the sign of the slope reasonable?
(c) From results in other parts of the country, the expert believed that for each additional ton of fertilizer, yield should increase by at least 50 bushels per acre over the relevant range. Is this prior belief confirmed at the $\alpha = 0.01$ level of significance?

15-19. A plant manager wants to estimate the relationship between the speed of an assembly line and the number of defective items produced from the assembly line. The manager will use these data to set the most appropriate assembly-line speed. From a series of 10 trials, the following data were found.

Speed of Assembly Line (inches per second)	Number of Defectives per Hour	Speed of Assembly Line (inches per second)	Number of Defectives per Hour
2.0	4.1	2.0	5.5
2.0	5.6	2.1	1.6
2.1	6.7	2.1	4.8
2.1	4.3	2.1	3.2
2.0	2.0	2.1	3.7

(a) Suggest problems that might exist with data gathered in this manner. (A graph may be helpful.) Does the slope value have the expected sign?
(b) Estimate a linear regression equation, and find a 95 percent confidence interval for the slope coefficient. Does this result confirm one's suspicions from part (a)?

15-20. A delivery service has gathered data for estimating the relationship between the annual repair costs and gasoline mileage for its fleet of trucks. Data for 10 trucks of similar size, make, model, and age are:

$x \equiv$ Annual Repair Expenditures	$y \equiv$ Miles per Gallon
752	11
411	15
1,060	8
847	10
621	13
995	9
708	9
323	18
1,146	7
1,002	6

(a) Estimate the regression line $y_c = a + bx$.

(b) Let a and b be the estimates from part (a). If the manager were to estimate $x_c = c + dy$, would the regression line remain the same with only the axes changed (that is, would $c = -a/b$ and $d = -1/b$)? Explain and demonstrate. Which equation is most useful? Why?

15-21.* A synthetic material manufactured by a chemical company is stored in a room in which the humidity changes. At random times, a quality control inspector checks the humidity of the room and measures the moisture absorbed by the chemical material. The results of 10 trials were as follows:

$x \equiv$ Humidity Reading	$y \equiv$ Moisture Absorbed (milligrams of water)
50	2.6
56	3.4
41	1.7
62	4.1
67	4.0
51	3.2
16	0.9
29	1.1
48	1.9
70	5.0

(a) Estimate a regression equation of the form $y_c = a + bx$.

(b) Define new variables $Q \equiv x - \bar{x}$ and $P \equiv y - \bar{y}$ and estimate the regression equation $P = c + dQ$. What is the relationship between the new slope and height coefficients and the old slope and height coefficients? Could you have predicted this result?

15-22. Fall enrollments at a major university depend critically upon the number of applications accepted the previous spring. Recent trustee mandates have set 6,000 as the ceiling on enrollment, but fall registration for the current academic year was 6,251 and brought complaints from both students and faculty. In order to estimate enrollment more accurately, the Registrar's office was asked to develop a regression equation that would relate spring acceptances to the number of new students enrolling the following fall. An assistant in the Registrar's office gathered the following data in order to develop a predictive equation:

Spring Acceptances	Fall Enrollment (New Students)	Spring Acceptances	Fall Enrollment (New Students)
3,072	2,765	3,391	2,971
3,146	2,782	3,116	2,768
3,221	2,846	3,202	2,826
3,056	2,765	3,218	2,819

(a) Estimate a linear regression equation.
(b) Test the slope coefficient for significance at the $\alpha = 0.05$ level. If the slope coefficient were found to be insignificant at the $\alpha = 0.05$ level, should the university continue to use the model? Why or why not?
(c) Suppose that 3,167 returning students are expected in the fall. How many applicants can be accepted this spring if the registrar wants to be at least 95 percent confident that the enrollment ceiling will not be exceeded?
(d) If acceptances of new students by all departments total 3,312, what is the probability that the enrollment ceiling will be exceeded, given that 3,167 returning students are expected?

15-23.* An engineer for a stereo-tape-cartridge manufacturer wants to estimate a relationship between the life of a certain kind of tape and the tracking force at which the tape is played. The engineer has equipment that can play tapes at several forces between 1 and 2 grams. For a $500 cost, the equipment can be modified to play at forces of from 1 to 3 grams. The engineer figures that for each trial, the cost of determining a tape life is approximately $10, and only $700 can be spent for the experiment. Therefore, if the manager elects to modify the equipment, only 20 tracking forces will be selected between 1 and 3 grams. These forces will be selected in either case according to random sampling using a uniform (rectangular)

REGRESSION ANALYSIS

distribution over the possible forces. The manager will elect not to modify the equipment only if an estimate of the slope coefficient for the regression equation can be obtained which has as little variance as the one that could be estimated by modifying the equipment. Given the $700 budget constraint, should the manager modify the equipment? Assume that regardless of this choice, the standard error of the estimate would be the same constant value over all tracking forces. Use the fact that

$$\sigma_b^2 = \frac{\sigma_e^2}{(n-1)\sigma^2}.$$

15-24. A large wholesale company is interested in the relationship between the price it charges for a particular item and the number of units purchased by customers. Over the past few years, the company has at random times altered the price of the item and noted the number of orders for the item at that price. Management concedes that other factors could be affecting demand, but they believe that fitting a regression equation to the data may yield important insights into the price–demand relationship. The following data represent the results of their experimentation.

Price per Unit	Average Demand (units per week)	Price per Unit	Average Demand (units per week)
$25.10	122.12	$25.60	119.62
25.20	121.43	25.70	119.16
25.30	121.22	25.80	118.51
25.40	120.45	25.90	118.13
25.50	120.41	26.00	117.71

(a) Estimate a regression of the form $y_c = a + bx$. Find s_e and s_b, and conduct an hypothesis test at the 0.05 significance level for the slope.
(b) Estimate a regression of the form $y = b/x$. Find s_e and s_b, and conduct an hypothesis test at the 0.05 significance level for the slope.
(c) Using the information in parts (a) and (b), which equation seems to fit the data better? Which equation should be chosen as the appropriate relationship? Why?

15-25. An insurance company wants to see if a relationship exists between the annual income of families and the amount of insurance coverage purchased by the families. Knowledge of this relationship would be extremely helpful in deciding where the company should concentrate its sales efforts. From a random sample of 20 current policyholders, the following data were obtained:

Annual Income (thousands)	Insurance Coverage (thousands)	Annual Income (thousands)	Insurance Coverage (thousands)
$11	$25	$ 9	$10
20	48	13	27
26	47	15	27
12	22	17	29
15	30	18	32
7	10	18	35
12	24	29	55
9	5	16	22
8	8	25	41
10	15	24	40

(a) Plot the data along with the least-squares regression line. Does a linear relationship appear to be a reasonable approximation?

(b) Using the following values for the independent variable, find 95 percent confidence intervals for both the individual values of the predicted dependent variable and the mean of the predicted dependent variable.

10, 12, 14, 16, 18, 20, 22, 24

Chapter **16**

Correlation Analysis

In regression analysis, we spent considerable effort developing a functional relationship between a single dependent variable (we denote it by y) and one independent variable (we denote it by x). We concentrated on linear relationships of the form $y = a + bx$. Our purposes, then, were to (1) predict values of y based on knowledge of x, (2) find independent variables whose changes cause, and hence explain, changes in the dependent variable, and (3) test the significance of the relationship between the two variables as expressed by the sample slope coefficient b.

In correlation analysis, the objective is somewhat different. What we hope to demonstrate is that two (or more) variables are related. There is no need in correlation analysis to distinguish between the dependent and independent variable because we are only concerned with whether they move together or not. We are tempted to ask whether correlation analysis is relevant to a regression problem since regression problems involve two (or more) variables. The answer is yes. Regression analysis and correlation analysis are related. A significant regression relationship (indicated by a significant slope value) implies a significant correlation between the two variables.

In fact, we know even more. If the slope coefficient, b, in a regression equation is significant and negative, it indicates that the dependent variable decreases as the independent variable increases and vice versa. The two variables are said to be *negatively correlated*; one increases when the other decreases. On the other hand, if the slope coefficient is significant and positive, then the dependent variable increases as the independent variable increases and vice versa. The two variables are said to be *positively correlated*. Finally, if the slope coefficient is not significantly different from zero, we conclude that the available data do not support a correlation between the two variables.

Correlation analysis will allow us to do more than just make the general types of observations concerning the co-movement of two (or more) variables alluded to in the previous paragraph. Correlation analysis also permits us to quantify, or measure the strength of the co-movement. It is our task in this chapter to derive the appropriate measures and to develop an intuitive feeling for what they tell us and why. Our task is more easily accomplished if we first study correlation in conjunction with a regression problem. This approach has the added advantage of relating back to our discussion and examples of regression analysis in Chapter 15.

16-1 Correlation Analysis in a Regression Problem

In a regression problem, a primary task is to fit the regression equation to a set of data points. We learned how to do this in Chapter 15. In Figure 16-1, we have reproduced the data from the example in Section 15-3, and we have drawn in the sample regression line as calculated using the sample of five observations given in Table 15-2. These data were as follows:

	\multicolumn{5}{c}{Sample Observation}	
	1	2	3	4	5	Average
x	4.0	5.0	6.0	7.0	8.0	6.0
y	205	250	260	300	315	266

The sample points do not fall precisely on the line. However, the line suggests that as the processing time increases, so does the expected ultimate life. In other words, if the processing time is above average, then we expect the ultimate life also to exceed its average. The average processing time in our sample is 6 hours, and the average ultimate life is 266 weeks. When we take a processing time of 7.0 hours, the actual life turned out to be 300 hours, which exceeds 266 hours. But we expected an ultimate life of 104 + 27(7), or 293, weeks as calculated using the regression equation. Why does the actual figure of 300 differ from the expected figure of 293? The reasons may lie in measurement error or in the impact of other variables, such as different ultimate users, which have not been included explicitly in the present example. (A multiple regression analysis is needed if more than one independent variable is to be considered.) The difference of 300 − 293, or 7, remains unexplained in our study. On the other hand, the difference of 293 − 266

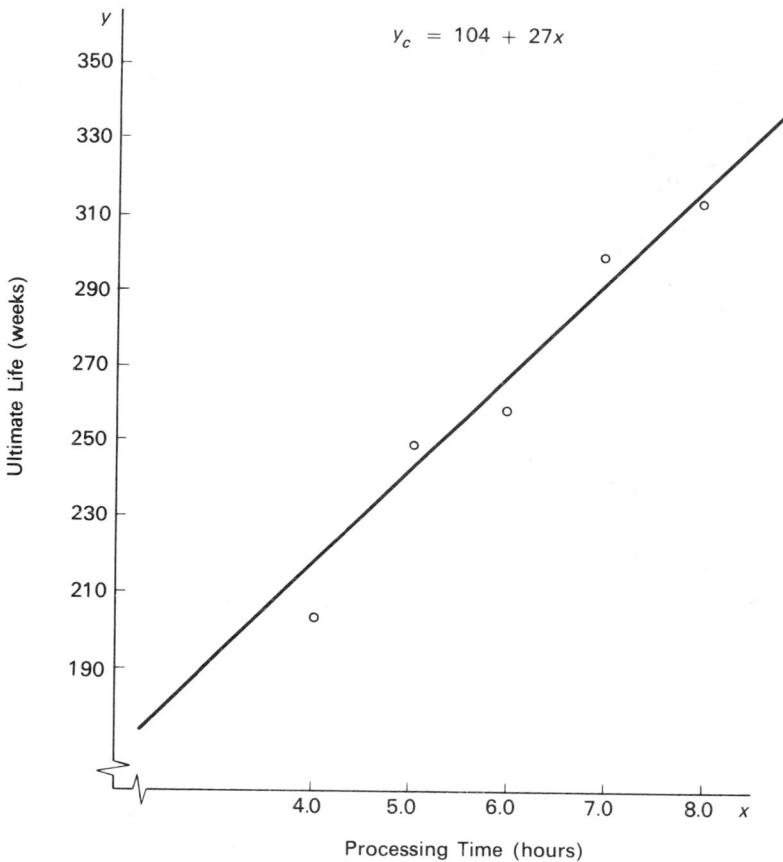

Figure 16-1 Sample: Processing Time and Ultimate Life

(that is, $y_c - \bar{y}$), or 27, was explained by the fact that when processing time exceeds the mean processing time of 6 by one unit, we expect the ultimate life to exceed its average of 266 by 27 units. Thus, of the $300 - 266 = 34$ total difference of y from $\bar{y}$, 27 units are explained by the line and 7 units are not. This relationship is illustrated for a single sample point, $x = 7.0$ and $y = 300$, in Figure 16-2. (An identical analysis would hold if $y = 286$ when $x = 7.0$.)

The correlation coefficient is related to this discussion of the explained and unexplained variation. First, we observe that the total variation is equal to the sum of the explained and unexplained variation.

$$\text{Total Variation} = \text{Unexplained Variation} + \text{Explained Variation}$$
$$y - \bar{y} = (y - y_c) + (y_c - \bar{y}). \qquad (16\text{-}1)$$

In computing the correlation coefficient, however, we measure the variation of each sample point in squared form and sum over all sample values. (The reason for using the squared variation of each sample point is the same as the reason for using squared deviations in computing the variance; namely, the sum of the deviations alone will cancel out.) Therefore, we compute (and set equal) the values given in formula (16-2) and not those given in (16-1). (The expression is mathematically valid although it is not an obvious extension of equation (16-1). The derivation is not given here.)

$$\underset{\text{measure of}}{\Sigma (y - \bar{y})^2} = \underset{\text{measure of}}{\Sigma (y - y_c)^2} + \underset{\text{measure of}}{\Sigma (y_c - \bar{y})^2} \quad (16\text{-}2)$$

total variation = unexplained variation + explained variation.

The summations are each taken over all points (x_i, y_i) in the sample. The y_c (calculated value of y) in each case is based on the x_i for the specific y_i value in question. After a brief bit of algebra, we can write

$$\frac{\text{sum of the squares of the explained variation}}{\text{sum of the squares of the total variation}}$$

$$= \frac{\left(\begin{array}{c}\text{sum of the squares of the}\\ \text{total variation}\end{array}\right) - \left(\begin{array}{c}\text{sum of the squares of the}\\ \text{unexplained variation}\end{array}\right)}{\text{sum of the squares of the total variation}}$$

$$= \frac{\Sigma (y - \bar{y})^2 - \Sigma (y - y_c)^2}{\Sigma (y - \bar{y})^2}$$

$$= 1 - \frac{\Sigma (y - y_c)^2}{\Sigma (y - \bar{y})^2}$$

$$= 1 - \frac{\text{sum of the squares of the unexplained variation}}{\text{sum of the squares of the total variation}}. \quad (16\text{-}3)$$

The various forms of expression (16-3) each give the ratio of a specific measure of the explained variation to the total variation. In intuitive terms, this ratio is the proportion of the total variation (measured in squared terms) in the dependent variable that is explained by the independent variable. Symbolically, we write r^2 for this concept when it is calculated from sample data. The name for r^2 is the *sample coefficient of determination*.

$$r^2 = 1 - \frac{\Sigma (y - y_c)^2}{\Sigma (y - \bar{y})^2}. \quad (16\text{-}4)$$

CORRELATION ANALYSIS

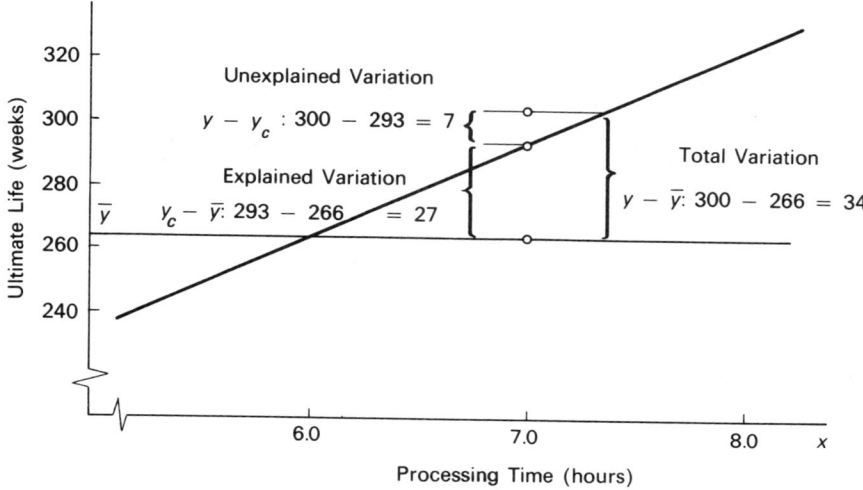

Figure 16-2 Explained, Unexplained, and Total Variation in y Illustrated for a Single Point

The square root of r^2, or r, is the *sample correlation coefficient*.

$$r = \sqrt{r^2} = \sqrt{1 - \frac{\Sigma (y - y_c)^2}{\Sigma (y - \bar{y})^2}}. \qquad (16\text{-}5)$$

The summations are again taken over all the sample points. The correlation coefficient is also a measure of the association between the two variables, but it lacks the intuitive interpretation that the coefficient of determination has (see Technical Note 16-1).

The corresponding parameter values are denoted by the Greek lowercase letter rho, ρ^2 and ρ.

New Symbol	Meaning
r^2	The sample coefficient of determination. It measures the proportion of the variation in y explained by x. It is computed from a sample.
r	The coefficient of correlation. It equals $\sqrt{r^2}$ and is a measure of the association between y and x. It is computed from a sample.
ρ^2	The universe coefficient of determination.
ρ	The universe coefficient of correlation.

Unfortunately, neither equation (16-4) nor (16-5) is easy to calculate without the aid of a computer if the sample is of even moderate size. One somewhat easier calculational formula is given by equation (16-6), which is obtained from (16-3) by substituting $y_c = a + bx$ and recognizing at the right point in the algebraic derivation that $\sum (\bar{y}^2)$ for a sample of size n equals $n\bar{y}^2$. We give the result now without showing the algebra needed to derive it:

$$r^2 = \frac{a \sum_i y_i + b \sum_i x_i y_i - n(\bar{y})^2}{\sum_i y_i^2 - n(\bar{y})^2}, \qquad (16\text{-}6)$$

where the summation is taken over all points in the sample. For our continuing problem and using the data in Table 15-2,

$$r^2 = \frac{104(1{,}330) + 27(8{,}250) - 5(266)^2}{361{,}350 - 5(266)^2},$$

$$= \frac{7{,}290}{7{,}570} \approx 0.96,$$

and hence,

$$r \approx 0.98.$$

This is a very high correlation (perfect correlation yields $r = 1.00$). Such high correlations are not often found among managerial variables considered in pairs.

We may also use another variation measure to compute r and r^2. The unexplained variability is the variability of the sample points around the regression line. A measure of this variability is given by s_e^2. This is the squared form of the standard error of the estimate. [The standard error of the estimate, s_e, was introduced in Section 15-5. A calculational formula is given by (15-9b).] Further, the total variability in y can be measured by

$$s_y^2 = \frac{\sum (y - \bar{y})^2}{n - 1}. \qquad (16\text{-}7)$$

For our problem $s_e^2 = 93.33$ as computed in Section 15-5 and $s_y^2 = 1{,}892$ using (16-7). Using s_e^2, the variance of the sample points around the sample regression line as a measure of the unexplained variation, and s_y^2, the variance of the y values as a measure of the total variation, we can estimate r^2 using

$$r_{\text{est}}^2 = 1 - \frac{s_e^2}{s_y^2} = 1 - \frac{93.33}{1{,}892} \approx 0.95. \qquad (16\text{-}8)$$

CORRELATION ANALYSIS 569

This estimate is known as the *adjusted coefficient of determination*, and for large sample sizes it is nearly identical to the r^2 value determined using equation (16-4) or (16-6).

Now that we know how to calculate the coefficient of correlation, let us examine the range of possible values it may assume. Equation (16-9), which generalizes equation (16-8) to the universe, will be used for this purpose.

$$\rho = \sqrt{1 - \frac{\sigma_e^2}{\sigma_Y^2}} = \sqrt{1 - \frac{\text{unexplained variance}}{\text{total variance}}}. \qquad (16\text{-}9)$$

Now if there is no association between the two variables, then none of the variation in Y is explained by X. In this case, the unexplained variance, σ_e^2, equals the total variance, σ_Y^2, and

$$\rho = \sqrt{1 - \frac{\sigma_Y^2}{\sigma_Y^2}} = \sqrt{1 - 1} = \sqrt{0} = 0.$$

This case is illustrated in Figure 16-3. In this case, the regression line is horizontal; the slope of the regression line is zero. Since sampling is an uncertain process, finding $r = 0$ does not ensure that $\rho = 0$. Nor does finding that $r \neq 0$ guarantee that $\rho \neq 0$. There is no way to remove all the uncertainty from the sampling process. But, as we shall see in Section 16-3, it is possible to test the hypothesis that $\rho = 0$.

The alternative extreme case is when all the points fall exactly on the regression line. In this case, illustrated in Figure 16-4, there is no residual variation of the points around the line. There is no unexplained variation. The regression line gives the exact value of the dependent variable for any

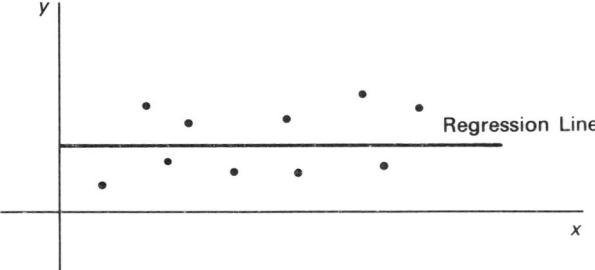

Figure 16-3 Case of No Correlation (ρ or $r = 0$)

Figure 16-4 Case of a Perfect Negative Correlation (ρ or $r = -1$)

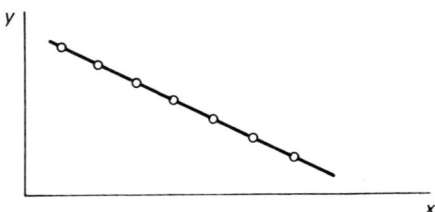

value of the independent variable. Hence, $\sigma_e^2 = 0$ and

$$\rho = \sqrt{1 - \frac{\sigma_e^2}{\sigma_Y^2}} = \sqrt{1 - \frac{0}{\sigma_Y^2}} = \sqrt{1} = \pm 1.$$

The only issue remaining is whether to use $\rho = +1$ or $\rho = -1$ (or $r = +1$ or -1, if we have sample data). The choice is based on whether y increases or decreases as x increases. If both y and x increase together, then the positive square root is used. If y decreases as x increases, the negative square root is used. The same convention is followed for all nonzero correlation coefficient calculations. Figure 16-4 illustrates the case where $\rho = -1$. Another way of saying this is to say that the correlation coefficient takes the same sign as the slope coefficient in the regression equation.

The correlation coefficient for the Dundee Uranium Dilatometer Company data was found to be 0.98 using equation (16-6). The correlation coefficient is positive since, as can be seen in Figure 16-1, the ultimate life increases as processing time increases.

S.S. A government regulatory agency is considering using the results of a personality test to predict job performance levels as measured by their superior's efficiency rating of the individual. The testing service reports that a correlation coefficient of 0.70 was obtained by several other organizations using the test.

(a) Interpret this figure for your nonstatistical superior.
(b) Is this a particularly high value?
(c) Would you recommend that the test be used?

Solution: The best way to explain the result is to use the squared form: $r^2 = 0.49$. This implies that just under 50 percent of the variability in performance, as measured by the efficiency rating, is explained by this test. The percentage is reasonably large in an absolute sense, but it still leaves

CORRELATION ANALYSIS 571

approximately 50 percent of the performance variability unexplained. Such tests generally do not reveal information about all the capacities and characteristics of the individual. Other measures are necessary in estimating success and some may not be easily quantified. Whether the test should be used depends on its cost, what improvement in predicting success will be obtained, and, more important, what value this improvement in predicting success will give the firm, and perhaps what long-run effects the testing procedure may have on employee attitudes. Alternatives should also be considered.

16-2 Correlation Analysis Without Regression

The coefficient of determination is typically used to express the proportion of the total variation in the dependent variable that is explained by the independent variable through the regression line. Regression, on the other hand, is used to predict values of the dependent variable for specific values of the independent variable given the regression equation. In a regression problem, it is hard to justify the predictive exercise unless some meaningful causal connection can be established from the dependent to the independent variable. This was the case for the DUD Company, where processing time is known (or at least suspected) by management to be an important variable in determining the ultimate life of a dilatometer.

But a correlation analysis can be done between any two (or more) variables, In this situation, we are interested in how closely one variable moves with the other. There need not be any way to establish a direction of causal effect between them. Consider an example. Ms. Soi Been is a speculator in the commodity futures market. Naturally, she is interested in the association between the movements in soybean prices and the price of leeks. Ms. Been has collected the sample data for 100 daily price changes. These data are summarized in Table 16-1.

Table 16-1 Price Changes of Soybeans and Leeks: Number of Occurrences

Change in Price of Soybeans: x	Change of Price of Leeks: y					
	-2%	-1%	0%	$+1\%$	$+2\%$	Total
-1%	0	3	4	5	8	20
0%	12	12	9	12	12	57
$+1\%$	9	6	3	4	1	23
Total	21	21	16	21	21	100

Ms. Been is interested in learning whether there is some association between the price movements of soybeans and leeks. The mixture of future contracts purchased by Ms. Been may depend on the association between the price changes of the two commodities. For example, if the price changes are negatively correlated, one increases when the other decreases, and a loss on holding one commodity will typically be balanced by a corresponding gain on the other. In this case, holding a mix of contracts would protect Ms. Been against large losses (and, unfortunately, also against large gains) and, hence, reduces her risk.

In solving Ms. Been's problem, we will approach the correlation coefficient from a new point of view. Consider the product sum (taken over all sample points):

$$\sum (x - \bar{x})(y - \bar{y}). \tag{16-10}$$

If when x exceeds (is less than) its mean, y also exceeds (is less than) its mean, this sum of products will be positive. Such a case is illustrated in Figure 16-1 for the data of the DUD Company. In this case, there is a positive relationship. On the other hand, if when x exceeds (is less than) its mean, y is less than (exceeds) its mean, the sum given by (16-10) will be negative. Hence, the sum given by (16-10) will suggest, in a general way, whether the two variables are correlated, and, if so, whether they move in the same or opposite direction. Figure 16-5a illustrates a case where the sum given by (16-10) is negative; $(x - \bar{x})$ is generally negative when $y - \bar{y}$ is positive, and vice versa. Figure 16-5b suggests a case where the points are scattered in such a way that the products, when summed, will involve both plus and minus terms, which tend to cancel. In Figure 16-5b, the sum given by (16-10) will be close to zero, indicating that movements in the two variables are not strongly associated.

There are problems in using the sum given by (16-10) as a measure of the association between the two variables. First, the size of this sum is a function

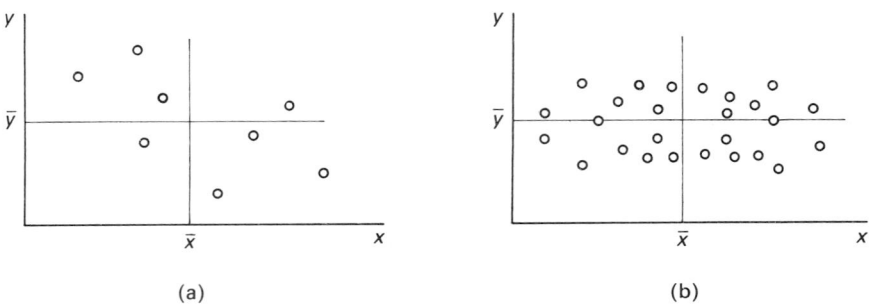

Figure 16-5 Two Situations Involving the Comovement of Two Variables

CORRELATION ANALYSIS

of the units used to measure the variables. For the data given in Table 16-1, the price changes are measured in percentages. The percentage changes are given by the integers -2, -1, 0, 1, and 2. Thus, the single observation of a price change of $+1$ percent for soybeans and $+2$ percent for leeks would enter the sum as $(1 - 0.03)(2 - 0)$, where $\bar{x} = +0.03$ percent and $\bar{y} = 0$ percent. But the decimal equivalent of percentages could also be used where 1 becomes 0.01. Then the price changes would be measured in units of -0.02, -0.01, 0.00, 0.01, and 0.02. In this case, the sum given by (16-10) would be only one ten-thousandth as large. (Using percentages where each number is 100 times as large as its decimal equivalent, yields a sum ten thousand times as large.) This just will not do!

Since a number that depends on the units of measurement is hard to interpret, we require a measure that provides the same value no matter what units are used to measure the variables. Such a measure can be obtained by dividing each term of the sum in (16-10) by its standard deviation; that is, by standardizing the two differences. This is done in the sum over all sample points given by (16-11).

$$\sum \left(\frac{x - \bar{x}}{s_x} \cdot \frac{y - \bar{y}}{s_y} \right) = \frac{\sum (x - \bar{x})(y - \bar{y})}{s_x \cdot s_y}, \qquad (16\text{-}11)$$

where s_y is given by (16-7) and s_x is an equivalent expression for the standard deviation of the x values.

The value of this sum for the DUD Company problem is about -0.42, and this result is independent of the units used to measure the two variables. (The calculations are not shown because they are both tedious and the result is only an intermediate step in the calculation of the correlation coefficient.) Dividing by s_x and s_y in (16-11), then, eliminates the effect of the particular measurement scale used.

The second difficulty that we must resolve relates to the sample size. The sum given by (16-11) is a function also of the number of observations we have, since $(x - \bar{x})(y - \bar{y})$ grows as the sample size increases while the standard deviations are not affected on the average. Both s_x and s_y contain the factor $\sqrt{1 \div (n - 1)}$ in their calculation. It is this averaging factor used in computing the standard deviation that prevents it from increasing as the sample size increases. Multiplying the denominator of (16-11) by $n - 1$ removes this averaging process and allows s_x and s_y to increase as n increases. Now, both the numerator and denominator of (16-11) increase in such a way as n increases that (16-12) is independent of n. Thus, our final measure of the correlation coefficient is

$$r = \frac{\sum (x - \bar{x})(y - \bar{y})}{s_x s_y (n - 1)}. \qquad (16\text{-}12)$$

By substituting

$$s_x = \sqrt{\frac{\sum (x - \bar{x})^2}{n - 1}} \quad \text{and} \quad s_y = \sqrt{\frac{\sum (y - \bar{y})^2}{n - 1}},$$

multiplying out, and simplifying, formula (16-12) can be reduced to a simpler form. Using several algebraic steps omitted here, we obtain

$$r = \frac{\sum (x - \bar{x})(y - \bar{y})}{\sqrt{\sum (x - \bar{x})^2 \sum (y - \bar{y})^2}} = \frac{(\sum_i x_i y_i) - n\bar{x}\bar{y}}{\sqrt{(\sum_i x_i^2 - n\bar{x}^2)(\sum_i y_i^2 - n\bar{y}^2)}}, \quad (16\text{-}13)$$

where again the sum is over all sample points. (To obtain the last expression, the facts $\bar{x} \sum \bar{y} = n\bar{x}\bar{y}$ and $\bar{x} \sum x = n\bar{x}^2$ are required.) Formula (16-13) is equivalent to formula (16-6). Both yield the same answer for a given sample. Formula (16-13) has two advantages over (16-6). First, it does not require the values of a and b from the regression equation. Formula (16-13) can be calculated directly from the sample data. Second, the sign of r results from the calculation. There is no need to refer to the sign of the slope coefficient, b. The calculations needed to obtain r using (16-13) for Ms. Been's problem are given in Table 16-2. Note that we have grouped the data. Thus, for example, the sum of the 100 x-values is given by the sum of the products obtained by multiplying each value of x by the frequency with which it occurs (see column 4 = $\sum$ [(col. 1) · (col. 3)].

Table 16-2 Data Required to Calculate the Correlation Coefficients: Soybean–Leek Example[a]

x	y	$f(xy)$	$xf(xy)$	$yf(xy)$	$xyf(xy)$	$x^2 f(xy)$	$y^2 f(xy)$
−1	−2	0	0	0	0	0	0
−1	−1	3	−3	−3	3	3	3
−1	0	4	−4	0	0	4	0
−1	+1	5	−5	5	−5	5	5
−1	+2	8	−8	16	−16	8	32
0	−2	12	0	−24	0	0	48
0	−1	12	0	−12	0	0	12
0	0	9	0	0	0	0	0
0	+1	12	0	+12	0	0	12
0	+2	12	0	+24	0	0	48
1	−2	9	9	−18	−18	9	36
1	−1	6	6	−6	−6	6	6
1	0	3	3	0	0	3	0
1	+1	4	4	4	4	4	4
1	+2	1	1	2	2	1	4
Total		$n = 100$	$\sum x = 3$	$\sum y = 0$	$\sum xy = -36$	$\sum x^2 = 43$	$\sum y^2 = 210$

[a] $f(xy) \equiv$ number of observations with the given values of x and y in columns 1 and 2.

CORRELATION ANALYSIS

Using the appropriate totals from Table 16-2, we have for the 100 data points,

$$\sum xy = -36 \qquad \bar{x} = \frac{\sum x}{n} = \frac{3}{100} = 0.03 \qquad \bar{y} = \frac{\sum y}{n} = \frac{0}{100} = 0$$

$$n = 100 \qquad \sum x^2 = 43 \qquad \sum y^2 = 210.$$

Hence,

$$r = \frac{-36 - 100(0.03)(0)}{\sqrt{[43 - 100(0.03)^2][210 - 100(0)^2]}}$$

$$= \frac{-36}{\sqrt{8,030}} = \frac{-36}{95.03}$$

$$\approx -0.379.$$

There is a negative correlation between the movement of soybean prices and leek prices. The relationship is not, however, a strong one (note $r^2 \approx 0.145$), but it may be sufficient to allow Ms. Been to "hedge" by buying both leeks and soybeans. However, before we advise her to use the relationship, it is worth finding out if the relationship is significant. In other words, how likely is it that the sample price data were drawn from a universe in which no actual association is present, and, hence, our sample value of r is just the result of chance? This is our task in the next section.

S.S. Could we have guessed the relationship between soybean prices and leek-price movements without doing a correlation analysis? If yes, then why bother with correlation analysis? Give two reasons.

Solution: An experienced speculator may have been able to predict that the price movements of these two commodities would be negatively correlated. A close inspection of Table 16-1 (but probably not Table 16-2) may also be sufficient to suggest the negative relationship. Notice, for example, that there are more occurrences of $(+1, -2)$ and $(-1, +2)$ than of $(-1, -2)$ and $(+1, +2)$, respectively. But even if we could know that a negative correlation exists, this does not tell us either how important the relationship is (as is suggested by the size of r or r^2) or whether the relationship is significant (the topic of Section 16-3).

16-3 Tests of Significance on the Correlation Coefficients

Probability statements concerning the correlation coefficient can only be made when x and y are random variables. In particular, this means that x cannot be predetermined or fixed in advance as was true in regression analysis.

Moreover, when the sample correlation coefficient, r, is used to form confidence intervals for ρ, it is necessary to assume further that the joint distribution of the two variables is bivariate normal. A discussion of the bivariate normal distribution is beyond the scope of this volume and, hence, we will not construct confidence intervals for ρ around the sample value of r.

A more common problem is to test whether or not ρ could be zero given that some nonzero value has been observed for r. This is an hypothesis-testing problem, one we encountered in the last section. We will consider this problem when it arises in regression analysis and when it arises in correlation analysis.

In regression analysis, as we noted in Section 15-6, a zero slope value indicates no relationship between the two variables. The situation is also illustrated in Figure 16-3 and discussed in Section 16-1. Hence, a significance test on the slope coefficient is equivalent to a significance test on the correlation coefficient. If we cannot reject the null hypothesis that $\beta = 0$ (or, for a one-sided test that, for example, $\beta \leq 0$), then we cannot reject the null hypothesis that $\rho = 0$ (alternatively, $\rho \leq 0$).

When, as is true in many cases, the hypothesized direction of the relationship can be prespecified, a more powerful one-tailed test of significance is favored. [In this regard, recall that the sign given to r when it is calculated using equation (16-6) is the sign taken on by the slope coefficient b in the estimated regression equation.]

In correlation analysis, the sign of r is established directly using equation (16-13). It would be possible also to calculate and test the slope coefficient for the related regression equation, but this seems somewhat out of place if the requirements for a regression analysis are not met; namely, if x and y are both random variables.

A test for the significance of r is based on the t distribution to test the null hypothesis

$$H_0: \rho = 0$$

versus

$$H_1: \rho \neq 0.$$

The t statistic for this test with $(n - 2)$ degrees of freedom is

$$t = \frac{r - 0}{s_r}, \tag{16-14}$$

where s_r is the standard deviation of the sample correlation coefficient. (There are $n - 2$ degrees of freedom since the data are used to calculate both $\bar{x}$ and $\bar{y}$ in obtaining r.) The value of s_r is calculated using

$$s_r = \sqrt{\frac{1 - r^2}{n - 2}}. \tag{16-15}$$

Correlation Analysis

For our data

$$s_r = \sqrt{\frac{1 - (-0.38)^2}{98}} \approx 0.093$$

and

$$t = \frac{-0.38 - 0}{0.093} \approx -4.09.$$

The -4.09 value of t is obtained using the sample data. It is tested using a given level of significance and the t value associated with this confidence level found in the t table. For example, using an 0.01 significance level and a two-tailed test, the appropriate t value for 98 degrees of freedom is approximated by the normal, $z_{0.495} = 2.576$. Since $-4.09 < -2.576$, the null hypothesis is rejected, and we conclude that $\rho \neq 0$. (If the direction of the relationship can be predetermined, a one-tailed test could be used.)

S.S. A study involving the estimation of direct labor time per unit of a standard product as a function of contract size in units yielded a correlation coefficient of -0.60 based on a sample size of 12 contracts. The firm wishes to test the significance of this value.

(a) Should a two-tailed test or a one-tailed test be used?

(b) State the hypothesis symbolically.

(c) Test the hypothesis.

Solution:

(a) Since labor time per unit should decline due to learning and other efficiencies obtained from larger-scale operations, the manager should expect a negative correlation (unless other variables interfere). Hence, a one-tailed test is in order.

(b) The null hypothesis would be $H_0: \rho \geq 0$. Now:

$$s_r = \sqrt{\frac{1 - 0.6^2}{10}} \approx 0.253$$

$$t = \frac{-0.60 - 0}{0.253} = -2.37.$$

(c) The value of r is significant at the 0.025 probability level since $t_{0.475} = 2.228$ for 10 degrees of freedom. It is not significant at the 0.01 probability level. If a two-tailed test were used, r would be significant at the 0.05 probability level, but not at the 0.025 probability level.

16-4 Correlation and Statistical Independence*

A question that often causes difficulty is whether statistical independence implies a correlation of zero and vice versa. There are two ways to address this problem: mathematically and graphically. The former is precise, the latter intuitive. We shall use both means and in the order listed.

Using an argument analogous to that which led up to equation (16-11), the universe correlation coefficient can be defined by

$$\rho = \frac{E[(X - \mu_X)(Y - \mu_Y)]}{\sigma_X \sigma_Y}, \qquad (16\text{-}16)$$

where the expectation symbol replaces the summation. Now if the dependent and independent variables are statistically independent, then

$E[(X - \mu_X)(Y - \mu_Y)]$
$= E(X - \mu_X) \cdot E(Y - \mu_Y)$ since $E(XY) = E(X)E(Y)$ if the variables are independent random variables
$= [E(X) - \mu_X][E(Y) - \mu_Y]$
$= (\mu_X - \mu_X)(\mu_Y - \mu_Y)$
$= 0$

and

$$\rho = \frac{0}{\sigma_X \sigma_Y} = 0.$$

Hence, statistical independence does imply that the two variables are uncorrelated, $\rho = 0$. But does it work the other way?

Consider Figure 16-6. In this case, if we were to require a linear relationship and hence compute r using equation (16-5), the regression line would have zero slope and r would equal zero. The correlation coefficient would also be zero if equation (16-13) were used. And yet the value of y changes as x changes in a way that, given the graph, is completely predictable. An example similar to that suggested in Figure 16-6 is the relationship between rainfall and crop yield. If there is no rain, there is no crop. And yet too much rain can also reduce the total yield.

Thus, we have a case of statistical dependence with a correlation of zero. Zero correlation does not necessarily imply statistical independence. (Independence and a zero correlation can occur simultaneously. If all the data

---

* This section can be omitted without loss of continuity.

CORRELATION ANALYSIS

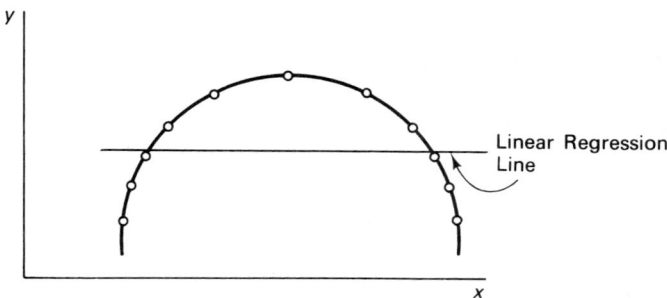

Figure 16-6 Correlation and Statistical Dependence

points fell on the horizontal regression line, then the variables would be both uncorrelated and statistically independent.)

In a manner of speaking, the result above is unfair since our calculations of the correlation coefficient assume explicitly [equation (16-6)] or implicitly [equation (16-13)] a linear relationship between the variables—possibly after transformation. In one sense, it is unfair to apply a linear relationship to the data in Figure 16-6 since the relationship appears to be nonlinear. Suppose that we recognize this and use the appropriate curvilinear function to estimate y_c. (Curvilinear regression can be done, although we avoid it here.) Then r, computed using equation (16-5) for the data plotted in Figure 16-6, would equal 1. Perfect correlation would reflect the statistical dependence present in the data.

We must be careful in correlation analysis not to impute to the relationship, through mathematical calculations, characteristics it does not possess. The old adage "garbage in–garbage out" applies. The manager must be careful to see that the relationship makes sense based on knowledge of the underlying economic phenomena before proceeding to use the results of a statistical analysis.

S.S. Consider the following variables: (1) traffic flow on major New York streets and (2) height of the tide along the Florida Coast.
 (a) Would you expect these two variables to be independent?
 (b) Would you expect these two variables to be correlated?

Solution: A clear dependence between the processes underlying these variables is not obvious, and yet data would show them to be correlated. The reason is that the tides are influenced by the moon, whose rise is related to the time of day and, hence, to traffic flow. The probability of various levels of traffic flow would not be the same depending on the height of the tide.

That is, knowing the current level of the tide would help us predict the traffic flow in the next hour. (Knowing the time of day would be better, however, since then a direct relationship can be determined.) The variables are dependent and they are therefore correlated.

16-5 A Warning

As was the case for regression analysis, most of the interesting behavioral, technical, and economic relationships involve several variables and not just two. The manager may be interested in the net explanatory power of several independent variables in a regression analysis as well as the overall tendency of the variables to explain the movements of a single dependent variable. This is the problem of multiple correlation analysis. Like multiple regression analysis, the problems are more complex and difficult. Indeed, whole books, and big ones at that, are devoted to multiple correlation and regression analyses. At some point you should pursue the topic of multiple relationships. Many of the basic ideas discussed in this and the preceding chapter, including prediction, hypothesis testing, and estimating the strength of relationships, carry over to the more complex multivariate case. An understanding of the single independent-variable case can help one understand the multivariate case that is often a result of a formal managerial application.

TECHNICAL NOTE 16-1: AN INTUITIVE APPROACH TO THE CORRELATION COEFFICIENT

The coefficient of determination was interpreted in regression analysis as the proportion of the total variability in the dependent variable that is explained by the independent variable(s). No such simple interpretation was given to the correlation coefficient. This note attempts to provide an intuitive means of conceptualizing the correlation coefficient.

In a regression analysis, the estimated slope coefficient measures the importance for the dependent variable of a change in the independent variable. It ignores, however, the likelihood of any particular change in the independent variable. Hence, in two situations described by linear functions with the same slope coefficient, a change of, say, 2 units in the independent variable has the same effect on the dependent variable. However, it may be that such a change is more likely in one case than the other.

This situation is akin to the problem of finding a measure of correlation that is independent of the units used to measure the variables. This problem was discussed in Section 16-2 in connection with formula (16-11). The solu-

Correlation Analysis

tion here is the same as the one we used then. What we can do is to fit the regression line to the standardized variables, that is, to

$$\frac{x - \mu_X}{\sigma_X} \quad \text{and} \quad \frac{y - \mu_Y}{\sigma_Y} \quad \text{for the universe,}$$

or to

$$\frac{x - \bar{x}}{s_x} \quad \text{and} \quad \frac{y - \bar{y}}{s_y} \quad \text{for the sample data,}$$

where s_x and s_y are the sample standard deviations of the sample values.

When this transformation of the data is made and the regression line is fitted to the standardized variables, the new slope coefficient tells the change in y in standardized units for a unit standard deviation change in x. It turns out that, once both variables are standardized, the slope coefficient is the correlation coefficient. That is, if we use the values

$$x' = \frac{x - \bar{x}}{s_x} \quad \text{and} \quad y' = \frac{y - \bar{y}}{s_y}$$

in place of the original x and y values, then

$$y'_c = a + rx',$$

where r is the sample correlation coefficient between x and y.

This is a particularly useful procedure when several independent variables are considered at one time. The slope coefficients, using the standardized form, allow for both the unit change effect and the likelihood of the change as measured by the standard deviation. These standardized regression coefficients are called beta coefficients. They assist us in comparing each independent variable's effect on the dependent variable by converting the slope measure to a unit-free measure. The value of the slope coefficient is not effected by the original units used to measure the independent variable. The value of β gives the standard deviation change in y for a unit standard deviation change in x. [Note that the β used here does not have the same interpretation as was given β in Chapter 15. In Chapter 15, β was the universe regression parameter (the universe slope value) for the variable, measured in the same units in which x is measured. Here, in this technical note, it is the sample slope coefficient when the regression equation is fitted to the standardized values of the variable(s).]

New Symbols

Symbol	Meaning
r^2	The sample coefficient of determination. A measure of the variability in the dependent variable explained by the independent variable.
ρ^2	The universe coefficient of determination.
r	The sample correlation coefficient, equals $\sqrt{r^2}$.
ρ	The universe correlation coefficient, equals $\sqrt{\rho^2}$.
s_r	The standard deviation of the correlation coefficient. Calculated from the sample. The universe concept would be represented by σ_r, which s_r estimates.

Key Formulas

Formula	Used to Compute:
$r^2 = \dfrac{a \sum_i y_i + b \sum_i x_i y_i - n(\bar{y})^2}{\sum_i y_i^2 - n(\bar{y})^2}$	The sample coefficient of determination. Calculational formula.
$r^2_{est} = 1 - \dfrac{s_e^2}{s_y^2}$	An estimate of the coefficient of determination.
$r = \dfrac{\sum_i x_i y_i - n\bar{x}\bar{y}}{\sqrt{(\sum_i x_i^2 - n\bar{x}^2)(\sum_i y_i^2 - n\bar{y}^2)}}$	The sample coefficient of correlation. Calculational formula that does not involve knowledge of the relevant regression equation.
$s_r = \sqrt{\dfrac{1 - r^2}{n - 2}}$	Standard deviation of the correlation coefficient.

PROBLEMS

16-1. How can one establish the sign of the correlation coefficient from a knowledge of the linear regression line?

16-2.* In correlation analysis, which of the following statements holds?
(a) Both variables are random variables.
(b) The independent variable only is a random variable.
(c) The dependent variable only is a random variable.
(d) Neither variable is a random variable.

Correlation Analysis

16-3. In regression analysis, which of the statements in Problem 16-2 is correct?

16-4.* If the correlation coefficient is -0.5, what proportion of the variation in the dependent variable is explained by the independent variable?

16-5. What is the difference in interpretation between r and ρ?

16-6.* The case of (a) zero correlation and (b) a correlation of 1, would appear in what form on a graph? Select the best choice for (a) and (b) from the four choices below.
  (1) All the data points fall on the regression line, which has a zero slope.
  (2) Some of the data points, at least, do not fall on the regression line, which has a nonzero slope.
  (3) All the data points fall on the regression line, which has a nonzero slope.
  (4) Some of the points, at least, do not fall on the regression line, which has a zero slope.

16-7.* Given the following data on cost and output, find the correlation coefficient and test it for significance.

Units of output, m	1	2	3
Dollars of cost, n	4	7	10

16-8.* A study of welfare expenditures involving a sample of 400 households in Tompkins County produced the following relationship:

$$w = 8.24 - 1.71x$$

$$r = -0.73, \quad s_e = 0.5, \quad s_b = 0.1,$$

where only welfare households were involved and $w \equiv$ yearly household welfare allocations in thousands of dollars, $x \equiv$ yearly household income in thousands of dollars:

$$\bar{w} = 2.768 \quad \text{and} \quad \bar{x} = 3.2.$$

(a) What percentage of welfare allocations are explained by income?
(b) Predict average welfare allocations for an income level of $4,500.
(c) Compute a 95 percent confidence interval for the average determined in part (b).

16-9. In a correlation-regression problem, describe a test of significance equivalent to testing the significance of r.

16-10.* For each of the following situations, indicate whether a correlation analysis or a regression analysis would be more appropriate.
  (a) In order to choose among various advertising media, an agency account executive is presently investigating the relationship between a woman's age and her annual expenditures on a client firm's cosmetics to see if they are related. The executive will also look at other demographic variables.
  (b) A trucker needs to establish a decision rule that will enable the firm to determine when to inspect and replace tires, based upon the number of miles driven.
  (c) A government agency wishes to identify which field offices of various sizes (in terms of the number of employees) are out of alignment with the prevailing pattern of working days lost due to sick leave.
  (d) A survey research firm conducts attitude surveys in two stages. The first stage involves identifying important variables, such as age and income. The second stage is more detailed, involving a separate study to predict values of one variable using the known values of those found associated with it in the initial stage.

16-11. An American Cancer Society study involving 800,000 men and women showed that the death rate from strokes was considerably higher, and from coronary heart disease generally higher, among persons who usually sleep nine to ten hours a night than among those who usually sleep seven hours. On the basis of these statistically significant results, is it reasonable to argue that longer sleep is the cause of deaths and coronary heart disease? Why or why not?

16-12. Determine the correlation coefficient for the following data:

x	y
8	6
24	18

16-13. A recent study of crime in all 100 wards of New York City produced the following:

$$C = 20 - 2X, \quad r = -0.6, \quad s_e = 1.0, \quad s_b = 0.3$$

where

$C \equiv$ number of major crimes per ward (a political subdivision) per month

$x \equiv$ average income of the families living in the ward in thousands ($\bar{x} = 5$).

CORRELATION ANALYSIS   585

(a) $C = 20 - 2x$ is
   (1) The universe relation between C and x.
   (2) The correlation equation.
   (3) The regression equation.
   (4) Nonlinear in x.
(b) The best explanation of the value 20 is that
   (1) It is the C intercept and allows values of C to be computed knowing x.
   (2) It provides a reasonable estimate of the number of major crimes per ward when average income is zero.
   (3) It gives the decrease in major crimes for an increase of $1,000 in average ward income.
   (4) It indicates that when average ward income exceeds $10,000, there will be no major crimes.
(c) How much of the variability in number of major crimes is explained by average ward income?
(d) Test the significance of the relationship between C and x. (Use a 0.01 probability level.)
(e) Give the major limitation that you see in the correlation regression analysis given in the problem.
(f) Estimate the *expected* number of major crimes per ward given an average income level of $6,000 and place a 99 percent confidence interval around this value.

Compute the correlation coefficient for the data in the problems indicated and test the coefficient for significance at the 0.05 probability level.

16-14.* Problem 15-6   16-15.   Problem 15-7
16-16.   Problem 15-8   16-17.* Problem 15-9
16-18.   Problem 15-11   16-19.   Problem 15-13
16-20.* Problem 15-14   16-21.   Problem 15-16
16-22.   Problem 15-19   16-23.* Problem 15-20
16-24.   Problem 15-21

16-25.   A credit manager for a large department store wants to determine whether their credit rating system for screening credit applicants is really capable of detecting risky accounts. From a random sample of 200 customers, the manager computed the following numbers, where x is the customer's initial credit rating and y is the past due amount in the customer's account.

$\sum x = 1,471$, $\sum y = 16,494$,

$\sum x^2 = 12,315$, $\sum y^2 = 1,581,677$, $\sum xy = 137,517$.

(a) Compute the sample correlation coefficient.
(b) Conduct an appropriate hypothesis test at the $\alpha = 0.05$ level of significance.
(c) How small could r be and still yield significant results at the 0.05 level?

16-26. For each of the following pairs of variables, suggest whether the correlation coefficient should be expected to be positive, negative, or zero:
(a) The amount of fertilizer applied to a plot of land and the subsequent crop yield.
(b) The monthly return on a share of stock traded on the New York Stock Exchange and the monthly return on the Standard and Poor's 500 market index.
(c) The number of inspectors stationed on an assembly line and the number of defective units shipped to customers.
(d) The age of machinery in a factory and the annual repair expenditures related to the machinery.
(e) Total salesmen's commissions and total sales for a wholesale distributor.
(f) Land value per acre and distance from the center of a city.
(g) The amount of service left in a battery and the number of hours it has been used.
(h) Pollen count and the sale of an anti-allergy drug.

16-27. The manager of a machine shop wants to determine whether there is a significant correlation between the age of a certain group of machines and the amount of time per week that the maintenance crew spends in servicing the machines. Each of the machines in the group was acquired at the same time, 5 years ago, and the manager decided to observe the maintenance time for each machine for a period of 1 month. Can the manager achieve meaningful results from a correlation analysis? Why or why not? Suppose that half the machines were acquired 4 years ago and half were acquired 5 years ago. Can the manager now achieve meaningful results? Explain.

16-28. In order to determine an efficiency rating for one appliance, a manager decided to conduct a series of experiments. The experiments consisted of running the appliance for a number of different lengths of time and noting the number of kilowatt-hours of electricity used. The results of the calculations indicated that the unexplained variance was twice as great as the explained variance. What is the sample correlation coefficient?

16-29. A credit manager for a company believes that the past due amount owed by customers is directly related to the size of the account. The manager selected a sample of 10 past-due accounts and found the following results:

Amount Past Due	Account Size	Amount Past Due	Account Size
$46	143	$47	128
55	158	93	251
48	151	27	111
81	227	74	206
41	136	66	194

What proportion of the variation in past-due amount is explained by account size?

Chapter **17**

Time-Series Analysis

The president of the Barnesville-Stumptown Water Ski Corporation, Joe Barstuwaski, wishes to forecast company sales. Mr. Barstuwaski's firm makes a complete line of water skis, including 25 different style-size combinations. He is interested in forecasting sales for several reasons. First, the firm is considering expanding its facilities so that it can produce water-ski accessories as well as the skis themselves. Mr. Barstuwaski wants a long-term (next 10 years) forecast of total dollar sales, so he can determine whether the firm will have sufficient funds (cash flow) to make all of the mortgage payments on the new facilities. If the total forecast of sales exceeds $100 million, he has determined that they should proceed with the expansion. Mr. Barstuwaski also needs an intermediate term (next 12 months) forecast of the total sales in units, so he can plan the amount of overall production for the next several months and establish an operating budget for the firm. Several options, including hiring, overtime, planned vacations, and other methods, can be implemented to meet the production plan. Finally, he needs a short-term (next month or two) forecast of each item's sales so he can plan to have sufficient inventory on hand to satisfy incoming orders.

Although the above reasons are sufficient, there are often additional uses for forecasts. For example, the intermediate-term forecast might be used in setting salesmen's goals, and the long-term forecast might be used in establishing a dividend policy.

Mr. Barstuwaski believes that he may be able to make these forecasts by examining past data. Accordingly, he has accumulated some information for his (and our) perusal and analysis. Table 17-1 contains total dollar sales in

Table 17-1 Total Sales (millions of dollars) for the Barnesville-Stumptown Water Ski Company, for the Period 1967–1976
(Average Selling Price of a Unit, as of January 1, 1977, is $94)

Year		Jan 1	Feb 2	Mar 3	Apr 4	May 5	June 6	July 7	Aug 8	Sept 9	Oct 10	Nov 11	Dec 12	Year's Total
(1967)	1	0.68	0.62	0.56	0.50	0.48	0.44	0.45	0.48	0.51	0.53	0.57	0.60	6.42
(1968)	2	0.69	0.59	0.53	0.51	0.45	0.45	0.42	0.49	0.46	0.54	0.58	0.60	6.31
(1969)	3	0.75	0.66	0.60	0.51	0.49	0.40	0.49	0.58	0.51	0.50	0.61	0.68	6.78
(1970)	4	0.72	0.71	0.65	0.47	0.55	0.42	0.47	0.62	0.50	0.55	0.63	0.75	7.04
(1971)	5	0.82	0.80	0.68	0.62	0.55	0.53	0.49	0.60	0.63	0.70	0.74	0.81	7.97
(1972)	6	0.85	0.87	0.75	0.61	0.56	0.52	0.54	0.60	0.56	0.68	0.71	0.80	8.05
(1973)	7	0.84	0.86	0.80	0.62	0.60	0.53	0.54	0.62	0.71	0.70	0.74	0.81	8.37
(1974)	8	0.90	0.86	0.79	0.63	0.58	0.52	0.52	0.61	0.66	0.67	0.76	0.84	8.34
(1975)	9	0.85	0.90	0.80	0.78	0.69	0.59	0.57	0.54	0.63	0.72	0.77	0.83	8.67
(1976)	10	1.01	0.89	0.81	0.80	0.72	0.58	0.63	0.59	0.60	0.74	0.81	0.82	9.00
Month's Total		8.11	7.76	6.97	6.05	5.67	4.98	5.12	5.73	5.77	6.33	6.92	7.54	(76.95)

millions, by month, for the last 10 years, 1967–1976. Table 17-2 gives the monthly sales for the past 3 years, in units, for one of the individual products.

Mr. Barstuwaski has given us two separate sets of time-series data. A *time series* is a set of values for one variable, with each value representing a different period of time. There are many examples of time-series data in addition to the two given above. In fact, most business and economic data take the form of a time series. For example, annual gross national product in the United States for the last 50 years would constitute a time series. Monthly rainfall in a specified area for the time period 1918–1920 would also constitute a time series. In this latter case, the data would consist of 36 numbers, one rainfall figure for each month; month one (January 1918) might have had 1.4 inches of rainfall, for example.

Table 17-2 Unit Sales for Item 1, Barnesville-Stumptown Water Ski Company
(Each Unit Sells for $112 as of January 1, 1977)

Year		Jan 1	Feb 2	Mar 3	Apr 4	May 5	June 6	July 7	Aug 8	Sept 9	Oct 10	Nov 11	Dec 12	Year's Total
(1974)	1	40	42	31	24	37	29	25	31	25	32	32	49	397
(1975)	2	51	49	40	42	24	30	26	26	30	36	44	45	443
(1976)	3	45	51	36	34	40	30	28	24	29	39	48	48	452
Month's Total		136	142	107	100	101	89	79	81	84	107	124	142	(1292)

We might study a GNP or rainfall time series to either describe what has happened or to forecast what will happen. In a management setting, we are usually concerned about forecasting, but understanding the past may help us to forecast. Mr. Barstuwaski is interested in obtaining three different forecasts, for three different purposes. The past data will prove useful in this endeavor.

In studying a time series, we try to find patterns that were present in the past, such as a tendency toward lower amounts of rainfall. We will not try to explain these patterns using other variables; we will merely try to find the patterns. Time-series analysis is based only on the time-series data, not on external factors. (However, as we will discuss at more length later, the intelligent use of a time-series analysis calls for the manager to use all available information, including subjective judgments.)

Mr. Barstuwaski has examined Table 17-1 and is prepared to make a forecast of the next 10 years' total sales. He has noticed something interesting about total sales for 1974, 1975, and 1976. (Since it is hard to digest the full set of data, Mr. Barstuwaski scanned it to try to pick out a pattern.) The three values are 8.34, 8.67, and 9.00. He points out that there has been an increase of 0.33 from 1974 to 1975 and from 1975 to 1976. He predicts that there will be a similar increase in each of the next 10 years. Thus, his yearly total forecasts are $9.33, $9.66, $9.99, $10.32, $10.65, $10.98, $11.31, $11.64, $11.97, and $12.30 million, for a total of $108.15 million. This is more than the $100 million Mr. Barstuwaski feels is necessary before approving the expansion plans, so he is prepared to proceed.

Mr. Barstuwaski is playing the familiar game: 2, 4, 6, 8, ... He is relying on a pattern based on only a few of the many data values available. Time-series analysis is a set of methods for finding patterns which are more complicated than noting that 10 completes the 2, 4, 6, 8, ... series. These methods are designed to use all the data the manager has and which he believes is relevant to the problem.

Consider some of the difficulties with the simplified approach first attempted by Mr. Barstuwaski. First, examine the following numbers, taken from a recent newspaper: 52, 53, 54, ... You might guess that the next number in the sequence is 55, when, in fact, it is 51. These numbers represent the lap times in a four-person relay (mile relay, each individual runs 440 yards), and the last runner is the fastest of the four. Knowing what the data represent, we would probably be unwilling to guess at the fourth value. If we did guess, the chosen value would be based on our knowledge of the strategy followed in sequencing the runners on a relay team. We must always consider the source of the data and our knowledge about it, in addition to studying the patterns.

The second example is Table 17-1. Looking at the overall pattern, we see that the last two yearly increases are larger than the increases just prior to that. (In fact, there was a decrease from 1973 to 1974.) This can be explained by the U.S. recession, centered around the end of 1974. Thus, it may be overly optimistic to forecast an $0.33 million increase for each of the next 10 years. However, some of the earlier increases are even larger than 0.33, so 0.33 might be pessimistic. It would seem that Mr. Barstuwaski (and the rest of us) have little choice but to read the rest of the chapter if he is to analyze his time-series data. It is too soon for him to make a decision regarding the facilities expansion.

17-1 General Time-Series Models

A time series is likely to have values affected by many external factors. For example, a company's total sales may be affected by general economic conditions, the weather, or their advertising expenditures. In addition, their sales may fluctuate because some of their large customers may periodically lower or raise their inventories. In regression analysis, we tried to isolate one (or, in multiple regression, a few) explanatory variables that explain the variability in the dependent variable. In time-series analysis, we examine only the time-series data itself in trying to explain the variability in the data.

A time series is usually assumed to contain four different components, *trend, seasonal variation, cyclical variation,* and *irregular variation*. The *trend* (or *long-term trend*) measures the change, either up or down, of the values over a long period of time. For example, an examination of the data in Table 17-1 indicates that total yearly sales has exhibited an upward trend.

Table 17-1 data also exhibits a seasonal variation. This can be seen by noting that the demand is consistently higher around December, January, and February than in June and July. A *seasonal variation* in general is a periodic pattern in the data that recurs at regular intervals, such as a year in Table 17-1. A seasonal variation can recur every week if, for example, we are examining grocery store sales; it is a systematic pattern, including high and low values, that is fairly consistent. Based on our knowledge of the variable expressed by the time series, we can be relatively sure how long a seasonal pattern takes to complete. In Table 17-1, the seasonal pattern covers 1 year or 12 monthly values. Figure 17-1 displays the Table 17-1 data graphically so that the trend and seasonal terms can be visualized. The seasonal peaks appear to recur every 12 months, and there does seem to be an upward trend in the data. (The trend line in Figure 17-1 is drawn by eye. Do you agree with it? If not, draw your own, and we will check it later.)

However, the data do not follow precisely the same seasonal pattern each year, and the trend does not necessarily follow the nice straight line we have drawn. These irregularities can be explained by both cyclical variation and irregular variation. *Cyclical variation* is a pattern in the data, much like a

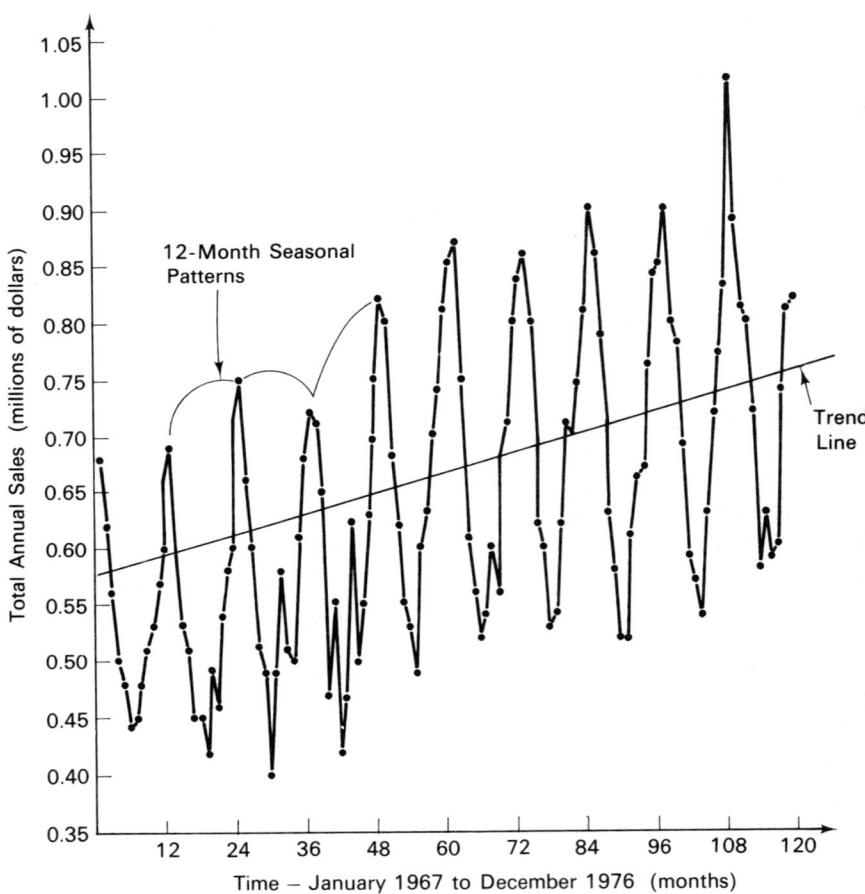

Figure 17-1 Pictorial Representation of Table 17-1 Data: Total Annual Sales (millions of dollars) for the Barnesville-Stumptown Water Ski Company

seasonal pattern, but one that does not reoccur every year (or week if we are using daily data). Examples include patterns that take 17 (or 29, or 7) months to repeat themselves. They are caused, for example, by swings in general economic conditions or by swings in aggregate customer inventories. Unfortunately, since there can be more than one cause, there can be more than one cyclical pattern in the data, making it difficult to determine and quantify the separate effects. In addition, the pattern may change through time. (An 18-month inventory cycle may, over time, become a 15-month inventory cycle.) Because of these difficulties the cyclical variation is frequently lumped into the irregular variation.

Time-Series Analysis

If all else fails, irregular variation can be used to explain data values. *Irregular variation* is a nonrecurring variation that is present to some degree in all time series, because the world is uncertain. While we cannot predict irregular variation, we will find ways to remove the other components from the series and observe how large the irregular variation tends to be. (This is similar to regression, where we were able to measure how much of the variability of the dependent variable could not be explained by the regression line.) We will usually not try to separate the cyclical and the irregular variation. These two sources will be treated as the "leftover" variation after the trend and seasonal have been removed.

The four components of a time series are commonly assumed to combine in one of two different models, a multiplicative model or an additive model.

time-series value: $y = T \cdot S \cdot C \cdot I$ (multiplicative model)

time-series value: $y = T + S + C + I$ (additive model),

where T, S, C, and I are the trend, seasonal, cyclical, and irregular terms, respectively.

New Symbol	Meaning
y (or y_i or y_t)	The time-series value (or the ith time-series value, y_i; or the time-series value in period t, y_t).
T	The trend term in the time series.
S	The seasonal term in the time series.
C	The cyclical term in the time series.
I	The irregular term in the time series.

If we knew all the individual terms, we could forecast the time-series value. That is why we will try to isolate and evaluate the rate at which the trend is changing and the amount of the seasonal effect in each period. We will not use any external facts to explain the individual terms; we will just try to explain the time series by examining the time series and its components. It is incumbent upon the user of time-series analysis to apply judgment, both before and after the analysis.

Since it is difficult to visualize the separate effects clearly in Figure 17-1, let's build a hypothetical, multiplicative time series from its components. Suppose the initial trend term is 200 and that it increases 25 units each month. That is, T equals 200, then 225, then 250, and so on. Suppose, further, that the seasonal term is 0.6, 0.8, 0.8, 1.2, 1.4, 0.8, 0.8, 0.6, 0.8, 1.2, 1.4, and 1.6 from month 1 through 12 in every year. Finally, assume there is an inventory cycle of 5 months' duration that repeats a pattern of 0.6, 1.0, 1.2, 1.2, and

1.0, starting with the first period. For now, we will assume there is no irregular variation, so $I = 1.0$ throughout. The calculations for the first 24 values of the multiplicative model are shown in Table 17-3.

Table 17-3 Time Series Based on a 12-Month Seasonal Pattern, 5-Month Cycle, and Trend

Period	Trend (T)	Seasonal (S)	Cyclical (C)	Irregular (I)	Time-Series Value ($y = TSCI$)
1	200	0.6	0.6	1.0	72
2	225	0.8	1.0	1.0	180
3	250	0.8	1.2	1.0	240
4	275	1.2	1.2	1.0	396
5	300	1.4	1.0	1.0	420
6	325	0.8	0.6	1.0	156
7	350	0.8	1.0	1.0	280
8	375	0.6	1.2	1.0	270
9	400	0.8	1.2	1.0	384
10	425	1.2	1.0	1.0	510
11	450	1.4	0.6	1.0	378
12	475	1.6	1.0	1.0	760
13	500	0.6	1.2	1.0	360
14	525	0.8	1.2	1.0	504
15	550	0.8	1.0	1.0	440
16	575	1.2	0.6	1.0	414
17	600	1.4	1.0	1.0	840
18	625	0.8	1.2	1.0	600
19	650	0.8	1.2	1.0	624
20	675	0.6	1.0	1.0	405
21	700	0.8	0.6	1.0	336
22	725	1.2	1.0	1.0	870
23	750	1.4	1.2	1.0	1,260
24	775	1.6	1.2	1.0	1,488
25	800	0.6	1.0	1.0	480
26	825	0.8	0.6	1.0	396
27	850	0.8	1.0	1.0	680
28	875	1.2	1.2	1.0	1,260
29	900	1.4	1.2	1.0	1,512
30	925	0.8	1.0	1.0	740

The Table 17-3 data are shown graphically in Figure 17-2. The correct trend line has been drawn in the graph. Examine the data representation and see what trend line you would have drawn in freehand. (Ours would have started lower and ended higher; it would have had a greater slope than the correct trend.) Now try to visualize the seasonal effect. There is no doubt that there are ups and downs, but it is not clear that they repeat on a 12-

TIME-SERIES ANALYSIS

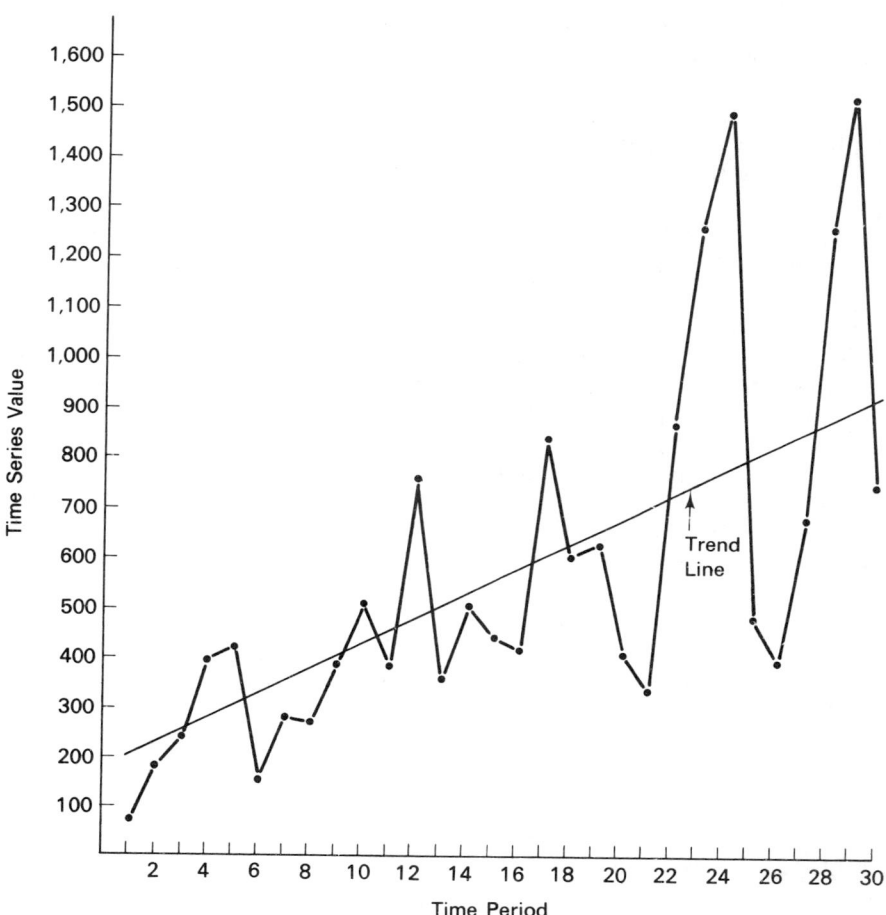

Figure 17-2 Time Series Described in Table 17-3

month basis. (For example, look at the data for periods 4, 16, and 28. Period 16 does not seem to fit the pattern that produces the values in periods 4 and 28. However, several other periods do follow a fairly consistent pattern.) The 5-month cycle is obscuring the 12-month seasonal pattern. Imagine how obscure it would look if irregular variation were included!

The 5-month cycle may be easier to spot if we look specifically at the two intervals between the high peaks. In fact, once we see that the first 5-period space between peaks (12 to 17) is followed by a 7-period space (to 24), it is easier to see that a 12-period seasonal pattern exists. The point is that the combined effect of all these factors makes a time series hard to analyze. We will not deal with all the many methods of analysis that exist. Instead, we will concentrate on analyzing the trend and seasonal factors. This is done in the next two sections.

S.S. Construct a time series using a trend term that stays constant at 500, a seasonal term identical to that in Table 17-3, a 5-period cyclical pattern identical to the one in Table 17-3, and an irregular variation term with the following values:

0.95, 0.95, 1.35, 1.25, 0.85, 0.85, 0.75, 0.65, 0.55, 0.75, 0.75, 0.95, 0.85, 1.45, 1.45, 1.35, 1.05, 0.85, 0.85, 1.25, 0.95, 1.35, 0.85, 0.85, 0.55, 1.15, 1.35, 0.55, 0.55, 0.95.

(These 30 values are based on random numbers to ensure their irregularity.) Once you have constructed six data values for this time series, graphically illustrate these values (plus the next 24 given below) and try to point out the trend, cycle, and seasonal effect.

Period	Trend (T)	Seasonal (S)	Cycle (C)	Irregular (I)	y = TSCI
1	500	0.6	0.6	0.95	
2	500	0.8	1.0	0.95	
3	500	0.8	1.2	1.35	
4	500	1.2	1.2	1.25	
5	500	1.4	1.0	0.85	
6	500	0.8	0.6	0.85	
7	500	0.8	1.0	0.75	300
8	500	0.6	1.2	0.65	234
9	500	0.8	1.2	0.55	264
10	500	1.2	1.0	0.75	450
11	500	1.4	0.6	0.75	315
12	500	1.6	1.0	0.95	760
13	500	0.6	1.2	0.85	306
14	500	0.8	1.2	1.45	696
15	500	0.8	1.0	1.45	580
16	500	1.2	0.6	1.35	486
17	500	1.4	1.0	1.05	735
18	500	0.8	1.2	0.85	408
19	500	0.8	1.2	0.85	408
20	500	0.6	1.0	1.25	375
21	500	0.8	0.6	0.95	228
22	500	1.2	1.0	1.35	810
23	500	1.4	1.2	0.85	714
24	500	1.6	1.2	0.85	816
25	500	0.6	1.0	0.55	165
26	500	0.8	0.6	1.15	276
27	500	0.8	1.0	1.35	540
28	500	1.2	1.2	0.55	396
29	500	1.4	1.2	0.55	462
30	500	0.8	1.0	0.95	380

TIME-SERIES ANALYSIS

Solution: The first six values are 171, 380, 648, 900, 595, and 204, respectively. The graph is given below in Figure 17-3. Looking at the graph we might guess that the trend line is flat, or nearly so. The 5-period cycle is hard to spot. Even the seasonal is hard to isolate, but we could convince ourselves that there is a seasonal by examining periods 1, 13, 25, and then 2, 14, 26, and so on. The pattern does not jump out of the page, although the seasonal effect can be discovered with sufficient diligence. We will study methods for finding these factors systematically in Section 17-3.

17-2 Estimating the Trend Line

Mr. Barstuwaski is interested in predicting total sales for the next 10 years. To do so, it would be helpful to know the long-term trend of sales. He need not concern himself with a seasonal pattern, since that will average

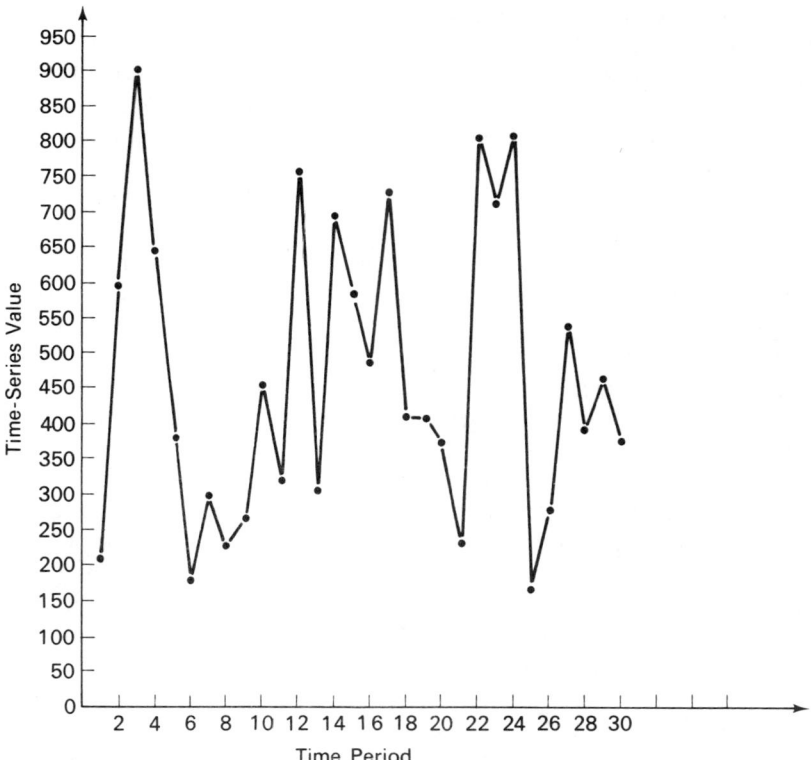

Figure 17-3 Time Series with a 12-Month Seasonal Pattern, a 5-Month Cycle, and Irregular Variation

out over the 10 years, and a cycle of duration less than 2 or 3 years will also have very little effect on the 10-year total. (A pronounced cycle of, say, 7 years could have a significant effect, but we will not attempt to predict cycles of this length.) Thus, Mr. Barstuwaski is most interested in the trend effect on sales, for the purpose of making this long-run forecast.

Long-term trend forecasting is used for strategic planning. Mr. Barstuwaski is interested in assessing whether long-run sales will support a facilities expansion. This is an important question for his company, and a good forecast is essential. He can ignore any short-term fluctuations in addressing the strategic question.

Governments, local, state, or federal, may be interested in predicting the trend in tax revenues under their current taxation method before they either enter into a major capital project (such as building a new school) or try to change the tax rates. Companies are interested in long-term sales of a product line both to assess their ability to finance new projects and to decide when new product lines will be necessary for continued growth. Thus, we will look for trends in past data to use in forecasting future values.

Unfortunately, there is more than one way in which a time series may manifest a long-term trend. The example of building a time series used in the previous section had an increase of 25 units in each time period. That is a linear trend; the time series was increasing by the same amount in each period. A time series can also exhibit increasing (or decreasing) rates of change. That is, the time series might increase 25 units the first period, then 30, 35, and so on. It might even grow (or decline) exponentially. Worse yet, it might change from one form to another during a given time period. In fact, we cannot list all the ways in which a time series might exhibit a trend. Figure 17-4 depicts the trend effects described in this paragraph and gives the equations that describe the trend curves.

There are many comments that can be made regarding Figure 17-4. First, there are many other possible underlying trend patterns, but these are the ones most commonly used. Second, the curves in Figure 17-4b and c have been given the names *parabolic* and *exponential*, referring to the equations used. The exponential growth curve results when a time series increases (or decreases) by a constant percent in each time period. The parabolic curve is a second-order polynomial, and is used sometimes to give a better fit. Third, the first three curves all have increasing time-series values as time progresses, but they can, by properly choosing the constant values, be made into decreasing-trend time series. The methods of time series analysis can be used to analyze either increasing or decreasing trends.

Finally, Figure 17-4d shows what is often termed a life-cycle curve. A new product may have sales following this curve in that initial growth is slow, then there is a period of quick growth, followed by a tapering off in the rate of increase, and eventually a decrease in sales. This curve is not easily modeled

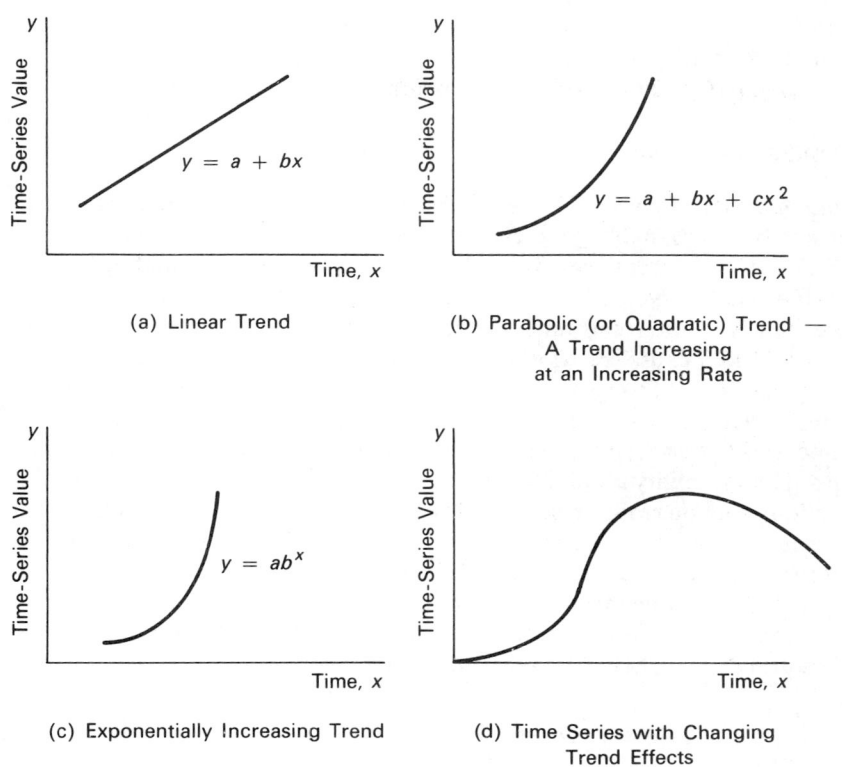

Figure 17-4 Four Representations of Trend Terms in a Time Series

by a single equation. Rather, we might think of two or more different trend equations underlying the time series. Any time a trend model is used for forecasting, we should be alert to the fact that the method we are using may become inappropriate at some point in the future. Thus, we take a leap of faith when making a long-term forecast; we assume (as we must) that the pattern that fits the past data fairly well will continue to fit the future data.

Before estimating a trend effect, we must choose the model to be used. There is no way to be sure of the choice, especially since we can never include all possible underlying models in the set from which the choice is made. The manager should, of course, utilize any special knowledge about the situation. One other step we should use in the process is to plot the data, and visually decide which one of the models (if any) best fits the data. If there is anything strange in the data, plotting will help us to find it. Next, we will briefly describe the computations used to determine the linear trend term when the underlying model is that of Figure 17-4a. The computations for the trend models of Figure 17-4b or c are not given here. [The formulas can be found in more advanced texts, such as: M. Hamburg, *Statistical Analysis for Decision Making* (New York: Harcourt Brace Jovanovich, 1970).]

LINEAR TREND

Looking back to Figure 17-1, we see that the linear trend line that was drawn seems to be a reasonably good expression of the trend. When we have reason to believe the trend is linear, we will use the least-squares method to find the best line. That is, we will use the normal equations from Chapter 15 to find the equation of the trend line. It is important to note here that we are not trying to find a causal relationship, as in regression analysis; we are simply finding the "best"-fitting linear line, the rate of growth over the past several years. In fact, the assumptions of regression analysis are usually not met in a time series. Successive y values are certainly not independent in a time series. The variability around the line as measured by the error term (irregular variation) may increase or decrease through time, rather than stay constant as assumed in regression analysis. In spite of this, we can use the fitted line to forecast in the same manner as in Chapter 15. We will take the equation $y = a + bx$, obtained using the normal equations, and forecast by plugging in the x value for future time periods. The constants a and b are found using equations (17-1) and (17-2), taken from Chapter 15.

$$b = \frac{\sum_{i=1}^{n} x_i y_i - n\bar{x}\bar{y}}{\sum_{i=1}^{n} x_i^2 - n\bar{x}^2} \tag{17-1}$$

$$a = \bar{y} - b\bar{x}, \tag{17-2}$$

where there are n data values, $i = 1, \ldots, n$.

We will not obtain variance measures and confidence intervals as in Chapter 15, since the statistical assumptions required are not met.

Since time is a continuous variable and we can start counting it whenever we wish, we can make $\bar{x} = 0$ by properly choosing the starting point. For example, if we have 5 years in a time series, 1892–1896, we can make $1892 = -2$, $1893 = -1$, $1894 = 0$, $1895 = +1$, and $1896 = +2$. If we are considering 4 years (an even number), such as 1685–1688, we can make the two middle years be -1 and $+1$, then count forward and backward in twos. Thus, $1685 = -3$, $1686 = -1$, $1687 = +1$, $1688 = +3$, and $\bar{x} = 0$. (Then the b value will be the estimated increase during one half of a year.) The advantage is that equations (17-1) and (17-2) are simplified, as follows:

$$b = \frac{\sum_{i=1}^{n} x_i y_i}{\sum_{i=1}^{n} x_i^2}, \quad \text{if } \bar{x} = 0 \tag{17-3}$$

$$a = \bar{y}, \quad \text{if } \bar{x} = 0. \tag{17-4}$$

As an example of finding a and b, we will use equations (17-1) and (17-2) and the Barnesville-Stumptown Water Ski Company sales data from Table 17-1. The time periods are numbered from 1 to 120, as shown in Figure 17-1. The calculations below show some of the required computations and the resulting values for

$$\sum_{i=1}^{n} x_i y_i, \quad \sum_{i=1}^{n} x_i^2, \quad \sum_{i=1}^{n} x_i, \quad \sum_{i=1}^{n} y_i.$$

x_i	y_i	$x_i y_i$	x_i^2
1	0.68	0.68	1
2	0.62	1.24	4
3	0.56	1.68	9
4	0.50	2.00	16
5	0.48	2.40	25
⋮	⋮	⋮	⋮
119	0.81	96.39	14,161
120	0.82	98.40	14,400
7,260	76.95	4,955.55	583,220

$$(\bar{x} = 60.5)(\bar{y} = 0.64125)$$

Thus, using equations (17-1) and (17-2),

$$b = \frac{\sum_{i=1}^{n} x_i y_i - n\bar{x}\bar{y}}{\sum_{i=1}^{n} x_i^2 - n\bar{x}^2} = \frac{4{,}955.55 - 120(60.5)(0.64125)}{583{,}220 - 120(60.5)^2}$$

$$= \frac{4.955.55 - 4{,}655.475}{583{,}220 - 439{,}230} = \frac{300.075}{143{,}990} = 0.002084$$

$$a = \bar{y} - b\bar{x} = 0.64125 - (0.002084)(60.5) = 0.5152.$$

The trend line is $y = a + bx = 0.5152 + 0.002084x$. Going back to Figure 17-1, we see that the line drawn by eye was less steep ($y = 0.575 + 0.00150x$). That is the reason we make the calculations rather than just drawing the line by eye.

To apply (17-3) and (17-4) we need to form a new set of x_i values. The new x_i values are $-119, -117, -115, \ldots, -1, +1, \ldots, 115, 117,$ and 119, to have 120 integer values that average to zero. The calculations using (17-3) and (17-4) are summarized on page 602.

x_i	y_i	x_iy_i	x_i^2
−119	0.68	−80.92	14,161
−117	0.62	−72.54	13,689
−115	0.56	−64.40	13,225
−113	0.50	−56.50	12,769
−111	0.48	−53.28	12,321
⋮	⋮	⋮	⋮
117	0.81	94.77	13,689
119	0.82	97.58	14,161
0	76.95	599.9	575,960

Now, applying (17-3) and (17-4), we obtain

$$a = \frac{76.95}{120} = 0.64125 \quad \text{and} \quad b = \frac{599.9}{575,960} = 0.001042.$$

The b value verifies the answer found above; b should be one half as large as before since it measures the increase in one half of a year.

To make a forecast using the trend line for any future time value, we simply plug in the appropriate value of time. These time values would be 121, 122, and so on when time is measured from 1 to 120, and they would be 121, 123, 125, and so on when time is measured from −119 to +119. We must remember that a forecast of the value of the time series must also include the seasonal variation. We can, however, forecast total sales for any number of full years, by adding up several of these yearly values, since the seasonal factor averages out over the year. The forecast of total sales for 1977, based on the trend line, is:

Forecast of total 1977 sales
$$= [a + b(121)] + [a + b(122)] + \cdots + [a + b(132)]$$
$$= 12a + b(121 + 122 + \cdots + 132) = 12a + 1518b$$
$$= 12(0.5152) + 1,518(0.002084) = 6.1824 + 3.1635 = 9.3459.$$

Similarly, we can make Mr. Barstuwaski's 10-year forecast:

Forecast of total sales from 1977 through 1986 inclusive
$$= 120a + b(121 + \cdots + 240) = 120a + b\left[\frac{121 + 240}{2} \, 120\right]$$
$$= 120a + 180.5(120)b = 120a + 21,660b$$
$$= 120(0.5152) + 21,660(0.002084)$$
$$= 61.824 + 45.139 = 106.96.$$

There are 120 values in the total, so $120a$ is part of the forecast. The 120 x values average $(121 + 240)/2$, so $120(121 + 240)/2$ is the total number of

Time-Series Analysis

b terms included. The forecast of $106.96 million is below the $108.15 million Mr. Barstuwaski predicted using rough figures. The $106.96 is probably the better forecast, but we cannot be sure of that. It is possible that Mr. Barstuwaski's three annual figures more accurately represent what will happen in the future than does the more complete history. In either case, more importantly, the forecast is above the $100 million necessary to justify the expansion plans. (There are, of course, many other factors, such as sales predictions for the new product line, that would have to be considered before beginning the expansion.)

Examining this calculational effort, you may wonder if there is not an easier way. There are several. First, computer programs are designed for this purpose, and we recommend them highly. Second, we can draw the line by eye, after plotting the data, but significant mistakes are likely. Third, we can get someone else to do the calculations (or talk to the computer) for us. This is entirely appropriate at a later stage in your career, but, alas, not at the current stage. Finally, since we have 10 full years of data, we could work with the yearly totals, reducing the number of data values to 10, and get a yearly trend equation. This is appropriate, and it will lead to similar (but not exactly the same) forecasts. It is preferable to work with full years in estimating trend in the method used here; otherwise the seasonal effect will introduce error into the estimates. It is possible, as the next section will show, to deseasonalize the data first, then estimate the trend using full or partial years.

S.S. In the Barnesville-Stumptown Water Ski Company data presented in Table 17-1, the last six yearly totals were 7.97, 8.05, 8.37, 8.34, 8.67, and 9.00, respectively. Using only those six yearly totals, instead of the data from Table 17-1, find the linear trend line. Make a forecast for the next annual total, 1977. (Be sure to number the years using $-5, -3, -1, 1, 3,$ and 5.)

Solution:

Year	x	y	xy	x^2
1971	−5	7.97	−39.85	25
1972	−3	8.05	−28.15	9
1973	−1	8.37	−8.37	1
1974	1	8.34	8.34	1
1975	3	8.67	26.01	9
1976	5	9.00	45.00	25
Sum	0	50.40	2.98	70

Linear trend: $a = \bar{y} - b\bar{x} = 8.40$,

$$b = \frac{\sum_{i=1}^{n} x_i y_i - n\bar{x}\bar{y}}{\sum_{i=1}^{n} x_i^2 - n\bar{x}^2} = \frac{2.98}{70} = 0.04257$$

For 1977, when $x = 7$,
$$forecast = a + b(7) = 8.40 + 0.04257(7) = 8.698.$$

[Notice that the b value is for one half of a year, so 1977 (numbered as 7) is 2 units away from 1976 (numbered as 5).]

17-3 Estimating the Seasonal Effect

In studying trends, in the previous section, we ignored seasonal effects because they averaged out over a year. If we are interested in forecasting for other than some integer number of years, we cannot ignore the seasonal factor. Mr. Barstuwaski wants a forecast of total sales for each of the next 12 months, so that he can plan the amount of overall production and establish the operating budget for the next several months. This is an important problem for his firm, and the seasonal effect must be the dominant consideration.

The main reason for estimating the seasonal pattern is to use it in intermediate-term forecasting. Seasonals are also useful in removing the effect of seasonal influences from (or *deseasonalizing*) the data. This allows easier estimation of the trend factor. In a similar manner, we might *detrend* the data (remove the trend effect) to allow easier analysis of the seasonal effects. We will return to these notions later.

In the general multiplicative time-series model, $Y = TSCI$, the seasonal variation indicates that a given month (or week or quarter) is equal to some constant times the long-term trend line, on the average. That is, in a retail store we might say that December's sales are, say, expected to be $1\frac{1}{2}$ times an average month's sales. January's sales are, say, expected to be 0.6 times an average month's sales. Our task is to estimate the 1.5, 0.6, and other values. Then we can use those values in forecasting. The seasonal need not occur by months. We can speak of a seasonal effect within the 5 days of a week, the 52 weeks of a year, or the four quarters of a year.

The basic idea of a multiplicative seasonal, then, is to find the average January (for example) demand, divided by the average monthly demand during the entire year. The easiest way to find the *seasonal index*, then, is to compute:

Seasonal index for January
$$\equiv S_{Jan} = \frac{\text{total January demand/number of Januarys}}{\text{total demand/number of months}}.$$

New Symbol	Meaning
$S_{Jan}, S_{Feb}, \ldots$	The seasonal index for January, February, and so on.

TIME-SERIES ANALYSIS

Other month's seasonal indexes would be computed similarly, and the deseasonalized value in any period is the original value divided by the seasonal factor. (This will be discussed further later in this section.) There is a weakness in this method, in that if the data are taken during full calendar years, the December values will be higher (given a positive trend) than January values, because of the trend term. The trend term would confuse the estimation of the seasonal indexes. However, the method is easy, and we will try it on the Barnesville-Stumptown Water Ski Company data before introducing the methods that allow us to correct for the trend effect.

Table 17-1 gives total sales by month and overall. Using those figures:

$$S_{Jan} = \frac{8.11/10}{76.95/120} = 1.265, \quad S_{Feb} = \frac{7.76/10}{76.95/120} = 1.210,$$

$S_{Mar} = 1.087$, $S_{Apr} = 0.943$, $S_{May} = 0.884$, $S_{June} = 0.777$, $S_{July} = 0.798$, $S_{Aug} = 0.894$, $S_{Sept} = 0.900$, $S_{Oct} = 0.987$, $S_{Nov} = 1.079$, $S_{Dec} = 1.176$.

These values will be checked against those obtained using the better (and harder) methods given later. We know that January's index (1.265) is low, since it includes less trend effect than later months. December's index is high because it contains more trend effect. One way of correcting this, if we already have done a linear trend analysis, is to subtract bx_i from each data value, thus eliminating the trend effect. Then the seasonal indexes can be computed using the detrended data. Even if no trend analysis has been done, we can use the ratio-to-moving-average method, given next. We will postpone a discussion of forecasting until after the ratio-to-moving-average method has been introduced.

RATIO-TO-MOVING-AVERAGE METHOD OF CALCULATING SEASONAL INDEXES

The *ratio-to-moving-average method* of calculating seasonal indexes computes the ratio of a time-series value to the average of a full year's data (or other period if appropriate) around that value. Centering the year's total around the value under consideration and then averaging by dividing by 12 should eliminate, or at least reduce, the trend effect. Unfortunately, when there is an even number of periods in a seasonal pattern, such as with monthly data, we cannot precisely center a year's data around one value. For example, January through December is not centered around either June or July. June has 5 months before it and 6 months after it, while the opposite is true for July. However, January through January, 13 data values, is centered around July. To avoid double counting the end month (January here), we count only half of each of the end values. That is, July's *centered moving average* consists of one half of the previous January value plus one half of the following

January value plus the 11 values between these two January values. This sum is then divided by 12. July is in the center since $5\frac{1}{2}$ months' values before and after are included in the average; and there is a total of 12 values. A formula for calculating the ratio to moving average using the centered moving average for monthly data is as follows (this ratio will then be used to estimate the seasonal indexes:

ratio to moving average for monthly data calculated using the centered moving average

$$\equiv \frac{y_i}{\text{centered moving average around } y_i}$$

$$\equiv \frac{y_i}{(\frac{1}{2}y_{i-6} + y_{i-5} + y_{i-4} + \cdots + y_{i+4} + y_{i+5} + \frac{1}{2}y_{i+6})/12}. \quad (17\text{-}5)$$

This equation can only be used with monthly data, and the value can be computed only 6 months after the beginning of the series and 6 months before the end of the series.

Given the results using this formula, the seasonal index is estimated by averaging this ratio for all the Januarys (giving S_{Jan}, the seasonal index for January). The same is done for February, and so on.

To do these calculations by hand (even with a calculator), it helps to know that we can generate successive centered moving averages by adding

$$\tfrac{1}{12}[-\tfrac{1}{2}(y_{i-6}) - \tfrac{1}{2}(y_{i-5}) + \tfrac{1}{2}(y_{i+6}) + \tfrac{1}{2}(y_{i+7})]$$
$$= \tfrac{1}{24}(-y_{i-6} - y_{i-5} + y_{i+6} + y_{i+7})$$

to the previous one. Although this looks complicated, it is easier than starting from scratch each time.

To show a sample of the calculations, we note that the first 14 data points from Table 17-1 were: 0.68, 0.62, 0.56, 0.50, 0.48, 0.44, 0.45, 0.48, 0.51, 0.53, 0.57, 0.60, 0.69, and 0.59. The first ratio we can compute is for period 7, July. Using (17-5):

July's ratio to moving average $= \dfrac{0.45}{[\frac{1}{2}(0.68) + 0.62 + \cdots + 0.60 + \frac{1}{2}(0.69)]/12}$

$$= \frac{0.45}{6.425/12} = \frac{0.45}{0.5354} = 0.840.$$

Next, August's centered moving average is obtained.

August's centered moving average
$$= [\tfrac{1}{2}(0.62) + 0.56 + \cdots + 0.69 + \tfrac{1}{2}(0.59)]/12$$
$$= 6.415/12 = 0.5346.$$

TIME-SERIES ANALYSIS

Or, using the computational method mentioned above:

August's centered moving average

$$= \text{July's centered moving average}$$
$$+ \tfrac{1}{24}(-y_1 - y_2 + y_{13} + y_{14})$$
$$= 6.425 + \tfrac{1}{24}(-0.68 - 0.62 + 0.69 + 0.59)$$
$$= 0.5354 + \tfrac{1}{24}(0.02) = 0.5346.$$

Then the ratio, computed using equation (17-5), is

$$\text{August's ratio to moving average} = \frac{0.48}{0.5346} = 0.898.$$

Table 17-4 summarizes the results of many calculations based on Mr. Barstuwaski's data. (Some calculations are omitted to keep the table from being even larger than it currently is.) The calculations of the seasonal indexes are given below the table.

January's seasonal index, S_{Jan}, equals the average of the ratio-to-moving-average figures for each January. The same method applies for each month. There are nine values of the ratio for each month. Thus,

$$S_{\text{Jan}} = (1.305 + 1.377 + 1.254 + 1.307 + 1.251 + 1.238$$
$$+ 1.287 + 1.173 + 1.483)/9 = 1.297$$
$$S_{\text{Feb}} = (1.117 + 1.219 + 1.235 + 1.275 + 1.423 + 1.266$$
$$+ 1.232 + 1.243 + 1.193)/9 = 1.245.$$

As the result of similar calculations, we obtain the other 10 seasonal indexes. Table 17-5 gives all 12 seasonal indexes both as computed earlier in the chapter using plain averages and as computed using the ratio-to-moving-average method. At that time we said that the 1.265 figure for the January seasonal index was too low, and the 1.176 December seasonal index was too high. This turns out to be true. With a few exceptions, the deviations follow the predicted pattern: average index $\leq$ ratio-to-moving-average index up to June and greater thereafter.

Table 17-5 also contains a final column, the adjusted ratio-to-moving-average index, which is adjusted so that the indexes average to exactly 1 (sum to 12). (If the indexes do not average to 1, they will cause forecasts using them to be biased.) This is accomplished by multiplying each value of the ratio-to-moving-average index by 12/12.039 (12.039 is the sum of the second column). Some rounding is necessary to keep each index to three decimals. This last column is the one we will use hereafter.

Table 17-4 Calculations of the Ratio-to-Moving-Average Values for the Barnesville-Stumptown Water Ski Company Data

Time Period		Sales	Centered Moving Average	Ratio to Moving Average	Time Period		Sales	Centered Moving Average	Ratio to Moving Average
1967	Jan.	0.68			1971	Jan	0.82	0.6275	1.307
	Feb.	0.62				Feb.	0.80	0.6275	1.275
	Mar.	0.56				Mar.	0.68	0.6321	1.076
	Apr.	0.50				⋮	⋮	⋮	⋮
	May	0.48			1972	Jan.	0.85	0.6795	1.251
	June	0.44				Feb.	0.87	0.6816	1.423
	July	0.45	0.5354	0.840		Mar.	0.75	0.6786	1.105
	Aug.	0.48	0.5346	0.898		⋮	⋮	⋮	⋮
	Sept.	0.51	0.5321	0.958	1973	Jan.	0.84	0.6783	1.238
	Oct.	0.53	0.5313	0.998		Feb.	0.86	0.6791	1.266
	Nov.	0.57	0.5305	1.074		Mar.	0.80	0.6862	1.166
	Dec.	0.60	0.5297	1.133		⋮	⋮	⋮	⋮
1968	Jan.	0.69	0.5289	1.305	1974	Jan.	0.90	0.6992	1.287
	Feb.	0.59	0.5280	1.117		Feb.	0.86	0.6980	1.232
	Mar.	0.53	0.5263	1.007		Mar.	0.79	0.6955	1.136
	Apr.	0.51	0.5246	0.972		⋮	⋮	⋮	⋮
	May	0.45	0.5254	0.856	1975	Jan.	0.85	0.7246	1.173
	June	0.45	0.5258	0.856		Feb.	0.90	0.7238	1.243
	July	0.42	0.5283	0.795		Mar.	0.80	0.7196	1.112
	Aug.	0.49	0.5337	0.918		⋮	⋮	⋮	⋮
	Sept.	0.46	0.5395	0.853	1971	Jan.	1.01	0.7417	1.483
	Oct.	0.54	0.5425	0.995		Feb.	0.89	0.7463	1.193
	Nov.	0.58	0.5442	1.066		Mar.	0.81	0.7471	1.084
	Dec.	0.60	0.5432	1.105		⋮	⋮	⋮	⋮
1969	Jan.	0.75	0.5446	1.377		Dec.	0.82		
	Feb.	0.66	0.5514	1.219					
	Mar.	0.60	0.5573	1.077					
	⋮	⋮	⋮	⋮					
1970	Jan.	0.72	0.5741	1.254					
	Feb.	0.71	0.5749	1.235					
	Mar.	0.65	0.5753	1.130					
	⋮	⋮	⋮	⋮					

The ratio-to-moving-average seasonal indexes are better estimates in that the trend effect has been averaged out in their computation. Both computations, however, are based on the assumption that there has been only one seasonal pattern during the entire 10-year period. If the pattern has changed (or is changing), the seasonal indexes will be an average of both the old and the new values. The values will to some degree not represent either the current or the upcoming situation.

Changes in seasonal patterns are usually slight over any short period of time, so we frequently ignore the error introduced by changing seasonals.

Table 17-5 Comparison of Seasonal Indexes Computed using Plain Averages and the Ratio-to-Moving-Average Method

Month	Plain Average Index	Ratio-to-Moving-Average Index	Adjusted Ratio-to-Moving-Average Index
January	1.265	1.297	1.293
February	1.210	1.245	1.241
March	1.087	1.099	1.095
April	0.943	0.947	0.944
May	0.884	0.886	0.883
June	0.777	0.777	0.775
July	0.798	0.791	0.788
August	0.894	0.904	0.901
September	0.900	0.902	0.899
October	0.987	0.971	0.968
November	1.079	1.059	1.056
December	1.176	1.161	1.157
Total	12.000	12.039	12.000

Examples of seasonal changes can be given to indicate how they occur. First, retail sales in the spring season follow a different pattern, depending on when spring holidays occur. (Easter, for example, occurs on the first Sunday after the first full moon of spring, and can vary in date from late March to late April, a period of about 5 weeks.) Thus, some forecasts are tied to so many weeks before Easter rather than just to the first week of April. Retail sales before the December holidays do not change pattern drastically from one year to the next, but Christmas retail sales have moved from almost entirely December (in the late 1940s) to November and December (in the 1960s), and even into October currently. This change is occurring slowly, and including old data will cause us to slightly misestimate the seasonal. There are other ways in which seasonals may change, but we will not discuss them here. The point is that seasonals do change, even drastically in some cases, and we should keep that possibility in mind.

The effect of a changing seasonal is important in forecasting. Forecasts frequently assume not only that the seasonal pattern has been stable, but that it will continue to be stable. A forecast is obtained by finding the trend and seasonal terms and forecasting that pattern into the future. This is done by forecasting the deseasonalized value using the trend equation, then multiplying that value by the seasonal index. For a linear trend equation, this means that a forecast for period t would be $(a + bt)$ times the appropriate seasonal factor for period t, S_t.

forecast for period t, using a linear trend and a seasonal index $= (a + bt)S_t$.

(17-6)

For examples, we can obtain the month-by-month forecast of units sold that Mr. Barstuwaski wants. Table 17-1 shows that the average price of a unit is $94 as of January 1, 1977. We can make the forecast in dollar terms, using the figures derived in Table 17-5 for the Table 17-1 data, and then convert them to unit sales forecasts using the $94 figure. To obtain the forecast, month by month, of total dollar sales, we use the linear trend equation found in the previous section ($0.5152 + 0.002084t$, where $t = 121$ for January 1977) and the seasonal factors given in the last column of Table 17-5. The calculations, using equation (17-6), are given in Table 17-6. The forecasts are made as of January 1977, after observing the December 1976 demand.

Table 17-6 Monthly Forecast of Total Dollar Sales and Unit Sales (both in millions) for 1977 for the Barnesville-Stumptown Water Ski Company

Month	t	$a + bt = 0.5152 + 0.002084t$	Seasonal Index: S_t	Forecast of Dollar Sales $(a + bt)S_t$	Forecast of Unit Sales = Dollar Sales/94
January	121	0.7674	1.293	0.9922	0.01056
February	122	0.7694	1.241	0.9548	0.01016
March	123	0.7715	1.095	0.8448	0.00899
April	124	0.7736	0.944	0.7303	0.00777
May	125	0.7757	0.883	0.6849	0.00729
June	126	0.7778	0.775	0.6028	0.00641
July	127	0.7799	0.788	0.6146	0.00654
August	128	0.7820	0.901	0.7046	0.00750
September	129	0.7840	0.899	0.7048	0.00750
October	130	0.7861	0.968	0.7609	0.00809
November	131	0.7882	1.056	0.8323	0.00885
December	132	0.7903	1.157	0.9144	0.00973

Table 17-6 gives Mr. Barstuwaski the forecasts he needs to plan production. As time progresses and more data values are available, the forecasts would be recomputed using the new data, and the production plan and operating budget could be revised.

S.S. A publishing house is interested in forecasting sales of its textbooks, by quarter (not by month), for the purpose of scheduling their reprinting (production process) and controlling their inventory. As an example, they have chosen an introductory psychology text. The data on sales, by quarter, are given for the last 3 years in Table 17-7.

Time-Series Analysis

Table 17-7 Sales (thousands) of an Introductory Psychology Text

Year	Quarter 1	Quarter 2	Quarter 3	Quarter 4	Total
1	4	14	10	7	35
2	4	11	8	6	29
3	6	15	8	6	35
Total	14	40	26	19	99

(a) Use the ratio-to-moving-average method to estimate the four seasonal indexes (Remember that 5 terms, not 13, will be involved in each centered moving average.)

(b) Using $8.25 + 0.0(t)$ as the trend line (that is, there is no trend), forecast the next four quarters' sales.

(c) In year 3, the book was revised. Does that pose any problems for the forecaster? Comment very briefly.

Solution:

(a) Seasonal index calculation:

Year	Quarter	Sales	Centered Moving Average	Ratio to Moving Average
1	1	4		
1	2	14		
1	3	10	$[\frac{1}{2}(4) + 14 + 10 + 7 + \frac{1}{2}(4)]/4 = 8.750$	$10/8.75 = 1.143$
1	4	7	$[\frac{1}{2}(14) + 10 + 7 + 4 + \frac{1}{2}(11)]/4 = 8.375$	$7/8.375 = 0.836$
2	1	4	$[\frac{1}{2}(10) + 7 + 4 + 11 + \frac{1}{2}(8)]/4 = 7.750$	$4/7.75 = 0.516$
2	2	11	$[\frac{1}{2}(7) + 4 + 11 + 8 + \frac{1}{2}(6)]/4 = 7.375$	$11/7.375 = 1.492$
2	3	8	$[\frac{1}{2}(4) + 11 + 8 + 6 + \frac{1}{2}(6)]/4 = 7.500$	$8/7.5 = 1.067$
2	4	6	$[\frac{1}{2}(11) + 8 + 6 + 6 + \frac{1}{2}(15)]/4 = 8.250$	$6/8.25 = 0.727$
3	1	6	$[\frac{1}{2}(8) + 6 + 6 + 15 + \frac{1}{2}(8)]/4 = 8.750$	$6/8.75 = 0.686$
3	2	15	$[\frac{1}{2}(6) + 6 + 15 + 8 + \frac{1}{2}(6)]/4 = 8.750$	$15/8.75 = 1.714$
3	3	8		
3	4	6		

Each seasonal index is the average of two of the ratio-to-moving-average values:

$$S_1 = \frac{0.516 + 0.686}{2} = 0.601, \quad S_2 = \frac{1.492 + 1.714}{2} = 1.603$$

$$S_3 = \frac{1.143 + 1.067}{2} = 1.105, \quad S_4 = \frac{0.836 + 0.727}{2} = 0.782.$$

These four values sum to 4.091, so they must all be multiplied by 4/4.091. The adjusted figures, and the answers, are:

$$S_1 = 0.588, \quad S_2 = 1.080, \quad S_3 = 1.567, \quad S_4 = 0.765.$$

(b) The forecasts are $(8.25 + 0t)S_t$, or 8.25 times the seasonal factor. The forecasts are:

quarter 1: $8.25(0.588) = 4.85$ quarter 2: $8.25(1.080) = 8.91$

quarter 3: $8.25(1.567) = 12.93$ quarter 4: $8.25(0.765) = 6.31$.

(c) A revision does cause a problem for the forecaster (and, thus, the manager). It introduces an extra factor that may confuse the effect of trend and/or seasonal effects. In particular, there may be a cyclic variation of the same length as the life of a revision (say 4 years). This cyclic effect might take the form of a jump to a higher sales level when a revision first appears, then a slow increase during the first 2 years, followed by a decline of increasing amount as the book moves into its third and fourth years. In such a case, the cyclic effect may be large and should be (and is) explicitly considered in making forecasts.

17-4 Exponential Smoothing Methods*

Sections 17-1, 17-2, and 17-3 assume that the same underlying pattern holds for the entire set of data and will continue to hold for the future. The implication of this assumption is that all historical data values are equally important. In many situations, however, the pattern is changing, either continuously or occasionally, and the recent data values are more relevant than the older ones.

Another observation concerning the methods of the previous sections is that a large amount of data must be available, perhaps saved in a computer-accessed file. Solving some forecasting problems using the time-series methods discussed so far would require an extensive, and therefore expensive, data storage capability. Consider, as an example, the problem of controlling inventories for a department store that stocks 10,000 different items. Analyzing 10,000 time series once a month (or even more frequently) using any complicated procedure may be prohibitively expensive.

Exponential smoothing methods provide quick methods of time-series analysis that require relatively little data storage. They are most useful when a firm faces many forecasting problems and cannot afford an expensive analysis. (A more detailed analysis may still be done for total sales, division

---

* This section may be omitted without loss of continuity.

Time-Series Analysis

sales, or product group sales.) In addition, exponential smoothing methods have the virtue of counting the more recent data values more heavily. This is an advantage if the underlying pattern is changing.

Exponential smoothing is used most frequently for short-term forecasts. As an example, Mr. Barstuwaski is interested in making short-term forecasts, for inventory control purposes, for each of his items. We will use the data in Table 17-2 to illustrate the new techniques.

In short-term forecasting we will be interested in what is happening now rather than in what occurred during the distant past. Since we are interested in short-term forecasting, trend effects are ignored in the first pass at these new methods. (Seasonal effects can be included separately, and this is discussed later.) Initially, an average of recent values is obtained, and this average is used as the short-term forecast.

One type of average that could be used is a moving average, where the last few (12 or 6 or 4 or some number) data values from the time series are averaged. This method is frequently used, and it performs well. However, it does not weigh the recent data values more heavily in the average. Further, it ignores older information that may have some value. Exponential smoothing methods avoid these problems, and it is easier computationally than the method of moving averages.

In exponential smoothing, a weighted average is computed by weighting the most recent time-series value by a given amount (say 0.05 or 0.10) and each previous value of the time series by a smaller amount. The weights must total 1. The basic formula is given by equation (17-7):

exponentially smoothed weighted average in period t,

$$ES_t = \alpha y_t + (1 - \alpha)ES_{t-1}, \qquad (17\text{-}7)$$

where

$y_t \equiv$ time-series value in period t,

$\alpha \equiv$ smoothing constant between 0 and 1 that is chosen by the forecaster (note that α, as used here, is not a significance level or the value of a regression parameter).

New Symbol	Meaning
ES_t	The exponentially smoothed, weighted average for period t.
α	The smoothing constant, $0 < \alpha < 1$.

For example, if $\alpha = 0.10$, $ES_{t-1} = 10$, and $y_t = 12$, then

$$ES_t = 0.1(12) + 0.9(10) = 10.2.$$

This value would be the forecast for all future periods, if (and only if) we can ignore trend and seasonal effects. This simplified form of exponential smoothing is frequently used, and if trend and seasonality are of minor importance, it performs well.

Several questions arise concerning the use of equation (17-7). First, how should the manager choose an α value? One method is to select a commonly used α value (0.05 or 0.10 for monthly data). A preferable approach is to examine past data and try different values of α to see which one would have worked best (resulted in the smallest forecast errors for the past data). In addition, α may be changed from time to time when conditions dictate. A high α value causes the smoothed average to react more quickly to a change, but it will also make larger errors in responding to random fluctuations. Hence, a low α is appropriate when the pattern of the time series is relatively stable. The α value may be increased when the manager has reason to believe something is changing or is going to change.

Second, equation (17-7) works well with a known value for ES_{t-1}, but where does the first value of ES come from? The first value is based on an average of some historical data, if there are any. If there are no such data, it is based on a guess. In any case, as we move forward in time, exponential smoothing "forgets" (gives less and less weight to) the original value, and the forgetting process is quicker for a higher α value.

Third, it is not obvious, looking at equation (17-7), that the data values are counted less and less as we move back into time. The most recent value, y_t, is weighted by α (say 0.10) in the average, and all other values, contained in ES_{t-1}, are weighted by a total of $1 - \alpha$ (say 0.90). But in ES_{t-1}, y_{t-1}, is weighted by α (say 0.10). Thus, believe it or not, y_{t-1} is weighted by α times $1 - \alpha$ (0.10 times 0.90) in calculating ES_t. In fact, the weights applied to the past data values are α, $\alpha(1 - \alpha)$, $\alpha(1 - \alpha)^2$, $\alpha(1 - \alpha)^3$, ... Using $0.1 = \alpha$, the weights applied are 0.1, 0.09, 0.081, 0.0729, ... to the values y_t, y_{t-1}, y_{t-2}, y_{t-3}, ... These numbers will sum to 1 if the weights are added back in time forever. In practice, the initial value or guess is given whatever weight is left to make the sum equal 1. This is not something to worry about; it just happens when ES_0 is assigned a value. [The mathematics of exponential smoothing is discussed at length in: R. G. Brown, *Smoothing, Forecasting and Prediction of Discrete Time Series* (Englewood Cliffs, N.J.: Prentice-Hall, Inc., 1963).]

To illustrate some of these points, consider the Table 17-2 data on sales for item 1 of the Barnesville-Stumptown Water Ski Company. The first 2 years of data are used to obtain the initial ES value. Thus, since the first 2 years' (first 24 months') totals are 397 and 443, respectively, the initial ES value is $(397 + 443)/24 = 35.00$. Table 17-8 shows the results of making exponential smoothing calculations using $\alpha = 0.1$ and $\alpha = 0.2$, assuming the 1976 data arrive one bit at a time. That is, these are the successive monthly forecasts for 1976.

Time-Series Analysis

Table 17-8 Exponential Smoothing of the 1976 Table 17-2 Data, where $ES_0 = 35.00$

Month	t	y_t	ES_t Using $\alpha = 0.1$	ES_t Using $\alpha = 0.2$
January	1	45	$0.1(45) + 0.9(35.00) = 36.00$	$0.2(45) + 0.8(35.00) = 37.00$
February	2	51	$0.1(51) + 0.9(36.00) = 37.50$	$0.2(51) + 0.8(37.00) = 39.80$
March	3	36	$0.1(36) + 0.9(37.50) = 37.35$	$0.2(36) + 0.8(39.80) = 39.04$
April	4	34	$0.1(34) + 0.9(37.35) = 37.02$	$0.2(34) + 0.8(39.04) = 38.03$
May	5	40	37.31	38.43
June	6	30	36.58	36.74
July	7	28	35.72	35.99
August	8	24	34.55	32.79
September	9	29	34.00	32.04
October	10	39	34.50	33.43
November	11	48	35.85	36.34
December	12	48	37.07	38.67

Each ES value is, in turn, used to compute the next one, and only the ES value needs to be saved for future calculations. The forecast for any future period, ignoring trend and seasonal effects, is the most recent ES value. The average is continuously updated because the average value may shift over time.

To include a seasonal effect, seasonal factors can be computed as in Section 17-3. Then each y_t value is divided by the appropriate seasonal factor, to deseasonalize the data, before applying (17-7). The seasonally adjusted forecast is made by multiplying ES_t by the seasonal factor for the period being forecast. (Other methods of dealing with seasonality are discussed in the book by Brown, referenced previously in this section.)

Trend effects can be estimated using a separate exponential smoothing formula to estimate the trend factor. (Only linear trend effects are considered here.) In addition, the basic formula, (17-7), must be modified to reflect the trend factor. These two formulas are given in (17-8) and (17-9); (17-10) gives the formula for computing a forecast for any future period when both linear trend and seasonal effects are considered.

When there is a trend and seasonal effect, the smoothed average is

$$ES_t = \alpha(y_t/S_t) + (1 - \alpha)(ES_{t-1} + b_{t-1}). \tag{17-8}$$

The estimate of the linear trend effect in period t is given by

$$b_t = \alpha(ES_t - ES_{t-1}) + (1 - \alpha)b_{t-1}. \tag{17-9}$$

The forecast, made in period t, for period $t + L$

$$= (ES_t + Lb_t)S_{t+L} \tag{17-10}$$

S_t and S_{t+L} are the seasonal indexes, and b_t is the average increase in the time series per period.

New Symbol	Meaning
b_t	The estimate of the linear-trend term. The expected amount of increase, per period, in the time series.

Including trend and seasonal effects adds complexity. Table 17-9 shows the calculations for the last year of the Table 17-2 data, using $\alpha = 0.2$. Using the first 2 years of data and the simplest computational method, the seasonal factors are: 1.30, 1.30, 1.01, 0.94, 0.87, 0.85, 0.73, 0.81, 0.79, 0.97, 1.09, and 1.34, respectively. Values for ES_0 and b_0 can be found using equations (17-1) and (17-2), where $b = b_0$ and $ES_0 = a + b(24)$, since 24 is the number of the last month used. Equations (17-1) and (17-2) give $b_0 = 0.08$ and $ES_0 = 36.85$.

Table 17-9 represents a lot of arithmetic. But it is a whole year's worth of arithmetic. Each month we would have to do only one line of Table 17-9.

We are now prepared to make forecasts for 1977 for item 1. To make each of the next 12 months' forecasts we will use $ES_{12} = 38.37$, $b_{12} = 0.17$, and the appropriate seasonal factor, since that is the most recent information available. The results, using equation (17-10), are summarized in Table 17-10.

Using Table 17-10, an inventory control scheme can be devised for item 1.

Table 17-9 Exponential Smoothing, Including Trend and Seasonal Effects, of the 1976 Table 17-2 Data

Month	t	y_t	S_t	y_t/S_t	ES_t	b_t
January	1	45	1.30	34.62	0.2(45/1.3) + 0.8(36.85 + 0.08) = 36.47	0.2(36.47 − 36.85) + 0.8(0.08) = −0.01
February	2	51	1.30	39.23	0.2(51/1.3) + 0.8(36.47 − 0.01) = 37.01	0.2(37.01 − 36.47) + 0.8(−0.01) = 0.10
March	3	36	1.01	35.64	0.2(36/1.01) + 0.8(37.01 + 0.10) = 36.82	0.2(36.82 − 37.01) + 0.8(0.10) = 0.03
April	4	34	0.94	36.17	0.2(34/0.94) + 0.8(36.82 + 0.03) = 36.75	0.2(36.75 − 36.82) + 0.8(0.03) = 0.01
May	5	40	0.88	45.45	38.50	0.36
June	6	30	0.85	35.29	38.15	0.22
July	7	28	0.73	38.36	38.37	0.22
August	8	24	0.81	29.63	36.80	−0.14
September	9	29	0.79	36.71	36.67	−0.14
October	10	39	0.97	40.21	37.27	0.01
November	11	48	1.08	44.44	38.71	0.30
December	12	48	1.34	35.82	38.37	0.17

Time-Series Analysis

Table 17-10   Forecasts of 1977 Sales, by Month, for Item 1 of the Barnesville-Stumptown Water Ski Company

Month	Forecast
January	[38.37 + 0.17](1.30) = 50.10
February	[38.37 + 2(0.17)](1.30) = 50.32
March	[38.37 + 3(0.17)](1.01) = 39.27
April	[38.37 + 4(0.17)](0.94) = 36.71
May	[38.37 + 5(0.17)](0.88) = 34.51
June	[38.37 + 6(0.17)](0.85) = 33.48
July	[38.37 + 7(0.17)](0.73) = 28.88
August	[38.37 + 8(0.17)](0.81) = 32.18
September	[38.37 + 9(0.17)](0.79) = 31.52
October	[38.37 + 10(0.17)](0.97) = 38.87
November	[38.37 + 11(0.17)](1.08) = 43.46
December	[38.37 + 12(0.17)](1.34) = 54.15

Mr. Barstuwaski will also make several updates (probably monthly) of the forecasts in Table 17-10 before finalizing inventory quantities, on a detailed, item-by-item basis, for the later months in 1977.

S.S.   In Table 17-7, data are given on book sales for an introductory psychology text. Based on the first 2 years of data the firm estimates $S_1 = 0.50$, $S_2 = 1.56$, $S_3 = 1.12$, $S_4 = 0.82$, $ES_0 = 8.0$, and $b_0 = 0.5$. Perform the exponential smoothing calculations that would have been made when receiving the third-year data, one quarter at a time, using $\alpha = 0.10$. Then use the last ES and b values to make forecasts for the next four quarters.

Solution:

ES_t	b_t
0.1(6/0.5) + 0.9(8.0 + 0.5) = 8.85	0.1(8.85 − 8.0) + 0.9(0.5) = 0.535
0.1(15/1.56) + 0.9(8.85 + 0.535) = 9.41	0.1(9.41 − 8.85) + 0.9(0.535) = 0.538
0.1(8/1.12) + 0.9(9.41 + 0.538) = 9.67	0.1(9.67 − 9.41) + 0.9(0.538) = 0.510
0.1(6/0.82) + 0.9(9.67 + 0.510) = 9.89	0.1(9.89 − 9.67) + 0.9(0.510) = 0.481

Quarter	Forecasts
1	(9.89 + 0.481)(0.5) = 5.19
2	[9.89 + 2(0.481)](1.56) = 16.93
3	[9.89 + 3(0.481)](1.12) = 12.69
4	[9.89 + 4(0.481)](0.82) = 9.69

17-5 A Warning

Time-series analysis is the process of analyzing historical data for a variable of interest (sales, for example) to find patterns of behavior. In this chapter we have discussed methods for isolating the trend and seasonal components of a time series. (Cyclical and irregular variation tend to make that task harder.) Once the historical patterns are estimated, the patterns can be used to make forecasts of the future time-series values.

There are many dangers involved in this process that can lead to poor forecasts. In time-series analysis, we should plot the data to see what kind of trend, seasonal or cyclical patterns there are in the data. These patterns must be estimated, and we must assume the patterns were stable in the past and will persist into the future. In fact, the pattern may have changed several times in the past, but we may not be able to discover this change since irregular variation may hide the true pattern. Worse, the pattern may be about to change, so that forecasts based on the analysis of the historical data will be in error. Another possibility is that the patterns may be stable, but we may have misspecified one or more of them. If any of these situations develop, the forecasts are likely to be in error. We should keep information on forecast performance so that the size of potential errors can be estimated. We cannot eliminate the errors, so we must learn to live with them.

Time-series analysis tries to explain a series using only the series itself. There are many factors that impact on sales, for example, such as advertising and average disposable income, but to some extent these are captured in trends, seasonals, and cycles that the time series has in common with the external factors. However, some external factors occur only occasionally. In investigating a time series, we should try to remove effects that are not systematic. In forecasting, we should make additions or subtractions for external factors that will have an impact. For example, if a new, enlarged advertising program is beginning, we should separately predict its impact and add an appropriate amount to the time-series forecast.

In managing a forecasting system that includes time-series analysis, a manager must be aware of exactly what time-series analysis will do and what it will not do. The manager must know what assumptions are made, and decide whether they are reasonable in a given situation or not. Time-series analysis can be best used when there are many forecasts to be made and when a mistake in one would not be too costly. Time-series analysis can isolate seasonal effects for use in intermediate-term planning, and the manager can consider whether the seasonal has any reason to be changing. In long-term planning, time-series analysis may be only one input, that management uses, together with other forecasts and judgments, in forming the long-term plan. In short, a forecasting system must be intelligently managed.

To manage a forecasting system, a manager should be aware of forecasting techniques other than the ones discussed here. There are behavioral techniques

Time-Series Analysis

for dealing with areas where few data exist. (An example is the "Delphi method.") There are more complicated quantitative techniques. There are survey methods. Finally, there are phone calls, salesmen's estimates, guesses, and hunches. All of these are important in certain situations. Unfortunately, we cannot discuss them all here. An article by J. Chambers *et al.* (How to Choose the Right Forecasting Technique, *Harvard Business Review*, July–August 1971) provides a good overview of many forecasting techniques.

New Symbols

Symbol	Meaning
y, y_i, y_t	y is the time-series value in the general time-series model; y_i or y_t represent the ith or period t value of a particular time series.
T	The trend term in the series.
S	The seasonal term in the series.
C	The cyclical term in the series.
I	The irregular term in the series.
$S_{\text{Jan}}, S_{\text{Feb}}, \ldots$	The seasonal index for January, February, and so on.
α	The smoothing constant, $0 < \alpha < 1$.

Key Formulas

Formula	Used to Compute:
$b = \sum_{i=1}^{n} x_i y_i \bigg/ \sum_{i=1}^{n} x_i^2$	The b term if $\bar{x} = 0$.
$a = \bar{y}$	The a term if $\bar{x} = 0$.
$\dfrac{y_i}{(\frac{1}{2}y_{i-6} + y_{i-5} + \cdots + y_{i+5} + \frac{1}{2}y_{i+6})/12}$	The ratio-to-moving-average value calculated as y_i over the centered moving average, for monthly data.
$(a + bt)S_t$	The forecast for period t when a linear-trend analysis and seasonal component are included in the time-series analysis.
$ES_t = \alpha y_t + (1 - \alpha)ES_{t-1}$	The exponentially smoothed, weighted average when there is no trend or seasonal term.
$ES_t = \alpha(y_t/S_t) + (1 - \alpha)(ES_{t-1} + b_{t-1})$	The exponentially smoothed, weighted average when there is a trend and seasonal term.
$b_t = \alpha(ES_t - ES_{t-1}) + (1 - \alpha)b_{t-1}$	The estimate of the trend term in an exponential smoothing analysis.
$(ES_t + Lb_t)S_{t+L}$	The forecast made in period t for period $t + L$ in an exponential smoothing analysis with a trend and seasonal term.

PROBLEMS

17-1.* Is a sequential list of the last 25 years' GNP figures a time series? Would it be wise to forecast the next value using only the time-series data?

17-2.* A production manager is interested in the time it takes to assemble a particular item, and asks one of the workers who is going to assemble the entire item to keep track of the time required for the first unit produced on each of the first five days of production. The times, in hours, are: 2.4, 1.6, 1.3, 1.2, and 1.2. Is this a time series? What would you forecast for the next value? Describe how you arrived at that forecast. Is the trend effect linear?

17-3. What are the four components of the general time-series model? Which of the four can be estimated using methods in this chapter?

17-4.* Any cycle that repeats within the year (when a year is the period of the seasonal) can be incorporated into the seasonal. Suppose we have a 3-month cycle that is: 0.8, 1.3, 0.9, and so on. Then the yearly seasonal would be: 0.8, 1.3, 0.9, 0.8, 1.3, 0.9, 0.8, 1.3, 0.9, 0.8, 1.3, and 0.9. Form a seasonal that is the result of multiplying the above 3-month cycle by the 4-month cyclical pattern: 0.6, 0.7, 1.1, and 1.6.

17-5. Form 10 values of a time series using the general additive model: $Y = T + S + C + I$. The time series represents daily sales of a government bond. The trend values begin with 10 and increase one unit each day. Beginning with the first value, the seasonal (weekly) pattern is: $-3, -3, 0, +1, +5$. There is no cyclical effect, so $C = 0$ for an additive model, and the irregular variation is: $-1, -1, +3, +1, +4, -2, 0, -2, -3, +1$.

17-6. Give an example of a time series that might exhibit linear trend (Figure 17-4a); exponentially increasing trend (Figure 17-4c). In your example of exponentially increasing trend, for how long would you expect that pattern to continue?

17-7.* A sporting goods distributor is interested in analyzing total tennis racquet sales. Total sales of tennis racquets, in thousands of racquets, for each of the last 10 years, is 24, 25, 27, 31, 39, 51, 59, 63, 65, 66.
(a) Find the linear trend line through the time series. Use equations (17-1) and (17-2) and number the years from 1 to 10.
(b) Discuss the validity of a linear trend. (A plot of the data will help to illustrate the trend.)

Time-Series Analysis

17-8.* A firm producing equipment used in mattress manufacturing has steadily lost sales on one product line. (Some of their other products have had increased sales.) They believe that the steady decline will continue over the next 5 years. For planning purposes, they would like a forecast of total sales over that period. The last 5 years of sales figures for this product line, in thousands of units, are: 4.61, 4.29, 4.31, 4.21, 4.01. Use equations (17-3) and (17-4) to obtain a linear-trend equation, and use that equation to make a forecast of total sales over the next 5 years.

17-9. Annual sales for a large clothing manufacturer are as follows:

Year	Sales (millions)	Year	Sales (millions)
1960	$4.30	1966	$0.94
1961	3.25	1967	0.82
1962	2.03	1968	0.78
1963	1.62	1969	0.84
1964	1.34	1970	0.80
1965	1.17	1971	0.76

Estimate a linear trend and forecast sales for the years 1972, 1973, 1974, and 1975.

17-10. A department store chain has experienced the following annual sales:

Year	Sales (millions)	Year	Sales (millions)
1965	$12.1	1971	$35.5
1966	13.5	1972	76.4
1967	18.6	1973	82.1
1968	22.7	1974	105.6
1969	37.1	1975	131.3
1970	38.4		

Find the linear trend and forecast sales for 1976.

17-11. What is the main reason for estimating a seasonal pattern? Give at least two examples of managerial problems in which the seasonal pattern might be important.

17-12. The sales manager for a shoe manufacturer has estimated a trend line based upon some historical data concerning monthly sales. The estimated line is $y = 20 + 1.5x + 0.12x^2$, where January 1974 and February 1974 correspond to x values of zero and 1, respectively.

The manager has also estimated seasonal indexes:

Jan.	Feb.	Mar.	Apr.	May	June	July	Aug.	Sept.	Oct.	Nov.	Dec.
0.72	0.85	0.81	1.07	1.14	1.20	1.15	1.12	0.86	0.74	1.14	1.20

Using these data, forecast sales for each month in 1975.

17-13. The sales manager for a textiles firm has estimated a trend equation for monthly sales as follows:

$$y = 52 + 1.46x + 0.91x^2.$$

An increment of 1 unit for x indicates a movement of 1 month with $x = 0$ at January 1974. An increment of 1 unit for y indicates an increase in sales of $1,000. The manager has developed the following seasonal indexes for each month:

Jan.	Feb.	Mar.	Apr.	May	June	July	Aug.	Sept.	Oct.	Nov.	Dec.
1.25	0.91	0.83	0.85	0.86	0.92	0.97	1.26	1.14	0.92	0.86	1.23

The president of the company recently congratulated the sales manager for a significant increase in first quarter 1975 sales over the same period in 1974, based on the following data:

First quarter sales, 1974 (thousands): $250.67

First quarter sales, 1975 (thousands): $401.32.

Based upon the time-series information, were the president's congratulations necessarily justified?

17-14.* A plant manager in the General Assembly Company (GAC) believes that the rate of output on the main assembly line is steadily increasing through time. However, there are daily (seasonal) variations in the week. (Monday and Friday tend to give slightly lower output.) Given the output figures below for the last four weeks, use the ratio-to-moving-average method to estimate the seasonal pattern within 1 week. (Notice that since there are 5 days in each "season," the centered moving average is an average of five values; no halves are used.)

Output figures: 45, 51, 54, 50, 41, 42, 55, 56, 57, 45, 47, 58, 54, 52, 48, 48, 55, 60, 59, 50.

17-15. Why is the ratio-to-moving-average method of determining seasonals preferred to simply computing the average of the period one (for example) demands and dividing it by the average of all periods?

17-16.* A company that manufactures and sells air conditioners has the following sales data (hundreds of units) for one of their lines:

Time-Series Analysis

	Quarter			
Year	1	2	3	4
1	29	19	8	15
2	34	16	7	19
3	31	15	8	20

The firm knows it has a seasonal, and that the first quarter is their largest sales quarter. (Retailers buy early to prepare for their own peak sales.) They want you to use the ratio-to-moving-average method to estimate the seasonal factor for each of the quarters.

17-17.* In Problem 17-14, the weekly output of an assembly line was examined. The data for 4 weeks' output is given there. The seasonal (daily) effect was estimated and found to be

$S_1 = 0.889$, $S_2 = 1.080$, $S_3 = 1.088$, $S_4 = 1.047$, $S_5 = 0.875$.

(These figures have been adjusted to average to 1.) The manager would like you to deseasonalize the data and use equations (17-1) and (17-2) to estimate the trend line for the deseasonalized data.

17-18.* In forecasting monthly sales, a sales manager examines 3 years of data. The results of the analysis are as follows:

The trend line is: $142.0 + 1.4t$

The seasonals are: 0.9, 1.1, 1.4, 1.5, 1.3, 1.0, 0.9, 0.8, 0.8, 0.7, 0.8, 0.8.

The next period is number 37, the beginning month of the seasonal pattern. Compute a forecast for periods 37, 39, 41, and 45.

17-19. Monthly data concerning orders received (thousands of dollars) at an electronic component manufacturing plant are as follows:

Month	1972	1973	1974
January	$41.6	$42.4	$43.0
February	43.4	46.1	42.6
March	47.8	47.3	49.5
April	50.6	51.2	47.6
May	54.3	53.9	50.8
June	58.2	55.7	56.2
July	65.7	61.9	61.7
August	62.4	62.7	63.4
September	61.5	60.8	60.9
October	59.3	58.8	60.6
November	52.7	57.1	60.2
December	48.4	49.0	51.7

(a) Find seasonal indexes using the ratio-to-moving-average method.
(b) The actual sales for January and July of 1975 were 44.3 and 64.7, respectively. How does the increase in sales between January and July compare with what could be expected on a purely seasonal basis? What assumption is necessary in answering this question?

17-20. The treasurer of a large company is trying to determine how much money to borrow for the coming year to meet the company's cash flow needs. The treasurer would like to have monthly forecasts of cash needs so that interest expenses could be held as low as possible. The treasurer has the following historical data available to aid in forecasting:

Cash Requirements (thousands)

Month	1972	1973	1974	1975
January	$14.6	$15.2	$15.7	$15.8
February	15.1	15.7	15.9	16.2
March	29.6	27.9	21.3	26.1
April	12.3	14.6	12.6	14.9
May	14.5	16.2	13.1	15.2
June	15.7	17.7	15.3	15.1
July	16.9	16.9	16.2	16.7
August	18.3	19.3	18.9	18.2
September	27.6	29.1	25.8	27.8
October	25.4	28.9	26.1	24.6
November	16.1	21.0	14.2	14.6
December	15.8	12.7	10.0	12.1

(a) Find an estimate of a linear-trend line.
(b) Use the ratio-to-moving-average method to estimate seasonal indexes.
(c) Forecast cash requirements for the next 12 months.

17-21. The following data represent quarterly sales for a sporting goods company:

Sales (millions)

	1971	1972	1973	1974	1975
Spring	$48	$49	$51	$43	$54
Summer	44	42	43	38	48
Winter	52	56	58	51	62
Fall	59	61	67	59	67

(a) Is there evidence of a seasonal pattern in the series?
(b) Is there evidence of movements other than seasonal ones?
(c) Estimate a linear-trend equation.
(d) Estimate seasonal indexes using the ratio-to-moving-average method.
(e) Forecast sales for each quarter of 1976.

17-22. (a) What are the main advantages of exponential smoothing?
(b) What advantage does a regression approach have over exponential smoothing?

17-23. (a) Describe how you can obtain initial estimates, ES_0 and b_0, for use in exponential smoothing.
(b) Describe how you can deal with data that has a strong seasonal, when using exponential smoothing.

17-24. In making forecasts for inventory control, an inventory manager has some items that have seasonality and trend and some items that are stable.
(a) For one of the stable items, $ES_0 = 14$. The manager believes there is no trend or seasonal for this item. What is the forecast for five periods hence?
(b) If the next demand is 17, and $\alpha = 0.2$ is used, what is the new ES-value, ES_1?

17-25.* An inventory control manager has an item that has both trend and seasonal effects. The monthly seasonals are:

0.7, 0.8, 1.2, 1.4, 1.0, 0.8, 0.7, 0.6, 0.9, 1.1, 1.4, 1.5.

The current period is March (period 3 in the above pattern), and $ES_0 = 21.2$ and $b_0 = 0.8$ after including March's demand.
(a) What is the forecast for April and for May?
(b) If April's demand is 27.0, what are the new ES and b values, and what is the revised forecast for May? Use $\alpha = 0.1$.

Chapter 18

Nonparametric Methods

In previous chapters we have investigated a number of statistical tests. In each such test we were compelled to make certain assumptions as preconditions for making any statistical inference. These assumptions were of two kinds: (1) assumptions concerning the form of the appropriate probability distribution and (2) assumptions concerning the parameters of the appropriate distribution.

For example, the use of the t distribution in tests of significance and the use of the F distribution in analysis of variance (Chapter 14) both require the analyst to assume that the underlying population is normally distributed. Although small departures from normality do not do major violence to the t test (no population is precisely normal), major deviations invalidate the test. It would, then, be useful to have testing methods that do not require assumptions concerning the underlying distribution. Such methods exist and some of the more important ones are discussed in this chapter. They are called *distribution-free methods* because they avoid the limitations introduced when the nature of the underlying probability distribution must be specified.

An alternative to distribution-free methods is to test the data to see if they meet the requirements of a specific probability law. Such tests are called *goodness-of-fit* tests. Several such tests are examined in this chapter.

Some statistical tests require knowledge of the parameters of a probability distribution. For example, statistical tests using the normal distribution are not precise if one must use the sample variance as an estimate of the population variance. This chapter describes testing methods that avoid such issues. These methods are called *nonparametric* tests. In common usage, the term "nonparametric" has come to be used to cover distribution-free as well as nonparametric methods, hence the title of this chapter.

The strength of measurement possible in a given situation is also relevant to a decision of whether or not to use nonparametric methods. In Chapter 1 we discussed four measurement scales: nominal, ordinal, interval, and ratio. The reader may find it useful to review that discussion at this point.

The tests we have described prior to this chapter (with one major exception) require measurement at least of interval strength. One of the advantages of nonparametric tests is that they can be used with nominal (class) and ordinal (rank) data. Nonparametric tests can also be used with interval and ratio data when the decision maker wishes to avoid one or more of the assumptions implied by the relevant parametric method. However, such use is not without cost. The more that may be properly assumed, the more powerful the statistical test that may be used. We saw this most forcefully when we considered one- and two-tailed tests of significance. One-tailed tests assume more and they are more powerful. Parametric tests assume more and, hence, are more powerful than nonparametric methods. On the other hand, the nonparametric methods typically possess considerable strength, often requiring only a 5 to 10 percent increase in the sample size to achieve the same power. And, as we have seen, they do not require certain assumptions that may prove untenable in a given problem situation. Finally, nonparametric tests often involve tests on the median, a more intuitive concept than the mean.

Let's summarize some of the advantages and disadvantages of nonparametric methods. Advantages of nonparametric methods:

1. They involve few restrictive assumptions.

2. They can be used with nominal and ordinal measurement (as well as with interval and ratio measures).

3. The tests often focus on medians which are more easily understood than are means on which parametric tests often center.

Disadvantages of nonparametric methods:

1. The techniques are slightly less powerful than parametric methods.

2. They may not use all of the data in cases involving interval and ratio data. The data will be treated as though it were taken from at most an ordinal scale.

3. Many nonparametric tests make experimental design and test selection complex.

4. Tables for many of the tests are not readily available and approximations are often inadequate for very small samples ($n < 10$ as a rule of thumb).

5. Measures of the strength of a relationship (in contrast to significance) are sometimes difficult to establish and exist only for two variables.

The remainder of this chapter is divided into sections which categorize and describe the available tests. These sections are:

18-1. Estimation Methods
18-2. Chi-Square Tests for Independence
18-3. Goodness-of-Fit Tests
18-4. The Runs Test for Randomness
18-5. The Signs Test of Significance
18-6. Rank Tests of Significance
18-7. Nonparametric Measures of Association

18-1 Estimation Methods

Often the issue in a statistical analysis is to estimate a population parameter rather than to make some statistical test. We have seen that the median, mode, and mean all measure the centrality of a distribution. Sometimes, with open-ended data classifications, for example, it may be impossible to compute the mean precisely.

One possibility would be to estimate the mean using the median or by averaging the first and third quartiles if the distribution is approximately symmetric:

$$\text{estimate of mean} = \frac{f_{0.25} + f_{0.75}}{2}. \qquad (18\text{-}1)$$

Shortcut methods are also available for estimating the sample standard deviation. One such method assumes the range of the data approximates three standard deviations on either side of the mean (99.7 percent of the data for a normal distribution). Thus, an estimate of the sample standard deviation is given by range/6. This estimate ignores the fact that the smaller the sample, the smaller the expected range. For small samples, a factor denoted by d_2 has been developed to reflect the sample size as well as the sample range in cases where the population may again be treated as roughly normal. Values of d_2 can be used to obtain improved estimates of the standard deviation.

$$\text{Estimated sample standard deviation} = \frac{\text{range}}{d_2}. \qquad (18\text{-}2)$$

New Symbol	Meaning
d_2	A factor used in estimating the sample standard deviation.

Nonparametric Methods

Suppose, for example, that a sample of size 10 yields a range of 18. The values of d_2 for various sample sizes are given in Table 18-1.

Table 18-1 Values of d_2 Used in Estimating the Sample Standard Deviation

n	d	n	d	n	d	n	d	n	d	n	d
2	1.13	4	2.06	6	2.53	8	8.25	10	3.08	12	3.26
3	1.69	5	2.33	7	2.70	9	2.97	11	3.18		

Hence, the sample standard deviation estimate would be $18/3.08 = 5.84$.

The two estimation procedures for the standard deviation we have just described both depend on normality and hence are really parametric estimation methods. We include them here because they are simple and easy to use.

A decision maker may have a set of data for which the mean and variance can be calculated but where it is not reasonable to assume normality. Yet the decision maker may wish to estimate the proportion of the population falling beyond some value. For example, suppose that the mean cost to perform a given task is $15.36 and the sample standard deviation is $1.20. Suppose, further, that the decision maker wishes to know the likelihood of the cost value (proportion of cost values) falling above $17.76, but the form of the underlying distribution is unknown.

The decision maker can use Tchebycheff's inequality to answer such questions provided it can be assumed that the underlying distribution has a finite mean and variance. The inequality states that the probability of a value being within k standard deviations of the mean is at least $1 - (1/k^2)$. Mathematically, using x for the specific value of the random variable,

$$P(|x - \mu| \leq k\sigma) \geq 1 - \frac{1}{k^2}. \tag{18-3}$$

For our example, we wishes to know the probability of being more than $(17.76 - 15.36)/1.20 = 2$ standard deviations above the mean. Formula (18-3) tells us that the probability of being within two standard deviations of the mean is at least $1 - (1/k^2) = 1 - (1/2^2) = 0.75$. This means that the probability of being beyond, below or above, two standard deviations from the mean is at most $1 - 0.75 = 0.25$. This is the most we can say unless we wish to assume something about the distribution. If we were willing to assume a symmetrical distribution, we could then estimate the upper bound on the probability of more than 17.76 to be $0.125 = \frac{1}{2}(0.25)$. Note that the 0.25 probability may be less, but it cannot be greater regardless of the actual probability distribution involved.

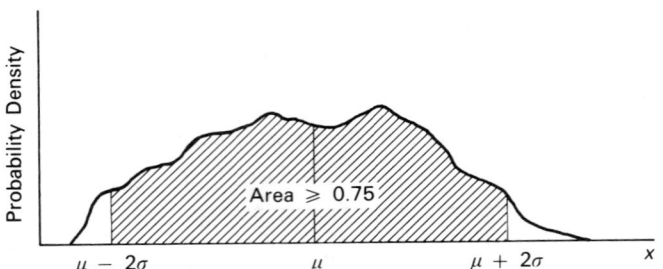

Figure 18-1 Area Referred to in Tchebycheff's Inequality for $k = 2$

Figure 18-1 illustrates Tchebycheff's inequality for our problem. The shaded area represents the probability of being within 2 standard deviations, which is at least 0.75 for any probability distribution with a finite mean and variance.

The statements that the inequality allows us to make are consistent with the weaker statements permitted by nonparametric methods. If we had been able to assume normality, then the probability of being within 2 standard deviations is 0.9544 instead of the 0.75 value given by the inequality.

S.S. 1. Estimate the variance for a sample of 8 with a range of 15.3. State any necessary assumptions.

2. What proportion of the data in a population are within one standard deviation of the mean according to Tchebycheff's inequality?

Solution:
1. If the population distribution can be assumed to be close to normal, then using the value of d_2, from Table 18-1, for a sample of size 8, we have

$$s^2 = \left(\frac{15.3}{2.85}\right)^2 = 28.82.$$

The variance is the square of the standard deviation.

2. Tchebycheff's inequality states that the proportion of observations within $k\sigma$ of the mean is at least $1 - (1/k^2)$. For our case, $k = 1$ and the inequality states that "at least zero percent of the data lies within one standard deviation of the mean." We already know this. Tchebycheff's inequality provides no information for $k \leq 1$.

18-2 Chi-Square Tests for Independence

The chi-square test deals with count data: for example, how many components failed, how many consumers prefer a given brand, how many

Nonparametric Methods

electors favor a given candidate. Count data are nominal and the chi-square methods provide a powerful test for determining the significance of count data.

Perhaps the most simple chi-square test is of the difference in two proportions. The null hypothesis is that they are equal: they represent two separate and independent estimates of a single proportion. Since some difference in the two separate estimates is inevitable, the question is whether this difference is more than could be explained by chance (random fluctuation).

Suppose, for example, that a firm is interested in whether male and female customers differ in their preference for its brand of coffee over the major competitive brand where the two brands sell equally well. The answer will be important to their promotional strategy. The data are given in Table 18-2 for a sample of 50 men and 50 women. The data do suggest a difference in preference but is it more than a chance occurrence? This is what we intend to test.

Table 18-2 Coffee Brand Preferences

	Brand	
Sex	Company	Competitor
Male	30	20
Female	22	28

To begin, what would we expect? If there were no difference in preference and given that the two brands sell equally well, we would expect half, 25 in this case, of each sex to prefer the company brand and half, again 25, of each sex to prefer the competitor's brand. The chi-square (written, as in Chapter 11, using the Greek letter χ as χ^2) test compares the observed and expected frequencies in the following formula:

$$\chi^2 = \sum_{i=1}^{m} \frac{(O_i - E_i)^2}{E_i}, \qquad (18\text{-}4)$$

where

$m \equiv$ number of cells, $i = 1, 2, \ldots, m$

$O_i \equiv$ observed frequency in cell i

$E_i \equiv$ expected frequency in cell i.

For our example: $m = 4$, $O_i = 30, 20, 22,$ and 28, $E_i = 25$ in each cell. Thus, for our problem, formula (18-4) yields

$$\chi^2 = \frac{(30-25)^2}{25} + \frac{(20-25)^2}{25} + \frac{(22-25)^2}{25} + \frac{(28-25)^2}{25} = \frac{68}{25} = 2.72.$$

Now if there were no difference between what we expected and what was observed, the computed value of chi-square would be zero. The further the observed values are from the expected values, the larger the value of chi-square. A graph of a chi-square distribution is given in Figure 18-2. The chi-square distribution, like the t distribution, is different for each different number of degrees of freedom.

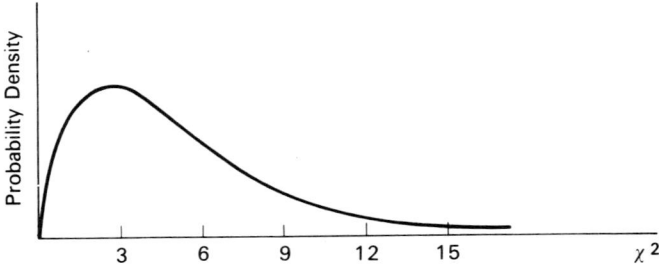

Figure 18-2 Chi-square Distribution for d.f. = 5

Before we can determine if our value of χ^2 equal to 2.72 is significant, we must establish the degrees of freedom present. One way is to ask how many numbers could be placed into the table before the rest of the numbers would be determined by the observed row and column totals. In other words, how much freedom do we have to specify the expected frequencies? For our example, Table 18-2 with only the observed totals listed would be as given below. There are four cells, as we already know. Only one of these can be filled in before it, together with the row and column totals, determines each of the other cell entries. Hence, the degrees of freedom for our brand-preference problem is one; d.f. = 1.

	Brand		
Sex	Company	Competitor	Total
Male			50
Female			50
Total	52	48	

NONPARAMETRIC METHODS

A mechanical rule that will yield the correct answer is to multiply the number of columns of data less one by the number of rows of data less one. Letting c be the number of columns and r be the number of rows, the rule states:

$$\chi^2_{\text{degrees of freedom}} = (c-1)(r-1). \tag{18-5}$$

For our problem $c = 2$ and $r = 2$, the number of subcategories of each variable, and d.f. $= (2-1)(2-1) = 1$. Hence, we write $\chi^2_{\text{d.f.}} = \chi^2_1 = 2.72$. Using the chi-square table in the appendix, we find $\chi^2_1 = 2.71$ for a significant level of 0.10 and 3.84 for a significance level of 0.05. Our value is significant at the 10 percent level but not at the 5 percent level. This is illustrated in Figure 18-3. Note the difference in the shape of the chi-square distribution for one degree of freedom.

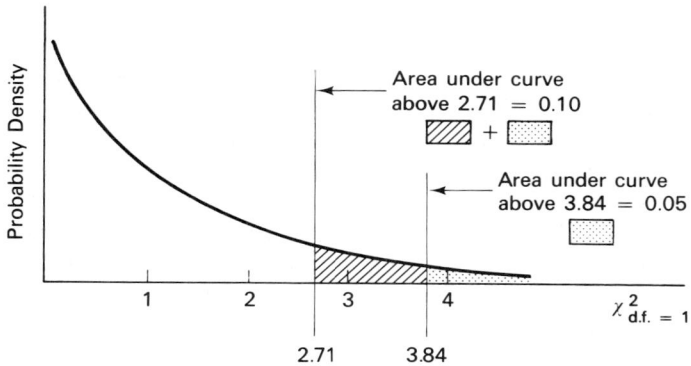

Figure 18-3 Chi-Square Test for Brand-Preference Problem d.f. = 1

If, as is appropriate in significance tests, the significance level (α) had been set prior to the test, a decision is indicated. Hence, if α had been set at 0.05, the data are evaluated as not significant in terms of a preference by sex between the two brands. It is possible, however, that if the decision is sufficiently important, a firm may elect to withhold judgment in this case and sample further before concluding that no preference of importance is present.

The test just made is equivalent to testing the difference in two proportions. The natural extension of such a test is to test the difference in several proportions. In other words, we can use the same approach to a problem in which each variable is divided into several subdivisions.

As an example, suppose four different treatments have been suggested for a particular medical condition. Patients are classified after a period of treatment as responding positively, not responding, or responding negatively.

The investigator wishes to know if the treatments have statistically different effects. The null hypothesis is that there is no difference in treatment effects. Suppose the data collected are as given in Table 18-3. Such tables are often called *contingency tables*.

Table 18-3 Response by Treatment

Response	T_1	T_2	T_3	T_4	Total
Positive	8	10	7	5	30
Neutral	9	6	13	12	40
Negative	3	14	10	3	30
Total	20	30	30	20	100

The first task is to establish the expected frequencies. These frequencies follow from the null hypothesis of independence. Recall from probability theory (see Chapters 5 and 6) that if two variables are independent, then the joint probability of a specific outcome must equal the product of the separate probabilities leading to that outcome: the product of the marginal probabilities. Hence, the probability of a positive response under treatment 1 is the probability a subject was given treatment 1 multiplied by the probability of a positive response. To obtain these joint probabilities, we first obtain the marginal probabilities. This is done in Table 18-4.

Table 18-4 Marginal Probabilities: Treatment Problem

Response	T_1	T_2	T_3	T_4	Total
Positive					$\frac{30}{100} = 0.3$
Neutral					$\frac{40}{100} = 0.4$
Negative					$\frac{30}{100} = 0.3$
Total	$\frac{20}{100} = 0.2$	$\frac{30}{100} = 0.3$	$\frac{30}{100} = 0.3$	$\frac{20}{100} = 0.2$	100

Under the null hypothesis of independence, the probabilities within the table are found by multiplying the marginal probabilities just established in Table 18-4. This is illustrated in Table 18-5.

NONPARAMETRIC METHODS

Table 18-5 Computation of Joint Probabilities: Treatment Problem

Response	T_1	T_2	T_3	T_4	Total
Positive	(0.3)(0.2) = 0.06	(0.3)(0.3) = 0.09	(0.3)(0.3) = 0.09	(0.3)(0.2) = 0.06	0.3
Neutral	(0.4)(0.2) = 0.08	(0.4)(0.3) = 0.12	(0.4)(0.3) = 0.12	(0.4)(0.2) = 0.08	0.4
Negative	(0.3)(0.2) = 0.06	(0.3)(0.3) = 0.09	(0.3)(0.3) = 0.09	(0.3)(0.2) = 0.06	0.3
Total	0.2	0.3	0.3	0.2	1.0

Finally, these joint probabilities can be used together with the total sample size of 100 to establish the expected frequencies. This is done in Table 18-6. Each value in Table 18-6 is obtained by multiplying the sample size of 100 by the joint probability under independence from Table 18-5.

Table 18-6 Expected Frequencies: Treatment Problem

Response	T_1	T_2	T_3	T_4	Total
Positive	6	9	9	6	30
Neutral	8	12	12	8	40
Negative	6	9	9	6	30
Total	20	30	30	20	100

Using the observed frequencies in Table 18-3 and the expected frequencies from Table 18-6, we may compute chi-squares:

$$\chi^2 = \frac{(8-6)^2}{6} + \frac{(10-9)^2}{9} + \frac{(7-9)^2}{9} + \frac{(5-6)^2}{6} + \frac{(9-8)^2}{8}$$
$$+ \frac{(6-12)^2}{12} + \frac{(13-12)^2}{12} + \frac{(12-8)^2}{8} + \frac{(3-6)^2}{6}$$
$$+ \frac{(14-9)^2}{9} + \frac{(10-9)^2}{9} + \frac{(3-6)^2}{6}$$
$$= 12.49.$$

The degrees of freedom are $(c-1)(r-1) = (4-1)(3-1) = 6$. We find, using the chi-square table in the appendix, that for $\alpha = 0.05$ that $\chi_6^2 = 12.59$. Thus, our computed value is not significant at the 0.05 probability level. The null hypothesis of no differences among treatments is not rejected at the 0.05 probability level.

The chi-square test breaks down when the expected frequency in any cell is zero and does not work well for expected frequencies less than 1. When this occurs two adjacent rows or columns must be combined until no expected cell frequency is less than 1. There is a resultant loss of degrees of freedom when this is done. Although some authors recommend using the test only when the expected frequency in each cell is at least 5, recent research suggests that when the expected values of each cell is at least one or more the test is satisfactory.

S.S. After sufficient practice, the steps illustrated in Tables 18-4 through 18-6 to obtain the expected frequencies can be abbreviated. The argument runs something like the following. Thirty percent, $30 \div 100$, of all subjects reacted positively. If there were no differences among treatments, then 30 percent of those subjects given each treatment should respond positively. Hence, the expected frequencies for the top row of the table are

$$\frac{30}{100}(20) = 6, \quad \frac{30}{100}(30) = 9, \quad \frac{30}{100}(30) = 9, \quad \frac{30}{100}(20) = 6.$$

Give the equivalent calculations for the other rows and put them in a table.

Solution:

Expected Frequencies: Treatment Problem

Response	T_1	T_2	T_3	T_4	Total
Positive	$\frac{30}{100}(20) = 6$	$\frac{30}{100}(30) = 9$	$\frac{30}{100}(30) = 9$	$\frac{30}{100}(20) = 6$	30
Neutral	$\frac{40}{100}(20) = 8$	$\frac{40}{100}(30) = 12$	$\frac{40}{100}(30) = 12$	$\frac{40}{100}(20) = 8$	40
Negative	$\frac{30}{100}(20) = 6$	$\frac{30}{100}(30) = 9$	$\frac{30}{100}(30) = 9$	$\frac{30}{100}(20) = 6$	30
Total	20	30	30	20	100

In practice it is common to put both the expected and observed frequencies in the same table. The value of chi-square can be calculated directly from such a single table.

18-3 Goodness-of-Fit Tests

In the previous section, we describe the chi-square test for independence. In making this test, observed frequencies are compared to expected frequencies. A related problem is the determination of how closely an observed

NONPARAMETRIC METHODS

(sample) frequency distribution fits an expected (proposed theoretical) probability distribution. It should not come as a surprise that the chi-square distribution can be used to make a goodness-of-fit test.

CHI-SQUARE TEST FOR GOODNESS OF FIT

Suppose that a hospital is considering a WATS (Wide Area Telephone Service) line. Based on phone-call data in similar hospitals, the belief is that incoming calls per minute arrive in accordance with a Poisson distribution with parameter $\lambda = 3.8$. Data on 100 randomly selected time periods, each 1 minute long, is given in Table 18-7.

Table 18-7   Observed Incoming Phone Calls per Minute

Number of Incoming Calls per Minute	Frequency Observed	Number of Incoming Calls per Minute	Frequency Observed
0	2	6	7
1	7	7	4
2	18	8	3
3	25	9	1
4	19	10	1
5	13		100

Based on the Poisson distribution with a λ of 3.8, the expected frequencies (taken from the Poisson table) are given in the second column of Table 18-8. The remainder of Table 18-8 is used to compute the value of chi-square. The farther the expected values from those observed values (the larger the χ^2 value is), the poorer the fit of the hypothesized distribution.

In computing the value of chi-square, the last two frequencies are combined because the expected frequency in the last cell is less than 1. For this test we are free to select eight of the nine expected frequencies before the last is determined, given a total of 100 and the fact that two cells were combined. Thus, we have $\chi_8^2 = 3.038$. This value is not significant even at the 0.10 probability level ($\chi_8^2 = 13.36$), and we conclude that our hypothesized Poisson distribution with a λ of 3.8 fits the data satisfactorily.

We should observe, however, that a Type II error is easily made. Thus, we would have found that a Poisson distribution with a $\lambda = 4.0$ could not be rejected, nor, more importantly, could a normal distribution with a mean of 3.5 and a standard deviation of 2, for example. This implies a great truth about statistical methods. What one knows about the underlying process is typically more important than the statistical manipulations. Both the manager and the statistician play an important role in this process. The statistician

Table 18-8 Expected Phone Calls and the Computation of Chi-Square

Number of Incoming Calls	Expected Frequency[a] $\lambda = 3.8; n = 100$	Observed Frequency	$(O_i - E_i)^2$	$\dfrac{(O_i - E_i)^2}{E_i}$
0	2.24	2	0.058	0.026
1	8.50	7	2.250	0.264
2	16.15	18	3.423	0.212
3	20.46	25	20.612	1.007
4	19.44	19	0.914	0.047
5	14.77	13	3.133	0.212
6	9.36	7	5.570	0.595
7	5.08	4	1.166	0.230
8	2.41	3	0.476	0.198
9	1.02 ⎫ 1.41	1 ⎫ 2	0.348	0.247
10	0.39 ⎭	1 ⎭		
	100	100		3.038

[a] Obtained using the values in the Poisson table.

may know that processes with the characteristics of incoming phone calls are typically Poisson in nature. The manager is in the best position to know whether the data from other hospitals are comparable and whether the day, the month, or the time of day are related to the number of incoming calls. Only if both the manager and the statistician work closely together will useful conclusions emerge.

KOLMOGROV–SMIRNOV TEST FOR GOODNESS OF FIT

The chi-square test requires a reasonably large sample. Another test, the Kolmogrov–Smirnov test, can be used for smaller samples. This test has the additional advantages that it is generally more powerful, easier to compute, and it does not make any stipulations concerning the minimum expected cell frequencies. The disadvantages of this test are that (1) it requires ordinal data, whereas the chi-square test requires only nominal data and (2) tables of the test statistic are not as available as χ^2 tables.

The test uses a statistic called D, which is computed as follows:

1. Compute the cumulative observed relative frequencies.
2. Compute the cumulative expected relative frequencies.
3. Compute the difference and drop the sign.
4. The value of D is the maximum difference.

$$D = \max_t \left| \sum_{i=0}^{t}(O_i/n) - \sum_{i=0}^{t}(E_i/n) \right|. \tag{18-6}$$

Nonparametric Methods

Table 18-9 Computation of the Kolmogrov–Smirnov Test Statistic D: Phone-Call Problem

Number of Incoming Calls	Cumulative Expected Relative Frequency $\lambda = 3.8; n = 100$	Cumulative Observed Relative Frequency	Cumulative Absolute Difference
0	0.0224	0.02	0.0024
1	0.1074	0.09	0.0174
2	0.2689	0.27	0.0011
3	0.4735	0.52	0.0465 = D
4	0.6679	0.71	0.0421
5	0.8156	0.84	0.0344
6	0.9092	0.91	0.0008
7	0.9610	0.95	0.0110
8	0.9851	0.98	0.0051
9	0.9953	0.99	0.0053
10	0.9992	1.00	0.0008

We illustrate the computation of D using the data in columns 2 and 3 of Table 18-8 to obtain Table 18-9.

The larger the value of D, the less likely the hypothesized distribution is suitable. Tables of the statistic D exist which give critical values of D for various probabilities and sample sizes. For our example, the value of D at a significance level of 0.10 (α) and a sample size of 100 is 0.122. Since our value of D is less than the critical value, it is not significant and we again accept the hypothesized distribution.

Tables of the critical values of D are not given in this book but may be found in: S. Siegel, *Nonparametric Statistics* (New York: McGraw-Hill Book Company, 1956), p. 251; L. Miller, "Tables of Percentage Points of Kolmogrov Statistics," *Journal of the American Statistical Association*, 1956, pp. 111–121; or D. Owen, *Handbook of Statistical Tables* (Reading, Mass.: Addison-Wesley Publishing Company, Inc., 1962).

S.S. A firm believes its sales of refrigerators are uniformly distributed over time. A count of last year's sales in units by month is as follows:

Month	Sales	Month	Sales	Month	Sales
January	22	May	27	September	19
February	27	June	26	October	23
March	33	July	22	November	21
April	32	August	20	December	16

Test the validity of the claim using the chi-square test of goodness of fit using an α level of 0.05. State the null hypothesis, the degrees of freedom, the value of chi-square, and whether the value of chi-square is significant.

Solution:

	\multicolumn{12}{c}{Month}
	J	F	M	A	M	J	J	A	S	O	N	D	Total
Observed	22	27	33	32	27	26	22	20	19	23	21	16	288
Expected[a]	24	24	24	24	24	24	24	24	24	24	24	24	288
$(O_i - E_i)^2/E_i$	0.17	0.38	3.38	2.67	0.38	0.17	0.17	0.67	1.04	0.04	0.38	2.67	12.12

[a] Expected frequency = $\frac{288}{12}$ = 24.

Null hypothesis: Sales are distributed uniformly through time. (Or we could say that monthly sales follow a uniform, a rectangular distribution.)

Degrees of freedom: 11

Value of chi-square: 12.12.

The value of chi-square is not significant at the 0.05 probability level since the chi-square value at that point is 19.68. This is true despite the apparent increased sales levels in the spring.

18-4 The Runs Test for Randomness

The runs test for randomness is based on the order or sequence of a set of observations. A *run* is defined as a succession of identical observations. If there are either too few or too many runs in a sequence of observations, nonrandomness is suspected. Several examples involving the output of a machine where each item is classified as either good, G, or bad, B, are as follows, where nonrandomness is likely.

Sequence 1: G, G, G, G, G, G, B, B, B, B, B, B

Sequence 2: B, G, B, G, B, G, B, G, B, G, B, G

Sequence 3: B, B, G, G, G, G, G, G, B, B, B, B.

In the first sequence the machine seems to have gone out of control after the sixth item. The second sequence might occur where items must be continuously monitored to retain control or it may be due to different in-

Nonparametric Methods 641

spectors. The third sequence is consistent with a process that is brought under control only to go out again after the eighth item.

Consider an example involving a series of 20 subcontracting decisions made by a major municipality classified by whether the contract was let to a big or a small contractor.

Total runs:
$\underline{b}, \underline{s, s}, \underline{b}, \underline{s}, \underline{b, b}, \underline{s}, \underline{b}, \underline{s}, \underline{b, b}, \underline{s, s}, \underline{b}, \underline{s}, \underline{b}, \underline{s}, \underline{b}, \underline{s}$
  1   2   3   4   5   6   7   8   9   10   11   12 13 14 15 16

Runs of b's:
  1   2   3   4   5   6   7   8

This series involves 16 total runs and 8 runs of b's as is indicated below the series. We wish to know whether 8 runs of b's is unusual, under a null hypothesis of randomness. We will focus on the runs of b's to obtain an answer. Let N_1 be the number of b's and N_2 the number of s's. Then $N_1 + N_2$ is the total sample size. We are interested here in the probability of 8 or more runs of b's given 20 items involving 10 b's and 10 s's. To determine this probability, we consider the general problem of constructing x runs of b's. Consider the 10 b's standing together. This is a single run of b's.

$$b \quad b \quad b \quad b \quad b \quad b \quad b \quad b \quad b \quad b$$

To produce additional runs of b's, we need to separate the b's into groups. We can do so by placing spaces at the appropriate places in the run of b's. There exist $9 = N_1 - 1$ possible choices for these spaces. We have 8 runs of b's in the original sequence. To obtain $8 = x$ runs of b's, we must select $8 - 1$ or 7 of the $9 = N_1 - 1$ spaces between the b's. This yields the 8 groups of b's. In general, we select $x - 1$ of the $N_1 - 1$ spaces. The number of ways to perform this task is given by the combination of $N_1 - 1$ things taken $x - 1$ at a time: $C_{x-1}^{N_1-1}$.

The 8 groups of b's are separated by inserting them between the s's. The 10 s's provide $11 = N_2 + 1$ places where the b's may be inserted:

$$\_\_s\_\_s\_\_s\_\_s\_\_s\_\_s\_\_s\_\_s\_\_s\_\_s\_\_$$

as indicated by the dashes among the s's. We must now select $8 = x$ of these places. This can be done in C_8^{11} ways. In general, this is given by the combination of $N_2 + 1$ things taken x at a time: $C_x^{N_2+1}$. The number of ways, then, to produce x runs of b's among $N_1 + N_2$ items is:

General: for x runs of b's in $N_1 + N_2$ items: $C_{x-1}^{N_1-1} \cdot C_x^{N_2+1}$

Specific: for 8 runs of b's in $10 + 10$ items: $C_7^9 \cdot C_8^{11}$.

The total number of distinguishable sequences is given by

$$\text{General: } C_{N_1}^{N_1+N_2}$$

$$\text{Specific: } C_{10}^{20}.$$

Hence, the probability we seek is

$$\text{General: } C_{x-1}^{N_1-1} \cdot C_x^{N_2+1} / C_{N_1}^{N_1+N_2}$$

$$= \frac{(N_1 - 1)!}{(x - 1)!(N_1 - x)!} \cdot \frac{(N_2 + 1)!}{x!(N_2 + 1 - x)!} \cdot \frac{(N_1)!(N_2)!}{(N_1 + N_2)!}$$

Specific: $C_7^9 \cdot C_8^{11} / C_{10}^{20} = 0.03215$.

It is a hypergeometric probability.

For $N_1 = N_2 = 10$, the probabilities for runs of b's of 7, 8, 9, and 10 are 0.15003, 0.03215, 0.00268, and 0.00006, respectively. Thus, the probability of 8 or more runs under a null hypothesis of randomness is

$$0.03215 + 0.00268 + 0.00006 = 0.03489.$$

Using an α significance level of 0.05, we could reject the hypothesis of randomness. Perhaps there is a conscious decision on the part of the municipality to balance contracts among large and small contractors over even short periods of time. In this case, the order analysis may give us information which is not suggested by the frequency of events: namely, that in addition to equality in the number of contracts let to large and small contractors, a large contract tends to be let after a small one, and vice versa.

In general, computation of the hypergeometric probabilities required is quite tedious. Tables do exist for small samples. One ($N \leq 20$) is contained in the book by Siegel referred to in Section 18-3. For samples over 20 when N_1 and N_2 are close to equal, a normal approximation may be used. For this approximation, let

$r \equiv$ number of runs in total for both kinds of items

$N_1 \equiv$ number of items of one kind

$N_2 \equiv$ number of items of the other kind

mean number of runs under the null hypothesis

$$\equiv \mu_r = \frac{2N_1 N_2}{N_1 + N_2} + 1 \tag{18-7}$$

NONPARAMETRIC METHODS

standard deviation of the number of runs under the null hypothesis

$$\equiv \sigma_r = \sqrt{\frac{2N_1 N_2 (2N_1 N_2 - N_1 - N_2)}{(N_1 + N_2)^2 (N_1 + N_2 - 1)}}. \tag{18-8}$$

The test is generally a two-tailed test.

S.S. 1. Apply the normal approximation to the data for contracting given earlier in this subsection using $\alpha = 0.05$.

2. The runs test can be used in testing for trends or seasonals in time-series data. Suppose that automobile sales in millions of units over the past 10 years were as follows:

6.83, 8.51, 7.31, 7.95, 8.50, 7.21, 10.73, 7.68, 9.13, 8.45.

Is there a trend in this yearly data? Use a 0.05 significance level. The test is made by constructing a series of plus and minus signs based on whether the value is above ($+$) or below ($-$) the median value. (Let N_1 be the number of $+$ signs.)

Solution:
1. $N_1 = 10 = N_2 : r = 16, \mu_r = [(2(10)10)/(10+10)] + 1 = 11$

$$\sigma_r = \sqrt{\frac{2(10)(10)[2(10)(10) - 10 - 10]}{(10+10)^2(10+10-1)}} = \sqrt{\frac{36,000}{7,600}} = 2.176$$

$$z = \frac{11 - 16}{2.176} = 2.30.$$

This value of z leaves 0.02 of the area in both tails and we again conclude that the data are significant at the 0.05 probability level.

2. The null hypothesis is that the data are random, and there is no trend. Median sales is any number between 7.95 and 8.45 inclusive. We select the midpoint of this range, 8.20, as our median. The run series is $-, +, -, -, +, -, +, -, +, +$. The series exhibits eight runs. $N_1 = 5 = N_2, x = 4$, and

$$C_3^4 \cdot C_4^6 / C_5^{10} = \frac{4!}{3!\,1!} \cdot \frac{6!}{4!\,2!} \bigg/ \frac{10!}{5!\,5!} = 0.24.$$

Theoretically, for a two-tail test, we would need to calculate the probability of four or less runs of one item, but since the probability of four runs already exceeds the significance level, we cannot reject the null hypothesis based on the data we have. This is true even though what we know about the economics of automobile sales might lead us to suspect a trend. In part the result is due to the fact that we have only a small sample. Furthermore, a small sample can be importantly influenced by the choice of year used to initiate the data series.

18-5 The Signs Test of Significance

Signs were used in the last section for a runs test of randomness. Signs can also be used to test for a significant difference between two populations. We have previously tested for the difference between two populations using small sample sizes (see Chapter 12). These tests required us to assume that the underlying populations were normally distributed and that they had the same variance. Often these assumptions are not realistic. The present signs test assumes only that the variable under consideration has a conceptually continuous distribution.

The test must be based on matched pairs of subjects in cases where it is possible to rank-order the two members of any pair on the variable of interest. It is implicitly assumed, and the decision maker must be careful here, that the pairs have been matched to eliminate the effect of any other variables that might explain the difference tested. (Sometimes this is done by random combination by which, it is hoped, systematic tendencies in other confounding variables are avoided.)

The null hypothesis is that there is no difference in the populations; more precisely, that their medians are identical. An advantage of the test is that it is based on the binomial distribution which we have studied previously and for which tables for small sample sizes are readily available. Moreover, approximations we have studied (Chapter 8) for large sample sizes are also available.

Let us try an example. The data in Table 18-10 represent return on investment figures for 10 randomly selected companies in 2 different years.

Table 18-10   Percentage Return on Investment for 2 Years

Company	Year 1	Year 2	Company	Year 1	Year 2
1	15.0	12.4	6	3.6	5.6
2	5.7	6.2	7	11.1	10.4
3	11.3	9.2	8	15.6	13.2
4	18.5	16.2	9	14.5	13.2
5	9.7	9.7	10	8.7	7.4

Can we say that the median return has declined in the second year using a 0.05 probability level? This is a one-tailed test and suggests that the direction of the change can be specified by the decision maker in advance.

Examining the signs of the change from year 1 to year 2, we find that 7 of the changes are negative, 2 are positive, and there is one tie. In using the binomial test, we must allow only two possibilities; hence ties are omitted.

The null hypothesis of equal medians implies that an equal number of pluses and minuses should occur. Thus, under the null hypothesis the probability of a plus is 0.5. We must find the binomial probability of 7 or more pluses in 9 trials when the probability of a plus on any given trial under the null hypothesis is 0.5. In symbols, we require

$$\sum_{x=7}^{9} b(x; 9, 0.5) = 1 - \sum_{x=0}^{6} b(x; 9, 0.5).$$

From the binomial tables, we find this probability to be 0.0899. For the data we have, we may not conclude significance at the 0.05 level. Even if we had been able to conclude significance, the test would not have suggested the reason for the change. Note, for example, that companies 2 and 6, the companies whose rate of return rose, had the two lowest returns. Perhaps they were both in the same industry, an industry affected by special economic conditions. Data like this often need additional analysis. We also observe that to avoid the parametric test assumptions, we have sacrificed data. The magnitudes of the differences have been suppressed and only the signs of the differences are used in the test.

Suppose, to carry the discussion one step further, that we had 30 companies in our sample and that there were three ties, 21 positive changes and 6 negative changes. This retains the same ratios we had before but gives us more data. Again, we could use the binomial and the tables at the end of this book to calculate

$$\sum_{x=21}^{27} b(x; 27, 0.5) = 1 - \sum_{x=0}^{20} b(x; 27, 0.5) \le 1 - \sum_{x=0}^{20} b(x; 28, 0.5)$$

$$= 1 - 0.9937 = 0.0063.$$

The data now imply significance at the 0.05 probability level. (Note that $n = 27$ is not in our tables but that the probability we desire is smaller than that found using the data for $n = 28$.)

Now let us try the normal approximation on the data. The normal works quite well even with small samples under a null hypothesis that sets the probability of a "success" close to 0.05. We obtain $\mu_+ = 0.5(27) = 13.5$, $x_+ = 21$, and $\sigma_+ = \sqrt{27(0.5)(0.5)} = 2.60$. Hence,

$$z = \frac{21 - 13.5}{2.60} = 2.88.$$

This leaves an area of 0.0020 in the tail. The value again leads us to conclude that the data are significant.

As a final point, the tests we have made in this section were one-tail tests. In effect, we tested the hypothesis that the (median) rate of return has declined versus the null hypothesis that it has not declined. This is not surprising since an investigator should have a feeling for the direction of change, if any. A two-tail test of the null hypothesis of no change in the (median) return versus the alternate hypothesis of a change, would require us to double the calculated probabilities. For our problem, the conclusions remain the same for $\alpha = 0.05$, regardless of which form of the null hypothesis is used. But this would not always be so.

S.S. The government is attempting to validate the claim of a major gasoline producer that their new nonleaded gasoline produces an increase in the miles per gallon obtained when compared to their regular leaded gasoline. To make this test 15 cars selected randomly are driven with a measured tank in normal driving with both types of gasoline. The data are given in Table 18-11. Do the data support the claim at the 0.05 probability level?

Table 18-11   Gas-Mileage Test Data

Car Pair	Miles per Gallon		Car Pair	Miles per Gallon	
	Leaded	Not Leaded		Leaded	Not Leaded
1	12.5	12.9	9	25.6	25.2
2	23.7	23.8	10	21.8	22.1
3	15.4	16.2	11	16.5	17.1
4	17.3	17.5	12	18.2	18.0
5	8.4	9.3	13	11.3	11.8
6	21.5	21.3	14	10.9	11.3
7	16.4	16.4	15	13.5	13.9
8	13.3	14.1			

Solution: The null hypothesis is that leaded gasoline does not lead to an increase in gas mileage. This is a one-tailed test since the claim of the manufacturer is for an increase, a directional change. Omitting the one tied value, there are 11 cases of an increase in 14 pairs. The binomial probability of 11 or more is given by

$$\sum_{x=11}^{14} b(x; 14, 0.5) = 1 - \sum_{x=0}^{10} b(x; 14, 0.5) = 1 - 0.9713 = 0.0287.$$

We conclude that the data are significant at the 0.05 probability level. In doing so, we assume that the test was run appropriately in terms of the gas used, the routes driven, that there was no disclosure of the type of gas used, and so on. We also note, however, that for the four cars achieving over 20 miles per gallon, two experienced a decline and one had an increase of only one tenth. Indeed, all the declines were in the high miles-per-gallon cars. Perhaps there is some factor at work that caused this leaded gasoline to work better for larger cars. Further analysis (and test data) is required if this possibility, suggested by the present data, is to be tested. The answer may be important to both the government and the producer.

18-6 Rank Tests of Significance

In this subsection we examine three tests. They are:

1. The Mann–Whitney U test. This test attempts to establish whether two samples come from populations with the same median. The test assumes only that the underlying populations are conceptually continuous and even minor violations here are not serious. The test is easy to apply. Tables for small sample sizes are available in books of statistical tables or books on nonparametric statistics. (This is also true for the next two tests.)

2. The Wilcoxon Matched-Pairs Signed-Rank Test. This test examines the more stringent hypothesis that two samples come from the same population. This test is also easy to make. Not all books contain the requisite tables for small sample sizes.

3. The Kruskal–Wallis One-Way Analysis-of-Variance Test. This test extends the Mann–Whitney test to more than two populations. As such it provides an alternative to the F test used in analysis of variance. (See Chapter 14.) The Kruskal–Wallis test avoids the limiting population assumptions, such as normality, required by the F test.

The tests described in this section all assume at least ordinal measurement. Hence, the numerical values may be used to assign ranks to the sample values once they have been pooled. The analysis tests whether the average ranks differ significantly from sample to sample.

MANN–WHITNEY U TEST

The procedure for making this test is as follows:

1. Pool all the observations from both samples and rank them in order of increasing size. In ranking, tied observations receive the average of the tied ranks.
2. Separate the ranked data back into the original two samples.

3. Compute:

$$T_1 = N_1 \cdot N_2 + \frac{N_1(N_1 + 1)}{2} - R_1 \qquad (18\text{-}9)$$

$$T_2 = N_1 \cdot N_2 - T_1, \qquad (18\text{-}10)$$

where

$N_1 \equiv$ number of cases in one sample
$N_2 \equiv$ number of cases in the other sample
$R_1 \equiv$ sum of the ranks for the sample corresponding to N_1.

4. Compute

$$U = \min \{T_1, T_2\}. \qquad (18\text{-}11)$$

Let us try an example. The use of accounting ratios in credit analysis has been well established. For example, the quick ratio, the ratio of cash and marketable securities to short-term payables, is an indication of short-term financial strength. A ratio near unity is usually considered a good sign. Too low a ratio implies future difficulties in meeting debts as they become due, while too high a ratio suggests a failure to use excess funds productively in the business. The data in Table 18-12 gives quick ratios for 9 railroads and 10 airline companies selected randomly in a recent year. For the year in question, we can use a Mann–Whitney test to see if a significant difference in the quick ratios of the two industries exists using a 0.10 significance level.

Table 18-12 Quick Ratios: Railroads and Airlines

Railroads		Airlines	
Quick Ratio	Rank	Quick Ratio	Rank
0.83	6	0.72	2
0.91	10	1.10	15
0.58	1	0.92	12
1.37	19	1.13	16
1.31	18	0.91	10
0.75	3	0.87	8
0.81	5	0.79	4
0.86	7	1.05	14
0.97	13	1.21	17
		0.91	10
Total	82		108

NONPARAMETRIC METHODS

We will, arbitrarily, select the railroad sample as the N_1 sample. Hence, $N_1 = 9$, $N_2 = 10$, $R_1 = 82$,

$$T_1 = 9(10) + \frac{9(10)}{2} - 82 = 53, \qquad T_2 = 90 - 53 = 37,$$

and

$$U = \min\{53, 37\} = 37.$$

(Note that the three tied 0.91 values are each given the rank of 10, which is the average of 9, 10, and 11, the ranks for which these values tied.)

Special tables are again required to test for the significance of this result when the sample sizes are as small as in the present case. One set of tables is contained in: S. Siegel, *Nonparametric Statistics* (New York: McGraw-Hill Book Company, 1956). Most handbooks of statistical tables also contain the necessary tables. Using the appropriate table, the critical value of U is 24. Since our value of U is 37, significance at the 0.10 level is not indicated. Using the tables, small values of U are required for significance; we reject the null hypothesis if U is less than or equal to the tabulated critical value.

When both samples are of size 20 or more, the sampling distribution of U is approximated closely by a normal distribution. (Indeed, most statisticians are content with the normality assumption when N_1 and N_2 both exceed 8; hence, we can use the normal approximation here.) The mean and standard deviation are estimated by

$$\mu_U = \frac{N_1 \cdot N_2}{2} \tag{18-12}$$

and

$$\sigma_U = \sqrt{\frac{N_1 \cdot N_2(N_1 + N_2 + 1)}{12}}. \tag{18-13}$$

For our problem $\mu_U = 45$ and $\sigma_U = \sqrt{9(10)(20)/12} = 12.24$. Thus, $z = (37 - 45)/12.24 = -0.65$. For a two-tail test a value of z in excess of 1.96 is required. Again no significant difference in the quick ratios of these two industries is suggested by the data with respect to medians.

S.S. Assuming that the 2 populations of quick ratios have the same median, test for a difference in variability by ranking the distance from the median observation to each data value. (The assumption is required for the test to be valid.) Use an α level of 0.10 and the normal approximation. Is the difference in variability between the two distributions significantly different at a 0.10 probability level? The median for the pooled sample is used to make the test.

Solution: First, the median of the data is 0.91 and, as before, $N_1 = 9$ and $N_2 = 10$. The ranking from the median is done in Table 18-13.

$$R_1 = 98; \quad T_1 = 90 + 45 - 98 = 37; \quad T_2 = 80 - 37 = 53;$$

$$U = 37; \quad \mu_U = 45; \quad \sigma_U = 12.24; \quad z = \frac{37 - 45}{12.24} = -0.65.$$

The difference between the variability of the 2 data sets is not significant.

Table 18-13 Quick-Ratio Data Used to Test Variability

	Railroads			Airlines	
Observation	Observation-Median	Rank	Observation	Observation-Median	Rank
0.83	0.08	8	0.72	0.19	13½
0.91	0.00	2	1.10	0.19	13½
0.58	0.33	17	0.92	0.01	4
1.37	0.46	19	1.13	0.22	15
1.31	0.40	18	0.91	0.00	2
0.75	0.16	12	0.87	0.04	5
0.81	0.10	9	0.79	0.12	10
0.85	0.06	6½	1.05	0.14	11
0.97	0.06	6½	1.21	0.30	16
			0.91	0.00	2
Total		98			92

WILCOXON MATCHED-PAIRS SIGNED-RANK TEST

The Wilcoxon test is based on dependent samples and hence can be used on the same type of data for which the sign test discussed in Section 18-5 was used. The Wilcoxon test is more powerful in that the relative magnitudes of the differences within pairs is used rather than just the direction of these differences. The assumption required to use the more powerful Wilcoxon test is that the strength of the effect between pairs is captured by the difference within pairs.

Let us assume that this is the case for the return-on-investment data in Table 18-10 and use the Wilcoxon test on these data. The assumption means that company 1 showed a greater decline in its return than did company 3

NONPARAMETRIC METHODS

since its return declined 2.6 percent while that for company 3 declined 2.0 percent.

The test procedure is as follows:

1. Compute the difference for each firm.
2. Rank the absolute value of the differences from the lowest to the highest. (Note that a difference of -1 is given a lower rank than either a difference of -2 or $+2$).
3. To each rank now affix the sign of the difference.
4. Sum the positive ranks and sum the negative ranks separately.
5. Let T be the absolute value of the smaller sum.
6. Ties: Any pairs with differences of zero are dropped from the analysis. If two or more differences are equal and nonzero, they are assigned the average rank of the ranks for which they tie.

Under a null hypothesis of no difference, we would expect the sum of the positive ranks to equal the (absolute) sum of the negative ranks. The Wilcoxon test tests this hypothesis.

Using the data from Table 18-10, we have constructed Table 18-14.

Table 18-14 Return Data: Wilcoxon Test

Company	Difference	Rank (with sign)	Company	Difference	Rank (with sign)
1	2.6	-9	6	2.0	$+5$
2	0.5	$+1$	7	0.7	-2
3	2.1	-6	8	2.4	-8
4	2.3	-7	9	1.3	-3.5
5	0		10	1.3	-3.5

Using the data in Table 18-14, T, the smaller sum, is equal to 6, the sum of the positive ranks. The sum of the negative ranks is 39. Again special tables are available in Siegel's book and in books of statistical tables for small sample sizes. Using this table, we find that the decline in the data is significant at the 0.025 probability level for a one-tail test. Recall that the decision maker asked whether the return on investment had declined.

Using the Wilcoxon test, we conclude significance although we were unable to do so using a sign test on the same data. The reason is that the Wilcoxon test uses more of the data and is, therefore, a more powerful test.

When the two sample sizes exceed 25 in total number, the distribution of T is approximately normal, with

$$\mu_T = \frac{N(N+1)}{4} \qquad (18\text{-}14)$$

$$\sigma_T = \sqrt{\frac{N(N+1)(2N+1)}{24}}, \qquad (18\text{-}15)$$

where N is the total number of observations.

To show how well this approximation works even for small samples, we use it for the nine observed differences (there was one tie which is omitted in testing) in Table 18-14. We obtain

$$\mu_T = \frac{9(10)}{4} = 22.5; \qquad \sigma_T = \sqrt{\frac{9(10)(19)}{24}} = 8.44.$$

Hence,

$$z = \frac{T - \mu_T}{\sigma_T} = \frac{6 - 22.5}{8.44} = -1.95.$$

This value of z is almost exactly equal to the critical value in a one-tail test at the 0.025 probability level (-1.96). [The table for T (for an N of 9) in Siegel's book gives a T of 6 as the critical value.]

S.S. 1. Could the Mann–Whitney U test be used to test the significance of these data? Why or why not?

2. Use the Wilcoxon test and assume normality to test the data on gasoline mileage given in Table 18-11 at the 0.01 probability level.

Solution:

1. Use of the Mann–Whitney U Test requires that the samples be independent. This is not true here. The two samples are matched by company. This is precisely the type of situation for which the Wilcoxon test was developed.

2. The 15 signed differences are given in the next line with the signed ranks indicated below:

Differences:

$+0.4, +0.1, +0.8, +0.2, +0.9, -0.2, 0, +0.8,$
$-0.4, +0.3, +0.6, -0.2, +0.5, +0.4, +0.4$

NONPARAMETRIC METHODS

Signed ranks:

+7.5, +1, +12.5, +3, +14, −3, no rank, +12.5,
−7.5, +5, +11, −3, +10, +7.5, +7.5

$$\mu_T = \frac{14(15)}{4} = 52.5$$

$$\sigma_T = \sqrt{\frac{14(15)(29)}{24}} = 15.93$$

$$T = |-3 - 7.5 - 3| = 13.5$$

$$z = \frac{13.5 - 52.5}{15.93} = -2.45$$

For a one-tail test with $\alpha = 0.01$, a value of $z = -2.33$ is required for significance. Since our value of z is beyond the critical value, we conclude a difference in the gasoline mileage exists.

KRUSKAL–WALLIS ONE-WAY ANALYSIS-OF-VARIANCE TEST

The Kruskal–Wallis test represents an extension of the Mann–Whitney test to more than two independent samples. The procedure is much the same.

1. Pool all the observations from all the samples and rank them in order of increasing size. In ranking, tied observations receive the average of the tied ranks.
2. Separate the ranked data by sample.
3. Compute

$$H = -3(N + 1) + \frac{12}{N(N + 1)} \sum_{j=1}^{k} \frac{R_j^2}{N_j}, \qquad (18\text{-}16)$$

where

$N \equiv$ number of cases in all samples combined

$k \equiv$ number of samples

$R_j \equiv$ sum of the ranks in the jth sample, $j = 1, 2, \ldots, k$

$N_j \equiv$ number of cases in the jth sample, $j = 1, 2, \ldots, k$.

When ties exist, the value of H must be divided by

$$1 - \frac{\sum t^*}{N^3 - N},$$

where

$$t^* = t^3 - t \quad (t \text{ is the number of tied observations in a group of scores})$$

$\sum t^*$ directs the summation over all tied groups.

The test assumes only that the variable under consideration has a conceptually continuous distribution and that ordinal measurement has been achieved. The null hypothesis is that the k samples all come from the same population. If the hypothesis is rejected, the method does not indicate which sample or samples are from different populations.

For an example, we expand the data in Table 18-12 to include 10 bus companies. The quick ratios and pooled ranks for all three industries are given in Table 18-15.

Table 18-15 Quick Ratios: Three Industries

Railroads		Airlines		Buses	
Quick Ratios	Rank	Quick Ratios	Rank	Quick Ratios	Rank
0.83	14	0.72	7.5	0.67	6
0.91	20	1.10	25	0.76	10
0.58	1	0.92	22	0.59	2
1.37	29	1.13	26	0.82	13
1.31	28	0.91	20	0.62	3
0.75	9	0.87	18	0.72	7.5
0.81	12	0.79	11	0.64	4
0.86	17	1.05	24	0.66	5
0.97	23	1.21	27	0.84	15
		0.91	20	0.85	16
Total	153		200.5		81.5

There are two sets of ties. Two returns of 0.72 are given an average rank of 7.5, and the three returns of 0.91 are given an average rank of 20.

Using the data in Table 18-15, we obtain

$$H = -3(29 + 1) + \frac{12}{29(30)} \left[\frac{(153)^2}{9} + \frac{(200.5)^2}{10} + \frac{(81.5)^2}{10} \right]$$
$$= 10.5,$$

and correcting for ties, we divide H by

$$1 - \frac{(2^3 - 2) + (3^3 - 3)}{29^3 - 29} = 0.9988.$$

NONPARAMETRIC METHODS

Thus, H corrected for ties equals 10.51. In most cases, as is true here, the correction for ties is negligible.

As long as there are more than five cases in each group, $N_j > 5$, H is closely approximated by a chi-square distribution under the null hypothesis with $k - 1$ degrees of freedom. Using a 0.10 significance level and $3 - 1$, or 2, degrees of freedom, we find the critical value of H to be 4.60. Since our value of H exceeds this critical value, we conclude that the differences in data are significant: the three samples do not all come from the same universe. Indeed, the differences in our data are significant at the 0.01 probability level.

S.S. 1. Since the only difference between the data examined here and that tested in the earlier part of this section using the Mann–Whitney U test is the addition of the data on bus companies, may we conclude that the bus company data comes from a different universe?

2. Is it appropriate to use these data to test the hypothesis that the three sets of data come from the same population with respect to variability?

Solution:

1. Recall that we did not find significance at the 0.10 probability level previously for the rails and airlines. The test we just made only tells us the three industries are different. Pair-wise tests would be required to test whether the bus-company data comes from a different universe. But further tests on the same data result in further losses of degrees of freedom and a conclusion of which industry or industries differ might better be made on the basis of a specific hypothesis and new data or, perhaps, on what the investigator knows about the economics of the situation. Moreover, it is not clear just what pair-wise tests should be made first or what significance level to use. (What conclusion would emerge if the pairwise comparisons turned out to be inconsistent?) Finally, it is always possible that what we have is merely an unusual data set from like populations. This will happen 10 percent of the time using our procedure with a 0.10 significance level.

2. Since we have already concluded that the samples come from different populations, it is inappropriate to test for a difference in variability. If, inappropriately, a Kruskal–Wallis test were made on the differences from the median, pooling the data and using a median of 0.84, then we would have obtained:

$$R_1 = 136, \quad R_2 = 161, \quad R_3 = 138$$

$$H = 0.367 \text{ (uncorrected for ties)}.$$

18-7 Nonparametric Measures of Association

Often we find that knowing that two samples come from different populations is not very informative. Instead, we wish to have some measure of the association between the populations, obtained using the sample data. When the data are at least interval in nature, we use correlation and regression techniques. With data that are not at least interval with respect to measurement, we must find new measures. Two are discussed in this section. The first is the contingency coefficient. The other is Spearman's rank correlation coefficient,.

As with other nonparametric methods, the present measures avoid some of the limiting assumptions made by the more powerful techniques. For example, to test the significance of a parametric correlation coefficient requires assumptions of normality. This assumption can be avoided using Spearman's rank correlation coefficient. For this reason the investigator may elect to use these measures even when interval or ratio measurement is attained, but the required test assumptions seem untenable.

CONTINGENCY COEFFICIENT

The contingency coefficient is a measure of association between two sets of sample data measured on a nominal scale. The coefficient is computed directly from a contingency table using the formula

$$C = \sqrt{\frac{\chi^2}{N + \chi^2}}, \qquad (18\text{-}17)$$

where

$N \equiv$ total number of observations

$\chi^2 \equiv$ chi-square value.

Using the data in Tables 18-3 and 18-6, we have $N = 100$ and $\chi^2 = 12.49$, a value we calculated for that example. Hence,

$$C = \sqrt{\frac{12.49}{100 + 12.49}} = 0.33.$$

A test of the significance of the contingency coefficient is not required since the chi-square test performed on the data is sufficient for this purpose. That is, if the χ^2 value is significant, the measure of association is also significant.

The contingency coefficient goes to zero for the case of a complete lack of association, since the chi-square value is zero in this case. However, perfect

NONPARAMETRIC METHODS

association is not characterized by $C = 1$. The upper limit to C depends on the number of classifications, equaling $\sqrt{1/2} = 0.707$ for a 2 × 2 table, $\sqrt{2/3} = 0.816$ for a 3 × 3 table, and increasing for larger-sized tables. Because of this situation, two contingency coefficients are comparable only if they come from tables with the same cell configuration.

S.S. Compute C for the data in Table 18-2. Is C significant at the 0.05 probability level?

Solution:

$$C = \sqrt{\frac{2.72}{100 + 2.72}} = 0.16.$$

As we discovered in Section 18-2, the chi-square value, and thus C, is not significant at the 0.05 probability level.

SPEARMAN'S RANK CORRELATION COEFFICIENT

Spearman's rank correlation coefficient requires at least ordinal data. It is based on the differences between two rankings of some phenomenon on two scales; hence, it deals with paired ranks. Examples include different evaluator ranks of performance for a group of workers and different consumer ranks for a set of products. There are two sets of ranks, call them X's and Y's, obtained independently. The value of Spearman's rank correlation coefficient is calculated using

$$r_s = 1 - \frac{6 \sum d^2}{N^3 - N}, \tag{18-18}$$

where

$d \equiv X - Y$, the difference in ranking for each item, and $\sum d^2$ is the sum of these differences

$N \equiv$ sample size.

Ties, if any, are assigned the average rank for which the observations tie. The effect of ties on r_s is typically negligible and, hence, no correction factor to r_s is given here. If major problems with ties exist, the reader is referred to Siegel's book.

Each week during the football season both the United Press International (UPI) and the Associated Press (AP) rank the top football teams. Let us examine the degree of association between these tests for 1 week. The data are given in Table 18-16. Only the names of the teams have been altered to protect the innocent.

Table 18-16 Ranks of Football Teams

Team	UPI Rank	AP Rank	$(UPI - AP)^2$
A	1	3	4
B	2	1	1
C	3	4	1
D	4	5	1
E	5	2	9
F	6	7	1
G	7	6	1
H	8	8	0
I	9	10	1
J	10	9	1
			$\sum d^2 = 20$

Using equation (18-16),

$$r_s = 1 - \frac{6(20)}{10^3 - 10} = 0.879.$$

If the set of items to be ranked is the same in each ranking, no problems arise. This is typically the case. With team ranks, however, one of the top 10 teams ranked by the UPI may not be ranked by the AP. This means that either the method must be abandoned here, or done over a common set of teams, ignoring those not ranked by both. (This produces an upward bias to this particular use of the test.) The measure, then, is not ideal for football rankings.

A test of significance for r_s is not required since the procedure described for the Wilcoxon test can be used. However, to save time, if the sample is 10 or more, the significance of r_s under a null hypothesis of no association can be tested using a t test with $(N - 2)$ degrees of freedom. The standard deviation of r_s is given by

$$S_{r_s} = \sqrt{\frac{1 - r_s^2}{N - 2}}. \tag{18-19}$$

Hence,

$$t = \frac{r_s - 0}{S_{r_s}} = r_s \sqrt{\frac{N - 2}{1 - r_s^2}}. \tag{18-20}$$

For our example,

$$t = 0.879 \sqrt{\frac{8}{1 - (0.879)^2}} = 5.214.$$

NONPARAMETRIC METHODS   659

This value of t is significant at the 0.001 probability level. We conclude that a significant relationship exists. Here we could use a one-tail test since the expected association is positive, not negative.

S.S. Calculate the value of Spearman's rank correlation coefficient for the data on investment returns given in Table 18-10. Test it for significance at the 0.01 probability level.

Solution: The ranks for year 1 are 8, 2, 6, 10, 4, 1, 5, 9, 7, 3 and for year 2: 7, 2, 5, 10, 4, 1, 6, 8.5, 8.5, and 3. $\sum d^2 = 5.50$ and

$$r_s = 1 - \frac{6(5.50)}{10^3 - 10} = 0.967$$

$$t = 0.967 \sqrt{\frac{8}{1 - (0.967)^2}} = 10.7.$$

This value is significant at the 0.01 probability level. We conclude that a significant relationship exists.

18-8 A Warning

Nonparametric methods are an important addition to our statistical weaponry. Where meaningful, the tests and measures described in this chapter are at least 75 percent as powerful as the related parametric test. That is, a sample size one-third larger is necessary (at most) to achieve equivalent power with a nonparametric test. Moreover, some tests such as the Mann–Whitney, the Wilcoxon, and the Kruskal–Wallis tests are 95 percent as powerful. Yet increased sampling does cost money, so if the stronger measurement and test assumptions are met, parametric methods are preferred.

Although many of the restrictive assumptions of the parametric approaches can be avoided, the basic requirement of randomness remains. All statistical tests rest on this critical assumption. The decision maker should be ever vigilant for violations.

We also observe that no nonparametric methods exist for testing interactions as can be done in the analysis-of-variance model. Nonparametric association measures, which capture the importance of relationships, exist only for the case of two variables. We have nothing to take the place of the correlation coefficient in a multiple correlation model. Further, tables for small sample sizes are often not easily available for some small-sample tests. (But see Siegal's book referenced earlier in this chapter.) For large sample sizes, approximations using the normal or χ^2 distributions can often be used.

Key Formulas

Formula	Used to Compute:
$P(\lvert x - \mu \rvert \leq k\sigma) \geq 1 - \dfrac{1}{k^2}$	Tchebycheff's unequality. The minimum probability of being within k standard deviations of the mean.
$\chi^2 = \sum_{i=1}^{m} \dfrac{(O_i - E_i)^2}{E_i}$	The chi-square statistic.
$D = \max_{t} \left\lvert \sum_{i=0}^{t} \left(\dfrac{O_i}{n}\right) - \sum_{i=0}^{t} \left(\dfrac{E_i}{n}\right) \right\rvert$	The Kolmogorov–Smirnov statistic.
$\mu_r = \dfrac{2N_1 N_2}{N_1 + N_2} + 1$	Expected number of runs under the null hypothesis: normal approximation.
$\sigma_r^2 = \dfrac{2N_1 N_2 (2N_1 N_2 - N_1 - N_2)}{(N_1 + N_2)^2 (N_1 + N_2 - 1)}$	Variance of the number of runs under the null hypothesis: normal approximation.
$T_1 = N_1 \cdot N_2 + \dfrac{N_1(N_1 + 1)}{2} - R_1$	Statistic used in the Mann–Whitney U test.
$T_2 = N_1 \cdot N_2 - T_1$	Statistic used in the Mann–Whitney U test.
$U = \min\{T_1, T_2\}$	Mann–Whitney U statistic.
$\mu_U = \dfrac{N_1 N_2}{2}$	Expected value of U: normal approximation: Mann–Whitney test.
$\sigma_U^2 = \dfrac{N_1 N_2 (N_1 + N_2 + 1)}{12}$	Variance of U: normal approximation: Mann–Whitney test.
$\mu_T = \dfrac{N(N + 1)}{4}$	Expected value of T: normal approximation: Wilcoxon test.
$\sigma_T^2 = \dfrac{N(N + 1)(2N + 1)}{24}$	Variance of T: normal approximation: Wilcoxon test.
$H = -3(N + 1) + \dfrac{12}{N(N + 1)} \sum_{j=1}^{k} \dfrac{R_j^2}{N_j}$	Kruskal–Wallis statistic, where $R_j \equiv$ sum of the ranks in the jth sample and $N_j \equiv$ number of observations in the jth sample.
$C = \sqrt{\dfrac{\chi^2}{N + \chi^2}}$	The contingency coefficient.
$r_s = 1 - \dfrac{6 \sum d^2}{N^3 - N}$	Spearman's rank correlation coefficient.
$s_{r_s}^2 = \dfrac{1 - r_s^2}{N - 2}$	Variance of Spearman's rank correlation coefficient used in a t test.

Nonparametric Methods

We close by noting that once we master the techniques in this book we will know as much about them as Sir Ronald Aylmar Fisher, often called the father of statistical methodology, knew about them. Of course, just to keep things in some sort of perspective, we also know everything William Shakespeare knew about the alphabet.

New Symbols

Symbol	Meaning
d_2	A factor used in estimating the sample standard deviation.
d	The difference in ranks for an item used in calculating Spearman's rank correlation coefficient.
T	Absolute value of smaller rank sum: Wilcoxon test.

PROBLEMS

18-1. Discuss the technical difference between "distribution-free" and "nonparametric" statistics. When using a z test (standard normal) to test an hypothesis, is the test either distribution-free or nonparametric? Why or why not?

18-2. Discuss the advantages and disadvantages of nonparametric methods, where nonparametric here is meant to include distribution-free methods.

18-3.* Is Tchebycheff's inequality a nonparametric method? Is it a distribution-free method?

18-4. A sales manager for a business equipment manufacturer has last year's total dollar sales for the firm's 12 sales representatives. The sales figures range from $150,000 to $2,400,000. Using this information, estimate the mean and standard deviation of the distribution. What assumptions, if any, are used? Discuss the validity of the estimates, and indicate how the mean and variance should be estimated in such a situation.

18-5. Sections 18-2 and 18-3 both use the chi-square distribution. Discuss the difference between the two tests. (In particular, explain how the "expected" numbers are derived.)

18-6. What advantage does the Kolmogorov–Smirnov test have over the chi-square test? What are the Kolmogorov–Smirnov test's relative disadvantages?

18-7.* The inventory and warehouse manager for a sporting goods distributor believes that tennis racquet sales, classified by grip size, follow a predictable distribution. (The manager believes that the same distribution has held for 15 years.) In particular, the manager thinks that for every three racquets sold with a $4\frac{3}{8}$-inch grip, five racquets are sold with a $4\frac{1}{2}$-inch grip and two with a $4\frac{5}{8}$-inch grip. If this pattern still holds, it would simplify the ordering decisions for restocking. Last month the firm sold 200 racquets, 42 with a $4\frac{3}{8}$-inch grip, 104 with a $4\frac{1}{2}$-inch grip and 54 with a $4\frac{5}{8}$-inch grip. At the $\alpha = 0.05$ level, test the null hypothesis that no change in the relative sales distribution has occurred.

18-8.* In Problem 18-7, we found, for sales of racquets by grip size:

	Grip		
	$4\frac{3}{8}$	$4\frac{1}{2}$	$4\frac{5}{8}$
O_i	42	104	54
E_i	60	100	40

Convert these data to relative frequencies and compute D, the value of the Kolmogorov–Smirnov test. For $\alpha = 0.05$, the critical value of D can be found to be $1.36/\sqrt{N} = 1.36/\sqrt{200} = 0.096$. Can we reject H_0 based on the Kolmogorov–Smirnov test?

18-9. In Section 18-4, a series of the sort B, G, B, G, B, G and one of the sort B, B, B, G, G, G were both described as nonrandom. One might say that both of these are random since the probability of a B is the same as the probability of a G. Why is this last statement incorrect and why might the two series be nonrandom?

18-10. The Automatic Widget Foundry Limited (AWF-L) is concerned about sales forecasts. The production manager believes that sales are random and, thus, unpredictable. The marketing manager believes there is a sales pattern. They both believe there is no trend. The last 24 months of sales (thousands of dollars) are as follows:

21, 25, 22, 20, 14, 12, 16, 17, 16, 20, 21, 25, 24, 23, 23, 14, 16, 17, 15, 19, 23, 24, 26, 27.

NONPARAMETRIC METHODS

(a) Form a + and − series based on the difference from a median of $20\frac{1}{2}$ and apply a runs test for randomness. What conclusion can you draw? (State it in words.)

(b) Why might the firm want to know whether their sales are random or not? What might they do next?

18-11. The signs test for significance is based on matched pairs. In Chapter 12, we discussed another test for distinguishing between two means, based on matched pairs. Discuss the assumptions made by the two tests.

18-12.* The yield from a particular manufacturing process depends on a key machine that is used in the process. A new machine has been suggested as a replacement for the existing machine. To compare the two, a test has been run to obtain the yields for 10 days of production. One batch of the product is produced by each machine on each day. The yields are:

Old machine:
98.1, 94.2, 93.0, 96.8, 91.7, 94.0, 91.9, 97.2, 95.5, 96.5

New machine:
91.2, 93.9, 95.8, 96.7, 99.0, 99.1, 98.4, 98.1, 99.5, 99.0

(a) Using a distribution-free test for matched pairs, at the 0.05 level, is the new machine superior to the old machine?

(b) Examine the data and decide (based only on visual observation) whether the new machine is superior or not. Give a reason for your answer.

18-13. Discuss the differences between a Mann–Whitney U test and a t test, both of which are used to compare measures of location for two samples.

18-14. What hypothesis does the Wilcoxon test consider? Is it analogous to a t test? Why or why not?

18-15. Compare the Kruskal–Wallis test to the F test.

18-16.* In Problem 18-12, data are given for yields of a process, using two different machines. Use the Mann–Whitney U test to determine if the new machine gives a higher median yield, at the 0.05 level. A one-tail test is appropriate, and the critical value for U is 27.

18-17. Read Problems 18-12 and 18-16 again. The signs test was unable to find a significant difference, while the Mann–Whitney test was. Explain why this happened. If either test is appropriate in a given situation, which would be preferable, and why? Why would the other test ever be used?

18-18.* (a) In Problem 18-12, data on yields for two machines are given. Apply the Wilcoxon test to that data to see if there is a difference between yields for the two machines. Use the normal approximation, and comment on whether it is appropriate to do so.
(b) Is the Wilcoxon test more or less powerful, in general, than the signs test? Why?

18-19.* A consumer testing organization wants to know if there are geographic differences with respect to how long individuals keep a certain type of car. Some data, for several individuals, are as follows:

Years of Ownership of Last Wheelie Whizzer

Individual Number	Region 1: Rural	Region 2: Suburban	Region 3: Urban	Region 4: Small Town
1	7	3	3	6
2	12	2	3	1
3	3	1	6	4
4	10	6	1	9
5	11	10	8	13
6	7	4	4	2
7			4	11
8			2	

Use a nonparametric test to see if there are geographic differences.

18-20.* (a) The contingency coefficient requires data measured on at least what type of scale? Give an example of that kind of measurement.
(b) Spearman's rank correlation coefficient requires data measured on at least what type of scale? Give an example of that kind of measurement.
(c) Would you ever choose to use the two measures of association above for data measured on a more powerful measurement scale? Why or why not?

18-21. Use a simple numerical example to demonstrate that $0.707 = C$ is the highest value possible for the 2×2 contingency-table case.

18-22.* A clothing store chain wants to see if different stores have the same size distribution of sales for sweaters. (The null hypothesis is that there is no association.) Last month's sales for one sweater line are as follows:

Sweater Size

Store	Small	Medium	Large	Total
1	40	80	50	170
2	20	80	80	180
Total	60	160	130	350

(a) Compute C, the contingency coefficient.
(b) Is the degree of association significantly different from zero? (Use $\alpha = 0.10$).

18-23.* A corporation is considering the validity of their procedures for filling executive vacancies. The procedure uses, among other things, a ranking of possible candidates by the executives familiar with the individual's ability. They would feel more comfortable about using these data if they knew that the rankings of different managers were consistent with each other. They have asked two managers to rank 10 candidates for a particular position:

Candidate	Executive 1	2
A	1	1
B	2	4
C	3	5
D	4	2
E	5	7
F	6	3
G	7	6
H	8	9
I	9	10
J	10	8

(a) Compute Spearman's rank correlation coefficient for these data.
(b) Test the significance of the value found in (a).

18-24. The Widgets of America Manufacturing Company (WAMCO!) is considering hiring unskilled workers. The assistant plant manager argues that these unskilled workers do not work effectively, and proposes the following test to verify the claim. Five unskilled workers

(A to E) and five skilled workers (F to J) performed the relevant task. The times are as follows:

Individual	A	B	C	D	E	F	G	H	I	J
Time (minutes)	4.99	5.42	5.68	5.91	6.42	4.91	5.62	6.99	8.12	9.43

The manager says that a ranking with the skilled workers F through J, respectively, in positions 1, 4, 5, 8, and 9 and unskilled workers A through E, respectively, in positions 2, 3, 6, 7, and 10 would show that the skilled were better than the unskilled. Thus, the manager wants to see if the actual ranking is significantly correlated with: $F, A, B, G, H, C, D, I, J, E$ [where the two sets of workers are (separately) ranked in order].

(a) Is the actual ranking significantly correlated with the manager's fabricated one? Use Spearman's rank correlated coefficient.

(b) The unskilled workers claim that the average performance is what counts and they say their average is better. They want the manager to use a one-tail t test, with $\alpha = 0.10$, to test the difference between the averages. (The two means are 5.684 and 7.014. The two standard deviations are 0.535 and 1.832, respectively. Also $s_{\bar{x}_1 - \bar{x}_2} = 0.854$.)

(c) What does it all mean?

18-25. Comment on the following:
        Question: Why is a t test like a distribution-free test?
        Response: Because of the Central Limit Theorem.

Appendix: Tables

Table I Areas for a Standard Normal Distribution

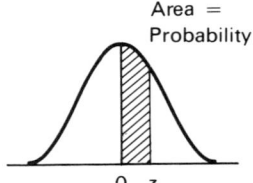

Area = Probability

Entries in the table represent the area under the curve between $z = 0$ and a positive value of z. Because of the symmetry of the curve, the area under the curve between $z = 0$ and a negative value of z would be found in a like manner.

z	.00	.01	.02	.03	.04	.05	.06	.07	.08	.09
0.0	.0000	.0040	.0080	.0120	.0160	.0199	.0239	.0279	.0319	.0359
0.1	.0398	.0438	.0478	.0517	.0557	.0596	.0636	.0675	.0714	.0753
0.2	.0793	.0832	.0871	.0910	.0948	.0987	.1026	.1064	.1103	.1141
0.3	.1179	.1217	.1255	.1293	.1331	.1368	.1406	.1443	.1480	.1517
0.4	.1554	.1591	.1628	.1664	.1700	.1736	.1772	.1808	.1844	.1879
0.5	.1915	.1950	.1985	.2019	.2054	.2088	.2123	.2157	.2190	.2224
0.6	.2257	.2291	.2324	.2357	.2389	.2422	.2454	.2486	.2518	.2549
0.7	.2580	.2612	.2642	.2673	.2704	.2734	.2764	.2794	.2823	.2852
0.8	.2881	.2910	.2939	.2967	.2995	.3023	.3051	.3078	.3106	.3133
0.9	.3159	.3186	.3212	.3238	.3264	.3289	.3315	.3340	.3365	.3389
1.0	.3413	.3438	.3461	.3485	.3508	.3531	.3554	.3577	.3599	.3621
1.1	.3643	.3665	.3686	.3708	.3729	.3749	.3770	.3790	.3810	.3830
1.2	.3849	.3869	.3888	.3907	.3925	.3944	.3962	.3980	.3997	.4015
1.3	.4032	.4049	.4066	.4082	.4099	.4115	.4131	.4147	.4162	.4177
1.4	.4192	.4207	.4222	.4236	.4251	.4265	.4279	.4292	.4306	.4319
1.5	.4332	.4345	.4357	.4370	.4382	.4394	.4406	.4418	.4429	.4441
1.6	.4452	.4463	.4474	.4484	.4495	.4505	.4515	.4525	.4535	.4545
1.7	.4554	.4564	.4573	.4582	.4591	.4599	.4608	.4616	.4625	.4633
1.8	.4641	.4649	.4656	.4664	.4671	.4678	.4686	.4693	.4699	.4706
1.9	.4713	.4719	.4726	.4732	.4738	.4744	.4750	.4756	.4761	.4767
2.0	.4772	.4778	.4783	.4788	.4793	.4798	.4803	.4808	.4812	.4817
2.1	.4821	.4826	.4830	.4834	.4838	.4842	.4846	.4850	.4854	.4857
2.2	.4861	.4864	.4868	.4871	.4875	.4878	.4881	.4884	.4887	.4890
2.3	.4893	.4896	.4898	.4901	.4904	.4906	.4909	.4911	.4913	.4916
2.4	.4918	.4920	.4922	.4925	.4927	.4929	.4931	.4932	.4934	.4936
2.5	.4938	.4940	.4941	.4943	.4945	.4946	.4948	.4949	.4951	.4952
2.6	.4953	.4955	.4956	.4957	.4959	.4960	.4961	.4962	.4963	.4964
2.7	.4965	.4966	.4967	.4968	.4969	.4970	.4971	.4972	.4973	.4974
2.8	.4974	.4975	.4976	.4977	.4977	.4978	.4979	.4979	.4980	.4981
2.9	.4981	.4982	.4982	.4983	.4984	.4984	.4985	.4985	.4986	.4986
3.0	.49865	.4987	.4987	.4988	.4988	.4989	.4989	.4989	.4990	.4990
4.0	.49997									

Table II Student t Distribution

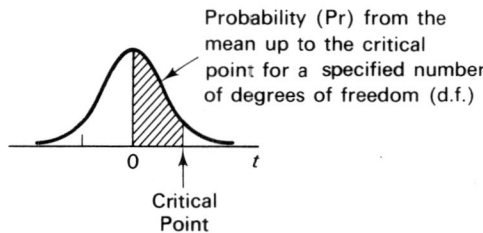

Probability (Pr) from the mean up to the critical point for a specified number of degrees of freedom (d.f.)

Critical Point

The following table provides the values of t_α that correspond to a given probability (area) between the mean and plus (or minus) t and a specified number of degrees of freedom.

Degrees of Freedom	0.10	0.25	0.40	0.45	0.475	0.49	0.495	0.4990
1	0.325	1.000	3.078	6.314	12.706	31.821	63.657	318.31
2	0.289	0.816	1.886	2.920	4.303	6.965	9.925	22.326
3	0.277	0.765	1.638	2.353	3.182	4.541	5.841	10.213
4	0.271	0.741	1.533	2.132	2.776	3.747	4.604	7.173
5	0.267	0.727	1.476	2.015	2.571	3.365	4.032	5.893
6	0.265	0.718	1.440	1.943	2.447	3.143	3.707	5.208
7	0.263	0.711	1.415	1.895	2.365	2.998	3.499	4.785
8	0.262	0.706	1.397	1.860	2.306	2.896	3.355	4.501
9	0.261	0.703	1.383	1.833	2.262	2.821	3.250	4.297
10	0.260	0.700	1.372	1.812	2.228	2.764	3.169	4.144
11	0.260	0.697	1.363	1.796	2.201	2.781	3.106	4.025
12	0.259	0.695	1.356	1.782	2.179	2.681	3.055	3.930
13	0.259	0.694	1.350	1.771	2.160	2.650	3.012	3.852
14	0.258	0.692	1.345	1.761	2.145	2.624	2.977	3.787
15	0.258	0.691	1.341	1.753	2.131	2.602	2.947	3.733
16	0.258	0.690	1.337	1.746	2.120	2.583	2.921	3.686
17	0.257	0.689	1.333	1.740	2.110	2.567	2.898	3.646
18	0.257	0.688	1.330	1.734	2.101	2.552	2.878	3.610
19	0.257	0.688	1.328	1.729	2.093	2.539	2.861	3.579
20	0.257	0.687	1.325	1.725	2.086	2.528	2.845	3.552
21	0.257	0.686	1.323	1.721	2.080	2.518	2.831	3.527
22	0.256	0.686	1.321	1.717	2.074	2.508	2.819	3.505
23	0.256	0.685	1.319	1.714	2.069	2.500	2.807	3.485
24	0.256	0.685	1.318	1.711	2.064	2.492	2.797	3.467
25	0.256	0.684	1.316	1.708	2.060	2.485	2.787	3.450
26	0.256	0.684	1.315	1.706	2.056	2.479	2.779	3.435
27	0.256	0.684	1.314	1.703	2.052	2.473	2.771	3.421
28	0.256	0.683	1.313	1.701	2.048	2.467	2.763	3.408
29	0.256	0.683	1.311	1.699	2.045	2.462	2.756	3.396
30	0.256	0.683	1.310	1.697	2.042	2.457	2.750	3.385
40	0.255	0.681	1.303	1.684	2.021	2.423	2.704	3.307
60	0.254	0.679	1.296	1.671	2.000	2.390	2.660	3.232
120	0.254	0.677	1.289	1.658	1.980	2.358	2.617	3.160
∞	0.253	0.674	1.282	1.645	1.960	2.236	2.576	3.090

SOURCE: E. S. Pearson and H. O. Hartley, *Biometrika Tables for Statisticians*, Vol. 1, 1966, London, by permission.

Table III Percentage Points of the χ^2 Distribution Function

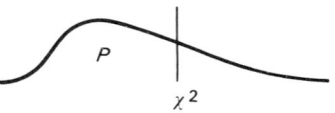

d.f.[a]	0.005	0.010	0.025	0.050	0.100	0.250	0.500
1	392704.10^{-10}	157088.10^{-9}	982069.10^{-9}	393214.10^{-8}	.0157908	.1015308	.454937
2	.0100251	.0201007	.0506356	.102587	.210720	.575364	1.38629
3	.0717212	.114832	.215795	.351846	.584375	1.212534	2.36597
4	.206990	.297110	.484419	.710721	1.063623	1.92255	3.35670
5	.411740	.554300	.831211	1.145476	1.61031	2.67460	4.35146
6	.675727	.872085	1.237347	1.63539	2.20413	3.45460	5.34812
7	.989265	1.239043	1.68987	2.16735	2.83311	4.25485	6.34581
8	1.344419	1.646482	2.17973	2.73264	3.48954	5.07064	7.34412
9	1.734926	2.087912	2.70039	3.32511	4.16816	5.89883	8.34283
10	2.15585	2.55821	3.24697	3.94030	4.86518	6.73720	9.34182
11	2.60321	3.05347	3.81575	4.57481	5.57779	7.58412	10.3410
12	3.07382	3.57056	4.40379	5.22603	6.30380	8.43842	11.3403
13	3.56503	4.10691	5.00874	5.89186	7.04150	9.29906	12.3398
14	4.07468	4.66043	5.62872	6.57063	7.78953	10.1653	13.3393
15	4.60094	5.22935	6.26214	7.26094	8.54675	11.0365	14.3389
16	5.14224	5.81221	6.90766	7.96164	9.31223	11.9122	15.3385
17	5.69724	6.40776	7.56418	8.67176	10.0852	12.7919	16.3381
18	6.26481	7.01491	8.23075	9.39046	10.8649	13.6753	17.3379
19	6.84398	7.63273	8.90655	10.1170	11.6509	14.5620	18.3376
20	7.43386	8.26040	9.59083	10.8508	12.4426	15.4518	19.3374
21	8.03366	8.89720	10.28293	11.5913	13.2396	16.3444	20.3372
22	8.64272	9.54249	10.9823	12.3380	14.0415	17.2396	21.3370
23	9.26042	10.19567	11.6885	13.0905	14.8479	18.1373	22.3369
24	9.88623	10.8564	12.4011	13.8484	15.6587	19.0372	23.3367
25	10.5197	11.5240	13.1197	14.6114	16.4734	19.9393	24.3366
26	11.1603	12.1981	13.8439	15.3791	17.2919	20.8434	25.3364
27	11.8076	12.8786	14.5733	16.1513	18.1138	21.7494	26.3363
28	12.4613	13.5648	15.3079	16.9279	18.9392	22.6572	27.3363
29	13.1211	14.2565	16.0471	17.7083	19.7677	23.5666	28.3362
30	13.7867	14.9535	16.7908	18.4926	20.5992	24.4776	29.3360
40	20.7065	22.1643	24.4331	26.5093	29.0505	33.6603	39.3354
50	27.9907	29.7067	32.3574	34.7642	37.6886	42.9421	49.3349
60	35.5346	37.4848	40.4817	43.1879	46.4589	52.2938	59.3347
70	43.2752	45.4418	48.7576	51.7393	55.3290	61.6983	69.3344
80	51.1720	53.5400	57.1532	60.3915	64.2778	71.1445	79.3343
90	59.1963	61.7541	65.6466	69.1260	73.2912	80.6247	89.3342
100	67.3276	70.0648	74.2219	77.9295	82.3581	90.1332	99.3341
X	-2.5758	-2.3263	-1.9600	-1.6449	-1.2816	-0.6745	0.0000

[a] The symbol d.f. stands for the number of degrees of freedom.

SOURCE: E. S. Pearson and H. O. Hartley, *Biometrika Tables for Statisticians* (Table 3), Vol. I with permission.

Table III (continued)

P

d.f.[a]	0.750	0.900	0.950	0.975	0.990	0.995	0.999
1	1.32330	2.70554	3.84146	5.02389	6.63490	7.87944	10.828
2	2.77259	4.60517	5.99147	7.37776	9.21034	10.5966	13.816
3	4.10835	6.25139	7.81473	9.34840	11.3449	12.8381	16.266
4	5.38527	7.77944	9.48773	11.1433	13.2767	14.8602	18.467
5	6.62568	9.23635	11.0705	12.8325	15.0863	16.7496	20.515
6	7.84080	10.6446	12.5916	14.4494	16.8119	18.5476	22.458
7	9.03715	12.0170	14.0671	16.0128	18.4753	20.2777	24.322
8	10.2188	13.3616	15.5073	17.5346	20.0902	21.9550	26.125
9	11.3887	14.6837	16.9190	19.0228	21.6660	23.5893	27.877
10	12.5489	15.9871	18.3070	20.4831	23.2093	25.1882	29.588
11	13.7007	17.2750	19.6751	21.9200	24.7250	26.7569	31.264
12	14.8454	18.5494	21.0261	23.3367	26.2170	28.2995	32.909
13	15.9839	19.8119	22.3621	24.7356	27.6883	29.8194	34.528
14	17.1170	21.0642	23.6848	26.1190	29.1413	31.3193	36.123
15	18.2451	22.3072	24.9958	27.4884	30.5779	32.8013	37.697
16	19.3688	23.5418	26.2962	28.8454	31.9999	34.2672	39.252
17	20.4887	24.7690	27.5871	30.1910	33.4087	35.7185	40.790
18	21.6049	25.9894	28.8693	31.5264	34.8053	37.1564	42.312
19	22.7178	27.2036	30.1435	32.8523	36.1908	38.5822	43.820
20	23.8277	28.4120	31.4104	34.1696	37.5662	39.9968	45.315
21	24.9348	29.6151	32.6705	35.4789	38.9321	41.4010	46.797
22	26.0393	30.8133	33.9244	36.7807	40.2894	42.7956	48.268
23	27.1413	32.0069	35.1725	38.0757	41.6384	44.1813	49.728
24	28.2412	33.1963	36.4151	39.3641	42.9798	45.5585	51.179
25	29.3389	34.3816	37.6525	40.6465	44.3141	46.9278	52.620
26	30.4345	35.5631	38.8852	41.9232	45.6417	48.2899	54.052
27	31.5284	36.7412	40.1133	43.1944	46.9630	49.6449	55.476
28	32.6205	37.9159	41.3372	44.4607	48.2782	50.9933	56.892
29	33.7109	39.0875	42.5569	45.7222	49.5879	52.3356	58.302
30	34.7998	40.2560	43.7729	46.9792	50.8922	53.6720	59.703
40	45.6160	51.8050	55.7585	59.3417	63.6907	66.7659	73.402
50	56.3336	63.1671	67.5048	71.4202	76.1539	79.4900	86.661
60	66.9814	74.3970	79.0819	83.2976	88.3794	91.9517	99.607
70	77.5766	85.5271	90.5312	95.0231	100.425	104.215	112.317
80	88.1303	96.5782	101.879	106.629	112.329	116.321	124.839
90	98.6499	107.565	113.145	118.136	124.116	128.299	137.208
100	109.141	118.498	124.342	129.561	135.807	140.169	149.449
X	+0.6745	+1.2816	+1.6449	+1.9600	+2.3263	+2.5758	+3.0902

[a] For d.f. $= v > 100$, take

$$\chi^2 = v\left(1 - \frac{2}{9v} + z\sqrt{\frac{2}{9v}}\right)^3 \quad \text{or} \quad \chi^2 = \tfrac{1}{2}(z + \sqrt{2v - 1})^2$$

according to the degree of accuracy required. z is the standardized normal deviate.

Table IV Poisson Probabilities

The table gives the probability of exactly x occurrences for given λ.

λ

x	0.005	0.01	0.02	0.03	0.04	0.05	0.06	0.07	0.08	0.09
0	.9950	.9900	.9802	.9704	.9608	.9512	.9418	.9324	.9231	.9139
1	.0050	.0099	.0192	.0291	.0384	.0476	.0565	.0653	.0738	.0823
2	.0000	.0000	.0002	.0004	.0008	.0012	.0017	.0023	.0030	.0037
3	.0000	.0000	.0000	.0000	.0000	.0000	.0000	.0001	.0001	.0001

λ

x	0.1	0.2	0.3	0.4	0.5	0.6	0.7	0.8	0.9	1.0
0	.9048	.8187	.7408	.6703	.6065	.5488	.4966	.4493	.4066	.3679
1	.0905	.1637	.2222	.2681	.3033	.3293	.3476	.3595	.3659	.3679
2	.0045	.0164	.0333	.0536	.0758	.0988	.1217	.1438	.1647	.1839
3	.0002	.0011	.0033	.0072	.0126	.0198	.0284	.0383	.0494	.0613
4	.0000	.0001	.0002	.0007	.0016	.0030	.0050	.0077	.0111	.0153
5	.0000	.0000	.0000	.0001	.0002	.0004	.0007	.0012	.0020	.0031
6	.0000	.0000	.0000	.0000	.0000	.0000	.0001	.0002	.0003	.0005
7	.0000	.0000	.0000	.0000	.0000	.0000	.0000	.0000	.0000	.0001

λ

x	1.1	1.2	1.3	1.4	1.5	1.6	1.7	1.8	1.9	2.0
0	.3329	.3012	.2725	.2466	.2231	.2019	.1827	.1653	.1496	.1353
1	.3662	.3614	.3543	.3452	.3347	.3230	.3106	.2975	.2842	.2707
2	.2014	.2169	.2303	.2417	.2510	.2584	.2640	.2678	.2700	.2707
3	.0738	.0867	.0998	.1128	.1255	.1378	.1496	.1607	.1710	.1804
4	.0203	.0260	.0324	.0395	.0471	.0551	.0636	.0723	.0812	.0902
5	.0045	.0062	.0084	.0111	.0141	.0176	.0216	.0260	.0309	.0361
6	.0008	.0012	.0018	.0026	.0035	.0047	.0061	.0078	.0098	.0120
7	.0001	.0002	.0003	.0005	.0008	.0011	.0015	.0020	.0027	.0034
8	.0000	.0000	.0001	.0001	.0001	.0002	.0003	.0005	.0006	.0009
9	.0000	.0000	.0000	.0000	.0000	.0000	.0001	.0001	.0001	.0002

λ

x	2.1	2.2	2.3	2.4	2.5	2.6	2.7	2.8	2.9	3.0
0	.1225	.1108	.1003	.0907	.0821	.0743	.0672	.0608	.0550	.0498
1	.2572	.2438	.2306	.2177	.2052	.1931	.1815	.1703	.1596	.1494
2	.2700	.2681	.2652	.2613	.2565	.2510	.2450	.2384	.2314	.2240
3	.1890	.1966	.2033	.2090	.2138	.2176	.2205	.2225	.2237	.2240
4	.0992	.1082	.1169	.1254	.1336	.1414	.1488	.1557	.1622	.1680
5	.0417	.0476	.0538	.0602	.0668	.0735	.0804	.0872	.0940	.1008
6	.0146	.0174	.0206	.0241	.0278	.0319	.0362	.0407	.0455	.0504
7	.0044	.0055	.0068	.0083	.0099	.0118	.0139	.0163	.0188	.0216
8	.0011	.0015	.0019	.0025	.0031	.0038	.0047	.0057	.0068	.0081
9	.0003	.0004	.0005	.0007	.0009	.0011	.0014	.0018	.0022	.0027
10	.0001	.0001	.0001	.0002	.0002	.0003	.0004	.0005	.0006	.0008
11	.0000	.0000	.0000	.0000	.0000	.0001	.0001	.0001	.0002	.0002
12	.0000	.0000	.0000	.0000	.0000	.0000	.0000	.0000	.0000	.0001

Table IV Poisson Probabilities (continued)

					λ					
x	3.1	3.2	3.3	3.4	3.5	3.6	3.7	3.8	3.9	4.0
0	.0450	.0408	.0369	.0334	.0302	.0273	.0247	.0224	.0202	.0183
1	.1397	.1304	.1217	.1135	.1057	.0984	.0915	.0850	.0789	.0733
2	.2165	.2087	.2008	.1929	.1850	.1771	.1692	.1615	.1539	.1465
3	.2237	.2226	.2209	.2186	.2158	.2125	.2087	.2046	.2001	.1954
4	.1734	.1781	.1823	.1858	.1888	.1912	.1931	.1944	.1951	.1954
5	.1075	.1140	.1203	.1264	.1322	.1377	.1429	.1477	.1522	.1563
6	.0555	.0608	.0662	.0716	.0771	.0826	.0881	.0936	.0989	.1042
7	.0246	.0278	.0312	.0348	.0385	.0425	.0466	.0508	.0551	.0595
8	.0095	.0111	.0129	.0148	.0169	.0191	.0215	.0241	.0269	.0298
9	.0033	.0040	.0047	.0056	.0066	.0076	.0089	.0102	.0116	.0132
10	.0010	.0013	.0016	.0019	.0023	.0028	.0033	.0039	.0045	.0053
11	.0003	.0004	.0005	.0006	.0007	.0009	.0011	.0013	.0016	.0019
12	.0001	.0001	.0001	.0002	.0002	.0003	.0003	.0004	.0005	.0006
13	.0000	.0000	.0000	.0000	.0001	.0001	.0001	.0001	.0002	.0002
14	.0000	.0000	.0000	.0000	.0000	.0000	.0000	.0000	.0000	.0001

					λ					
x	4.1	4.2	4.3	4.4	4.5	4.6	4.7	4.8	4.9	5.0
0	.0166	.0150	.0136	.0123	.0111	.0101	.0091	.0082	.0074	.0067
1	.0679	.0630	.0583	.0540	.0500	.0462	.0427	.0395	.0365	.0337
2	.1393	.1323	.1254	.1188	.1125	.1063	.1005	.0948	.0894	.0842
3	.1904	.1852	.1798	.1743	.1687	.1631	.1574	.1517	.1460	.1404
4	.1951	.1944	.1933	.1917	.1898	.1875	.1849	.1820	.1789	.1755
5	.1600	.1633	.1662	.1687	.1708	.1725	.1738	.1747	.1753	.1755
6	.1093	.1143	.1191	.1237	.1281	.1323	.1362	.1398	.1432	.1462
7	.0640	.0686	.0732	.0778	.0824	.0869	.0914	.0959	.1002	.1044
8	.0328	.0360	.0393	.0428	.0463	.0500	.0537	.0575	.0614	.0653
9	.0150	.0168	.0188	.0209	.0232	.0255	.0280	.0307	.0334	.0363
10	.0061	.0071	.0081	.0092	.0104	.0118	.0132	.0147	.0164	.0181
11	.0023	.0027	.0032	.0037	.0043	.0049	.0056	.0064	.0073	.0082
12	.0008	.0009	.0011	.0014	.0016	.0019	.0022	.0026	.0030	.0034
13	.0002	.0003	.0004	.0005	.0006	.0007	.0008	.0009	.0011	.0013
14	.0001	.0001	.0001	.0001	.0002	.0002	.0003	.0003	.0004	.0005
15	.0000	.0000	.0000	.0000	.0001	.0001	.0001	.0001	.0001	.0002

Table IV Poisson Probabilities (continued)

λ

x	5.1	5.2	5.3	5.4	5.5	5.6	6.7	5.8	5.9	6.0
0	.0061	.0055	.0050	.0045	.0041	.0037	.0033	.0030	.0027	.0025
1	.0311	.0287	.0265	.0244	.0225	.0207	.0191	.0176	.0162	.0149
2	.0793	.0746	.0701	.0659	.0618	.0580	.0544	.0509	.0477	.0446
3	.1348	.1293	.1239	.1185	.1133	.1082	.1033	.0985	.0938	.0892
4	.1719	.1681	.1641	.1600	.1558	.1515	.1472	.1428	.1383	.1339
5	.1753	.1748	.1740	.1728	.1714	.1697	.1678	.1656	.1632	.1606
6	.1490	.1515	.1537	.1555	.1571	.1584	.1594	.1601	.1605	.1606
7	.1086	.1125	.1163	.1200	.1234	.1267	.1298	.1326	.1353	.1377
8	.0692	.0731	.0771	.0810	.0849	.0887	.0925	.0962	.0998	.1033
9	.0392	.0423	.0454	.0486	.0519	.0552	.0586	.0620	.0654	.0688
10	.0200	.0220	.0241	.0262	.0285	.0309	.0334	.0359	.0386	.0413
11	.0093	.0104	.0116	.0129	.0143	.0157	.0173	.0190	.0207	.0225
12	.0039	.0045	.0051	.0058	.0065	.0073	.0082	.0092	.0102	.0113
13	.0015	.0018	.0021	.0024	.0028	.0032	.0036	.0041	.0046	.0052
14	.0006	.0007	.0008	.0009	.0011	.0013	.0015	.0017	.0019	.0022
15	.0002	.0002	.0003	.0003	.0004	.0005	.0006	.0007	.0008	.0009
16	.0001	.0001	.0001	.0001	.0001	.0002	.0002	.0002	.0003	.0003
17	.0000	.0000	.0000	.0000	.0000	.0001	.0001	.0001	.0001	.0001

λ

x	6.1	6.2	6.3	6.4	6.5	6.6	6.7	6.8	6.9	7.0
0	.0022	.0020	.0018	.0017	.0015	.0014	.0012	.0011	.0010	.0009
1	.0137	.0126	.0116	.0106	.0098	.0090	.0082	.0076	.0070	.0064
2	.0417	.0390	.0364	.0340	.0318	.0296	.0276	.0258	.0240	.0223
3	.0848	.0806	.0765	.0726	.0688	.0652	.0617	.0584	.0552	.0521
4	.1294	.1249	.1205	.1162	.1118	.1076	.1034	.0992	.0952	.0912
5	.1579	.1549	.1519	.1487	.1454	.1420	.1385	.1349	.1314	.1277
6	.1605	.1601	.1595	.1586	.1575	.1562	.1546	.1529	.1511	.1490
7	.1399	.1418	.1435	.1450	.1462	.1472	.1480	.1486	.1489	.1490
8	.1066	.1099	.1130	.1160	.1188	.1215	.1240	.1263	.1284	.1304
9	.0723	.0757	.0791	.0825	.0858	.0891	.0923	.0954	.0985	.1014
10	.0441	.0469	.0498	.0528	.0558	.0588	.0618	.0649	.0679	.0710
11	.0245	.0265	.0285	.0307	.0330	.0353	.0377	.0401	.0426	.0452
12	.0124	.0137	.0150	.0164	.0179	.0194	.0210	.0227	.0245	.0264
13	.0058	.0065	.0073	.0081	.0089	.0098	.0108	.0119	.0130	.0142
14	.0025	.0029	.0033	.0037	.0041	.0046	.0052	.0058	.0064	.0071
15	.0010	.0012	.0014	.0016	.0018	.0020	.0023	.0026	.0029	.0033
16	.0004	.0005	.0005	.0006	.0007	.0008	.0010	.0011	.0013	.0014
17	.0001	.0002	.0002	.0002	.0003	.0003	.0004	.0004	.0005	.0006
18	.0000	.0001	.0001	.0001	.0001	.0001	.0001	.0002	.0002	.0002
19	.0000	.0000	.0000	.0000	.0000	.0000	.0000	.0001	.0001	.0001

Table IV Poisson Probabilities (continued)

χ	λ=7.1	7.2	7.3	7.4	7.5	7.6	7.7	7.8	7.9	8.0
0	.0008	.0007	.0007	.0006	.0006	.0005	.0005	.0004	.0004	.0003
1	.0059	.0054	.0049	.0045	.0041	.0038	.0035	.0032	.0029	.0027
2	.0208	.0194	.0180	.0167	.0156	.0145	.0134	.0125	.0116	.0107
3	.0492	.0464	.0438	.0413	.0389	.0366	.0345	.0324	.0305	.0286
4	.0874	.0836	.0799	.0764	.0729	.0696	.0663	.0632	.0602	.0573
5	.1241	.1204	.1167	.1130	.1094	.1057	.1021	.0986	.0951	.0916
6	.1468	.1445	.1420	.1394	.1367	.1339	.1311	.1282	.1252	.1221
7	.1489	.1486	.1481	.1474	.1465	.1454	.1442	.1428	.1413	.1396
8	.1321	.1337	.1351	.1363	.1373	.1382	.1388	.1392	.1395	.1396
9	.1042	.1070	.1096	.1121	.1144	.1167	.1187	.1207	.1224	.1241
10	.0740	.0770	.0800	.0829	.0858	.0887	.0914	.0941	.0967	.0993
11	.0478	.0504	.0531	.0558	.0585	.0613	.0640	.0667	.0695	.0722
12	.0283	.0303	.0323	.0344	.0366	.0388	.0411	.0434	.0457	.0481
13	.0154	.0168	.0181	.0196	.0211	.0227	.0243	.0260	.0278	.0296
14	.0078	.0086	.0095	.0104	.0113	.0123	.0134	.0145	.0157	.0169
15	.0037	.0041	.0046	.0051	.0057	.0062	.0069	.0075	.0083	.0090
16	.0016	.0019	.0021	.0024	.0026	.0030	.0033	.0037	.0041	.0045
17	.0007	.0008	.0009	.0010	.0012	.0013	.0015	.0017	.0019	.0021
18	.0003	.0003	.0004	.0004	.0005	.0006	.0006	.0007	.0008	.0009
19	.0001	.0001	.0001	.0002	.0002	.0002	.0003	.0003	.0003	.0004
20	.0000	.0000	.0001	.0001	.0001	.0001	.0001	.0001	.0001	.0002
21	.0000	.0000	.0000	.0000	.0000	.0000	.0000	.0000	.0001	.0001

χ	λ=8.1	8.2	8.3	8.4	8.5	8.6	8.7	8.8	8.9	9.0
0	.0003	.0003	.0002	.0002	.0002	.0002	.0002	.0002	.0001	.0001
1	.0025	.0023	.0021	.0019	.0017	.0016	.0014	.0013	.0012	.0011
2	.0100	.0092	.0086	.0079	.0074	.0068	.0063	.0058	.0054	.0050
3	.0269	.0252	.0237	.0222	.0208	.0195	.0183	.0171	.0160	.0150
4	.0544	.0517	.0491	.0466	.0443	.0420	.0398	.0377	.0357	.0337
5	.0882	.0849	.0816	.0784	.0752	.0722	.0692	.0663	.0635	.0607
6	.1191	.1160	.1128	.1097	.1066	.1034	.1003	.0972	.0941	.0911
7	.1378	.1358	.1338	.1317	.1294	.1271	.1247	.1222	.1197	.1171
8	.1395	.1392	.1388	.1382	.1375	.1366	.1356	.1344	.1332	.1318
9	.1256	.1269	.1280	.1290	.1299	.1306	.1311	.1315	.1317	.1318
10	.1017	.1040	.1063	.1084	.1104	.1123	.1140	.1157	.1172	.1186
11	.0749	.0776	.0802	.0828	.0853	.0878	.0902	.0925	.0948	.0970
12	.0505	.0530	.0555	.0579	.0604	.0629	.0654	.0679	.0703	.0728
13	.0315	.0334	.0354	.0374	.0395	.0416	.0438	.0459	.0481	.0504
14	.0182	.0196	.0210	.0225	.0240	.0256	.0272	.0289	.0306	.0324
15	.0098	.0107	.0116	.0126	.0136	.0147	.0158	.0169	.0182	.0194
16	.0050	.0055	.0060	.0066	.0072	.0079	.0086	.0093	.0101	.0109
17	.0024	.0026	.0029	.0033	.0036	.0040	.0044	.0048	.0053	.0058
18	.0011	.0012	.0014	.0015	.0017	.0019	.0021	.0024	.0026	.0029
19	.0005	.0005	.0006	.0007	.0008	.0009	.0010	.0011	.0012	.0014
20	.0002	.0002	.0002	.0003	.0003	.0004	.0004	.0005	.0005	.0006
21	.0001	.0001	.0001	.0001	.0001	.0002	.0002	.0002	.0002	.0003
22	.0000	.0000	.0000	.0000	.0001	.0001	.0001	.0001	.0001	.0001

Table IV Poisson Probabilities (continued)

λ

χ	9.1	9.2	9.3	9.4	9.5	9.6	9.7	9.8	9.9	10.0
0	.0001	.0001	.0001	.0001	.0001	.0001	.0001	.0001	.0001	.0000
1	.0010	.0009	.0009	.0008	.0007	.0007	.0006	.0005	.0005	.0005
2	.0046	.0043	.0040	.0037	.0034	.0031	.0029	.0027	.0025	.0023
3	.0140	.0131	.0123	.0115	.0107	.0100	.0093	.0087	.0081	.0076
4	.0319	.0302	.0285	.0269	.0254	.0240	.0226	.0213	.0201	.0189
5	.0581	.0555	.0530	.0506	.0483	.0460	.0439	.0418	.0398	.0378
6	.0881	.0851	.0822	.0793	.0764	.0736	.0709	.0682	.0656	.0631
7	.1145	.1118	.1091	.1064	.1037	.1010	.0982	.0955	.0928	.0901
8	.1302	.1286	.1269	.1251	.1232	.1212	.1191	.1170	.1148	.1126
9	.1317	.1315	.1311	.1306	.1300	.1293	.1284	.1274	.1263	.1251
10	.1198	.1210	.1219	.1228	.1235	.1241	.1245	.1249	.1250	.1251
11	.0991	.1012	.1031	.1049	.1067	.1083	.1098	.1112	.1125	.1137
12	.0752	.0776	.0799	.0822	.0844	.0866	.0888	.0908	.0928	.0948
13	.0526	.0549	.0572	.0594	.0617	.0640	.0662	.0685	.0707	.0729
14	.0342	.0361	.0380	.0399	.0419	.0439	.0459	.0479	.0500	.0521
15	.0208	.0221	.0235	.0250	.0265	.0281	.0297	.0313	.0330	.0347
16	.0118	.0127	.0137	.0147	.0157	.0168	.0180	.0192	.0204	.0217
17	.0063	.0069	.0075	.0081	.0088	.0095	.0103	.0111	.0119	.0128
18	.0032	.0035	.0039	.0042	.0046	.0051	.0055	.0060	.0065	.0071
19	.0015	.0017	.0019	.0021	.0023	.0026	.0028	.0031	.0034	.0037
20	.0007	.0008	.0009	.0010	.0011	.0012	.0014	.0015	.0017	.0019
21	.0003	.0003	.0004	.0004	.0005	.0006	.0006	.0007	.0008	.0009
22	.0001	.0001	.0002	.0002	.0002	.0002	.0003	.0003	.0004	.0004
23	.0000	.0001	.0001	.0001	.0001	.0001	.0001	.0001	.0002	.0002
24	.0000	.0000	.0000	.0000	.0000	.0000	.0000	.0001	.0001	.0001

Table V Binomial Distribution Function (Probability of x or Fewer Successes in n Trials, Where p Is the Probability on One Trial)

n	x	p=.01	.02	.03	.04	.05	.06	.07	.08	.09
2	0	.9801	.9604	.9409	.9216	.9025	.8836	.8649	.8464	.8281
2	1	.9999	.9996	.9991	.9984	.9975	.9964	.9951	.9936	.9919
2	2	1.0000	1.0000	1.0000	1.0000	1.0000	1.0000	1.0000	1.0000	1.0000
3	0	.9703	.9411	.9126	.8847	.8573	.8305	.8043	.7786	.7535
3	1	.9997	.9988	.9973	.9953	.9927	.9896	.9859	.9818	.9771
3	2	1.0000	.9999	.9999	.9999	.9998	.9997	.9996	.9994	.9992
3	3	1.0000	1.0000	1.0000	1.0000	1.0000	1.0000	1.0000	1.0000	1.0000
4	0	.9606	.9223	.8852	.8493	.8145	.7807	.7480	.7163	.6857
4	1	.9994	.9976	.9948	.9909	.9859	.9800	.9732	.9655	.9570
4	2	1.0000	.9999	.9998	.9997	.9995	.9991	.9987	.9980	.9972
4	3	1.0000	1.0000	1.0000	1.0000	.9999	.9999	.9999	.9999	.9999
4	4	1.0000	1.0000	1.0000	1.0000	1.0000	1.0000	1.0000	1.0000	1.0000
5	0	.9509	.9039	.8587	.8153	.7737	.7339	.6956	.6590	.6240
5	1	.9990	.9961	.9915	.9852	.9774	.9681	.9575	.9456	.9326
5	2	.9999	.9999	.9997	.9994	.9988	.9980	.9969	.9954	.9936
5	3	1.0000	1.0000	1.0000	.9999	.9999	.9999	.9998	.9998	.9997
5	4	1.0000	1.0000	1.0000	1.0000	1.0000	1.0000	1.0000	1.0000	.9999
5	5	1.0000	1.0000	1.0000	1.0000	1.0000	1.0000	1.0000	1.0000	1.0000
6	0	.9414	.8858	.8329	.7827	.7350	.6898	.6469	.6063	.5678
6	1	.9985	.9943	.9875	.9784	.9672	.9540	.9391	.9227	.9048
6	2	.9999	.9998	.9995	.9988	.9977	.9962	.9941	.9914	.9881
6	3	1.0000	1.0000	.9999	.9999	.9999	.9998	.9996	.9994	.9991
6	4	1.0000	1.0000	1.0000	1.0000	1.0000	1.0000	.9999	.9999	.9999
6	5	1.0000	1.0000	1.0000	1.0000	1.0000	1.0000	1.0000	1.0000	1.0000
6	6	1.0000	1.0000	1.0000	1.0000	1.0000	1.0000	1.0000	1.0000	1.0000
7	0	.9320	.8681	.8079	.7514	.6983	.6484	.6017	.5578	.5167
7	1	.9979	.9921	.9829	.9706	.9556	.9382	.9187	.8974	.8745
7	2	.9999	.9997	.9991	.9980	.9962	.9937	.9903	.9859	.9806
7	3	1.0000	.9999	.9999	.9999	.9998	.9996	.9992	.9988	.9981
7	4	1.0000	1.0000	1.0000	1.0000	.9999	.9999	.9999	.9999	.9998
7	5	1.0000	1.0000	1.0000	1.0000	1.0000	1.0000	1.0000	1.0000	1.0000
7	6	1.0000	1.0000	1.0000	1.0000	1.0000	1.0000	1.0000	1.0000	1.0000
7	7	1.0000	1.0000	1.0000	1.0000	1.0000	1.0000	1.0000	1.0000	1.0000
8	0	.9227	.8507	.7837	.7213	.6634	.6095	.5595	.5132	.4702
8	1	.9973	.9896	.9776	.9618	.9427	.9208	.8965	.8702	.8423
8	2	.9999	.9995	.9986	.9969	.9942	.9903	.9853	.9789	.9711
8	3	1.0000	.9999	.9999	.9998	.9996	.9992	.9986	.9978	.9965
8	4	1.0000	1.0000	1.0000	.9999	.9999	.9999	.9999	.9998	.9997
8	5	1.0000	1.0000	1.0000	1.0000	1.0000	1.0000	1.0000	.9999	.9999
8	6	1.0000	1.0000	1.0000	1.0000	1.0000	1.0000	1.0000	1.0000	1.0000
8	7	1.0000	1.0000	1.0000	1.0000	1.0000	1.0000	1.0000	1.0000	1.0000
8	8	1.0000	1.0000	1.0000	1.0000	1.0000	1.0000	1.0000	1.0000	1.0000
9	0	.9135	.8337	.7602	.6925	.6302	.5729	.5204	.4721	.4279
9	1	.9965	.9868	.9718	.9522	.9287	.9021	.8729	.8416	.8088
9	2	.9999	.9993	.9980	.9955	.9916	.9862	.9790	.9702	.9595
9	3	1.0000	.9999	.9999	.9997	.9993	.9987	.9977	.9962	.9943
9	4	1.0000	1.0000	1.0000	.9999	.9999	.9999	.9998	.9996	.9994
9	5	1.0000	1.0000	1.0000	1.0000	1.0000	1.0000	.9999	.9999	.9999
9	6	1.0000	1.0000	1.0000	1.0000	1.0000	1.0000	1.0000	1.0000	1.0000
9	7	1.0000	1.0000	1.0000	1.0000	1.0000	1.0000	1.0000	1.0000	1.0000
9	8	1.0000	1.0000	1.0000	1.0000	1.0000	1.0000	1.0000	1.0000	1.0000
9	9	1.0000	1.0000	1.0000	1.0000	1.0000	1.0000	1.0000	1.0000	1.0000

Table V (continued)

n	x	p=.01	.02	.03	.04	.05	.06	.07	.08	.09
10	0	.9043	.8170	.7374	.6648	.5987	.5386	.4839	.4343	.3894
10	1	.9957	.9838	.9654	.9418	.9138	.8824	.8482	.8121	.7745
10	2	.9998	.9991	.9972	.9937	.9885	.9811	.9716	.9599	.9459
10	3	1.0000	.9999	.9998	.9995	.9989	.9979	.9964	.9942	.9911
10	4	1.0000	1.0000	.9999	.9999	.9999	.9998	.9996	.9994	.9989
10	5	1.0000	1.0000	1.0000	1.0000	1.0000	.9999	.9999	.9999	.9999
10	6	1.0000	1.0000	1.0000	1.0000	1.0000	1.0000	1.0000	1.0000	1.0000
10	7	1.0000	1.0000	1.0000	1.0000	1.0000	1.0000	1.0000	1.0000	1.0000
11	0	.8953	.8007	.7153	.6382	.5688	.5063	.4501	.3996	.3543
11	1	.9948	.9804	.9586	.9307	.8981	.8617	.8227	.7819	.7398
11	2	.9998	.9988	.9962	.9917	.9847	.9752	.9630	.9481	.9305
11	3	1.0000	.9999	.9997	.9993	.9984	.9969	.9946	.9914	.9871
11	4	1.0000	1.0000	.9999	.9999	.9998	.9997	.9994	.9990	.9982
11	5	1.0000	1.0000	1.0000	1.0000	.9999	.9999	.9999	.9999	.9998
11	6	1.0000	1.0000	1.0000	1.0000	1.0000	1.0000	1.0000	.9999	.9999
11	7	1.0000	1.0000	1.0000	1.0000	1.0000	1.0000	1.0000	1.0000	1.0000
12	0	.8863	.7847	.6938	.6127	.5403	.4759	.4186	.3676	.3224
12	1	.9938	.9768	.9513	.9190	.8816	.8404	.7966	.7513	.7051
12	2	.9997	.9984	.9951	.9892	.9804	.9684	.9532	.9348	.9133
12	3	1.0000	.9999	.9996	.9990	.9977	.9956	.9924	.9879	.9820
12	4	1.0000	1.0000	.9999	.9999	.9998	.9995	.9991	.9983	.9972
12	5	1.0000	1.0000	1.0000	1.0000	.9999	.9999	.9999	.9998	.9997
12	6	1.0000	1.0000	1.0000	1.0000	1.0000	1.0000	1.0000	.9999	.9999
12	7	1.0000	1.0000	1.0000	1.0000	1.0000	1.0000	1.0000	1.0000	1.0000
13	0	.8775	.7690	.6730	.5882	.5133	.4473	.3892	.3382	.2934
13	1	.9927	.9730	.9436	.9068	.8645	.8185	.7702	.7206	.6707
13	2	.9997	.9980	.9938	.9864	.9754	.9607	.9422	.9201	.8946
13	3	.9999	.9999	.9995	.9986	.9969	.9940	.9897	.9837	.9758
13	4	1.0000	1.0000	.9999	.9999	.9997	.9993	.9986	.9975	.9959
13	5	1.0000	1.0000	1.0000	.9999	.9999	.9999	.9998	.9997	.9994
13	6	1.0000	1.0000	1.0000	1.0000	1.0000	1.0000	.9999	.9999	.9999
13	7	1.0000	1.0000	1.0000	1.0000	1.0000	1.0000	1.0000	1.0000	1.0000
14	0	.8687	.7536	.6528	.5646	.4876	.4205	.3620	.3111	.2670
14	1	.9916	.9689	.9355	.8940	.8470	.7963	.7435	.6900	.6367
14	2	.9996	.9975	.9923	.9832	.9699	.9522	.9302	.9041	.8744
14	3	.9999	.9998	.9993	.9981	.9958	.9920	.9864	.9786	.9685
14	4	1.0000	.9999	.9999	.9998	.9995	.9990	.9980	.9964	.9941
14	5	1.0000	1.0000	1.0000	.9999	.9999	.9999	.9997	.9995	.9991
14	6	1.0000	1.0000	1.0000	1.0000	1.0000	1.0000	.9999	.9999	.9999
14	7	1.0000	1.0000	1.0000	1.0000	1.0000	1.0000	1.0000	1.0000	.9999
15	0	.8600	.7385	.6332	.5420	.4632	.3952	.3367	.2863	.2430
15	1	.9903	.9646	.9270	.8808	.8290	.7737	.7168	.6597	.6035
15	2	.9995	.9969	.9906	.9797	.9638	.9428	.9171	.8870	.8531
15	3	.9999	.9998	.9991	.9975	.9945	.9896	.9824	.9726	.9600
15	4	1.0000	.9999	.9999	.9997	.9993	.9986	.9972	.9950	.9918
15	5	1.0000	1.0000	1.0000	.9999	.9999	.9998	.9996	.9993	.9987
15	6	1.0000	1.0000	1.0000	1.0000	1.0000	.9999	.9999	.9999	.9998
15	7	1.0000	1.0000	1.0000	1.0000	1.0000	1.0000	1.0000	.9999	.9999
15	8	1.0000	1.0000	1.0000	1.0000	1.0000	1.0000	1.0000	1.0000	1.0000
16	0	.8514	.7238	.6142	.5204	.4401	.3715	.3131	.2633	.2211
16	1	.9890	.9601	.9182	.8673	.8107	.7510	.6902	.6298	.5710
16	2	.9994	.9963	.9887	.9757	.9570	.9327	.9031	.8688	.8306
16	3	.9999	.9997	.9989	.9968	.9930	.9868	.9778	.9658	.9504
16	4	1.0000	.9999	.9999	.9996	.9991	.9980	.9961	.9932	.9889
16	5	1.0000	1.0000	1.0000	.9999	.9999	.9997	.9994	.9989	.9980
16	6	1.0000	1.0000	1.0000	1.0000	.9999	.9999	.9999	.9998	.9997
16	7	1.0000	1.0000	1.0000	1.0000	1.0000	1.0000	1.0000	.9999	.9999
16	8	1.0000	1.0000	1.0000	1.0000	1.0000	1.0000	1.0000	1.0000	1.0000

Table V (continued)

n	x	p=.01	.02	.03	.04	.05	.06	.07	.08	.09
17	0	.8429	.7093	.5958	.4995	.4181	.3492	.2912	.2423	.2012
17	1	.9876	.9554	.9091	.8534	.7922	.7282	.6638	.6005	.5395
17	2	.9993	.9955	.9866	.9714	.9497	.9218	.8882	.8497	.8072
17	3	.9999	.9996	.9985	.9959	.9912	.9835	.9726	.9580	.9396
17	4	1.0000	.9999	.9998	.9995	.9988	.9973	.9949	.9910	.9854
17	5	1.0000	1.0000	.9999	.9999	.9998	.9996	.9992	.9985	.9972
17	6	1.0000	1.0000	1.0000	1.0000	.9999	.9999	.9999	.9998	.9995
17	7	1.0000	1.0000	1.0000	1.0000	1.0000	1.0000	.9999	.9999	.9999
17	8	1.0000	1.0000	1.0000	1.0000	1.0000	1.0000	1.0000	1.0000	1.0000
18	0	.8345	.6951	.5779	.4796	.3972	.3283	.2708	.2229	.1831
18	1	.9862	.9504	.8997	.8393	.7735	.7055	.6377	.5718	.5091
18	2	.9992	.9947	.9842	.9667	.9418	.9102	.8725	.8298	.7831
18	3	.9999	.9996	.9982	.9950	.9891	.9798	.9667	.9494	.9277
18	4	1.0000	.9999	.9998	.9994	.9984	.9965	.9933	.9884	.9813
18	5	1.0000	1.0000	.9999	.9999	.9998	.9995	.9989	.9979	.9962
18	6	1.0000	1.0000	1.0000	1.0000	.9999	.9999	.9998	.9997	.9993
18	7	1.0000	1.0000	1.0000	1.0000	1.0000	1.0000	.9999	.9999	.9999
18	8	1.0000	1.0000	1.0000	1.0000	1.0000	1.0000	1.0000	1.0000	.9999
19	0	.8261	.6812	.5606	.4604	.3773	.3086	.2518	.2051	.1666
19	1	.9847	.9453	.8900	.8249	.7547	.6829	.6120	.5439	.4797
19	2	.9991	.9939	.9817	.9616	.9334	.8979	.8560	.8091	.7585
19	3	.9999	.9995	.9978	.9938	.9867	.9757	.9601	.9398	.9147
19	4	1.0000	.9999	.9998	.9992	.9979	.9955	.9914	.9852	.9765
19	5	1.0000	1.0000	.9999	.9999	.9997	.9993	.9985	.9971	.9948
19	6	1.0000	1.0000	1.0000	.9999	.9999	.9999	.9998	.9995	.9990
19	7	1.0000	1.0000	1.0000	1.0000	1.0000	.9999	.9999	.9999	.9998
19	8	1.0000	1.0000	1.0000	1.0000	1.0000	1.0000	1.0000	.9999	.9999
20	0	.8179	.6676	.5437	.4420	.3584	.2901	.2342	.1886	.1516
20	1	.9831	.9401	.8801	.8103	.7358	.6604	.5868	.5168	.4516
20	2	.9990	.9929	.9789	.9561	.9245	.8850	.8390	.7879	.7334
20	3	.9999	.9994	.9973	.9925	.9841	.9710	.9528	.9293	.9006
20	4	1.0000	.9999	.9997	.9990	.9974	.9943	.9892	.9816	.9709
20	5	1.0000	1.0000	.9999	.9999	.9996	.9991	.9980	.9962	.9932
20	6	1.0000	1.0000	1.0000	.9999	.9999	.9998	.9997	.9993	.9987
20	7	1.0000	1.0000	1.0000	1.0000	1.0000	.9999	.9999	.9999	.9998
20	8	1.0000	1.0000	1.0000	1.0000	1.0000	1.0000	1.0000	.9999	.9999
22	0	.8016	.6411	.5116	.4073	.3235	.2563	.2025	.1597	.1255
22	1	.9797	.9290	.8597	.7807	.6981	.6163	.5380	.4652	.3988
22	2	.9986	.9907	.9728	.9441	.9051	.8575	.8032	.7442	.6825
22	3	.9999	.9991	.9961	.9895	.9778	.9602	.9362	.9059	.8696
22	4	1.0000	.9999	.9995	.9984	.9959	.9913	.9838	.9727	.9575
22	5	1.0000	1.0000	.9999	.9998	.9994	.9984	.9967	.9936	.9888
22	6	1.0000	1.0000	1.0000	.9999	.9999	.9997	.9994	.9987	.9975
22	7	1.0000	1.0000	1.0000	1.0000	.9999	.9999	.9999	.9998	.9995
22	8	1.0000	1.0000	1.0000	1.0000	1.0000	1.0000	.9999	.9999	.9999
24	0	.7856	.6157	.4814	.3754	.2919	.2265	.1752	.1351	.1039
24	1	.9761	.9173	.8387	.7508	.6608	.5734	.4917	.4172	.3508
24	2	.9982	.9881	.9658	.9307	.8840	.8281	.7657	.6994	.6315
24	3	.9999	.9987	.9946	.9856	.9702	.9474	.9169	.8793	.8351
24	4	1.0000	.9999	.9993	.9977	.9940	.9873	.9767	.9614	.9409
24	5	1.0000	.9999	.9999	.9997	.9990	.9975	.9947	.9899	.9827
24	6	1.0000	1.0000	1.0000	.9999	.9998	.9996	.9990	.9978	.9958
24	7	1.0000	1.0000	1.0000	1.0000	.9999	.9999	.9998	.9996	.9991
24	8	1.0000	1.0000	1.0000	1.0000	1.0000	.9999	.9999	.9999	.9998
26	0	.7700	.5914	.4529	.3459	.2635	.2001	.1515	.1144	.0861
26	1	.9722	.9052	.8172	.7207	.6241	.5322	.4481	.3730	.3075

Table V (continued)

n	x	p=.01	.02	.03	.04	.05	.06	.07	.08	.09
26	2	.9978	.9852	.9580	.9160	.8613	.7972	.7271	.6542	.5813
26	3	.9998	.9983	.9928	.9810	.9612	.9326	.8952	.8498	.7979
26	4	.9999	.9998	.9990	.9966	.9914	.9822	.9679	.9476	.9210
26	5	1.0000	.9999	.9999	.9995	.9984	.9962	.9920	.9850	.9746
26	6	1.0000	1.0000	.9999	.9999	.9997	.9993	.9983	.9964	.9932
26	7	1.0000	1.0000	1.0000	.9999	.9999	.9999	.9997	.9993	.9984
26	8	1.0000	1.0000	1.0000	1.0000	1.0000	.9999	.9999	.9998	.9997
28	0	.7547	.5679	.4262	.3188	.2378	.1768	.1310	.0968	.0713
28	1	.9681	.8925	.7952	.6908	.5883	.4928	.4073	.3326	.2687
28	2	.9972	.9819	.9493	.9001	.8373	.7652	.6880	.6094	.5324
28	3	.9998	.9977	.9906	.9756	.9509	.9159	.8711	.8180	.7584
28	4	.9999	.9997	.9986	.9953	.9882	.9760	.9572	.9313	.8981
28	5	1.0000	.9999	.9998	.9992	.9977	.9944	.9883	.9787	.9644
28	6	1.0000	1.0000	.9999	.9999	.9996	.9989	.9973	.9944	.9896
28	7	1.0000	1.0000	1.0000	.9999	.9999	.9998	.9995	.9987	.9974
28	8	1.0000	1.0000	1.0000	1.0000	.9999	.9999	.9999	.9997	.9994
30	0	.7397	.5454	.4010	.2938	.2146	.1562	.1133	.0819	.0590
30	1	.9638	.8794	.7730	.6611	.5535	.4554	.3693	.2957	.2342
30	2	.9966	.9782	.9399	.8831	.8122	.7324	.6487	.5654	.4855
30	3	.9997	.9971	.9881	.9694	.9392	.8973	.8450	.7842	.7174
30	4	.9999	.9997	.9981	.9936	.9843	.9684	.9447	.9126	.8723
30	5	1.0000	.9999	.9997	.9989	.9967	.9920	.9837	.9707	.9519
30	6	1.0000	1.0000	.9999	.9998	.9994	.9983	.9960	.9917	.9847
30	7	1.0000	1.0000	1.0000	.9999	.9999	.9997	.9991	.9980	.9958
30	8	1.0000	1.0000	1.0000	1.0000	.9999	.9999	.9998	.9995	.9990
50	0	.6050	.3641	.2180	.1298	.0769	.0453	.0265	.0154	.0089
50	1	.9105	.7357	.5552	.4004	.2794	.1900	.1264	.0827	.0532
50	2	.9861	.9215	.8108	.6767	.5405	.4162	.3107	.2259	.1605
50	3	.9984	.9822	.9372	.8608	.7604	.6473	.5327	.4253	.3303
50	4	.9998	.9967	.9831	.9510	.8963	.8206	.7290	.6289	.5276
50	5	.9999	.9995	.9962	.9855	.9622	.9223	.8649	.7918	.7071
50	6	1.0000	.9999	.9993	.9963	.9882	.9710	.9416	.8981	.8403
50	7	1.0000	.9999	.9998	.9992	.9968	.9906	.9779	.9562	.9231
50	8	1.0000	1.0000	.9999	.9998	.9992	.9973	.9926	.9833	.9671
50	9	1.0000	1.0000	1.0000	.9999	.9998	.9993	.9978	.9943	.9874
50	10	1.0000	1.0000	1.0000	1.0000	.9999	.9998	.9994	.9982	.9957
50	11	1.0000	1.0000	1.0000	1.0000	.9999	.9999	.9998	.9995	.9986
50	12	1.0000	1.0000	1.0000	1.0000	1.0000	.9999	.9999	.9998	.9996
50	13	1.0000	1.0000	1.0000	1.0000	1.0000	1.0000	.9999	.9999	.9999
50	14	1.0000	1.0000	1.0000	1.0000	1.0000	1.0000	1.0000	.9999	.9999
50	15	1.0000	1.0000	1.0000	1.0000	1.0000	1.0000	1.0000	1.0000	1.0000

n	x	p=.01	.02	.03	.04	.05	.06	.07	.08	.09	.10
100	0	.3660	.1326	.0476	.0169	.0059	.0021	.0007	.0002	.0001	.0000
100	1	.7358	.4033	.1946	.0872	.0371	.0152	.0060	.0023	.0009	.0003
100	2	.9206	.6767	.4198	.2321	.1183	.0566	.0258	.0113	.0048	.0019
100	3	.9816	.8590	.6472	.4295	.2578	.1430	.0744	.0367	.0173	.0078
100	4	.9966	.9492	.8179	.6289	.4360	.2768	.1632	.0902	.0474	.0237
100	5	.9995	.9845	.9192	.7884	.6160	.4407	.2914	.1799	.1045	.0576
100	6	.9999	.9959	.9688	.8936	.7660	.6064	.4443	.3032	.1940	.1172
100	7	1.0000	.9991	.9894	.9525	.8720	.7483	.5988	.4471	.3128	.2061
100	8	1.0000	.9998	.9968	.9810	.9369	.8537	.7340	.5926	.4494	.3209
100	9	1.0000	1.0000	.9991	.9932	.9718	.9225	.8380	.7220	.5875	.4513
100	10	1.0000	1.0000	.9998	.9978	.9885	.9624	.9092	.8243	.7118	.5832
100	11	1.0000	1.0000	1.0000	.9993	.9957	.9832	.9531	.8972	.8124	.7030
100	12	1.0000	1.0000	1.0000	.9998	.9985	.9931	.9776	.9441	.8862	.8018

Table V (continued)

n	x	p=.01	.02	.03	.04	.05	.06	.07	.08	.09	.10
100	13	1.0000	1.0000	1.0000	1.0000	.9995	.9974	.9901	.9718	.9355	.8761
100	14	1.0000	1.0000	1.0000	1.0000	.9999	.9991	.9959	.9867	.9659	.9274
100	15	1.0000	1.0000	1.0000	1.0000	1.0000	.9997	.9984	.9942	.9831	.9601
100	16	1.0000	1.0000	1.0000	1.0000	1.0000	.9999	.9994	.9976	.9922	.9794
100	17	1.0000	1.0000	1.0000	1.0000	1.0000	1.0000	.9998	.9991	.9966	.9900
100	18	1.0000	1.0000	1.0000	1.0000	1.0000	1.0000	.9999	.9997	.9986	.9954
100	19	1.0000	1.0000	1.0000	1.0000	1.0000	1.0000	1.0000	.9999	.9995	.9980
100	20	1.0000	1.0000	1.0000	1.0000	1.0000	1.0000	1.0000	1.0000	.9998	.9992
100	21	1.0000	1.0000	1.0000	1.0000	1.0000	1.0000	1.0000	1.0000	.9999	.9997
100	22	1.0000	1.0000	1.0000	1.0000	1.0000	1.0000	1.0000	1.0000	1.0000	.9999
100	23	1.0000	1.0000	1.0000	1.0000	1.0000	1.0000	1.0000	1.0000	1.0000	1.0000

n	x	p=.10	.15	.20	.25	.30	.35	.40	.45	.50
2	0	.8100	.7225	.6400	.5625	.4900	.4225	.3600	.3025	.2500
2	1	.9900	.9775	.9600	.9375	.9100	.8775	.8400	.7975	.7500
2	2	1.0000	1.0000	1.0000	1.0000	1.0000	1.0000	1.0000	1.0000	1.0000
3	0	.7290	.6141	.5120	.4218	.3430	.2746	.2160	.1663	.1250
3	1	.9720	.9392	.8960	.8437	.7840	.7182	.6480	.5747	.5000
3	2	.9990	.9966	.9920	.9843	.9730	.9571	.9360	.9088	.8750
3	3	1.0000	1.0000	1.0000	1.0000	1.0000	1.0000	1.0000	1.0000	1.0000
4	0	.6561	.5220	.4096	.3164	.2401	.1785	.1296	.0915	.0625
4	1	.9477	.8904	.8192	.7382	.6517	.5629	.4752	.3909	.3125
4	2	.9963	.9880	.9728	.9492	.9163	.8735	.8208	.7585	.6875
4	3	.9999	.9994	.9984	.9960	.9919	.9849	.9744	.9589	.9375
4	4	1.0000	1.0000	1.0000	1.0000	1.0000	1.0000	1.0000	1.0000	1.0000
5	0	.5904	.4437	.3276	.2373	.1680	.1160	.0777	.0503	.0312
5	1	.9185	.8352	.7372	.6328	.5282	.4284	.3369	.2562	.1875
5	2	.9914	.9733	.9420	.8964	.8369	.7648	.6825	.5931	.5000
5	3	.9995	.9977	.9932	.9843	.9692	.9459	.9129	.8687	.8125
5	4	.9999	.9999	.9996	.9990	.9975	.9947	.9897	.9815	.9687
5	5	1.0000	1.0000	1.0000	1.0000	1.0000	1.0000	1.0000	1.0000	1.0000
6	0	.5314	.3771	.2621	.1779	.1176	.0754	.0466	.0276	.0156
6	1	.8857	.7764	.6553	.5339	.4201	.3190	.2332	.1635	.1093
6	2	.9841	.9526	.9011	.8305	.7443	.6470	.5443	.4415	.3437
6	3	.9987	.9941	.9830	.9624	.9295	.8825	.8208	.7447	.6562
6	4	.9999	.9996	.9984	.9953	.9890	.9776	.9590	.9308	.8906
6	5	1.0000	.9999	.9999	.9997	.9992	.9981	.9959	.9917	.9843
6	6	1.0000	1.0000	1.0000	1.0000	1.0000	1.0000	1.0000	1.0000	1.0000
7	0	.4783	.3205	.2097	.1334	.0823	.0490	.0279	.0152	.0078
7	1	.8503	.7165	.5767	.4449	.3294	.2338	.1586	.1024	.0625
7	2	.9743	.9262	.8519	.7564	.6470	.5322	.4199	.3164	.2265
7	3	.9972	.9879	.9666	.9294	.8739	.8001	.7102	.6082	.5000
7	4	.9998	.9987	.9953	.9871	.9712	.9443	.9037	.8470	.7734
7	5	.9999	.9999	.9996	.9986	.9962	.9909	.9811	.9642	.9375
7	6	1.0000	1.0000	.9999	.9999	.9997	.9993	.9983	.9962	.9921
7	7	1.0000	1.0000	1.0000	1.0000	1.0000	1.0000	1.0000	1.0000	1.0000
8	0	.4304	.2724	.1677	.1001	.0576	.0318	.0168	.0083	.0039
8	1	.8131	.6571	.5033	.3670	.2553	.1691	.1063	.0631	.0351
8	2	.9619	.8947	.7969	.6785	.5517	.4278	.3153	.2201	.1445
8	3	.9949	.9786	.9437	.8861	.8059	.7064	.5940	.4769	.3632
8	4	.9995	.9971	.9895	.9727	.9420	.8939	.8263	.7396	.6367
8	5	.9999	.9997	.9987	.9957	.9887	.9746	.9501	.9115	.8554
8	6	1.0000	.9999	.9999	.9996	.9987	.9964	.9914	.9818	.9648
8	7	1.0000	1.0000	1.0000	.9999	.9999	.9997	.9993	.9983	.9960
8	8	1.0000	1.0000	1.0000	1.0000	1.0000	1.0000	1.0000	1.0000	1.0000

Table V (continued)

n	x	p=.10	.15	.20	.25	.30	.35	.40	.45	.50
9	0	.3874	.2316	.1342	.0750	.0403	.0207	.0100	.0046	.0019
9	1	.7748	.5994	.4362	.3003	.1960	.1210	.0705	.0385	.0195
9	2	.9470	.8591	.7382	.6006	.4628	.3372	.2317	.1495	.0898
9	3	.9916	.9660	.9143	.8342	.7296	.6088	.4826	.3613	.2539
9	4	.9991	.9943	.9804	.9510	.9011	.8282	.7334	.6214	.5000
9	5	.9999	.9993	.9969	.9900	.9747	.9464	.9006	.8341	.7460
9	6	1.0000	.9999	.9996	.9986	.9957	.9888	.9749	.9502	.9101
9	7	1.0000	1.0000	.9999	.9998	.9995	.9986	.9962	.9909	.9804
9	8	1.0000	1.0000	1.0000	1.0000	.9999	.9999	.9997	.9992	.9980
9	9	1.0000	1.0000	1.0000	1.0000	1.0000	1.0000	1.0000	1.0000	1.0000
10	0	.3486	.1968	.1073	.0563	.0282	.0134	.0060	.0025	.0009
10	1	.7361	.5443	.3758	.2440	.1493	.0859	.0463	.0232	.0107
10	2	.9298	.8202	.6778	.5255	.3827	.2616	.1672	.0995	.0546
10	3	.9872	.9500	.8791	.7758	.6496	.5138	.3822	.2660	.1718
10	4	.9983	.9901	.9672	.9218	.8497	.7515	.6331	.5044	.3769
10	5	.9998	.9986	.9936	.9802	.9526	.9050	.8337	.7384	.6230
10	6	.9999	.9998	.9991	.9964	.9894	.9739	.9452	.8980	.8281
10	7	1.0000	.9999	.9999	.9995	.9984	.9951	.9877	.9726	.9453
10	8	1.0000	1.0000	1.0000	.9999	.9998	.9994	.9983	.9955	.9892
10	9	1.0000	1.0000	1.0000	1.0000	.9999	.9999	.9999	.9996	.9990
10	10	1.0000	1.0000	1.0000	1.0000	1.0000	1.0000	1.0000	1.0000	1.0000
11	0	.3138	.1673	.0859	.0422	.0197	.0087	.0036	.0013	.0004
11	1	.6973	.4921	.3221	.1971	.1129	.0605	.0302	.0139	.0058
11	2	.9104	.7788	.6174	.4552	.3127	.2001	.1189	.0652	.0327
11	3	.9814	.9305	.8388	.7133	.5695	.4255	.2962	.1911	.1132
11	4	.9972	.9841	.9495	.8853	.7897	.6683	.5327	.3971	.2744
11	5	.9997	.9973	.9883	.9656	.9217	.8513	.7535	.6331	.5000
11	6	.9999	.9996	.9980	.9924	.9783	.9498	.9006	.8262	.7255
11	7	1.0000	.9999	.9997	.9988	.9957	.9877	.9707	.9390	.8867
11	8	1.0000	1.0000	.9999	.9998	.9994	.9979	.9940	.9852	.9672
11	9	1.0000	1.0000	1.0000	.9999	.9999	.9997	.9992	.9977	.9941
11	10	1.0000	1.0000	1.0000	1.0000	1.0000	.9999	.9999	.9998	.9995
11	11	1.0000	1.0000	1.0000	1.0000	1.0000	1.0000	1.0000	1.0000	1.0000
12	0	.2824	.1422	.0687	.0316	.0138	.0056	.0021	.0007	.0002
12	1	.6590	.4434	.2748	.1583	.0850	.0424	.0195	.0082	.0031
12	2	.8891	.7358	.5583	.3906	.2528	.1512	.0834	.0421	.0192
12	3	.9743	.9077	.7945	.6487	.4925	.3466	.2253	.1344	.0730
12	4	.9956	.9760	.9274	.8423	.7236	.5833	.4381	.3044	.1938
12	5	.9994	.9953	.9805	.9456	.8821	.7872	.6652	.5269	.3872
12	6	.9999	.9993	.9961	.9857	.9614	.9153	.8417	.7393	.6127
12	7	1.0000	.9999	.9994	.9972	.9905	.9744	.9426	.8882	.8061
12	8	1.0000	1.0000	.9999	.9996	.9983	.9943	.9847	.9644	.9270
12	9	1.0000	1.0000	1.0000	.9999	.9997	.9991	.9971	.9921	.9807
12	10	1.0000	1.0000	1.0000	1.0000	.9999	.9999	.9996	.9989	.9968
12	11	1.0000	1.0000	1.0000	1.0000	1.0000	1.0000	.9999	.9999	.9997
12	12	1.0000	1.0000	1.0000	1.0000	1.0000	1.0000	1.0000	1.0000	1.0000
13	0	.2541	.1209	.0549	.0237	.0096	.0037	.0013	.0004	.0001
13	1	.6213	.3982	.2336	.1267	.0636	.0295	.0126	.0049	.0017
13	2	.8661	.6919	.5016	.3326	.2024	.1131	.0579	.0269	.0112
13	3	.9658	.8820	.7473	.5842	.4206	.2782	.1685	.0929	.0461
13	4	.9935	.9658	.9008	.7939	.6543	.5005	.3530	.2279	.1334
13	5	.9990	.9924	.9699	.9197	.8346	.7158	.5744	.4268	.2905
13	6	.9999	.9987	.9930	.9757	.9376	.8705	.7711	.6437	.5000
13	7	.9999	.9998	.9987	.9943	.9817	.9538	.9023	.8212	.7094
13	8	1.0000	.9999	.9998	.9990	.9959	.9874	.9679	.9301	.8665
13	9	1.0000	1.0000	.9999	.9998	.9993	.9974	.9922	.9796	.9538
13	10	1.0000	1.0000	1.0000	.9999	.9999	.9996	.9986	.9958	.9887
13	11	1.0000	1.0000	1.0000	1.0000	.9999	.9999	.9998	.9994	.9982
13	12	1.0000	1.0000	1.0000	1.0000	1.0000	1.0000	.9999	.9999	.9998
13	13	1.0000	1.0000	1.0000	1.0000	1.0000	1.0000	1.0000	1.0000	1.0000

Table V (continued)

n	x	p=.10	.15	.20	.25	.30	.35	.40	.45	.50
14	0	.2287	.1027	.0439	.0178	.0067	.0024	.0007	.0002	.0000
14	1	.5846	.3566	.1979	.1009	.0474	.0205	.0081	.0028	.0009
14	2	.8416	.6479	.4480	.2811	.1608	.0839	.0397	.0170	.0064
14	3	.9558	.8534	.6981	.5213	.3551	.2205	.1243	.0632	.0286
14	4	.9907	.9533	.8701	.7415	.5842	.4227	.2792	.1671	.0897
14	5	.9985	.9884	.9561	.8883	.7805	.6405	.4858	.3373	.2119
14	6	.9998	.9977	.9883	.9617	.9067	.8164	.6924	.5461	.3953
14	7	.9999	.9996	.9976	.9897	.9685	.9247	.8499	.7414	.6047
14	8	1.0000	.9999	.9996	.9978	.9917	.9757	.9417	.8811	.7880
14	9	1.0000	1.0000	.9999	.9997	.9983	.9940	.9825	.9574	.9102
14	10	1.0000	1.0000	1.0000	.9999	.9998	.9989	.9961	.9886	.9713
14	11	1.0000	1.0000	1.0000	1.0000	.9999	.9999	.9994	.9978	.9935
14	12	1.0000	1.0000	1.0000	1.0000	1.0000	.9999	.9999	.9997	.9990
14	13	1.0000	1.0000	1.0000	1.0000	1.0000	1.0000	1.0000	.9999	.9999
14	14	1.0000	1.0000	1.0000	1.0000	1.0000	1.0000	1.0000	1.0000	1.0000
15	0	.2059	.0874	.0352	.0134	.0048	.0016	.0005	.0002	.0000
15	1	.5490	.3186	.1671	.0802	.0353	.0142	.0052	.0017	.0005
15	2	.8159	.6042	.3980	.2361	.1268	.0617	.0271	.0107	.0037
15	3	.9444	.8227	.6482	.4613	.2969	.1727	.0905	.0424	.0176
15	4	.9873	.9383	.8358	.6865	.5155	.3520	.2173	.1204	.0592
15	5	.9977	.9832	.9389	.8516	.7216	.5643	.4033	.2608	.1509
15	6	.9996	.9964	.9819	.9434	.8688	.7548	.6099	.4522	.3036
15	7	.9999	.9994	.9958	.9827	.9500	.8868	.7869	.6536	.5001
15	8	1.0000	.9999	.9992	.9958	.9848	.9578	.9050	.8182	.6964
15	9	1.0000	1.0000	.9999	.9992	.9963	.9876	.9662	.9231	.8491
15	10	1.0000	1.0000	1.0000	.9999	.9994	.9972	.9907	.9745	.9408
15	11	1.0000	1.0000	1.0000	1.0000	.9999	.9995	.9981	.9937	.9824
15	12	1.0000	1.0000	1.0000	1.0000	1.0000	.9999	.9997	.9989	.9963
15	13	1.0000	1.0000	1.0000	1.0000	1.0000	1.0000	1.0000	.9999	.9995
15	14	1.0000	1.0000	1.0000	1.0000	1.0000	1.0000	1.0000	1.0000	1.0000
15	15	1.0000	1.0000	1.0000	1.0000	1.0000	1.0000	1.0000	1.0000	1.0000
16	0	.1853	.0743	.0281	.0100	.0033	.0010	.0003	.0001	.0000
16	1	.5147	.2839	.1407	.0635	.0261	.0098	.0033	.0010	.0003
16	2	.7893	.5614	.3518	.1971	.0994	.0451	.0183	.0066	.0021
16	3	.9316	.7899	.5982	.4050	.2459	.1339	.0651	.0281	.0106
16	4	.9830	.9209	.7983	.6302	.4499	.2892	.1666	.0853	.0384
16	5	.9967	.9765	.9183	.8104	.6598	.4900	.3288	.1976	.1051
16	6	.9995	.9944	.9733	.9204	.8247	.6881	.5272	.3660	.2272
16	7	.9999	.9989	.9930	.9729	.9256	.8406	.7161	.5629	.4018
16	8	1.0000	.9998	.9985	.9925	.9744	.9329	.8577	.7442	.5982
16	9	1.0000	1.0000	.9998	.9984	.9929	.9771	.9417	.8759	.7728
16	10	1.0000	1.0000	1.0000	.9997	.9984	.9938	.9809	.9514	.8949
16	11	1.0000	1.0000	1.0000	1.0000	.9997	.9988	.9951	.9851	.9616
16	12	1.0000	1.0000	1.0000	1.0000	.9999	.9998	.9991	.9965	.9894
16	13	1.0000	1.0000	1.0000	1.0000	1.0000	1.0000	.9999	.9994	.9979
16	14	1.0000	1.0000	1.0000	1.0000	1.0000	1.0000	1.0000	.9999	.9997
16	15	1.0000	1.0000	1.0000	1.0000	1.0000	1.0000	1.0000	1.0000	1.0000
16	16	1.0000	1.0000	1.0000	1.0000	1.0000	1.0000	1.0000	1.0000	1.0000
18	0	.1501	.0536	.0180	.0056	.0016	.0004	.0001	.0000	.0000
18	1	.4503	.2241	.0991	.0395	.0142	.0046	.0013	.0003	.0001
18	2	.7338	.4797	.2713	.1353	.0600	.0236	.0082	.0025	.0007
18	3	.9018	.7202	.5010	.3057	.1646	.0783	.0328	.0120	.0038
18	4	.9718	.8794	.7164	.5187	.3327	.1886	.0942	.0411	.0154
18	5	.9936	.9581	.8671	.7175	.5344	.3550	.2088	.1077	.0481
18	6	.9988	.9882	.9487	.8610	.7217	.5491	.3743	.2258	.1189
18	7	.9998	.9973	.9837	.9431	.8593	.7283	.5634	.3915	.2403
18	8	1.0000	.9995	.9957	.9807	.9404	.8609	.7368	.5778	.4073
18	9	1.0000	.9999	.9991	.9946	.9790	.9403	.8653	.7473	.5927
18	10	1.0000	1.0000	.9998	.9988	.9939	.9788	.9424	.8721	.7597
18	11	1.0000	1.0000	1.0000	.9998	.9986	.9938	.9797	.9463	.8811

Table V (continued)

n	x	p=.10	.15	.20	.25	.30	.35	.40	.45	.50
18	12	1.0000	1.0000	1.0000	.9999	.9997	.9985	.9942	.9817	.9518
18	13	1.0000	1.0000	1.0000	1.0000	.9999	.9997	.9987	.9950	.9845
18	14	1.0000	1.0000	1.0000	1.0000	1.0000	.9999	.9997	.9990	.9962
18	15	1.0000	1.0000	1.0000	1.0000	1.0000	1.0000	.9999	.9998	.9993
18	16	1.0000	1.0000	1.0000	1.0000	1.0000	1.0000	1.0000	.9999	.9999
18	17	1.0000	1.0000	1.0000	1.0000	1.0000	1.0000	1.0000	1.0000	1.0000
18	18	1.0000	1.0000	1.0000	1.0000	1.0000	1.0000	1.0000	1.0000	1.0000
20	0	.1215	.0387	.0115	.0031	.0008	.0001	.0000	.0000	.0000
20	1	.3917	.1755	.0691	.0243	.0076	.0021	.0005	.0001	.0000
20	2	.6769	.4049	.2060	.0912	.0354	.0121	.0036	.0009	.0002
20	3	.8670	.6477	.4114	.2251	.1070	.0443	.0159	.0049	.0012
20	4	.9568	.8298	.6296	.4148	.2375	.1182	.0509	.0188	.0059
20	5	.9887	.9326	.8042	.6171	.4163	.2454	.1256	.0553	.0206
20	6	.9976	.9780	.9133	.7857	.6080	.4166	.2500	.1299	.0576
20	7	.9995	.9940	.9678	.8981	.7722	.6010	.4158	.2520	.1315
20	8	.9999	.9986	.9900	.9590	.8866	.7623	.5956	.4143	.2517
20	9	.9999	.9997	.9974	.9861	.9520	.8782	.7553	.5913	.4119
20	10	1.0000	.9999	.9994	.9960	.9828	.9468	.8724	.7507	.5881
20	11	1.0000	.9999	.9999	.9990	.9948	.9804	.9434	.8692	.7482
20	12	1.0000	1.0000	.9999	.9998	.9987	.9939	.9789	.9419	.8684
20	13	1.0000	1.0000	1.0000	.9999	.9997	.9984	.9935	.9785	.9423
20	14	1.0000	1.0000	1.0000	1.0000	.9999	.9996	.9983	.9935	.9793
20	15	1.0000	1.0000	1.0000	1.0000	.9999	.9999	.9996	.9984	.9940
20	16	1.0000	1.0000	1.0000	1.0000	1.0000	.9999	.9999	.9997	.9987
20	17	1.0000	1.0000	1.0000	1.0000	1.0000	1.0000	.9999	.9999	.9998
20	18	1.0000	1.0000	1.0000	1.0000	1.0000	1.0000	1.0000	1.0000	.9999
20	19	1.0000	1.0000	1.0000	1.0000	1.0000	1.0000	1.0000	1.0000	1.0000
20	20	1.0000	1.0000	1.0000	1.0000	1.0000	1.0000	1.0000	1.0000	1.0000
22	0	.0984	.0280	.0073	.0017	.0003	.0000	.0000	.0000	.0000
22	1	.3392	.1367	.0479	.0148	.0040	.0009	.0002	.0000	.0000
22	2	.6200	.3381	.1544	.0606	.0206	.0061	.0015	.0003	.0000
22	3	.8280	.5751	.3320	.1623	.0680	.0245	.0075	.0019	.0004
22	4	.9378	.7738	.5428	.3234	.1645	.0716	.0265	.0083	.0021
22	5	.9817	.9000	.7326	.5168	.3134	.1629	.0722	.0270	.0084
22	6	.9956	.9631	.8670	.6993	.4941	.3021	.1584	.0705	.0262
22	7	.9991	.9886	.9438	.8384	.6712	.4735	.2898	.1517	.0669
22	8	.9998	.9970	.9798	.9254	.8135	.6466	.4540	.2763	.1431
22	9	.9999	.9993	.9938	.9704	.9084	.7915	.6243	.4350	.2617
22	10	1.0000	.9998	.9984	.9900	.9612	.8930	.7719	.6037	.4159
22	11	1.0000	.9999	.9996	.9971	.9859	.9526	.8792	.7543	.5840
22	12	1.0000	1.0000	.9999	.9993	.9956	.9820	.9443	.8672	.7382
22	13	1.0000	1.0000	.9999	.9998	.9988	.9942	.9785	.9383	.8568
22	14	1.0000	1.0000	1.0000	.9999	.9997	.9984	.9929	.9757	.9331
22	15	1.0000	1.0000	1.0000	1.0000	.9999	.9996	.9980	.9920	.9737
22	16	1.0000	1.0000	1.0000	1.0000	.9999	.9999	.9995	.9978	.9915
22	17	1.0000	1.0000	1.0000	1.0000	1.0000	.9999	.9999	.9995	.9978
22	18	1.0000	1.0000	1.0000	1.0000	1.0000	1.0000	.9999	.9999	.9995
22	19	1.0000	1.0000	1.0000	1.0000	1.0000	1.0000	1.0000	.9999	.9999
22	20	1.0000	1.0000	1.0000	1.0000	1.0000	1.0000	1.0000	1.0000	.9999
22	21	1.0000	1.0000	1.0000	1.0000	1.0000	1.0000	1.0000	1.0000	1.0000
22	22	1.0000	1.0000	1.0000	1.0000	1.0000	1.0000	1.0000	1.0000	1.0000
24	0	.0797	.0202	.0047	.0010	.0001	.0000	.0000	.0000	.0000
24	1	.2924	.1059	.0330	.0090	.0021	.0004	.0000	.0000	.0000
24	2	.5642	.2798	.1145	.0398	.0118	.0030	.0006	.0001	.0000
24	3	.7857	.5048	.2638	.1150	.0424	.0132	.0035	.0007	.0001
24	4	.9149	.7133	.4598	.2466	.1110	.0421	.0134	.0035	.0007
24	5	.9723	.8605	.6558	.4221	.2288	.1044	.0399	.0127	.0033
24	6	.9925	.9428	.8110	.6074	.3885	.2105	.0959	.0364	.0113
24	7	.9983	.9801	.9108	.7662	.5646	.3575	.1919	.0862	.0319
24	8	.9996	.9941	.9638	.8786	.7250	.5256	.3279	.1730	.0757
24	9	.9999	.9985	.9873	.9453	.8472	.6866	.4890	.2991	.1537

Table V (continued)

n	x	p=.10	.15	.20	.25	.30	.35	.40	.45	.50
24	10	.9999	.9996	.9962	.9786	.9257	.8166	.6502	.4539	.2706
24	11	1.0000	.9999	.9990	.9928	.9686	.9057	.7869	.6151	.4194
24	12	1.0000	.9999	.9997	.9979	.9885	.9577	.8857	.7579	.5805
24	13	1.0000	1.0000	.9999	.9994	.9963	.9835	.9465	.8658	.7293
24	14	1.0000	1.0000	.9999	.9998	.9990	.9945	.9783	.9352	.8462
24	15	1.0000	1.0000	1.0000	.9999	.9997	.9984	.9924	.9730	.9242
24	16	1.0000	1.0000	1.0000	1.0000	.9999	.9996	.9978	.9904	.9680
24	17	1.0000	1.0000	1.0000	1.0000	.9999	.9999	.9994	.9971	.9886
24	18	1.0000	1.0000	1.0000	1.0000	1.0000	.9999	.9998	.9993	.9966
24	19	1.0000	1.0000	1.0000	1.0000	1.0000	1.0000	.9999	.9998	.9992
24	20	1.0000	1.0000	1.0000	1.0000	1.0000	1.0000	1.0000	.9999	.9998
24	21	1.0000	1.0000	1.0000	1.0000	1.0000	1.0000	1.0000	1.0000	.9999
24	22	1.0000	1.0000	1.0000	1.0000	1.0000	1.0000	1.0000	1.0000	1.0000
24	23	1.0000	1.0000	1.0000	1.0000	1.0000	1.0000	1.0000	1.0000	1.0000
24	24	1.0000	1.0000	1.0000	1.0000	1.0000	1.0000	1.0000	1.0000	1.0000
26	0	.0646	.0146	.0030	.0005	.0000	.0000	.0000	.0000	.0000
26	1	.2512	.0816	.0226	.0054	.0011	.0002	.0000	.0000	.0000
26	2	.5105	.2296	.0840	.0258	.0067	.0014	.0002	.0000	.0000
26	3	.7409	.4385	.2068	.0801	.0259	.0070	.0015	.0003	.0000
26	4	.8881	.6504	.3833	.1843	.0733	.0242	.0066	.0014	.0002
26	5	.9601	.8150	.5774	.3371	.1625	.0649	.0214	.0057	.0012
26	6	.9881	.9166	.7473	.5153	.2965	.1416	.0558	.0180	.0046
26	7	.9970	.9679	.8687	.6851	.4604	.2596	.1215	.0467	.0144
26	8	.9993	.9894	.9407	.8195	.6274	.4105	.2255	.1024	.0377
26	9	.9998	.9969	.9767	.9091	.7704	.5730	.3641	.1936	.0843
26	10	.9999	.9992	.9920	.9599	.8747	.7218	.5213	.3204	.1634
26	11	1.0000	.9998	.9976	.9845	.9396	.8383	.6736	.4713	.2786
26	12	1.0000	.9999	.9994	.9947	.9745	.9168	.8006	.6256	.4225
26	13	1.0000	1.0000	.9998	.9984	.9905	.9622	.8918	.7616	.5774
26	14	1.0000	1.0000	.9999	.9996	.9969	.9850	.9482	.8650	.7214
26	15	1.0000	1.0000	1.0000	.9999	.9991	.9948	.9783	.9326	.8365
26	16	1.0000	1.0000	1.0000	.9999	.9997	.9984	.9921	.9706	.9156
26	17	1.0000	1.0000	1.0000	1.0000	.9999	.9996	.9975	.9889	.9622
26	18	1.0000	1.0000	1.0000	1.0000	.9999	.9999	.9993	.9964	.9855
26	19	1.0000	1.0000	1.0000	1.0000	1.0000	.9999	.9998	.9990	.9953
26	20	1.0000	1.0000	1.0000	1.0000	1.0000	1.0000	.9999	.9997	.9987
26	21	1.0000	1.0000	1.0000	1.0000	1.0000	1.0000	1.0000	.9999	.9997
26	22	1.0000	1.0000	1.0000	1.0000	1.0000	1.0000	1.0000	.9999	.9999
26	23	1.0000	1.0000	1.0000	1.0000	1.0000	1.0000	1.0000	1.0000	.9999
26	24	1.0000	1.0000	1.0000	1.0000	1.0000	1.0000	1.0000	1.0000	1.0000
26	25	1.0000	1.0000	1.0000	1.0000	1.0000	1.0000	1.0000	1.0000	1.0000
26	26	1.0000	1.0000	1.0000	1.0000	1.0000	1.0000	1.0000	1.0000	1.0000
28	0	.0523	.0105	.0019	.0003	.0000	.0000	.0000	.0000	.0000
28	1	.2151	.0627	.0154	.0032	.0006	.0000	.0000	.0000	.0000
28	2	.4593	.1870	.0611	.0166	.0037	.0007	.0001	.0000	.0000
28	3	.6945	.3772	.1601	.0551	.0156	.0036	.0007	.0001	.0000
28	4	.8578	.5869	.3148	.1353	.0474	.0136	.0031	.0006	.0000
28	5	.9449	.7645	.5005	.2637	.1127	.0393	.0111	.0025	.0004
28	6	.9820	.8847	.6784	.4278	.2201	.0923	.0314	.0086	.0018
28	7	.9950	.9514	.8182	.5997	.3648	.1820	.0740	.0242	.0062
28	8	.9988	.9822	.9099	.7501	.5275	.3089	.1484	.0577	.0178
28	9	.9997	.9944	.9609	.8615	.6824	.4607	.2588	.1187	.0435
28	10	.9999	.9984	.9851	.9321	.8086	.6160	.3985	.2135	.0924
28	11	.9999	.9996	.9950	.9705	.8971	.7528	.5510	.3403	.1724
28	12	1.0000	.9999	.9985	.9887	.9509	.8572	.6950	.4874	.2857
28	13	1.0000	.9999	.9996	.9962	.9792	.9264	.8131	.6355	.4252
28	14	1.0000	1.0000	.9999	.9988	.9922	.9663	.8975	.7653	.5747
28	15	1.0000	1.0000	.9999	.9997	.9974	.9863	.9500	.8645	.7142
28	16	1.0000	1.0000	1.0000	.9999	.9992	.9951	.9784	.9304	.8275
28	17	1.0000	1.0000	1.0000	.9999	.9998	.9985	.9918	.9685	.9075
28	18	1.0000	1.0000	1.0000	1.0000	.9999	.9996	.9973	.9875	.9564

Table V (continued)

n	x	p = .10	.15	.20	.25	.30	.35	.40	.45	.50
28	19	1.0000	1.0000	1.0000	1.0000	.9999	.9999	.9992	.9957	.9821
28	20	1.0000	1.0000	1.0000	1.0000	1.0000	.9999	.9998	.9987	.9937
28	21	1.0000	1.0000	1.0000	1.0000	1.0000	1.0000	.9999	.9996	.9981
28	22	1.0000	1.0000	1.0000	1.0000	1.0000	1.0000	.9999	.9999	.9995
28	23	1.0000	1.0000	1.0000	1.0000	1.0000	1.0000	1.0000	.9999	.9999
28	24	1.0000	1.0000	1.0000	1.0000	1.0000	1.0000	1.0000	1.0000	.9999
28	25	1.0000	1.0000	1.0000	1.0000	1.0000	1.0000	1.0000	1.0000	1.0000
28	26	1.0000	1.0000	1.0000	1.0000	1.0000	1.0000	1.0000	1.0000	1.0000
28	27	1.0000	1.0000	1.0000	1.0000	1.0000	1.0000	1.0000	1.0000	1.0000
28	28	1.0000	1.0000	1.0000	1.0000	1.0000	1.0000	1.0000	1.0000	1.0000
30	0	.0423	.0076	.0012	.0001	.0000	.0000	.0000	.0000	.0000
30	1	.1837	.0480	.0105	.0019	.0003	.0000	.0000	.0000	.0000
30	2	.4113	.1514	.0441	.0106	.0021	.0003	.0000	.0000	.0000
30	3	.6474	.3216	.1227	.0374	.0093	.0019	.0003	.0000	.0000
30	4	.8245	.5244	.2552	.0978	.0301	.0075	.0015	.0002	.0000
30	5	.9268	.7105	.4275	.2026	.0765	.0232	.0056	.0010	.0001
30	6	.9741	.8474	.6069	.3480	.1595	.0585	.0171	.0039	.0007
30	7	.9922	.9302	.7607	.5142	.2813	.1237	.0435	.0121	.0026
30	8	.9979	.9722	.8713	.6736	.4315	.2247	.0940	.0312	.0080
30	9	.9995	.9903	.9389	.8034	.5888	.3575	.1762	.0694	.0213
30	10	.9999	.9970	.9743	.8942	.7303	.5077	.2914	.1350	.0493
30	11	.9999	.9992	.9905	.9493	.8406	.6548	.4310	.2326	.1002
30	12	1.0000	.9998	.9968	.9784	.9155	.7802	.5784	.3591	.1808
30	13	1.0000	.9999	.9991	.9918	.9599	.8736	.7145	.5024	.2923
30	14	1.0000	.9999	.9997	.9972	.9830	.9348	.8246	.6448	.4277
30	15	1.0000	1.0000	.9999	.9991	.9936	.9699	.9029	.7690	.5722
30	16	1.0000	1.0000	.9999	.9997	.9978	.9876	.9518	.8644	.7076
30	17	1.0000	1.0000	1.0000	.9999	.9993	.9955	.9787	.9286	.8192
30	18	1.0000	1.0000	1.0000	.9999	.9998	.9985	.9917	.9665	.8997
30	19	1.0000	1.0000	1.0000	1.0000	.9999	.9995	.9971	.9861	.9506
30	20	1.0000	1.0000	1.0000	1.0000	.9999	.9999	.9991	.9949	.9786
30	21	1.0000	1.0000	1.0000	1.0000	1.0000	.9999	.9997	.9984	.9919
30	22	1.0000	1.0000	1.0000	1.0000	1.0000	1.0000	.9999	.9995	.9973
30	23	1.0000	1.0000	1.0000	1.0000	1.0000	1.0000	.9999	.9999	.9992
30	24	1.0000	1.0000	1.0000	1.0000	1.0000	1.0000	1.0000	.9999	.9998
30	25	1.0000	1.0000	1.0000	1.0000	1.0000	1.0000	1.0000	1.0000	.9999
30	26	1.0000	1.0000	1.0000	1.0000	1.0000	1.0000	1.0000	1.0000	1.0000
30	27	1.0000	1.0000	1.0000	1.0000	1.0000	1.0000	1.0000	1.0000	1.0000
30	28	1.0000	1.0000	1.0000	1.0000	1.0000	1.0000	1.0000	1.0000	1.0000
30	29	1.0000	1.0000	1.0000	1.0000	1.0000	1.0000	1.0000	1.0000	1.0000
30	30	1.0000	1.0000	1.0000	1.0000	1.0000	1.0000	1.0000	1.0000	1.0000
50	0	.0051	.0003	.0000	.0000	.0000	.0000	.0000	.0000	.0000
50	1	.0337	.0029	.0001	.0000	.0000	.0000	.0000	.0000	.0000
50	2	.1117	.0141	.0012	.0000	.0000	.0000	.0000	.0000	.0000
50	3	.2502	.0460	.0056	.0005	.0000	.0000	.0000	.0000	.0000
50	4	.4312	.1121	.0185	.0021	.0001	.0000	.0000	.0000	.0000
50	5	.6161	.2193	.0480	.0070	.0007	.0000	.0000	.0000	.0000
50	6	.7702	.3613	.1034	.0193	.0024	.0002	.0000	.0000	.0000
50	7	.8778	.5187	.1904	.0452	.0072	.0008	.0000	.0000	.0000
50	8	.9421	.6681	.3073	.0916	.0182	.0024	.0002	.0000	.0000
50	9	.9754	.7910	.4437	.1636	.0402	.0067	.0007	.0000	.0000
50	10	.9906	.8800	.5835	.2622	.0788	.0160	.0022	.0002	.0000
50	11	.9967	.9371	.7106	.3816	.1390	.0342	.0056	.0006	.0000
50	12	.9990	.9699	.8139	.5109	.2228	.0661	.0132	.0017	.0001
50	13	.9997	.9868	.8894	.6370	.3278	.1163	.0279	.0044	.0004
50	14	.9999	.9947	.9392	.7480	.4468	.1877	.0539	.0103	.0013
50	15	.9999	.9980	.9692	.8369	.5691	.2801	.0955	.0219	.0033
50	16	1.0000	.9993	.9855	.9016	.6838	.3888	.1560	.0426	.0076
50	17	1.0000	.9997	.9937	.9448	.7821	.5059	.2368	.0765	.0164
50	18	1.0000	.9999	.9974	.9712	.8594	.6215	.3356	.1273	.0324
50	19	1.0000	.9999	.9990	.9860	.9152	.7264	.4464	.1973	.0594

Table V (continued)

n	x	p=.10	.15	.20	.25	.30	.35	.40	.45	.50
50	20	1.0000	1.0000	.9996	.9937	.9522	.8139	.5610	.2861	.1013
50	21	1.0000	1.0000	.9999	.9973	.9749	.8812	.6701	.3899	.1611
50	22	1.0000	1.0000	.9999	.9989	.9877	.9290	.7660	.5019	.2399
50	23	1.0000	1.0000	.9999	.9996	.9944	.9603	.8438	.6134	.3359
50	24	1.0000	1.0000	1.0000	.9998	.9976	.9793	.9021	.7160	.4438
50	25	1.0000	1.0000	1.0000	.9999	.9990	.9899	.9426	.8033	.5561
50	26	1.0000	1.0000	1.0000	.9999	.9996	.9954	.9685	.8720	.6640
50	27	1.0000	1.0000	1.0000	1.0000	.9998	.9980	.9839	.9220	.7600
50	28	1.0000	1.0000	1.0000	1.0000	.9999	.9992	.9923	.9556	.8388
50	29	1.0000	1.0000	1.0000	1.0000	.9999	.9997	.9966	.9764	.8986
50	30	1.0000	1.0000	1.0000	1.0000	1.0000	.9999	.9986	.9884	.9405
50	31	1.0000	1.0000	1.0000	1.0000	1.0000	.9999	.9994	.9947	.9675
50	32	1.0000	1.0000	1.0000	1.0000	1.0000	.9999	.9998	.9977	.9835
50	33	1.0000	1.0000	1.0000	1.0000	1.0000	1.0000	.9999	.9991	.9923
50	34	1.0000	1.0000	1.0000	1.0000	1.0000	1.0000	.9999	.9996	.9967
50	35	1.0000	1.0000	1.0000	1.0000	1.0000	1.0000	1.0000	.9999	.9987
50	36	1.0000	1.0000	1.0000	1.0000	1.0000	1.0000	1.0000	.9999	.9995
50	37	1.0000	1.0000	1.0000	1.0000	1.0000	1.0000	1.0000	.9999	.9998
50	38	1.0000	1.0000	1.0000	1.0000	1.0000	1.0000	1.0000	1.0000	.9999
50	39	1.0000	1.0000	1.0000	1.0000	1.0000	1.0000	1.0000	1.0000	.9999
50	40	1.0000	1.0000	1.0000	1.0000	1.0000	1.0000	1.0000	1.0000	1.0000

n	x	p=.15	.20	.25	.30	.35	.40	.45	.50
100	0	.0000	.0000	.0000	.0000	.0000	.0000	.0000	.0000
100	1	.0000	.0000	.0000	.0000	.0000	.0000	.0000	.0000
100	2	.0000	.0000	.0000	.0000	.0000	.0000	.0000	.0000
100	3	.0001	.0000	.0000	.0000	.0000	.0000	.0000	.0000
100	4	.0004	.0000	.0000	.0000	.0000	.0000	.0000	.0000
100	5	.0016	.0000	.0000	.0000	.0000	.0000	.0000	.0000
100	6	.0047	.0001	.0000	.0000	.0000	.0000	.0000	.0000
100	7	.0122	.0003	.0000	.0000	.0000	.0000	.0000	.0000
100	8	.0275	.0009	.0000	.0000	.0000	.0000	.0000	.0000
100	9	.0551	.0023	.0000	.0000	.0000	.0000	.0000	.0000
100	10	.0994	.0057	.0001	.0000	.0000	.0000	.0000	.0000
100	11	.1635	.0126	.0004	.0000	.0000	.0000	.0000	.0000
100	12	.2473	.0253	.0010	.0000	.0000	.0000	.0000	.0000
100	13	.3474	.0469	.0025	.0001	.0000	.0000	.0000	.0000
100	14	.4572	.0804	.0054	.0002	.0000	.0000	.0000	.0000
100	15	.5683	.1285	.0111	.0004	.0000	.0000	.0000	.0000
100	16	.6725	.1923	.0211	.0010	.0000	.0000	.0000	.0000
100	17	.7633	.2712	.0376	.0022	.0001	.0000	.0000	.0000
100	18	.8372	.3621	.0630	.0045	.0001	.0000	.0000	.0000
100	19	.8935	.4602	.0995	.0089	.0003	.0000	.0000	.0000
100	20	.9337	.5595	.1488	.0165	.0008	.0000	.0000	.0000
100	21	.9607	.6540	.2114	.0288	.0017	.0000	.0000	.0000
100	22	.9779	.7389	.2864	.0479	.0034	.0001	.0000	.0000
100	23	.9881	.8109	.3711	.0755	.0066	.0003	.0000	.0000
100	24	.9939	.8686	.4617	.1136	.0121	.0006	.0000	.0000
100	25	.9970	.9125	.5535	.1631	.0211	.0012	.0000	.0000
100	26	.9986	.9442	.6417	.2244	.0351	.0024	.0001	.0000
100	27	.9994	.9658	.7224	.2964	.0558	.0046	.0002	.0000
100	28	.9997	.9800	.7925	.3768	.0848	.0084	.0004	.0000
100	29	.9999	.9888	.8505	.4623	.1236	.0148	.0008	.0000
100	30	1.0000	.9939	.8962	.5491	.1730	.0248	.0015	.0000
100	31	1.0000	.9969	.9307	.6331	.2331	.0398	.0030	.0001
100	32	1.0000	.9984	.9554	.7107	.3029	.0615	.0055	.0002
100	33	1.0000	.9993	.9724	.7793	.3803	.0913	.0098	.0004
100	34	1.0000	.9997	.9836	.8371	.4624	.1303	.0166	.0009
100	35	1.0000	.9999	.9906	.8839	.5458	.1795	.0272	.0018

Table V (continued)

n	x	p=.15	.20	.25	.30	.35	.40	.45	.50
100	36	1.0000	.9999	.9948	.9201	.6269	.2386	.0429	.0033
100	37	1.0000	1.0000	.9973	.9470	.7024	.3068	.0651	.0060
100	38	1.0000	1.0000	.9986	.9660	.7699	.3822	.0951	.0105
100	39	1.0000	1.0000	.9993	.9790	.8276	.4621	.1343	.0176
100	40	1.0000	1.0000	.9997	.9875	.8750	.5433	.1831	.0284
100	41	1.0000	1.0000	.9999	.9928	.9123	.6225	.2415	.0443
100	42	1.0000	1.0000	.9999	.9960	.9406	.6967	.3087	.0666
100	43	1.0000	1.0000	1.0000	.9979	.9611	.7635	.3828	.0967
100	44	1.0000	1.0000	1.0000	.9989	.9754	.8211	.4613	.1356
100	45	1.0000	1.0000	1.0000	.9995	.9850	.8689	.5413	.1841
100	46	1.0000	1.0000	1.0000	.9997	.9912	.9070	.6196	.2421
100	47	1.0000	1.0000	1.0000	.9999	.9950	.9362	.6931	.3086
100	48	1.0000	1.0000	1.0000	.9999	.9973	.9577	.7596	.3822
100	49	1.0000	1.0000	1.0000	1.0000	.9985	.9729	.8173	.4602
100	50	1.0000	1.0000	1.0000	1.0000	.9993	.9832	.8654	.5398
100	51	1.0000	1.0000	1.0000	1.0000	.9996	.9900	.9040	.6178
100	52	1.0000	1.0000	1.0000	1.0000	.9998	.9942	.9338	.6914
100	53	1.0000	1.0000	1.0000	1.0000	.9999	.9968	.9559	.7579
100	54	1.0000	1.0000	1.0000	1.0000	1.0000	.9983	.9716	.8159
100	55	1.0000	1.0000	1.0000	1.0000	1.0000	.9991	.9824	.8644
100	56	1.0000	1.0000	1.0000	1.0000	1.0000	.9996	.9894	.9033
100	57	1.0000	1.0000	1.0000	1.0000	1.0000	.9998	.9939	.9334
100	58	1.0000	1.0000	1.0000	1.0000	1.0000	.9999	.9966	.9557
100	59	1.0000	1.0000	1.0000	1.0000	1.0000	1.0000	.9982	.9716
100	60	1.0000	1.0000	1.0000	1.0000	1.0000	1.0000	.9991	.9824
100	61	1.0000	1.0000	1.0000	1.0000	1.0000	1.0000	.9995	.9895
100	62	1.0000	1.0000	1.0000	1.0000	1.0000	1.0000	.9998	.9940
100	63	1.0000	1.0000	1.0000	1.0000	1.0000	1.0000	.9999	.9967
100	64	1.0000	1.0000	1.0000	1.0000	1.0000	1.0000	1.0000	.9982
100	65	1.0000	1.0000	1.0000	1.0000	1.0000	1.0000	1.0000	.9991
100	66	1.0000	1.0000	1.0000	1.0000	1.0000	1.0000	1.0000	.9996
100	67	1.0000	1.0000	1.0000	1.0000	1.0000	1.0000	1.0000	.9998
100	68	1.0000	1.0000	1.0000	1.0000	1.0000	1.0000	1.0000	.9999
100	69	1.0000	1.0000	1.0000	1.0000	1.0000	1.0000	1.0000	1.0000

Table VI The F Distribution

This table gives the F value such that there is a 0.05 or 0.01 probability of exceeding that value, for the 2 degrees of freedom indicated. The boldface value uses $\alpha = 0.01$, and the regular type value uses $\alpha = 0.05$. For example, if there are 5 degrees of freedom in the numerator and 20 degrees of freedom in the denominator, $F_{0.01} = 4.10$ and $F_{0.05} = 2.71$.

Values of F
Right tail of the distribution for $P = 0.05$ (light-face type), 0.01 (bold-face type)

	Degrees of Freedom for Numerator											
	1	2	3	4	5	6	7	8	9	10	11	12
1	161 **4,052**	200 **4,999**	216 **5,403**	225 **5,625**	230 **5,764**	234 **5,859**	237 **5,928**	239 **5,981**	241 **6,022**	242 **6,056**	243 **6,082**	244 **6,106**
2	18.51 **98.49**	19.00 **99.01**	19.16 **99.17**	19.25 **99.25**	19.30 **99.30**	19.33 **99.33**	19.36 **99.34**	19.37 **99.36**	19.38 **99.38**	19.39 **99.40**	19.40 **99.41**	19.41 **99.42**
3	10.13 **34.12**	9.55 **30.81**	9.28 **29.46**	9.12 **28.71**	9.01 **28.24**	8.94 **27.91**	8.88 **27.67**	8.84 **27.49**	8.81 **27.34**	8.78 **27.23**	8.76 **27.13**	8.74 **27.05**
4	7.71 **21.20**	6.94 **18.00**	6.59 **16.69**	6.39 **15.98**	6.26 **15.52**	6.16 **15.21**	6.09 **14.98**	6.04 **14.80**	6.00 **14.66**	5.96 **14.54**	5.93 **14.45**	5.91 **14.37**
5	6.61 **16.26**	5.79 **13.27**	5.41 **12.06**	5.19 **11.39**	5.05 **10.97**	4.95 **10.67**	4.88 **10.45**	4.82 **10.27**	4.78 **10.15**	4.74 **10.05**	4.70 **9.96**	4.68 **9.89**
6	5.99 **13.74**	5.14 **10.92**	4.76 **9.78**	4.53 **9.15**	4.39 **8.75**	4.28 **8.47**	4.21 **8.26**	4.15 **8.10**	4.10 **7.98**	4.06 **7.87**	4.03 **7.79**	4.00 **7.72**
7	5.59 **12.25**	4.74 **9.55**	4.35 **8.45**	4.12 **7.85**	3.97 **7.46**	3.87 **7.19**	3.79 **7.00**	3.73 **6.84**	3.68 **6.71**	3.63 **6.62**	3.60 **6.54**	3.57 **6.47**
8	5.32 **11.26**	4.46 **8.65**	4.07 **7.59**	3.84 **7.01**	3.69 **6.63**	3.58 **6.37**	3.50 **6.19**	3.44 **6.03**	3.39 **5.91**	3.34 **5.82**	3.31 **5.74**	3.28 **5.67**
9	5.12 **10.56**	4.26 **8.02**	3.86 **6.99**	3.63 **6.42**	3.48 **6.06**	3.37 **5.80**	3.29 **5.62**	3.23 **5.47**	3.18 **5.35**	3.13 **5.26**	3.10 **5.18**	3.07 **5.11**
10	4.96 **10.04**	4.10 **7.56**	3.71 **6.55**	3.48 **5.99**	3.33 **5.64**	3.22 **5.39**	3.14 **5.21**	3.07 **5.06**	3.02 **4.95**	2.97 **4.85**	2.94 **4.78**	2.91 **4.71**
11	4.84 **9.65**	3.98 **7.20**	3.59 **6.22**	3.36 **5.67**	3.20 **5.32**	3.09 **5.07**	3.01 **4.88**	2.95 **4.74**	2.90 **4.63**	2.86 **4.54**	2.82 **4.46**	2.79 **4.40**
12	4.75 **9.33**	3.88 **6.93**	3.49 **5.95**	3.26 **5.41**	3.11 **5.06**	3.00 **4.82**	2.92 **4.65**	2.85 **4.50**	2.80 **4.39**	2.76 **4.30**	2.72 **4.22**	2.69 **4.16**
13	4.67 **9.07**	3.80 **6.70**	3.41 **5.74**	3.18 **5.20**	3.02 **4.86**	2.92 **4.62**	2.84 **4.44**	2.77 **4.30**	2.72 **4.19**	2.67 **4.10**	2.63 **4.02**	2.60 **3.96**
14	4.60 **8.86**	3.74 **6.51**	3.34 **5.56**	3.11 **5.03**	2.96 **4.69**	2.85 **4.46**	2.77 **4.28**	2.70 **4.14**	2.65 **4.03**	2.60 **3.94**	2.56 **3.86**	2.53 **3.80**
15	4.54 **8.68**	3.68 **6.36**	3.29 **5.42**	3.06 **4.89**	2.90 **4.56**	2.79 **4.32**	2.70 **4.14**	2.64 **4.00**	2.59 **3.89**	2.55 **3.80**	2.51 **3.73**	2.48 **3.67**
16	4.49 **8.53**	3.63 **6.23**	3.24 **5.29**	3.01 **4.77**	2.85 **4.44**	2.74 **4.20**	2.66 **4.03**	2.59 **3.89**	2.54 **3.78**	2.49 **3.69**	2.45 **3.61**	2.42 **3.55**
17	4.45 **8.40**	3.59 **6.11**	3.20 **5.18**	2.96 **4.67**	2.81 **4.34**	2.70 **4.10**	2.62 **3.93**	2.55 **3.79**	2.50 **3.68**	2.45 **3.59**	2.41 **3.52**	2.38 **3.45**
18	4.41 **8.28**	3.55 **6.01**	3.16 **5.09**	2.93 **4.58**	2.77 **4.25**	2.66 **4.01**	2.58 **3.85**	2.51 **3.71**	2.46 **3.60**	2.41 **3.51**	2.37 **3.44**	2.34 **3.37**
19	4.38 **8.18**	3.52 **5.93**	3.13 **5.01**	2.90 **4.50**	2.74 **4.17**	2.63 **3.94**	2.55 **3.77**	2.48 **3.63**	2.43 **3.52**	2.38 **3.43**	2.34 **3.36**	2.31 **3.30**
20	4.35 **8.10**	3.49 **5.85**	3.10 **4.94**	2.87 **4.43**	2.71 **4.10**	2.60 **3.87**	2.52 **3.71**	2.45 **3.56**	2.40 **3.45**	2.35 **3.37**	2.31 **3.30**	2.28 **3.23**
21	4.32 **8.02**	3.47 **5.78**	3.07 **4.87**	2.84 **4.37**	2.68 **4.04**	2.57 **3.81**	2.49 **3.65**	2.42 **3.51**	2.37 **3.40**	2.32 **3.31**	2.28 **3.24**	2.25 **3.17**
22	4.30 **7.94**	3.44 **5.72**	3.05 **4.82**	2.82 **4.31**	2.66 **3.99**	2.55 **3.76**	2.47 **3.59**	2.40 **3.45**	2.35 **3.35**	2.30 **3.26**	2.26 **3.18**	2.23 **3.12**
23	4.28 **7.88**	3.42 **5.66**	3.03 **4.76**	2.80 **4.26**	2.64 **3.94**	2.53 **3.71**	2.45 **3.54**	2.38 **3.41**	2.32 **3.30**	2.28 **3.21**	2.24 **3.14**	2.20 **3.07**
24	4.26 **7.82**	3.40 **5.61**	3.01 **4.72**	2.78 **4.22**	2.62 **3.90**	2.51 **3.67**	2.43 **3.50**	2.36 **3.36**	2.30 **3.25**	2.26 **3.17**	2.22 **3.09**	2.18 **3.03**
25	4.24 **7.77**	3.38 **5.57**	2.99 **4.68**	2.76 **4.18**	2.60 **3.86**	2.49 **3.63**	2.41 **3.46**	2.34 **3.32**	2.28 **3.21**	2.24 **3.13**	2.20 **3.05**	2.16 **2.99**
26	4.22 **7.72**	3.37 **5.53**	2.98 **4.64**	2.74 **4.14**	2.59 **3.82**	2.47 **3.59**	2.39 **3.42**	2.32 **3.29**	2.27 **3.17**	2.22 **3.09**	2.18 **3.02**	2.15 **2.96**

SOURCE: Adapted by permission from George W. Snedecor and William G. Cochran, *Statistical Methods*, 6th ed. © 1967 by The Iowa State University Press, Ames, Iowa.

Table VI The F Distribution (continued)

Degrees of Freedom for Denominator		Degrees of Freedom for Numerator											
		1	2	3	4	5	6	7	8	9	10	11	12
	27	4.21	3.35	2.96	2.73	2.57	2.46	2.37	2.30	2.25	2.20	2.16	2.13
		7.68	**5.49**	**4.60**	**4.11**	**3.79**	**3.56**	**3.39**	**3.26**	**3.14**	**3.06**	**2.98**	**2.93**
	28	4.20	3.34	2.95	2.71	2.56	2.44	2.36	2.29	2.24	2.19	2.15	2.12
		7.64	**5.45**	**4.57**	**4.07**	**3.76**	**3.53**	**3.36**	**3.23**	**3.11**	**3.03**	**2.95**	**2.90**
	29	4.18	3.33	2.93	2.70	2.54	2.43	2.35	2.28	2.22	2.18	2.14	2.10
		7.60	**5.42**	**4.54**	**4.04**	**3.73**	**3.50**	**3.33**	**3.20**	**3.08**	**3.00**	**2.92**	**2.87**
	30	4.17	3.32	2.92	2.69	2.53	2.42	2.34	2.27	2.21	2.16	2.12	2.09
		7.56	**5.39**	**4.51**	**4.02**	**3.70**	**3.47**	**3.30**	**3.17**	**3.06**	**2.98**	**2.90**	**2.84**
	32	4.15	3.30	2.90	2.67	2.51	2.40	2.32	2.25	2.19	2.14	2.10	2.07
		7.50	**5.34**	**4.46**	**3.97**	**3.66**	**3.42**	**3.25**	**3.12**	**3.01**	**2.94**	**2.86**	**2.80**
	34	4.13	3.28	2.88	2.65	2.49	2.38	2.30	2.23	2.17	2.12	2.08	2.05
		7.44	**5.29**	**4.42**	**3.93**	**3.61**	**3.38**	**3.21**	**3.08**	**2.97**	**2.89**	**2.82**	**2.76**
	36	4.11	3.26	2.86	2.63	2.48	2.36	2.28	2.21	2.15	2.10	2.06	2.03
		7.39	**5.25**	**4.38**	**3.89**	**3.58**	**3.35**	**3.18**	**3.04**	**2.94**	**2.86**	**2.78**	**2.72**
	38	4.10	3.25	2.85	2.62	2.46	2.35	2.26	2.19	2.14	2.09	2.05	2.02
		7.35	**5.21**	**4.34**	**3.86**	**3.54**	**3.32**	**3.15**	**3.02**	**2.91**	**2.82**	**2.75**	**2.69**
	40	4.08	3.23	2.84	2.61	2.45	2.34	2.25	2.18	2.12	2.07	2.04	2.00
		7.31	**5.18**	**4.31**	**3.83**	**3.51**	**3.29**	**3.12**	**2.99**	**2.88**	**2.80**	**2.73**	**2.66**
	42	4.07	3.22	2.83	2.59	2.44	2.32	2.24	2.17	2.11	2.06	2.02	1.99
		7.27	**5.15**	**4.29**	**3.80**	**3.49**	**3.26**	**3.10**	**2.96**	**2.86**	**2.77**	**2.70**	**2.64**
	44	4.06	3.21	2.82	2.58	2.43	2.31	2.23	2.16	2.10	2.05	2.01	1.98
		7.24	**5.12**	**4.26**	**3.78**	**3.46**	**3.24**	**3.07**	**2.94**	**2.84**	**2.75**	**2.68**	**2.62**
	46	4.05	3.20	2.81	2.57	2.42	2.30	2.22	2.14	2.09	2.04	2.00	1.97
		7.21	**5.10**	**4.24**	**3.76**	**3.44**	**3.22**	**3.05**	**2.92**	**2.82**	**2.73**	**2.66**	**2.60**
	48	4.04	3.19	2.80	2.56	2.41	2.30	2.21	2.14	2.08	2.03	1.99	1.96
		7.19	**5.08**	**4.22**	**3.74**	**3.42**	**3.20**	**3.04**	**2.90**	**2.80**	**2.71**	**2.64**	**2.58**
	50	4.03	3.18	2.79	2.56	2.40	2.29	2.20	2.13	2.07	2.02	1.98	1.95
		7.17	**5.06**	**4.20**	**3.72**	**3.41**	**3.18**	**3.02**	**2.88**	**2.78**	**2.70**	**2.62**	**2.56**
	55	4.02	3.17	2.78	2.54	2.38	2.27	2.18	2.11	2.05	2.00	1.97	1.93
		7.12	**5.01**	**4.16**	**3.68**	**3.37**	**3.15**	**2.98**	**2.85**	**2.75**	**2.66**	**2.59**	**2.53**
	60	4.00	3.15	2.76	2.52	2.37	2.25	2.17	2.10	2.04	1.99	1.95	1.92
		7.08	**4.98**	**4.13**	**3.65**	**3.34**	**3.12**	**2.95**	**2.82**	**2.72**	**2.63**	**2.56**	**2.50**
	65	3.99	3.14	2.75	2.51	2.36	2.24	2.15	2.08	2.02	1.98	1.94	1.90
		7.04	**4.95**	**4.10**	**3.62**	**3.31**	**3.09**	**2.93**	**2.79**	**2.70**	**2.61**	**2.54**	**2.47**
	70	3.98	3.13	2.74	2.50	2.35	2.23	2.14	2.07	2.01	1.97	1.93	1.89
		7.01	**4.92**	**4.08**	**3.60**	**3.29**	**3.07**	**2.91**	**2.77**	**2.67**	**2.59**	**2.51**	**2.45**
	80	3.96	3.11	2.72	2.48	2.33	2.21	2.12	2.05	1.99	1.95	1.91	1.88
		6.96	**4.88**	**4.04**	**3.56**	**3.25**	**3.04**	**2.87**	**2.74**	**2.64**	**2.55**	**2.48**	**2.41**
	100	3.94	3.09	2.70	2.46	2.30	2.19	2.10	2.03	1.97	1.92	1.88	1.85
		6.90	**4.82**	**3.98**	**3.51**	**3.20**	**2.99**	**2.82**	**2.69**	**2.59**	**2.51**	**2.43**	**2.36**
	125	3.92	3.07	2.68	2.44	2.29	2.17	2.08	2.01	1.95	1.90	1.86	1.83
		6.84	**4.78**	**3.94**	**3.47**	**3.17**	**2.95**	**2.79**	**2.65**	**2.56**	**2.47**	**2.40**	**2.33**
	150	3.91	3.06	2.67	2.43	2.27	2.16	2.07	2.00	1.94	1.89	1.85	1.82
		6.81	**4.75**	**3.91**	**3.44**	**3.14**	**2.92**	**2.76**	**2.62**	**2.53**	**2.44**	**2.37**	**2.30**
	200	3.89	3.04	2.65	2.41	2.26	2.14	2.05	1.98	1.92	1.87	1.83	1.80
		6.76	**4.71**	**3.88**	**3.41**	**3.11**	**2.90**	**2.73**	**2.60**	**2.50**	**2.41**	**2.34**	**2.28**
	400	3.86	3.02	2.62	2.39	2.23	2.12	2.03	1.96	1.90	1.85	1.81	1.78
		6.70	**4.66**	**3.83**	**3.36**	**3.06**	**2.85**	**2.69**	**2.55**	**2.46**	**2.37**	**2.29**	**2.23**
	1,000	3.85	3.00	2.61	2.38	2.22	2.10	2.02	1.95	1.89	1.84	1.80	1.76
		6.66	**4.62**	**3.80**	**3.34**	**3.04**	**2.82**	**2.66**	**2.53**	**2.43**	**2.34**	**2.26**	**2.20**
	∞	3.84	2.99	2.60	2.37	2.21	2.09	2.01	1.94	1.88	1.83	1.79	1.75
		6.64	**4.60**	**3.78**	**3.32**	**3.02**	**2.80**	**2.64**	**2.51**	**2.41**	**2.32**	**2.24**	**2.18**

Table VI The F Distribution (continued)

Degrees of Freedom for Denominator		14	16	20	24	30	40	50	75	100	200	500	∞
		\multicolumn{12}{c}{Degrees of Freedom for Numerator}
	1	245 **6,142**	246 **6,169**	248 **6,208**	249 **6,234**	250 **6,258**	251 **6,286**	252 **6,302**	253 **6,323**	253 **6,334**	254 **6,352**	254 **6,361**	254 **6,366**
	2	19.42 **99.43**	19.43 **99.44**	19.44 **99.45**	19.45 **99.46**	19.46 **99.47**	19.47 **99.48**	19.47 **99.48**	19.48 **99.49**	19.49 **99.49**	19.49 **99.49**	19.50 **99.50**	19.50 **99.50**
	3	8.71 **26.92**	8.69 **26.83**	8.66 **26.69**	8.64 **26.60**	8.62 **26.50**	8.60 **26.41**	8.58 **26.35**	8.57 **26.27**	8.56 **26.23**	8.54 **26.18**	8.54 **26.14**	8.53 **26.12**
	4	5.87 **14.24**	5.84 **14.15**	5.80 **14.02**	5.77 **13.93**	5.74 **13.83**	5.71 **13.74**	5.70 **13.69**	5.68 **13.61**	5.66 **13.57**	5.65 **13.52**	5.64 **13.48**	5.63 **13.46**
	5	4.64 **9.77**	4.60 **9.68**	4.56 **9.55**	4.53 **9.47**	4.50 **9.38**	4.46 **9.29**	4.44 **9.24**	4.42 **9.17**	4.40 **9.13**	4.38 **9.07**	4.37 **9.04**	4.36 **9.02**
	6	3.96 **7.60**	3.92 **7.52**	3.87 **7.39**	3.84 **7.31**	3.81 **7.23**	3.77 **7.14**	3.75 **7.09**	3.72 **7.02**	3.71 **6.99**	3.69 **6.94**	3.68 **6.90**	3.67 **6.88**
	7	3.52 **6.35**	3.49 **6.27**	3.44 **6.15**	3.41 **6.07**	3.38 **5.98**	3.34 **5.90**	3.32 **5.85**	3.29 **5.78**	3.28 **5.75**	3.25 **5.70**	3.24 **5.67**	3.23 **5.65**
	8	3.23 **5.56**	3.20 **5.48**	3.15 **5.36**	3.12 **5.28**	3.08 **5.20**	3.05 **5.11**	3.03 **5.06**	3.00 **5.00**	2.98 **4.96**	2.96 **4.91**	2.94 **4.88**	2.93 **4.86**
	9	3.02 **5.00**	2.98 **4.92**	2.93 **4.80**	2.90 **4.73**	2.86 **4.64**	2.82 **4.56**	2.80 **4.51**	2.77 **4.45**	2.76 **4.41**	2.73 **4.36**	2.72 **4.33**	2.71 **4.31**
	10	2.86 **4.60**	2.82 **4.52**	2.77 **4.41**	2.74 **4.33**	2.70 **4.25**	2.67 **4.17**	2.64 **4.12**	2.61 **4.05**	2.59 **4.01**	2.56 **3.96**	2.55 **3.93**	2.54 **3.91**
	11	2.74 **4.29**	2.70 **4.21**	2.65 **4.10**	2.61 **4.02**	2.57 **3.94**	2.53 **3.86**	2.50 **3.80**	2.47 **3.74**	2.45 **3.70**	2.42 **3.66**	2.41 **3.62**	2.40 **3.60**
	12	2.64 **4.05**	2.60 **3.98**	2.54 **3.86**	2.50 **3.78**	2.46 **3.70**	2.42 **3.61**	2.40 **3.56**	2.36 **3.49**	2.35 **3.46**	2.32 **3.41**	2.31 **3.38**	2.30 **3.36**
	13	2.55 **3.85**	2.51 **3.78**	2.46 **3.67**	2.42 **3.59**	2.38 **3.51**	2.34 **3.42**	2.32 **3.37**	2.28 **3.30**	2.26 **3.27**	2.24 **3.21**	2.22 **3.18**	2.21 **3.16**
	14	2.48 **3.70**	2.44 **3.62**	2.39 **3.51**	2.35 **3.43**	2.31 **3.34**	2.27 **3.26**	2.24 **3.21**	2.21 **3.14**	2.19 **3.11**	2.16 **3.06**	2.14 **3.02**	2.13 **3.00**
	15	2.43 **3.56**	2.39 **3.48**	2.33 **3.36**	2.29 **3.29**	2.25 **3.20**	2.21 **3.12**	2.18 **3.07**	2.15 **3.00**	2.12 **2.97**	2.10 **2.92**	2.08 **2.89**	2.07 **2.87**
	16	2.37 **3.45**	2.33 **3.37**	2.28 **3.25**	2.24 **3.18**	2.20 **3.10**	2.16 **3.01**	2.13 **2.96**	2.09 **2.89**	2.07 **2.86**	2.04 **2.80**	2.02 **2.77**	2.01 **2.75**
	17	2.33 **3.35**	2.29 **3.27**	2.23 **3.16**	2.19 **3.08**	2.15 **3.00**	2.11 **2.92**	2.08 **2.86**	2.04 **2.79**	2.02 **2.76**	1.99 **2.70**	1.97 **2.67**	1.96 **2.65**
	18	2.29 **3.27**	2.25 **3.19**	2.19 **3.07**	2.15 **3.00**	2.11 **2.91**	2.07 **2.83**	2.04 **2.78**	2.00 **2.71**	1.98 **2.68**	1.95 **2.62**	1.93 **2.59**	1.92 **2.57**
	19	2.26 **3.19**	2.21 **3.12**	2.15 **3.00**	2.11 **2.92**	2.07 **2.84**	2.02 **2.76**	2.00 **2.70**	1.96 **2.63**	1.94 **2.60**	1.91 **2.54**	1.90 **2.51**	1.88 **2.49**
	20	2.23 **3.13**	2.18 **3.05**	2.12 **2.94**	2.08 **2.86**	2.04 **2.77**	1.99 **2.69**	1.96 **2.63**	1.92 **2.56**	1.90 **2.53**	1.87 **2.47**	1.85 **2.44**	1.84 **2.42**
	21	2.20 **3.07**	2.15 **2.99**	2.09 **2.88**	2.05 **2.80**	2.00 **2.72**	1.96 **2.63**	1.93 **2.58**	1.89 **2.51**	1.87 **2.47**	1.84 **2.42**	1.82 **2.38**	1.81 **2.36**
	22	2.18 **3.02**	2.13 **2.94**	2.07 **2.83**	2.03 **2.75**	1.98 **2.67**	1.93 **2.58**	1.91 **2.53**	1.87 **2.46**	1.84 **2.42**	1.81 **2.37**	1.80 **2.33**	1.78 **2.31**
	23	2.14 **2.97**	2.10 **2.89**	2.04 **2.78**	2.00 **2.70**	1.96 **2.62**	1.91 **2.53**	1.88 **2.48**	1.84 **2.41**	1.82 **2.37**	1.79 **2.32**	1.77 **2.28**	1.76 **2.26**
	24	2.13 **2.93**	2.09 **2.85**	2.02 **2.74**	1.98 **2.66**	1.94 **2.58**	1.89 **2.49**	1.86 **2.44**	1.82 **2.36**	1.80 **2.33**	1.76 **2.27**	1.74 **2.23**	1.73 **2.21**
	25	2.11 **2.89**	2.06 **2.81**	2.00 **2.70**	1.96 **2.62**	1.92 **2.54**	1.87 **2.45**	1.84 **2.40**	1.80 **2.32**	1.77 **2.29**	1.74 **2.23**	1.72 **2.19**	1.71 **2.17**
	26	2.10 **2.86**	2.05 **2.77**	1.99 **2.66**	1.95 **2.58**	1.90 **2.50**	1.85 **2.41**	1.82 **2.36**	1.78 **2.28**	1.76 **2.25**	1.72 **2.19**	1.70 **2.15**	1.69 **2.13**

Table VI The *F* Distribution (continued)

		Degrees of Freedom for Numerator										
		14	16	20	24	30	40	50	75	100	200	500	∞
Degrees of Freedom for Denominator	27	2.08 **2.83**	2.03 **2.74**	1.97 **2.63**	1.93 **2.55**	1.88 **2.47**	1.84 **2.38**	1.80 **2.33**	1.76 **2.25**	1.74 **2.21**	1.71 **2.16**	1.68 **2.12**	1.67 **2.10**
	28	2.06 **2.80**	2.02 **2.71**	1.96 **2.60**	1.91 **2.52**	1.87 **2.44**	1.81 **2.35**	1.78 **2.30**	1.75 **2.22**	1.72 **2.18**	1.69 **2.13**	1.67 **2.09**	1.65 **2.06**
	29	2.05 **2.77**	2.00 **2.68**	1.94 **2.57**	1.90 **2.49**	1.85 **2.41**	1.80 **2.32**	1.77 **2.27**	1.73 **2.19**	1.71 **2.15**	1.68 **2.10**	1.65 **2.06**	1.64 **2.03**
	30	2.04 **2.74**	1.99 **2.66**	1.93 **2.55**	1.89 **2.47**	1.84 **2.38**	1.79 **2.29**	1.76 **2.24**	1.72 **2.16**	1.69 **2.13**	1.66 **2.07**	1.64 **2.03**	1.62 **2.01**
	32	2.02 **2.70**	1.97 **2.62**	1.91 **2.51**	1.86 **2.42**	1.82 **2.34**	1.76 **2.25**	1.74 **2.20**	1.69 **2.12**	1.67 **2.08**	1.64 **2.02**	1.61 **1.98**	1.59 **1.96**
	34	2.00 **2.66**	1.95 **2.58**	1.89 **2.47**	1.84 **2.38**	1.80 **2.30**	1.74 **2.21**	1.71 **2.15**	1.67 **2.08**	1.64 **2.04**	1.61 **1.98**	1.59 **1.94**	1.57 **1.91**
	36	1.98 **2.62**	1.93 **2.54**	1.87 **2.43**	1.82 **2.35**	1.78 **2.26**	1.72 **2.17**	1.69 **2.12**	1.65 **2.04**	1.62 **2.00**	1.59 **1.94**	1.56 **1.90**	1.55 **1.87**
	38	1.96 **2.59**	1.92 **2.51**	1.85 **2.40**	1.80 **2.32**	1.76 **2.22**	1.71 **2.14**	1.67 **2.08**	1.63 **2.00**	1.60 **1.97**	1.57 **1.90**	1.54 **1.86**	1.53 **1.84**
	40	1.95 **2.56**	1.90 **2.49**	1.84 **2.37**	1.79 **2.29**	1.74 **2.20**	1.69 **2.11**	1.66 **2.05**	1.61 **1.97**	1.59 **1.94**	1.55 **1.88**	1.53 **1.84**	1.51 **1.81**
	42	1.94 **2.54**	1.89 **2.46**	1.82 **2.35**	1.78 **2.26**	1.73 **2.17**	1.68 **2.08**	1.64 **2.02**	1.60 **1.94**	1.57 **1.91**	1.54 **1.85**	1.51 **1.80**	1.49 **1.78**
	44	1.92 **2.52**	1.88 **2.44**	1.81 **2.32**	1.76 **2.24**	1.72 **2.15**	1.66 **2.06**	1.63 **2.00**	1.58 **1.92**	1.56 **1.88**	1.52 **1.82**	1.50 **1.78**	1.48 **1.75**
	46	1.91 **2.50**	1.87 **2.42**	1.80 **2.30**	1.75 **2.22**	1.71 **2.13**	1.65 **2.04**	1.62 **1.98**	1.57 **1.90**	1.54 **1.86**	1.51 **1.80**	1.48 **1.76**	1.46 **1.72**
	48	1.90 **2.48**	1.86 **2.40**	1.79 **2.28**	1.74 **2.20**	1.70 **2.11**	1.64 **2.02**	1.61 **1.96**	1.56 **1.88**	1.53 **1.84**	1.50 **1.78**	1.47 **1.73**	1.45 **1.70**
	50	1.90 **2.46**	1.85 **2.39**	1.78 **2.26**	1.74 **2.18**	1.69 **2.10**	1.63 **2.00**	1.60 **1.94**	1.55 **1.86**	1.52 **1.82**	1.48 **1.76**	1.46 **1.71**	1.44 **1.68**
	55	1.88 **2.43**	1.83 **2.35**	1.76 **2.23**	1.72 **2.15**	1.67 **2.06**	1.61 **1.96**	1.58 **1.90**	1.52 **1.82**	1.50 **1.78**	1.46 **1.71**	1.43 **1.66**	1.41 **1.64**
	60	1.86 **2.40**	1.81 **2.32**	1.75 **2.20**	1.70 **2.12**	1.65 **2.03**	1.59 **1.93**	1.56 **1.87**	1.50 **1.79**	1.48 **1.74**	1.44 **1.68**	1.41 **1.63**	1.39 **1.60**
	65	1.85 **2.37**	1.80 **2.30**	1.73 **2.18**	1.68 **2.09**	1.63 **2.00**	1.57 **1.90**	1.54 **1.84**	1.49 **1.76**	1.46 **1.71**	1.42 **1.64**	1.39 **1.60**	1.37 **1.56**
	70	1.84 **2.35**	1.79 **2.28**	1.72 **2.15**	1.67 **2.07**	1.62 **1.98**	1.56 **1.88**	1.53 **1.82**	1.47 **1.74**	1.45 **1.69**	1.40 **1.62**	1.37 **1.56**	1.35 **1.53**
	80	1.82 **2.32**	1.77 **2.24**	1.70 **2.11**	1.65 **2.03**	1.60 **1.94**	1.54 **1.84**	1.51 **1.78**	1.45 **1.70**	1.42 **1.65**	1.38 **1.57**	1.35 **1.52**	1.32 **1.49**
	100	1.79 **2.26**	1.75 **2.19**	1.68 **2.06**	1.63 **1.98**	1.57 **1.89**	1.51 **1.79**	1.48 **1.73**	1.42 **1.64**	1.39 **1.59**	1.34 **1.51**	1.30 **1.46**	1.28 **1.43**
	125	1.77 **2.23**	1.72 **2.15**	1.65 **2.03**	1.60 **1.94**	1.55 **1.85**	1.49 **1.75**	1.45 **1.68**	1.39 **1.59**	1.36 **1.54**	1.31 **1.46**	1.27 **1.40**	1.25 **1.37**
	150	1.76 **2.20**	1.71 **2.12**	1.64 **2.00**	1.59 **1.91**	1.54 **1.83**	1.47 **1.72**	1.44 **1.66**	1.37 **1.56**	1.34 **1.51**	1.29 **1.43**	1.25 **1.37**	1.22 **1.33**
	200	1.74 **2.17**	1.69 **2.09**	1.62 **1.97**	1.57 **1.88**	1.52 **1.79**	1.45 **1.69**	1.42 **1.62**	1.35 **1.53**	1.32 **1.48**	1.26 **1.39**	1.22 **1.33**	1.19 **1.28**
	400	1.72 **2.12**	1.67 **2.04**	1.60 **1.92**	1.54 **1.84**	1.49 **1.74**	1.42 **1.64**	1.38 **1.57**	1.32 **1.47**	1.28 **1.42**	1.22 **1.32**	1.16 **1.24**	1.13 **1.19**
	1,000	1.70 **2.09**	1.65 **2.01**	1.58 **1.89**	1.53 **1.81**	1.47 **1.71**	1.41 **1.61**	1.36 **1.54**	1.30 **1.44**	1.26 **1.38**	1.19 **1.28**	1.13 **1.19**	1.08 **1.11**
	∞	1.69 **2.07**	1.64 **1.99**	1.57 **1.87**	1.52 **1.79**	1.46 **1.69**	1.40 **1.59**	1.35 **1.52**	1.28 **1.41**	1.24 **1.36**	1.17 **1.25**	1.11 **1.15**	1.00 **1.00**

Table VII Random Numbers

```
09 18 82 00 97   32 82 53 95 27   04 22 08 63 04   83 38 98 73 74   64 27 85 80 44
90 04 58 54 97   51 98 15 06 54   94 93 88 19 97   91 87 07 61 50   68 47 66 46 59
73 18 95 02 07   47 67 72 62 69   62 29 06 44 64   27 12 46 70 18   41 36 18 27 60
75 76 87 64 90   20 97 18 17 49   90 42 91 22 72   95 37 50 58 71   93 82 34 31 78
54 01 64 40 56   66 28 13 10 03   00 68 22 73 98   20 71 45 32 95   07 70 61 78 13

08 35 86 99 10   78 54 24 27 85   13 66 15 88 73   04 61 89 75 53   31 22 30 84 20
28 30 60 32 64   81 33 31 05 91   40 51 00 78 93   32 60 46 04 75   94 11 90 18 40
53 84 08 62 33   81 59 41 36 28   51 21 59 02 90   28 46 66 87 95   77 76 22 07 91
91 75 75 37 41   61 61 36 22 69   50 26 39 02 12   55 78 17 65 14   83 48 34 70 55
89 41 59 26 94   00 39 75 83 91   12 50 71 76 46   48 94 97 23 06   94 54 13 74 08

77 51 30 38 20   86 83 42 99 01   68 41 48 27 74   51 90 81 39 80   72 89 35 55 07
19 50 23 71 74   69 97 92 02 88   55 21 02 97 73   74 28 77 52 51   65 34 46 74 15
21 81 85 93 13   93 27 88 17 57   05 68 67 31 56   07 08 28 50 46   31 85 33 84 52
51 47 46 64 99   68 10 72 36 21   94 04 99 13 45   42 83 60 91 91   08 00 74 54 49
99 55 96 83 31   62 53 52 41 70   69 77 71 28 30   74 81 97 81 42   43 86 07 28 34

33 71 34 80 07   93 58 47 28 69   51 92 66 47 21   58 30 32 98 22   93 17 49 39 72
85 27 48 68 93   11 30 32 92 70   28 83 43 41 37   73 51 59 04 00   71 14 84 36 43
84 13 38 96 40   44 03 55 21 66   73 85 27 00 91   61 22 26 05 61   62 32 71 84 23
56 73 21 62 34   17 39 59 61 31   10 12 39 16 22   85 49 65 75 60   81 60 41 88 80
65 13 85 68 06   87 64 88 52 61   34 31 36 58 61   45 87 52 10 69   85 64 44 72 77

38 00 10 21 76   81 71 91 17 11   71 60 29 29 37   74 21 96 40 49   65 58 44 96 98
37 40 29 63 97   01 30 47 75 86   56 27 11 00 86   47 32 46 26 05   40 03 03 74 38
97 12 54 03 48   87 08 33 14 17   21 81 53 92 50   75 23 76 20 47   15 50 12 95 78
21 82 64 11 34   47 14 33 40 72   64 63 88 59 02   49 13 90 64 41   03 85 65 45 52
73 13 54 27 42   95 71 90 90 35   85 79 47 42 96   08 78 98 81 56   64 69 11 92 02

07 63 87 79 29   03 06 11 80 72   96 20 74 41 56   23 82 19 95 38   04 71 36 69 94
60 52 88 34 41   07 95 41 98 14   59 17 52 06 95   05 53 35 21 39   61 21 20 64 55
83 59 63 56 55   06 95 89 29 83   05 12 80 97 19   77 43 35 37 83   92 30 15 04 98
10 85 06 27 46   99 59 91 05 07   13 49 90 63 19   53 07 57 18 39   06 41 01 93 62
39 82 09 89 52   43 62 26 31 47   64 42 18 08 14   43 80 00 93 51   31 02 47 31 67

59 58 00 64 78   75 56 97 88 00   88 83 55 44 86   23 76 80 61 56   04 11 10 84 08
38 50 80 73 41   23 79 34 87 63   90 82 29 70 22   17 71 90 42 07   95 95 44 99 53
30 69 27 06 68   94 68 81 61 27   56 19 68 00 91   82 06 76 34 00   05 46 26 92 00
65 44 39 56 59   18 28 82 74 37   49 63 22 40 41   08 33 76 56 76   96 29 99 08 36
27 26 75 02 64   13 19 27 22 94   07 47 74 46 06   17 98 54 89 11   97 34 13 03 58

91 30 70 69 91   19 07 22 42 10   36 69 95 37 28   28 82 53 57 93   28 97 66 62 52
68 43 49 46 88   84 47 31 36 22   62 12 69 84 08   12 84 38 25 90   09 81 59 31 46
48 90 81 58 77   54 74 52 45 91   35 70 00 47 54   83 82 45 26 92   54 13 05 51 60
06 91 34 51 97   42 67 27 86 01   11 83 30 95 28   63 01 19 89 01   14 97 44 03 44
10 45 51 60 19   14 21 03 37 12   91 34 23 78 21   88 32 58 08 51   43 66 77 08 83

12 88 39 73 43   65 02 76 11 84   04 28 50 13 92   17 97 41 50 77   90 71 22 67 69
21 77 83 09 76   38 80 73 69 61   31 64 94 20 96   63 28 10 20 23   08 81 64 74 49
19 52 35 95 15   65 12 25 96 59   86 28 36 82 58   69 57 21 37 98   16 43 59 15 29
67 24 55 26 70   35 58 31 65 63   79 24 68 66 86   76 46 33 42 22   26 65 59 08 02
60 58 44 73 77   07 50 03 79 92   45 13 42 65 29   26 76 08 36 37   41 32 64 43 44

53 85 34 13 77   36 06 69 48 50   58 83 87 38 59   49 36 47 33 31   96 24 04 36 42
24 63 73 87 36   74 38 48 93 42   52 62 30 79 92   12 36 91 86 01   03 74 28 38 73
83 08 01 24 51   38 99 22 28 15   07 75 95 17 77   97 37 72 75 85   51 97 23 78 67
16 44 42 43 34   36 15 19 90 73   27 49 37 09 39   85 13 03 25 52   54 84 65 47 59
60 79 01 81 57   57 17 86 57 62   11 16 17 85 76   45 81 95 29 79   65 13 00 48 60
```

SOURCE: Reproduced by permission from tables of the RAND Corporation in *A Million Random Digits with* 100,000 *Normal Deviates,* New York, The Free Press, 1955.

Solutions to Starred Problems

Chapter 1

1-2. Cost or time may argue against complete enumeration.
1-4. Problem identification and solution implementation.
1-5. It ignores uncertainty.
1-9. It ignores a possible relationship between family size and income.
1-11. In simplification, we may lose the essence of the problem.
1-14. Accurate data. Determining when a sale is made. The largest problem is defining an industry and determining what firms belong to each.
1-15. Fewer hours worked per week or higher taxes or higher withholding accompanying larger fringe benefits.
1-19. Perhaps, but other factors instituted at the same time could account for the result; also, the result may be a chance phenomenon.
1-20. The statement is meaningless since no basis for comparison is given.
1-25. (a) Ratio
     (b) Nominal
     (c) Nominal
     (d) Ratio
     (e) Ratio

Chapter 2

2-5. (a) Ogive
    (b) Histogram or frequency polygon
    (c) Frequency polygon

Solutions to Starred Problems

2-7. (a) Classes are not exhaustive
(b) Classes overlap
(c) 25 included twice

2-8. Data do not support the conclusion. 90 percent of the stockholders own less than 100 shares. The size of large holdings (if any) is not given.

2-10. 11(10.9); 3.

2-14.

Volume	Frequency	Relative Frequency
84–94	2	0.04
95–105	9	0.18
106–116	13	0.26
117–127	11	0.22
128–138	11	0.22
139–149	1	0.02
150–160	3	0.06

2-16.

Output	Frequency	Relative Frequency
55–62	2	0.067
63–70	0	0.000
71–78	3	0.100
79–86	13	0.433
87–94	7	0.233
95–102	5	0.167

2-20.

Class	Class Limit	Frequency	Relative Frequency	Less Than	More Than
			Before		
	$16\frac{1}{2}$			0	30
17–23	$23\frac{1}{2}$	6	0.200	6	24
24–30	$30\frac{1}{2}$	2	0.067	8	22
31–37	$37\frac{1}{2}$	3	0.100	11	19
38–44	$44\frac{1}{2}$	7	0.233	18	12
45–51	$51\frac{1}{2}$	8	0.267	26	4
52–58	$58\frac{1}{2}$	4	0.133	30	0
			After		
	$5\frac{1}{2}$			0	30
6–13	$13\frac{1}{2}$	4	0.133	4	26
14–21	$21\frac{1}{2}$	11	0.367	15	15
22–29	$29\frac{1}{2}$	12	0.400	27	3
30–37	$37\frac{1}{2}$	2	0.067	29	1
38–45	$45\frac{1}{2}$	0	0.000	29	1
46–53	$53\frac{1}{2}$	1	0.033	30	0

2-23. No. The information that is lost is crucial.

2-25. It is a good table. We might wonder, however, whether it is based on 1972 deaths or is a forecast for persons living in 1972.

Chapter 3

3-1. All coffee drinkers in the United States; those from whom opinions are obtained.

3-4. $\bar{x} = 0.55$; $M_0 = 0$; $M = 0$; all three have some information value.

3-7. $\bar{x} = 15$; $M = 16$; MAD $= 2.4$; $s = 2.83$. The statistics do not indicate the trend in the values.

3-8. 22.9; 22.0 (with grouped data). The second is lower since the actual data values fall toward the top of the classes.

3-10. $\bar{x} = 7.5$; $M = 4$; $M_0 = 3$. Range $= 2$ to 24; MAD $= 5.5$; $M_2 = s^2 = 43.75$; $s = 6.614$. Sk $= 1.588$; $M_4 = 6872.8$; $a = 3.59$. There may be marketing reasons for maintaining books below the break-even level.

3-12. (a) 15.9
 (b) 15.95
 (c) 0.083

3-16. (a) Foreign (coefficient of variation $= 4.60/25 = 0.184$, whereas American has a coefficient of variation $= 2.07/14 = 0.15$).
 (b) Variance is unchanged at 4.29.
 (c) Variance is 100 times greater.

3-17. (a) 17,960,000
 (b) Mean $= 35,920$; median $= 35,400$; modal class is 30 to less than 40.

3-19. (a) 3.6
 (b) 4.66

3-23. (a) 23.5; 23.25; modal class is 22–23.
 (b) $s^2 = 3.8$

Chapter 4

4-2. If the value of your house increases as the CPI increases, the CPI overstates the effect.

4-4. Weighted index: Weights average change by the proportion of the item in the mix.
 Unweighted index: Averages the individual changes.
 The CPI is a weighted index.

Solutions to Starred Problems

4-5. (a) 1.380; CPI change = 1.378.
 (b) Lower salary levels, which have risen less than 1.38, comprise more than $\frac{1}{4}$ of the individuals.

4-8. (a) 1.231 and 1.321
 (b) 1.205 and 1.308
 (c) 1.305

4-12. (a) 104.9
 (b) 100, 98.9, 104.9
 (c) 100, 98.9, 104.9
 (d) 100, 98.9, 104.9

4-13. (a) No; the data simply indicate that consumer prices have risen faster in New York City.
 (b) Chicago experienced the smallest price rise; New York, the greatest.

4-15. (a) Year: 1964 1965 1966 1967 1968 1969
 Index: 89.9 91.6 93.2 95.0 100.0 110.9
 (b) Year: 1965 1966 1967 1968 1969 1970 1971
 Index: 75.6 81.0 83.5 91.1 95.3 100.0 108.9

4-19. After adjusting for price changes, the "real" cost has declined.

4-20. No; workers earn 20 percent less; $\left(\dfrac{1 + 0.20}{1 + 0.50} - 1\right) 100 = 80$.

4-25. (a) 31.12
 (b) Such shares are more risky.

Chapter 5

5-3. {acceptable, unacceptable}

5-4. The statement ignores the number of flights.

5-5. {f}, {o}, {n}, {f, o}, {f, n}, {o, n}, {f, o, n}, $\emptyset$

5-6. 3, 4

5-7. {f, n}

5-9. $\frac{17}{60}$

5-10. $P(\text{tack } X) > 0.5$ $P(\text{tack } Y) < 0.5$

5-11. $\frac{1}{12}$ to $\frac{3}{4}$; no

5-15. (a) mutually exclusive
 (b), (c), (d) not mutually exclusive

5-16. The odds are 2 to 3 that Britelite will not purchase Buy Me.

Chapter 6

6-1. $10^3 = 1{,}000$, using the fundamental principle of counting.

6-2. $(\frac{1}{6})^8 \approx 0.0000006$

6-4. (a) $\dfrac{123{,}552}{2{,}598{,}960} \approx 0.048$

(b) $\dfrac{54{,}912}{2{,}598{,}960} \approx 0.021$

(c) $\dfrac{624}{2{,}598{,}960} \approx 0.00024$

6-6. (a) $\dfrac{1}{12{,}000{,}000}$

(b) The assumption is not justified.

6-9. $r \geq 23$

6-11. (a) No
(b) 4
(c) Yes

6-14. (a) 0.46
(b) 0.46

(c) $\dfrac{0.30}{0.46} \approx 0.65$

6-20. 8,580

6-22. (a) 0.80
(b) 4/7

6-26. (a) $f(x) \geq 0$ for all x and area $= \frac{1}{2}(1)(2) = 1$
(b) $P(x < 1) = \frac{1}{4} = P(x \leq 1)$
$P(x > 1) = \frac{3}{4}$
$P(x = 1) = 0$

Chapter 7

7-4. 3.5; no

7-5. (a) 5
(b) 13
(c) Cannot say
(d) Cannot say

Solutions to Starred Problems

7-7. 80,000; they need to determine if the potential $200,000 loss would ruin them. If not, the product looks good.

7-8. $\sigma^2 = 1{,}376 \times 10^8$, so $\sigma = 37.1 \times 10^4$.
The variance is not as meaningful as knowing there is a 40 percent chance of a $200,000 loss.

7-12. $437.50

7-15. 1.0

7-16. 0.7; 0.21

7-17. 160; 17.436

7-21. 35,910 (based on 3.94 expected new franchises); 2,961,900

7-25. (a) 2 or 3 cakes
(b) 3 cakes
(Both answers require a lot of arithmetic.)

Chapter 8

8-1. 0.0547

8-3. 0.1353

8-5. (a) 0.5
(b) 0.023
(c) 0.954

8-9. 0.3193; binomial approximation gives $b(0; 5, 0.2) = 0.3276$.

8-15. 34.46 percent

8-17. 0.0691

8-18. $P(a) = 0.1073 > P(b) = 0.0105$

8-20. (a) 0.0067
(b) 0.0404

8-21. $0.3 = 1 - (1 - e^{-1.2})$

8-22. (a) 0.0197
(b) 0.5002

Chapter 9

9-1. The first 100 may tend to be either good or bad. A better sample would include items throughout the week.

9-2. Cost and feasibility

SOLUTIONS TO STARRED PROBLEMS

9-7. Yes; no

9-10. (a) Binomial (with replacement); hypergeometric (without replacement)
(b) 0.1398
(c) It would sample throughout the run, avoiding any bunching of good or bad items.

9-11. Serial numbers: 62675-30, 62675-8, 62675-28, 62675-2, and 62675-12.

9-14. (a) Hands 1 and 2 each have a pair. Hand 3 has three fours and is best.
(b) Hand 3 is still best.

9-18. $\bar{x} = 5.755$ and $s_{\bar{x}}^2 = 0.01375$

9-19. (a) Stratified sampling is reasonable, but they should use more strata and sample according to population size.
(b) $P(\bar{x} \geq 5.755 \mid \mu = 5.5, \sigma_{\bar{x}} = 0.11726) = P(z \geq 2.17) = 0.015$.

9-23. Income and occupation are two possibilities.

9-24. No; we could use 335,000,000 (straight average), or we might go with 307,500,000 (average from the two more experienced individuals).

Chapter 10

10-1. Central Limit Theorem

10-2. (a) Binomial
(b) 0.1183

10-5. (a) 2.016 and 0.02
(b) 0.2119
(c) 0; 0.0548

10-8. (a) 0.2525
(b) Cannot be obtained, owing to the seeming nonnormality of the original universe. If normality is assumed, the probability is: $P(z \geq 3.67) \approx 0$.
(c) Sk = 1.70; it is skewed to the right.

10-11. $\sigma_{\bar{x}} = 0.2238$ (use finite population correction factor)
$P(\bar{x} > 32) = 0$

10-12. $P(z \leq -10) = 0$

10-15. 0.9962 (using a normal approximation)

10-19. (a) 0.0013
(b) 0.2473
(c) 0.9974

Solutions to Starred Problems

10-24. (a) 0.4778
(b) 0.7264

10-28. (a) 3
(b) 5

Chapter 11

11-4. The one with $n - 1$ in the denominator; both; factor of $\frac{100}{99}$.

11-5. 1,000,000; 160,000

11-7. (a) $P(19.888 \leq \mu \leq 19.912) = 0.95$
(b) 0.0475
(c) Depends on the costs.

11-9. (a) 0.8
(b) $0.72 < p < 0.88$; $0.73 < p < 0.87$; $0.77 < p < 0.83$

11-12. $P(8.043 < \mu < 8.357) = 0.95$

11-18. $P(0.6102 < p < 0.7898) = 0.95$, using the normal approximation.

11-19. (a) $P(10{,}448 < \mu < 12{,}664) = 0.99$
(b) $n = 1{,}104$

11-22. $P(8.549 < \mu < 11.851) = 0.95$

11-26. $P(-0.1808 < p < -0.0192) = 0.95$

11-30. $P(1.0145 < \mu < 1.0255) = 0.90$ and $P(0.0001686 < \sigma^2 < 0.0004436) = 0.90$

Chapter 12

12-2. Based on costs; H_0 is chosen so that falsely accepting it is less expensive than falsely accepting H_1. The choice is frequently unclear.

12-5. One-tailed; when we do not know, a priori, the direction of any change that might occur.

12-12. 0.95 occurs for μ just above 25 (for example, 25.00001). The new test will be more powerful, but the maximum type II error is still 0.95.

12-13. (a) $H_0: \mu = 0.500$ and $H_1: \mu \neq 0.500$
(b) H_0 is rejected ($\bar{x}_{c,\text{upper}} = 0.503$).

12-14. (a) 0; 0.0475; 0.50; 0.0475
(b) Several; for example, how is "lasting" defined?

12-18. (a) No ($p' = 0.85 < p'_c = 0.93$).
(b) No; the sample size is too small.

12-22. Assuming that $\bar{x}$ is normally distributed, $\bar{x}_c = 3{,}788 > \bar{x} = 3{,}467$. H_1 cannot be accepted.

12-23. $\bar{x}_1 - \bar{x}_2 = 2.55 > \bar{x}_c = 0.7728$. Accept H_1.

12-24. Use a two-tailed test. Now $\bar{x}_{c,\text{upper}} = 0.843$. Accept H_1.

12-27. $H_0: p_1 = p_2$ and $H_1: p_1 \neq p_2$. $p'_1 - p'_2 = 0.025 < p'_{c,\text{upper}} = 0.0798$. ($p'_{c,\text{lower}} = -0.0798$.) We cannot accept H_1.

Chapter 13

13-1. (a) Maximin: stock zero; minimax: stock two.
(b) Stock two.
(c) No

13-4. 3

13-5. Job 1

13-6. 5/6

13-8. (a) $\sigma \approx 0.7407$
(b) 0.053
(c) 0.8438

13-10. a_2

13-11. (a) Expected Utility of investing $= 11.7 > 9.0 =$ Expected Utility of present position
(b) $468.27

13-13. (a) $E_0(\mu) = 2{,}000 < 2{,}080 = \mu_b$; no
(b) No: $\bar{x}$ and $E_0(\mu)$ are both less than μ_b.
(c) $E_1(\mu) = 2{,}090 > 2{,}080 = \mu_b$; yes.

13-16. If she can stock 6 only, minimax directs her to stock 4. If she can stock 7 units, minimax directs her to stock 5. She could have stocked 5 initially.

13-25. Select $X = 11$; EVPI $= \$1.80$.

Chapter 14

14-3. Ratio of two sample variances, such as within cells and between factor levels; the basic data are assumed to be normally distributed.

14-5. Yes; no (the two tabulated F values are 3.06 and 4.89).

Solutions to Starred Problems 703

14-6. $MS_B = 363{,}302$; $MS_W = 285{,}678$; $F = 1.27 < 5.14$ (from table). Accept H_0.

14-10. $MS_B = 4.376$; $MS_W = 59.686$; $F = 0.0733 < 3.55$ (from table). Accept H_0.

14-13. $MS_B = 1{,}826$; $MS_W = 268$; $F = 6.81$. The values from the F table for 2 and 12 degrees of freedom are 6.91 for $\alpha = 0.01$ and 3.89 for $\alpha = 0.05$. With $\alpha = 0.01$, H_1 cannot be accepted; with $\alpha = 0.05$, H_1 is accepted.

14-16. $MS_B = 20.067$; $MS_W = 4.633$; $F = 4.33 > 3.89$ (from table). Accept H_1.

14-19. $MS_B(1) = 19.59$; $MS_B(2) = 55.61$; $MS_I = 14.48$; $MS_W = 3.61$. To test factor 1, $F = 5.42 > 2.52$ (from table). Reject H_0.

14-22. $MS_B(1) = 1.635$; $MS_B(2) = 0.3016$; $MS_I = 0.1726$; $MS_W = 0.0383$. The three F values are 7.875, 42.69, and 4.506. Each is significant at the 0.05 level.

14-25. (a) $MS_B(1) = 4.0055$; $MS_B(2) = 114.16$; $MS_I = 0.1588$; $MS_W = 0.8633$; $F = 4.6398 > 3.88$ (from table). The dog foods are different.
(b) This is tested by examining the interaction effect, which is not significant ($F = 0.1839$).

14-27. $MS_B(1) = 7.6915$; $MS_B(2) = 11.7893$; $MS_I = 2.3181$; $MS_W = 1.0174$.
(a) $F = 7.56 > 6.01$. Accept H_1.
(b) $F = 11.59 > 6.01$. Accept H_1.
(c) $F = 2.28 < 4.58$. H_1 cannot be accepted.

Chapter 15

15-1. Reject H_0.

15-3. (a) $n = 1 + 3m$
(b) Cost increases by $3 for every unit increase in output. Cost is $1 at zero output; an estimate of fixed cost.

15-5. (a) Yes, the slope is significant.
(b) 96.802 to 133.798

15-6. (a) $y_c = 2.1 + 0.85x$
(b) Yes
(c) $s_e = 1.07$
(d) $s_b = 0.239$
(e) 0.3 to 1.4
(f) H_1 is accepted ($t = 3.556$).

15-12. (a) $t = 3.42$; H_0 is rejected.
   (b) Yes; $H_0: \beta \leq 0$. Yes; reject H_0.
   (c) 8.094 ± 4.356

15-16. (a) $y_c = -139.5732 + 2.6602x$
   (b) $y_c = 38.6602$
   (c) No

15-17. (a) Zero
   (b) $1.46 per cc.

15-18. (a) $y_c = -2.2724 + 0.5016x$
   (b) Yes
   (c) $t = 0.02187$; no

15-21. (a) $y_c = 0.983 + 0.077x$
   (b) $P = 0.077Q$

15-23. Yes

Chapter 16

16-2. (a)

16-4. 25 percent

16-6. (a) (4)
   (b) (3)

16-7. $r = 1$, $s_r = 0$. The coefficient is significant at any α level.

16-8. (a) 0.53
   (b) 0.54
   (c) 0.32 to 0.76 thousand

16-10. Regression, (c) and (b); correlation, (a); both, (d).

16-14. $r = 0.783$, $t = 3.56$; accept H_1.

16-17. $r = -0.455$, $t = -1.44$; accept H_0.

16-20. $r = 0.993$, $t = 22.07$; accept H_1.

16-23. $r = 0.925$, $t = 7.28$; accept H_1.

Chapter 17

17-1. Yes. No, predictive variables should be used.

17-2. Yes. A reasonable forecast would be 1.2 or 1.1. The values are stabilizing.

Solutions to Starred Problems

17-4. 0.48, 0.91, 0.99, 1.28, 0.78, 0.63, 0.88, 2.08, 0.54, 0.56, 1.43, 1.44.

17-7. (a) $13.87 + 5.66t$
(b) The curve appears to be nonlinear.

17-8. $a = 4.286$, $b = -0.128$, and the 5-year total forecast $= 18.23$.

17-14. Adjusted values are: $S_1 = 0.889$; $S_2 = 1.080$; $S_3 = 1.088$; $S_4 = 1.047$; $S_5 = 0.875$.

17-16. Adjusted values are: $S_1 = 1.808$; $S_2 = 0.848$; $S_3 = 0.409$; $S_4 = 0.935$.

17-17. $47.38 + 0.401x$. The deseasonalized values are 50.62, 47.22, and so on.

17-18. 177.42; 275.24; 259.22; 164.00.

17-25. (a) 30.8; 22.8
(b) $ES_1 = 21.0$; $b_1 = 0.70$; the forecast is 21.70.

Chapter 18

18-3. No; yes.

18-7. $\chi^2 = 10.46 > \chi^2_{0.05} = 5.99$ for d.f. $= 2$. Reject H_0.

18-8. $D = 0.09 < 0.096$. No, accept H_0.

18-12. $\sum_{x=7}^{10} b(x; 10, 0.5) = 0.1719$. Accept H_0.

The new machine did better as time passed. Early results may reflect learning, and the new machine may be best.

18-16. $U = 24.5 < 27$. Reject H_0.

18-18. $z = 1.58$. Accept H_0. The normal is not appropriate due to the small sample size. The Wilcoxon test is more powerful since it uses more of the information in the data.

18-19. $H = 5.63$ (or 5.69 if corrected for ties) $< \chi^2_{0.10}$ for 3 d.f. Accept H_0.

18-20. (a) Nominal
(b) Ordinal
(c) Yes, because the assumptions of the more powerful method may not be met.

18-22. (a) $C = 0.19$. Reject H_0.
(b) $\chi^2 = 13.43 > \chi^2_{0.10} = 4.61$ for 2 d.f.

18-23. (a) $r_s = 0.806$
(b) $t = 3.85 > t_{0.01} = 2.90$ for 8 d.f. and a one-tail test.

Index

A

Addition rule of probability, 141–44
Adjusted coefficient of determination, 569
Alternate hypothesis, 370
Analysis of variance, 472–504
  experimental design in, 502
  fixed-effects model, 501
  one-factor, 474–84
    computational formulas, 481–84
    F distribution, 476–81
  random-effects model, 501
  two-factor, 486–97
    computational formulas, 492–97
    interaction effect, 488
    main effects, 488
  See also Nonparametric methods
Analysis-of-variance table, 482
Arithmetic average value, *see* Mean
Association, nonparametric measures of, 656–59
  contingency coefficient, 656–57
  Spearman's rank correlation coefficient, 657–59
Average squared deviation, *see* Variance

B

Bayesian decision rule, 420–21, 424–25
Bayes' Theorem, 182–88
Bernoulli, James, 136

Biased estimator, 336–37
Binomial probability mass function, 240–48
Bounded classes, 22
Brown, R. G., 614

C

Cardano, Gerolamo, 137
Causation, regression and, 535
Central Limit Theorem, 315–17, 325–26
Chi square distribution, 233, 348–52
Chi-square tests:
  for goodness-of-fit, 637–38
  for independence, 630–36
Classification schemes, 21–26
Class interval, 22, 25
Class limits, 22
Class midvalue, 22
Close-ended classes, 22
Cluster sampling, 295–96
Cochran, W. G., 481
Coefficient of variation, 63
Combinations, 155, 159–61
Complement events, 129
Composite hypothesis, 369
Compound events, 124
Conditional probability, 154, 175–82
  defined, 178
  finding, 178–79
  multiplication law of probability, 179–81
  statistical independence, 180–81

707

Confidence intervals, 340–52
  defined, 341
  for standard deviation, 347–52
  for universe mean, 341–45
  for universe proportion, 345–47
  for variance, 347–52
Consistent estimator, 337–38
Consumer Price Index (CPI), 86, 92, 94–95
"Consumer Price Index Pricing and Calculation Procedures, The" (Rothwell), 94
Contingency coefficient, 656–57
Continuous classes, 23, 24–25
Continuous probability functions:
  binomial, 240–48
  exponential, 257–59
  hypergeometric, 249–51
  normal, 259–75
  Poisson, 252–55
Continuous sample spaces, 121–22, 166, 168–71
Continuous state variable, 455–56
Convenience sampling, 289
Correlation analysis, 563–82
  intuitive approach to correlation coefficients, 580–81
  in regression problem, 564–72
  without regression, 571–75
  statistical independence and, 578–79
  tests of significance on correlation coefficients, 575–77
Counting techniques, 154–65
  classifying sets of objects, 161–62
  combinations, 155, 159–61
  fundamental principle of, 155–57, 163–64
  permutations, 155, 157–59
  using, 162–65
Cox, G. M., 481
Critical value, 374–76
Cross-section data in regression analysis, 543–45
Cumulative probability functions, 171–73, 188–91
Cumulative relative frequency, 32
Current ratio, 96–97
Cyclical variation, in time series:
  defined, 591–92
  in multiplicative model, 593–95

D

Data-generating process, 309–11
Decision Analysis (Raiffa), 12, 118
Decision, defined, 417
Decision maker, identifying, 11–12
Decision to obtain more information, 433–41
  expected value of perfect information, 434–36
  revision of probabilities based on sample information, 436–41
Decision rules, 374, 417–21
  Bayesian, 420–21, 424–25
  maximum probability, 418
  minimax, 419–20, 424
  most favorable payoff value, 418–19
Decision table, 415, 432
Decision theory, *see* Statistical decision theory
Decision tree:
  for multistage problems, 427–29
  for single stage problem, 412–15
Degrees of freedom, 321, 323, 324, 476–78, 480
DeMorgan, Augustus, 136
Dependent variable, 451–52
Descriptive statistics, concept of, 2–3
Discrete probability functions, 166–68, 171, 239–55
Discrete sample spaces, 121–22, 165–68, 171–73
Discrete state variable, 454–55
Disjoint events, 127–28
Distribution-free methods, 626
Dominance, concept of, 417–18
Dow Jones Index, 86

E

Efficient estimator, 336–37
Empty event, 125
Error characteristic curve, 382–84
Estimation, 334–57
  confidence intervals, 340–52
    defined, 341
    for standard deviation, 347–52
    for universe mean, 341–45
    for universe proportion, 345–47
    for variance, 347–52
  estimators:
    characteristics of, 336–39
    defined, 335
  nonparametric methods, 628–30
  sample size determination, 352–55
  sample standard deviation, 628–29
Events, 124–33
  defined, 124
  determining possible outcomes, 131–33
  mutually exclusive, 127–28, 141–43
  occurrence of, 125

INDEX   709

Events (cont.)
  Venn diagrams and, 126–30
  See also Probabilities
Expectations, 200–223
  examples, 213–16
  general form of, 214
  involving probability density functions, 220–23
  of linear functions, 208–11
  mean, 201–4
  random variables, 217–20
  standard deviation, 205–7
  variance, 205–7
Expected frequencies, 634–36
Expected payoffs, 213–14
Expected value of perfect information, 434–36
Experimental design, 502
Experimental Designs (Cochran and Cox), 481
Exponential curve, 598, 599
Exponential probability density function, 257–59
Exponential smoothing methods, 612–17
Extrapolation, 533–34

F

F distribution, 476–81
  table, 689–92
Feller, W., 453
Finite population correction factor, 314, 326–27
First moment, 70
Fisher's ideal index, 106
Fixed-effects model, 501
Fixed-weight aggregative index, 91
Fractile rule, 454–56
Fractiles, 58–59
Frame, 286
Frequency curves, 27–31
Frequency distribution, 19–39
  classification schemes, 21–26
  defined, 20
  graphical presentation of, 27–34
    frequency polygon, 27–31
    histogram, 27, 33–34
    ogive, 30, 32–33
  relative frequency, 24–25
  tabular presentation of, 36–38
Frequency polygon, 27–31
Frisch, Ragnar, 103
Fundamental events, 124
Fundamental principle of counting, 155–57, 163–64

G

Geis, Irving, 33, 286
Geometric average, 101
Goodness-of-fit, 636–39
  chi-square test for, 637–38
  defined, 626
  Kolmogrov-Smirnov test for, 638–39
Gosset, W. S., 320
Graphical presentation, 27–34
  frequency polygon, 27–31
  histogram, 27, 33–34
  ogive, 30, 32–33
Gross National Product Price Deflator (GNPPD), 86
Grouped data, describing, 64–69
  interquartile range, 66
  mean, 65
  mean absolute deviation, 66–67
  median, 65
  mode, 66
  range, 66
  standard deviation, 67–68
  variance, 67–68

H

Handbook of Statistical Tables (Owen), 639
Heteroscedasticity, 537
Histogram, 27, 33–34
Homoscedasticity, 537
How to Lie with Statistics (Huff and Geis), 33, 286
Huff, Darrell, 33, 286
Hypergeometric probability mass function, 249–51
Hypothesis testing, 366–69
  correlation analysis and, 575–77
  definitions, 369
  examples, 366–68
  incorrect decisions, 370–72
    error characteristic curve, 382–84
    level of significance, 377–79
    probability of making, 381–84
    Type I errors, defined, 371–72
    Type II errors, defined, 371
  null hypothesis:
    critical value, 374–76
    decision rule, 374
    defined, 369–70
    selection of, 370–76
  one-tailed tests, 389, 541–42
  power of, 384
  regression analysis and, 541–43

Hypothesis testing (cont.)
   tests involving means and proportions, 386–96
   two-tailed tests, 389–91

Interval measurement, 8
Introduction to Probability Theory and Its Applications, An (Feller), 453
Intuitive probabilities, 136
Irregular variation, in time series, 593

I

Implementation, 12
Independence:
   chi-square tests for, 630–36
   statistical, 180–81, 578–79
Independent variable, 522
Index numbers, 85–106
   applications of, 94–98
   base period:
      selection of, 98–99, 103–4
      shifting of, 99–100
   changes in quality of items and, 103
   Consumer Price Index (CPI), 86, 92, 94–95
   current ratio, 96–97
   economic utility and, 104–6
   Fisher's ideal index, 106
   fixed-weight aggregative index, 91
   Index of Industrial Production (IIP), 86, 95–96
   Laspeyres index, 91, 92, 93, 104, 105
   nature of, 85–87
   Paasche index, 91–93, 104, 105
   parity ratio, 97–98
   quick ratio, 86–87
   sampling, 101–2
   splicing series of, 100–1
   unweighted, 87–90
   weighted, 90–93
   Wholesale Price Index, 86, 95
Index of Consumer Confidence, 96
Index of Industrial Production (IIP), 86, 95–96
Index of Spot Market Prices, 90
Inferential statistics:
   concept of, 3–6
   defined, 52
Interaction effect, 488
Intercept, 523
Interpolation, 534
Interquartile range:
   defined, 59
   for grouped data, 66
Intersection of two events, 128
Intersection of two sets, 147–48
Interval estimates:
   defined, 335
   See also Confidence intervals

J

Joint probabilities, 177, 634–35
Journal of the American Statistical Association, 15, 639
Judgment sampling, 289, 290

K

Kelvin, Lord, 2
Kolmogrov-Smirnov test for goodness-of-fit, 638–39
Kruskal-Wallis one-way analysis-of-variance test, 501, 647, 653–55
Kurtosis, 69–70, 73–74

L

Laspeyres index, 91, 92, 93, 104, 105
Level of significance of hypothesis testing, 377–79
Linear functions, expectations of, 208–11
Location, measures of, 53–56
   mean, *see* Mean
   median, 54–55, 65
   mode, 55–56, 66

M

Main effects, 488
Mann-Whitney U test, 647–49
Marginal probabilities, 177, 634
Maximum decision rule, 419
Maximum probability decision rule, 418
Mean:
   sample, 53–55, 335
      defined, 202
      distribution of, *see* Sampling distributions
      as estimator, 336, 337–38
      for grouped data, 65
      moments around, 70
   universe, 201–4, 308–9

Mean (cont.)
    confidence intervals for, 341–45
    defined, 202
    hypotheses involving, see Hypothesis testing
    point estimates for, 336, 337–38
    of probability density functions, 261, 263, 265, 269, 272
    of probability mass functions, 244, 245, 250, 253–54
Mean absolute deviation, 59–60, 66–67
Mean rule, 456–61
Mean sum of squares, 482–84, 489–97
Measurement scales, 6–10
Median, 54–55, 65
Middle value, see Median
Midvalue, class, 22
Miller, L., 639
Minimal sufficient statistic, 339
Minimax decision rule, 419–20, 424
Minimax-loss decision rule, 424
Minimax-regret decision rule, 424
Mode, 55–56, 66
Modern Probability Theory and Its Applications (Parzen), 254
Moments of samples, 69–70
Most common value, see Mode
Most favorable payoff value decision rule, 418–19
Multiple factor analysis of variance, see Two-factor analysis of variance
Multiple regression analysis, 546–47
Multiple-stage sampling, 296
Multiplication law of probability, 179–81
Multistage problems, 426–32
    decision tree for, 427–29
    roll-back procedure, 429–32
Mutually exclusive events, 127–28, 141–43

N

Negatively correlated variables, 563
New York Stock Exchange Index, 86
Nominal class limits, 22, 23
Nominal measurement, 6–7
Nonequivalent state-action problems, 456–61
Nonlinear probability functions, 188–91
Nonparametric methods, 626–61
    advantages of, 627
    of association, 656–59
        contingency coefficient, 656–57
        Spearman's rank correlation coefficient, 657–59

Nonparametric methods (cont.)
    chi-square tests:
        for goodness-of-fit, 637–38
        for independence, 630–36
    defined, 626–27
    disadvantages of, 627
    estimation methods, 628–30
    Kolmogrov-Smirnov test for goodness-of-fit, 638–39
    rank tests of significance, 647–55
        Kruskal-Wallis one-way analysis-of-variance test, 50, 647, 653–55
        Mann-Whitney U test, 647–49
        Wilcoxon matched-pairs signed-rank test, 647, 650–52
    runs test for randomness, 640–43
    signs test of significance, 644–46
Nonparametric Statistics (Siegel), 639
Nonrespondent bias, 287
Nonsampling error, 287
Normal equations, 531–32, 549
Normal probability density function, 259–75
Null event, 125
Null hypothesis:
    critical value, 374–76
    decision rule, 374
    defined, 369–70
    selection of, 370–76
Null set, 147

O

Occurrence of events, 125–30
Odds, 137–38
Ogive, 30, 32–33
One-factor analysis of variance, 474–84
    computational formulas, 481–84
    F distribution, 476–81
One Million Random Digits, 291
One-tailed tests, 389, 541–42
Open-ended classes, 22
Opportunity loss, 422–26
Ordered set, 146–47
Ordinal measurement, 7–8
Outcome, defined, 119
Overlapping events, 128
Owen, D., 639

P

Paasche index, 91–92, 104, 105
Parabolic curve, 598, 599

Parameters, 239–40
Parity Index, 86
Parity ratio, 97–98
Parzen, E., 254
Payoff values:
　evaluating, 442–49
　　modification of monetary values, 443–45
　　nonmonetary payoffs, 446–49
　　most favorable decision rule, 418–19
Percentiles, 58
Permutation, 155, 157–59
Personal probabilities, 136
Point estimates, 335–39
Poisson probability mass function, 252–55
　table, 672–87
Poisson, Siméon D., 135
Population:
　defined, 5, 51
　identification of, 52
Positively correlated variables, 563
Possible outcomes, determining, 131–32
Power of hypothesis test, 384
Predictions:
　placing confidence intervals around, 536–40
　regression equation for, 533–35
Price indexes, computation of, 87–95
Prior probabilities, 436–38, 451–54
Probabilities, 115–91
　addition rule of, 141–44
　algebra of sets and, 145–48
　conditional, 154, 175–82
　　defined, 178
　　finding, 178–79
　　multiplication law of probability, 179–81
　　statistical independence, 180–81
　counting techniques, 154–65
　　classifying sets of objects, 161–62
　　combinations, 155, 159–61
　　fundamental principle of, 155–57, 163–64
　　permutations, 155, 157–59
　　using, 162–65
　defined, 116
　events, 124–33
　　defined, 124
　　determining possible outcomes, 131–33
　　mutually exclusive, 127–28, 141–43
　　occurrence of, 125
　　Venn diagrams and, 126–30
　numerical measures for, 133–34
　odds and, 137–38
　prior, 436–38, 451–54
　relative frequency and, 135–36

Probabilities (cont.)
　revision of, 182–86, 436–41
　sample spaces, 119–24
　　continuous, 121–22, 166, 168–71
　　data and, 120–22
　　defined, 119
　　discrete, 121–22, 165–68, 171–73
　　numbers and real-world phenomena, 122–23
　　Venn diagrams and, 126–30
　uncertainty, 115–18
　unique cases and, 138–39
　zero or 1, 139–40
　See also Expectations; Probability distributions
Probability density functions, 168–71, 256–69
　expectations involving, 220–23
　exponential, 257–59
　normal, 259–75
　triangular, 170
　uniform, 169
Probability distributions, 231–75
　chi square distribution, 233
　nonlinear probability functions, 188–91
　probability density functions, see Probability density functions
　probability mass functions, see Probability mass functions
　random variables, 233–38
Probability mass functions, 166–68, 171, 239–55
　binomial, 240–48
　expectations involving, 205–7
　hypergeometric, 249–51
　parameters, 239–40
　Poisson, 252–55, 672–87
Problem analysis, 10–13
Proportion:
　sample:
　　as estimator, 339
　　Tchebycheff's inequality and, 629–30
　universe:
　　confidence intervals for, 345–47
　　hypotheses involving, 386–96
　　point estimates for, 339

Q

Qualitative data, 121
Quantitative data, 121
Quartiles, 58–59
Quick ratio, 86–87
Quota sampling, 289

INDEX

R

Raiffa, Howard, 12, 118, 137
Ramsey, Frank, 138
Random-effects model, 501
Random numbers, table, 693
Random sampling, 289–96, 544
  cluster sampling, 295–96
  defined, 289
  multi-stage sampling, 296
  process of, 290–91
  without replacement, 251, 290
  with replacement, 290, 314
  simple, 293
  stratified sampling, 293–95
  systematic sampling, 295
Random variables, 217–20, 233–38
Range, 57, 66
Rank tests of significance, 647–55
  Kruskal-Wallis one-way analysis-of-variance test, 501, 647, 653–55
  Mann-Whitney U test, 647–49
  Wilcoxon matched-pairs signed-rank test, 647, 650–52
Ratio measurement, 8–9
Ratio-to-moving-average method, 605–10
Regression analysis, 521–50
  cross-section data in, 543–45
  limitation of linearity, 545
  multiple relationships, 546–47
  normal equations, 531–32, 549
  objectives of, 521–25
  population regression line, estimation of, 529–32
  predictions:
    placing confidence intervals around, 536–40
    regression equation for, 533–35
  regression population, 525–28
  testing hypothesis, 541–43
  time-series data in, 543–45
Relative frequency, 24–25, 135–36
Relative kurtosis, 74
Relative skewness, 72–73
Roberts, Harry V., 117
Roll-back procedure, 429–32
Rothwell, Doris P., 94
Runs test for randomness, 640–43

S

Sample:
  defined, 5, 51, 285
  identification of, 52
  moments of, 69–70

Sample coefficient of determination, 566, 567
Sample correlation coefficient, 567
Sample size, determining, 352–55
Sample spaces, 119–24
  continuous, 121–22, 166, 168–71
  data and, 120–22
  defined, 119
  discrete, 121–22, 165–68, 171–73
  numbers and real-world phenomena, 122–23
  Venn diagrams and, 126–30
Sampling, 283–300
  definitions, 285–87
  index numbers, 101–2
  judgment, 289, 290
  potential problems, 287–88
  random, 289–96, 544
    cluster sampling, 295–96
    defined, 289
    multiple-stage sampling, 296
    process of, 290–91
    without replacement, 251, 290
    with replacement, 290, 314
    simple, 293
    stratified sampling, 293–95
    systematic sampling, 295
Sampling distributions, 308–27
  data-generating process, 309–11
  finite population correction factor, 314, 326–27
  sample means, 312–26
    Central Limit Theorem, 315–17, 325–26
    unknown universe standard deviation and, 318–24
Sampling error, 287
Seasonal variation, in time series:
  analysis of, 604–10, 615–16
  defined, 591
  in multiplicative model, 593–95
Second moment, 70
Sets, algebra of, 145–48
Siegel, S., 639, 649
Significance, tests of, *see* Tests of significance
Significance level of hypothesis testing, 377–79
Signs test of significance, 644–46
Simple hypothesis, 369
Simple random sampling, 293
Skewness, 69–73
Smoothing, Forecasting and Prediction of Discrete Time Series (Brown), 614
Spearman's rank correlation coefficient, 657–59

Standard deviation:
  sample, 60–62, 335
    estimation of, 628–29
    for grouped data, 67–68
  universe, 205–7, 314
    confidence intervals for, 347–52
    point estimates for, 336–37
    of probability density functions, 261, 262, 265, 272
Standard error of the estimate, 536–37
Standard normal probability density function, 265–69, 272–75
Statistical decision theory, 409–61
  decision to obtain more information, 433–41
    expected value of perfect information, 434–36
    revision of probabilities based on sample information, 436–41
  decision rules, 417–21
    Bayesian, 420–21, 424–25
    maximum probability, 418
    minimax, 419–20, 424
    most favorable payoff value, 418–19
  decision-theory-model, 412–16
  evaluating payoff values, 442–49
    modification of monetary values, 443–45
    nonmonetary payoffs, 446–49
  fractile rule, 454–56
  mean rule, 456–61
  multistage problems, 426–32
    decision tree for, 427–29
    roll-back procedure, 429–32
  opportunity loss, 422–26
  prior probabilities, 436–38, 451–54
Statistical descriptions, 49–77
  describing grouped data, 64–69
    interquartile range, 66
    mean, 65
    mean absolute deviation, 66–67
    median, 65
    mode, 66
    range, 66
    standard deviation, 67–68
    variance, 67–68
  kurtosis, 69–70, 73–74
  measures of location, 53–56
    mean, see Mean
    median, 54–55
    mode, 55–56
  measures of variation, 56–64
    coefficient of variation, 63
    fractiles, 58–59
    mean absolute deviation, 59–60
    range, 57
    standard deviation, see Standard deviation

Statistical descriptions (*cont.*)
  variance, see Variance
  populations, 51–52
  samples, 51–52
  skewness, 69–73
Statistical independence, 180–81, 578–79
Statistical inference, concept of, 3–6
Statistical Principles in Experimental Design (Winer), 481
Statistics: A New Approach (Wallis and Roberts), 117
Statistics, defined, 1, 239
Stratified sampling, 293–95
Student's t distribution, see t distribution
Sturges' Rule, 23–25
Sufficient estimator, 339
Sum of squares, 481–84, 489–97
Sun Tze, 137
Systematic sampling, 295

T

t distribution, 233, 320–24
  table, 669
Tabular presentation, 36–38
Tchebycheff's inequality, 629–30
Tests of significance:
  on correlation coefficients, 575–77
  rank, 647–55
    Kruskal-Wallis one-way analysis-of-variance test, 501, 647, 653–55
    Mann-Whitney U test, 647–49
    Wilcoxon matched-pairs signed-rank test, 647, 650–52
  signs, 644–46
Time-series analysis, 588–619
  cyclical variation:
    defined, 591–92
    in multiplicative model, 593–95
  exponential smoothing methods, 612–17
  general-time series models, 591–96
  irregular variation, 593
  seasonal variation:
    analysis of, 604–10, 615–16
    defined, 591
    in multiplicative model, 593–95
  time series, defined, 589
  trend:
    analysis of, 597–603, 615–16
    defined, 591
    in multiplicative model, 593–95
Trend, in time series:
  analysis of, 597–603, 615–16
  defined, 591
  in multiplicative model, 593–95

INDEX

Triangular probability density function, 170
True class limits, 23
True-factor analysis of variance, 486–97
 computational formulas, 492–97
 interaction effect, 488
 main effects, 488
Two-stage cluster sample, 295
Two-tailed tests, 389–91
Type I and II errors in hypothesis testing, see Hypothesis testing

U

Unbiased estimator, 336–37
Uncertainty, 115–18
Uniform probability density function, 169
Union of several events, 128–29
Union of two sets, 147–48
Unique cases, 138–39
Universe, 5, 51
Universe set, 147
Unweighted index numbers, 87–90
Utility theory, 445–49

V

Variables:
 defined, 5
 random, 217–20, 233–38
Variance:
 analysis of, see Analysis of variance
 sample, 59–62

Variance (cont.)
 distribution of sample means and, 312–14
 as estimator, 338
 for grouped data, 67–68
 universe, 205–7, 312–15
  confidence intervals for, 347–52, 536–40
  point estimates for, 338
 of probability density functions, 258
 of probability mass functions, 244–45, 250, 253–54
Variance ratio, 477–78, 480, 481–82
Variation, measures of, 56–64
 coefficient of variation, 63
 fractiles, 58–59
 mean absolute deviation, 59–60
 range, 57
 standard deviation, see Standard deviation
 variance, see Variance
Venn, John, 126
Venn diagrams, 126–30

W

Wallis, W. Allen, 15, 117
Ward, Artemus, 27
Weighted index numbers, 90–93
Wholesale Price Index (WPI), 86, 95
Wilcoxon matched-pairs signed-rank test, 647, 650–52
Winer, B. J., 481